Lecture Notes in Computer Scie

T0238578

Commenced Publication in 1973
Founding and Former Series Editors:
Gerhard Goos, Juris Hartmanis, and Jan van Leeuwen

Stefan Berghofer   Tobias Nipkow
Christian Urban   Makarius Wenzel (Eds.)

Theorem Proving
in Higher Order Logics

22nd International Conference, TPHOLs 2009
Munich, Germany, August 17-20, 2009
Proceedings

 Springer

Volume Editors

Stefan Berghofer
Tobias Nipkow
Christian Urban
Makarius Wenzel

Technische Universität München
Institut für Informatik
Boltzmannstraße 3
85748, Garching, Germany

E-mail: {berghofe,nipkow,urbanc,wenzelm}@in.tum.de

Library of Congress Control Number: 2009931594

CR Subject Classification (1998): F.4, F.3, F.1, D.2.4, B.6.3, B.6.1, D.4.5, G.4, I.2.2

LNCS Sublibrary: SL 1 – Theoretical Computer Science and General Issues

ISSN   0302-9743
ISBN-10   3-642-03358-X Springer Berlin Heidelberg New York
ISBN-13   978-3-642-03358-2 Springer Berlin Heidelberg New York

springer.com

© Springer-Verlag Berlin Heidelberg 2009
Printed in Germany

Typesetting: Camera-ready by author, data conversion by Scientific Publishing Services, Chennai, India
Printed on acid-free paper   SPIN: 12727186   06/3180   5 4 3 2 1 0

Preface

This volume constitutes the proceedings of the 22nd International Conference on Theorem Proving in Higher Order Logics (TPHOLs 2009), which was held during August 17-20, 2009 in Munich, Germany. TPHOLs covers all aspects of theorem proving in higher order logics as well as related topics in theorem proving and verification.

There were 55 papers submitted to TPHOLs 2009 in the full research category, each of which was refereed by at least three reviewers selected by the Program Committee. Of these submissions, 26 research papers and 1 proof pearl were accepted for presentation at the conference and publication in this volume. In keeping with longstanding tradition, TPHOLs 2009 also offered a venue for the presentation of emerging trends, where researchers invited discussion by means of a brief introductory talk and then discussed their work at a poster session. A supplementary proceedings volume was published as a 2009 technical report of the Technische Universität München.

The organizers are grateful to David Basin, John Harrison and Wolfram Schulte for agreeing to give invited talks. We also invited four tool developers to give tutorials about their systems. The following speakers kindly accepted our invitation and we are grateful to them: John Harrison (HOL Light), Adam Naumowicz (Mizar), Ulf Norell (Agda) and Carsten Schürmann (Twelf).

The TPHOLs conference traditionally changes continents each year to maximize the chances that researchers around the world can attend. TPHOLs started in 1998 in the University of Cambridge as an informal users' meeting for the HOL system. Since 1993, the proceedings of TPHOLs have been published in the *Springer Lecture Notes in Computer Science* series:

1993 (Canada)	Vol. 780	2001 (UK)	Vol. 2152
1994 (Malta)	Vol. 859	2002 (USA)	Vol. 2410
1995 (USA)	Vol. 971	2003 (Italy)	Vol. 2758
1996 (Finland)	Vol. 1125	2004 (USA)	Vol. 3223
1197 (USA)	Vol. 1275	2005 (UK)	Vol. 3603
1998 (Australia)	Vol. 1479	2006 (USA)	Vol. 4130
1999 (France)	Vol. 1690	2007 (Germany)	Vol. 4732
2000 (USA)	Vol. 1869	2008 (Canada)	Vol. 5170

We thank our sponsors: Microsoft Research Redmond, Galois, Verisoft XT, Validas AG and the DFG doctorate programme Puma, for their support.

Finally, we are grateful to Andrei Voronkov. His EasyChair tool greatly eased the task of reviewing the submissions and of generating these proceedings. He also helped us with the finer details of EasyChair.

Next year, in 2010, TPHOLs will change its name to ITP, Interactive Theorem Proving. This is not a change in direction but merely reflects the fact better that

TPHOLs is the premier forum for interactive theorem proving. ITP 2010 will be part of the Federated Logic Conference, FLoC, in Edinburgh.

June 2009   Stefan Berghofer
                                                         Tobias Nipkow
                                                         Christian Urban
                                                         Makarius Wenzel

Organisation

Programme Chairs

Tobias Nipkow   TU München, Germany
Christian Urban   TU München, Germany

Programme Committee

Thorsten Altenkirch	David Aspinall	Jeremy Avigad
Gilles Barthe	Christoph Benzmüller	Peter Dybjer
Jean-Christophe Filliâtre	Georges Gonthier	Mike Gordon
Jim Grundy	Joe Hurd	Reiner Hähnle
Gerwin Klein	Xavier Leroy	Pete Manolios
César Muñoz	Michael Norrish	Sam Owre
Larry Paulson	Frank Pfenning	Randy Pollack
Sofiène Tahar	Laurent Théry	Freek Wiedijk

Local Organisation

Stefan Berghofer
Makarius Wenzel

External Reviewers

Naeem Abbasi	Martin Giese	Zhaohui Luo
Behzad Akbarpour	Alwyn Goodloe	Kenneth MacKenzie
Knut Akesson	Thomas Göthel	Jeff Maddalon
June Andronick	Osman Hasan	Lionel Mamane
Bob Atkey	Daniel Hedin	Conor McBride
Stefan Berghofer	Hugo Herbelin	James McKinna
Yves Bertot	Brian Huffman	Russell O'Connor
Johannes Borgstrom	Clment Hurlin	Steven Obua
Ana Bove	Ullrich Hustadt	Anne Pacalet
Cristiano Calcagno	Rafal Kolanski	Florian Rabe
Harsh Raju Chamarthi	Alexander Krauss	Bernhard Reus
Benjamin Chambers	Sava Krstic	Norbert Schirmer
Nils Anders Danielsson	Cesar Kunz	Stefan Schwoon
William Denman	Stphane Lescuyer	Jaroslav Sevcik
Peter Dillinger	Rebekah Leslie	Thomas Sewell
Bruno Dutertre	Pierre Letouzey	Natarajan Shankar

Table of Contents

Let's Get Physical: Models and Methods for Real-World Security Protocols

David Basin, Srdjan Capkun, Patrick Schaller, and Benedikt Schmidt

ETH Zurich, 8092 Zurich, Switzerland

Abstract. Traditional security protocols are mainly concerned with key establishment and principal authentication and rely on predistributed keys and properties of cryptographic operators. In contrast, new application areas are emerging that establish and rely on properties of the physical world. Examples include protocols for secure localization, distance bounding, and device pairing.

We present a formal model that extends inductive, trace-based approaches in two directions. First, we refine the standard Dolev-Yao model to account for network topology, transmission delays, and node positions. This results in a distributed intruder with restricted, but more realistic, communication capabilities. Second, we develop an abstract message theory that formalizes protocol-independent facts about messages, which hold for all instances. When verifying protocols, we instantiate the abstract message theory, modeling the properties of the cryptographic operators under consideration. We have formalized this model in Isabelle/HOL and used it to verify distance bounding protocols where the concrete message theory includes exclusive-or.

1 Introduction

Situating Adversaries in the Physical World. There are now over three decades of research on symbolic models and associated formal methods for security protocol verification. The models developed represent messages as terms rather than bit strings, take an idealized view of cryptography, and focus on the communication of agents over a network controlled by an active intruder. The standard intruder model used, the Dolev-Yao model, captures the above aspects. Noteworthy for our work is that this model abstracts away all aspects of the physical environment, such as the location of principals and the speed of the communication medium used. This is understandable: the Dolev-Yao model was developed for authentication and key-exchange protocols whose correctness is independent of the principals' physical environment. Abstracting away these details, effectively by identifying the network with the intruder, results in a simpler model that is adequate for verifying such protocols.

With the emergence of wireless networks, protocols have been developed whose security goals and assumptions differ from those in traditional wireline networks. A prominent example is distance bounding [1,2,3,4,5], where one device must determine an upper bound on its *physical* distance to another, potentially untrusted, device. The goal of distance bounding is neither message secrecy nor

S. Berghofer et al. (Eds.): TPHOLs 2009, LNCS 5674, pp. 1–22, 2009.

authentication, but rather to establish a physical property. To achieve this, distance bounding protocols typically combine cryptographic guarantees, such as message-origin authentication, with properties of the physical (communication) layer, for example that attackers cannot relay messages between locations faster than the speed of light. Other examples of "physical protocols" include secure time synchronization, wormhole and neighborhood detection, secure localization, broadcast authentication, and device pairing.

In [6], we presented the first formal model that is capable of modeling and reasoning about a wide class of physical protocols and their properties. The key idea is to reflect relevant aspects of the physical world in the model, namely network topology, transmission delays, and node positions. In particular, all agents are modeled as network nodes. This includes the intruder, who is no longer a single entity but instead is distributed and therefore corresponds to a set of nodes. Communication between nodes is subject to restrictions reflecting the nodes' physical environment and communication capabilities. For example, not all nodes can communicate and communication takes time determined by the network topology and the propagation delays of the communication technologies used. Hence, nodes require time to share their knowledge and information cannot travel at speeds faster than the speed of light. Possible communication histories are formalized as traces and the resulting model is an inductively-defined, symbolic, trace-based model, along the lines of Paulson's *Inductive Approach* [7].

In [6], we formalized this model in Isabelle/HOL [8] and verified the security properties of three physical protocols: an authenticated ranging protocol [9], a protocol for distance bounding using ultrasound [5], and a broadcast-authentication protocol based on delayed key disclosure [10].

Verifying distance bounding protocols. Our starting point in this paper is a family of distance bounding protocols proposed by Meadows [4]. The family is defined by a protocol pattern containing a function variable F, where different instances of F result in different protocols. We present two security properties, which distinguish between the cases of honest and dishonest participants. For each property, we reduce the security of a protocol defined by an instance of F to conditions on F. Afterwards, we analyze several instances of F, either showing that the conditions are fulfilled or presenting counterexamples to the security properties.

This protocol family is interesting as a practically-relevant case study in applying our framework to formalize and reason about nontrivial physical protocols. Moreover, it also illustrates how we can extend our framework (originally defined over a free term algebra) to handle protocols involving equationally-defined operators on messages and how this can be done in a general way. Altogether, we have worked with five different protocols and two different message theories. To support this, we have used Isabelle's locales construct to formalize an abstract message theory and a general theory of protocols. Within the locales, we prove general, protocol-independent facts about (abstract) messages, which hold when we subsequently instantiate the locales with our different concrete message theories and protocols.

Contributions. First, we show that our framework for modeling physical security protocols can be extended to handle protocols involving equationally-defined operators. This results in a message theory extended with an XOR operator and a zero element, consisting of equivalence classes of messages with respect to the equational theory of XOR. We use normalized terms here as the representatives of the equivalence classes. With this extension, we substantially widen the scope of our approach. Note that this extension is actually independent of our "physical" refinement of communication and also could be used in protocol models based on the standard Dolev-Yao intruder.

Second, we show how such extensions can be made in a generic, modular way. Noteworthy here is that we could formulate a collection of message-independent and protocol-independent facts that hold for a large class of intended extensions. An example of such a fact is that the minimal message-transmission time between two agents A and B determines a lower bound on the time difference between A creating a fresh nonce and B learning it.

Finally, physical protocols often contain time-critical steps, which must be optimized to reduce computation and communication time. As a result, these steps typically employ low-level operations like XOR, in contrast to more conventional protocols where nanosecond time differences are unimportant. Our experience indicates that the use of such low-level, equationally-defined operators results in substantial additional complexity in reasoning about protocols in comparison to the standard Dolev-Yao model. Moreover, the complexity is also higher because security properties are topology dependent and so are attacks. Attacks now depend not only on what the attackers know, but also their own physical properties, i.e., the possible constellations of the distributed intruders. Due to this complexity, pencil-and-paper proofs quickly reach their limits. Our work highlights the important role that Formal Methods can play in the systematic development and analysis of physical protocols.

Organization. In Section 2, we provide background on Isabelle/HOL and the distance bounding protocols that we analyze in this paper. In Section 3, we present our formal model of physical protocols, which we apply in Section 4. Finally, in Section 5, we discuss related work and draw conclusions.

2   Background

2.1   Isabelle/HOL

Isabelle [8] is a generic theorem prover with a specialization for higher-order logic (HOL). We will avoid Isabelle-specific details in this paper as far as possible or explain them in context, as needed.

We briefly review two aspects of Isabelle/HOL that are central to our work. First, Isabelle supports the definition of (parameterized) inductively-defined sets. An inductively-defined set is defined by sets of rules and denotes the least set closed under the rules. Given an inductive definition, Isabelle generates a rule for proof by induction.

Second, Isabelle provides a mechanism, called *locales* [11] that can be used to structure generic developments, which can later be specialized. A locale can be seen as either a general kind of proof context or, alternatively, as a kind of parameterized module. A locale declaration contains:

- a name, so that the locale can be referenced and used,
- typed parameters, e.g., ranging over relations or functions,
- assumptions about the parameters (the module axioms), and
- functions defined using the parameters.

In the context of a locale, one can make definitions and prove theorems that depend on the locale's assumptions and parameters. Finally, a locale can be interpreted by instantiating its parameters so that the assumptions are theorems. After interpretation, not only can the assumptions be used for the instance, but also all theorems proved and definitions made in the locale's context.

2.2 Distance Bounding Protocols

Distance bounding protocols are two-party protocols involving a *verifier* who must establish a bound on his distance to a *prover*. These protocols were originally introduced in [1] to prevent a man-in-the-middle attack called Mafia Fraud. Suppose, for example, that an attacker possesses a fake automated teller machine (ATM). When a user uses his banking card to authenticate himself to the fake ATM, the attacker simply forwards the authenticating information to a real ATM. After successful authentication, the attacker can plunder the user's account. Distance bounding protocols prevent this attack by determining an upper bound on the distance between the ATM and the banking card. The ATM is the verifier and checks that the card, acting as the prover, is sufficiently close by to rule out the man-in-the-middle.

The idea behind distance bounding is simple. The verifier starts by sending a challenge to the prover. The prover's reply contains an authenticated message involving the challenge, which shows that it has been received by the prover. After receiving the reply, the verifier knows that the challenge has traveled back and forth between him and the prover. Assuming that the signal encoding the challenge travels with a known speed, the verifier can compute an upper bound on the distance to the prover by multiplying the measured round-trip time of his challenge by the signal's velocity.

For distance bounding to yield accurate results, the verifier's round-trip time measurement should correspond as close as possible to the physical distance between the prover and verifier. This is achieved by having the prover generate his response as quickly as possible. Expensive cryptographic operations such as digital signatures should therefore be avoided. A distance bounding protocol can typically be decomposed into three phases: a *setup phase*, a *measurement phase*, and a *validation phase*. Only the measurement phase is time-critical. The prover makes computationally inexpensive operations, such as XOR, during this phase and may use more sophisticated cryptographic algorithms, such as commitment schemes and message-authentication codes, in the other phases.

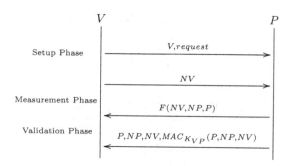

Fig. 1. Pattern for Distance Bounding Protocols

In [4], Meadows et al. present a suite of distance bounding protocols, following the pattern shown in Diagram 1. Here, V denotes the verifier and P is the prover. Both parties initially create nonces NV and NP. V then sends a request to P, followed by a nonce NV. Upon receiving NV, P replies as quickly as possible with $F(NV, NP, P)$, where F is instantiated with an appropriate function. Finally P uses a key K_{VP} shared with V to create a message-authentication code (MAC). This proves that the nonce NP originated with P and binds the reply in the measurement phase to P's identity.

This protocol description is schematic in F. [4] provides four examples of instantiations of $F(NV, NP, P)$ built from different combinations of concatenation, exclusive-or, and hashing, e.g. $(NV \oplus P, NP)$ or, even simpler, (NV, NP, P). Each instantiation uses only simple cryptographic operations, which could even be implemented in hardware to further reduce their computation time.

The security property we want to prove is: "If V has successfully finished a protocol run with P, then V's conclusion about the distance to P is an upper bound on the physical distance between the two nodes." We will formalize this property, along with associated provisos, in subsequent sections.

3   Formal Model

In this section, we present our model of physical protocols. To support the verification of multiple protocols, we use locales to parameterize our model both with respect to the concrete protocol and message theory. Figure 2 depicts the theories we formalized in Isabelle and their dependencies. Some of these theories are concrete to begin with (e.g. *Geometric Properties of* $\mathbb{R}^3$) whereas other theories consist of locales or their interpretations. For example, the *Abstract Message Theory* contains a locale describing message theories, which is interpreted in our two concrete message theories (*Free* and *XOR*). In the theory *Parametrized Communication Systems*, we abstractly define the set of valid traces as a set of (parametric) inductive rules. In formalizations of concrete protocols using either of the two concrete message theories, we can therefore use both message-theory independent and message-theory specific facts by importing the required theories.

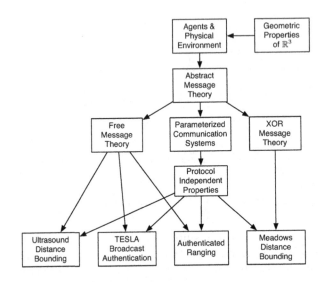

Fig. 2. Dependency Graph of our Isabelle Theory Files

3.1   Agents and Environment

Agents are either honest agents or dishonest intruders. We model each kind using the natural numbers *nat*. Hence there are infinitely many agents of each kind.

$$\textbf{datatype } \mathit{agent} = \textsf{Honest } \mathit{nat} \mid \textsf{Intruder } \mathit{nat}$$

We refer to agents using capital letters like A and B. We also write H_A and H_B for honest agents and I_A and I_B for intruders, when we require this distinction. In contrast to the Dolev-Yao setting, agents' communication abilities are subject to the network topology and physical laws. Therefore, we cannot reduce a set of dishonest users at different locations to a single one.

Location and Physical Distance. To support reasoning about physical protocols, we associate every node A with a location loc_A. We define $loc : agent \rightarrow \mathbb{R}^3$ as an uninterpreted function constant. Protocol-specific assumptions about the position of nodes can be added as local assumptions to the corresponding theorems or using Isabelle/HOL's specification mechanism.[1] We use the standard Euclidean metric on $\mathbb{R}^3$ to define the physical distance between two agents A and B as $\mid loc_A - loc_B \mid$.

Taking the straight-line distance between the locations of the agents A and B in $\mathbb{R}^3$ as the shortest path (taken for example by electromagnetic waves

---

[1] Definition by specification allows us to assert properties of an uninterpreted function. It uses Hilbert's ϵ-operator and requires a proof that a function with the required properties exists.

when there are no obstacles), we define the line-of-sight communication distance $cdist_{LoS} : agent \times agent \to \mathbb{R}$ as

$$cdist_{LoS}(A, B) = \frac{|\,loc_A - loc_B\,|}{c},$$

where c is the speed of light. Note that $cdist_{LoS}$ depends only on A and B's location and is independent of the network topology.

Transmitters, Receivers, and Communication Distance. To distinguish communication technologies with different characteristics, we equip each agent with an indexed set of transmitters.

$$\textbf{datatype } transmitter = \mathsf{Tx}\ agent\ nat$$

The constructor Tx returns a transmitter, given an agent A and an index i, denoted Tx_A^i. Receivers are formalized analogously.

$$\textbf{datatype } receiver = \mathsf{Rx}\ agent\ nat$$

We model the network topology using the uninterpreted function constant $cdist_{Net} : transmitter \times receiver \to \mathbb{R}_{\geq 0} \cup \{\bot\}$. We use $cdist_{Net}(Tx_A^i, Rx_B^j) = \bot$ to denote that Rx_B^j cannot receive transmissions from Tx_A^i. In contrast, $cdist_{Net}(Tx_A^i, Rx_B^j) = t$, where $t \neq \bot$, describes that Rx_B^j can receive signals (messages) emitted by Tx_A^i after a delay of at least t time units. This function models the minimal signal-transmission time for the given configuration. This time reflects environmental factors such as the communication medium used by the given transceivers and obstacles between transmitters and receivers. Since we assume that information cannot travel faster than the speed of light, we always require that $cdist_{LoS}(A, B) \leq cdist_{Net}(Tx_A^i, Rx_B^j)$ using the specification mechanism.

We use the formalization of real numbers and vectors provided in Isabelle's standard library for time and location. Additionally, we use the formalization of the Cauchy-Schwarz inequality [12] to establish that $cdist_{LoS}$ is a pseudometric.

3.2 Messages

Instead of restricting our model to a concrete message theory, we first define a locale that specifies a collection of message operators and their properties. In the context of this locale, we prove a number of properties independent of the protocol and message theory. For example, $cdist_{LoS}(A, B)$ is a lower bound on the time required for a nonce freshly created by A to become known by another agent B, since the nonce must be transmitted. For the results in [6], we have instantiated the locale with a message theory similar to Paulson's [7], modeling a free term algebra. In Section 3.6, we describe the instantiation with a message theory that includes the algebraic properties of the XOR operator, which we use in Section 4.

The theory of keys is shared by all concrete message theories and reuses Paulson's formalization. Keys are represented by natural numbers. The function

$inv : key \rightarrow key$ partitions the set of keys into symmetric keys, where $inv\ k = k$, and asymmetric keys. We model key distributions as functions from agents to keys, e.g. the theory assumes that K_{AB} returns a shared symmetric key for a pair of agents A and B.

Abstract Message Theory. Our MESSAGE_THEORY locale is parametric in the message type $'msg$ and consists of the following function constants.

$Nonce : agent \rightarrow nat \rightarrow 'msg$	$Key : key \rightarrow 'msg$
$Int : int \rightarrow 'msg$	$Real : real \rightarrow 'msg$
$Hash : 'msg \rightarrow 'msg$	$Crypt : key \rightarrow 'msg \rightarrow 'msg$
$MPair : 'msg \rightarrow 'msg \rightarrow 'msg$	$parts : 'msg\ set \rightarrow 'msg\ set$
$subterms : 'msg\ set \rightarrow 'msg\ set$	$dm : agent \rightarrow 'msg\ set \rightarrow 'msg\ set$

This formalizes that every interpretation of the MESSAGE_THEORY locale defines the seven given message construction functions and three functions on message sets. A *Nonce* is tagged with a unique identifier and the name of the agent who created it. This ensures that independently created nonces never collide. Indeed, even colluding intruders must communicate to share a nonce. The constructor *Crypt* denotes signing, asymmetric, or symmetric encryption, depending on the key used. We also require that functions for pairing (*MPair*), hashing (*Hash*), integers (*Int*), and reals (*Real*) are defined. We use the abbreviations $\langle A, B \rangle$ for *MPair A B* and $\{m\}_k$ for *Crypt k m*. Moreover, we define $MAC_k(m) = Hash\langle Key\ k, m\rangle$ as the keyed MAC of the message m and $MACM_k(m) = \langle MAC_k(m), m\rangle$ as the pair consisting of m and its *MAC*. Additionally, every interpretation of MESSAGE_THEORY must define the functions *subterms*, *parts*, and *dm*. These respectively formalize the notions of subterms, extractable subterms, and the set of messages derivable from a set of known messages by a given agent. In the free message theory, *subterms* corresponds to syntactic subterms, for example $x \in subterms(\{Hash\ x\})$ while $x \notin parts(\{Hash\ x\})$.

We assume that the following properties hold for any interpretation of *parts*.

$$\frac{X \in H}{X \in parts(H)} \qquad \frac{X \in parts(H)}{\exists Y \in H.X \in parts(\{Y\})} \qquad \frac{G \subseteq H}{parts(G) \subseteq parts(H)}$$

$$parts(parts(H)) = parts(H) \qquad parts(H) \subseteq subterms(H)$$

These properties allow us to derive most of the lemmas about *parts* from Paulson's formalization [7] in our abstract setting. For example,

$$parts(G) \cup parts(H) = parts(G \cup H).$$

Similar properties are assumed to hold for the *subterms* function.

We also assume properties of the message-derivation operator *dm* that state that no agent can guess another agent's nonces or keys, or forge encryptions or *MAC*s. These assumptions are reasonable for message theories formalizing idealized encryption.

$$\frac{Nonce\ B\ N_B \in subterms(dm\ A\ H) \quad A \neq B}{Nonce\ B\ N_B \in subterms(H)} \qquad \frac{Key\ k \in parts(dm\ A\ H)}{Key\ k \in parts(H)}$$

$$\frac{\{m\}_k \in subterms(dm\ A\ H)}{\{m\}_k \in subterms(H) \vee Key\ k \in parts(H)}$$

$$\frac{MAC_k(m) \in subterms(dm\ A\ H)}{MAC_k(m) \in subterms(H) \vee Key\ k \in parts(H)}$$

3.3   Events and Traces

We distinguish between three types of events: an agent sending a message, receiving a message, or making a claim. We use a polymorphic data type to model these different message types.

> **datatype** *'msg event* = Send *transmitter 'msg ('msg list)*
> | Recv *receiver 'msg* | Claim *agent 'msg*

A *trace* is a list of timed events, where a timed event $(t, e) \in real \times event$ pairs a time-stamp with an event.

A timed event $(t^S, Send\ Tx_A^i\ m\ L)$ denotes that the agent A has sent the message m using his transmitter Tx_A^i at time t^S and has associated the protocol data L with the event. The list of messages L models local state information and contains the messages used to construct m. The sender may require these messages in subsequent protocol steps. Storing L with the *Send* event is necessary since we support non-free message construction functions like XOR where a function's arguments cannot be recovered from the function's image alone.

A send event like the above may result in multiple timed *Recv*-events of the form $(t^R, Recv\ Rx_B^j\ m)$, where the time-stamps t^R and the receivers Rx_B^j must be consistent with the network topology. Note that the protocol data stored in L when sending the message does not affect the events on the receiver's side.

A *Claim*-event models a belief or conclusion made by a protocol participant, formalized as a message. For example, after successfully completing a run of a distance bounding protocol with a prover P, the verifier V concludes at time t that d is an upper bound on the distance to P. We model this by adding the timed event $(t, Claim\ V\ \langle P, Real\ d\rangle)$ to the trace. The protocol is secure if the conclusion holds for all traces containing this claim event.

Note that the time-stamps used in traces and the rules use the notion of absolute time. However, agents' clocks may deviate arbitrarily from absolute time. We must therefore translate the absolute time-stamps to model the local views of agents. We describe this translation in Section 3.4.

$$\frac{\begin{array}{l} tr \in Tr \qquad t^R \geq maxtime(tr) \\ (t^S, Send\ Tx_A^i\ m\ L) \in tr \\ cdist_{Net}(Tx_A^i, Rx_B^j) = t_{AB} \\ t_{AB} \neq \bot \quad t^R \geq t^S + t_{AB} \end{array}}{tr.(t^R, Recv\ Rx_B^j\ m) \in Tr}\ \text{NET} \qquad \frac{\begin{array}{l} tr \in Tr \qquad t \geq maxtime(tr) \\ m \in dm_{I_A}(knows_{I_A}(tr)) \end{array}}{tr.(t, Send\ Tx_{I_A}^k\ m\ []) \in Tr}\ \text{FAKE}$$

$$\frac{}{[] \in Tr}\ \text{NIL} \qquad \frac{\begin{array}{c} tr \in Tr \qquad t \geq maxtime(tr) \qquad step \in proto \\ (act, m) \in step(view(H_A, tr), H_A, ctime(H_A, t)) \\ m \in dm_{H_A}(knows_{H_A}(tr)) \end{array}}{tr.(t, translateEv(H_A, act, m)) \in Tr}\ \text{PROTO}$$

Fig. 3. Rules for Tr

Knowledge and Used Messages. Each agent A initially possesses some knowledge, denoted $initKnows_A$, which depends on the protocol executed. We use locales to underspecify the initial knowledge. We define a locale INITKNOWS that only includes the constant $initKnows : agent \rightarrow\ 'msg\ set$. Different key distributions are specified by locales extending INITKNOWS with additional assumptions. For example, the locale INITKNOWS_SHARED assumes that any two agents A and B share a secret key $Key\ K_{AB}$. In a system run with trace tr, A's knowledge consists of all messages he received together with his initial knowledge.

$$knows_A(tr) = \{m\ |\exists\ k\ t.(t,\ Recv\ Tx_A^k\ m) \in tr\} \cup\ initKnows_A$$

Each agent can derive all messages in the set $dm_A(knows_A(tr))$ by applying the derivation operator to the set of known messages. We use the *subterms* function to define the set of messages used in a trace tr.

$$used(tr) = \{n\ |\ \exists\ A\ k\ t\ m\ L.(t, Send\ Tx_A^k\ m\ L) \in tr\ \land\ n \in subterms(\{m\})\}$$

A nonce is *fresh* for a trace tr if it is not in $used(tr)$. Note that since a nonce is not fresh if its hash has been sent, we cannot use *parts* instead of *subterms* in the above definition.

3.4 Network, Intruder, and Protocols

We now describe the rules used to inductively define the set of traces Tr for a system parameterized by a protocol *proto*, an initial knowledge function *initKnows*, and the parameters from the abstract message theory. The base case, modeled by the NIL rule in Figure 3, states that the empty trace is a valid trace for all protocols. The other rules describe how valid traces can be extended. The rules model the network behavior, the possible actions of the intruders, and the actions taken by honest agents following the protocol steps.

Network Rule. The NET-rule models message transmission from transmitters to receivers, constrained by the network topology as given by $cdist_{Net}$. A *Send*-event from a transmitter may induce a *Recv*-event at a receiver only if the receiver can receive messages from the transmitter as specified by $cdist_{Net}$. The time delay between these events is bounded below by the communication distance between the transmitter and the receiver.

If there is a *Send*-event in the trace tr and the NET-rule's premises are fulfilled, a corresponding *Recv*-event is appended (denoted by $xs.x$) to the trace. The restriction on connectivity and transmission delay are ensured by $t_{AB} \neq \bot$ and $t^R \geq t^S + t_{AB}$. Here, t_{AB} is the communication distance between the receiver and transmitter, t^S is the sending time, and t^R is the receiving time.

Note that one *Send*-event can result in multiple *Recv*-events at the same receiver at different times. This is because $cdist_{Net}$ models the minimal communication distance and messages may also arrive later, for example due to the reflection of the signal carrying the message. Moreover, a *Send*-event can result in multiple *Recv*-events at different receivers, modeling for example broadcast communication. Finally, note that transmission failures and jamming by an intruder, resulting in message loss, are captured by not applying the NET-rule for a given *Send*-event and receiver, even if all premises are fulfilled.

The time-stamps associated with *Send*-events and *Recv*-events denote the starting times of message transmission and reception. Thus, our network rule captures the latency of links, but not the message-transmission time, which also depends on the message's size and the transmission speed of the transmitter and the receiver. Some implementation-specific attacks, such as those described in [13,5], are therefore not captured in our model.

The premise $t \geq maxtime(tr)$, included in every rule (except NIL), ensures that time-stamps increase monotonically within each trace. Here t denotes the time-stamp associated with the new event and $maxtime(tr)$ denotes the latest time-stamp in the trace tr. This premise guarantees that the partial order on events induced by their time-stamps (note that events can happen simultaneously) is consistent with the order of events in the list representing the trace.

Intruder Rule. The FAKE-rule in Figure 3 describes the intruders' behavior. An intruder can always send any message m derivable from his knowledge. Intruders do not need any protocol state since they behave arbitrarily.

Since knowledge is distributed, we use explicit *Send*-events and *Recv*-events to model the exchange of information between colluding intruders. With an appropriate $cdist_{Net}$ function, it is possible to model an environment where the intruders are connected by high-speed links, allowing them to carry out wormhole attacks. Restrictions on the degree of cooperation between intruders can be modeled as predicates on traces. Internal and external attackers are both captured since they differ only in their initial knowledge (or associated transceivers).

Protocols. In contrast to intruders who can send arbitrary derivable messages, honest agents follow the protocol. A protocol is defined by a set of step functions.

Each step function takes the local view and time of an agent as input and returns all possible actions consistent with the protocol specification.

There are two types of possible actions, which model an agent either sending a message with a given transmitter id and storing the associated protocol data or making a claim.

$$\textbf{datatype } \textit{'msg action} = \ \mathsf{SendA} \ \textit{nat} \ (\textit{'msg} \ \textit{list} \ ) \ | \ \mathsf{ClaimA}$$

Note that message reception has already been modeled by the NET-rule.

An *action* associated with an agent and a message can be translated into the corresponding trace event using the *translateEv* function.

$$translateEv(A, \mathsf{SendA} \ k \ L, m) = Send \ Tx_A^k \ m \ L$$
$$translateEv(A, \mathsf{ClaimA} \ , m) = Claim \ A \ m$$

A protocol *step* is therefore of type *agent* $\times$ *trace* $\times$ *real* $\rightarrow$ (*action* $\times$ *msg*) *set*. Since our protocol rule PROTO (described below) is parameterized by the protocol, we define a locale PROTOCOL that defines a constant *proto* of type *step set* and inductively define *Tr* in the context of this locale.

Since the actions of an agent A only depend on his own previous actions and observations, we define A's view of a trace tr as the projection of tr on those events involving A. For this purpose, we introduce the function *occursAt*, which maps events to associated agents, e.g. $occursAt(Send \ Tx_A^i \ m \ L) = A$.

$$view(A, tr) = [(ctime(A, t), ev) \ |(t, ev) \in tr \wedge occursAt(ev) = A]$$

Since the time-stamps of trace events refer to absolute time, the *view* function accounts for the offset of A's clock by translating times using the *ctime* function. Given an agent and an absolute time-stamp, the uninterpreted function *ctime* : *agent* $\times$ *real* $\rightarrow$ *real* returns the corresponding time-stamp for the agent's clock.

Using the above definitions, we define the PROTO-rule in Figure 3. For a given protocol, specified as a set of the step functions, the PROTO rule describes all possible actions of honest agents, given their local views of a valid trace tr at a given time t. If all premises are met, the PROTO-rule appends the translated event to the trace. Note that agents' behavior, modeled by the function *step*, is based only on the local clocks of the agents, i.e., agents cannot access the global time. Moreover, the restriction that all messages must be in $dm_{H_A}(knows_{H_A}(tr))$ ensures that agents only send messages derivable from their knowledge.

3.5   Protocol-Independent Results

The set of all possible traces Tr is inductively defined by the rules NIL, NET, FAKE, and PROTO in the context of the MESSAGE_THEORY, INITKNOWS, and PROTOCOL locales. To verify a concrete protocol, we instantiate these locales thereby defining the concrete set of traces for the given protocol, initial knowledge, and message theory. Additional requirements are specified by defining new locales that extend PROTOCOL and INITKNOWS.

Our first lemma specifies a lower bound on the time between when an agent first uses a nonce and another agent later uses the same nonce. The lemma holds whenever the initial knowledge of all agents does not contain any nonces.

Lemma 3.1. *Let A be an arbitrary (honest or dishonest) agent, N an arbitrary nonce, and $(t_A^S, Send \ \ Tx_A^i \ m_A \ L_A)$ the first event in a trace tr where $N \in$ subterms $\{m_A\}$. If tr contains an event $(t, Send \ \ Tx_B^j \ m_B \ L_B)$ or $(t, Recv \ Rx_B^j \ m_B)$ where $A \neq B$ and $N \in$ subterms $\{m_B\}$, then $t - t_A^S \geq cdist_{LoS}(A, B)$.*

Our next lemma holds whenever agents' keys are not *parts* of protocol messages and concerns when MACs can be created. Note that we need the notion of extractable subterms here since protocols use keys in MACs, but never send them in extractable positions.

Lemma 3.2. *Let A and B be honest agents and C a different possibly dishonest agent. Furthermore let $(t_C^S, Send \ \ Tx_C^i \ m_C \ L_C)$ be an event in the trace tr where $MAC_{K_{AB}}(m) \in$ subterms $\{m_C\}$ for some message m and a shared secret key K_{AB}. Then, for E either equal to A or B, there is a send event $(t_E^S, Send \ \ Tx_E^j \ m_E \ L_E) \in tr$ where $MAC_{K_{AB}}(m) \in$ subterms $\{m_E\}$ and $t_C^S - t_E^S \geq cdist_{LoS}(E, C)$.*

Note that the lemmas are similar to the axioms presented in [4]. The proofs of these lemmas can be found in our Isabelle/HOL formalization [14].

3.6 XOR Message Theory

In this section, we present a message theory including XOR, which instantiates the message-theory locale introduced in Section 3.2.

The Free Message Type. We first define the free term algebra of messages. Messages are built from agent names, integers, reals, nonces, keys, hashes, pairs, encryption, exclusive-or, and zero.

$$
\begin{aligned}
\textbf{datatype } fmsg = \ &\mathsf{AGENT} \ agent \mid \mathsf{INT} \ int \ \mid \mathsf{REAL} \ real \\
&\mid \mathsf{NONCE} \ agent \ nat \ \mid \mathsf{KEY} \ key \ \mid \mathsf{HASH} \ fmsg \\
&\mid \mathsf{MPAIR} \ fmsg \ fmsg \ \mid \mathsf{CRYPT} \ key \ fmsg \\
&\mid fmsg \ \bar{\oplus} \ fmsg \ \mid \mathsf{ZERO}
\end{aligned}
$$

To faithfully model $\bar{\oplus}$, we require the following set of equations E:

$$(x \bar{\oplus} y) \bar{\oplus} z \approx x \bar{\oplus} (y \bar{\oplus} z) \ \ \text{(A)} \qquad\qquad x \bar{\oplus} y \approx y \bar{\oplus} x \ \ \text{(C)}$$
$$x \bar{\oplus} \mathsf{ZERO} \approx x \ \ \text{(U)} \qquad\qquad x \bar{\oplus} x \approx \mathsf{ZERO} \ \ \text{(N)}$$

We define the corresponding equivalence relation $=_E$ as the reflexive, symmetric, transitive, and congruent closure of E. We also define the reduction relation $\rightarrow_E$ as the reflexive, transitive, and congruent closure of E, where the cancellation rules (U) and (N) are directed from left to right and (A) and (C) can be used in both directions. Note that $x \rightarrow_E y$ implies $x =_E y$, for all x and y.

$$\frac{}{reduced\ (\mathsf{AGENT}\ \ a)}\ \text{AGENT} \qquad \frac{}{reduced\ (\mathsf{INT}\ \ i)}\ \text{INT} \qquad \frac{}{reduced\ (\mathsf{REAL}\ \ i)}\ \text{REAL}$$

$$\frac{}{reduced\ (\mathsf{NONCE}\ \ a\ na)}\ \text{NONCE} \qquad \frac{reduced\ h}{reduced\ (\mathsf{HASH}\ \ h)}\ \text{HASH}$$

$$\frac{reduced\ a \quad reduced\ b}{reduced\ (\mathsf{MPAIR}\ \ a\ b)}\ \text{MPAIR} \qquad \frac{reduced\ m}{reduced\ (\mathsf{CRYPT}\ \ k\ m)}\ \text{CRYPT}$$

$$\frac{reduced\ a \quad reduced\ b \quad standard\ a \quad a < first\ b \quad b \neq \mathsf{ZERO}}{reduced\ (a\ \bar{\oplus}\ b)}\ \text{XOR}$$

Fig. 4. Rules for *reduced*

Reduced Messages and the Reduction Function. We define the predicate *standard* on *fmsg* that returns true for all messages where the outermost constructor is neither equal to $\bar{\oplus}$ nor *ZERO*. We also define the projection function *first*, where *first* x equals a when $x = a\ \bar{\oplus}\ b$ for some a and b and equals x otherwise. We use both functions to define *reduced* messages. We show below that every equivalence class with respect to $=_E$ contains exactly one *reduced* message, used as the classes canonical representative. To handle commutativity, we define a linear order on *fmsg* using the underlying orders on *nat*, *int*, and *agent*. A message is *reduced* if $\bar{\oplus}$ messages are right-associated, ordered, and all cancellation rules have been applied. This is captured by the inductive definition in Figure 4.

To obtain a decision procedure for $x =_E y$, we define a reduction function $\downarrow$ on *fmsg* that reduces a message, that is $x\downarrow$ is reduced and $(x\downarrow) =_E x$. We begin with the definition of an auxiliary function $\oplus^?$: *fmsg* $\rightarrow$ *fmsg*: $a \oplus^? b =$ if $b =$ ZERO then a else $a\ \bar{\oplus}\ b$.

We now define the main part of the reduction: the function $\oplus^\downarrow$: *fmsg* $\rightarrow$ *fmsg* $\rightarrow$ *fmsg* presented in Figure 5 combines two reduced messages a and b, yielding $(a\ \bar{\oplus}\ b)\downarrow$. Note that the order of the equations is relevant: given overlapping patterns, the first applicable equation is used. The algorithm is similar to a merge-sort on lists. The first two cases are straightforward and correspond to the application of the (U) rule. The other cases are justified by combinations of all four rules.

The definition of $(\cdot)\downarrow$: *fmsg* $\rightarrow$ *fmsg* is straightforward:

$$
\begin{aligned}
(\mathsf{HASH}\ \ m)\downarrow &= \mathsf{HASH}\ \ m\downarrow \\
(\mathsf{MPAIR}\ \ a\ b)\downarrow &= \mathsf{MPAIR}\ \ (a\downarrow)\ (b\downarrow) \\
(\mathsf{CRYPT}\ \ k\ m)\downarrow &= \mathsf{CRYPT}\ \ k\ (m\downarrow) \\
(a\ \bar{\oplus}\ b)\downarrow &= (a\downarrow) \oplus^\downarrow (b\downarrow) \\
x\downarrow &= x
\end{aligned}
$$

$$x \oplus^{\downarrow} \text{ZERO} \qquad\qquad = x \tag{1}$$

$$\text{ZERO} \oplus^{\downarrow} x \qquad\qquad = x \tag{2}$$

$$(a_1 \bar{\oplus} a_2) \oplus^{\downarrow} (b_1 \bar{\oplus} b_2) = \text{if } a_1 = b_1 \text{ then } a_2 \oplus^{\downarrow} b_2 \tag{3}$$

$$\text{else if } a_1 < b_1 \text{ then } a_1 \oplus^{?} (a_2 \oplus^{\downarrow} (b_1 \bar{\oplus} b_2)) \tag{4}$$

$$\text{else } b_1 \oplus^{?} ((a_1 \bar{\oplus} a_2) \oplus^{\downarrow} b_2) \tag{5}$$

$$(a_1 \bar{\oplus} a_2) \oplus^{\downarrow} b \qquad = \text{if } a_1 = b \text{ then } a_2 \tag{6}$$

$$\text{else if } a_1 < b \text{ then } a_1 \oplus^{?} (a_2 \oplus^{\downarrow} b) \tag{7}$$

$$\text{else } b \bar{\oplus} (a_1 \bar{\oplus} a_2) \tag{8}$$

$$a \oplus^{\downarrow} (b_1 \bar{\oplus} b_2) \qquad = (b_1 \bar{\oplus} b_2) \oplus^{\downarrow} a \tag{9}$$

$$a \oplus^{\downarrow} b \qquad\qquad = \text{if } a = b \text{ then } \text{ZERO} \tag{10}$$

$$\text{else if } a < b \text{ then } a \bar{\oplus} b \text{ else } b \bar{\oplus} a \tag{11}$$

Fig. 5. Definition of $\oplus^{\downarrow}$

We have proved the following facts about reduction: (1) if *reduced* x then $(x\downarrow) = x$, (2) *reduced* $(x\downarrow)$, and (3) $x \to_E (x\downarrow)$. Using these facts we establish:

Lemma 3.3. *For all messages x and y, $x =_E y$ iff (if and only if) $(x\downarrow) = (y\downarrow)$. Furthermore, if reduced x and reduced y, then $x =_E y$ iff $x = y$.*

The Message Type, Parts, and dm. Given the above lemma, we use the function $\downarrow$ and the predicate *reduced* to characterize $=_E$. Isabelle's *typedef* mechanism allows us to define the quotient type *msg* with $\{m \mid reduced\ m\}$ as the representing set. This defines a new type *msg* with a bijection between the representing set in *fmsg* and *msg* given by the function $Abs\_msg : fmsg \to msg$ and its inverse $Rep\_msg : msg \to fmsg$. Note that $=_E$ on *fmsg* corresponds to object-logic equality on *msg*. This is reflected in the following lemma.

Lemma 3.4. *For all messages x and y, $x = y$ iff $Rep\_msg(x) =_E Rep\_msg(y)$.*

We define functions on *msg* by using the corresponding definitions on *fmsg* and the embedding and projection functions. That is, we lift the message constructors to *msg* using the $\downarrow$ function. For example:

$$Nonce\ a\ n = Abs\_msg(\text{NONCE } a\ n)$$
$$MPair\ a\ b = Abs\_msg(\text{MPAIR } (Rep\_msg(a))\ (Rep\_msg(b)))$$
$$Hash\ m \quad = Abs\_msg(\text{HASH } (Rep\_msg(m)))$$
$$Xor\ a\ b \quad = Abs\_msg((Rep\_msg(a) \bar{\oplus} Rep\_msg(b))\downarrow)$$
$$Zero \quad = Abs\_msg(\text{ZERO })$$

In the following, we write 0 for *Zero* and $x \oplus y$ for *Xor* x y. We define a function *fparts* on *fmsg* that returns all extractable subterms of a given message, e.g. $m \in fparts(\{CRYPT\ k\ m\})$, but $m \notin fparts(\{HASH\ m\})$. The function *parts*

$$\frac{m \in M}{m \in dm_A(M)} \text{ INJ} \qquad \frac{}{0 \in dm_A(M)} \text{ ZERO} \qquad \frac{m \in dm_A(M)}{Hash\ m \in dm_A(M)} \text{ HASH}$$

$$\frac{\langle m, n \rangle \in dm_A(M)}{m \in dm_A(M)} \text{ FST} \qquad \frac{\langle m, n \rangle \in dm_A(M)}{n \in dm_A(M)} \text{ SND}$$

$$\frac{m \in dm_A(M) \qquad n \in dm_A(M)}{\langle m, n \rangle \in dm_A(M)} \text{ PAIR} \qquad \frac{m \in dm_A(M) \qquad n \in dm_A(M)}{m \oplus n \in dm_A(M)} \text{ XOR}$$

$$\frac{m \in dm_A(M) \qquad Key\ k \in DM_A(M)}{\{m\}_k \in dm_A(M)} \text{ ENC} \qquad \frac{}{Nonce\ A\ n \in dm_A(M)} \text{ NONCE}$$

$$\frac{\{m\}_k \in dm_A(M) \qquad (Key\ k)^{-1} \in dm_A(M)}{m \in dm_A(M)} \text{ DEC} \qquad \frac{}{Agent\ a \in dm_A(M)} \text{ AGENT}$$

$$\frac{}{Int\ n \in dm_A(M)} \text{ INT} \qquad \frac{}{Real\ n \in dm_A(M)} \text{ REAL}$$

Fig. 6. Rules for $dm_A(M)$

on *msg* that is used to instantiate the function of the same name in the message-theory locale is then defined as

$$parts(H) = \{ Abs\_msg(m) \mid m \in fparts\ \{Rep\_msg(x) \mid x \in H\}\}.$$

This defines the *parts* of a message m in the equivalence class represented by $m{\downarrow}$ as the *fparts* of $m{\downarrow}$. For example $parts(\{X \oplus X\}) = \{0\}$. The function *subterms* is defined similarly, but returns all subterms and not just the extractable ones. We give the rules for the inductively-defined message-derivation operator $dm : agent \rightarrow msg\ set \rightarrow msg\ set$ in Figure 6. The rules specify message decryption, projection on pairs, pairing, encryption, signing, hashing, *XOR*ing, and the generation of integers, reals, agent names, and nonces. For example, the *Dec*-rule states that if an agent A can derive the ciphertext $\{m\}_k$ and the decryption key $(Key\ k)^{-1}$, then he can also derive the plaintext m. When $Key\ k$ is used as a signing key, A uses the verification key $(Key\ k)^{-1}$ to verify the signature. The XOR rule uses the constructor *Xor*, which ensures that the resulting message is reduced. We can now interpret the locale MESSAGE_THEORY by proving that the defined operators have the required properties.

4   Protocol Correctness

In this section, we present highlights from our verification of instances of the protocol pattern in Figure 1. Complete details are provided in [14]. We first formalize the protocol pattern and its desired security properties. Afterwards, we reduce the security of pattern instances to properties of the rapid-response function used. Finally, we consider several concrete rapid-response functions proposed in [4] and analyze the security of the corresponding instantiations.

4.1 Protocol Rules

We conduct our security analysis using our XOR message theory. We first define the concrete initial knowledge as $initKnows_A = \bigcup_B \{Key\, K_{AB}\}$, which we interpret as an instance of the INITKNOWS_SHARED locale. Next, we instantiate the PROTOCOL locale by defining the protocol pattern in Figure 1 as $proto = \{mdb_1, mdb_2, mdb_3, mdb_4\}$. Each step function $mdb_i(A, tr, t)$ yields the possible actions of the agent A executing the protocol step i with his view of the trace tr at the local time t.

Our distance bounding protocol definition uses the uninterpreted rapid-response function $F : msg \times msg \times msg \rightarrow msg$ and consists of the following steps.[2]

Start: An honest verifier V can start a protocol run by sending a fresh nonce using his radio transmitter r at any local time t.

$$\frac{Nonce\, V\, NV \notin used(tr)}{(SendA\ r\ [\,],\, Nonce\, V\, NV) \in mdb_1(V, tr, t)}$$

Rapid-Response: If a prover P receives a nonce NV, he may continue the protocol by replying with the message $F(NV, NP, P)$ and storing the protocol data $[NV, Nonce\, P\, NP]$. This information must be stored since it is needed in the authentication step and cannot, in general, be reconstructed from $F(NV, NP, P)$.

$$\frac{(t_P^R, Recv\ Rx_P^r\ NV) \in tr \qquad Nonce\, P\, NP \notin used(tr)}{(SendA\ r\ [NV, Nonce\, P\, NP],\, F(NV, Nonce\, P\, NP, Agent\, P)) \in mdb_2(P, tr, t)}$$

Authentication: After a prover P has answered a verifier's challenge with a rapid-response, he authenticates the response with the corresponding MAC.

$$\frac{(t_P^R, Recv\ Rx_P^r\ NV) \in tr \quad (t_P^S, Send\ Tx_P^r\ F(NV, Nonce\, P\, NP, Agent\, P)\ [NV, Nonce\, P\, NP]) \in tr}{(SendA\ r\ [\,],\, MACM_{K_{VP}}(NV, Nonce\, P\, NP, Agent\ P)) \in mdb_3(P, tr, t)}$$

Claim: Suppose the verifier receives a rapid-response in the measurement phase at time t_1^R and the corresponding MAC in the validation phase, both involving the nonce that he initially sent at time t_1^S. The verifier therefore concludes that $(t_1^R - t_1^S) * c/2$ is an upper bound on the distance to the prover P, where c denotes the speed of light.

$$\frac{\begin{array}{c}(t_1^S, Send\ Tx_V^r\ (Nonce\, V\, NV)\ [\,]) \in tr \\ (t_1^R, Recv\ (Rx_V^r)\ F(Nonce\, V\, NV, NP, Agent\ P)) \in tr \\ (t_2^R, Recv\ Rx_V^r\ MACM_{K_{VP}}(Nonce\, V\, NV, NP, Agent\, P)) \in tr\end{array}}{(ClaimA, (Agent\ P, Real\ (t_1^R - t_1^S) * c/2)) \in mdb_4(V, tr, t)}$$

---

[2] We have formalized each step in Isabelle/HOL using set comprehension, but present the steps here as rules for readability. For each rule r, the set we define by comprehension is equivalent to the set defined inductively by the rule r.

The set of traces Tr is inductively defined by the rules NIL, FAKE, NET, and PROTO. Note that the same set of traces can be inductively defined by the NIL, FAKE, and NET rules along with rules describing the individual protocol steps. See [6] for more details on these different representations.

4.2 Security Properties

In this section, we present properties of distance bounding protocols instantiating the protocol pattern from Figure 1. First, we note that we only consider honest verifiers. Since successful protocol execution leads to claims on the verifier's side, it makes no sense to consider dishonest verifiers. However, we distinguish between honest and dishonest provers. For honest provers, we require that the claimed distance after a successful protocol execution denoted by a *Claim* event is an upper bound on the physical distance between the prover and the verifier.

Definition 4.1. *A distance bounding protocol is secure for honest provers (hp-secure) iff whenever Claim V $\langle P, Real\,d \rangle \in tr$, then $d \geq |\,loc_V - loc_P\,|$.*

In the case of a dishonest prover, it is impossible to distinguish between different intruders who exchange their keys. Hence, our weaker property must accommodate for attacks where one intruder pretends to be another intruder. We therefore require that the claimed distance is an upper bound on the distance between the verifier and some intruder.

Definition 4.2. *A distance bounding protocol is secure for dishonest provers (dp-secure) iff whenever Claim V $\langle P, Real\,d \rangle \in tr$ for an intruder P, then there is an intruder P' such that $d \geq |\,loc_V - loc_{P'}\,|$.*

4.3 Security Proofs Based on Properties of F

In order to prove security properties of an instance of the protocol pattern, we show how security properties of the protocol can be reduced to properties of F. To prove the security of a concrete instantiation, we must then only prove that it has the required properties.

Definition 4.3. *In the following, let X, Y, and Z, be messages, m an atomic message or a MAC, A, B, and C, agents, and NA, NB, and NC nonce tags. We define the following properties for a function $F : msg \times msg \times msg \rightarrow msg$:*

(P0)
    (a) If $X \in H$, then $F(X, Nonce\,A\,NA, Agent\,A) \in dm_A\,H$.
    (b) If $m \in parts(\{F(X, Y, Z)\})$, then $m \in parts(\{X, Y, Z\})$.
    (c) $F(X, Nonce\,A\,NA, Agent\,A) \neq Nonce\,C\,NC$.
    (d) $F(X, Nonce\,A\,NA, Agent\,A) \neq MACM_{K_{BC}}(Y, Nonce\,C\,NC, Agent\,C)$.
(P1) *Nonce A $NA \in subterms(F(Nonce\,A\,NA, X, Agent\,B))$*
(P2) *Nonce B $NB \in subterms(F(X, Nonce\,B\,NB, Agent\,B))$*
(P3) *Agent $B \in subterms(F(Nonce\,A\,NA, X, Agent\,B))$*

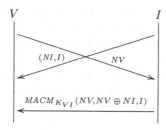

Fig. 7. Jumping the Gun Attack

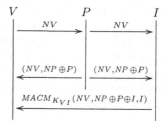

Fig. 8. Impersonation Attack

The property (P0) specifies well-formedness conditions on F: (a) F can be computed from the challenge, (b) F neither introduces new atomic messages nor MACs, which are not among its arguments, and (c)–(d) F cannot be confused with the other protocol messages. Properties (P1)–(P3) state that arguments in certain positions are always subterms of F's result as long as the remaining arguments have the required types. Using these properties, we prove the following lemmas that give sufficient conditions on F for *hp-security* and *dp-security*.

Theorem 4.4. *For every function F with properties (P0)–(P2), the resulting instance of the protocol pattern is an hp-secure distance bounding protocol.*

In the proof of Theorem 4.4, (P1) is used to ensure that the nonce sent as a challenge by the verifier must be involved in the computation of the prover's response. Analogously, (P2) ensures that the response can only be computed if the fresh nonce contributed by the prover is known.

For the case of a dishonest prover, we additionally require (P3) to prevent a dishonest prover from taking credit for a response sent by an honest one.

Theorem 4.5. *For every function F with properties (P0)–(P3), the resulting instance of the protocol pattern is a dp-secure distance bounding protocol.*

We have proved that $\langle NV \oplus P, NP \rangle$ and $\langle NV, NP, P \rangle$ have properties (P0)–(P3) and therefore the corresponding protocol instances are both *hp-secure* and *dp-secure*.

The function $F_1(NV, NP, P) = \langle NV \oplus NP, P \rangle$ lacks (P1) and is not *dp-secure*. To see that (P1) fails, consider $F_1(NV, NV, P) = \langle 0, P \rangle$, which does not contain NV as a subterm. Remember that we have defined the subterms of a message t as the subterms of $t{\downarrow}$. A dishonest prover I can use this to execute the "jumping the gun" attack given in Figure 7. The attack uses the equivalence $F_1(NV, NV \oplus NI, I) = \langle NV \oplus(NV \oplus NI), I \rangle = \langle NI, I \rangle$.

In contrast, the function $F_2(NV, NP, P) = \langle NV, NP \oplus P \rangle$ lacks property (P3) and is therefore *hp-secure* but not *dp-secure*. To see that (P3) fails, consider $F_2(NV, NI \oplus Agent\ P, Agent\ P) = \langle NV, NI \rangle$, which does not contain $Agent\ P$ as a subterm. This leads to the impersonation attack depicted in Figure 8 violating the *dp-security* property. This attack uses the equivalence $F_2(NV, NP, P) = \langle NV, NP \oplus P \rangle = F_2(NV, NP \oplus I \oplus P, I)$.

Overall, proving the properties (P0)–(P4) for a given function and applying Theorems 4.4 and 4.5 is much simpler than the corresponding direct proofs. However, finding the correct properties and proving these theorems for the XOR message theory turned out to be considerably harder than proofs for comparable theorems about a fixed protocol in the free message theory. This additional complexity mainly stems from the untyped protocol formalization necessary to realistically model the XOR operator.

5   Related Work and Conclusions

Our model of physical security protocols extends the Dolev-Yao model with dense time, network topology, and node location. Each of these properties has been handled individually in other approaches. For example, discrete and dense time have been used to reason about security protocols involving timestamps or timing issues like timeouts and retransmissions [15,16]. Models encompassing network topology have been studied in the context of secure routing protocols for ad hoc networks [17,18]. Node location and relative distance have been considered in [4,5]. In [6], we compare our work with these approaches in more detail. While these approaches address individual physical properties, to the best of our knowledge our model is the first that provides a general foundation for reasoning about all three of these properties and their interplay.

The protocol pattern we study comes from Meadows et al. [4]. The authors give a condition on instances of F (called "simply stable") under which the resulting protocol is correct in the case of honest provers. They also investigate the two function instances F_1 and F_2 that we presented above. They give the attack on F_1. However, as they do not consider the possibility of dishonest provers in their proof, they classify F_2 as secure. Their correctness proofs are based on a specialized authentication logic, tailored for distance-bounding protocols, that is presented axiomatically. While they do not provide a semantics, we note that their axioms can be suitably interpreted and derived within our setting.

From the specification side, our work builds on several strands of research. The first is the modeling of security protocols as inductively-defined sets of traces. Our work is not only inspired by Paulson's inductive method [7], we were able to reuse some of his theories, in particular his key theory and much of his free message algebra.

The second strand is research on formalizing equational theories in theorem provers. Courant and Monin [19] formalize a message algebra including XOR in Coq, which they use to verify security APIs. They introduce an uninterpreted normalization function with axiomatically-defined properties, which in turn is used to define their equivalence relation. Since they do not use a quotient type to represent equivalence classes, they must account for different representations of equivalent messages. Paulson's work on defining functions on equivalence classes [20] uses a quotient type construction in Isabelle/HOL that is similar to ours, but represents equivalence classes as sets instead of canonical elements.

The final strand concerns developing reusable results by proving theorems about *families* of inductively-defined sets. In our work, we give generic formalizations of sets of messages and traces, including lemmas, which hold for different instances. The key idea is to use one parameterized protocol rule, instead of a collection of individual rules, for the protocol steps. The inductive definition is then packaged in a locale, where the locale parameter is the rule parameter, and locale assumptions formalize constraints on the protocol steps (such as well-formedness). Note that this idea is related to earlier work on structuring (meta)theory [21,22,23] using parameterized inductively-defined sets, where the theorems themselves directly formalize the families of sets. Overall, our work constitutes a substantial case study in using locales to structure reusable theories about protocols. Another case study, in the domain of linear arithmetic, is that of [24].

In conclusion, our model has enabled us to formalize protocols, security properties, and environmental assumptions that are not amenable to formal analysis using other existing approaches. As future work, we plan to extend our model to capture additional properties of wireless security protocols. We also intend to refine our model to capture message sizes and transmission rate, rapid bit exchange, and online guessing attacks, which would allow us to analyze protocols such as those presented in [1].

References

1. Brands, S., Chaum, D.: Distance-bounding protocols. In: Helleseth, T. (ed.) EUROCRYPT 1993. LNCS, vol. 765, pp. 344–359. Springer, Heidelberg (1994)
2. Capkun, S., Buttyan, L., Hubaux, J.P.: SECTOR: secure tracking of node encounters in multi-hop wireless networks. In: SASN 2003: Proceedings of the 1st ACM Workshop on Security of Ad Hoc and Sensor Networks, pp. 21–32. ACM Press, New York (2003)
3. Hancke, G.P., Kuhn, M.G.: An RFID distance bounding protocol. In: SECURECOMM 2005: Proceedings of the 1st International Conference on Security and Privacy for Emerging Areas in Communications Networks, Washington, DC, USA, pp. 67–73. IEEE Computer Society, Los Alamitos (2005)
4. Meadows, C., Poovendran, R., Pavlovic, D., Chang, L., Syverson, P.: Distance bounding protocols: Authentication logic analysis and collusion attacks. In: Secure Localization and Time Synchronization for Wireless Sensor and Ad Hoc Networks, pp. 279–298. Springer, Heidelberg (2006)
5. Sastry, N., Shankar, U., Wagner, D.: Secure verification of location claims. In: WiSe 2003: Proceedings of the 2003 ACM workshop on Wireless security, pp. 1–10. ACM Press, New York (2003)
6. Schaller, P., Schmidt, B., Basin, D., Capkun, S.: Modeling and verifying physical properties of security protocols for wireless networks. In: CSF-22: 22nd IEEE Computer Security Foundations Symposium (to appear, 2009)
7. Paulson, L.C.: The inductive approach to verifying cryptographic protocols. Journal of Computer Security 6, 85–128 (1998)
8. Nipkow, T., Paulson, L., Wenzel, M.: Isabelle/HOL. LNCS, vol. 2283. Springer, Heidelberg (2002)
9. Capkun, S., Hubaux, J.P.: Secure positioning of wireless devices with application to sensor networks. In: INFOCOM, pp. 1917–1928. IEEE, Los Alamitos (2005)

10. Perrig, A., Tygar, J.D.: Secure Broadcast Communication in Wired and Wireless Networks. Kluwer Academic Publishers, Norwell (2002)
11. Ballarin, C.: Interpretation of locales in Isabelle: Theories and proof contexts. In: Borwein, J.M., Farmer, W.M. (eds.) MKM 2006. LNCS (LNAI), vol. 4108, pp. 31–43. Springer, Heidelberg (2006)
12. Porter, B.: Cauchy's mean theorem and the cauchy-schwarz inequality. The Archive of Formal Proofs, Formal proof development (March 2006)
13. Clulow, J., Hancke, G.P., Kuhn, M.G., Moore, T.: So near and yet so far: Distance-bounding attacks in wireless networks. In: Buttyán, L., Gligor, V.D., Westhoff, D. (eds.) ESAS 2006. LNCS, vol. 4357, pp. 83–97. Springer, Heidelberg (2006)
14. Schmidt, B., Schaller, P.: Isabelle Theory Files: Modeling and Verifying Physical Properties of Security Protocols for Wireless Networks, http://people.inf.ethz.ch/benschmi/ProtoVeriPhy/
15. Delzanno, G., Ganty, P.: Automatic Verification of Time Sensitive Cryptographic Protocols. In: Jensen, K., Podelski, A. (eds.) TACAS 2004. LNCS, vol. 2988, pp. 342–356. Springer, Heidelberg (2004)
16. Evans, N., Schneider, S.: Analysing Time Dependent Security Properties in CSP Using PVS. In: Cuppens, F., Deswarte, Y., Gollmann, D., Waidner, M. (eds.) ESORICS 2000. LNCS, vol. 1895, pp. 222–237. Springer, Heidelberg (2000)
17. Acs, G., Buttyan, L., Vajda, I.: Provably Secure On-Demand Source Routing in Mobile Ad Hoc Networks. IEEE Transactions on Mobile Computing 5(11), 1533–1546 (2006)
18. Yang, S., Baras, J.S.: Modeling vulnerabilities of ad hoc routing protocols. In: SASN 2003: Proceedings of the 1st ACM Workshop on Security of Ad Hoc and Sensor Networks, pp. 12–20. ACM, New York (2003)
19. Courant, J., Monin, J.: Defending the bank with a proof assistant. In: Proceedings of the 6th International Workshop on Issues in the Theory of Security (WITS 2006), pp. 87–98 (2006)
20. Paulson, L.: Defining functions on equivalence classes. ACM Transactions on Computational Logic 7(4), 658–675 (2006)
21. Basin, D., Constable, R.: Metalogical frameworks. In: Huet, G., Plotkin, G. (eds.) Logical Environments, pp. 1–29. Cambridge University Press, Cambridge (1993); Also available as Technical Report MPI-I-92-205
22. Basin, D., Matthews, S.: Logical frameworks. In: Gabbay, D., Guenthner, F. (eds.) Handbook of Philosophical Logic, 2nd edn., vol. 9, pp. 89–164. Kluwer Academic Publishers, Dordrecht (2002)
23. Basin, D., Matthews, S.: Structuring metatheory on inductive definitions. Information and Computation 162(1–2) (October/November 2000)
24. Nipkow, T.: Reflecting quantifier elimination for linear arithmetic. Formal Logical Methods for System Security and Correctness, 245 (2008)

VCC: A Practical System
for Verifying Concurrent C

Ernie Cohen[1], Markus Dahlweid[2], Mark Hillebrand[3], Dirk Leinenbach[3],
Michał Moskal[2], Thomas Santen[2], Wolfram Schulte[4], and Stephan Tobies[2]

[1] Microsoft Corporation, Redmond, WA, USA
ernie.cohen@microsoft.com
[2] European Microsoft Innovation Center, Aachen, Germany
{markus.dahlweid,michal.moskal,thomas.santen,
stephan.tobies}@microsoft.com
[3] German Research Center for Artificial Intelligence (DFKI), Saarbrücken, Germany
{mah,dirk.leinenbach}@dfki.de
[4] Microsoft Research, Redmond, WA, USA
schulte@microsoft.com

Abstract. VCC is an industrial-strength verification environment for
low-level concurrent system code written in C. VCC takes a program
(annotated with function contracts, state assertions, and type invariants)
and attempts to prove the correctness of these annotations. It includes
tools for monitoring proof attempts and constructing partial counterexample executions for failed proofs. This paper motivates VCC, describes
our verification methodology, describes the architecture of VCC, and
reports on our experience using VCC to verify the Microsoft Hyper-V
hypervisor.[1]

1 Introduction

The mission of the Hypervisor Verification Project (part of Verisoft XT [1]) is to
develop an industrially viable tool-supported process for the sound verification
of functional correctness properties of commercial, off-the-shelf, system software,
and to use this process to verify the Microsoft Hyper-V hypervisor. In this paper,
we describe the proof methodology and tools developed in pursuit of this mission.

Our methodology and tool design has been driven by the following challenges:

Reasoning Engine. In an industrial process, developers and testers must drive
the verification process (even if more specialized verification engineers architect
global aspects of the verification, such as invariants on types). Thus, verification should be primarily driven by assertions stated at the level of code itself,
rather than by guidance provided to interactive theorem provers. The need for
mostly automatic reasoning led us to generate verification conditions that could

---

[1] Work partially funded by the German Federal Ministry of Education and Research
(BMBF) in the framework of the Verisoft XT project under grant 01 IS 07 008.

S. Berghofer et al. (Eds.): TPHOLs 2009, LNCS 5674, pp. 23–42, 2009.

be discharged automatically by an SMT (first-order satisfiability modulo theories) solver. The determination to stick to first-order methods means that the only form of induction available is computational induction, which required developing methodological workarounds for inductive data structures.

Moreover, to allow users to understand failed verification attempts, we try whenever possible to reflect information from the underlying reasoning engine back to the level of the program. For example, countrexamples generated by failed proofs in the SMT solver are reflected to the user as (partial) counterexample traces by the VCC Model Viewer.

Weak Typing. Almost all critical system software today is written in C (or C++). C has only a weak, easily circumvented type system and explicit memory (de)allocation, so memory safety has to be explicitly verified. Moreover, address arithmetic enables many nasty programming tricks that are absent from typesafe code.

Still, most code in a well-written C system adheres to a much stricter type discipline. The VCC memory model [2] leverages this by maintaining in ghost memory a *typestate* that tracks where the "valid" typed memory objects are. On each memory reference and pointer dereference, there is an implicit assertion that resulting object is in the typestate. System invariants guarantee that valid objects do not overlap in any state, so valid objects behave like objects in a modern (typesafe) OO system. Well-behaved programs incur little additional annotation overhead, but nasty code (e.g., inside of the memory allocator) may require explicit manipulation of the typestate[2].

While C is flexible enough to be used in a very low-level way, we still want program annotations to take advantage of the meaningful structure provided by well-written code. Because C structs are commonly used to group semantically related data, we use them by default like objects in OO verification methodologies (e.g., as the container of invariants). Users can introduce additional (ghost) levels of structure to reflect additional semantic structure.

Concurrency. Most modern system software is concurrent. Indeed, the architecturally visible caching of modern processors means that even uniprocessor operating systems are effectively concurrent programs. Moreover, real system code makes heavy use of lock-free synchronization. However, typical modular and thread-local approaches to verifying concurrent programs (e.g., [3]) are based on locks or monitors, and forbid direct concurrent access to memory.

As in some other concurrency methodologies (e.g., [4]), we use an ownership discipline that allows a thread to perform sequential writes only to data that it owns, and sequential reads only to data that it owns or can prove is not changing. But in addition, we allow concurrent access to data that is marked as volatile in the typestate (using operations guaranteed by the compiler to be atomic on the

---

[2] Our slogan is "It's no harder to functionally verify a typesafe program written in an unsafe language than one written in a safe language." Thus, verification actually makes unsafe languages more attractive.

given platform), leveraging the observation that a correct concurrent program typically can only race on volatile data (to prevent an optimizing compiler from changing the program behavior). Volatile updates are required to preserve invariants but are otherwise unconstrained, and such updates do not have to be reported in the framing of a function specification.

Cross-Object Invariants. A challenge in modular verification is to how to make sure that updates don't break invariants that are out of scope. This is usually accomplished by restricting invariants to data within the object, or mild relaxations based on the ownership hierarchy. However, sophisticated implementations often require coordination between unrelated objects.

Instead, we allow invariants to mention arbitrary parts of the state. To keep modular verification sound, VCC checks that no object invariant can be broken by invariant-preserving changes to other objects. This *admissibility* check is done based on the type declarations alone.

Simulation. A typical way to prove properties of a concurrent data type is to show that it simulates some simpler type. In existing methodologies, simulation is typically expressed as a theorem about a program, e.g., by introducing an existentially quantified variable representing the simulated state. This is acceptable when verifying an abstract program (expressed, say, with a transition relation), but is awkward when updates are scattered throughout the codebase, and it violates our principle of keeping the annotations tightly integrated with the code.

Instead, we prove concurrent simulation in the code itself, by representing the abstract target with ghost state. The coupling invariant is expressed as an ordinary (single state) invariant linking the concrete and ghost state, and the specification of the simulated system is expressed with a two-state invariant on the ghost state. These invariants imply a (forward) simulation, with updates to the ghost state providing the needed existential witnesses.

Claims. Concurrent programs implicitly deal with chunks of knowledge about the state. For example, a program attempting to acquire a spin lock must "know" that the spin lock hasn't been destroyed. But such knowledge is ephemeral – it could be broken by any write that is not thread local – so passing knowledge to a function in the form of a precondition is too weak. Instead, we package the knowledge as the invariant of a ghost object; these knowledge-containing objects are called *claims*. Because claims are first-class objects, they can be passed in and out of functions and stored in data structures. They form a critical part of our verification methodology.

C and Assembly Code. System software requires interaction between C and assembly code. This involves subtleties such as the semantics of calls between C and assembly (in each direction), and consideration of which hardware resources (e.g., general purpose registers) can be silently manipulated by compiled C code. Assembly verification in VCC is discussed in [5].

Weak Memory Models. Concurrent program reasoning methods usually tacitly assume sequentially consistent memory. However, system software has to run on modern processors which, in a concurrent setting, do not provide an efficient implementation of sequentially consistent memory (primarily because of architecturally visible store buffering). Additional proof obligations are needed to guarantee that sequentially consistent reasoning is sound. We have developed a suitable set of conditions for x64 memory, but VCC does not yet enforce the corresponding verification conditions.

Content. Section 2 gives an overview of Hyper-V. Section 3 introduces the VCC methodology. Section 4 looks at the main components of VCC's tool suite. Section 5 reflects on the past year's experience on using VCC for verifying Hyper-V. Section 6 concludes with related work.

2   The Microsoft Hypervisor

The development of our verification environment is driven by the verification of the Microsoft Hyper-V hypervisor, which is an ongoing collaborative research project between the European Microsoft Innovation Center, Microsoft Research, the German Research Center for Artificial Intelligence, and Saarland University in the Verisoft XT project. The hypervisor is a relatively thin layer of software (100KLOC of C, 5KLOC of assembly) that runs directly on x64 hardware. The hypervisor turns a single real multi-processor x64 machine (with AMD SVM [6] or Intel VMX [7] virtualization extensions) into a number of virtual multi-processor x64 machines. (These virtual machines include additional machine instructions to create and manage other virtual machines.)

The hypervisor was not written with formal verification in mind. Verification requires substantial annotations to the code, but these annotations are structured so that they can be easily removed by macro preprocessing, so the annotated code can still be compiled by the standard C compiler. Our goal is that the annotations will eventually be integrated into the codebase and maintained by the software developers, evolving along with the code.

The hypervisor code consists of about 20 hierarchical layers, with essentially no up-calls except to pure functions. The functions and data of each layer is separated into public and private parts. These visibility properties are ensured statically using compiler and preprocessor hacks, but the soundness of the verification does not depend on these properties. These layers are divided into two strata. The lower layers form the *kernel stratum* which is a small multi-processor operating system, complete with hardware abstraction layer, kernel, memory manager, and scheduler (but no device drivers). The *virtualization stratum* runs in each thread an "application" that simulates an x64 machine without the virtualization features, but with some additional machine instructions, and running under an additional level of memory address translation (so that each machine can see 0-based, contiguous physical memory).

For the most part, a virtual machine is simulated by simply running the real hardware; the extra level of virtual address translation is accomplished by using *shadow page tables* (SPTs). The SPTs, along with the hardware *translation lookaside buffers* (TLBs) (which asynchronously gather and cache virtual to physical address translations), implement a virtual TLB. This simulation is subtle for two reasons. First, the hardware TLB is architecturally visible, because (1) translations are not automatically flushed in response to edits to page tables stored in memory, and (2) translations are gathered asynchronously and nonatomically (requiring multiple reads and writes to traverse the page tables), creating races with system code that operates on the page tables. Even the semantics of TLBs are subtle, and the hypervisor verification required constructing the first accurate formal models of the x86/x64 TLBs. Second, the TLB simulation is the most important factor in system performance; simple SPT algorithms, even with substantial optimization, can introduce virtualization overheads of 50% or more for some workloads. The hypervisor therefore uses a very large and complex SPT algorithm, with dozens of tricky optimizations, many of which leverage the freedoms allowed by the weak TLB semantics.

3   VCC Methodology

VCC extends C with annotations giving function pre- and post-conditions, assertions, type invariants, and ghost code. Many of these annotations are similar to those found in ESC/Java [8], Spec# [9], or Havoc [10]. With contracts in place, VCC performs a static modular analysis, in which each function is verified in isolation, using only the contracts of functions that it calls and invariants of types used in its code. But unlike the aforementioned systems, VCC is geared towards sound verification of functional properties of low-level concurrent C code.

We show VCC's use by specifying hypervisor partitions; the data structure which keeps state to execute a guest operating system. Listing 1 shows a much simplified but annotated definition of the data structure. (The actual struct has 98 fields.)

```
typedef enum { Undefined, Initialized, Active, Terminating } LifeState;

typedef struct _Partition {
  bool signaled;
  LifeState lifeState;
  invariant(lifeState == Initialized || lifeState == Active ||
            lifeState == Terminating)
  invariant(signaled ==> lifeState == Active)
} Partition;

void part_send_signal(Partition *part)
  requires(part->lifeState == Active)
  ensures(part->signaled)
  maintains(wrapped(part))
  writes(part)
{
  unwrap(part);
  part->signaled = 1;
  wrap(part);
}
```

Listing 1. Sequential partition

Function Contracts. Every function can have a specification, consisting of four kinds of clauses. Preconditions, introduced by **requires** clauses, declare under which condition the function is allowed to be called. Postconditions, introduced by **ensures** clauses, declare under which condition the function is allowed to return. A **maintains** clause combines a precondition followed by a postcondition with the same predicate. Frame conditions, introduced by **writes** clauses, limit the part of the program's state that the function is allowed to modify.

So part_send_signal of Listing 1 is allowed to be called if the actual parameter's lifeState is Active. When the function terminates it guarantees (1) that the formal parameter's signaled bit has been set and (2) that it modified at most the passed partition object. We will discuss the notion of a wrapped object, which is mentioned in the **maintains** clause, in Sect. 3.1.

Type Invariants. Type definitions can have type invariants, which are one- or two-state predicates on data. Other specifications can refer to the invariant of object o as **inv**(o) (or **inv2**(o)). VCC implicitly uses invariants at various locations, as will be explained in the following subsections.

The invariant of the *Partition* struct of Listing 1 says that lifeState must be one of the valid ones defined for a partition, and that if the signaled bit is set, lifeState is Active.

Ghosts. A crucial concept in the VCC specification language is the division into operational code and ghost code. Ghost code is seen only by the static verifier, not the regular compiler. Ghost code comes in various forms: *Ghost type definitions* are types, which can either be regular C types, or special types for verification purposes like maps and claims (see Sect. 3.2). *Ghost fields* of arbitrary type can be introduced as specially marked fields in operational types. These fields do not interfere with the size and ordering of the operational fields. Likewise, static or automatic *ghost variables* of arbitrary type are supported. Like ghost fields, they are marked special and do not interfere with operational variables. *Ghost parameters* of arbitrary type can pass additional ghost state information in and out of the called function. *Ghost state updates* perform operations on only the ghost memory state. Any flow of data from the ghost state to the operational state of the software is forbidden.

One application of ghost code is maintaining *shadow copies* of implementation data of the operational software. Shadow copies usually introduce abstractions, e.g., representing a list as a set. They are also introduced to allow for atomic update of the shadow, even if the underlying data structure cannot be updated atomically. The atomic updates are required to enforce protocols on the overall system using two-state invariants.

3.1   Verifying Sequential Programs

Ownership and Invariants. To deal with high-level aliasing, VCC implements a Spec#-style [9] ownership model: The heap is organized into a forest of tree

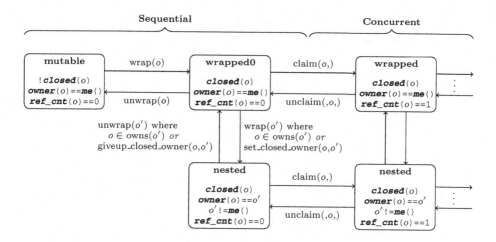

Fig. 1. Objects states, transitions, and access permissions

structures. The edges of the trees indicate ownership, that is, an aggregate / sub-object relation. The roots of trees in the ownership forest are objects representing threads of execution. The set of objects directly or transitively owned by an object is called the ownership domain of that object.

We couple ownership and type invariants. Intuitively, a type invariant can depend only on state in its ownership domain. We later relax this notion. Of course ownership relationships change over time and type invariants cannot always hold. We thus track the status for each object o in meta-state: **owner**(o) denotes the owner of an object, **owns**(o) specifies the set of objects owned by o (the methodology ensures that **owner**() and **owns**() stay in sync), **closed**(o) guarantees that o's invariant holds.

Figure 1 discusses the possible meta-states of an object (The *Concurrent* part of this figure will be explained in Sect. 3.2):

- **mutable**(o) holds if o is not closed and is owned by the current thread (henceforth called **me**). Allocated objects are always mutable and *fresh* (i.e., not previously present in the owns-set of the current thread).
- **wrapped**(o) holds if o is closed and owned by **me**. Non-volatile fields of wrapped objects cannot change. (**wrapped0**(o) abbreviates **wrapped**(o) and **ref_cnt**(o)==0).
- **nested**(o) holds if o is closed and owned by an object.

Ghost operations, like **wrap**, **unwrap**, etc. update the state as depicted in Fig. 1; note that unwrapping an object moves its owned object from nested to wrapped, wrapping the object moves them back.

The function `part_send_signal()` from Listing 1 respects this meta-state protocol. The function precondition requires `part` to be wrapped, i.e., `part`'s invariant holds. The function body first unwraps `part`, which suspends its

```
typedef struct vcc(dynamic_owns) _PartitionDB {
  Partition *partitions[MAXPART];
  invariant(forall(unsigned i; i < MAXPART;
                  partitions[i] != NULL ==> set_in(partitions[i], owns(this))))
} PartitionDB;

void db_send_signal(PartitionDB *db, unsigned idx)
  requires(idx < MAXPART)
  maintains(wrapped(db))
  ensures((db->partitions[idx] != NULL) && (db->partitions[idx]->lifeState == Active)
         ==> db->partitions[idx]->signaled)
  writes(db)
{
  unwrap(db);
  if ((db->partitions[idx] != NULL) && (db->partitions[idx]->lifeState == Active)) {
    part_send_signal(db->partitions[idx]);
  }
  wrap(db);
}
```

Listing 2. Sequential partition database

invariant, next its fields are written to. To establish the postcondition, part is wrapped again; at this point, VCC checks that all invariants of part hold.

The write clauses work accordingly: write access to the root of an ownership domain enables writing to the entire ownership domain. In our example, **writes**(part) gives the part_send_signal function write access to all fields of part (and the objects part owns), and tells the caller, that state updates are confined to the ownership domain of part. Additionally, one can always write to objects that are fresh.

Conditional Ownership. The actual hypervisor implementation does not use partition pointers but abstract partition identifiers to refer to partitions. This is because partitions can be created and destroyed anytime during the operation of the hypervisor, which might lead to dangling pointers. Listing 2 simulates the hypervisor solution: before any operation on a partition can take place, the pointer to the partition is retrieved from a partition database using the partition identifier. The *PartitionDB* type contains an array partitions of MAXPART entries of pointers to *Partition*. The index in partitions serves as the partition identifier. The partition database invariant states that all (non-null) elements of partitions are owned by the partition database.

The function db_send_signal() in Listing 2 attempts to send a signal to the partition with index idx of the partition database db. Is uses the function part_send_signal() from Listing 1, so we need to ensure that the preconditions of that function are met: part->lifeState is Active follows from the condition of the if-statement; **wrapped**(part) holds since part is contained in the owns-set of db; unwrapping db transitions the partition from nested into the wrapped state. It also makes the partition writable as it has not previously been owned by the current thread and thus is also considered fresh from the point of view of the current thread.

```
#define ISSET(n, v) (((v) & (1ULL << (n))) != 0)
typedef Partition *PPartition;

typedef struct vcc(claimable) _PartitionDB {
  volatile uint64_t allSignaled;
  volatile PPartition partitions[MAXPART];
  invariant(forall(unsigned i; i < MAXPART;
                   unchanged(partitions[i]) ||
                   old(partitions[i]) == NULL || !closed(old(partitions[i])))))
  invariant(forall(unsigned i; i < MAXPART;
                   unchanged(ISSET(i, allSignaled)) ||
                   inv2(partitions[i]))))
} PartitionDB;
```

Listing 3. Concurrent partition database

3.2   Verifying Concurrent Programs

We now proceed with a concurrent version of the partition database structure
from the previous example (cf. Listing 3). The array of partitions is declared as
volatile to mark the intent of allowing arbitrary threads to add and remove
partitions without unwrapping the partition database. The partitions are also
no longer owned by the database, instead we imagine that the thread currently
executing a partition owns it.[3] The first two-state invariant of the database
prevents removal of closed partitions. The meaning of the invariant is: for any
two consecutive states of the machine either partitions[i] stays the same
(**unchanged**(x) is defined as **old**(x)==(x)), or it was NULL in the first state, or
the object pointed to by partitions[i] in the first state is open. VCC enforces
this two-state invariant on every write to the database. Thus, if one has a closed
partition at index i, one can rely on it staying there.

In the concurrent database, the individual signaled fields from the sequential
version have been collected into a bit mask in the database. This allows taking
an atomic snapshot of partitions currently being signaled. On the other hand,
the details of how these bits can change logically belong with the individual
partitions. This is stated by the second database invariant, saying that whenever
the i-th bit of allSignaled is changed, the invariant of the i-th partition shall
be preserved.

Listing 4 shows the updated version of the partition. The lifeState field
remains the same. Since it is not marked **volatile**, the partition must be unwrapped before changing its life state. Because we want to keep the signature of
the part_send_signal() function, the partition now needs to hold a pointer
to the database and its index. An invariant enforces that the current partition
is indeed stored at that index. This makes the invariant of the partition depend
on fields of the database; so without further precaution, we would need to check
the invariants of partitions when updating the database – but this would make
reasoning about invariants non-modular. Instead, VCC requires that invariants
are *admissible*, as described below.

---

[3] In reality, if one takes a reference to a partition from the database, the database
needs to provide some guarantees that the partition will stick around long enough.
This is achieved using rundowns, but for brevity we skip it here.

```
typedef struct vcc(claimable) _Partition {
  LifeState lifeState;
  invariant(lifeState == Initialized || lifeState == Active ||
            lifeState == Terminating)

  struct _PartitionDB *db;
  unsigned idx;
  invariant(idx < MAXPART && db->partitions[idx] == this)

  spec(volatile bool signaled;)
  invariant(signaled <==> ISSET(idx, db->allSignaled))
  invariant(signaled ==> lifeState == Active)

  spec(claim_t db_claim;)
  invariant(keeps(db_claim) && claims_obj(db_claim, db))
} Partition;
```

Listing 4. Admissibility, volatile fields, shadow fields

Admissibility. A state transition is *legal* iff, for every object *o* that is closed in the transition's prestate or poststate, if any field of *o* is updated (including the "field" indicating closedness) the two-state invariant of *o* is preserved. An invariant of an object *o* is *admissible* iff it is satisfied by every legal state transition. Stated differently, an invariant is admissible if it is preserved by every transition that preserves invariants of all modified objects. Note that admissibility depends only on type definitions (not function specs or code), and is monotonic (i.e., if an invariant has been proved admissible, the addition of further types or invariants cannot make it inadmissible). VCC checks that all invariants are admissible. Thus, when checking that a state update doesn't break any invariant, VCC has to check only the invariants of the updated objects.

Some forms of invariants are trivially admissible. In particular, an invariant in object *o* that mentions only fields of *o* is admissible. This applies to idx < MAXPART. For db->partitions[idx]==**this**, let us assume that db->partitions[idx] changes across a transition (other components of that expression could not change). We know db->partitions[idx] was **this** in the prestate. Assume for a moment, that we know db was closed in both the prestate and the poststate. Then we know db->partitions[idx] was unchanged, it was NULL in the prestate (but **this** != NULL), or **this** was open in the poststate: all three cases are contradictory. But if we knew that db stays closed, then the invariant would be admissible.

Claims. The required knowledge is provided by the *claim* object, owned by the partition and stored in the ghost field db_claim. A claim, as it is used here, can be thought of as a handle that keeps its claimed object from opening. If an object *o* has a type which is marked with **vcc(claimable)** the field **ref_cnt**(*o*) tracks the number of outstanding claims that claim *o*. An object cannot be unwrapped if this count is positive, and a claim can only be created when the object is closed. Thus, when a claim to an object exists, the object is known to be closed.[4]

---

[4] Claims can actually be implemented using admissible two-state invariants. We decided to build them into the annotation language for convenience.

```
void part_send_signal(Partition *part spec(claim_t c))
  requires(wrapped(c) && claims_obj(c, part))
{
  PartitionDB *db = part->db;
  uint64_t idx = part->idx;

  if (part->lifeState != Active) return;

  bv_lemma(forall(int i, j; uint64_t v; 0 <= i && i < 64 && 0 <= j && j < 64 ==>
    i != j ==> (ISSET(j, v) <==> ISSET(j, v | (1ULL << i)))));

  atomic(part, db, c) {
    speconly(part->signaled = true;)
    InterlockedBitSet(&db->allSignaled, idx);
  }
}
```

Listing 5. Atomic operation

More generally, a claim is created by giving an invariant and a set of claimed (claimable) objects on which the claim depends. At the point at which the claim is created, the claimed objects must all be closed and the invariant must hold. Moreover, the claim invariant, conjoined with the invariants of the claimed objects, must imply that the claim invariant cannot be falsified without opening one of the claimed objects.

Pointers to claims are often passed as ghost arguments to functions (most often with the precondition that the claim is wrapped). In this capacity, the claim serves as a stronger kind of precondition. Whereas an ordinary precondition can only usefully constrain state local to the thread, a claim can constrain volatile (shared) state. Moreover, unlike a precondition, the claim invariant is guaranteed to hold until the claim is destroyed. In a function specification, the macro **always**(o, P) means that the function maintains **wrapped**(o) and that o points to a valid claim, and that the invariant of the claim o implies the predicate P. Thus, this contract guarantees that P holds throughout the function call (both to the function and to its callers).

Atomic Blocks. Listing 5 shows how objects can be concurrently updated. The signaling function now only needs a claim to the partition, passed as a ghost parameter, and does not need to list the partition in its writes clause. In fact, the writes clause of the signaling function is empty, reflecting the fact that from the caller perspective, the actions could have been performed by another thread, without the current thread calling any function. A thread can read its own non-volatile state; it can also read non-volatile fields of closed objects, in particular objects for which it holds claims. On the other hand, the volatile fields can only be read and written inside of atomic blocks. Such a block identifies the objects that will be read or written, as well as claims that are needed to establish closedness of those objects. It can contain at most one physical state update or read, which is assumed to be performed atomically by the underlying hardware. In our example, we set the idx-th bit of allSignaled field, using a dedicated CPU instruction (it also returns the old value, but we ignore it). On top of that, the atomic block can perform any number of updates of the ghost state. Both

```
#define Write(state) ((state)&0x80000000)
#define Readers(state) ((state)&0x7FFFFFFF)

typedef struct vcc(claimable) vcc(volatile_owns) _LOCK {
  volatile long state;
  spec(obj_t protected_obj;)
  spec(volatile bool initialized, writing;)
  spec(volatile claim_t self_claim;)

  invariant(old(initialized) ==> initialized && unchanged(self_claim))
  invariant(initialized ==>
    is_claimable(protected_obj) && is_non_primitive_ptr(protected_obj) &&
    set_in(self_claim,owns(this)) && claims_obj(self_claim, this) &&
    protected_obj != self_claim)
  invariant(initialized && !writing ==>
    set_in(protected_obj,owns(this)) &&
    ref_cnt(protected_obj) == (unsigned) Readers(state) &&
    ref_cnt(self_claim) == (unsigned)(Readers(state) + (Write(state)!=0)))
  invariant(initialized && old(Write(state)) ==>
    Readers(state) <= old(Readers(state)) && (!Write(state) ==> old(writing)))
  invariant(initialized && writing ==>
    Readers(state) == 0 && Write(state) && ref_cnt(self_claim) == 0)
} LOCK;
```

Listing 6. Annotated lock data structure

physical and ghost updates can only be performed on objects listed in the header of the atomic block. The resulting state transition is checked for legality, i.e., we check the two-state invariants of updated objects across the atomic block. The beginning of the atomic block is the only place where we simulate actions of other threads; technically this is done by forgetting everything we used to know about volatile state. The only other possible state updates are performed on mutable (and thus open) objects and thus are automatically legal.

3.3   Verification of Concurrency Primitives

In VCC, concurrency primitives (other than atomic operations) are verified (or just specified), rather than being built in. As an example we present the acquisition of a reader-writer lock in exclusive (i.e., writing) mode.[5] In this example, claims are used to capture not only closedness of objects but also properties of their fields.

The data structure LOCK (cf. Listing 6) contains a single volatile implementation variable called state. Its most significant bit holds the write flag that is set when a client requests exclusive access. The remaining bits hold the number of readers. Both values can be updated atomically using interlocked operations.

Acquiring a lock in exclusive mode proceeds in two phases. First, we spin on setting the write bit of the lock atomically. After the write bit has been set, no new shared locks may be taken. Second, we spin until the number of readers reaches zero. This protocol is formalized using lock ghost fields and invariants.

The lock contains four ghost variables: a pointer protected_obj identifying the object protected by the lock, a flag initialized that is set after

---

[5] For details and full annotated source code see [11].

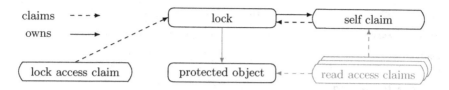

Fig. 2. Ownership and claims structure (shared and exclusive access)

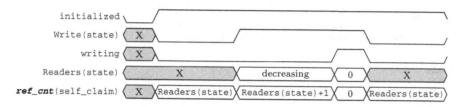

Fig. 3. Relation of lock implementation and ghost variables

initialization, a flag `writing` that is one when exclusive access has been granted (and no reader holds the lock), and a claim `self_claim`. The use of `self_claim` is twofold. First, we tie its reference count to the implementation variables of the lock. This allows restricting changes of these variables by maintaining claims on `self_claim`. Second, it is used to claim lock properties, serving as a proxy between the claimant and the lock. For this purpose it claims the lock and is owned by it. It thus becomes writable and claimable in atomic operations on the lock without requiring it or the lock to be listed in function writes clauses.

Figures 2 and 3 contain a graphical representation of the lock invariants. Figure 2 shows the setup of ownership and claims. The lock access claim is created after initialization. It ensures that the lock remains initialized and allocated, and clients use it (or a derived claim) when calling lock functions. During non-exclusive access each reader holds a read access claim on the protected object and the lock, and the lock owns the protected object, as indicated in gray. During exclusive access the protected object is owned by the client and there may be no readers. Figure 3 depicts the dynamic relation between implementation and ghost variables. As long as the write bit is zero, shared locks may be acquired and released, as indicated by the number of readers. The write bit is set when the acquisition of an exclusive lock starts. In this phase the number of readers must decrease. When it reaches zero, exclusive lock acquisition can complete by activating the `writing` flag. For each reader and each request for write access (which is at most one) there is a reference on `self_claim`.

Listing 7 shows the annotated code for acquisition of an exclusive lock. The macro **`claimp`** wrapped around the parameter `lock_access_claim` means that `lock_access_claim` is a ghost pointer to a wrapped claim; the **`always`** clause says that this claim is wrapped, is not destroyed by the function, and that its invariant implies that the lock is closed and initialized (and hence, will remain

```
void AcquireExclusive(LOCK *lock claimp(lock_access_claim))
  always(lock_access_claim, closed(lock) && lock->initialized)
  ensures(wrapped0(lock->protected_obj) && is_fresh(lock->protected_obj))
{
  bool done;
  spec(claim_t write_bit_claim;)

  bv_lemma(forall(long i; Write(i|0x80000000) &&
    Readers(i|0x80000000) == Readers(i)));

  do
    atomic (lock, lock_access_claim) {
      done = !Write(InterlockedOr(&lock->state, 0x80000000));
      speconly(if (done) {
        write_bit_claim = claim(lock->self_claim, lock->initialized &&
          stays_unchanged(lock->self_claim) && Write(lock->state));
      })
    }
  while (!done);
  do
    invariant(wrapped0(write_bit_claim))
    atomic (lock, write_bit_claim) {
      done = Readers(lock->state)==0;
      speconly(if (done) {
        giveup_closed_owner(lock->protected_obj, lock);
        unclaim(write_bit_claim, lock->self_claim);
        lock->writing = 1;
      })
    }
  while (!done);
}
```

Listing 7. Acquisition of an exclusive lock

so during the function call). After the function returns it guarantees that the protected object is unreferenced, wrapped, and fresh (and thus, writable).

In the first loop of the implementation we spin until the write bit could be atomically set (via the InterlockedOr intrinsic), i.e., in an atomic block the write bit has been seen as zero and then set to one. In the terminating loop case we create a temporary claim write_bit_claim, which references the self claim and states that the lock stays initialized, that the self claim stays, and that the write bit of the lock has been set. VCC checks that the claimed property holds initially and is stable against interference. The former is true by the passed-in lock access claim and the state seen and updated in the atomic operation; the latter is true because as long as there remains a reference to the self claim, the writing flag cannot be activated and the write bit cannot be reset. Also, the atomic update satisfies the lock invariant.

The second loop waits for the readers to disappear. If the number of readers has been seen as zero, we remove the protected object from the ownership of the lock, discard the temporary claim, and set the writing flag. All of this can be justified by the claimed property and the lock's invariant. Setting the writing flag is allowed because the write bit is known to be set. Furthermore, the writing flag is known to be zero in the pre-state of the atomic operation because the reference count of the self claim, which is referenced by write_bit_claim, cannot be zero. This justifies the remaining operations.

4   VCC Tool Suite

VCC reuses the Spec# tool chain [9], which has allowed developing a comprehensive C verifier with limited effort. In addition we developed auxiliary tools to support the process of verification engineering in a real-world effort.

4.1   The Static Verifier

We base our verification methodology on inline annotations in the, otherwise unaltered, source code of the implementation. The C preprocessor is used to eliminate these annotations for normal C compilation. For verification, the output of the preprocessor (with annotations still intact) is fed to the VCC compiler.

CCI. The VCC compiler is build using Microsoft Research's Common Compiler Infrastructure (CCI) libraries [12]. VCC reads annotated C and turns the input into CCI's internal representation to perform name resolution, type and error check as any normal C compiler would do.

Source Transformations and Plugins. Next, the fully resolved input program undergoes multiple source-to-source transformations. These transformations first simplify the source, and then add proof obligations stemming from the methodology. The last transformation generates the Boogie source.

VCC provides a plugin interface, where users can insert and remove transformations, including the final translation. Currently two plugins have been implemented: to generate contracts for assembly functions from their C correspondents; and to build a new methodology based on separation logic [13].

Boogie. Once the source code has been analyzed and found to be valid, it is translated into a Boogie program that encodes the input program according to our formalization of C. Boogie [14] is an intermediate language that is used by a number of software verification tools including Spec# and HAVOC. Boogie adds minimal imperative control flow, procedural and functional abstractions, and types on top of first order predicated logic. The translation from annotated C to Boogie encodes both static information about the input program, like types and their invariants, and dynamic information like the control flow of the program and the corresponding state updates. Additionally, a fixed axiomatization of C memory, object ownership, type state, and arithmetic operations (the *prelude*) is added. The resulting program is fed to the Boogie program verifier, which translates it into a sequence of verification conditions. Usually, these are then passed to an automated theorem prover to be proved or refuted. Alternatively, they can be discharged interactively. The HOL-Boogie tool [15] provides support for this approach based on the Isabelle interactive theorem prover.

Z3. Our use of Boogie targets Z3 [16], a state-of-the art first order theorem prover that supports satisfiability modulo theories (SMT). VCC makes heavy use of Z3's fast decision procedures for linear arithmetic and uses the slower fixed-length bit vector arithmetic only when explicitly invoked by VCC's **bv_lemma**() mechanism (see Listing 5 for an example). These lemmas are typically used when reasoning for overflowing arithmetic or bitwise operations.

4.2 Static Debugging

In the ideal case, the work flow described above is all there is to running VCC: an annotated program is translated via Boogie into a sequence of verification conditions that are successfully proved by Z3. Unfortunately, this ideal situation is encountered only seldomly during the process of verification engineering, where most time is spent debugging failed verification attempts. Due to the undecidability of the underlying problem, these failures can either be caused by a genuine error in either the code or the annotations, or by the inability of the SMT solver to prove or refute a verification condition within available resources like computer memory, time, or verification engineer's patience.

VCC Model Viewer. In case of a refutation, Z3 constructs a counterexample that VCC and Boogie can tie back to a location in the original source code. However that is not that easy, since these counterexamples contain many artifacts of the underlying axiomatization, and so are not well-suited for direct inspection. The VCC Model Viewer translates the counterexample into a representation that allows inspecting the sequence of program states that led to the failure, including the value of local and global variables and the heap state.

Z3 Inspector. A different kind of verification failure occurs when the prover takes an excessive amount of time to come up with a proof or refutation for a verification condition. To counter this problem, we provide the Z3 Inspector, a tool that allows to monitor the progress of Z3 tied to the annotations in the source code. This allows to pinpoint those verification conditions that take excessively long to be processed. There can be two causes for this: either the verification condition is valid and the prover requires a long time to find the proof, or the verification condition is invalid and the search for a counterexample takes a very long time. In the latter case, the Z3 Inspector helps identifying the problematic assertion quickly.

Z3 Axiom Profiler. In the former case a closer inspection of the quantifier instantiation pattern can help to determine inefficiencies in the underlying axiomatization of C or the program annotations. This is facilitated by the Z3 Axiom Profiler, which allows to analyze the quantifier instantiation patters to detect, e.g., triggering cycles.

Visual Studio. All of this functionality is directly accessible from within the Visual Studio IDE, including verifying only individual functions. We have found that this combination of tools enables the verification engineer to efficiently develop and debug the annotations required to prove correctness of the scrutinized codebase.

5   VCC Experience

The methodology presented in this paper was implemented in VCC in late 2008. Since this methodology differs significantly from earlier approaches, the annotation of the hypervisor codebase had to start from scratch. As of June 2009, fourteen verification engineers are working on annotating the codebase and verifying functions. Since November 2008 approx. 13 500 lines of annotations have been added to the hypervisor codebase. About 350 functions have been successfully verified resulting in an average of two verified functions per day. Additionally, invariants for most public and private data types (consisting of about 150 structures or groups) have been specified and proved admissable. This means that currently about 20% of the hypervisor codebase has been successfully verified using our methodology.

A major milestone in the verification effort is having the specifications of all public functions from all layers so that the verification of the different layers require no interaction of the verification engineers, since all required information has been captured in the contracts and invariants. This milestone has been reached or will be reached soon for seven modules. Also for three modules already more than 50% of the functions have been successfully verified.

We have found that having acceptable turnaround times for verify-and-fix cycles is crucial to maintain productivity of the verification engineers. Currently VCC verifies most functions in 0.5 to 500 seconds with an average of about 25 seconds. The longest running function needs ca. 2 000 seconds to be verified.

The all-time high was around 50 000 seconds for a successful proof attempt. In general failing proof attempts tend to take longer than successfully verifying a function. A dedicated test suite has been created to constantly monitor verification performance. Performance has improved by one to two orders of magnitude. Many changes have contributed to these improvements, ranging from changes in our methodology, the encoding of type state, our approach to invariant checking, the support of out parameters, to updates in the underlying tools Boogie and Z3. With these changes, we have, for example, reduced the verification time for the 50 000s function down to under 1 000s.

Still, in many cases the verification performance is unacceptable. Empirically, we have found that verification times of over a minute start having an impact on the productivity of the verification engineer, and that functions that require one hour or longer are essential intractable. We are currently working on many levels to alleviate these problems: improvements in the methodology, grid-style distribution of verification condition checking, parallelization of proof search for a single verification condition, and other improvements of SMT solver technology.

6   Related Work

Methodology. The approach of Owicki and Gries [17] requires annotations to be stable with respect to *every* atomic action of the other threads, i.e., that the assertions are interference free. This dependency on the other threads makes the

Owicki-Gries method non-modular and the number of interference tests grows quadratically with the number of atomic actions. Ashcroft [18] proposed to just use a single big state invariant to verify concurrent programs. This gets rid of the interference check, and makes verification thread-modular.

Jones developed the more modular rely/guarantee method [19] which abstracts the possible interleavings with other threads to rely on and guarantee assertions. Now, it suffices to check that each thread respects these assertions locally and that the rely and guarantee assertions of all threads fit together. Still, their approach (and also the original approach of Owicki and Gries) do not support data modularity: there is no hiding mechanism, a single bit change requires the guarantees of *all* threads to be checked.

Flanagan et al. [3] describe a rely/guarantee based prover for multi-threaded Java programs. They present the verification of three synchronization primitives but do not report on larger verification examples. The approach is thread modular (as it is based on rely/guarantee) but not function modular (they simulate function calls by inlining).

In contrast to rely/guarantee, concurrent separation logic exploits that large portions of the program state may be operated on mutually exclusive [20, 21]. Thus, like in our approach, interference is restricted to critical regions and verification can be completely sequential elsewhere. Originally, concurrent separation logic was restricted to exclusive access (and atomic update) of shared resources. Later, Bornat proposed a fractional ownership scheme to allow for shared read-only state also [22]. Recently, Vafeiadis and Parkinson [23] have worked on combining the ideas of concurrent separation logic with rely/guarantee reasoning.

Our ownership model, with uniform treatment of objects and threads, is very similar to the one employed in Concurrent Spec# [4]. Consequently, the visible specifications of locks, being the basis of Concurrent Spec# methodology, is essentially the same. We however do not treat locks as primitives, and allow for verifying implementation of various concurrency primitives. The Spec# ideas have also permeated into recent work by Leino and Mueller [24]. They use dynamic frames and fractional permissions for verifying fine grained locking. History invariants [25] are two-state object invariants, requiring admissibility check similar to ours. These invariants are however restricted to be transitive and are only used in the sequential context.

Systems Verification. Klein [26] provides a comprehensive overview of the history and current state of the art in operating systems verification, which is supplemented by a recent special issue of the Journal of Automated Reasoning on operating system verification [27]. The VFiasco [28] project, followed by the Robin projects attempted the verification of a micro kernel, based on a translation of C++ code into its corresponding semantics in the theorem prover PVS. While these projects have been successful in providing a semantic model for C++, no significant portions of the kernel implementation has been verified. Recent related projects in this area include the project L4.verified [29] (verification of an industrial microkernel), the FLINT project [30] (verification of an assembly kernel), and the Verisoft project [31] (the predecessor project of Verisoft XT focusing on the pervasive verification of hardware-software systems). All three projects are based on interactive theorem proving (with Isabelle or Coq). Our hypervisor verification attempt is significantly more ambitious, both with respect to size (ca. 100KLOC of C) and complexity (industrial code for a modern multiprocessor architecture with a weak memory model).

Acknowledgments. Thanks to everyone in the project: Artem Alekhin, Eyad Alkassar, Mike Barnett, Nikolaj Bjørner, Sebastian Bogan, Sascha Böhme, Matko Botinĉan, Vladimir Boyarinov, Ulan Degenbaev, Lieven Desmet, Sebastian Fillinger, Tom In der Rieden, Bruno Langenstein, K. Rustan M. Leino, Wolfgang Manousek, Stefan Maus, Leonardo de Moura, Andreas Nonnengart, Steven Obua, Wolfgang Paul, Hristo Pentchev, Elena Petrova, Norbert Schirmer, Sabine Schmaltz, Peter-Michael Seidel, Andrey Shadrin, Alexandra Tsyban, Sergey Tverdyshev, Herman Venter, and Burkhart Wolff.

References

1. Verisoft XT: The Verisoft XT project (2007), http://www.verisoftxt.de
2. Cohen, E., Moskal, M., Schulte, W., Tobies, S.: A precise yet efficient memory model for C. In: SSV 2009. ENTCS. Elsevier Science B.V., Amsterdam (2009)
3. Flanagan, C., Freund, S.N., Qadeer, S.: Thread-modular verification for shared-memory programs. In: Le Métayer, D. (ed.) ESOP 2002. LNCS, vol. 2305, pp. 262–277. Springer, Heidelberg (2002)
4. Jacobs, B., Piessens, F., Leino, K.R.M., Schulte, W.: Safe concurrency for aggregate objects with invariants. In: Aichernig, B.K., Beckert, B. (eds.) SEFM 2005, pp. 137–147. IEEE, Los Alamitos (2005)
5. Maus, S., Moskal, M., Schulte, W.: Vx86: x86 assembler simulated in C powered by automated theorem proving. In: Meseguer, J., Roşu, G. (eds.) AMAST 2008. LNCS, vol. 5140, pp. 284–298. Springer, Heidelberg (2008)
6. Advanced Micro Devices (AMD), Inc.: AMD64 Architecture Programmer's Manual: Vol. 1-3 (2006)
7. Intel Corporation: Intel 64 and IA-32 Architectures Software Developer's Manual: Vol. 1-3b (2006)
8. Flanagan, C., Leino, K.R.M., Lillibridge, M., Nelson, G., Saxe, J.B., Stata, R.: Extended static checking for Java. SIGPLAN Notices 37(5), 234–245 (2002)
9. Barnett, M., Leino, K.R.M., Schulte, W.: The Spec# programming system: An overview. In: Barthe, G., Burdy, L., Huisman, M., Lanet, J.-L., Muntean, T. (eds.) CASSIS 2004. LNCS, vol. 3362, pp. 49–69. Springer, Heidelberg (2005)
10. Microsoft Research: The HAVOC property checker, http://research.microsoft.com/projects/havoc
11. Hillebrand, M.A., Leinenbach, D.C.: Formal verification of a reader-writer lock implementation in C. In: SSV 2009. ENTCS, Elsevier Science B.V., Amsterdam (2009); Source code, http://www.verisoftxt.de/PublicationPage.html
12. Microsoft Research: Common compiler infrastructure, http://ccimetadata.codeplex.com/
13. Botinĉan, M., Parkinson, M., Schulte, W.: Separation logic verification of C programs with an SMT solver. In: SSV 2009. ENTCS. Elsevier Science B.V., Amsterdam (2009)

14. Barnett, M., Chang, B.Y.E., Deline, R., Jacobs, B., Leino, K.R.M.: Boogie: A modular reusable verifier for object-oriented programs. In: de Boer, F.S., Bonsangue, M.M., Graf, S., de Roever, W.-P. (eds.) FMCO 2005. LNCS, vol. 4111, pp. 364–387. Springer, Heidelberg (2006)
15. Böhme, S., Moskal, M., Schulte, W., Wolff, B.: HOL-Boogie: An interactive prover-backend for the Verifiying C Compiler. Journal of Automated Reasoning (to appear, 2009)
16. de Moura, L., Bjørner, N.: Z3: An efficient SMT solver. In: Ramakrishnan, C.R., Rehof, J. (eds.) TACAS 2008. LNCS, vol. 4963, pp. 337–340. Springer, Heidelberg (2008)
17. Owicki, S., Gries, D.: Verifying properties of parallel programs: An axiomatic approach. Communications of the ACM 19(5), 279–285 (1976)
18. Ashcroft, E.A.: Proving assertions about parallel programs. Journal of Computer and System Sciences 10(1), 110–135 (1975)
19. Jones, C.B.: Tentative steps toward a development method for interfering programs. ACM Transactions on Programming Languages and Systems 5(4), 596–619 (1983)
20. O'Hearn, P.W.: Resources, concurrency, and local reasoning. Theoretical Computer Science 375(1-3), 271–307 (2007)
21. Reynolds, J.C.: Separation logic: A logic for shared mutable data structures. In: LICS 2002, pp. 55–74. IEEE, Los Alamitos (2002)
22. Bornat, R., Calcagno, C., O'Hearn, P.W., Parkinson, M.J.: Permission accounting in separation logic. In: Palsberg, J., Abadi, M. (eds.) POPL 2005, pp. 259–270. ACM, New York (2005)
23. Vafeiadis, V., Parkinson, M.J.: A marriage of rely/guarantee and separation logic. In: Caires, L., Vasconcelos, V.T. (eds.) CONCUR 2007. LNCS, vol. 4703, pp. 256–271. Springer, Heidelberg (2007)
24. Leino, K.R.M., Müller, P.: A basis for verifying multi-threaded programs. In: Castagna, G. (ed.) ESOP 2009. LNCS, vol. 5502, pp. 378–393. Springer, Heidelberg (2009)
25. Leino, K.R.M., Schulte, W.: Using history invariants to verify observers. In: De Nicola, R. (ed.) ESOP 2007. LNCS, vol. 4421, pp. 80–94. Springer, Heidelberg (2007)
26. Klein, G.: Operating system verification – An overview. Sādhanā: Academy Proceedings in Engineering Sciences 34(1), 27–69 (2009)
27. Journal of Automated Reasoning: Operating System Verification 42(2–4) (2009)
28. Hohmuth, M., Tews, H.: The VFiasco approach for a verified operating system. In: 2nd ECOOP Workshop in Programming Languages and Operating Systems (2005)
29. Heiser, G., Elphinstone, K., Kuz, I., Klein, G., Petters, S.M.: Towards trustworthy computing systems: Taking microkernels to the next level. SIGOPS Oper. Syst. Rev. 41(4), 3–11 (2007)
30. Ni, Z., Yu, D., Shao, Z.: Using XCAP to certify realistic systems code: Machine context management. In: Schneider, K., Brandt, J. (eds.) TPHOLs 2007. LNCS, vol. 4732, pp. 189–206. Springer, Heidelberg (2007)
31. Alkassar, E., Hillebrand, M.A., Leinenbach, D.C., Schirmer, N.W., Starostin, A., Tsyban, A.: Balancing the load: Leveraging a semantics stack for systems verification. Journal of Automated Reasoning: Operating System Verification 27, 389–454

Without Loss of Generality

John Harrison

Intel Corporation, JF1-13
2111 NE 25th Avenue, Hillsboro OR 97124, USA
johnh@ichips.intel.com

Abstract. One sometimes reads in a mathematical proof that a certain assumption can be made 'without loss of generality' (WLOG). In other words, it is claimed that considering what first appears only a special case does nevertheless suffice to prove the general result. Typically the intuitive justification for this is that one can exploit symmetry in the problem. We examine how to formalize such 'WLOG' arguments in a mechanical theorem prover. Geometric reasoning is particularly rich in examples and we pay special attention to this area.

1 Introduction

Mathematical proofs sometimes state that a certain assumption can be made 'without loss of generality', often abbreviated to 'WLOG'. The phase suggest that although making the assumption at first sight only proves the theorem in a more restricted case, this does nevertheless justify the theorem in full generality. What is the intuitive justification for this sort of reasoning? Occasionally the phrase covers situations where we neglect special cases that are obviously trivial for other reasons. But more usually it suggests the exploitation of symmetry in the problem. For example, consider Schur's inequality, which asserts that for any nonnegative real numbers a, b and c and integer $k \geq 0$ one has $0 \leq a^k(a-b)(b-c) + b^k(b-a)(b-c) + c^k(c-a)(c-b)$. A typical proof might begin:

> Without loss of generality, let $a \leq b \leq c$.

If asked to spell this out in more detail, we might say something like:

> Since $\leq$ is a total order, the three numbers must be ordered somehow, i.e. we must have (at least) one of $a \leq b \leq c$, $a \leq c \leq b$, $b \leq a \leq c$, $b \leq c \leq a$, $c \leq a \leq b$ or $c \leq b \leq a$. But the theorem is completely symmetric between a, b and c, so each of these cases is just a version of the other with a change of variables, and we may as well just consider one of them.

Suppose that we are interested in formalizing mathematics in a mechanical theorem prover. Generally speaking, for an experienced formalizer it's rather routine to take an existing proof and construct a formal counterpart, even though it may require a great deal of work to get things just right and encourage the proof assistant check all the details. But with such 'without loss of generality' constructs, it's not immediately obvious what the formal counterpart should be. We can plausibly suggest two possible formalizations:

S. Berghofer et al. (Eds.): TPHOLs 2009, LNCS 5674, pp. 43–59, 2009.

- The phrase may be an informal shorthand saying 'we should really do 6 very similar proofs here, but if we do one, all the others are exactly analogous and can be left to the reader'.
- The phrase may be asserting that 'by a general logical principle, the apparently more general case and the special WLOG case are in fact equivalent (or at least the special case implies the general one)'.

The former point of view can be quite natural in a computer proof assistant. If we have a proof script covering one of the 6 cases, we might simply perform a 6-way case-split and for each case use a duplicate of the initial script, changing the names of variables systematically in an editor. Indeed, if we have a programmable proof assistant, it would be more elegant to write a general parametrized proof script that we could use for all 6 cases with different parameters. This sort of programming is exactly the kind of thing that LCF-style systems [3] like HOL [2] are designed to make easy via their 'metalanguage' ML, and sometimes its convenience makes it irresistible. However, this approach is open to criticism on at least three grounds:

- Ugly/clumsy
- Inefficient
- Not faithful to the informal proof.

Indeed, it seems unnatural, even with the improvement of using a parametrized script, to perform essentially the same proof 6 different times, and if each proof takes a while to run, it could waste computer resources. And it is arguably *not* what the phrase 'without loss of generality' is meant to conjure up. If the book had intended that interpretation, it would probably have said something like 'the other cases are similar and are left to the reader'. So let us turn to how we might formalize and use a general logical principle.

2   A HOL Light Proof of Schur's Inequality

In fact, in HOL Light there is already a standard theorem with an analogous principle for a property of *two* real numbers:

```
REAL_WLOG_LE =
 |- (∀x y. P x y ⇔ P y x) ∧
    (∀x y. x <= y ⇒ P x y)
    ⇒ (∀x y. P x y)
```

This asserts that for any property P of two real numbers, if the property is symmetric between those two numbers ($\forall x\ y.\ P\ x\ y \Leftrightarrow P\ y\ x$) and assuming $x \leq y$ the property holds ($\forall x\ y.\ x \leq y \Rightarrow P\ x\ y$), then we can conclude that it holds for all real numbers ($\forall x\ y.\ P\ x\ y$). In order to tackle the Schur inequality we will prove a version for *three* variables. Our chosen formulation is quite analogous, but using a more minimal formulation of symmetry between all three variables:

```
REAL_WLOG_3_LE =
 |- (∀x y z. P x y z ⇒ P y x z ∧ P x z y) ∧
    (∀x y z. x <= y ∧ y <= z ⇒ P x y z)
    ⇒ (∀x y z. P x y z)
```

The proof is relatively straightforward following the informal intuition: we observe that one of the six possible ordering sequences must occur, and in each case we can deduce the general case from the more limited one and symmetry. The following is the tactic script to prove `REAL_WLOG_3_LE`:

```
REPEAT STRIP_TAC THEN (STRIP_ASSUME_TAC o REAL_ARITH)
 `x <= y /\ y <= z \/ x <= z /\ z <= y \/ y <= x /\ x <= z \/
  y <= z /\ z <= x \/ z <= x /\ x <= y \/ z <= y /\ y <= x` THEN
ASM_MESON_TAC[]
```

Now let us see how to use this to prove Schur's inequality in HOL Light, which we formulate as follows:

```
|- ∀k a b c. &0 <= a /\ &0 <= b /\ &0 <= c
    ⇒ &0 <= a pow k * (a - b) * (a - c) +
            b pow k * (b - a) * (b - c) +
            c pow k * (c - a) * (c - b)
```

The first step in the proof is to strip off the additional variable k (which will not play a role in the symmetry argument), use backwards chaining with the WLOG theorem `REAL_WLOG_3_LE`, and then break the resulting goal into two subgoals, one corresponding to the symmetry and the other to the special case.

```
GEN_TAC THEN MATCH_MP_TAC REAL_WLOG_3_LE THEN CONJ_TAC
```

The first subgoal, corresponding to symmetry of the problem, is the following:

```
`∀a b c. (&0 <= a /\ &0 <= b /\ &0 <= c
    ⇒ &0 <= a pow k * (a - b) * (a - c) +
            b pow k * (b - a) * (b - c) +
            c pow k * (c - a) * (c - b))
    ⇒ (&0 <= b /\ &0 <= a /\ &0 <= c
        ⇒ &0 <= b pow k * (b - a) * (b - c) +
                a pow k * (a - b) * (a - c) +
                c pow k * (c - b) * (c - a)) /\
      (&0 <= a /\ &0 <= c /\ &0 <= b
        ⇒ &0 <= a pow k * (a - c) * (a - b) +
                c pow k * (c - a) * (c - b) +
                b pow k * (b - a) * (b - c))`
```

Although this looks rather large, the proof simply exploits the fact that addition and multiplication are associative and commutative via routine logical reasoning, so we can solve it by:

```
MESON_TAC[REAL_ADD_AC; REAL_MUL_AC]
```

We have now succeeded in reducing the original goal to the special case:

```
`∀a b c. a <= b ∧ b <= c
        ⇒ &0 <= a ∧ &0 <= b ∧ &0 <= c
          ⇒ &0 <= a pow k * (a - b) * (a - c) +
                  b pow k * (b - a) * (b - c) +
                  c pow k * (c - a) * (c - b)`
```

and so we can claim that the foregoing proof steps correspond almost exactly to the informal WLOG principle. We now rewrite the expression into a more convenient form:

```
REPEAT STRIP_TAC THEN ONCE_REWRITE_TAC[REAL_ARITH
`a pow k * (a - b) * (a - c) +
 b pow k * (b - a) * (b - c) +
 c pow k * (c - a) * (c - b) =
 (c - b) * (c pow k * (c - a) - b pow k * (b - a)) +
 a pow k * (c - a) * (b - a)`]
```

The form of this expression is now congenial, so we can simply proceed by repeatedly chaining through various monotonicity theorems and then use linear arithmetic reasoning to finish the proof:

```
REPEAT(FIRST(map MATCH_MP_TAC
             [REAL_LE_ADD; REAL_LE_MUL; REAL_LE_MUL2]) THEN
       ASM_SIMP_TAC[REAL_POW_LE2; REAL_POW_LE; REAL_SUB_LE] THEN
       REPEAT CONJ_TAC) THEN
ASM_REAL_ARITH_TAC
```

We have therefore succeeded in deploying WLOG reasoning in a natural way and following a standard textbook proof quite closely. However, a remaining weak spot is the proof of the required symmetry for the particular problem. In this case, we were just able to use standard first-order automation (MESON_TAC) to deduce this symmetry from the associativity and commutativity of the two main operations involved (real addition and multiplication). However, we can well imagine that in more complicated situations, this kind of crude method might be tedious or impractical. We will investigate how to approach this more systematically using reasoning from a somewhat different domain.

3   WLOG Reasoning in Geometry

Geometry is particularly rich in WLOG principles, perhaps reflecting the fundamental importance in geometry of property-preserving transformations. The modern view of geometry has been heavily influenced by Klein's "Erlanger Programm" [7], which emphasizes the role of transformations and invariance under classes of transformations, while modern physical theories usually regard conservation laws as manifestations of invariance properties: the conservation of angular momentum arises from invariance under rotations, while conservation of energy arises from invariance under shifts in time, and so on [8].

One of the most important ways in which such invariances are used in proofs is to make a convenient choice of coordinate system. In our formulation of Euclidean space in HOL Light [6], geometric concepts are all defined in analytic terms using vectors, which in turn are expressed with respect to a standard coordinate basis. For example, the angle formed by three points is defined in terms of the angle between two vectors:

```
|- angle(a,b,c) = vector_angle (a - b) (c - b)
```

which is defined in terms of norms and dot products using the inverse cosine function acs (degenerating to $\pi/2$ if either vector is zero):

```
|- vector_angle x y =
        if x = vec 0 ∨ y = vec 0 then pi / &2
        else acs((x dot y) / (norm x * norm y))
```

where norms are defined in terms of dot products:

```
|- ∀x. norm x = sqrt(x dot x)
```

and finally dot products in $\mathbb{R}^N$ are defined in terms of the N components in the usual way as $x \cdot y = \sum_{i=1}^{N} x_i y_i$, or in HOL Light:

```
|- x dot y = sum(1..dimindex(:N)) (ı. x$i * y$i)
```

This means that whenever we state geometric theorems, most of the concepts ultimately rest on a *particular* choice of coordinate system and standard basis vectors. When we are performing high-level reasoning, we can often reason about geometric concepts directly using lemmas established earlier without ever dropping down to the ultimate representation with respect to the standard basis. But when we *do* need to reason algebraically in terms of coordinates, we often find that a different choice of coordinate system would make the reasoning much more tractable.

The simplest example is probably choosing the origin of the coordinate system. If a proposition $\forall x.\ P[x]$ is invariant under spatial translation, i.e. changing x to any $a + x$, then it suffices to prove the special case $P[0]$, or in other words, to assume without loss of generality that x is the origin. The reasoning is essentially trivial: if we have $P[0]$ and also $\forall a\ x.\ P[x] \Rightarrow P[a + x]$, then we can deduce $P[x + 0]$ and so $P[x]$. In HOL Light we can state this as the following general theorem, asserting that if P is invariant under translation and we have the special case $P[0]$, then we can conclude $\forall x.\ P[x]$:

```
WLOG_ORIGIN =
   |- (∀a x. P(a + x) ⇔ P x) ∧ P(vec 0) ⇒ (∀x. P x)
```

Thus, when confronted with a goal, we can simply rearrange the universally quantified variables so that the one we want to take as the origin is at the outside, then apply this theorem, giving us the special case $P[0]$ together with the invariance of the goal under translation. For example, suppose we want to prove that the angles of a triangle that is not completely degenerate all add up to π radians (180 degrees):

```
`∀A B C.  ˜(A = B ∧ B = C ∧ A = C)
        ⇒ angle(B,A,C) + angle(A,B,C) + angle(B,C,A) = pi`
```

If we apply our theorem by MATCH_MP_TAC WLOG_ORIGIN and split the resulting goal into two conjuncts, we get one subgoal corresponding to the special case when A is the origin:

```
`∀B C.
   ˜(vec 0 = B ∧ B = C ∧ vec 0 = C)
 ⇒ angle(B,vec 0,C) + angle(vec 0,B,C) + angle(B,C,vec 0) =
    pi`
```

and another goal for the invariance of the property under translation by a:

```
`∀a A.  (∀B C.
             ˜(a + A = B ∧ B = C ∧ a + A = C)
           ⇒ angle(B,a + A,C) +
               angle(a + A,B,C) + angle(B,C,a + A) = pi) ⇔
        (∀B C.
             ˜(A = B ∧ B = C ∧ A = C)
           ⇒ angle(B,A,C) + angle(A,B,C) + angle(B,C,A) = pi)`
```

We will not dwell more on the detailed proof of the theorem in the special case where A is the origin, but will instead focus on the invariance proof. In contrast to the case of Schur's inequality, this is somewhat less easy and can't obviously be deferred to basic first-order automation. So how do we prove it?

At first sight, things don't look right: it seems that we ought to have translated not just A but *all* the variables A, B and C together. However, note that for any given a the translation mapping $x \mapsto a + x$ is surjective: for any y there is an x such that $a + x = y$ (namely $x = y - a$). That means that we can replace universal quantifiers over vectors, and even existential ones too, by translated versions. This general principle can be embodied in the following HOL theorem, easily proven automatically by MESON_TAC:

```
QUANTIFY_SURJECTION_THM =
 |- ∀f:A->B.
          (∀y. ∃x. f x = y)
          ⇒ (∀P. (∀x. P x) ⇔ (∀x. P (f x))) ∧
             (∀P. (∃x. P x) ⇔ (∃x. P (f x))) ∧
```

We can apply it with a bit of instantiation and higher-order rewriting to all the universally quantified variables on the left-hand side of the equivalence in the goal and obtain:

```
'∀a A.
   (∀B C.
       ˜(a + A = a + B ∧ a + B = a + C ∧ a + A = a + C)
       ⇒ angle(a + B,a + A,a + C) +
           angle(a + A,a + B,a + C) +
           angle(a + B,a + C,a + A) = pi) ⇔
   (∀B C.
       ˜(A = B ∧ B = C ∧ A = C)
       ⇒ angle(B,A,C) + angle(A,B,C) + angle(B,C,A) = pi)'
```

Now things are becoming better. First of all, it is clear that $a + A = a + B \Leftrightarrow A = B$ etc. just by general properties of vector addition. As for angles, recall that angle(x,y,z) is defined as the vector angle between the two differences $x - y$ and $z - y$. Because it is defined in terms of such differences, it again follows from basic properties of vector addition that, for example, $(a + B) - (a + A) = A - B$, and so we can deduce the invariance property that we seek.

This is all very well, but the process is quite laborious. We have to carefully apply translation to all the quantified variables just once so that we don't get into an infinite loop, and then we have to appeal to suitable basic invariance theorems for pretty much *all* the concepts that appear in our theorems. Even in this case, doing so is not entirely trivial, and for more involved theorems it can be worse, as Hales [5] notes:

> [...] formal proofs by symmetry are much harder than anticipated. It was necessary to give a total of nearly a hundred lemmas, showing that the symmetries preserve all of the relevant structures, all the way back to the foundations.

Indeed, this process seems unpleasant enough that we should consider automating it, and for geometric invariants this is just what we have done.

4   Tactics Using Invariance under Translation

Our WLOG tactic for choosing the origin is based on a list of theorems asserting invariance under translation for various geometric concepts, stored in a reference variable invariant_under_translation. The vision is that each time a new geometric concept (angle, collinear, etc.) is defined, one proves a corresponding invariance theorem and adds it to this list, so that thereafter the invariance will be exploitable automatically by the WLOG tactic. For example, the entry corresponding to angle is

```
|- ∀a b c d. angle(a + b,a + c,a + d) = angle(b,c,d)
```

While we usually aim to prove that numerical functions of vectors (e.g. distances or angles) or predicates on vectors (e.g. collinearity) are completely invariant under translation, for operations returning more vectors, we normally want to prove that the translation can be 'pulled outside', e.g.

```
|- ∀a x y. midpoint(a + x,a + y) = a + midpoint (x,y)
```

Then a translated formula can be systematically mapped into its untranslated form by applying these transformations in a bottom-up fashion, pulling the translation up through vector-producing functions like midpoint and then systematically eliminating them when they reach the level of predicates or numerical functions of vectors.

Our setup is somewhat more ambitious in that it applies not only to properties of vectors but also to properties of *sets* of vectors, many of which are also invariant under translation. For example, recall that a set is convex if whenever it contains the points x and y it also contains each intermediate point between x and y, i.e. each $ux + vy$ where $0 \le u, v$ and $u + v = 1$:

```
|- ∀s. convex s ⇔
      (∀x y u v.
           x IN s ∧ y IN s ∧ &0 <= u ∧ &0 <= v ∧ u + v = &1
           ⇒ u % x + v % y IN s)
```

This is invariant under translation in the following sense:

```
|- ∀a s. convex (IMAGE (λx. a + x) s) ⇔ convex s
```

as are many other geometric or topological predicates (bounded, closed, compact, path-connected, ...) and numerical functions on sets such as measure (area, volume etc. depending on dimension):

```
|- ∀a s. measure (IMAGE (λx. a + x) s) = measure s
```

As with points, for functions that return other sets of vectors, our theorems state rather that the 'image under translation' operation can be pulled up through the function, e.g.

```
|- ∀a s. convex hull IMAGE (λx. a + x) s =
         IMAGE (λx. a + x) (convex hull s)
```

We include in the list other theorems of the same type for the basic set operations, so that they can be handled as well, e.g.

```
|- ∀a s t. IMAGE (λy. a + y) s UNION IMAGE (λy. a + y) t =
           IMAGE (λy. a + y) (s UNION t)

|- ∀a s t. IMAGE (λy. a + y) s SUBSET IMAGE (λy. a + y) t ⇔
           s SUBSET t
```

Our conversion (GEOM_ORIGIN_CONV) and corresponding tactic that is defined in terms of it (GEOM_ORIGIN_TAC) work by automating the process sketched in the special case in the previous section. First, they apply the basic reduction so that we need to prove equivalence when one nominated variable is translated. Then the other quantifiers are modified to apply similar translation to the other variables, even if quantification is nested in a complicated way. We use an enhanced version of the theorem

`QUANTIFY_SURJECTION_THM` which applies a similarly systematic modification to quantifiers over sets of vectors and set comprehensions such as $\{x \mid \text{angle}(a, b, x) = \pi/3\}$:

```
|- ∀f. (∀y. ∃x. f x = y)
       ⇒ (∀P. (∀x. P x) ⇔ (∀x. P (f x))) ∧
         (∀P. (∃x. P x) ⇔ (∃x. P (f x))) ∧
         (∀Q. (∀s. Q s) ⇔ (∀s. Q (IMAGE f s))) ∧
         (∀Q. (∃s. Q s) ⇔ (∃s. Q (IMAGE f s))) ∧
         (∀P. {x | P x} = IMAGE f {x | P (f x)})
```

With this done, it remains only to rewrite with the invariance theorems taken from the list `invariant_under_translation` in a bottom-up sweep. If the intended result uses only these properties in a suitable fashion, then this should automatically reduce the invariance goal to triviality. The user does not even see it, but is presented instead with the special case. (If the process of rewriting does not solve the invariance goal, then that is returned as an additional subgoal so that the user can either help the proof along manually or perhaps observe that a concept is used for which no invariance theorem has yet been stored.) For example, if we set out to prove the formula for the volume of a ball:

```
`∀z:real^3 r. &0 <= r
             ⇒ measure(cball(z,r)) = &4 / &3 * pi * r pow 3`
```

a simple application of `GEOM_ORIGIN_TAC` `z:real^3` reduces it to the special case when the ball is centered at the origin:

```
`∀r. &0 <= r
    ⇒ measure(cball(vec 0,r)) = &4 / &3 * pi * r pow 3`
```

Here is an example with a more complicated quantifier structure and a mix of sets and points. We want to prove that for any point a and nonempty closed set s there is a closest point of s to a. (A set is closed if it contains all its limit points, i.e. all points that can be approached arbitrarily closely by a member of the set.) We set up the goal:

```
g `∀s a:real^N.
      closed s ∧ ~(s = {})
      ⇒ ∃x. x IN s ∧
            (∀y. y IN s ⇒ dist(a,x) <= dist(a,y))`;;
```

and with a single application of our tactic, we can suppose the point in question is the origin:

```
# e(GEOM_ORIGIN_TAC `a:real^N`);;
val it : goalstack = 1 subgoal (1 total)

`∀s. closed s ∧ ~(s = {})
    ⇒ (∃x. x IN s ∧
           (∀y. y IN s ⇒ dist(vec 0,x) <= dist(vec 0,y)))`
```

5 Tactics Using Invariance under Orthogonal Transformations

This is just one of several analogous tactics that we have defined. Many other tactics also exploit the invariance of many properties under *orthogonal transformations*. These are essentially maps $f : \mathbb{R}^N \to \mathbb{R}^N$ that are linear and preserve dot products:

```
|- ∀f. orthogonal_transformation f ⇔
       linear f ∧ (∀v w. f v dot f w = v dot w)
```

where linearity of a function $f : \mathbb{R}^M \to \mathbb{R}^N$ is defined as

```
|- ∀f. linear f ⇔
       (∀x y. f(x + y) = f x + f y) ∧
       (∀c x. f(c % x) = c % f x)
```

Orthogonal transformations can be characterized in various other ways. For example, a linear map f is an orthogonal transformation iff its corresponding matrix is an orthogonal matrix:

```
|- orthogonal_transformation f ⇔
   linear f ∧ orthogonal_matrix(matrix f)
```

where an $N \times N$ matrix Q is orthogonal if its transpose is also its inverse, i.e. $Q^T Q = QQ^T = 1$:

```
|- orthogonal_matrix(Q) ⇔
       transp(Q) ** Q = mat 1 ∧ Q ** transp(Q) = mat 1`;;
```

It is easy to prove that the determinant of an orthogonal matrix is either 1 or -1, and this gives a classification of orthogonal transformations into 'rotations', where the matrix has determinant 1 and 'rotoinversions' where the matrix has determinant -1. Intuitively, rotations do indeed correspond to rotation about the origin in n-dimensional space, while rotoinversions involve additional reflections. For example, in two dimensions, each rotation matrix is of the form

$$\begin{bmatrix} \cos(\theta) & -\sin(\theta) \\ \sin(\theta) & \cos(\theta) \end{bmatrix}$$

where θ is the (anticlockwise) angle of rotation. Invariance under orthogonal transformation is used in several tactics that allow us to transform a particular nonzero vector into another more convenient one of the same magnitude. The following theorem guarantees us that given any two vectors a and b in $\mathbb{R}^N$ of the same magnitude, there exists an orthogonal transformation that maps one into the other:

```
|- ∀a b:real^N.
       norm(a) = norm(b)
       ⇒ ∃f. orthogonal_transformation f ∧ f a = b
```

If we furthermore want $f : \mathbb{R}^N \to \mathbb{R}^N$ be a rotation, then *almost* the same theorem is true, except that we need the dimension to be at least 2. (An orthogonal transformation taking a vector into its negation in $\mathbb{R}^1$ must have a matrix with determinant -1.)

```
|- ∀a b:real^N.
       2 <= dimindex(:N) ∧ norm(a) = norm(b)
     ⇒ ∃f. orthogonal_transformation f ∧
             det(matrix f) = &1 ∧
             f a = b
```

Just as a reference variable `invariant_under_translation` is used to store theorems asserting the invariance of various concepts under translation, we use a second reference variable `invariant_under_linear` to store analogous theorems for invariance under linear transformations. These in general apply to slightly different classes of linear transformations, almost all of which are more general than orthogonal transformations. For each concept we try to use the most general natural class of linear mappings. Some theorems apply to all linear maps, e.g. the one for convex hulls:

```
|- ∀f s. linear f
         ⇒ convex hull IMAGE f s = IMAGE f (convex hull s)
```

Some apply to all injective linear maps, e.g. those for closedness of a set:

```
|- ∀f s. linear f ∧ (∀x y. f x = f y ⇒ x = y)
         ⇒ (closed (IMAGE f s) ⇔ closed s)
```

Some apply to all bijective (injective and surjective) linear maps, e.g. those for openness of a set:

```
|- ∀f s. linear f ∧
         (∀x y. f x = f y ⇒ x = y) ∧ (∀y. ∃x. f x = y)
         ⇒ (open (IMAGE f s) ⇔ open s)
```

Some apply to all norm-preserving linear maps, e.g. those for angles:

```
|- ∀f a b c. linear f ∧ (∀x. norm(f x) = norm x)
             ⇒ angle(f a,f b,f c) = angle(a,b,c)
```

Note that a norm-preserving linear map is also injective, so this property also suffices for all those requiring injectivity. For a function $f : \mathbb{R}^N \to \mathbb{R}^N$ this property is precisely equivalent to being an orthogonal transformation:

```
|- ∀f:real^N->real^N.
       orthogonal_transformation f ⇔
       linear f ∧ (∀v. norm(f v) = norm v)
```

However, it is important for some other related applications (an example is below) that we make theorems applicable to maps where the dimensions of the domain and codomain spaces are not necessarily the same.

Finally, the most restrictive requirement applies to just one theorem, the one for the vector cross product. This has a kind of chirality, so may have its sign changed by a general orthogonal transformation. Its invariance theorem requires a *rotation* of type $\mathbb{R}^3 \to \mathbb{R}^3$:

```
|- ∀f x y. linear f ∧
             (∀x. norm(f x) = norm x) ∧ det(matrix f) = &1
             ⇒ (f x) cross (f y) = f(x cross y)
```

We actually store the theorem in a slightly peculiar form, which makes it easier to apply uniformly in a framework where we can assume a transformation is a rotation except in dimension 1:

```
|- ∀f x y. linear f ∧ (∀x. norm(f x) = norm x) ∧
             (2 <= dimindex(:3) ⇒ det(matrix f) = &1)
             ⇒ (f x) cross (f y) = f(x cross y)
```

We can implement various tactics that exploit our invariance theorems to make various simplifying transformations without loss of generality:

- GEOM_BASIS_MULTIPLE_TAC chooses an orthogonal transformation or rotation to transform a vector into a nonnegative multiple of a chosen basis vector.
- GEOM_HORIZONTAL_PLANE_TAC chooses a combination of a translation and orthogonal transformation to transform a plane p in $\mathbb{R}^3$ into a 'horizontal' one $\{(x, y, z) \mid z = 0\}$.
- PAD2D3D_TAC transforms a problem in $\mathbb{R}^3$ where all points have zero third coordinate into a corresponding problem in $\mathbb{R}^2$.

The first two work in much the same way as the earlier tactic for choosing the origin. We apply the general theorem, modify all the other quantified variables and then rewrite with invariance theorems. We can profitably think of the basic processes in such cases as instances of general HOL theorems, though this is not actually how they are implemented. For example, we might say that if for each x we can find a 'transform' (e.g. translation, or orthogonal transformation) f such that $f(x)$ is 'nice' (e.g. is zero, or a multiple of some basis vector), and can also deduce for any 'transform' that $P(f(x)) \Leftrightarrow P(x)$, then proving $P(x)$ for all x is equivalent to proving it for 'nice' x. (The theorem that follows is automatically proved by MESON.)

```
|- ∀P. (∀x. ∃f. transform(f) ∧ nice(f x)) ∧
        (∀f x. transform(f) ⇒ (P(f x) ⇔ P x))
        ⇒ ((∀x. P x) ⇔ (∀x. nice(x) ⇒ P(x)))
```

However, in some more general situations we don't exactly want to show that $P(f(x)) \Leftrightarrow P(x)$, but rather that $P(f(x)) \Leftrightarrow P'(x)$ for some related but not identical property

P', for example if we want to transfer a property to a different type. For this reason, it is actually more convenient to observe that we can choose a 'transform' *from* a 'nice' value rather than *to* it, i.e. rely on the following:

```
|- ∀P P'. (∀x. ∃f y. transform(f) ∧ nice(y) ∧ f y = x) ∧
          (∀f x. transform(f) ∧ nice x ⇒ (P(f x) ⇔ P' x))
   ⇒ ((∀x. P x) ⇔ (∀y. nice(y) ⇒ P'(y)))`
```

The advantage of this is that in our approach based on rewriting by applying invariance theorems, the new property P' can emerge naturally from the rewriting of $P(f(x))$, instead of requiring extra code for its computation. Even in cases where the generality is not needed, we typically use this structure, i.e. choose our mapping *from* a 'nice' value.

6   An Extended Example

Let us see a variety of our tactics at work on a problem that was, in fact, the original motivation for most of the work described here.

```
`∀u1:real^3 u2 p a b.
    ~(u1 = u2) ∧
    plane p ∧
    {u1,u2} SUBSET p ∧
    dist(u1,u2) <= a + b ∧
    abs(a - b) < dist(u1,u2) ∧
    &0 <= a ∧
    &0 <= b
    ⇒ (∃d1 d2. {d1,d2} SUBSET p ∧
               &1 / &2 % (d1 + d2) IN affine hull {u1, u2} ∧
               dist(d1,u1) = a ∧
               dist(d1,u2) = b ∧
               dist(d2,u1) = a ∧
               dist(d2,u2) = b)`
```

The first step is to assume without loss of generality that the plane p is $\{(x, y, z) \mid z = 0\}$, i.e. the set of points whose third coordinate is zero, following which we manually massage the goal so that the quantifiers over u_1, u_2, d_1 and d_2 carry explicit restrictions:

```
# e(GEOM_HORIZONTAL_PLANE_TAC `p:real^3->bool` THEN
    ONCE_REWRITE_TAC[TAUT
      `a ∧ b ∧ c ∧ d ⇒ e ⇔ c ∧ a ∧ b ∧ d ⇒ e`] THEN
    REWRITE_TAC[INSERT_SUBSET; EMPTY_SUBSET] THEN
    REWRITE_TAC[IMP_CONJ; RIGHT_FORALL_IMP_THM] THEN
    REWRITE_TAC[GSYM CONJ_ASSOC; RIGHT_EXISTS_AND_THM] THEN
    REWRITE_TAC[IN_ELIM_THM]);;
```

which produces the result:

```
`∀u1. u1$3 = &0
    ⇒ (∀u2. u2$3 = &0
              ⇒ ~(u1 = u2)
              ⇒ plane {z | z$3 = &0}
              ⇒ (∀a b.
                        dist(u1,u2) <= a + b
                    ⇒ abs(a - b) < dist(u1,u2)
                    ⇒ &0 <= a
                    ⇒ &0 <= b
                    ⇒ (∃d1. d1$3 = &0 ∧
                                (∃d2. d2$3 = &0 ∧
                                        &1 / &2 % (d1 + d2) IN
                                        affine hull {u1, u2} ∧
                                        dist(d1,u1) = a ∧
                                        dist(d1,u2) = b ∧
                                        dist(d2,u1) = a ∧
                                        dist(d2,u2) = b))))`
```

Now we apply another WLOG tactic to reduce the problem from $\mathbb{R}^3$ to $\mathbb{R}^2$, and again make a few superficial rearrangements:

```
# e(PAD2D3D_TAC THEN
     SIMP_TAC[RIGHT_IMP_FORALL_THM; IMP_IMP; GSYM CONJ_ASSOC]);;
```

resulting in:

```
`∀u1 u2 a b.
    ~(u1 = u2) ∧
    plane {z | z$3 = &0} ∧
    dist(u1,u2) <= a + b ∧
    abs(a - b) < dist(u1,u2) ∧
    &0 <= a ∧
    &0 <= b
    ⇒ (∃d1 d2.
                &1 / &2 % (d1 + d2) IN affine hull {u1, u2} ∧
                dist(d1,u1) = a ∧
                dist(d1,u2) = b ∧
                dist(d2,u1) = a ∧
                dist(d2,u2) = b)`
```

Although HOL Light does not by default show the types, all the vector variables are now in $\mathbb{R}^2$ instead of $\mathbb{R}^3$ (except for the bound variable z in the residual planarity hypothesis, which is no longer useful anyway). Having collapsed the problem from 3 dimensions to 2 in this way, we finally choose u_1 as the origin:

```
# e(GEOM_ORIGIN_TAC `u1:real^2`);;
val it : goalstack = 1 subgoal (1 total)

`∀u2 a b.
    ~(vec 0 = u2) ∧
    plane {z | z$3 = &0} ∧
    dist(vec 0,u2) <= a + b ∧
    abs(a - b) < dist(vec 0,u2) ∧
    &0 <= a ∧
    &0 <= b
    ⇒ (∃d1 d2.
            &1 / &2 % (d1 + d2) IN affine hull {vec 0, u2} ∧
            dist(d1,vec 0) = a ∧
            dist(d1,u2) = b ∧
            dist(d2,vec 0) = a ∧
            dist(d2,u2) = b)`
```

and now u_2 as a multiple of the first standard basis vector:

```
# e(GEOM_BASIS_MULTIPLE_TAC 1 `u2:real^2`);;
val it : goalstack = 1 subgoal (1 total)

`∀u2. &0 <= u2
      ⇒ (∀a b.
              ~(vec 0 = u2 % basis 1) ∧
              plane {z | z$3 = &0} ∧
              dist(vec 0,u2 % basis 1) <= a + b ∧
              abs(a - b) < dist(vec 0,u2 % basis 1) ∧
              &0 <= a ∧
              &0 <= b
              ⇒ (∃d1 d2.
                      &1 / &2 % (d1 + d2) IN
                      affine hull {vec 0, u2 % basis 1} ∧
                      dist(d1,vec 0) = a ∧
                      dist(d1,u2 % basis 1) = b ∧
                      dist(d2,vec 0) = a ∧
                      dist(d2,u2 % basis 1) = b))`
```

We have thus reduced the original problem to a nicely oriented situation where the points we consider live in 2-dimensional space and are of the form $(0,0)$ and $(0, u_2)$. The final coordinate geometry is now relatively straightforward.

7   Future Work

Our battery of tactics so far is already a great help in proving geometric theorems. There are several possible avenues for improvement and further development.

One is to make use of still broader classes of transformations when handling theorems about correspondingly narrower classes of concepts. For example, some geometric properties, e.g. those involving collinearity and incidence but not distances and angles, are invariant under still broader classes of transformations, such as *shearing*, and this can be of use in choosing an even more convenient coordinate system — see for example the proof of Pappus's theorem given by Chou [1]. Other classes of theorems behave nicely under scaling, so we can freely turn some point $(0, a) \neq (0, 0)$ into just $(0, 1)$ and so eliminate another variable. Indeed, for still more restricted propositions, e.g. those involving only topological properties, we can consider continuous maps that may not be linear.

It would also be potentially interesting to extend the process to additional 'higher-order' properties. To some extent, we already do this with our support for sets of vectors, but we could take it much further, e.g. considering properties of sequences and series and their limits. A nice example where we would like to exploit a higher-order invariance arises in proving that every polygon has a triangulation. The proof given in [4] says: 'Pick the coordinate axis so that no two vertices have the same y coordinate'. It should not be difficult to extend the methods here to prove invariance of notions like 'triangulation of', and we could then pick a suitable orthogonal transformation to force the required property (there are only finitely many vertices but uncountably many angles of rotation to choose).

Another interesting idea would be to reformulate the process in a more 'metalogical' or 'reflective' fashion, by formalizing the class of problems for which our transformations suffice once and for all, instead of rewriting with the current selection of theorems and then either succeeding or failing. From a practical point of view, we think our current approach is usually better. It is actually appealing *not* to delimit the class of permissible geometric properties, but have that class expand automatically as new invariance theorems are added. Moreover, to use the reflective approach we would need to map into some formal syntax, which needs similar transformations anyway. However, there may be some situations where it would be easier to prove general properties in a metatheoretic fashion. For example, a first-order assertion over vectors with M vector variables, even if the pattern of quantification is involved, can be reduced to spaces of dimension $\leq M$ [9]. It should be feasible to handle important special cases (e.g. purely universal formulas) within our existing framework, but exploiting the full result might be a good use for metatheory.

Acknowledgements

The author is grateful to Truong Nguyen, whose stimulating questions on the Flyspeck project mailing list were the inspiration for most of this work.

References

1. Chou, S.-C.: Proving elementary geometry theorems using Wu's algorithm. In: Bledsoe, W.W., Loveland, D.W. (eds.) Automated Theorem Proving: After 25 Years. Contemporary Mathematics, vol. 29, pp. 243–286. American Mathematical Society, Providence (1984)

2. Gordon, M.J.C., Melham, T.F.: Introduction to HOL: a theorem proving environment for higher order logic. Cambridge University Press, Cambridge (1993)
3. Gordon, M., Wadsworth, C.P., Milner, R.: Edinburgh LCF. LNCS, vol. 78. Springer, Heidelberg (1979)
4. Hales, T.C.: Easy pieces in geometry (2007),
   `http://www.math.pitt.edu/~thales/papers/`
5. Hales, T.C.: The Jordan curve theorem, formally and informally. The American Mathematical Monthly 114, 882–894 (2007)
6. Harrison, J.: A HOL theory of Euclidean space. In: Hurd, J., Melham, T. (eds.) TPHOLs 2005. LNCS, vol. 3603, pp. 114–129. Springer, Heidelberg (2005)
7. Klein, F.: Vergleichende Betrachtungen ber neuere geometrische Forschungen. Mathematische Annalen 43, 63–100 (1893); Based on the speech given on admission to the faculty of the Univerity of Erlang in 1872. English translation "A comparative review of recent researches in geometry" in Bulletin of the New York Mathematical Society 2, 460–497 (1892-1893)
8. Noether, E.: Invariante Variationsprobleme. Nachrichten von der Königlichen Gesellschaft der Wissenschaften zu Gttingen: Mathematisch-physikalische Klasse, 235–257 (1918); English translation "Invariant variation problems" by M.A. Travel in 'Transport Theory and Statistical Physics', 1, 183–207 (1971)
9. Solovay, R.M., Arthan, R., Harrison, J.: Some new results on decidability for elementary algebra and geometry. ArXiV preprint 0904.3482 (2009); submitted to Annals of Pure and Applied Logic, `http://arxiv.org/PS_cache/arxiv/pdf/0904/0904.3482v1.pdf`

HOL Light: An Overview

John Harrison

Intel Corporation, JF1-13
2111 NE 25th Avenue
Hillsboro OR 97124
johnh@ichips.intel.com

Abstract. HOL Light is an interactive proof assistant for classical higher-order logic, intended as a clean and simplified version of Mike Gordon's original HOL system. Theorem provers in this family use a version of ML as both the implementation and interaction language; in HOL Light's case this is Objective CAML (OCaml). Thanks to its adherence to the so-called 'LCF approach', the system can be extended with new inference rules without compromising soundness. While retaining this reliability and programmability from earlier HOL systems, HOL Light is distinguished by its clean and simple design and extremely small logical kernel. Despite this, it provides powerful proof tools and has been applied to some non-trivial tasks in the formalization of mathematics and industrial formal verification.

1 LCF, HOL and HOL Light

Both HOL Light and its implementation language OCaml can trace their origins back to Edinburgh LCF, developed by Milner and his research assistants in the 1970s [6]. The LCF approach to theorem proving involves two key ideas:

- All proofs are ultimately performed in terms of a small set of primitive inferences, so provided this small logical 'kernel' is correct the results should be reliable.
- The entire system is embedded inside a powerful functional programming language, which can be used to program new inference rules. The type discipline of the programming language is used to ensure that these ultimately reduce to the primitives.

The original Edinburgh LCF was a theorem prover for Scott's Logic of Computable Functions [16], hence the name LCF. But as emphasized by Gordon [4], the basic LCF approach is applicable to any logic, and now there are descendents implementing a variety of higher order logics, set theories and constructive type theories. In particular, members of the HOL family [5] implement a version of classical higher order logic, hence the name HOL. They take the LCF approach a step further in that all theory developments are pursued 'definitionally'. New mathematical structures, such as the real numbers, may be defined only by exhibiting a model for them in the existing theories (say as Dedekind

S. Berghofer et al. (Eds.): TPHOLs 2009, LNCS 5674, pp. 60–66, 2009.
© Springer-Verlag Berlin Heidelberg 2009

cuts of rationals). New constants may only be introduced by definitional extension (roughly speaking, merely being a shorthand for an expression in the existing theory). This fits naturally with the LCF style, since it ensures that all extensions, whether of the deductive system or the mathematical theories, are consistent per construction.

2   HOL Light's Logical Foundations

HOL Light's logic is simple type theory [1,2] with polymorphic type variables. The terms of the logic are those of simply typed lambda calculus, with formulas being terms of boolean type, rather than a separate category. Every term has a single welldefined type, but each constant with polymorphic type gives rise to an infinite family of constant terms. There are just two primitive types: bool (boolean) and ind (individuals), and given any two types σ and τ one can form the function type $\sigma \to \tau$.[1]

For the core HOL logic, there is essentially only one predefined logical constant, equality ($=$) with polymorphic type $\alpha \to \alpha \to$ bool. However to state one of the mathematical axioms we also include another constant $\varepsilon : (\alpha \to$ bool$) \to \alpha$, explained further below. For equations, we use the conventional concrete syntax $s = t$, but this is just surface syntax for the λ-calculus term $((=)s)t$, where juxtaposition represents function application. For equations between boolean terms we often use $s \Leftrightarrow t$, but this again is just surface syntax.

The HOL Light deductive system governs the deducibility of one-sided sequents $\Gamma \vdash p$ where p is a term of boolean type and Γ is a set (possibly empty) of terms of boolean type. There are ten primitive rules of inference, rather similar to those for the internal logic of a topos [14].

$$\frac{}{\vdash t = t} \text{ REFL}$$

$$\frac{\Gamma \vdash s = t \quad \Delta \vdash t = u}{\Gamma \cup \Delta \vdash s = u} \text{ TRANS}$$

$$\frac{\Gamma \vdash s = t \quad \Delta \vdash u = v}{\Gamma \cup \Delta \vdash s(u) = t(v)} \text{ MK\_COMB}$$

$$\frac{\Gamma \vdash s = t}{\Gamma \vdash (\lambda x.\, s) = (\lambda x.\, t)} \text{ ABS}$$

$$\frac{}{\vdash (\lambda x.\, t)x = t} \text{ BETA}$$

$$\frac{}{\{p\} \vdash p} \text{ ASSUME}$$

---

[1] In Church's original notation, also used by Andrews, these are written o, ι and $\tau\sigma$ respectively. Of course the particular concrete syntax has no logical significance.

$$\frac{\Gamma \vdash p \Leftrightarrow q \quad \Delta \vdash p}{\Gamma \cup \Delta \vdash q} \text{ EQ\_MP}$$

$$\frac{\Gamma \vdash p \quad \Delta \vdash q}{(\Gamma - \{q\}) \cup (\Delta - \{p\}) \vdash p \Leftrightarrow q} \text{ DEDUCT\_ANTISYM\_RULE}$$

$$\frac{\Gamma[x_1, \ldots, x_n] \vdash p[x_1, \ldots, x_n]}{\Gamma[t_1, \ldots, t_n] \vdash p[t_1, \ldots, t_n]} \text{ INST}$$

$$\frac{\Gamma[\alpha_1, \ldots, \alpha_n] \vdash p[\alpha_1, \ldots, \alpha_n]}{\Gamma[\gamma_1, \ldots, \gamma_n] \vdash p[\gamma_1, \ldots, \gamma_n]} \text{ INST\_TYPE}$$

In MK_COMB it is necessary for the types to agree so that the composite terms are well-typed, and in ABS it is required that the variable x not be free in any of the assumptions Γ, while our notation for term and type instantiation assumes capture-avoiding substitution. All the usual logical constants are defined in terms of equality. The conventional syntax $\forall x.\, P[x]$ for quantifiers is surface syntax for $(\forall)(\lambda x.\, P[x])$, and we also use this 'binder' notation for the ε operator.

$$\top =_{def} (\lambda p.\, p) = (\lambda p.\, p)$$
$$\wedge =_{def} \lambda p.\, \lambda q.\, (\lambda f.\, f\, p\, q) = (\lambda f.\, f\, \top\, \top)$$
$$\Longrightarrow =_{def} \lambda p.\, \lambda q.\, p \wedge q \Leftrightarrow p$$
$$\forall =_{def} \lambda P.\, P = \lambda x.\, \top$$
$$\exists =_{def} \lambda P.\, \forall q.\, (\forall x.\, P(x) \Longrightarrow q) \Longrightarrow q$$
$$\vee =_{def} \lambda p.\, \lambda q.\, \forall r.\, (p \Longrightarrow r) \Longrightarrow (q \Longrightarrow r) \Longrightarrow r$$
$$\bot =_{def} \forall p.\, p$$
$$\neg =_{def} \lambda p.\, p \Longrightarrow \bot$$
$$\exists! =_{def} \lambda P.\, \exists P \wedge \forall x.\, \forall y.\, P\, x \wedge P\, y \Longrightarrow x = y$$

These definitions allow us to derive all the usual (intuitionistic) natural deduction rules for the connectives in terms of the primitive rules above. All of the core 'logic' is derived in this way. But then we add three mathematical axioms:

- The axiom of extensionality, in the form of an eta-conversion axiom ETA_AX: $\vdash (\lambda x.t\ x) = t$. We could have considered this as part of the core logic rather than a mathematical axiom; this is largely a question of taste.
- The axiom of choice SELECT_AX, asserting that the Hilbert operator ε is a choice operator: $\vdash P\ x \Longrightarrow P((\varepsilon)P)$. It is only from this axiom that we can deduce that the HOL logic is classical [3].
- The axiom of infinity INFINITY_AX, which implies that the type ind is infinite.

In addition, HOL Light includes two principles of definition, which allow one to extend the set of constants and the set of types in a way guaranteed to preserve consistency. The rule of constant definition allows one to introduce a new constant c and an axiom $\vdash c = t$, subject to some conditions on free variables and polymorphic types in t, and provided no previous definition for c has been introduced. All the definitions of the logical connectives above are introduced in this way. Note that this is 'object-level' definition: the constant and its defining axiom exists in the object logic. Nevertheless, the definitional principles are designed so that they always give a conservative (in particular consistency-preserving) extension of the logic.

3   The HOL Light Implementation

Like other LCF provers, HOL Light is in essence simply a large ML program that defines data structures to represent logical entities, together with a suite of functions to manipulate them in a way guaranteeing soundness. The most important data structures belong to one of the datatypes `hol_type`, `term` and `thm`, which represent types, terms (including formulas) and theorems respectively. The user can write arbitrary programs to manipulate these objects, and it is by creating new objects of type `thm` that one proves theorems. HOL's notion of an 'inference rule' is simply a function with return type `thm`.

In order to guarantee logical soundness, however, all these types are encapsulated as abstract types. In particular, the only way of creating objects of type `thm` is to apply one of the 10 very simple inference rules listed above or to make a new term or type definition. Thus, whatever the circuitous route by which one arrives at it, the validity of any object of type `thm` rests only on the correctness of the rather simple primitive rules (and of course the correctness of OCaml's type checking etc.).

To illustrate how inference rules are represented as functions in OCaml, suppose that two theorems of the form $\Gamma \vdash s = t$ and $\Delta \vdash t = u$ have already been proved and bound to the OCaml variables `th1` and `th2` respectively. In abstract logical terms, the rule TRANS ensures that the theorem $\Gamma \cup \Delta \vdash s = u$ is derivable. In terms of the HOL implementation, one can apply the OCaml function TRANS, of type `thm -> thm -> thm`, to these two theorems as arguments, and hence bind name `th3` to that theorem $\Gamma \cup \Delta \vdash s = u$:

```
let th3 = TRANS th1 th2;;
```

One doesn't normally use such low-level rules much, but instead interacts with HOL via a series of higher-level derived rules, using built-in parsers and printers to read and write terms in a more natural syntax. For example, if one wants to bind the name `th6` to the theorem of real arithmetic that when $|c - a| < e$ and $|b| \leq d$ then $|(a + b) - c| < d + e$, one simply does:

```
let th6 = REAL_ARITH
  `abs(c - a) < e /\ abs(b) <= d ==> abs((a + b) - c) < d + e`;;
```

If the purported fact in quotations turns out not to be true, then the rule will fail by raising an exception. Similarly, any bug in the derived rule (which represents several dozen pages of code written by the present author) would lead to an exception.[2] But we can be rather confident in the truth of any theorem that is returned, since it must have been created via applications of primitive rules, even though the precise choreographing of these rules is automatic and of no concern to the user. What's more, users can write their own special-purpose proof rules in the same style when the standard ones seem inadequate — HOL is fully programmable, yet retains its logical trustworthiness when extended by ordinary users.

Among the facilities provided by HOL is the ability to organize proofs in a mixture of forward and backward steps, which users often find more congenial. The user invokes so-called *tactics* to break down the goal into more manageable subgoals. For example, in HOL's inbuilt foundations of number theory, the proof that addition of natural numbers is commutative is written as follows (the symbol $\forall$ means 'for all'):

```
let ADD_SYM = prove
 (`∀m n. m + n = n + m`,
  INDUCT_TAC THEN
  ASM_REWRITE_TAC[ADD_CLAUSES]);;
```

The tactic INDUCT_TAC uses mathematical induction to break the original goal down into two separate goals, one for $m = 0$ and one for $m + 1$ on the assumption that the goal holds for m. Both of these are disposed of quickly simply by repeated rewriting with the current assumptions and a previous, even more elementary, theorem about the addition operator. The identifier THEN is a so-called *tactical*, i.e. a function that takes two tactics and produces another tactic, which applies the first tactic then applies the second to any resulting subgoals (there are two in this case).

For another example, we can prove that there is a unique x such that $x = f(g(x))$ if and only if there is a unique y with $y = g(f(y))$ using a single standard tactic MESON_TAC, which performs model elimination [15] to prove theorems about first order logic with equality. As usual, the actual proof under the surface happens by the standard primitive inference rules.

```
let WISHNU = prove
 (`(∃!x. x = f (g x)) ⇔ (∃!y. y = g(f y))`,
  MESON_TAC[]);;
```

These and similar higher-level rules certainly make the construction of proofs manageable whereas it would be almost unbearable in terms of the primitive rules alone. Nevertheless, we want to dispel any false impression given by the simple examples above: proofs often require long and complicated sequences of rules. The

---

[2] Or possibly to a true but different theorem being returned, but this is easily guarded against by inserting sanity checks in the rules.

construction of these proofs often requires considerable persistence. Moreover, the resulting proof scripts can be quite hard to read, and in some cases hard to modify to prove a slightly different theorem. One source of these difficulties is that the proof scripts are highly procedural — they are, ultimately, OCaml programs, albeit of a fairly stylized form. There are arguments in favour of a more declarative style for proof scripts, but the procedural approach has its merits too, particularly in applications using specialized derived inference rules [9].

4   HOL Light Applications

Over the years, HOL Light has been used for a wide range of applications, and in concert with this its library of pre-proved formalized mathematics and its stock of more powerful derived inference rules have both been expanded. As well as the usual battery of automated techniques like first-order reasoning and linear arithmetic, HOL Light has been used to explore and apply unusual and novel decision procedures [12,17].

In verification, HOL Light has been used at Intel to verify a number of complex floating-point algorithms including division, square root and transcendental functions [11]. HOL Light seems well-suited to applications like this. It has a substantial library of formalized real analysis, which is used incessantly when justifying the correctness of such algorithms. The flexibility and programmability that the LCF approach affords are also important here since one can write custom derived rules for special tasks like accumulating bounds on rounding errors or enumerating the solutions to Diophantine equations of special kinds.

As for the formalization of mathematics, HOL Light has from the very beginning had a useful formalization of real analysis [10]. More recently this has been substantially developed to cover multivariate analysis in Euclidean space and complex analysis. As well as the miscellany of theorems noted in the list at http://www.cs.ru.nl/~freek/100/, HOL Light has been used to formalize some particularly significant results such as the Jordan Curve Theorem [8] and the Prime Number Theorem [13]. HOL Light is also heavily used in the Flyspeck Project [7] to formalize the proof of the Kepler sphere-packing conjecture, possibly the most ambitious formalization project to date.

References

1. Andrews, P.B.: An Introduction to Mathematical Logic and Type Theory: To Truth Through Proof. Academic Press, London (1986)
2. Church, A.: A formulation of the Simple Theory of Types. Journal of Symbolic Logic 5, 56–68 (1940)
3. Diaconescu, R.: Axiom of choice and complementation. Proceedings of the American Mathematical Society 51, 176–178 (1975)
4. Gordon, M.J.C.: Representing a logic in the LCF metalanguage. In: Néel, D. (ed.) Tools and notions for program construction: an advanced course, pp. 163–185. Cambridge University Press, Cambridge (1982)

5. Gordon, M.J.C., Melham, T.F.: Introduction to HOL: a theorem proving environment for higher order logic. Cambridge University Press, Cambridge (1993)
6. Gordon, M.J.C., Milner, R., Wadsworth, C.P.: Edinburgh LCF. LNCS, vol. 78. Springer, Heidelberg (1979)
7. Hales, T.C.: Introduction to the Flyspeck project. In: Coquand, T., Lombardi, H., Roy, M.-F. (eds.) Mathematics, Algorithms, Proofs. Dagstuhl Seminar Proceedings, vol. 05021. Internationales Begegnungs- und Forschungszentrum fuer Informatik (IBFI), Schloss Dagstuhl, Germany (2006)
8. Hales, T.C.: The Jordan curve theorem, formally and informally. The American Mathematical Monthly 114, 882–894 (2007)
9. Harrison, J.: Proof style. In: Giménez, E., Paulin-Mohring, C. (eds.) TYPES 1996. LNCS, vol. 1512, pp. 154–172. Springer, Heidelberg (1998)
10. Harrison, J.: Theorem Proving with the Real Numbers. Springer, Heidelberg (1998); Revised version of author's PhD thesis
11. Harrison, J.: Floating-point verification using theorem proving. In: Bernardo, M., Cimatti, A. (eds.) SFM 2006. LNCS, vol. 3965, pp. 211–242. Springer, Heidelberg (2006)
12. Harrison, J.: Verifying nonlinear real formulas via sums of squares. In: Schneider, K., Brandt, J. (eds.) TPHOLs 2007. LNCS, vol. 4732, pp. 102–118. Springer, Heidelberg (2007)
13. Harrison, J.: Formalizing an analytic proof of the Prime Number Theorem (dedicated to Mike Gordon on the occasion of his 60th birthday). Journal of Automated Reasoning (to appear, 2009)
14. Lambek, J., Scott, P.J.: Introduction to higher order categorical logic. Cambridge studies in advanced mathematics, vol. 7. Cambridge University Press, Cambridge (1986)
15. Loveland, D.W.: Mechanical theorem-proving by model elimination. Journal of the ACM 15, 236–251 (1968)
16. Scott, D.: A type-theoretical alternative to ISWIM, CUCH, OWHY. Theoretical Computer Science 121, 411–440 (1993); Annotated version of a 1969 manuscript
17. Solovay, R.M., Arthan, R., Harrison, J.: Some new results on decidability for elementary algebra and geometry. ArXiV preprint 0904.3482 (2009); submitted to Annals of Pure and Applied Logic,
   http://arxiv.org/PS_cache/arxiv/pdf/0904/0904.3482v1.pdf

A Brief Overview of MIZAR

Adam Naumowicz and Artur Korniłowicz

Institute of Informatics
University of Białystok, Poland
{adamn,arturk}@math.uwb.edu.pl

Abstract. MIZAR is the name of a formal language derived from informal mathematics and computer software that enables proof-checking of texts written in that language. The system has been actively developed since 1970s, growing into a popular proof assistant accompanied with a huge repository of formalized mathematical knowledge. In this short overview, we give an outline of the key features of the MIZAR language, the ideas and theory behind the system, its main applications, and current development.

1 Introduction

The original goal of MIZAR [8], as conceived by its inventor, Andrzej Trybulec in the early 1970s, was to construct a formal language close to the mathematical jargon used in publications, but at the same time simple enough to enable computerized processing, in particular verification of full logical correctness. The historical description of the first 30 years of MIZAR presented in [7] outlines the evolution of the project, from its relatively modest initial implementations constrained by the capabilities of computers available at that time, to the current proof assistant system successfully used for practical formalization of mathematics.

In late 1980s MIZAR developers started to systematically collect formalizations, which gave rise to the Mizar Mathematical Library - MML. Since then the development of MML has been the central activity in the MIZAR project, as it has been believed that only substantial experience may help in improving the system. When in 1993 there emerged the QED initiative [12] to devise a computer-based database of all mathematical knowledge, strictly formalized and with all proofs having been checked automatically, MIZAR was ready to actively implement that ideology. Although the QED project has not been continued, to some extent the development of MIZAR is still driven in the spirit of its main goals.

Nowadays, when it has been demonstrated by MIZAR and numerous other systems that the computer mechanization of mathematics can be done in practice, the important field for research is how to do it in a relatively easy and comfortable way. Therefore useful constructs that occur in informal mathematics are still being incorporated into the linguistic layer to extend the expressiveness of the MIZAR language, and at the same time the efforts of MIZAR developers

S. Berghofer et al. (Eds.): TPHOLs 2009, LNCS 5674, pp. 67–72, 2009.

concentrate on strengthening the computational power and providing more automation on the side of the verifying software. Both directions in the development are intended to support and intensify further growth of MML.

2   The MIZAR Language

The idea of the MIZAR language being as close as possible to the language used in mathematical papers and being automatically verifiable is achieved by selecting a set of English words and phrases which occur most often in informal mathematics. In fact, MIZAR is intended to be close to the mathematical vernacular on the semantic level even more than on the level of the actual grammar. Therefore the syntax of MIZAR is much simplified compared to the natural language, stylistic variants are not distinguished and instead of English words in some cases their abbreviations are used.

The MIZAR language includes the standard set of first order logical connectives and quantifiers for forming formulas and also provides means for using free second order variables for forming schemes of theorems (infinite families of theorems, e.g. the induction scheme). The rest of MIZAR syntactic constructs is used for writing proofs and defining new mathematical objects.

By its design, MIZAR supports writing proofs in a declarative way (i.e mostly forward reasoning), resembling the standard mathematical practice. The proofs written in MIZAR are constructed according to the rules of the Jaśkowski style of natural deduction [6], or similar systems developed independently by F.B. Fitch [4] or K. Ono [11]. It is this part of the MIZAR language that has had the biggest influence on other systems and became the inspiration to develop similar proof layers on top of several procedural systems. To name the most important ones, there was the system Declare by D. Syme [13], the MIZAR mode for HOL by J. Harrison [5], the Isar language for Isabelle by M. Wenzel [16], MIZAR-light for HOL-light by F. Wiedijk [18] and most recently the declarative proof language (DPL) for Coq by P. Corbineau [3]. The MIZAR way of writing proofs was also the model for the notion of 'formal proof sketches' developed by F. Wiedijk [17].

Following the mathematical practice, MIZAR offers a number of definitional facilities to enable introducing notions of various linguistic categories. Each MIZAR definition defines a new constructor later used in syntactic constructions, and gives its syntax and meaning. In MIZAR terminology, predicates are constructors of (atomic) formulas, modes are constructors of types, functors are constructors of terms, and attributes are constructors of adjectives. The syntactic format of a constructor specifies the symbol of the constructor and the place and number of arguments. The format of a constructor together with information about the types of arguments is called a pattern. The formats are used for parsing and the patterns for identifying constructors.

A constructor may be represented by different patterns as synonyms and antonyms are allowed. The language allows to define prefix, postfix, infix, and also circumfix (for various kinds of brackets) operators. Moreover, MIZAR supports operator overloading to enable using the same 'natural' symbols with a different meaning in different contexts.

3   The MIZAR Proof-Checker

The checker of MIZAR is a disprover based on classical logic. An inference of the form $\frac{\alpha^1,\ldots,\alpha^k}{\beta}$ is transformed to $\frac{\alpha^1,\ldots,\alpha^k,\neg\beta}{\bot}$. A disjunctive normal form (DNF) of the premises is then created and the system tries to refute it

$$\frac{([\neg]\alpha^{1,1} \wedge \ldots \wedge [\neg]\alpha^{1,k_1}) \vee \ldots \vee ([\neg]\alpha^{n,1} \wedge \ldots \wedge [\neg]\alpha^{n,k_n})}{\bot}$$

where $\alpha^{i,j}$ are atomic or universal sentences (negated or not) - for the inference to be accepted, all disjuncts must be refuted.

Internally, all MIZAR formulas are expressed in a simplified "canonical" form (semantic correlates) using only the *verum* constant, negation, conjunction and universal quantifier together with atomic formulas. Thanks to that all inferences valid on the grounds of classical propositional reasoning are automatically accepted by the checker.

The checker is not based on a set of inference rules, but rather on M. Davis's concept of "obviousness w.r.t an algorithm". Its deductive power is still being strengthened by adding new computation mechanisms [10]. The algorithm includes processing of a hierarchy of dependent types ordered by a widening relation and extended by adjectives as type modifiers. MIZAR also supports a notion of polymorphic structure types to facilitate abstract developments in the Bourbaki style.

The MIZAR proof-checking software and a suite of additional utilities can be freely downloaded from the project's web site in a pre-compiled form for most popular hardware and OS combinations. Currently the supported platforms include Intel-based Linux, Solaris, FreeBSD, Darwin/Mac OS X and Microsoft Windows, and also Darwin/Mac OS X and Linux on PowerPC. There are also test releases available for palmtops running Linux on ARM processors. Technically, MIZAR processes its input files as a batch-like compiler. Its output contains marks of unaccepted fragments of the source text, so the user may proceed filling the gaps in reasoning until no errors are flagged. Essentially the system can be run from the command line, but the preferred method for interactive use is Josef Urban's Emacs-lisp Mizar mode that provides a fully functional interface to the system.

4   MIZAR Mathematical Library

When the systematic collecting of MIZAR formalizations started around 1989, there were plans to build several libraries with different axiomatics, e.g. based on various set theories or on the Peano arithmetic. However, when it became apparent that simultaneous maintaining several such libraries was a non-trivial task, the decision was made to support only one centralized repository, which has evolved into today's MML.

Since then all MIZAR developments have been created in a steady fashion on top of the chosen axiomatics and previously formalized data. MML is today a collection of interrelated texts (articles) fully checked for correctness by the MIZAR checker, based on the axioms of the Tarski-Grothendieck set theory, which basically is the Zermelo-Fraenkel set theory extended with Tarski's axiom that provides the existence of arbitrarily large strongly inaccessible cardinals [14].

The most recent distribution of MML (version 4.117.1046) includes 1047 articles written by 219 authors and comprises 48199 theorems, 9262 definitions (using 7001 different symbols), 757 schemes and 8573 registrations (statements about relations between adjectives that can be processed automatically). Of course the facts collected as 'theorems' vary in their importance. From a point of view of a mathematician, most of them are rather simple lemmas. However, several fields of mathematics are relatively well-developed and significant mathematical results have been formalized and included in MML. For example, the library contains many proofs of advanced topological theorems, like the Jordan Curve Theorem, the Brouwer Fixed Point Theorem, Urysohn's Lemma, the Tichonov Theorem, or the Tietze Extension Theorem, and also such fundamental statements in other domains like the Gödel Completeness Theorem, the Fundamental Theorem of Arithmetic, the Fundamental Theorem of Algebra, or Sylow's Theorems to name just a few.

MML is subject to continuous revisions performed by the Library Committee or Development Committee – two agendas of the Association of Mizar Users working on maintaining and optimizing the contents of the repository. The revisions are most often yielded by the strengthening of the MIZAR checker, reformulations of statements using new syntactic constructs, elimination of repetitions or 'weaker' statements, better solutions or improved ways of formalization, enhancement of proofs, or reorganization of items among articles. Whenever an article or a sequence of articles is revised, the rest of articles must always be checked, and improved if necessary, to keep the whole repository coherent.

5 Main Applications of MIZAR

Apart from the long-term goal of developing MML into a database for mathematics, the most important applications of MIZAR today are playing the role of a proof assistant to support creating rigorous mathematics, in mathematics education and in software and hardware verification.

To facilitate the whole process of writing formal mathematics, several external systems have been developed that complement the MIZAR proof checker. For example, effective semantic-based information retrieval, i.e., searching, browsing and presentation of MML can be done with the MML Query system developed by G. Bancerek [1]. Several sites provide an on-line MIZAR processor, writing proofs may also be assisted by the systems MoMM (a matching and interreduction tool) and the MIZAR Proof Advisor developed by J. Urban. The contents of MML as well as newly created documents can be presented in various userfriendly formats, including a semantically-linked XML-based web pages [15] or an automatically generated translation into English in the form of an electronic and printed journal, *Formalized Mathematics*.

For several decades Mizar has been used for educational purposes on various levels: from secondary school to doctoral studies. Usually the teaching was organized as Mizar-aided courses, most typically on introduction to logic, topology, lattice theory, general and universal algebra, category theory, etc. Recent applications in regular university-level courses being part of the obligatory curriculum for CS students at the University of Białystok are presented in [2,9].

Mizar has been used to define mathematical models of computers and prove properties of their programs. One approach which is well-developed in MML is based on the theory of random access Turing machines. There are also other formalized attempts to model and analyze standalone algorithms. Numerous MML articles are also devoted to the construction and analysis of gates and digital circuits.

6 Current Development

Despite its origins and initial implementations in 1970s, Mizar is still being actively developed. The development concerns both the language and the proof-checking software. The evolution of the Mizar language goes into the direction of best possible expressiveness, and still new useful language constructs are identified in mathematical texts and transformed into the formal setting of Mizar. Much work in this area has been concentrated on the processing of attributes, which in the most recent implementation can be expressed with their own visible arguments (e.g. n-dimensional, X-valued, etc.) in much the same way types have been constructed. As the Mizar type checking mechanism uses quite powerful automation techniques based on adjectives, the change makes it possible to formalize many concepts in a more natural and, what is maybe even more important, automatic way.

The capabilities of the proof-checker has recently been strengthened by providing means for a more complete adjective processing and the use of global choice (selecting unique representatives of types) to enable eliminating the so called 'permissive' definitions. The system has also been equipped with an efficient method of identifying semantically equivalent operations defined in different contexts, e.g the addition of numbers and the corresponding operation in the field of real numbers. The system has also been extended with more powerful automation of numerical computations.

Between the planned and currently considered future enhancements there are several forms of ellipsis (the ubiquitous '...' notation) and a syntactic extension to support binding operators like the sum, product or integral.

7 Miscellanea

More information on Mizar can be found on the project's web page [8] or its several mirrors. The site contains information on the Mizar language (e.g. the formal syntax, available manuals and other bibliographic links) and provides downloading of the system and its library. There are also pointers to other

MIZAR-related facilities, e.g. Mizar-Forum (a general-purpose MIZAR mailing list), Mizar User Service (an e-mail-based troubleshooting helpdesk), MIZAR TWiki (a collaboration platform) or the Association of Mizar Users (an international organization of active users and developers).

References

1. Bancerek, G., Rudnicki, P.: Information retrieval in MML. In: Asperti, A., Buchberger, B., Davenport, J.H. (eds.) MKM 2003. LNCS, vol. 2594, pp. 119–132. Springer, Heidelberg (2003)
2. Borak, E., Zalewska, A.: Mizar course in logic and set theory. In: Kauers, M., Kerber, M., Miner, R., Windsteiger, W. (eds.) MKM/CALCULEMUS 2007. LNCS, vol. 4573, pp. 191–204. Springer, Heidelberg (2007)
3. Corbineau, P.: A declarative language for the Coq proof assistant. In: Miculan, M., Scagnetto, I., Honsell, F. (eds.) TYPES 2007. LNCS, vol. 4941, pp. 69–84. Springer, Heidelberg (2008)
4. Fitch, F.B.: Symbolic Logic. An Introduction. The Ronald Press Company (1952)
5. Harrison, J.: A Mizar Mode for HOL. In: von Wright, J., Harrison, J., Grundy, J. (eds.) TPHOLs 1996. LNCS, vol. 1125, pp. 203–220. Springer, Heidelberg (1996)
6. Jaśkowski, S.: On the rules of supposition in formal logic. Studia Logica 1 (1934)
7. Matuszewski, R., Rudnicki, P.: Mizar: the first 30 years. Mechanized Mathematics and Its Applications 4(1), 3–24 (2005)
8. Mizar home page: http://mizar.org
9. Naumowicz, A.: Teaching How to Write a Proof. In: Formed 2008: Formal Methods in Computer Science Education, pp. 91–100 (2008)
10. Naumowicz, A., Byliński, C.: Improving Mizar texts with properties and requirements. In: Asperti, A., Bancerek, G., Trybulec, A. (eds.) MKM 2004. LNCS, vol. 3119, pp. 290–301. Springer, Heidelberg (2004)
11. Ono, K.: On a practical way of describing formal deductions. Nagoya Mathematical Journal 21 (1962)
12. QED Manifesto: http://www.rbjones.com/rbjpub/logic/qedres00.htm
13. Syme, D.: Three tactic theorem proving. In: Bertot, Y., Dowek, G., Hirschowitz, A., Paulin, C., Théry, L. (eds.) TPHOLs 1999. LNCS, vol. 1690, pp. 203–220. Springer, Heidelberg (1999)
14. Trybulec, A.: Tarski Grothendieck set theory. Formalized Mathematics 1(1), 9–11 (1990)
15. Urban, J.: XML-izing Mizar: Making Semantic Processing and Presentation of MML Easy. In: Kohlhase, M. (ed.) MKM 2005. LNCS, vol. 3863, pp. 346–360. Springer, Heidelberg (2006)
16. Wenzel, M., Wiedijk, F.: A comparison of Mizar and Isar. Journal of Automated Reasoning 29(3-4), 389–411 (2002)
17. Wiedijk, F.: Formal Proof Sketches. In: Berardi, S., Coppo, M., Damiani, F. (eds.) TYPES 2003. LNCS, vol. 3085, pp. 378–393. Springer, Heidelberg (2004)
18. Wiedijk, F.: Mizar Light for HOL Light. In: Boulton, R.J., Jackson, P.B. (eds.) TPHOLs 2001. LNCS, vol. 2152, pp. 378–393. Springer, Heidelberg (2001)

A Brief Overview of Agda –
A Functional Language with Dependent Types

Ana Bove, Peter Dybjer, and Ulf Norell

Chalmers University of Technology, Gothenburg, Sweden
{bove,peterd,ulfn}@chalmers.se

Abstract. We give an overview of Agda, the latest in a series of dependently typed programming languages developed in Gothenburg. Agda is based on Martin-Löf's intuitionistic type theory but extends it with numerous programming language features. It supports a wide range of inductive data types, including inductive families and inductive-recursive types, with associated flexible pattern-matching. Unlike other proof assistants, Agda is not tactic-based. Instead it has an Emacs-based interface which allows programming by gradual refinement of incomplete type-correct terms.

1 Introduction

A dependently typed programming language and proof assistant. Agda is a functional programming language with dependent types. It is an extension of Martin-Löf's intuitionistic type theory [12,13] with numerous features which are useful for practical programming. Agda is also a proof assistant. By the Curry-Howard identification, we can represent logical propositions by types. A proposition is proved by writing a program of the corresponding type. However, Agda is primarily being developed as a programming language and not as a proof assistant.

Agda is the latest in a series of implementations of intensional type theory which have been developed in Gothenburg (beginning with the ALF-system) since 1990. The current version (Agda 2) has been designed and implemented by Ulf Norell and is a complete redesign of the original Agda system. Like its predecessors, the current Agda supports a wide range of inductive data types, pattern matching, termination checking, and comes with an interface for programming and proving by direct manipulation of proof terms. On the other hand, the new Agda goes beyond the earlier systems in several respects: flexibility of pattern-matching, more powerful module system, flexible and attractive concrete syntax (using unicode), etc.

A system for functional programmers. A programmer familiar with a standard functional language such as Haskell or OCaml will find it easy to get started with Agda. Like in ordinary functional languages, programming (and proving) consists of defining data types and recursive functions. Moreover, users familiar with Haskell's generalised algebraic data types (GADTs) will find it easy to use Agda's inductive families [5].

S. Berghofer et al. (Eds.): TPHOLs 2009, LNCS 5674, pp. 73–78, 2009.

The Agda wiki. More information about Agda can be found on the Agda wiki [1]. There are tutorials [3,15], a guide to editing, type checking, and compiling Agda code, a link to the standard library, and much else. There is also a link to Norell's PhD thesis [14] with a language definition and detailed discussions of the features of Agda.

2   Agda Features

We begin by listing the logically significant parts of Agda.

Logical framework. The core of Agda is Martin-Löf's logical framework [13] which gives us the type Set and dependent function types $(x : A) \to B$ (using Agda's syntax). Agda's logical framework also provides record types and a countable sequence of larger universes $Set = Set_0, Set_1, Set_2, \ldots$.

Data type definitions. Agda supports a rich family of strictly positive inductive and inductive-recursive data types and families. Agda checks that the data type definitions are well-formed according to a discipline similar to that in [6,7].

Recursive function definitions. One of Agda's main features is its flexible pattern matching for inductive families. A coverage checker makes sure the patterns cover all possible cases. As in Martin-Löf type theory, all functions definable in Agda must terminate, which is ensured by the termination checker.

Codata. The current version of Agda also provides coinductive data types. This feature is however somewhat experimental and not yet stable.

Agda also provides several features to make it useful in practice:

Concrete syntax. The concrete syntax of Agda is much inspired by Haskell, but also contains a few distinctive features such as mixfix operators and full support for unicode identifiers and keywords.

Implicit arguments. The mechanism for implicit arguments allows the omission of parts of the programs that can be inferred by the typechecker.

Module system. Agda's module system supports separate compilation and allows parametrised modules. Together with Agda's record types, the module system provides a powerful mechanism for structuring larger developments.

Compilation. There is a simple compiler that compiles Agda programs via Haskell and allows Haskell functions to be called from within Agda.

Emacs interface. Using Agda's Emacs interface, programs can be developed incrementally, leaving parts of the program unfinished. By type checking the unfinished program, the programmer can get useful information on how to fill in the missing parts. The Emacs interface also provides syntax highlighting and code navigation facilities.

3   Agda and Some Related Languages

Agda and Martin-Löf type theory. Agda is an extension of Martin-Löf's type theory. An implementation of the latter in Agda can be found on the Agda wiki [1]. Meaning explanations of foundational interest for type theory have been provided by Martin-Löf [11,12], and all constructions in Agda (except codata) are intended to satisfy them. Agda is thus a predicative theory.

Agda and Coq. The most well-known system with dependent types which is based on the Curry-Howard identification is Coq [2]. Coq is an implementation of the Calculus of Inductive Constructions, an extension of the Calculus of Constructions [4] with inductive (but not inductive-recursive) types. Unlike Agda, Coq has an impredicative universe Prop. Moreover, for the purpose of program extraction, there is a distinction between Prop and Set in Coq which is not present in Agda. There are many other differences between Agda and Coq. For example, Agda's pattern matching for inductive families is more flexible than Coq's. On the other hand, Coq supports tactical theorem proving in the tradition of LCF [10], but Agda does not.

Agda and Haskell. Haskell has GADTs, a feature which mimics inductive families by representing them by type-indexed types. A fundamental difference is that Haskell allows partial general recursive functions and non-strictly positive data types. Hence, logic cannot be obtained by the Curry-Howard correspondence.

Other languages with dependent types. There are nowadays a number of functional languages with dependent types (some with and some without general recursion). Among these McBride's Epigram [8] is closest in spirit to Agda.

4   Example: Equational Proofs in Commutative Monoids

We will now show some of the code for a module which decides equality in commutative monoids. This is an example of reflection, a technique which makes it possible to program and use efficient verified decision procedure inside the system. Reflection was for example used extensively by Gonthier in his proof of the four colour theorem [9].

An example of a commutative monoid is the natural numbers with addition. Thus our decision procedure can automatically prove arithmetic equations such as

$\forall\ n\ m\ \rightarrow\ (n\ +\ m)\ +\ n\ \equiv\ m\ +\ (n\ +\ n).$

The above is a valid type in Agda syntax. To prove it in Agda we create a file `Example.agda` with the following content:

```
module Example where

open import Data.Nat
open import Relation.Binary.PropositionalEquality
```

```
prf : ∀ n m → (n + m) + n ≡ m + (n + n)
prf n m = ?
```

Natural numbers and propositional equality are imported from the standard library and opened to make their content available. Finally, we declare a proof object prf, the type of which represents the proposition to be proved; here $\forall x \to B$ is an abbreviation of $(x : A) \to B$ which does not explicitly mention the argument type. The final line is the incomplete definition of prf: it is a function of two arguments, but we do not yet know how to build a proof of the equation so we leave a "?" in the right hand side. The "?" is a placeholder that can be stepwise refined to obtain a complete proof.

In this way we can manually build a proof of the equation from associativity and commutativity of +, and basic properties of equality which can be found in the standard library. Manual equational reasoning however can become tedious for complex equations. We shall therefore write a general procedure for equational reasoning in commutative monoids, and show how to use it for proving the equation above.

Decision procedure for commutative monoids. First we define monoid expressions as an inductive family indexed by the number of variables:

```
data Expr n : Set where
  var  : Fin n → Expr n
  _⊕_  : Expr n → Expr n → Expr n
  zero : Expr n
```

Fin n is a finite set with n elements; there are at most n variables. Note that infix (and mixfix) operators are declared by using underscores to indicate where the arguments should go.

To decide whether two monoid expressions are equal we normalise them and compare the results. The normalisation function is

```
norm : ∀ {n} → Expr n → Expr n
```

Note that the first argument (the number of variables) is enclosed in braces, which signifies that it is implicit. To define this function we employ normalisation by evaluation, that is, we first interpret the expressions in a domain of "values", and then reify these values into normal expressions. Below, we omit the definitions of eval and reify and give only their types:

```
norm = reify ∘ eval

eval  : ∀ {n} → Expr n → NF n
reify : ∀ {n} → NF n → Expr n
```

The values in NF n are vectors recording the number of occurrences of each variable:

```
NF : ℕ → Set
NF n = Vec ℕ n
```

Next we define the type of equations between monoid expressions:

```
data Eqn n : Set where
  _==_ : Expr n → Expr n → Eqn n
```

We can define our arithmetic equation above as follows:

```
eqn₁ : Eqn 2
eqn₁ = build 2 λ a b → a ⊕ b ⊕ a == b ⊕ (a ⊕ a)
```

where we have used an auxiliary function `build` which builds an equation in `Eqn n` from an n-place curried function by applying it to variables.

Equations will be proved by normalising both sides:

```
simpl : ∀ {n} → Eqn n → Eqn n
simpl (e₁ == e₂) = norm e₁ == norm e₂
```

We are now ready to define a general decision procedure for arbitrary commutative monoids (the complete definition is given later):

```
prove : ∀ {n} (eqn : Eqn n) ρ → Prf (simpl eqn) ρ → Prf eqn ρ
```

The function takes an equation and an environment in which to interpret it, and builds a proof of the equation given a proof of its normal form. The definition of `Prf` will be given below.

We can instantiate this procedure to the commutative monoid of natural numbers and apply it to our equation, an environment with the two variables, and a proof of the normalised equation. Since the two sides of the equation will be equal after normalisation we prove it by reflexivity:

```
prf : ∀ n m → (n + m) + n ≡ m + (n + n)
prf n m = prove eqn₁ (n :: m :: []) ≡-refl
```

The `prove` function is defined in a module `Semantics` which is parametrised over an arbitrary commutative monoid

```
module Semantics (CM : CommutativeMonoid) where

  open CommutativeMonoid CM renaming (carrier to C)
```

Opening the `CommutativeMonoid` module brings into scope the carrier `C` with its equality relation `_≈_` and the monoid operations `_•_` and ε. A monoid expression is interpreted as a function from an environment containing values for the variables to an element of `C`.

```
Env : ℕ → Set
Env n = Vec C n

[_] : ∀ {n} → Expr n → Env n → C
[ var i ] ρ = lookup i ρ
[ a ⊕ b ] ρ = [ a ] ρ • [ b ] ρ
[ zero ] ρ = ε
```

Equations are also interpreted:

```
Prf : ∀ {n} → Eqn n → Env n → Set
Prf (e₁ == e₂) ρ = ⟦ e₁ ⟧ ρ ≈ ⟦ e₂ ⟧ ρ
```

One can prove that the normalisation function is sound in the sense that the normal form is semantically equal to the original expression in any environment:

```
sound : ∀ {n} (e : Expr n) ρ → ⟦ e ⟧ ρ ≈ ⟦ norm e ⟧ ρ
```

Hence, to prove an equation it suffices to prove the normalised version. The proof uses the module `Relation.Binary.EqReasoning` from the standard library for the equational reasoning:

```
prove : ∀ {n} (eqn : Eqn n) ρ → Prf (simpl eqn) ρ → Prf eqn ρ
prove (e₁ == e₂) ρ prf =
  begin ⟦ e₁ ⟧ ρ        ≈⟨ sound e₁ ρ ⟩
        ⟦ norm e₁ ⟧ ρ   ≈⟨ prf ⟩
        ⟦ norm e₂ ⟧ ρ   ≈⟨ sym (sound e₂ ρ) ⟩
        ⟦ e₂ ⟧ ρ
```
□

The complete code is available on the Agda wiki [1].

References

1. Agda wiki page, http://wiki.portal.chalmers.se/agda/
2. Bertot, Y., Castéran, P.: Interactive Theorem Proving and Program Development. In: Coq'Art: The Calculus of Inductive Constructions. Springer, Heidelberg (2004)
3. Bove, A., Dybjer, P.: Dependent types at work. In: Barbosa, L., Bove, A., Pardo, A., Pinto, J.S. (eds.) LerNet ALFA Summer School 2008. LNCS, vol. 5520, pp. 57–99. Springer, Heidelberg (to appear, 2009)
4. Coquand, T., Huet, G.: The calculus of constructions. Information and Computation 76, 95–120 (1988)
5. Dybjer, P.: Inductive families. Formal Aspects of Computing 6, 440–465 (1994)
6. Dybjer, P.: A general formulation of simultaneous inductive-recursive definitions in type theory. Journal of Symbolic Logic 65(2) (June 2000)
7. Dybjer, P., Setzer, A.: Indexed induction-recursion. Journal of Logic and Algebraic Programming 66(1), 1–49 (2006)
8. Epigram homepage, http://www.e-pig.org
9. Gonthier, G.: The four colour theorem: Engineering of a formal proof. In: Kapur, D. (ed.) ASCM 2007. LNCS, vol. 5081, p. 333. Springer, Heidelberg (2008)
10. Gordon, M., Milner, R., Wadsworth, C.: Edinburgh LCF. In: Kahn, G. (ed.) Semantics of Concurrent Computation. LNCS, vol. 70. Springer, Heidelberg (1979)
11. Martin-Löf, P.: Constructive mathematics and computer programming. In: Logic, Methodology and Philosophy of Science, VI, 1979, pp. 153–175. North-Holland, Amsterdam (1982)
12. Martin-Löf, P.: Intuitionistic Type Theory. Bibliopolis, Napoli (1984)
13. Nordström, B., Petersson, K., Smith, J.M.: Programming in Martin-Löf's Type Theory. An Introduction. Oxford University Press, Oxford (1990)
14. Norell, U.: Towards a practical programming language based on dependent type theory. PhD thesis, Chalmers University of Technology (2007)
15. Norell, U.: Dependently typed programming in Agda. In: Lecture Notes from the Summer School in Advanced Functional Programming (2008) (to appear)

The Twelf Proof Assistant*

Carsten Schürmann

IT University of Copenhagen
carsten@itu.dk

Logical framework research is based on the philosophical point of view that it should be possible to capture mathematical concepts such as proofs, logics, and meaning in a formal system — directly, adequately (in the sense that there are no spurious or exotic witnesses), and without having to commit to a particular logical theory. Instead of working with one general purpose representation language, we design special purpose logical frameworks for capturing reoccurring concepts for special domains, such as, for example, variable renaming, substitution application, and resource management for programming language theory. Most logical frameworks are based on constructive type theories, such as Isabelle (on the simply-typed λ-calculus), LF [HHP93] (on the dependently typed λ-calculus), and LLF (on a linearly typed λ-calculus). The representational strength of the logical framework stems from the choice of definitional equality on terms. For example, α-conversion models the tacit renaming of variables, β-contraction models substitution application, and η-expansion guarantees the adequacy of encodings.

The Twelf system [PS99] is an implementation of the logical framework LF and its equational theory based on $\alpha\beta\eta$. It was originally released by Pfenning and Schürmann in 1999 and supersedes the Elf system. Twelf is a proof assistant, also called a meta-logical framework, that excels at representing deductive systems with side conditions (such as the Eigen variable condition) and judgments with contexts that have the usual intuitionistic properties, such as weakening, contraction, and exchange. Twelf provides a logic programming interpreter to experiment with encodings and a reasoning engine to verify their meta-theoretic properties, such as, for example, cut-elimination, semantical equivalence, type soundness. Furthermore, it provides a module system to organize large developments.

Since its release, the Twelf system has become a popular tool for reasoning about the design and properties of modern programming languages and logics. It has been used, for example, to verify the soundness of typed assembly language [Cra03] and Standard ML [LCH07], for checking cut-elimination proofs for intuitionistic and classical logic [Pfe95], and for specifying and validating logic morphisms, for example, between HOL and Nuprl [SS06].

In this paper, we illustrate a few of Twelf's features by example. In particular, we show how to use Twelf to represent deductive systems in Section 1, how to reason with Twelf in Section 2, and how Twelf supports the modular design of code in Section 3.

---

* This work is supported in part by NABITT grant 2106-07-0019 of the Danish Strategic Research Council.

1   Representation

In order to use Twelf, one has to subscribe to the judgments-as-types representation paradigm, which means that one should appreciate that *judgments*, such as, for example, "$A ::= p \mid A_1 \supset A_2 \mid \neg A$ is a formula" and "A is derivable in a logic written as $\vdash A$" are best represented as *types* or *families of types*. Type families may then be declared in Twelf as constant o : type and |- : o -> type, respectively. Consider the following Twelf signature that gives an adequate representation of a natural deduction calculus for the fragment of propositional logic defined by implication and negation.

```
%sig IL = {
    o   : type.                              %name o A.
    |-  : o -> type.                         %prefix 9 |-.
    =>  : o -> o -> o.                       %infix right 10 =>.
    ~   : o -> o.
    =>I : (|- A -> |- B) -> |- A => B.
    =>E : |- A => B -> |- A -> |- B.
    ~I  : ({p:o} |- A -> |- p) -> |- ~ A.
    ~E  : |- ~ A -> |- A -> |- B.
    n   = [p:o] ~ (~ p).
    nI  : |- A -> |- n A = [D] ~I [p:o] [u: |- ~ A] ~E u D.
}.
```

Twelf is an implementation of LF type theory, and therefore certain syntactical connectives are primitive. type stands for the kind *type*, .->. for the non-dependent function arrow, {.}. for the dependent function arrow, and [.]. for λ-abstraction. Twelf has a powerful type inference algorithm based on higher-order unification that can often reconstruct omitted arguments and type labels. Working with Twelf corresponds to declaring and defining object and type level constants interleaved by the occasional Twelf specific instruction to check meta theoretic properties, record extra logical information, or execute queries. Examples of these instructions include %name, %prefix, and %infix that record the user's preference for name and fixity information. We will encounter a few more instructions below.

Returning to the signature above, the third and fourth line declare the connectives => for ⊃, and ~ for ¬. The fifth to eighth line declare the respective standard introduction and elimination rules (whose mathematical depiction we omit in the interest of space). We point out that every capitalized variable, such as A or B, should be thought of as implicitly Π-quantified. Furthermore, note how the LF function arrow is used to encode the hypothetical derivations in the premiss of =>I and ~I. It shamelessly uses the LF context to represent hypotheses. This particular technique of representation illustrates the true strength of the logical framework. It is sound because hypotheses in our logic enjoy the same weakening, contraction, exchange, and substitution properties as the context in LF. Moreover, LF's equational theory provides us with a free substitution principle for hypotheses. As we rest the representation on the higher-order features of LF, this technique is also called *higher-order abstract syntax*.

The bottom two definitions (as opposed to declarations) in the Twelf signature above introduce n as an abbreviation for the double negation ~~ and nI as a derived rule of inference stating soundness.

2   Reasoning

Besides being able to reconstruct and check LF types, Twelf is designed as a proof assistant that allows Twelf users to reason about the meta-theoretic properties of their encodings. In Twelf we separate cleanly the logical framework LF for representation from a meta-logic $\mathcal{M}_\omega$ for reasoning. It is well-known that in LF every term reduces to a unique β-normal η-long form that is also called *canonical form*. These forms are inductively defined and give rise to induction principles that are built into the meta-logic $\mathcal{M}_\omega$. These principles allow the Twelf user to reason about LF encodings even though they might be defined using higher-order abstract syntax.

If we restrict ourselves to the Π_2-fragment of $\mathcal{M}_\omega$, meta proofs can thankfully be encoded as relations in LF — with the only caveat being that we need to check that those relations behave as total functions (when executed on the Twelf logic programming engine). For every well-typed input, those functions must compute well-typed outputs, which means that computation must be terminating and may not get stuck. As an illustrative example of such a meta proof, consider the following signature that defines the Hilbert calculus and gives a proof of the deduction theorem.

```
%sig HILBERT = {
  o   : type.                              %name o A.
  |-  : o -> type.                         %prefix 9 |-.
  =>  : o -> o -> o.                        %infix right 10 =>.
  K   : |- A => B => A.
  S   : |- (A => B => C) => (A => B) => A => C.
  MP  : |- A => B -> |- A -> |- B.

  ded : (|- A -> |- B) -> |- A => B -> type.    %mode ded +D -E.
  aK  : ded ([x] K) (MP K K).
  aS  : ded ([x] S) (MP K S).
  aID : ded ([x] x) (MP (MP S (K : |- A => (A => A) => A)) K).
  aMP : ded ([x] MP (D x) (E x)) (MP (MP S D') E')
          <- ded ([x] D x) D' <- ded ([x] E x) E'.
  %worlds () (ded _ _).
  %total D (ded D _).
}.
```

The first six lines in this signature define the syntax of formulas, and the Hilbert calculus for the implicational fragment of propositional logic. The remainder of this signature defines the meta-proof as a relation between hypothetical derivations $\vdash B$ under the assumption $\vdash A$ and the internalized version $\vdash A \supset B$. The input is represented as a function from $|- A \to |- B$, and because every term of this type has a canonical form, there are only four cases that one needs to consider: `[x] K`, `[x] S`, `[x] x` and `[x] MP (D x) (E x)`, where `D` and `E` are canonical as well. No matter which input of type $|- A \to |- B$ is given, Twelf's operational semantic will always find a case that matches.

Twelf offers various tools to inspect the properties of this relation, first and foremost a mode checker `%mode` that checks that for ground[1] inputs one can expect ground outputs, a world checker `%worlds` that checks that the implicit LF context is regularly built[2] and a totality checker `%total` that checks that the relation is indeed a total function, i.e. terminating and all cases are covered. In this example, the world is empty as indicated by () and the termination uses the subterm ordering on the first argument of `ded`, which is named `D`.

3   Organization

Last but not least, Twelf offers a deceptively simple but useful module system. The module system provides the user with the ability to manage name spaces but does not extend LF. In fact, every Twelf development that contains module system features can be elaborated into an equivalent and pure LF signature. Using *structures*, one may embed one signature into another, and using *views* one may define maps from one signature to another.

Recall the definition of intuitionistic logic from Section 1. The following Twelf development defines classical logic as an extension of intuitionistic logic by the law of the excluded middle.

```
%sig CL = {
    %struct IL : IL %open o |- => =>I =>E ~ n ~I ~E nI.
    exm : |- ~ (~ A) => A.
}.
```

`%struct` imports IL into CL, and the `%open` allows the user to refer unqualified to the subsequent list of constant names.

Next, we give the Kolmogorov translation from classical logic into intuitionistic logic. To get this to work, we need to think of the usual turnstyle as $\vdash A$ as $\vdash \neg\neg A$. This is possible by defining a view in two steps. First, we define a view from IL to IL.

```
%view KOLMIL : IL -> IL = {
    o    := o.
    |-   := [x] |- n x.
    =>   := [x][y] (n x) => (n y).
    ~    := [x] ~ x.
```

---

[1] A term is ground if it doesn't contain free logic variables.

[2] Recall that terms may be open when using higher-order abstract syntax.

```
=>I := [A][B][D] nI (=>I D).
=>E := [A][B][D][E] ~I [p][u] ~E D (~I [q][v] ~E (=>E v E) u).
~I  := [A][D] ~I [q][u] ~E (D (~ A) u) u.
~E  := [A][C][D][E] ~I [p][u] ~E D E .
}.
```

For example, =>I maps to a term representing a derivation that under the assumption that ⊢ ¬¬A implies ⊢ ¬¬B the following holds: ⊢ ¬¬(A ⊃ B). Second, we extend this view to to a view for classical logic from CL to IL.

```
%view KOLM : CL -> IL = {
  %struct IL := KOLMIL.
  exm := [A] ~I [p] [u] ~E u (=>I [u] ~I [p] [v]
                ~E u (~I [q][w] ~E w v)).
}.
```

In this view, the substructure IL is mapped to the view KOLMIL, and the law of the excluded middle to a term representing a derivation of ⊢ ¬¬(¬¬A ⊃ A).

The Twelf system and documentation can be accessed from our homepage at http://www.twelf.org. More information about the module system is available from http://www.twelf.org/~mod.

References

[Cra03]   Crary, K.: Toward a foundational typed assembly language. In: Morrisett, G. (ed.) Proceedings of the 30th ACM Symposium on Principles of Programming Languages, New Orleans, Louisiana. SIGPLAN Notices, vol. 38(1), pp. 198–212. ACM Press, New York (2003)

[HHP93]   Harper, R., Honsell, F., Plotkin, G.: A framework for defining logics. Journal of the Association for Computing Machinery 40(1), 143–184 (1993)

[LCH07]   Lee, D.K., Crary, K., Harper, R.: Towards a mechanized metatheory of standard ML. In: Proceedings of the 34th Annual Symposium on Principles of Programming Languages, pp. 173–184. ACM Press, New York (2007)

[Pfe95]   Pfenning, F.: Structural cut elimination. In: Kozen, D. (ed.) Proceedings of the Tenth Annual Symposium on Logic in Computer Science, San Diego, California, pp. 156–166. IEEE Computer Society Press, Los Alamitos (1995)

[PS99]   Pfenning, F., Schürmann, C.: System description: Twelf — a meta-logical framework for deductive systems. In: Ganzinger, H. (ed.) CADE 1999. LNCS (LNAI), vol. 1632, pp. 202–206. Springer, Heidelberg (1999)

[SS06]   Schürmann, C., Stehr, M.O.: An executable formalization of the HOL/Nuprl connection in the meta-logical framework Twelf. In: Hermann, M., Voronkov, A. (eds.) LPAR 2006. LNCS, vol. 4246, pp. 150–166. Springer, Heidelberg (2006)

Hints in Unification

Andrea Asperti, Wilmer Ricciotti, Claudio Sacerdoti Coen, and Enrico Tassi

Department of Computer Science, University of Bologna
Mura Anteo Zamboni, 7 — 40127 Bologna, Italy
{asperti,ricciott,sacerdot,tassi}@cs.unibo.it

Abstract. Several mechanisms such as Canonical Structures [14], Type Classes [13,16], or Pullbacks [10] have been recently introduced with the aim to improve the power and flexibility of the type inference algorithm for interactive theorem provers. We claim that all these mechanisms are particular instances of a simpler and more general technique, just consisting in providing suitable hints to the unification procedure underlying type inference. This allows a simple, modular and not intrusive implementation of all the above mentioned techniques, opening at the same time innovative and unexpected perspectives on its possible applications.

1   Introduction

Mathematical objects commonly have multiple, isomorphic representations or can be seen at different levels of an algebraic hierarchy, according to the kind or amount of information we wish to expose or emphasise. This richness is a major tool in mathematics, allowing to implicitly pass from one representation to another depending on the user needs. This operation is much more difficult for machines, and many works have been devoted to the problem of adding syntactic facilities to mimic the *abus de notation* so typical of the mathematical language. The point is not only to free the user by the need of typing redundant information, but to switch to a more flexible linkage model, by combining, for instance, resolution of overloaded methods, or supporting multiple views of a same component.

All these operations, in systems based on type theories, are traditionally performed during type-inference, by a module that we call "refiner". The refiner is not only responsible for inferring types that have not been explicitly declared: it must synthesise or constrain terms omitted by the user; it must adjust the formula, for instance by inserting functions to pass from one representation to another one; it may help the user in identifying the minimal algebraic structure providing a meaning to the formula.

From the user point of view, the refiner is the primary source of "intelligence" of the system: the more effective it is, the easier becomes the communication with the system. Thus, a natural trend in the development of proof assistants consists in constantly improving the functionalities of this component, and in particular to move towards a tighter integration between the refiner and the modules in charge of proof automation.

S. Berghofer et al. (Eds.): TPHOLs 2009, LNCS 5674, pp. 84–98, 2009.
© Springer-Verlag Berlin Heidelberg 2009

Among the mechanisms which have been recently introduced in the literature with the aim to improve the power and flexibility of the refiner, we recall, in Section 2, Canonical Structures [14], Type Classes [13], and Pullbacks [10]. Our claim is that all these mechanisms are particular instances of a simpler and more general technique presented in Section 3, just consisting in providing suitable hints to the unification procedure underlying the type inference algorithm. This simple observation paves the way to a light, modular and not intrusive implementation of all the above mentioned techniques, and looks suitable to interesting generalisations as discussed in Section 4.

In the rest of the paper we shall use the notation $\equiv$ to express the type equivalence relation of the given calculus. A unification problem will be expressed as $A \overset{?}{\equiv} B$, resulting in a substitution σ such that $A\sigma \equiv B\sigma$. Metavariables will be denoted with $?_i$, and substitutions are described as lists of assignments of the form $?_i := t$.

2   Type Inference Heuristics

In this section, we recall some heuristics for type refinement already described in the literature and implemented in interactive provers like Coq, Isabelle and Matita.

2.1   Canonical Structures

A canonical structure is declaration of a particular instance of a record to be used by the type checker to solve unification problems. For example, consider the record type of groups, and its particular instance over integers ($\mathbb{Z}$).

$$\mathcal{Z} : \text{Group} := \{\text{gcarr} := \mathbb{Z}; \text{ gunit} := 0; \text{ gop} := Zplus; \ldots\}$$

The user inputs the following formula, where 0 is of type $\mathbb{Z}$.

$$0 + x = x \tag{1}$$

Suppose that the notation $(x + y)$ is associated with gop $?$ x y where gop is the projection of the group operation with type:

$$\text{gop} : \forall g : \text{Group.gcarr } g \to \text{gcarr } g \to \text{gcarr } g$$

and gcarr is of type Group $\to$ *Type* (i.e. the projection of the group carrier). After notation expansion equation (1) becomes

$$\text{gop } ?_1 \ 0 \ x = x$$

where $?_1$ is a metavariable. For (1) to be well typed the arguments of gop have to be of type gcarr g for some group g. In particular, the first user provided argument 0 is of type $\mathbb{Z}$, generating the following unification problem:

$$\text{gcarr } ?_1 \overset{?}{\equiv} \mathbb{Z}$$

If the user declared $\mathcal{Z}$ as the canonical group structure over $\mathbb{Z}$, the system finds the solution $?_1 := \mathcal{Z}$. This heuristic is triggered only when the unification problem involves a record projection π_i applied to a metavariable versus a constant c. Canonical structures $S := \{c_1; \ldots; c_n\}$ can be easily indexed using as keys all the pairs of the form $\langle \pi_i, c_i \rangle$.

This device was introduced by A.Saibi in the Coq system [14] and is extensively used in the formalisation of finite group theory by Gonthier et al. [2,6].

2.2 Type Classes

Type classes were introduced in the context of programming languages to properly handle symbol overloading in [7,15], and they have been later adopted in interactive provers [13,16].

In a programming language with explicit polymorphism, dispatching an overloaded method amounts to suitably instantiate a type variable. This generalises canonical structures exploiting a Prolog-like mechanism to search for type class instances.

For instance we show how to define the group theoretical construction of the Cartesian product using a simplification of the Coq syntax.

```
Class Group (A : Type) := { unit : A; gop : A → A → A; ...}
Instance Z : Group Z := { unit := 0; gop := Zplus; ...}
Instance _ × _ (A,B: Type) (G: Group A) (H: Group B) : Group (A × B) := {
    unit := ⟨unit G, unit H⟩;
    gop ⟨x1,x2⟩ ⟨y1,y2⟩ := ⟨gop G x1 y1, gop H x2 y2⟩;
    ...
}
```

With this device a slightly more complicated formula than (1) can be accepted by the system, such as:

$$\langle 0, 0 \rangle + x = x$$

Unfolding the $\_ + \_$ notation we obtain

$$\text{gop } ?_1 \ ?_2 \ \langle 0, 0 \rangle \ x = x$$

where the type of gop and the type of $?_2$ are:

$$\text{gop} : \forall T : \text{Type}.\forall g : \text{Group } T.T \to T \to T$$
$$?_2 : \text{Group } ?_1$$

After $?_1$ is instantiated with $\mathbb{Z} \times \mathbb{Z}$ proof automation is used to inhabit $?_2$ whose type has become Group $(\mathbb{Z} \times \mathbb{Z})$. Automation is limited to a Prolog-like search whose clauses are the user declared instances. Notice that the user has not defined a type class instance (i.e. a canonical structure) over the group $\mathcal{Z} \times \mathcal{Z}$.

2.3   Coercions Pullback

The coercions pullback device was introduced as part of the manifesting coercions technique by Sacerdoti Coen and Tassi in [10] to ease the encoding of algebraic structures in type theory (see [11] for a formalisation explicating that technique).

This devices comes to play in a setting with a hierarchy of structures, some of which are built combining together simpler structures. The carrier projection is very frequently declared as a coercion [8], allowing the user to type formulas like $\forall g : Group.\forall x : g.P(x)$ omitting to apply gcarr to g (i.e. the system is able to insert the application of coercions when needed [12]).

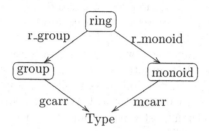

The algebraic structure of rings is composed by a multiplicative monoid and an additive group, respectively projected out by the coercions r_group and r_monoid so that a ring can be automatically seen by the system as a monoid or a group. The ring structure can be built when the carriers of the two structures are compatible (that in intensional type theories can require some non trivial efforts, see [10] for a detailed explanation).

When the operations of both structures are used in the same formula, the system has to solve a particular kind of unification problems. For example, consider the usual distributivity law of the ring structure:

$$x * (y + z) = x * y + x * z$$

Expanding the notation we obtain as the left hand side the following

$$\text{mop } ?_1\, x\, (\text{gop } ?_2\, y\, z)$$

The second argument of mop has type gcarr $?_2$ but is expected to have type mcarr $?_1$, corresponding to the unification problem:

$$\text{gcarr } ?_2 \overset{?}{\equiv} \text{mcarr } ?_1$$

The system should infer the minimal algebraic structure in which the formula can be interpreted, and the coercions pullback devices amounts to the calculation of the pullback (in categorical sense) of the coercions graph for the arrows gcarr and mcarr. The solution, in our example, is the following substitution:

$$?_2 := \text{r\_group } ?_3 \qquad ?_1 := \text{r\_monoid } ?_3$$

The solution is correct since the carriers of the structures composing the ring structure are compatible w.r.t. equivalence (i.e. the two paths in the coercions graph commute), that corresponds to the following property: for every ring r

$$\text{gcarr (r\_group } r) \equiv \text{mcarr (r\_monoid } r)$$

3   A Unifying Framework: Unification Hints

In higher order logic, or also in first order logic modulo sufficiently powerful rewriting, unification $\mathcal{U}$ is undecidable. To avoid divergence and to manage the complexity of the problem, theorem provers usually implement a simplified, decidable unification algorithm $\mathcal{U}_o$, essentially based on first order logic, sometimes extended to cope with reduction (two terms t_1 and t_2 are unifiable if they have reducts t_1' and t_2'' - usually computed w.r.t. a given reduction strategy - which are first order unifiable). Unification hints provide a way to easily extend the system's unification algorithm $\mathcal{U}_o$ (towards $\mathcal{U}$) with heuristics to choose solutions which can be less than most general, but nevertheless constitute a sensible default instantiation according to the user.

The general structure of a hint is

$$\frac{\vec{?_x} := \vec{H}}{P \equiv Q} \text{ myhint}$$

where $P \equiv Q$ is a *linear* pattern with free variable $FV(P,Q) = \vec{?_v}$, $\vec{?_x} \subseteq \vec{?_v}$, all variables in $\vec{?_x}$ are distinct and H_i cannot depend on $?_{x_i}, \ldots, ?_{x_n}$. A hint is *acceptable* if $P[\vec{H} / \vec{?_x}] \equiv Q[\vec{H} / \vec{?_x}]$, i.e. if the two terms obtained by telescopic substitution, are *convertible*. Since convertibility is (typically) a decidable relation, the system is able to discriminate acceptable hints.

Hints are supposed to be declared by the user, or automatically generated by the systems in peculiar situation. Formally a unification hint induces a schematic unification rule over the schematic variables $\vec{?_v}$ to reduce unification problems to simpler ones:

$$\frac{\vec{?_x} \overset{?}{\equiv} \vec{H}}{P \overset{?}{\equiv} Q} \text{ myhint}$$

Since $\vec{?_x}$ are schematic variables, when the rule is instantiated, the unification problems $\vec{?_x} \overset{?}{\equiv} \vec{H}$ become non trivial.

When a hint is acceptable, the corresponding schematic rule for unification is sound (proof: a solution to $\vec{?_x} \overset{?}{\equiv} \vec{H}$ is a substitution σ such that $\vec{?_x}\sigma \equiv \vec{H}\sigma$ and thus $P\sigma \equiv P[\vec{H}/\vec{?_x}]\sigma \equiv Q[\vec{H}/\vec{?_x}]\sigma \equiv Q\sigma$; hence σ is also a solution to $P \overset{?}{\equiv} Q$).

From the user perspective, the intuitive reading is that, having a unification problem of the kind $P \overset{?}{\equiv} Q$, then the "hinted" solution is $\overrightarrow{?_x} := \overrightarrow{H}$.

The intended use of hints is upon failure of the basic unification algorithm $\mathcal{U}_o$: the recursive definition unif that implements $\mathcal{U}_o$

```
let rec unif m n = body
```

is meant to be simply replaced by

```
let rec unif m n =
  try body
  with failure -> try_hints m n
```

The function try_hints simply matches the two terms m and n against the hints patterns (in a fixed order decided by the user) and returns the first solution found:

```
and try_hints m n =
  match m,n with
  | ...
  | P,Q when unif(x,H) as sigma -> sigma (* myhint *)
  | ...
```

This simple integration excludes the possibility of backtracking on hints, but is already expressive enough to cover, as we shall see in the next Section, all the cases discussed in Section 2.

Due to the lack of backtracking, hints are particularly useful when they are *invertible*, in the sense that the hinted solution is also unique, or at least "canonical" from the user point of view. However, even when hints are not canonical, they provide a strict and sound extension to the basic unification algorithm.

Hints may be easily indexed with efficient data structures investigated in the field of automatic theorem proving, like discrimination trees.

3.1 Implementing Canonical Structures

Every canonical structure declaration that declares T as the canonical solution for a unification problem $\pi_i \ ?_S \overset{?}{\equiv} t \mapsto ?_S := T$ can be simply turned in the corresponding unification hint:

$$\frac{?_S := T}{\pi_i \ ?_S \equiv t}$$

3.2 Implementing Type Classes

Like canonical structures, type classes are used to solve problems like $\pi_i \ ? \overset{?}{\equiv} t$, where π_i is a projection for a record type R. This kind of unification problem can be seen as inhabitation problems of the form "$? : R$ with $\pi_i := t$". Because of the lack of the with construction in the Calculus of Inductive Constructions, Sozeau encodes the problem abstracting the record type over t, thus reducing the problem to the inhabitation of the type $R\ t$. Since the the structure of t is explicit in the type, parametric type class instances like the Cartesian product described in Section 2.2 can be effectively used as Prolog-like clauses to solve the inhabitation problem. This approach forces a particular encoding of algebraic structures, where all the fields that are used to drive inhabitation search have to be abstracted out. This practice has a nasty impact on the modularity of non trivial algebraic hierarchies, as already observed in [9,10].

Unification hints can be employed to implement type classes without requiring an ad-hoc representation of algebraic structures. The following hint schema

$$\frac{?_R := \{\pi_1 := ?_1 \ \ldots\ \pi_i := ?_i \ \ldots\ \pi_n := ?_n\}}{\pi_i\ ?_R \equiv ?_i}\ \text{h-struct-i}$$

allows to reduce unification problems of the form $\pi_i\ ? \overset{?}{\equiv} t$ to the inhabitation of the fields $?_1 \ldots ?_n$. Moreover, if we dispose of canonical inhabitants for these fields we may already express them in the hint. Note that the user is not required to explicitly declare classes and instances.

Unification hints are flexible enough to also support a different approach that does not rely on inhabitation but reduces the unification problem to simpler problems of the same kind.

For example, the unification problem

$$\text{gcarr}\ ?_1 \overset{?}{\equiv} \mathbb{Z} \times \mathbb{Z}$$

can be solved by the following hint:

$$\frac{?_1 := \text{gcarr}\ ?_3 \qquad ?_2 := \text{gcarr}\ ?_4 \qquad ?_0 := ?_3 \times ?_4}{\text{gcarr}\ ?_0 \equiv ?_1 \times ?_2}\ \text{h-prod}$$

Intuitively, the hint says that, if the carrier of a group $?_0$ is a product $?_1 \times ?_2$, where $?_1$ is the carrier of a group $?_3$ and $?_2$ is the carrier of a group $?_4$ then we *may guess* that $?_0$ is the group product of $?_3$ and $?_4$. This is not the only possible solution but, in lack of alternatives, it is a case worth to be explored.

3.3   Implementing Coercions Pullback

Coercions are usually represented as arrows between type schemes in a DAG. A type scheme is a type that can contain metavariables. So, for instance, it is possible to declare a coercion from the type scheme Vect $?_A$ to the type scheme List $?_A$. Since coercions form a DAG, there may exist multiple paths between two nodes, i.e. alternative ways to map inhabitants of one type to inhabitants of another type. Since an arc in the graph is a function, a path corresponds to the functional composition of its arcs. A coercion graph is coherent [8] when every two paths, seen as composed functions p_1 and p_2, are equivalent, i.e. $p_1 \equiv p_2$.

Table 1. Unification problems solved by coercion hints

Problem	Solution
gcarr $?_1 \overset{?}{\equiv}$ mcarr $?_2$	$?_1 := $ r_group $?_3$, $?_2 := $ r_monoid $?_3$
gcarr $?_1 \overset{?}{\equiv}$ mcarr (r_monoid $?_2$)	$?_1 := $ r_group $?_2$
gcarr (r_group $?_1$) $\overset{?}{\equiv}$ mcarr $?_2$	$?_2 := $ r_monoid $?_1$
gcarr (r_group $?_1$) $\overset{?}{\equiv}$ mcarr (r_monoid $?_2$)	$?_2 := ?_1$

In a coherent dag, any pair of cofinal coercions defines a hint pattern, and the corresponding pullback projections (if they exist) are the hinted solution.

Consider again the example given in Sect. 2.3. The generated hint is

$$\frac{?_1 := \text{r\_group } ?_3 \qquad ?_2 := \text{r\_monoid } ?_3}{\text{gcarr } ?_1 \equiv \text{mcarr } ?_2}$$

This hint is enough to solve all the unification problems listed in Table 1, that occur often when formalising algebraic structures (e.g. in [11]).

4   Extensions

All the previous examples are essentially based on simple conversions involving records and projections. A natural idea is to extend the approach to more complex cases involving arbitrary, possibly recursive functions.

As we already observed, the natural use of hints is in presence of invertible reductions, where we may infer part of the structure of a term from its reduct.

A couple of typical situations borrowed from arithmetics could be the following, where *plus* and *times* are defined be recursion on the first argument, in the obvious way:

$$\frac{?_1 := 0 \qquad ?_2 := 0}{?_1 + ?_2 \equiv 0} \text{ plus0} \qquad \frac{?_1 := 1 \qquad ?_2 := 1}{?_1 * ?_2 \equiv 1} \text{ times1}$$

To understand the possible use of these hints, suppose for instance to have the goal

$$1 \leq a * b$$

under the assumptions $1 \leq a$ and $1 \leq b$; we may directly apply the monotonicity of times

$$\forall x, y, w, z. x \leq w \rightarrow y \leq z \rightarrow x * y \leq w * z$$

that will succeed unifying (by means of the hint) both x and y with 1, w with a and z with b.

Even when patterns do not admit a unique solution we may nevertheless identify an "intended" hint.

Consider for instance the unification problem

$$?_n + ?_m \overset{?}{\equiv} S ?_p$$

In this case there are two possible solutions:

$$1)\ ?_n := 0\quad \text{and}\quad ?_m := S\ ?_p$$
$$2)\ ?_n := S\ ?_q\quad \text{and}\quad ?_p :=?_q+?_m$$

however, the first one can be considered as somewhat degenerate, suggesting to keep the second one as a possible hint.

$$\frac{?_n := S\ ?_q \qquad ?_p :=?_q+?_m}{?_n+?_m \equiv S\ ?_p}\ \text{plus-S}$$

This would for instance allow to apply the lemma le_plus : $\forall x, y : \mathbb{N}.x \le y + x$ to prove that $m \le S(n+m)$.

The hint can also be used recursively: the unification problem

$$?_j + ?_i \overset{?}{\equiv} S(S(n+m))$$

will result in two subgoals,

$$\frac{?_j \overset{?}{\equiv} S\ ?_q \qquad S(n+m) \overset{?}{\equiv}?_q+?_i}{?_j+?_i \overset{?}{\equiv} S(S(n+m))}\ \text{plus-S}$$

and the second one will recursively call the hint, resulting in the instantiation $?_j := S(S\ n)$ and $?_i := m$ (other possible solutions, not captured by the hint, would instantiate $?_j$ with 0, 1 and 2).

4.1 Simple Reflexive Tactics Implementation

Reflexive tactics [1,3] are characterised by an initial phase in which the problem to be processed is interpreted in an abstract syntax, that is later fed to a normalisation function on the abstract syntax that is defined inside the logic. This step needs to be performed outside the logic, since there is no way to perform pattern matching on the primitive CIC constructors (i.e. the λ-calculus application).

Let us consider a simple reflexive tactic performing simplification in a semigroup structure (that amounts to eliminating all parentheses thanks to the associativity property).

The abstract syntax that will represent the input of the reflexive tactic is encoded by the following inductive type, where EOp represents the binary semigroup operation and EVar a semi-group expression that is opaque (that will be treated as a black box by the reflexive tactic).

```
inductive Expr (S : semigroup) : Type :=
 | EVar : sgcarr S → Expr S
 | EOp : Expr S → Expr S → Expr S.
```

We call sgcarr the projection extracting the carrier of the semi-group structure, and semigroup the record type representing the algebraic structure under analysis. Associated to that abstract syntax there is an interpretation function $[\![\cdot]\!]_S$ mapping an abstract term of type Expr S to a concrete one of type sgcarr S.

```
let rec [[e : Expr S]](S:semigroup) : sgcarr S :=
  match e with
  [ EVar x ⇒ x
  | Eop x y ⇒ sgop S [[x]]S [[y]]S
  ].
```

The normalisation function simpl is given the following type and is proved sound:

```
let rec simpl (e: Expr S) : Expr S := ...
lemma soundness:
  ∀S:semigroup.∀ P:sgcarr S → Prop.∀x:Expr S. P [[simpl x]]S →P [[x]]S
```

Given the following sample goal, imagine the user applies the soundness lemma (where P is instantiated with $\lambda x.x = d$).

$$a + (b + c) = d$$

yielding the unification problem

$$[[?_1]]_g \overset{?}{\equiv} a + (b + c) \tag{2}$$

This is exactly what the extra-logical initial phase of every reflexive tactic has to do: interpret a given concrete term into an abstract syntax.

We now show how the unification problem is solved declaring the two following hints, where h-add is declared with higher precedence.

$$\frac{?_a := \text{Eop } ?_S ?_x ?_y \qquad ?_m := [[?_x]]_{?_S} \qquad ?_n := [[?_y]]_{?_S}}{[[?_a]]_{?_S} \equiv ?_m + ?_n} \text{ h-add}$$

$$\frac{?_a := \text{EVar } ?_S ?_z}{[[?_a]]_{?_S} \equiv ?_z} \text{ h-base}$$

Hint h-add can be applied to problem (2), yielding three new recursive unification problems. H-base is the only hint that can be applied to the second problem, while the third one is matched by h-add, yielding three more problems whose last two can be solved by h-base:

$$\frac{?_1 \overset{?}{\equiv} \text{Eop } g ?_x ?_y \quad \frac{?_x \overset{?}{\equiv} \text{EVar } g \, a}{a \overset{?}{\equiv} [[?_x]]_g} \text{ h-base} \quad \frac{?_y \overset{?}{\equiv} \text{Eop } g ?_x ?_y \quad b \overset{?}{\equiv} [[?_x]]_g \quad c \overset{?}{\equiv} [[?_y]]_g}{b + c \overset{?}{\equiv} [[?_y]]_g} \text{ h-add}}{[[?_1]]_g \overset{?}{\equiv} a + b + c}$$

The leaves of the tree are all trivial instantiations of metavariables that together form a substitution that instantiates $?_1$ with the following expected term:

$$\text{Eop } g \text{ (EVar } g \, a) \text{ (Eop } g \text{ (EVar } g \, b) \text{ (EVar } g \, c))$$

4.2 Advanced Reflexive Tactic Implementation

The reflexive tactic to put a semi-group expression in canonical form is made easy by the fact that the mathematical property on which it is based has linear variable occurrences on both sides of the equation:

$$\forall g : \text{semigroup}.\forall a, b, c : \text{sgcarr } g. a + (b + c) = (a + b) + c$$

If we consider a richer structure, like groups, we immediately have properties that are characterised by non linear variable occurrences, for example

$$\forall g : \text{group}.\forall x : \text{gcarr } g. x * x^{-1} = 1$$

To apply the simplification rule above, the data type for abstract terms must support a decidable comparison function. We represent concrete terms external to the group signature by pointers (De Bruijn indexes) to a heap (represented as a context Γ). Thanks to the heap, we can share convertible concrete terms so that the test for equality is reduced to testing equality of pointers.

```
record group : Type :={
  gcarr : Type;
  1 : gcarr;
  _*_ : gcarr → gcarr → gcarr;
  _⁻¹ : gcarr → gcarr
}.
```

The abstract syntax for expressions is encoded in the following inductive type:

```
inductive Expr : Type :=
  | Eunit : Expr
  | Emult : Expr → Expr → Expr
  | Eopp : Expr → Expr
  | Evar : ℕ → Expr.
```

The interpretation function takes an additional argument that is the heap Γ. Lookup in Γ is written $\Gamma(m)$ and returns a dummy value when m is a dandling pointer.

```
let rec [e : Expr; Γ : list (gcarr g)]₍g:group₎ on e : gcarr g :=
  match e with
  [ Eunit ⇒ 1
  | Emult x y ⇒ [x; Γ]_g * [y; Γ]_g
  | Eopp x ⇒ [x; Γ]_g⁻¹
  | Evar n ⇒ Γ(n) ].
```

For example:

$$[\text{Emult (Evar } O) (\text{Emult (Eopp (Evar } O)) (\text{Evar } (S\ O)))); [x; y]]_g \equiv x * (x^{-1} * y)$$

The unification problem generated by the application of the reflexive tactic is of the form

$$[\![?_1; \ ?_2]\!]_{?_3} \stackrel{?}{\equiv} x * (x^{-1} * y)$$

and admits multiple solutions (corresponding to permutations of elements in the heap).

To be able to interpret the whole concrete syntax of groups in the abstract syntax described by the Expr type, we need the following hints:

$$\frac{?_a := \text{Emult } ?_x \ ?_y \qquad ?_m := [\![?_x; \ ?_r]\!]_{?_g} \qquad ?_n := [\![?_y; \ ?_r]\!]_{?_g}}{[\![?_a; \ ?_r]\!]_{?_g} \equiv ?_m * ?_n} \text{ h-times}$$

$$\frac{?_a := \text{Eunit}}{[\![?_a; \ ?_r]\!]_{?_g} \equiv 1} \text{ h-unit} \qquad \frac{?_a := \text{Eopp } ?_z \qquad ?_o := [\![?_z; \ ?_r]\!]_{?_g}}{[\![?_a; \ ?_r]\!]_{?_g} \equiv ?_o^{-1}} \text{ h-opp}$$

To identify equal variables, and give them the same abstract representation, we need two hints, implementing the lookup operation in the heap (or better, the generation of a duplicate free heap by means of explicit sharing).

$$\frac{?_a := \text{Evar } 0 \qquad ?_r := ?_r :: ?_\Theta}{[\![?_a; \ ?_r]\!]_{?_g} \equiv ?_r} \text{ h-var-base}$$

$$\frac{?_a := \text{Evar } (S \ ?_p) \qquad ?_r := ?_s :: ?_\Delta \qquad ?_q := [\![\text{Evar } ?_p; \ ?_\Delta]\!]_{?_g}}{[\![?_a; \ ?_r]\!]_{?_g} \equiv ?_q} \text{ h-var-rec}$$

To understand the former rule, consider the following unification problem:

$$[\![\text{Evar } 0; \ ?_t :: ?_r]\!]_{?_g} \stackrel{?}{\equiv} x$$

Since the first context item is a metavariable, unification (unfolding and computing the definition of $[\![\text{Evar } 0; \ ?_t :: ?_r]\!]_{?_g}$ to $?_t$) instantiates $?_t$ with x, that amounts to reserving the first heap position for the concrete term x.

In case the first context item has been already reserved for a different variable, unification falls back to hint h-var-rec, skipping that context item, and possibly instantiating the tail of the context $?_r$ with $x :: ?_\Delta$ for some fresh metavariable $?_\Delta$.

We now go back to our initial example $[\![?_1; \ ?_2]\!]_{?_3} \stackrel{?}{\equiv} x * (x^{-1} * y)$ and follow step by step how unification is able to find a solution for $?_1$ and $?_2$ using hints. The algorithm starts by applying the hint h-times, yielding one trivial and two non trivial recursive unification problems:

$$\frac{?_1 \stackrel{?}{\equiv} \text{Emult } ?_x \ ?_y \qquad x \stackrel{?}{\equiv} [\![?_x; \ ?_2]\!]_{?_g} \qquad x^{-1} * y \stackrel{?}{\equiv} [\![?_y; \ ?_2]\!]_{?_g}}{[\![?_1; \ ?_2]\!]_{?_g} \stackrel{?}{\equiv} x * (x^{-1} * y)} \text{ h-times}$$

The second recursive unification problem can be solved applying hint h-var-base:

$$\frac{?_x \stackrel{?}{\equiv} \text{Evar } 0 \qquad ?_2 \stackrel{?}{\equiv} x :: ?_\Theta}{x \stackrel{?}{\equiv} [\![?_x; \ ?_2]\!]} \text{ h-var-base}$$

The application of the hint h-var-base forces the instantiation of $?_2$ with $x ::$ $?_\Theta$, thus fixing the first entry of the context to x, but still allowing the free instantiation of the following elements.

Under the latter instantiation, the third unification problem to be solved becomes $x^{-1} * y \stackrel{?}{\equiv} [\![?_y; \ x ::?_\Theta]\!]$ that requires another application of hint h-times followed by h-opp on the first premise.

$$\frac{?_y \stackrel{?}{\equiv} \text{Emult (Evar 0) } ?_y \qquad x^{-1} \stackrel{?}{\equiv} [\![?_{x'}; \ x ::?_\Theta]\!]_{?_g} \qquad y \stackrel{?}{\equiv} [\![?_{y'}; \ x ::?_\Theta]\!]_{?_g}}{x^{-1} * y \stackrel{?}{\equiv} [\![?_y; \ x ::?_\Theta]\!]_{?_g}} \text{ h-times}$$

The first non-trivial recursive unification problem is $x^{-1} \stackrel{?}{\equiv} [\![?_{x'}; \ x ::?_\Theta]\!]_{?_g}$ and can be solved applying hint h-opp first and then h-var-base. The second problem is more interesting, since it requires an application of h-var-rec:

$$\frac{?_{y'} \stackrel{?}{\equiv} \text{Evar } (S \ ?_p) \qquad x ::?_\Theta \stackrel{?}{\equiv} ?_s ::?_\Delta \qquad y \stackrel{?}{\equiv} [\![\text{Evar } ?_p; \ ?_\Delta]\!]_{?_g}}{y \stackrel{?}{\equiv} [\![?_{y'}; \ x ::?_\Theta]\!]_{?_g}} \text{ h-var-rec}$$

The two unification problems on the left are easy to solve and lead to the following instantiation

$$?_{y'} := \text{Evar } (S \ ?_p) \qquad ?_s := x; \qquad ?_\Delta :=?_\Theta$$

The unification problem left is thus $y \stackrel{?}{\equiv} [\![\text{Evar } ?_p; \ ?_\Theta]\!]_{?_g}$ and can be solved using hint h-var-base. It leads to the instantiation

$$?_p := \text{Evar } 0 \qquad ?_\Theta := y ::?_{\Theta'}$$

for a fresh metavariable $?_{\Theta'}$. Note that hint h-var-base was not applicable in place of h-var-rec since it leads to an unsolvable unification problem that requires the first item of the context to be equal to both x and y:

$$\frac{?_{y'} \stackrel{?}{\equiv} \text{Evar } 0 \qquad x ::?_\Theta \stackrel{?}{\equiv} y ::?_\Theta}{y \equiv [\![?_{y'}; \ x ::?_\Theta]\!]_{?_g}} \text{ h-var-base}$$

The solution found for the initial unification problem is thus:

$$?_1 := \text{Emult (Evar } O) \text{ (Emult (Eopp (Evar } O)) \text{ (Evar } (S \ O))))$$
$$?_2 := x :: y ::?_{\Theta'}$$

Note that $?_{\Theta'}$ is still not instantiated, since the solution for $?_1$ is valid for every context that extends $x :: y ::?_{\Theta'}$. The user has to choose one of them, the empty one being the obvious choice.

All problems obtained by the application of the soundness lemma are of the form $[\![?_1; \ ?_2]\!]_{?_3} \stackrel{?}{\equiv} t$. If t contains no metavariables, hints cannot cause divergence since: h-opp, h-unit and h-times are used a finite number of times since they consume t; every other problem recursively generated has the form $[\![?_1; \ \vec{s} :: ?_r]\!]_{?_3} \stackrel{?}{\equiv} r$ where r is outside the group signature. To solve each goal, h-var-rec can be applied at most $| \vec{s} | + 1$ times and eventually h-var-base will succeed.

5   Conclusions

In a higher order setting, unification problems of the kind $f \; ?_i \overset{?}{=} o$ and $?_f \; i \overset{?}{=} o$ are extremely complex. In the latter case, one can do little better than using generate-and-test techniques; in the first case, the search can be partially driven by the structure of the function, but still the operation is very expensive. Moreover, higher order unification does not admit most general unifiers, so both problems above usually have several different solutions, and it is hard to guide the procedure towards the intended solution.

On the other side, it is simple to hint solutions to the unification algorithm, since the system has merely to check their correctness. By adding suitable hints in a controlled way, we can restrict to a first order setting keeping interesting higher-order inferences. In particular, we proved that hints are expressive enough to mimic some interesting ad-hoc unification heuristics like canonical structures, type classes and coercion pullbacks. It also seems that system provided unification errors in case of error-free formulae can be used to suggest to the user the need for a missing hint, in the spirit of "productive use of failure" [4].

Unification hints can be efficiently indexed using data structures for first order terms like discrimination trees. Their integration with the general flow of the unification algorithm is less intrusive than the previously cited ad-hoc techniques.

We have also shown an interesting example of application of unification hints to the implementation of reflexive tactics. In particular, we instruct the unification procedure to automatically infer a syntactic representation S of a term t such that $[\![S]\!] \equiv t$, introducing sharing in the process. This operation previously had to be done by writing a small extra-logical program in the programming language used to write the system, or in some ad-hoc language for customisation, like $\mathcal{L}$-tac [5]. Our proposal is superior since the refiner itself becomes able to solve such unification problems, that can be triggered in situations where the external language is not accessible, like during semantic analysis of formulae.

A possible extension consists in adding backtracking to the management of hints. This would require a more intrusive reimplementation of the unification algorithm; moreover it is not clear that this is the right development direction since the point is not to just add expressive power to the unification algorithm, but to get the right balance between expressiveness and effectiveness, especially in case of failure.

Another possible extension is to relax the linearity constraint on patterns with the aim to capture more invertible rules, like in the following cases:

$$\frac{?_x := 0}{?_x + ?_y \equiv ?_y} \; \text{plus-0} \qquad\qquad \frac{?_x := S \; ?_z}{?_x + ?_y \equiv S \; (?_z + ?_y)} \; \text{plus-S}$$

It seems natural to enlarge the matching relation allowing the recursive use of hints, at least when they are invertible. For instance, to solve the unification problem $?_1 + (?_2 + x) \overset{?}{=} x$ we need to apply hint plus-0 but matching the hint pattern requires a recursive application of hint plus-0 (hence it is not matching in the usual sense, since $?_2$ has to be instantiated with 0). The properties of this "matching" relation need a proper investigation that we leave for future work.

References

1. Barthe, G., Ruys, M., Barendregt, H.: A two-level approach towards lean proof-checking. In: Berardi, S., Coppo, M. (eds.) TYPES 1995. LNCS, vol. 1158, pp. 16–35. Springer, Heidelberg (1996)
2. Bertot, Y., Gonthier, G., Ould Biha, S., Pasca, I.: Canonical big operators. In: Mohamed, O.A., Muñoz, C., Tahar, S. (eds.) TPHOLs 2008. LNCS, vol. 5170, pp. 86–101. Springer, Heidelberg (2008)
3. Boutin, S.: Using reflection to build efficient and certified decision procedures. In: Ito, T., Abadi, M. (eds.) TACS 1997. LNCS, vol. 1281, pp. 515–529. Springer, Heidelberg (1997)
4. Bundy, A., Basin, D., Hutter, D., Ireland, A.: Rippling: meta-level guidance for mathematical reasoning. Cambridge University Press, New York (2005)
5. Delahaye, D.: A Tactic Language for the System Coq. In: Parigot, M., Voronkov, A. (eds.) LPAR 2000. LNCS, vol. 1955, pp. 85–95. Springer, Heidelberg (2000)
6. Gonthier, G., Mahboubi, A., Rideau, L., Tassi, E., Thery, L.: A Modular Formalisation of Finite Group Theory. In: Schneider, K., Brandt, J. (eds.) TPHOLs 2007. LNCS, vol. 4732, pp. 86–101. Springer, Heidelberg (2007)
7. Hall, C., Hammond, K., Jones, S.P., Wadler, P.: Type classes in haskell. ACM Transactions on Programming Languages and Systems 18, 241–256 (1996)
8. Luo, Z.: Coercive subtyping. J. Logic and Computation 9(1), 105–130 (1999)
9. Luo, Z.: Manifest fields and module mechanisms in intensional type theory. In: Miculan, M., Scagnetto, I., Honsell, F. (eds.) TYPES 2007. LNCS, vol. 4941. Springer, Heidelberg (2008)
10. Sacerdoti Coen, C., Tassi, E.: Working with mathematical structures in type theory. In: Miculan, M., Scagnetto, I., Honsell, F. (eds.) TYPES 2007. LNCS, vol. 4941, pp. 157–172. Springer, Heidelberg (2008)
11. Sacerdoti Coen, C., Tassi, E.: A constructive and formal proof of Lebesgue's dominated convergence theorem in the interactive theorem prover Matita. Journal of Formalized Reasoning 1, 51–89 (2008)
12. Saibi, A.: Typing algorithm in type theory with inheritance. In: The 24th Annual ACM SIGPLAN - SIGACT Symposium on Principle of Programming Language (POPL) (1997)
13. Sozeau, M., Oury, N.: First-class type classes. In: Mohamed, O.A., Muñoz, C., Tahar, S. (eds.) TPHOLs 2008. LNCS, vol. 5170, pp. 278–293. Springer, Heidelberg (2008)
14. The Coq Development Team. The Coq proof assistant reference manual (2005), http://coq.inria.fr/doc/main.html
15. Wadler, P., Blott, S.: How to make ad-hoc polymorphism less ad hoc. In: POPL 1989: Proceedings of the 16th ACM SIGPLAN-SIGACT symposium on Principles of programming languages, pp. 60–76. ACM, New York (1989)
16. Wenzel, M.: Type classes and overloading in higher-order logic. In: Gunter, E.L., Felty, A.P. (eds.) TPHOLs 1997. LNCS, vol. 1275, pp. 307–322. Springer, Heidelberg (1997)

Psi-calculi in Isabelle

Jesper Bengtson and Joachim Parrow

Dept. of Information Technology, Uppsala University, Sweden

Abstract. Psi-calculi are extensions of the pi-calculus, accommodating arbitrary nominal datatypes to represent not only data but also communication channels, assertions and conditions, giving it an expressive power beyond the applied pi-calculus and the concurrent constraint pi-calculus.

We have formalised psi-calculi in the interactive theorem prover Isabelle using its nominal datatype package. One distinctive feature is that the framework needs to treat binding sequences, as opposed to single binders, in an efficient way. While different methods for formalising single binder calculi have been proposed over the last decades, representations for such binding sequences are not very well explored.

The main effort in the formalisation is to keep the machine checked proofs as close to their pen-and-paper counterparts as possible. We discuss two approaches to reasoning about binding sequences along with their strengths and weaknesses. We also cover custom induction rules to remove the bulk of manual alpha-conversions.

1 Introduction

There are today several formalisms to describe the behaviour of computer systems. Some of them, like the lambda-calculus and the pi-calculus, are intended to explore fundamental principles of computing and consequently contain as few and basic primitives as possible. Other are more tailored to application areas and include many constructions for modeling convenience. Such formalisms are now being developed en masse. While this is not necessarily a bad thing there is a danger in developing complicated theories too quickly. The proofs (for example of compositionality properties) become gruesome with very many cases to check and the temptation to resort to formulations such as "by analogy with ..." or "is easily seen..." can be overwhelming. For examples in point, both the applied pi-calculus [1] and the concurrent constraint pi-calculus [8] have recently been discovered to have flaws or incompletenesses in the sense that the claimed compositionality results do not hold [5].

Since such proofs often require stamina and attention to detail rather than ingenuity and complicated new constructions they should be amenable to proof mechanisation. Our contribution in this paper is to implement a family of application oriented calculi in Isabelle [12]. The calculi we consider are the so called psi-calculi [5], obtained by extending the basic untyped pi-calculus with the following parameters: (1) a set of data terms, which can function as both

S. Berghofer et al. (Eds.): TPHOLs 2009, LNCS 5674, pp. 99–114, 2009.

communication channels and communicated objects, (2) a set of conditions, for use in conditional constructs such as **if** statements, (3) a set of assertions, used to express e.g. constraints or aliases. We base our exposition on nominal data types and these accommodate e.g. alpha-equivalence classes of terms with binders. For example, we can use a higher-order logic for assertions and conditions, and higher-order formalisms such as the lambda calculus for data terms and channels.

The main difficulty in representing calculi such as the lambda-, pi- or psi-calculi is to find an efficient treatment of binders. Informal proofs often use the Barendregt variable convention [4], that everything bound is unique. This convention provides a tractable abstraction when doing proofs involving binders, but it has recently been proven to be unsound in the general case [16]. Theorem provers have commonly used approaches based on de Bruijn indices [9], higher order abstract syntax, or nominal logic [13]. We use the nominal datatype package in Isabelle [15], and its strategy for dealing with single binders. Recent work by Aydemir et. al. introduce the locally nameless framework [2] which might be an improvement since the infrastructure is small and elegant.

One of our main contributions in the present paper is to extend the strategy to finite sequences of binders. Though it is possible to recurse over such sequences and treat each binder individually the resulting proofs would then become morasses of details with no counterpart in the informal proofs. To overcome this difficulty we introduce the notion of a *binding sequence*, which simultaneously binds arbitrarily finitely many names, and show how it can be implemented in Isabelle. We use such binding sequences to formulate and establish induction and inversion rules for the semantics of psi-calculi. The rules have been used to formally establish compositionality properties of strong bisimilarity. The proofs are close to their informal counterparts.

We are not aware of any other work on implementing calculi of this calibre in a proof assistant such as Isabelle. The closest related work are implementations of the basic pi-calculus, by ourselves [6] and also by others [10,11,14]. Neither are we aware of any other general technique for multiple binders, other than the yet unpublished work by Berghofer and Urban which we describe in Section 3.

The rest of the paper is structured as follows. In Section 2 we give a brief account of psi-calculi and how they use nominal data types. Section 3 treats implementation issues related to binding sequences and alpha-conversion. In Section 4 we show how these are used to create our formalisation. In Section 5 we report on the current status of the effort and ideas for further work.

2   Psi-calculi

This section is a brief recapitulation of psi-calculi and nominal data types; for a more extensive treatment including motivations and examples see [5].

2.1   Nominal Data Types

We assume a countably infinite set of atomic *names* $\mathcal{N}$ ranged over by $a, b, \ldots, z$. Intuitively, names will represent the symbols that can be statically scoped, and also represent symbols acting as variables in the sense that they can be subject to substitution. A *nominal set* [13] is a set equipped with *name swapping* functions written $(a\ b)$, for any names a, b. An intuition is that $(a\ b) \cdot X$ is X with a replaced by b and b replaced by a. A sequence of swappings is called a permutation, often denoted p, where $p \cdot X$ means the term X with the permutation p applied to it. We write p^- for the reverse of p. The *support* of X, written $n(X)$, is the least set of names A such that $(a\ b) \cdot X = X$ for all a, b not in A. We write $a\#X$, pronounced "a is fresh for X", for $a \notin n(X)$. If A is a set of names we write $A\#X$ to mean $\forall a \in A \,.\, a\#X$. We require all elements to have finite support, i.e., $n(X)$ is finite for all X. A function f is *equivariant* if $(a\ b) \cdot f(X) = f((a\ b) \cdot X)$ holds for all X, and similarly for functions and relations of any arity. Intuitively, this means that all names are treated equally.

2.2   Agents

A psi-calculus is defined by instantiating three nominal data types and four operators:

Definition 1 (Psi-calculus parameters). *A psi-calculus requires the three (not necessarily disjoint) nominal data types:*

$\qquad$ **T** *the (data) terms, ranged over by* M, N
$\qquad$ **C** *the conditions, ranged over by* φ
$\qquad$ **A** *the assertions, ranged over by* Ψ

and the four equivariant operators:

$\qquad \leftrightarrow: \mathbf{T} \times \mathbf{T} \to \mathbf{C}$ $\quad$ *Channel Equivalence*
$\qquad \otimes : \mathbf{A} \times \mathbf{A} \to \mathbf{A}$ $\quad$ *Composition*
$\qquad \mathbf{1} : \mathbf{A}$ $\qquad\qquad$ *Unit*
$\qquad \vdash\, \subseteq \mathbf{A} \times \mathbf{C}$ $\qquad$ *Entailment*

We require the existence of a substitution function for **T**, **C** and **A**. When X is a term, condition or assertion we write $X[\tilde{a} := \tilde{T}]$ to mean the simultaneous substitution of the names $\tilde{a}$ for the terms $\tilde{T}$ in X. The exact requisites of this function will be covered in Section 4.

The binary functions above will be written in infix. Thus, if M and N are terms then $M \leftrightarrow N$ is a condition, pronounced "M and N are channel equivalent" and if Ψ and Ψ' are assertions then so is $\Psi \otimes \Psi'$. Also we write $\Psi \vdash \varphi$, "Ψ entails φ", for $(\Psi, \varphi) \in \vdash$.

We say that two assertions are equivalent if they entail the same conditions:

Definition 2 (assertion equivalence). *Two assertions are equivalent, written* $\Psi \simeq \Psi'$, *if for all* φ *we have that* $\Psi \vdash \varphi \Leftrightarrow \Psi' \vdash \varphi$.

Channel equivalence must be symmetric and transitive, $\otimes$ must be compositional with regard to $\simeq$, and the assertions with $(\otimes, \mathbf{1})$ form an abelian monoid.

In the following $\tilde{a}$ means a finite (possibly empty) sequence of names, $a_1, \ldots, a_n$. The empty sequence is written ϵ and the concatenation of $\tilde{a}$ and $\tilde{b}$ is written $\tilde{a}\tilde{b}$.

When occurring as an operand of a set operator, $\tilde{a}$ means the corresponding set of names $\{a_1, \ldots, a_n\}$. We also use sequences of terms, conditions, assertions etc. in the same way.

A *frame* can intuitively be thought of as an assertion with local names:

Definition 3 (Frame). *A frame F is a pair $\langle B_F, \Psi_F \rangle$ where B_F is a sequence of names that bind into the assertion Ψ_F. We use F, G to range over frames.*

Name swapping on a frame just distributes to its two components. We identify alpha equivalent frames, so $n(F) = n(\Psi_F) - n(B_F)$. We overload $\mathbf{1}$ to also mean the least informative frame $\langle \epsilon, \mathbf{1} \rangle$ and $\otimes$ to mean composition on frames defined by $\langle B_1, \Psi_1 \rangle \otimes \langle B_2, \Psi_2 \rangle = \langle B_1 B_2, \Psi_1 \otimes \Psi_2 \rangle$ where B_1 is disjoint from $n(B_2, \Psi_2)$ and vice versa. We also write $\Psi \otimes F$ to mean $\langle \epsilon, \Psi \rangle \otimes F$, and $(\nu b)F$ to mean $\langle b B_F, \Psi_F \rangle$.

Definition 4 (Equivalence of frames). *We define $F \vdash \varphi$ to mean that there exist B_F and Ψ_F such that $F = \langle B_F, \Psi_F \rangle$, $B_F \# \varphi$, and $\Psi_F \vdash \varphi$. We also define $F \simeq G$ to mean that for all φ it holds that $F \vdash \varphi$ iff $G \vdash \varphi$.*

Intuitively a condition is entailed by a frame if it is entailed by the assertion and does not contain any names bound by the frame. Two frames are equivalent if they entail the same conditions.

Definition 5 (psi-calculus agents). *Given valid psi-calculus parameters as in Definition 1, the psi-calculus agents, ranged over by $P, Q, \ldots$, are of the following forms.*

$\overline{M} \, N.P$	Output
$\underline{M}(\lambda \tilde{x})N.P$	Input
case $\varphi_1 : P_1 \; [] \; \cdots \; [] \; \varphi_n : P_n$	Case
$(\nu a)P$	Restriction
$P \mid Q$	Parallel
$!P$	Replication
$(\!|\Psi|\!)$	Assertion

In the Input $\underline{M}(\lambda \tilde{x})N.P$ we require that $\tilde{x} \subseteq n(N)$ is a sequence without duplicates, and here any name in $\tilde{x}$ binds its occurrences in both N and P. Restriction binds a in P. An assertion is guarded *if it is a subterm of an Input or Output . In a replication $!P$ there may be no unguarded assertions in P, and in* **case** $\varphi_1 : P_1 \; [] \; \cdots \; [] \; \varphi_n : P_n$ *there may be no unguarded assertion in any P_i.*

Formally, we define name swapping on agents by distributing it over all constructors, and substitution on agents by distributing it and avoiding captures by binders through alpha-conversion in the usual way. We identify alpha-equivalent agents; in that way we get a nominal data type of agents where the support $n(P)$ of P is the union of the supports of the components of P, removing the names bound by λ and ν, and corresponds to the names with a free occurrence in P.

Table 1. Structured operational semantics. Symmetric versions of COM and PAR are elided. In the rule COM we assume that $\mathcal{F}(P) = \langle B_P, \Psi_P \rangle$ and $\mathcal{F}(Q) = \langle B_Q, \Psi_Q \rangle$ where B_P is fresh for all of Ψ, B_Q, Q, M and P, and that B_Q is similarly fresh. In the rule PAR we assume that $\mathcal{F}(Q) = \langle B_Q, \Psi_Q \rangle$ where B_Q is fresh for Ψ, P and α. In OPEN the expression $\nu\tilde{a} \cup \{b\}$ means the sequence $\tilde{a}$ with b inserted anywhere.

$$\text{IN} \; \frac{\Psi \vdash M \overset{\cdot}{\leftrightarrow} K}{\Psi \triangleright \underline{M}(\lambda\tilde{y})N.P \xrightarrow{K\,N[\tilde{y}:=\tilde{L}]} P[\tilde{y} := \tilde{L}]} \qquad \text{OUT} \; \frac{\Psi \vdash M \overset{\cdot}{\leftrightarrow} K}{\Psi \triangleright \overline{M}\,N.P \xrightarrow{\overline{K}\,N} P}$$

$$\text{CASE} \; \frac{\Psi \triangleright P_i \xrightarrow{\alpha} P' \qquad \Psi \vdash \varphi_i}{\Psi \triangleright \mathbf{case}\ \tilde{\varphi} : \tilde{P} \xrightarrow{\alpha} P'}$$

$$\text{COM} \; \frac{\Psi_Q \otimes \Psi \triangleright P \xrightarrow{\overline{M}\,(\nu\tilde{a})N} P' \qquad \Psi_P \otimes \Psi \triangleright Q \xrightarrow{K\,N} Q' \qquad \Psi \otimes \Psi_P \otimes \Psi_Q \vdash M \overset{\cdot}{\leftrightarrow} K}{\Psi \triangleright P \,|\, Q \xrightarrow{\tau} (\nu\tilde{a})(P' \,|\, Q')} \; \tilde{a}\#Q$$

$$\text{PAR} \; \frac{\Psi_Q \otimes \Psi \triangleright P \xrightarrow{\alpha} P'}{\Psi \triangleright P|Q \xrightarrow{\alpha} P'|Q} \; \mathrm{bn}(\alpha)\#Q \qquad \text{SCOPE} \; \frac{\Psi \triangleright P \xrightarrow{\alpha} P'}{\Psi \triangleright (\nu b)P \xrightarrow{\alpha} (\nu b)P'} \; b\#\alpha, \Psi$$

$$\text{OPEN} \; \frac{\Psi \triangleright P \xrightarrow{\overline{M}\,(\nu\tilde{a})N} P'}{\Psi \triangleright (\nu b)P \xrightarrow{\overline{M}\,(\nu\tilde{a}\cup\{b\})N} P'} \; \begin{matrix} b\#\tilde{a}, \Psi, M \\ b \in \mathrm{n}(N) \end{matrix} \qquad \text{REP} \; \frac{\Psi \triangleright P\,|\,!P \xrightarrow{\alpha} P'}{\Psi \triangleright\, !P \xrightarrow{\alpha} P'}$$

Definition 6 (Frame of an agent). *The* frame $\mathcal{F}(P)$ *of an agent* P *is defined inductively as follows:*

$$\mathcal{F}(\underline{M}(\lambda\tilde{x})N.P) = \mathcal{F}(\overline{M}\,N.P) = \mathcal{F}(\mathbf{case}\ \tilde{\varphi} : \tilde{P}) = \mathcal{F}(!P) = 1$$
$$\mathcal{F}(\!(\!|\Psi|\!)\!) = \langle \epsilon, \Psi \rangle$$
$$\mathcal{F}(P \,|\, Q) = \mathcal{F}(P) \otimes \mathcal{F}(Q)$$
$$\mathcal{F}((\nu b)P) = (\nu b)\mathcal{F}(P)$$

2.3 Operational Semantics

The *actions* ranged over by α, β are of the following three kinds: Output $\overline{M}(\nu\tilde{a})N$, Input $\underline{M}\,N$, and Silent τ. Here we refer to M as the *subject* and N as the *object*. We define $\mathrm{bn}(\overline{M}\,(\nu\tilde{a})N) = \tilde{a}$, and $\mathrm{bn}(\alpha) = \emptyset$ if α is an input or τ.

Definition 7 (Transitions). *A transition is of the kind* $\Psi \triangleright P \xrightarrow{\alpha} P'$, *meaning that in the environment* Ψ *the agent* P *can do an* α *to become* P'. *The transitions are defined inductively in Table 1.*

3 Binding Sequences

The main difficulty when formalising any calculus with binders is to handle alpha-equivalence. The techniques that have been used thus far by theorem provers share the trait that they only reason about single binders. This works well for many calculi, but psi-calculi require binding sequences of arbitrary length. For our psi-calculus datatype (Def. 5), a binding sequence is needed in the Input-case where the term $\underline{M}(\lambda \widetilde{x})N.P$ has the sequence $\widetilde{x}$ binding into N and P. The second place sequences are needed is when defining frames (Def 3). Frames are derived from processes (Def. 6) and as agents can have an arbitrary number of binders, so can the frames. The third occurrence of binding sequences can be found in the operational semantics (Table 1). In the transition $\Psi \rhd P \xrightarrow{\overline{M}(\nu\widetilde{a})N} P'$, the sequence $\widetilde{a}$ represents the bound names in P which occur in the object N.

In order to formalise these types of calculi efficiently in a theorem prover, libraries with support for sequences of binders have to be added. In the next sections we will discuss two approaches that have been made in this area, first one by us, which we call explicit binding sequences, and then one by Berghofer and Urban which we in this paper will call implicit binding sequences. They both build on the existing nominal representation of alpha-equivalence classes where a binding occurrence of the name a in the term T is written $[a].T$, and the support of $[a].T$ is the support of T with a removed. From this definition, creating a term with the binding sequence $\widetilde{a}$ in the term T, written $[\widetilde{a}].T$, can easily be done by recursion over $\widetilde{a}$. The proof that the support of $[\widetilde{a}].T$ is equal to the support of T with the names of $\widetilde{a}$ removed is trivial. Similarly, the notion of freshness needs to be expanded to handle sequences. The expression $\widetilde{a}\#T$ is defined as: $\forall x \in \mathsf{set}\ \widetilde{a}.\ x\#T$. This expression is overloaded for when $\widetilde{a}$ is either a list or a set.

3.1   Explicit Binding Sequences

Our approach is to scale the existing single binder setting to sequences. Isabelle has native support for generating fresh names, i.e. given any finite context of names $\mathcal{C}$, Isabelle can generate a name fresh for that context. There is also a distinctness predicate, written $\mathsf{distinct}\ \widetilde{a}$ which states that $\widetilde{a}$ contains no duplicates. From these we can generate a finite sequence $\widetilde{a}$ of arbitrary length n where $\mathsf{length}\ \widetilde{a} = n$, $\widetilde{a}\#\mathcal{C}$ and $\mathsf{distinct}\ \widetilde{a}$ by induction on n.

The term $[a].T$ can be alpha-converted into the term $[b].(a\ b)\cdot T$ if $b\#T$, where we call $(a\ b)$ an alpha-converting swapping. In order to mimic this behaviour with sequences, we lift name swapping to sequence swapping by pairwise composing the elements of two sequences to create an alpha-converting permutation. We will write $(\widetilde{a}\ \widetilde{b})$ for such a composition defined in the following manner:

Definition 8
$$([]\ []) = []$$
$$((x :: xs)\ (y :: ys)) = (x, y) :: (xs\ ys)$$

All theories that construct permutations using this function will ensure that the length of the sequences are equal.

We can now lift alpha-equivalence to support sequences.

Lemma 1. *If* length $\tilde{x}$ = length $\tilde{y}$, distinct $\tilde{y}$, $\tilde{x}\#\tilde{y}$ *and* $\tilde{y}\#T$
then $[\tilde{x}].T = [\tilde{y}].(\tilde{x}\ \tilde{y}) \cdot T$.

Proof. By induction on the length of $\tilde{x}$ and $\tilde{y}$.

The distinctness property is a bit stronger than strictly necessary; we only need that the names in $\tilde{x}$ that actually occur in T have a unique corresponding member in $\tilde{y}$. Describing this property formally would be cumbersome and distinctness is sufficient and easy to work with.

Long proofs tend to introduce alpha-converting permutations and it is therefor important to have a strategy for cancelling these. If a term T has been alpha-converted using the swapping $(a\ b)$, becoming $(a\ b) \cdot T$, it is possible to apply the same swapping to the expression where $(a\ b) \cdot T$ occurs. Using equivariance properties, the swapping can be distributed over the expression, and when it reaches $(a\ b) \cdot T$, it will cancel out since $(a\ b) \cdot (a\ b) \cdot T = T$. It can also be cancelled from any remaining term U in the expression, as long as $a\#U$ and $b\#U$. This technique is also applicable when dealing with sequences, where the alpha-converted term has the form $(\tilde{a}\ \tilde{b}) \cdot T$, with one important observation. Even though $(a\ b) \cdot (a\ b) \cdot T = T$, it is not generally the case that $p \cdot p \cdot T = T$. To cancel a permutation on a term, its inverse must be applied, i.e. $p^- \cdot p \cdot T = T$. By applying $(\tilde{a}\ \tilde{b})^-$ to the expression, the alpha-converting permutation will cancel out. The permutation will also be cancelled from any remaining term U as long as $\tilde{a}\#U$ and $\tilde{b}\#U$ since $\tilde{a}\#U$ and $\tilde{b}\#U$ implies $(\tilde{a}\ \tilde{b}) \cdot U = U$ and $(\tilde{a}\ \tilde{b})^- \cdot U = U$.

In this setting we are able to fully formalise our theories using binding sequences. The disadvantage is that facts regarding lengths of sequences and distinctness need to be maintained throughout the proofs.

3.2 Implicit Binding Sequences

Parallel to our work, Berghofer and Urban developed an alternative theory for binding sequences which is also being included in the nominal package. Their approach is to generate the alpha-converting permutation directly using the following lemma:

Lemma 2. *There exists a permutation p s.t.* set $p \subseteq$ set $\tilde{x} \times$ set$(p \cdot \tilde{x})$ *and* $(p \cdot \tilde{x})\#C$.

The intuition is that instead of creating a fresh sequence, a permutation is created which when applied to a sequence ensures the needed freshness conditions. The following corollary makes it possible to discard permutations which are sufficiently fresh:

Corollary 1. *If* $\tilde{x}\#T$, $\tilde{y}\#T$ *and* set $p \subseteq$ set $\tilde{x} \times$ set $\tilde{y}$ *then* $p \cdot T = T$.

From this, a corollary to perform alpha-conversions can be created.

Corollary 2. *If* set $p \subseteq$ set $\tilde{x} \times$ set$(p \cdot \tilde{x})$ *and* $(p \cdot \tilde{x})\#T$ *then* $[\tilde{x}].T = [p \cdot \tilde{x}].p \cdot T$.

Proof. since $\tilde{x}\#[\tilde{x}].T$ and $(p \cdot \tilde{x})\#T$ we have by Cor. 1 that $[\tilde{x}].T = p \cdot [\tilde{x}].T$ and hence by equivariance that $[\tilde{x}].T = [p \cdot \tilde{x}].p \cdot T$.

This method has the problem that when cancelling alpha-converting permutations as in section 3.1, the freshness conditions we use to cancel the permutation from the remaining terms are lost since $(p \cdot \tilde{x}) \# U$ does not imply $(p^- \cdot \tilde{x}) \# U$. We define the following predicate to fix this.

Definition 9. `distinctPerm` $p \equiv$ `distinct((map fst` p`)@(map snd` p`))`

Intuitively, the `distinctPerm` predicate ensures that all names in a permutation are distinct.

Corollary 3. *If* `distinctPerm` p *then* $p \cdot p \cdot T = T$

Proof. By induction on p.

Thus, by extending Lemma 2 with the condition `distintPerm` p we get permutations p which can be cancelled by applying p again rather than its inverse.

In general, proofs are easier if we know that the binding sequences are distinct. The following corollary helps.

Corollary 4. *If* $\tilde{a} \# C$ *then there exists an* $\tilde{b}$ *s.t.* $[\tilde{a}].T = [\tilde{b}].T$ *and* `distinct` $\tilde{b}$ *and* $\tilde{b} \# C$.

Proof. Since each name in $\tilde{a}$ can only bind once in T we can construct $\tilde{b}$ by replacing any duplicate name in $\tilde{a}$ with a sufficiently fresh name.

The advantage of implicit alpha-conversions is that facts about length and distinctness of sequences do not need to be maintained through the proofs. The freshness conditions are the ones needed for the single binder case and the distinctness properties are only needed when cancelling permutations. For most cases, this method is more convenient to work with. There are disadvantages regarding inversion rules, and alpha-equivalence properties that will be discussed in the next section.

3.3   Alpha-Equivalence

When reasoning with single binders, the nominal approach to alpha-equivalence is quite straightforward. Two terms $[a].T$ and $[b].U$ are equal if and only if either $a = b$ and $T = U$ or $a \neq b$, $a \# U$ and $U = (a\ b) \cdot T$. Reasoning about binding sequences is more difficult. Exactly what does it mean for two terms $[\tilde{a}].T$ and $[\tilde{b}].U$ to be equal? As long as T and U cannot themselves have binding sequences on a top level we know that `length` $\tilde{a} = $ `length` $\tilde{b}$, but the problem with the general case is what happens when $\tilde{a}$ and $\tilde{b}$ partially share names. As it turns out, this case is not important in order to reason about these types of equalities, but special heuristics are required.

The times where we actually get assumptions such as $[\tilde{a}].T = [\tilde{b}].U$ in our proofs are when we do induction or inversion over a term with binders. Typically, $[\tilde{b}].U$ is the term we start with, and $[\tilde{a}].T$ is the term that appears in the induction or inversion rule. These rules are designed in such a way that any bound names appearing in the rules can be assumed to be sufficiently fresh. More precisely, we can ensure that $\widetilde{a}\#\widetilde{b}$ and $\widetilde{a}\#U$. If we are working with explicit binding sequences we can also know that $\widetilde{a}$ is distinct. In this case, the heuristic is straightforward. Using the information provided by the induction rule we know using Lemma 1 that $[\widetilde{b}].U = [\widetilde{a}].(\widetilde{a}\ \widetilde{b}) \cdot U$ and hence that $T = (\widetilde{a}\ \widetilde{b}) \cdot U$. From here we continue with the proofs similarly to the single binder case.

When working with implicit sequences the problem is a bit more delicate. These rules have been generated using a permutation designed to ensure freshness conditions and we do not know exactly how $\widetilde{a}$ and $\widetilde{b}$ originally related to each other. We do know that the terms are alpha-equivalent and as such, there is a permutation which equates them. We first prove the following corollary:

Corollary 5. *If $[a].T = [b].U$ then $a \in \mathsf{supp}\ T = b \in \mathsf{supp}\ U$ and $a\#T = b\#U$.*

Proof. By the definition of alpha-equivalence on terms.

We can now prove the following lemma:

Lemma 3. *If $[\widetilde{a}].T = [\widetilde{b}].U$ and $\widetilde{a}\#\widetilde{b}$ then there exists a permutation p s.t.*
$$\mathsf{set}\ p \subseteq \mathsf{set}\ \widetilde{a} \times \mathsf{set}\ \widetilde{b},\ \widetilde{a}\#U\ and\ T = p \cdot U$$

Proof. The intuition here is to construct p by using Cor. 5 to filter out the pairs of names from $\widetilde{a}$ and $\widetilde{b}$ that do not occur in T and U respectively and pairing together the rest. The proof is done by induction on the length of $\widetilde{a}$ and $\widetilde{b}$.

The problem with this approach is that we do not know how $\widetilde{a}$ and $\widetilde{b}$ are related. If we know that they are both distinct then we can construct p such that $\widetilde{a} = p \cdot \widetilde{b}$ but generally we do not know this. The problematic cases are the ones dealing with inversion, in which case we resort to explicit binding sequences, but for the majority of our proofs Lemma 3 is enough.

4   Formalisation

Psi-calculi are parametric calculi. A specific instance is created by instantiating the framework with dataterms for the terms, assertions and conditions of the calculus. We also require an entailment relation, a notion of channel equality and composition of assertions. Isabelle has good support for reasoning about parametric systems through the use of locales [3].

4.1   Substitution Properties

We require a substitution function on agents. Since terms, assertions and conditions of psi-calculi are parameters, a locale is created to ensure that a set of substitution properties hold.

Definition 10. *A term M of type α is a $\mathsf{substType}$ if there is a substitution function $\mathsf{subst} :: \alpha \Rightarrow \mathsf{name\ list} \Rightarrow \beta\ \mathsf{list} \Rightarrow \alpha$ which meets the following constraints, where $\mathsf{length}\ \widetilde{x} = \mathsf{length}\ \widetilde{T}$ and $\mathsf{distinct}\ \widetilde{x}$*

Equivariance: $\quad p \cdot (M[\tilde{x} := \tilde{T}]) = (p \cdot M)[(p \cdot \tilde{x}) := (p \cdot \tilde{T})]$

Freshness: $\quad$ if $a\#M[\tilde{x} := \tilde{T}]$ and $a\#\tilde{x}$ then $a\#M$

$\qquad\qquad$ if $a\#M$ and $a\#\tilde{T}$ then $a\#M[\tilde{x} := \tilde{T}]$

$\qquad\qquad$ if set $\tilde{x} \subseteq$ supp M and $a\#M[\tilde{x} := \tilde{T}]$ then $a\#\tilde{T}$

$\qquad\qquad$ if $\tilde{x}\#\tilde{M}$ then $M[\tilde{x} := \tilde{T}] = M$

$\qquad\qquad$ if $\tilde{x}\#\tilde{y}$ and $\tilde{y}\#\tilde{T}$ then $M[\tilde{x}\tilde{y} := \tilde{T}\tilde{U}] = (M[\tilde{x} := \tilde{T}])[\tilde{y} := \tilde{U}]$

Alpha-equivalence: if set $p \subseteq$ set $\tilde{x} \times$ set$(p \cdot \tilde{x})$ and $(p \cdot \tilde{x})\#M$ then

$\qquad\qquad M[\tilde{x} := \tilde{T}] = (p \cdot M)[(p \cdot \tilde{x}) := \tilde{T}]$

The intuition is that subst is a simultaneous substitution function which replaces all occurrences of the names in $\tilde{x}$ in M with the corresponding dataterm in $\tilde{T}$. All that the locale dictates is that there is a function of the correct type which satisfies the constraints. Exactly how it works needs only be specified when creating an instance of the calculus in order to prove that the constraints are satisfied.

These constraints are the ones we need for the formalisation but we have not proven that they are strictly minimal. We leave this for future work.

4.2   The Psi Datatype

Nominal Isabelle does not support datatypes with binding sequences or nested datatypes. The two cases that are problematic when formalising psi-calculi are the Input case, which requires a binding sequence, and the Case case which requires a list of assertions and processes. The required datatype can be encoded using mutual recursion in the following way.

Definition 11. *The psi-calculi datatype has three type variables for terms, assertions and conditions respectively. In the* Res *and the* Bind *cases,* name *is a binding occurrence.*

nominal_datatype (α, β, γ) psi $=$ Output α α (α, β, γ) psi

$\qquad\qquad\qquad\qquad$ | Input α (α, β, γ) input

$\qquad\qquad\qquad\qquad$ | Case (α, β, γ) case

$\qquad\qquad\qquad\qquad$ | Par $((\alpha, \beta, \gamma)$ psi$)$ $((\alpha, \beta, \gamma)$ psi$)$

$\qquad\qquad\qquad\qquad$ | Res $\ll$ name $\gg$ $((\alpha, \beta, \gamma)$ psi$)$

$\qquad\qquad\qquad\qquad$ | Assert β

$\qquad\qquad\qquad\qquad$ | Bang (α, β, γ) psi

and (α, β, γ) input $\qquad\quad=$ Term α (α, β, γ) psi

$\qquad\qquad\qquad\qquad$ | Bind $\ll$ name $\gg$ $((\alpha, \beta, \gamma)$ input$)$

and (α, β, γ) case $\qquad\quad=$ EmptyCase

$\qquad\qquad\qquad\qquad$ | Cond γ $((\alpha, \beta, \gamma)$ psi$)$ $((\alpha, \beta, \gamma)$ case$)$

In order to create a substitution function for $(\alpha\ \beta\ \gamma)$ psi we create a locale with the following three substitution functions as substTypes.

```
substTerm    :: α ⇒ name list ⇒ α list ⇒ α
substAssert :: β ⇒ name list ⇒ α list ⇒ β
substCond    :: γ ⇒ name list ⇒ α list ⇒ γ
```

These functions will handle substitutions on terms, assertions and conditions respectively. Note that we always substitute names for terms.

The substitution function for `psi` can now be defined in the standard way where the substitutions are pushed through the datatype avoiding the binders. The axioms for `substType` can then be proven for the `psi` substitution function where the axioms themselves are used when the proofs reaches the terms, assertions and conditions.

4.3 Frames

The four nominal morphisms from Def. 1 are also encoded using locales along with their equivariance properties. From this definition, implementing Def. 2 and a locale for our requirements on assertion equivalence $\simeq$ is straightforward. To implement frames, the following nominal datatype is created:

Definition 12
```
nominal_datatype β frame = Assertion β
                         | FStep ≪ name ≫ (β frame)
```

In order to overload the $\otimes$ operator to work on frames as described in Def. 3 we create the following two nominal functions.

Definition 13
$$\text{insertAssertion} (\text{Assertion}\, \Psi)\, \Psi' = \text{Assertion}(\Psi' \otimes \Psi)$$
$$x\#\Psi' \Rightarrow \text{insertAssertion} (\text{FStep}\, x\, F)\, \Psi = \text{FStep}\, x\, (\text{insertAssertion}\, F\, \Psi')$$

$$(\text{Assertion}\, \Psi) \otimes G = \text{insertAssertion}\, G\, \Psi$$
$$x\#G \Rightarrow (\text{FStep}\, x\, F) \otimes G = \text{FStep}\, x\, (F \otimes G)$$

The following lemma is then derivable:

Lemma 4. *If $B_P \# B_Q$, $B_P \# \Psi_Q$ and $B_Q \# \Psi_P$*
  then $\langle B_P, \Psi_P \rangle \otimes \langle B_Q, \Psi_Q \rangle = \langle B_P @ B_Q, \Psi_P \otimes \Psi_Q \rangle$.

The implementations of Defs. 4 and 6 are then straightforward.

4.4 Operational Semantics

The operational semantics in Def. 7 is formalised in a similar manner to [6]. Since the actions on the labels can contain bound names which bind into the derivative of the transition, a residual datatype needs to be created which combines the actions with their derivatives. Since a bound output can contain an arbitrary number of bound names, binding sequences must be used here in a similar manner to `psi` and `frame`.

```
nominal_datatype (α, β, γ) boundOutput =
    Output α (α, β, γ) psi
    | BStep ≪ name ≫ (α, β, γ) boundOutput
```

datatype α action = Input α α
                     | Tau

datatype (α, β, γ) residual = Free $(\alpha$ action$)$ $((\alpha, \beta, \gamma)$ psi$)$
                               | Bound α $((\alpha, \beta, \gamma)$ boundOutput$)$

We will use the notation $(\nu\tilde{a})N \prec P$ for a term of type boundOutput which has the binding sequence $\tilde{a}$ into N and P. We can also write $\Psi \triangleright P \longmapsto M(\nu\tilde{a})N \prec P'$ for $\Psi \triangleright P \xrightarrow{M(\nu\tilde{a})N} P'$ and similarly for input and tau transitions.

As usual, the operational semantics is defined using an inductively defined predicate. As in [6] rules which can have either free or bound residuals are split into these two cases. We also saturate our rules with freshness conditions to ensure that the bound names are fresh for for all terms outside their scope. This is done to satisfy the vc-property described in [16] so that Isabelle can automatically infer an induction rule, but also to give us as much freshness information as possible when doing induction on transitions. Moreover, all frames are required to have distinct binding sequences. The introduction rules in Table 1 only include the freshness conditions which are strictly necessary and frames with non distinct binding sequences. These can be inferred from our inductive definition using regular alpha converting techniques and Cor. 4.

We will not cover the complete semantics here, just two rules to demonstrate some differences to the pen-and-paper formalisation.

The transition rule PAR has the implicit assumption that $\mathcal{F}(Q) = \langle B_Q, \Psi_Q \rangle$. When formalising the semantics, one inductive case will look as follows:

$$\text{PAR} \frac{\Psi_Q \otimes \Psi \triangleright P \xrightarrow{\alpha} P' \qquad \mathcal{F}(Q) = \langle B_Q, \Psi_Q \rangle \quad B_Q \# \Psi, P, \alpha, P', Q}{\Psi \triangleright P|Q \xrightarrow{\alpha} P'|Q} \quad \text{distinct } B_Q$$

Inferring the transition for P means selecting a specific alpha-variant of $\mathcal{F}(Q)$ as Ψ_Q appears without binders in the inference of the transition. Freshness conditions for B_Q are central for the proofs to hold.

Next consider the rule OPEN. We want the binding sequence on the transition to behave like a set in that we must not depend on the order of its names. Our formalisation solves this by explicitly splitting the binding sequence in two and placing the opened name in between. By creating a rule which holds for all such splits, we mimic the effect of a set.

$$\text{OPEN} \frac{\Psi \triangleright P \xrightarrow{\overline{M}(\nu\tilde{a}\tilde{c})N} P' \qquad \begin{array}{c} b\#\tilde{a},\tilde{c},\Psi,M \\ b \in n(N) \end{array}}{\Psi \triangleright (\nu b)P \xrightarrow{\overline{M}(\nu\tilde{a}b\tilde{c})N} P' \qquad \begin{array}{c} \tilde{a}\#\Psi,P,M,\tilde{c} \\ \tilde{c}\#\Psi,P,M \end{array}}$$

4.5   Induction Rules

At the core of any nominal formalisation is the need to create custom induction rules which allow the introduced bound names to be fresh for any given context. Without these, the user is forced to do manual alpha-conversions throughout the proofs and such proofs will differ significantly from their pen and paper counterparts, where freshness is just assumed. An in depth description can be found in [16]. Very recent additions to the nominal package generate induction rules where the user is allowed to choose a set of name which can be arbitrarily fresh for each inductive case. In most cases, this set will be the set of binders present in the rule.

Standard induction. Isabelle will automatically create a rule for doing induction on transitions of the form $\Psi \rhd P \longmapsto Rs$, where Rs is a residual. In nominal induction the predicate to be proven has the extra argument $\mathcal{C}$, such that all bound names introduced by the induction rule are fresh for $\mathcal{C}$. Thus, the predicate has the form $\texttt{Prop}\ \mathcal{C}\ \Psi\ P\ Rs$. This induction rule is useful for very general proofs about transitions, but we often need proofs which are specialised for input, output, or tau transitions. We create the following custom induction rules:

Lemma 5.

$$
\cfrac{\Psi \rhd P \xrightarrow{M\,N} P'}{\vdots}{\texttt{Prop}\,\mathcal{C}\,\Psi\,P\,M\,N\,P'}
\qquad\Bigg|\qquad
\cfrac{\Psi \rhd P \xrightarrow{\overline{M}\,(\nu\widetilde{a})N} P'}{\vdots}{\texttt{Prop}\,\mathcal{C}\,\Psi\,P\,M\,((\nu\widetilde{a})(N \prec P'))}
\qquad\Bigg|\qquad
\cfrac{\Psi \rhd P \xrightarrow{\tau} P'}{\vdots}{\texttt{Prop}\,\mathcal{C}\,\Psi\,P\,P'}
$$

Proof. Follows immediately from the induction rule generated by Isabelle.

The inductive steps for each rule have been left out as they are instances of the ones from the automatically generated induction rule, but with the predicates changed to match the corresponding transition.

These induction rules work well only as long as the predicate to be proven does not depend on anything under the scope of a binder. Trying to prove the following lemma illustrates the problem.

Lemma 6. *If* $\Psi \rhd P \xrightarrow{\overline{M}\,(\nu\widetilde{a})N} P'$, $x\#P$ *and* $x\#\widetilde{a}$ *then* $x\#N$ *and* $x\#P'$

Proof. By induction over the transitions of the form $\Psi \rhd P \xrightarrow{\overline{M}\,(\nu\widetilde{a})N} P'$.

The problem is that none of the induction rules we have will prove this lemma in a satisfactory way. Every applicable case in the induction rule will introduce its own bound output term $(\nu\widetilde{b})N' \prec P''$ where we know that $(\nu\widetilde{b})N' \prec P'' = (\nu\widetilde{a})N \prec P'$. What we need to prove relates to the term P', what the inductive hypotheses will give us is something related to the term P'' where all we know is that they are part of alpha-equivalent terms.

Proving this lemma on its own is not too difficult but in every step of every proof of this type, manual alpha-conversions and equivariance properties are needed. The following induction rule solves this problem.

$$\forall \Psi\ P\ M\ \tilde{a}\ N\ P'\ \tilde{b}\ p\ \mathcal{C}. \quad \frac{\Psi \vartriangleright P \xrightarrow{\overline{M}\,(\nu\tilde{a})N} P'}{\left(\begin{array}{l} \tilde{a}\#\tilde{b}, \Psi, P, M, \mathcal{C} \wedge \tilde{b}\#N, P' \wedge \\ \text{set } p \subseteq \text{set } \tilde{a} \times \text{set } \tilde{b} \wedge \\ \text{Prop } \mathcal{C}\ \Psi\ P\ M\ \tilde{a}\ N\ P' \longrightarrow \\ \text{Prop } \mathcal{C}\ \Psi\ P\ M\ \tilde{b}\ (p \cdot N)\ (p \cdot P') \end{array}\right)}$$

$$\vdots$$

$$\overline{\text{Prop } \mathcal{C}\ \Psi\ P\ M\ \tilde{a}\ N\ P'}$$

The difference between this rule and the output rule in Lemma 5 is that the predicate in Lemma 5 takes a residual $(\nu\tilde{a})N \prec P'$ as one argument and the predicate in this rule takes $\tilde{a}$, N and P' as three separate ones. By disassociating the binding sequence from the residual in this manner we have lost the ability to alpha-convert the residual, but we have gained the ability to reason about terms under the binding sequence. The extra added case in the induction rule above (beginning with $\forall \Psi\ P\ M\ \ldots$) is designed to allow the predicate to mimic the alpha-conversion abilities we have lost. When proving this induction rule, Lemma 3 is used in each step to generate the alpha-converting permutation, Prop is proven in the standard way and then alpha-converted using the new inductive case.

With this lemma, we must prove that the predicate we are trying to prove can respect alpha-conversions. The advantage is that it only has to be done once for each proof. Moreover, the case is very general and does not require the processes or actions to be of a specific form.

Using this induction rule will not allow us to prove lemmas which reason directly about the binding sequence $\tilde{a}$. The new inductive case swaps a sequence $\tilde{a}$ for $\tilde{b}$ but as in Lemma 3, we do not know exactly how these sequences relate to each other.

Induction with frames. A very common proof strategy in the psi-calculus is to do induction on a transition of a process which has a specific frame. Trying to prove the following lemma illustrates this.

Lemma 7. *If* $\Psi \vartriangleright P \xrightarrow{\overline{M}\,N} P'$, $\mathcal{F}(P) = \langle B_P, \Psi_P \rangle$, $X\#P$ *and* $B_P\#X, \Psi, P, M$ *then there exists a* K. *s.t.* $\Psi \otimes \Psi_P \vdash M \leftrightarrow K$ *and* $X\#K$.

Proof. By induction on the transition $\Psi \vartriangleright P \xrightarrow{\overline{M}\,N} P'$. The intuition of the proof is that K is the subject in the process P.

This lemma suffers from the same problem as Lemma 6 – every inductive step will generate a frame alpha-equivalent to $\langle B_P, \Psi_P \rangle$ and many tedious alpha-conversions have to be done to prove the lemma. Moreover, some of our lemmas need to directly reason about the binding sequence of the frame. A similar induction rule as for output transitions can be created to solve the problem.

$$\forall \Psi \; P \; M \; N \; P' \; B_P \; \Psi_P \; p \; \mathcal{C}. \quad
\begin{array}{c}
\Psi \rhd P \xrightarrow{M \, N} P' \\
\mathcal{F}(P) = \langle B_P, \, \Psi_P \rangle \\
\text{distinct } B_P \\
\begin{pmatrix}
(p \cdot B_P) \# \Psi, P, M, \mathcal{C}, N, P', B_P \; \wedge B_P \# \Psi_P \; \wedge \\
\text{set } p \subseteq \text{set } B_P \times \text{set}(p \cdot B_P) \; \wedge \\
\text{Prop } \mathcal{C} \; \Psi \; P \; M \; N \; P' \; B_P \; \Psi_P \longrightarrow \\
\text{Prop } \mathcal{C} \; \Psi \; P \; M \; N \; P' \; (p \cdot B_P) \; (p \cdot \Psi_P)
\end{pmatrix} \\
\vdots
\end{array}$$

$$\overline{\text{Prop } \mathcal{C} \; \Psi \; P \; M \; N \; P' \; B_P \; \Psi_P}$$

This lemma requires that the binding sequence B_P is distinct. This added requirement allows the alpha converting case to relate the sequence B_P to $p \cdot B_P$ allowing for a larger class of lemmas to be proven. Our semantics require all frames to have distinct binding sequences making this added requirement unproblematic.

A corresponding lemma has to be created for output transitions as well, but since frames only affect subjects as far as input and output transitions are concerned, this induction rule does not have to use the same mechanism for the bound names in the residual as for the ones in the frame.

After introducing these custom induction rules, we were able to remove thousands of lines of code which were only dealing with alpha-conversions.

5   Conclusions and Future Work

Nominal Isabelle has proven to be a very potent tool when doing this formalisation. Its support for locales has made the formalisation of parametric calculi such as psi-calculi feasible and the nominal datatype package handles binders elegantly.

Psi-calculi require substantially more infrastructure than the pi-calculus [6]. The reason for this is mainly that binding sequences are a very new addition to the nominal package, and many of the automatic rules are not fully developed. Extending the support for binding sequences will require a fair bit of work, but we believe that the custom induction rules that we have designed can be created automatically as they do not use any intrinsic properties of psi-calculi.

We are currently working on extending our framework to include weak bisimulation and barbs. We also plan to work on typed psi calculi where we aim to make the type system as general and parametric as psi calculi themselves.

The source files for this formalisation can be found at:
http://www.it.uu.se/katalog/jesperb/psi.tar.gz

Acknowledgments. We want to convey our sincere thanks to Stefan Berghofer for his hard work on expanding the nominal package to include the features we have needed for this formalisation.

References

1. Abadi, M., Fournet, C.: Mobile values, new names, and secure communication. In: Proceedings of POPL 2001, pp. 104–115. ACM, New York (2001)
2. Aydemir, B., Charguéraud, A., Pierce, B.C., Pollack, R., Weirich, S.: Engineering formal metatheory. In: POPL 2008: Proceedings of the 35th annual ACM SIGPLAN-SIGACT symposium on Principles of programming languages, pp. 3–15. ACM, New York (2008)
3. Ballarin, C.: Locales and locale expressions in isabelle/isar. In: Berardi, S., Coppo, M., Damiani, F. (eds.) TYPES 2003. LNCS, vol. 3085, pp. 34–50. Springer, Heidelberg (2004)
4. Barendregt, H.P.: The Lambda Calculus – Its Syntax and Semantics. Studies in Logic and the Foundations of Mathematics, vol. 103. North-Holland, Amsterdam (1984)
5. Bengtson, J., Johansson, M., Parrow, J., Victor, B.: Psi-calculi: Mobile processes, nominal data, and logic. Technical report, Uppsala University (2009); (submitted), http://user.it.uu.se/~joachim/psi.pdf
6. Bengtson, J., Parrow, J.: Formalising the pi-calculus using nominal logic. In: Seidl, H. (ed.) FOSSACS 2007. LNCS, vol. 4423, pp. 63–77. Springer, Heidelberg (2007)
7. Berghofer, S., Urban, C.: Nominal Inversion Principles. In: Mohamed, O.A., Muñoz, C., Tahar, S. (eds.) TPHOLs 2008. LNCS, vol. 5170, pp. 71–85. Springer, Heidelberg (2008)
8. Buscemi, M.G., Montanari, U.: Open bisimulation for the concurrent constraint π-calculus. In: Drossopoulou, S. (ed.) ESOP 2008. LNCS, vol. 4960, pp. 254–268. Springer, Heidelberg (2008)
9. de Bruijn, N.G.: Lambda calculus notation with nameless dummies. a tool for automatic formula manipulation with application to the church-rosser theorem. Indagationes Mathematicae 34, 381–392 (1972)
10. Hirschkoff, D.: A full formalisation of π-calculus theory in the calculus of constructions. In: Gunter, E.L., Felty, A.P. (eds.) TPHOLs 1997. LNCS, vol. 1275, pp. 153–169. Springer, Heidelberg (1997)
11. Honsell, F., Miculan, M., Scagnetto, I.: π-calculus in (co)inductive type theory. Theoretical Comput. Sci. 253(2), 239–285 (2001)
12. Nipkow, T., Paulson, L.C., Wenzel, M.: Isabelle/HOL. LNCS, vol. 2283. Springer, Heidelberg (2002)
13. Pitts, A.M.: Nominal logic, a first order theory of names and binding. Information and Computation 186, 165–193 (2003)
14. Röckl, C., Hirschkoff, D.: A fully adequate shallow embedding of the π-calculus in Isabelle/HOL with mechanized syntax analysis. J. Funct. Program. 13(2), 415–451 (2003)
15. Urban, C.: Nominal techniques in Isabelle/HOL. Journal of Automated Reasoning 40(4), 327–356 (2008)
16. Urban, C., Berghofer, S., Norrish, M.: Barendregt's variable convention in rule inductions. In: Pfenning, F. (ed.) CADE 2007. LNCS, vol. 4603, pp. 35–50. Springer, Heidelberg (2007)

Some Domain Theory and Denotational Semantics in Coq

Nick Benton[1], Andrew Kennedy[1], and Carsten Varming[2,*]

[1] Microsoft Research, Cambridge, UK
{nick,akenn}@microsoft.com
[2] Carnegie-Mellon University, Pittsburgh, USA
varming@cmu.edu

Abstract. We present a Coq formalization of constructive ω-cpos (extending earlier work by Paulin-Mohring) up to and including the inverse-limit construction of solutions to mixed-variance recursive domain equations, and the existence of invariant relations on those solutions. We then define operational and denotational semantics for both a simply-typed CBV language with recursion and an untyped CBV language, and establish soundness and adequacy results in each case.

1   Introduction

The use of proof assistants in formalizing language metatheory and implementing certified tools has grown enormously over the last five years or so. Most current work on mechanizing language definitions and type soundness results, certified compilation, proof carrying code, and so on has been based on operational semantics. But in our work on both certified compilation and on the semantics of languages with state, we have often found ourselves wanting a Coq formalization of the kind of denotational semantics that we have grown accustomed to working with on paper.

Mechanizing domain theory and denotational semantics has an illustrious history. Provers such as HOL, Isabelle/HOL and Coq can all trace an ancestral line back to Milner's LCF [16], which was a proof checker for Scott's PPλ logic of cpos, continuous functions and admissible predicates. And although later systems were built on less domain-specific foundations, there have subsequently been dozens of formalizations of different notions of domains and bits of semantics, with examples in all the major provers. Few, however, have really gone far enough to be useful. This paper describes our Coq formalization of ω-cpos and the denotational semantics of both typed and untyped versions of a simple functional language, going considerably further than previous work. A companion paper [8] describes a non-trivial compiler correctness theorem that has been formalized and proved using one of these denotational models.

---

* Research supported in part by National Science Foundation Grants CCF-0541021, CCF-0429505.

S. Berghofer et al. (Eds.): TPHOLs 2009, LNCS 5674, pp. 115–130, 2009.
© Springer-Verlag Berlin Heidelberg 2009

Our formalization is based on a Coq library for constructive pointed ω-cpos and continuous functions written by Paulin-Mohring [20] as a basis for a semantics of Kahn networks, and of probabilistic programs [6]. Section 2 describes our slight generalization of Paulin-Mohring's library to treat predomains and a general lift monad. In Section 3, we then define a simply-typed call-by-value functional language, give it a denotational semantics using our predomains and prove the standard soundness and adequacy theorems, establishing the correspondence between the operational and denotational semantics. These results seem not to have been previously mechanized for a higher-order language.

Section 4 is about solving recursive domain equations. We formalize Scott's inverse limit construction along the lines of work by Freyd [11,12] and Pitts [22,23]. This approach characterizes the solutions as minimal invariants, yielding reasoning principles that allow one to construct and work with recursively-defined predicates and relations over the recursively-defined domains. In Section 5, we define the semantics of an untyped call-by-value language using a particular recursive domain, and use the associated reasoning principles to again establish soundness and adequacy theorems.

2 Basic Domain Theory

This first part of the development is essentially unchanged from the earlier work of Paulin-Mohring [20]. The main difference is that Paulin-Mohring formalized *pointed* cpos and continuous maps, with a special-case construction of flat cpos (those that arise from adding a bottom element under all elements of an otherwise discretely ordered set), whereas we use potentially bottomless cpos ('predomains') and formalize a general constructive lift monad.

2.1 Complete Partial Orders

We start by defining the type of preorders, comprising a carrier type *tord* (to which :> means we can implicitly coerce), a binary relation *Ole* (written infix as $\sqsubseteq$), and proofs that *Ole* is reflexive and transitive:

```
Record ord:= mk_ord
  {tord :> Type;
   Ole : tord → tord → Prop;
   Ole_refl : ∀ x : tord, Ole x x;
   Ole_trans : ∀ x y z : tord, Ole x y → Ole y z → Ole x z}.
Infix "⊑" := Ole.
```

The equivalence relation == is then defined to be the symmetrisation of $\sqsubseteq$:

```
Definition Oeq (O : ord) (x y : O) := x ⊑ y ∧ y ⊑ x.
Infix "==" := Oeq (at level 70).
```

Both == and $\sqsubseteq$ are declared as parametric Setoid relations, with $\sqsubseteq$ being a partial order modulo ==. Most of the constructions that follow are proved and declared to be morphisms with respect to these relations, which then allows convenient (in)equational rewriting in proofs.

The type of monotone functions between partial orders is a parameterized record type, comprising a function between the underlying types of the two order parameters and a proof that that function preserves order:

Definition *monotonic* $(O_1\ O_2 : ord)\ (f : O_1 \to O_2) := \forall\ x\ y,\ x \sqsubseteq y \to f\ x \sqsubseteq f\ y.$
Record *fmono* $(O_1\ O_2 : ord) := mk\_fmono$
  $\{fmonot :> O_1 \to O_2;$
  fmonotonic: *monotonic fmonot*$\}.$

For any $O_1\ O_2 : ord$, the monotonic function space $O_1 \to_m O_2 : ord$ is defined by equipping *fmono* $O_1\ O_2$ with the order inherited from the codomain, $f \sqsubseteq g$ iff $f\ x \sqsubseteq g\ x$ for all x.

We define *natO* : *ord* by equipping the set of natural numbers, *nat*, with the usual 'vertical' order, $\leq$. If $c : natO \to_m O$ for some $O : ord$, we call c a *chain* in O. Now a complete partial order is defined as a dependent record comprising an underlying order, *tord*, a function $\bigsqcup$ for computing the least upper bound of any chain in *tord*, and proofs that this *is* both an upper bound (*le_lub*), and less than or equal to any other upper bound (*lub_le*):

Record *cpo*:= *mk_cpo*
  $\{tcpo :> ord;$
  $\bigsqcup : (natO \to_m tcpo) \to tcpo;$
  le_lub : $\forall\ (c : natO \to_m tcpo)\ (n : nat),\ c\ n \sqsubseteq \bigsqcup c;$
  lub_le : $\forall\ (c : natO \to_m tcpo)\ (x : tcpo),\ (\forall\ n,\ c\ n \sqsubseteq x) \to \bigsqcup c \sqsubseteq x\}.$

This definition of a complete partial order is constructive in the sense that we require least upper bounds of chains not only to exist, but to be computable in Coq's logic of total functions.

A monotone function f between two *cpos*, D_1 and D_2, is *continuous* if it preserves (up to ==) least upper bounds. One direction of this is already a consequence of monotonicity, so we just have to specify the other:

Definition *continuous* $(D_1\ D_2 : cpo)\ (f : D_1 \to_m D_2) :=$
  $\forall\ c : natO \to_m D_1,\ f\ (\bigsqcup c) \sqsubseteq \bigsqcup (f \circ c).$
Record *fconti* $(D_1\ D_2 : cpo) := mk\_fconti$
  $\{fcontit : D_1 \to_m D_2;$
  fcontinuous : *continuous fcontit*$\}.$

For any $D_1\ D_2 : cpo$, the continuous function space $D_1 \to_c D_2 : ord$ is defined by equipping the type *fconti* $D_1\ D_2$ with the pointwise order inherited from D_2. We then define $D_1 \Rightarrow_c D_2 : cpo$ by equipping $D_1 \to_c D_2$ with least upper bounds computed pointwise: if $c : natO \to_m (D_1 \to_c D_2)$ is a chain, then $\bigsqcup c : (D_1 \to_c D_2)$ is $\lambda d_1.\bigsqcup(\lambda n.c\ n\ d_1)$.

If $D : cpo$, write *ID D* : $D \to_c D$ for the continuous identity function on D. If $f : D \to_c E$ and $g : E \to_c F$ write $g \circ f : D \to_c F$ for their composition. Composition of continuous maps is associative, with *ID* as a unit.

Discrete cpos. If $X : $ **Type** then equipping X with the order $x_1 \sqsubseteq x_2$ iff $x_1 = x_2$ (i.e. Leibniz equality) yields a cpo that we write *Discrete X*.

Finite products. Write **1** for the one-point cpo, *Discrete unit*, which is terminal, in that for any $f\ g : D \to_c \mathbf{1},\ f == g$. If $D_1\ D_2 : cpo$ then equipping the usual product of the underlying types of their underlying orders with the pointwise ordering yields a product order. Equipping that order with a pointwise least upper bound operation $\bigsqcup c = (\bigsqcup(fst \circ c), \bigsqcup(snd \circ c))$ for $c \to_m D_1 \times D_2$ yields a product cpo $D_1 \times D_2$ with continuous $\pi_i : D_1 \times D_2 \to_c D_i$. We write $\langle f, g \rangle$ for the unique (up to $==$) continuous function such that $f == \pi_1 \circ \langle f, g \rangle$ and $g == \pi_2 \circ \langle f, g \rangle$.

Closed structure. We can define operations $curry : (D \times E \to_c F) \to (D \to_c E \Rightarrow_c F)$ and $ev : (E \Rightarrow_c D) \times E \to_c D$ such that for any $f : D \times E \to_c F$, $curry(f)$ is the unique continuous map such that $f == ev \circ \langle curry\, f \circ \pi_1, \pi_2 \rangle$. We define $uncurry : (D \Rightarrow_c E \Rightarrow_c F) \to_c D \times E \Rightarrow_c F$ by $uncurry = curry(ev \circ \langle ev \circ \langle \pi_1, \pi_1 \circ \pi_2 \rangle, \pi_2 \circ \pi_2 \rangle)$ and we check that $uncurry(curry(f)) == f$ and $curry(uncurry(h)) == h$ for all f and h.

So our internal category $\mathbb{CPO}$ of *cpos* and continuous maps is Cartesian closed. We elide the details of other constructions, including finite coproducts, strict function spaces and general indexed products, that are in the formalization. Although our cpos are not required to have least elements, those that do are of special interest. We use Coq's typeclass mechanism to capture them:

Class $Pointed(D : cpo) := \{\ \bot : D;\ Pleast : \forall\, d : D,\ \bot \sqsubseteq d\ \}$.
Instance $DOne\_pointed : Pointed\,\mathbf{1}$.
Instance $prod\_pointed\ A\ B\ \{\ pa : Pointed\,A\}\ \{pb : Pointed\,B\} : Pointed(A \times B)$.
Instance $fun\_pointed\ A\ B\ \{pb : Pointed\,B\} : Pointed(A \Rightarrow_c B)$.

Now if D is *Pointed*, and $f : D \to_c D$ then we can define *fixp f*, the least fixed point of f in the usual way, as the least upper bound of the chain of iterates of f starting at $\bot$. We define $FIXP : (D \Rightarrow_c D) \to_c D$ to be the 'internalised' version of *fixp*.

If $D : cpo$ and $P : D \to Prop$, then P is *admissible* if for all chains $c : natO \to_m D$ such that $(\forall n.\ P(c_n))$, one has $P(\bigsqcup c)$. In such a case, the subset type $\{d : D \mid P(d)\}$ with the order and lubs inherited from D is a cpo. We can also prove the standard fixed point induction principle:

Definition $fixp\_ind\ D\ \{\ pd : Pointed\,D\} : \forall\ (F: D \to_m D)(P : D \to \mathtt{Prop})$,
    $admissible\ P \to P\ \bot \to (\forall\ x,\ P\ x \to P\ (F\ x)) \to P\ (fixp\ F)$.

The main technical complexity in this part of the formalization is simply the layering of definitions, with (for example) cpos being built on *ords*, and $D \Rightarrow_c E$ being built on $D \to_c E$, which is built on $D \to_m E$, which is built on $D \to E$. Definitions have to be built up in multiple staged versions and there are many implicit coercions and hints for Coq's **auto** tactic, which are tricky to get right. There is also much boilerplate associated with morphism declarations supporting setoid rewriting, and there is some tension between the elementwise and 'point-free' styles of working.

2.2 The Lift Monad

The basic order theory of the previous section goes through essentially as it does when working classically on paper. In particular, the definitions of lubs in products and function spaces are already constructive. But lifting will allow us to express general partial recursive functions, which, in Coq's logic of total functions, is clearly going to involve some work. Our solution is a slight generalization of Paulin-Mohring's treatment of the particular case of flat cpos, which in turn builds on work of Capretta [9] on general recursion in type theory. We exploit Coq's support for coinductive datatypes [10], defining lifting in terms of a type *Stream* of potentially infinite streams:

Variable D : *cpo*.
CoInductive *Stream* := *Eps* : *Stream* → *Stream* | *Val* : D → *Stream*.

An element of *Stream* is (classically) either the infinite $Eps(Eps(Eps(\dots)))$, or some finite sequence of *Eps* steps, terminated by *Val d* for some $d : D$, $Eps(Eps(\dots Eps(Val\ d)\ \dots))$. One can think of *Stream* as defining a resumptions monad, which we will subsequently quotient to define lifting. For $x : Stream$ and $n : nat$, *pred_nth x n* is the stream that results from removing the first n *Eps* steps from x. The order on *Stream* is coinductively defined by

CoInductive *DLle* : *Stream*→ *Stream*→ **Prop** :=
| *DLleEps* : ∀ $x\ y$, *DLle x y* → *DLle* (*Eps x*) (*Eps y*)
| *DLleEpsVal* : ∀ $x\ d$, *DLle x* (*Val d*) → *DLle* (*Eps x*) (*Val d*)
| *DLleVal* : ∀ $d\ d'\ n\ y$, *pred_nth y n* = *Val d'* → $d \sqsubseteq d'$ → *DLle* (*Val d*) y.

which satisfies the following coinduction principle:

Lemma *DLle_rec* : ∀ R : *Stream*→ *Stream*→ **Prop**,
  (∀ $x\ y$, R (*Eps x*) (*Eps y*) → $R\ x\ y$) →
  (∀ $x\ d$, R (*Eps x*) (*Val d*) → $R\ x$ (*Val d*)) →
  (∀ $d\ y$, R (*Val d*) y → ∃ n, ∃ d', *pred_nth y n* = *Val d'* ∧ $d \sqsubseteq d'$)
  → ∀ $x\ y$, $R\ x\ y$ → *DLle x y*.

The coinduction principle is used to show that *DLle* is reflexive and transitive, allowing us to construct a preorder *DL_ord* : *ord* (and we now write the usual $\sqsubseteq$ for the order). The infinite stream of *Eps*'s, Ω, is the least element of the order.

Constructing a *cpo* from *DL_ord* is slightly subtle. We need to define a function that maps chains c : ($natO \rightarrow_m DL\_ord$) to their lubs in *DL_ord*. An important observation is that if some c_n is non-Ω, i.e. there exists a d_n such that $c_n == Val d_n$, then for any $m \geq n$, there is a d_m such that $c_m == Val d_m$ and that moreover, the sequence $d_n, d_{n+1}, \dots$, forms a chain in D. Classically, the idea is that we look for such a c_n; if we find one, then we can construct a chain in D and return *Val* applied to the least upper bound of that chain. If there's no such chain then the least upper bound is Ω. But we cannot simply test whether a particular c_n is Ω or not: we can only examine finite prefixes. So we make a 'parallel' corecursive search through all the c_ns, which may be pictured something like this:[1]

$$
\begin{array}{cccc}
c_0 = & Eps & Eps & Eps & \cdots \\
c_1 = & Eps & Eps & ? & ? \\
c_2 = & Eps & ? & ? & ? \\
\end{array}
$$

---

[1] In reality, the output stream 'ticks' less frequently than the picture would suggest.

The output we are trying to produce is an element of DL_ord. Each time our interleaving search finds an Eps, we produce an Eps on the output. So if every element of the chain is Ω, we will end up producing Ω on the output. But should we find a $Vald$ after outputting some finite number of Eps s, then we know all later elements of the chain are also non-Ω, so we go ahead and build the chain in D that they form and compute its least upper bound using the lub operation of D. The details of this construction, and the proof that it does indeed yield the least upper bound of the chain c, involve interesting bits of constructive reasoning: going from knowing that there *is* a chain in D to actually having that chain in one's hand so as to take its lub uses (a provable form of) constructive indefinite description, for example. But at the end of the day, we end up with a constructive definition of $D_\perp : cpo$, which is clearly *Pointed*.

Lifting gives a strong monad [17] on $\mathbb{CPO}$. The unit $\eta : D \to_c D_\perp$ applies the Val constructor. If $f : D \to_c E_\perp$ define $kleisli\ f : D_\perp \to_c E_\perp$ to be the map

$$cofix\ kl\ (d : D_\perp) : E_\perp\ :=\ match\ d\ with\ Eps\ dl\ \Rightarrow\ Eps\ (kl\ dl)\ |\ Val\ d' \Rightarrow f\ d'$$

Thinking operationally, the way in which $kleisli$ sequences computations is very intuitive. To run $kleisli\ f\ d$, we start by running d. Every time d takes an Eps step, we do too, so if d diverges so does $kleisli\ f\ d$. Should d yield a value d', however, the remaining steps are those of $f\ d'$. We prove that $kleisli\ f$ actually is a continuous function and, amongst other things, satisfies all the equations making $(-_\perp, \eta, kleisli(-))$ a Kleisli triple on $\mathbb{CPO}$. It is also convenient to have 'parameterized' versions of the Kleisli operators $Kleislir\ D\ E\ :\ (D \times E \to_c F_\perp) \to (D \times E_\perp \to_c F_\perp)$ defined by composing $kleisli$ with the evident strength $\tau : D \times E_\perp \to_c (D \times E)_\perp$.

3   A Simply-Typed Functional Language

Our first application of the domain theory formalization is mechanize the denotational semantics of PCF_v, a simply-typed, call-by-value functional language with recursion. Types in PCF_v consist of integer, boolean, functions and products; we represent typing environments by a list of types.

Inductive $Ty := Int\ |\ Bool\ |\ Arrow\ (\tau_1\ \tau_2 : Ty)\ |\ Prod\ (\tau_1\ \tau_2 : Ty)$.
Infix " -> " := $Arrow$. Infix " * " := $Prod$ (at $level$ 55).
Definition $Env := list\ Ty$.

We separate syntactic values v from general expressions e, and restrict the syntax to ANF (administrative normal form), with explicit sequencing of evaluation by LET and inclusion of values into expressions by VAL. As usual, there immediately arises the question of how to represent binders. Our first attempt used de Bruijn indices of type nat in the syntax representation, and a separate type for typing judgments:

Inductive $Value := VAR : nat \to Value\ |\ FIX : Ty \to Ty \to Exp \to Value\ |\ldots$
Inductive $VTy\ (\Gamma : Env)\ (\tau : Ty) : Value \to$ Type :=
| $TVAR : \forall\ m\ ,\ nth\_error\ \Gamma m = Some\ \tau \to VTy\ \Gamma\ (VAR\ m)\ \tau$
| $TFIX : \forall\ e\ \tau_1\ \tau_2, (\tau = \tau_1\ ->\ \tau_2) \to ETy\ (\tau_1 :: \tau_1\ ->\ \tau_2 :: \Gamma)\ e\ \tau_2 \to VTy\ \Gamma\ (FIX\ \tau_1\ \tau_2\ e)\ \tau \ldots$

The major drawback of the above is that typing judgments contain proof objects: simple equalities between types, as in *TFIX*, and the existence of a variable in the environment, as in *TVAR*. It's necessary to prove (at some length) that any two typings of the same term are equal, whilst definitions and theorems are hedged with well-formedness side-conditions.

We recently switched to a strongly-typed term representation in which variable and term types are indexed by *Ty* and *Env*, ensuring that terms are well-typed by construction. Definitions and theorems become more natural and much more concise, and the problems with equality proofs go away.[2] Here is the complete definition of well-typed terms:

Inductive *Var* : *Env* → *Ty* → **Type** :=
| *ZVAR* : ∀ Γ τ, *Var* (τ :: Γ) τ | *SVAR* : ∀ Γ τ τ', *Var* Γ τ → *Var* (τ' :: Γ) τ.

Inductive *Value* : *Env* → *Ty* → **Type** :=
| *TINT* : ∀ Γ, *nat* → *Value* Γ *Int* | *TBOOL* : ∀ Γ, *bool* → *Value* Γ *Bool*
| *TVAR* : ∀ Γ τ, *Var* Γ τ → *Value* Γ τ
| *TFIX* : ∀ Γ τ_1 τ_2, *Exp* (τ_1 :: τ_1 -> τ_2 :: Γ) τ_2 → *Value* Γ (τ_1 -> τ_2)
| *TPAIR* : ∀ Γ τ_1 τ_2, *Value* Γ τ_1 → *Value* Γ τ_2 → *Value* Γ (τ_1 * τ_2)
with *Exp* : *Env* → *Ty* → **Type** :=
| *TFST* : ∀ Γ τ_1 τ_2, *Value* Γ (τ_1 * τ_2) → *Exp* Γ τ_1
| *TSND* : ∀ Γ τ_1 τ_2, *Value* Γ (τ_1 * τ_2) → *Exp* Γ τ_2
| *TOP* : ∀ Γ, (*nat* → *nat* → *nat*) → *Value* Γ *Int* → *Value* Γ *Int* → *Exp* Γ *Int*
| *TGT* : ∀ Γ, *Value* Γ *Int* → *Value* Γ *Int* → *Exp* Γ *Bool*
| *TVAL* : ∀ Γ τ, *Value* Γ τ → *Exp* Γ τ
| *TLET* : ∀ Γ τ_1 τ_2, *Exp* Γ τ_1 → *Exp* (τ_1 :: Γ) τ_2 → *Exp* Γ τ_2
| *TAPP* : ∀ Γ τ_1 τ_2, *Value* Γ (τ_1 -> τ_2) → *Value* Γ τ_1 → *Exp* Γ τ_2
| *TIF* : ∀ Γ τ, *Value* Γ *Bool* → *Exp* Γ τ → *Exp* Γ τ → *Exp* Γ τ.

Definition *CExp* τ := *Exp nil* τ. **Definition** *CValue* τ := *Value nil* τ.

Variables of type *Var* Γ τ are represented by a "typed" de Bruijn index that is in essence a proof that τ lives at that index in Γ. The typing rule associated with each term constructor can be read directly off its definition: for example, *TLET* takes an expression typed as τ_1 under Γ, and another expression typed as τ_2 under Γ extended with a new variable of type τ_1; its whole type is then τ_2 under Γ. The abbreviations *CExp* and *CValue* define closed terms.

Now the operational semantics can be presented very directly:

Inductive *Ev* : ∀ τ, *CExp* τ → *CValue* τ → **Prop** :=
| *e_Val* : ∀ τ (*v* : *CValue* τ), *TVAL v* ⇓ *v*
| *e_Op* : ∀ *op* n_1 n_2, *TOP op* (*TINT* n_1) (*TINT* n_2) ⇓ *TINT* (*op* n_1 n_2)
| *e_Gt* : ∀ n_1 n_2, *TGT* (*TINT* n_1) (*TINT* n_2) ⇓ *TBOOL* (*ble_nat* n_2 n_1)
| *e_Fst* : ∀ τ_1 τ_2 (v_1 : *CValue* τ_1) (v_2 : *CValue* τ_2), *TFST* (*TPAIR* v_1 v_2) ⇓ v_1
| *e_Snd* : ∀ τ_1 τ_2 (v_1 : *CValue* τ_1) (v_2 : *CValue* τ_2), *TSND* (*TPAIR* v_1 v_2) ⇓ v_2
| *e_App* : ∀ τ_1 τ_2 *e* (v_1 : *CValue* τ_1) (v_2 : *CValue* τ_2),
     substExp (*doubleSubst* v_1 (*TFIX e*)) *e* ⇓ v_2 → *TAPP* (*TFIX e*) v_1 ⇓ v_2
| *e_Let* : ∀ τ_1 τ_2 e_1 e_2 (v_1 : *CValue* τ_1) (v_2 : *CValue* τ_2),
     e_1 ⇓ v_1 → *substExp* (*singleSubst* v_1) e_2 ⇓ v_2 → *TLET* e_1 e_2 ⇓ v_2
| *e_IfTrue* : ∀ τ (e_1 e_2 : *CExp* τ) *v*, e_1 ⇓ *v* → *TIF* (*TBOOL true*) e_1 e_2 ⇓ *v*

---

[2] The new **Program** and **dependent destruction** tactics in Coq 8.2 are invaluable for working with this kind of strongly dependent representation.

| $e\_IfFalse : \forall \tau \, (e_1 \, e_2 : CExp \, \tau) \, v, e_2 \Downarrow v \to TIF \, (TBOOL \, false) \, e_1 \, e_2 \Downarrow v$
where $"e \, '\Downarrow' \, v" := (Ev \, e \, v)$.

Substitutions are typed maps from variables to values:

Definition $Subst \, \Gamma \, \Gamma' := \forall \tau, Var \, \Gamma \, \tau \to Value \, \Gamma' \, \tau$.

Definition $hdSubst \, \Gamma \, \Gamma' \, \tau : Subst \, (\tau::\Gamma) \, \Gamma' \to Value \, \Gamma' \, \tau := \ldots$.

Definition $tlSubst \, \Gamma \, \Gamma' \, \tau : Subst \, (\tau::\Gamma) \, \Gamma' \to Subst \, \Gamma \, \Gamma' := \ldots$.

To apply a substitution on de Bruijn terms (functions $substVal$ and $subst$-Exp), one would conventionally define a *shift* operator, but the full dependent type of this operator (namely $Val(\Gamma ++ \Gamma') \, \tau \to Val(\Gamma ++ [\tau'] ++ \Gamma') \, \tau$) is hard to work with. Instead, we first define general *renamings* (maps from variables to variables), and then bootstrap substitution on terms, defining *shift* in terms of renaming [5,15,1]. Definitions and lemmas regarding composition must be similarly bootstrapped: first composition of renamings is defined, then composition of substitution with renaming, and finally composition of substitutions.

3.1 Denotational Semantics

The semantics of types is inductive, using product of cpo's to interpret products and continuous functions into a lifted cpo to represent call-by-value functions.

Fixpoint $SemTy \, \tau :=$ **match** τ **with**
| $Int \Rightarrow Discrete \, nat$ | $Bool \Rightarrow Discrete \, bool$
| $\tau_1 \to \tau_2 \Rightarrow SemTy \, \tau_1 \Rightarrow_c (SemTy \, \tau_2)_\perp$
| $\tau_1 * \tau_2 \Rightarrow SemTy \, \tau_1 \times SemTy \, \tau_2$ **end**.
Fixpoint $SemEnv \, \Gamma :=$ **match** Γ **with** $nil \Rightarrow 1 \,|\, \tau :: \Gamma' \Rightarrow SemEnv \, \Gamma' \times SemTy \, \tau$
end.

We interpret $Value \, \Gamma \, \tau$ in $SemEnv \, \Gamma \to_c SemTy \, \tau$. Expressions are similar, except that the range is a lifted cpo. Note how we have used a 'point-free' style, with no explicit mention of value environments.

Fixpoint $SemVar \, \Gamma \, \tau \, (var : Var \, \Gamma \, \tau) : SemEnv \, \Gamma \to_c SemTy \, \tau :=$
match var **with** $ZVAR \_\_ \Rightarrow \pi_2$ | $SVAR \_\_\_ v \Rightarrow SemVar \, v \circ \pi_1$ **end**.
Fixpoint $SemExp \, \Gamma \, \tau \, (e : Exp \, \Gamma \, \tau) : SemEnv \, \Gamma \to_c (SemTy \, \tau)_\perp :=$
match e **with**
| $TOP \_ op \, v_1 \, v_2 \Rightarrow \eta \circ uncurry \, (SimpleOp2 \, op) \circ \langle SemVal \, v_1 , SemVal \, v_2 \rangle$
| $TGT \_ v_1 \, v_2 \Rightarrow \eta \circ uncurry \, (SimpleOp2 \, ble\_nat) \circ \langle SemVal \, v_2 , SemVal \, v_1 \rangle$
| $TAPP \_\_\_ v_1 \, v_2 \Rightarrow ev \circ \langle SemVal \, v_1 , SemVal \, v_2 \rangle$
| $TVAL \_\_ v \Rightarrow \eta \circ SemVal \, v$
| $TLET \_\_\_ e_1 \, e_2 \Rightarrow Kleislir(SemExp \, e_2) \circ \langle ID, SemExp \, e_1 \rangle$
| $TIF \_\_ v \, e_1 \, e_2 \Rightarrow (choose \_ @3\_ (SemExp \, e_1)) \, (SemExp \, e_2) \, (SemVal \, v)$
| $TFST \_\_\_ v \Rightarrow \eta \circ \pi_1 \circ SemVal \, v$
| $TSND \_\_\_ v \Rightarrow \eta \circ \pi_2 \circ SemVal \, v$
end with $SemVal \, \Gamma \, \tau \, (v : Value \, \Gamma \, \tau) : SemEnv \, \Gamma \to_c SemTy \, \tau :=$
match v **with**
| $TINT \_ n \Rightarrow K \_ (n : Discrete \, nat)$
| $TBOOL \_ b \Rightarrow K \_ (b : Discrete \, bool)$
| $TVAR \_\_ i \Rightarrow SemVar \, i$
| $TFIX \_\_\_ e \Rightarrow FIXP \circ curry \, (curry \, (SemExp \, e))$
| $TPAIR \_\_\_ v_1 \, v_2 \Rightarrow \langle SemVal \, v_1 , SemVal \, v_2 \rangle$ **end**.

In the above *SimpleOp2* lifts binary Coq functions to continuous maps on discrete cpos, *K* is the usual combinator and *choose* is a continuous conditional.

3.2 Soundness and Adequacy

We first prove *soundness*, showing that if an expression e evaluates to a value v, then the denotation of e is indeed the denotation of v. This requires that substitution commutes with the semantic meaning function. We define the 'semantics' of a substitution $s : Subst\ \Gamma'\ \Gamma$ to be a map in $SemEnv\ \Gamma \to_c SemEnv\ \Gamma'$.

Fixpoint *SemSubst* $\Gamma\ \Gamma'$: *Subst* $\Gamma'\ \Gamma \to SemEnv\ \Gamma \to_c SemEnv\ \Gamma'$:=
  match Γ' **with**
  | *nil* $\Rightarrow$ **fun** $s \Rightarrow K\ \_\ (tt : \mathbf{1})$
  | $\_ :: \_ \Rightarrow$ **fun** $s \Rightarrow \langle\ SemSubst\ (tlSubst\ s)\ ,\ SemVal\ (hdSubst\ s)\ \rangle$ **end**.

This is then used to prove the substitution lemma, which in turn is used in the *e_App* and *e_Let* cases of the soundness proof.

Lemma *SemCommutesWithSubst*:
  $(\forall\ \Gamma\ \tau\ (v : Value\ \Gamma\ \tau)\ \Gamma'\ (s : Subst\ \Gamma\ \Gamma'),$
  $SemVal\ v \circ SemSubst\ s == SemVal\ (substVal\ s\ v))$
$\wedge\ (\forall\ \Gamma\ \tau\ (e : Exp\ \Gamma\ \tau)\ \Gamma'\ (s : Subst\ \Gamma\ \Gamma'),$
  $SemExp\ e \circ SemSubst\ s == SemExp\ (substExp\ s\ e)).$

Theorem *Soundness*: $\forall\ \tau\ (e : CExp\ \tau)\ v,\ e \Downarrow v \to SemExp\ e == \eta \circ SemVal\ v.$

We now prove *adequacy*, showing that if the denotation of a closed expression e is some (lifted) element, then e converges to a value. The proof uses a logical relation between syntax and semantics. We start by defining a *liftRel* operation that takes a relation between a cpo and values and lifts it to a relation between a lifted cpo and expressions, then use this to define *relExp* in terms of *relVal*.

Definition *liftRel* τ $(R : SemTy\ \tau \to CValue\ \tau \to \mathbf{Prop})$:=
  fun $d\ e \Rightarrow \forall\ d',\ d == Val\ d' \to \exists\ v,\ e \Downarrow v \wedge R\ d'\ v.$

Fixpoint *relVal* τ : $SemTy\ \tau \to CValue\ \tau \to \mathbf{Prop}$:= **match** τ **with**
| *Int* $\Rightarrow$ **fun** $d\ v \Rightarrow v = TINT\ d$
| *Bool* $\Rightarrow$ **fun** $d\ v \Rightarrow v = TBOOL\ d$
| $\tau_1 \to \tau_2 \Rightarrow$ **fun** $d\ v \Rightarrow \exists\ e,\ v = TFIX\ e \wedge \forall\ d_1\ v_1,\ relVal\ \tau_1\ d_1\ v_1 \to liftRel\ (relVal$
$\tau_2)\ (d\ d_1)\ (substExp\ (doubleSubst\ v_1\ v\ )\ e)$
| $\tau_1 * \tau_2 \Rightarrow$ **fun** $d\ v \Rightarrow \exists\ v_1,\ \exists\ v_2,\ v = TPAIR\ v_1\ v_2 \wedge relVal\ \tau_1\ (\pi_1 d)\ v_1 \wedge relVal\ \tau_2$
$(\pi_2 d)\ v_2$ **end**.

Fixpoint *relEnv* Γ : $SemEnv\ \Gamma \to Subst\ \Gamma\ nil \to \mathbf{Prop}$:= **match** Γ **with**
  | *nil* $\Rightarrow$ **fun** $\_\ \_ \Rightarrow True$
  | $\tau :: \Gamma \Rightarrow$ **fun** $d\ s \Rightarrow relVal\ \tau\ (\pi_2 d)\ (hdSubst\ s) \wedge relEnv\ \Gamma\ (\pi_1 d)\ (tlSubst\ s)$ **end**.

Definition *relExp* τ := *liftRel* $(relVal\ \tau)$.

The logical relation reflects == and is admissible:

Lemma *relVal_lower*: $\forall\ \tau\ d\ d'\ v,\ d \sqsubseteq d' \to relVal\ \tau\ d'\ v \to relVal\ \tau\ d\ v.$
Lemma *relVal_admissible*: $\forall\ \tau\ v,$ admissible $(\mathbf{fun}\ d \Rightarrow relVal\ \tau\ d\ v).$

These lemmas are then used in the proof of the Fundamental Theorem for the logical relation, which is proved by induction on the structure of terms.

Theorem FT: $(\forall\ \Gamma\ \tau\ v\ \rho\ s,\ relEnv\ \Gamma\ \rho\ s \rightarrow relVal\ \tau\ (SemVal\ v\ \rho)\ (substVal\ s\ v))$
$\qquad\qquad \wedge\ (\forall\ \Gamma\ \tau\ e\ \rho\ s,\ relEnv\ \Gamma\ \rho\ s \rightarrow relExp\ \tau\ (SemExp\ e\ \rho)\ (substExp\ s\ e)).$

Now we instantiate the fundamental theorem with closed expressions to obtain

Corollary $Adequacy$: $\forall\ \tau\ (e : CExp\ \tau)\ d,\ SemExp\ e\ tt == Val\ d \rightarrow \exists\ v,\ e \Downarrow v.$

4   Recursive Domain Equations

We now outline our formalization of the solution of mixed-variance recursive domain equations, such as arise in modelling untyped higher-order languages, languages with higher-typed store or languages with general recursive types.

The basic technology for solving domain equations is Scott's inverse limit construction, our formalization of which follows an approach due to Freyd [11,12] and Pitts [23]. A key idea is to separate the positive and negative occurences, specifying recursive domains as fixed points of locally continuous bi-functors $F : \mathbb{CPO}^{op} \times \mathbb{CPO} \rightarrow \mathbb{CPO}$, i.e. domains D such that such that $F(D, D) \simeq D$.

The type of mixed variance locally-continuous bifunctors on $\mathbb{CPO}$ is defined as the type of records comprising an action on pairs of objects (ob), a continuous action on pairs of morphisms (mor), contravariant in the first argument and covariant in the second, together with proofs that mor respects both composition ($morph\_comp$) and identities ($morph\_id$): **Record** $BiFunctor := mk\_functor$
$\quad \{\ ob : cpo \rightarrow cpo \rightarrow cpo;$
$\qquad mor: \forall\ (A\ B\ C\ D : cpo),\ (B \Rightarrow_c A) \times (C \Rightarrow_c D) \Rightarrow_c (ob\ A\ C \Rightarrow_c ob\ B\ D)\ ;$
$\qquad morph\_comp: \forall\ A\ B\ C\ D\ E\ F\ f\ g\ h\ k,$
$\qquad\qquad\qquad morB\ E\ D\ F\ (f, g) \circ morA\ B\ C\ D\ (h, k)$
$\qquad\qquad\qquad == mor\_\ \_\ \_\ \_\ (h \circ f, g \circ k)\ ;$
$\qquad morph\_id : \forall\ A\ B,\ mor\_\ \_\ \_\ \_\ (IDA\ ,\ IDB) == ID\_$

We single out the *strict* bifunctors, taking pointed cpos to pointed cpos:
Definition $FStrict : BiFunctor \rightarrow$ **Type** :=
$\quad$ **fun** $BF \Rightarrow \forall\ D\ E,\ PointedD \rightarrow PointedE \rightarrow (Pointed(ob\ BF\ D\ E)).$

We build interesting bifunctors from a few primitive ones in a combinatory style. If $D : cpo$ then $BiConst\ D : BiFunctor$ on objects is constantly D and on morphisms is constantly $ID\ D$. This is strict if D is pointed. $BiArrow : BiFunctor$ on objects takes (D, E) to $D \Rightarrow_c E$ with the action on morphisms given by conjugation. This is strict. If $F : BiFunctor$ then $BiLift\ F : BiFunctor$ on objects takes (D, E) to $(ob\ F\ (D, E))_\perp$ and on morphisms composes $mor\ F$ with the morphism part of the lift functor. This is always strict. If $F_1\ F_2 : BiFunctor$ then $BiPair\ F_1\ F_2 : BiFunctor$ on objects takes (D, E) to $(ob\ F_1\ (D, E)) \times (ob\ F_2\ (D, E))$ with the evident action on morphisms. This is strict if both F_1 and F_2 are. The definition of $BiSum\ F_1\ F_2 : BiFunctor$ is similar, though this is not generally strict as our coproduct is a separated sum.

A pair $(f : D \rightarrow_c E, g : E \rightarrow_c D)$ is an embedding-projection (e-p) pair if $g \circ f == id_D$ and $f \circ g \sqsubseteq id_E$. If F is a bifunctor and (f, g) an e-p pair, then $(mor\ F\ (g, f),\ mor\ F\ (f, g))$ is an e-p pair.

Now let $F : BiFunctor$ and $FS : FStrict\ F$. We define $\langle D_i \rangle$ to be the sequence of cpos defined by $D_0 = \mathbf{1}$ and $D_{n+1} = ob\ F\ (D_n, D_n)$. We then define a sequence of e-p pairs:

$$e_0 = \bot : D_0 \to_c D_1 \qquad\qquad p_0 = \bot : D_1 \to_c D_0$$
$$e_{n+1} = mor\,F\,(p_n, e_n) : D_{n+1} \to_c D_{n+2} \quad p_{n+1} = mor\,F\,(e_n, p_n) : D_{n+2} \to_c D_{n+1}.$$

Let $\pi_i : \Pi_j D_j \to_c D_i$ be the projections from the product of all the D_js. The predicate $P : \Pi_j D_j \to Prop$ defined by $P\,d := \forall i,\ \pi_i\ d == p_n(\pi_{i+1}\ d)$ is admissible, so we can define the sub-cpo D_∞ to be $\{d \mid P\,d\}$ with order and lubs inherited from the indexed product. D_∞ will be the *cpo* we seek, so we now need to construct the required isomorphism.

Define $t_n : D_n \to D_\infty$ to be the map that for $i < n$ projects D_n to D_i via $p_i \circ \cdots \circ p_{n-1}$ and for $i > n$ embeds D_n in D_i via $e_n \circ \cdots \circ e_{i-1}$. Then $mor\,F\,(t_i, \pi_i) : ob\,F\,(D_\infty, D_\infty) \to_c ob\,F\,(D_i, D_i) = D_{i+1}$, so $t_{i+1} \circ mor\,F\,(t_i, \pi_i) : ob\,F\,(D_\infty, D_\infty) \to_c D_\infty$, and $mor\,F\,(\pi_i, t_i) \circ \pi_{1+1} : D_\infty \to_c ob\,F\,(D_\infty, D_\infty)$. We then define

$$UNROLL := \bigsqcup_i (mor\,F\,(\pi_i, t_i) \circ \pi_{i+1}) : D_\infty \to_c ob\,F\,(D_\infty, D_\infty)$$

$$ROLL := \bigsqcup_i (t_{i+1} \circ mor\,F\,(t_i, \pi_i)) : ob\,F\,(D_\infty, D_\infty) \to_c D_\infty$$

Some calculation shows that $ROLL \circ UNROLL == ID\,D_\infty$ and $UNROLL \circ ROLL == ID(ob\,F\,(D_\infty, D_\infty))$, so we have constructed the desired isomorphism.

In order to do anything useful with recursively defined domains, we really need some general reasoning principles that allow us to avoid unpicking all the complex details of the construction above every time we want to prove something. One 'partially' abstract interface to the construction reveals that D_∞ comes equipped with a chain of retractions $\rho_i : D_\infty \to_c D_\infty$ such that $\bigsqcup_i \rho_i == IDD_\infty$; concretely, ρ_i can be taken to be $t_i \circ \pi_i$. A more abstract and useful principle is given by Pitts's [23] characterization of the solution as a *minimal invariant*, which is how we will establish the existence of a recursively defined logical relation in Section 5.1. Let $\delta : (D_\infty \Rightarrow_c D_\infty) \to_c (D_\infty \Rightarrow_c D_\infty)$ be given by

$$\delta\,e\ =\ ROLL \circ mor\,F\,(e, e) \circ UNROLL$$

The minimal invariance property is then the assertion that IDD_∞ is equal to *fix*(δ), which we prove via a pointwise comparison of the chain of retractions whose lub we know to be the identity function with the chain whose lub gives the least fixed point of δ.

5   A Uni-typed Lambda Calculus

We now apply the technology of the previous section to formalize the denotational semantics of an uni-typed (untyped) CBV lambda calculus with constants. This time the values are variables, numeric constants, and λ abstractions; expressions are again in ANF with *LET* and *VAL* constructs, together with function application, numeric operations, and a zero-test conditional. For binding, we use de Bruijn indices and separate well-formedness judgments *ETyping* and *VTyping*. The evaluation relation is as follows:

Inductive *Evaluation* : *Exp* → *Value* → **Type** :=
| *e_Value* : ∀ *v*, *VAL v* ⇓ *v*
| *e_App* : ∀ *e* v_2 *v*, *substExp* [v_2] *e* ⇓ *v* → (*APP* (*LAMBDA e*) v_2) ⇓ *v*
| *e_Let* : ∀ e_1 v_1 e_2 v_2, e_1 ⇓ v_1 → *substExp* [v_1] e_2 ⇓ v_2 → (*LET* e_1 e_2) ⇓ v_2
| *e_Ifz1* : ∀ e_1 e_2 v_1, e_1 ⇓ v_1 → *IFZ* (*NUM O*) e_1 e_2 ⇓ v_1
| *e_Ifz2* : ∀ e_1 e_2 v_2 *n*, e_2 ⇓ v_2 → *IFZ* (*NUM* (*S n*)) e_1 e_2 ⇓ v_2
| *e_Op* : ∀ *op* v_1 *n1* v_2 *n2*, *OP op* (*NUM n1*) (*NUM n2*) ⇓ *NUM* (*op n1 n2*)
where "*e* '⇓' *v*" := (*Evaluation e v*).
Inductive *Converges e* : **Prop** := *e_Conv* : ∀ *v* (_ : *Evaluation e v*), *Converges e*.
Notation "*e* '⇓'" := (*Converges e*).

Note that *Evaluation* is here in **Type** rather than **Prop**, which is a knock-on effect of separating the definition of terms from that of well-formedness.[3] We plan instead to make untyped terms well-scoped by construction, using the same techniques as we did for the typed language in Section 3.

5.1   Semantic Model

We interpret the unityped language in a solution for the recursive domain equation $D \simeq (nat + (D \rightarrow_c D))_\perp$, following the intuition that a computation either diverges or produces a value which is a number or a function. This is not the 'tightest' domain equation one could use for CBV: one could make function space strict, or equivalently make the argument of the function space be a domain of values rather than computations. But this equation still gives an adequate model. The construction in Coq is an instantiation of results from the previous section. First we build the strict bifunctor $F(D, E) = (nat + (D \rightarrow_c E))_\perp$:

Definition *FS* := *BiLift_strict* (*BiSum* (*BiConst* (*Discrete nat*)) *BiArrow*).

And then we construct the solution, defining domains D_∞ for computations and V_∞ for values:

Definition D_∞ := D_∞ *FS*.
Definition V_∞ := *Dsum* (*Discrete nat*) ($D_\infty \rightarrow_c D_\infty$).
Definition *Roll* : (V_∞)$_\perp \rightarrow_c D_\infty$:= *ROLL FS*.
Definition *Unroll* : $D_\infty \rightarrow_c$ (V_∞)$_\perp$:= *UNROLL FS*.
Definition *UR_iso* : *Unroll* ∘ *Roll* == *ID* _ := *DIso_ur FS*.
Definition *RU_iso* : *Roll* ∘ *Unroll* == *ID* _ := *DIso_ru FS*.

For environments we define the *n*-ary product of V_∞ and projection function.

Fixpoint *SemEnv n* : *cpo* := **match** *n* **with** *O* ⇒ **1**| *S n* ⇒ *SemEnv n* × V_∞ **end**.
Fixpoint *projenv* (*m n* : *nat*) : (*m* < *n*) → *SemEnv n* $\rightarrow_c$ V_∞ :=
match *m, n* **with**
| *m, O* ⇒ **fun** *inconsistent* ⇒ **match** (*lt_n_O m inconsistent*) **with end**
| *O, S n* ⇒ **fun** _ ⇒ π_2
| *S m, S n* ⇒ **fun** *h* ⇒ *projenv* (*lt_S_n* _ _ *h*) ∘ π_1 **end**.

---

[3] We induce over evaluations to construct well-formedness derivations when showing well-formedness preservation, and well-formedness derivations are themselves in **Type** so that we can use them to inductively define the denotational semantics.

We define a lifting operator $Dlift : (V_\infty \to_c D_\infty) \to_c D_\infty \to_c D_\infty$ and an operator $Dapp : V_\infty \times V_\infty \to_c (V_\infty)_\perp$ that applies the first component of a pair to the second, returning $\perp$ in the when the first component is not a function. (Coproducts are introduced with INL and INR, and eliminated with $[\cdot, \cdot]$.)

Definition $Dlift : (V_\infty \to_c D_\infty) \to_c D_\infty \to_c D_\infty :=$
$\quad curry \ (Roll \circ ev \circ \langle kleisli \circ (Unroll \circ -) \circ \pi_1, Unroll \circ \pi_2 \rangle).$
Definition $Dapp : V_\infty \times V_\infty \to_c (V_\infty)_\perp :=$
$\quad ev \circ \langle [\perp : Discrete \ nat \to_c D_\infty \to_c (V_\infty)_\perp, (Unroll \circ -)] \circ \pi_1, Roll \circ \eta \circ \pi_2 \rangle.$

We can then define the semantics of the unityped language:

Fixpoint $SemVal \ v \ n \ (vt : VTyping \ n \ v) : SemEnv \ n \to_c V_\infty :=$
match vt **with**
$| \ TNUM \ n \Rightarrow INL \ \_ \ \_ \circ (@K \ \_ \ (Discrete \ nat) \ n)$
$| \ TVAR \ m \ nthm \Rightarrow projenv \ nthm$
$| \ TLAMBDA \ t \ b \Rightarrow INR \ \_ \ \_ \circ Dlift \circ curry \ (Roll \circ SemExp \ b)$
end with $SemExp \ e \ n \ (et : ETyping \ n \ e) : SemEnv \ n \to_c (V_\infty)_\perp :=$
match et **with**
$| \ TAPP \ \_ \ \_ \ v_1 \ v_2 \Rightarrow Dapp \circ \langle SemVal \ v_1, SemVal \ v_2 \rangle$
$| \ TVAL \ \_ \ v \Rightarrow \eta \circ SemVal \ v$
$| \ TLET \ \_ \ \_ \ e_1 \ e_2 \Rightarrow ev \circ \langle curry(Kleislir(SemExp \ e_2)), SemExp \ e_1 \rangle$
$| \ TOP \ op \ \_ \ \_ \ v_1 \ v_2 \Rightarrow$
$\qquad uncurry(Operator2 \ (\eta \circ INL \ \_ \ \_ \circ uncurry(SimpleOp2 \ op))) \circ$
$\qquad \langle [\eta, \perp : (D_\infty \to_c D_\infty) \to_c (Discrete \ nat)_\perp] \circ SemVal \ v_1,$
$\qquad \ [\eta, \perp : (D_\infty \to_c D_\infty) \to_c (Discrete \ nat)_\perp] \circ SemVal \ v_2 \rangle$
$| \ TIFZ \ \_ \ \_ \ \_ \ v \ e_1 \ e_2 \Rightarrow ev \circ$
$\qquad \langle [[K \ \_ \ (SemExp \ e_1), K \ \_ \ (SemExp \ e_2)] \circ zeroCase,$
$\qquad \ \perp : (D_\infty \to_c D_\infty) \to_c SemEnv \ n \to_c (V_\infty)_\perp] \circ (SemVal \ v), ID \ \_ \rangle$ **end.**

5.2   Soundness and Adequacy

As with the typed language, the proof of soundness makes use of a substitution lemma, and in addition uses the isomorphism of the domain D_∞ in the case for *APP*. The proof then proceeds by induction, using equational reasoning to show that evaluation preserves semantics:

Lemma *Soundness*: $\forall \ e \ v \ (et : ETyping \ 0 \ e) \ (vt : VTyping \ 0 \ v),$
$\qquad e \Downarrow v \to SemExp \ et == \eta \circ SemVal \ vt.$

We again use a logical relation between syntax and semantics to prove adequacy, but now cannot define the relation by induction on types. Instead we have a recursive specification of a logical relation over our recursively defined domain, but it is not at all clear that such a relation exists: because of the mixed variance of the function space, the operator on relations whose fixpoint we seek is not monotone. Following Pitts [23], however, we again use the technique of separating positive and negative occurrences, defining a monotone operator in the complete lattice of *pairs* of relations, with the superset order in the first component and the subset order in the second. A fixed point of that operator is then constructed by Knaster-Tarski.

We first define a notion of *admissibility* on ==-respecting relations between elements of our domain of values V_∞ and closed syntactic values in *Value* and show that this is closed under intersection, so admissible relations form a complete lattice.

We then define a relational action corresponding to the bifunctor used in defining our recursive domain. This action, $RelV$, maps a pair of relations R, S on $(V_\infty)_\perp \times Value$ to a new relation that relates $(inl\ m)$ to $(NUM\ m)$ for all $m : nat$, and relates $(inr\ f)$ to $(LAMBDA\ e)$ just when $f : D_\infty \to_c D_\infty$ is strict and satisfies the 'logical' property

$$\forall(d, v) \in R, \forall d', Unroll(f(Roll(Val\ d))) == Val\ d'$$
$$\to \quad \exists v', substExp\ v\ e \Downarrow v' \wedge (d', v') \in S$$

We then show that $RelV$ maps admissible relations to admissible relations and is contravariant in its first argument and covariant in its second. Hence the function $\lambda R : RelAdm^{op}. \lambda S : RelAdm. (RelV\ S\ R, RelV\ R\ S)$ is monotone on the complete lattice $RelAdm^{op} \times RelAdm$. Thus it has a least fixed point (Δ^-, Δ^+). By applying the minimal invariant property from the previous section, we prove that in fact $\Delta^- == \Delta^+$, so we have found a fixed point, Δ of $RelV$, which is the logical relation we need to prove adequacy.

We extend Δ to Δ_e, a relation on $(V_\infty)_\perp \times Exp$, by $(d, e) \in \Delta_e$ if and only if for all d' if $d == Val\ d'$ then there exists a value v and a derivation $e \Downarrow v$ such that $(d', v) \in \Delta$.

The fundamental theorem for this relation is that for any environment env, derivations of $VTyping\ n\ v$ and $ETyping\ n\ e$, and any list vs of n closed values such that $nth\_error\ vs\ i = Some\ v$ implies $(projenv\ i\ env, v) \in \Delta$ for all i and v, $(SemVal\ v\ env, substVal\ vs\ v) \in \Delta$ and $(SemExp\ e\ env, substExp\ vs\ e) \in \Delta_e$.

Adequacy is then a corollary of the fundamental theorem:

Theorem *Adequacy*: $\forall e\ (te : ETyping\ 0\ e)\ d, SemExp\ te\ tt == Val\ d \to e \Downarrow$.

6   Discussion

As we noted in the introduction, there have been many mechanized treatments of different aspects of domain theory and denotational semantics. One rough division of this previous work is between axiomatic approaches and those in which definitions and proofs of basic results about cpos, continuous functions and so on are made explicitly with the prover's logic. LCF falls into the first category, as does Reus's work on synthetic domain theory in LEGO [25]. In the second category, HOLCF, originally due to Regensburger [24] and later reworked by Müller et al [18], uses Isabelle's axiomatic type class mechanism to define and prove properties of domain-theoretic structures within higher order logic. HOLCPO [2,4] was an extension of HOL with similar goals, and basic definitions have also been formalized in PVS [7]. Coq's library includes a formalization by Kahn of some general theory of dcpos [14].

HOLCF is probably the most developed of these systems, and has been used to prove interesting results [19,26], but working in a richer dependent type theory gives us some advantages. We can express the semantics of a typed language as a dependently typed map from syntax to semantics, rather than only being able to do shallow embeddings – this is clearly necessary if one wishes to prove theorems like adequacy or do compiler correctness. Secondly, it seems one really needs dependent types to work conveniently with monads and logical relations, or to formalize the inverse limit construction.[4]

The constructive nature of our formalization and the coinductive treatment of lifting has both benefits and drawbacks. On the minus side, some of the proofs and constructions are much more complex than they would be classically and one does sometimes have to pay attention to which of two classically-equivalent forms of definition one works with. Worse, some constructions do not seem to be possible, such as the smash product of pointed domains; not being able to define $\otimes$ was one motivation for moving from Paulin-Mohring's pointed cpos to our unpointed ones. One benefit that we have not yet seriously investigated, however, is that it is possible to extract actual executable code from the denotational semantics. Indeed, the lift monad can be seen as a kind of syntax-free operational semantics, not entirely unlike game semantics; this perspective, and possible connections with step-indexing, seem to merit further study.

The Coq development is of a reasonable size. The domain theory library, including the theory of recursive domain equations, is around 7000 lines. The formalization of the typed language and its soundness and adequacy proofs are around 1700 lines and the untyped language takes around 2500. Although all the theorems go through (with no axioms), we have to admit that the development is currently rather 'rough'. Nevertheless, we have already used it as the basis of a non-trivial formalization of some new research [8] and our intention is to develop the formalization into something that is more widely useful. Apart from general polishing, we plan to abstract some of the structure of our category of domains to make it convenient to work simultaneously with different categories, including categories of algebras. We would also like to provide better support for 'diagrammatic' rewriting in monoidal (multi)categories. It is convenient to use Setoid rewriting for pointfree equational reasoning, direct translating the normal categorical commuting diagrams. But dealing with all the structural morphisms is still awkward, and it should be possible to support something more like the diagrammatic proofs one can do with 'string diagrams' [13].

References

1. Adams, R.: Formalized metatheory with terms represented by an indexed family of types. In: Filliâtre, J.-C., Paulin-Mohring, C., Werner, B. (eds.) TYPES 2004. LNCS, vol. 3839, pp. 1–16. Springer, Heidelberg (2006)
2. Agerholm, S.: Domain theory in HOL. In: Joyce, J.J., Seger, C.-J.H. (eds.) HUG 1993. LNCS, vol. 780. Springer, Heidelberg (1994)

---

[4] Agerholm [3] formalized the construction of a model of the untyped lambda calculus using HOL-ST, a version of HOL that supports ZF-like set theory; this is elegant but HOL-ST is not widely used and no denotational semantics seems to have been done with the model. Petersen [21] formalized a reflexive cpo based on $P\omega$ in HOL, though this also appears not to have been developed far enough to be useful.

3. Agerholm, S.: Formalizing a model of the lambda calculus in HOL-ST. Technical Report 354, University of Cambridge Computer Laboratory (1994)
4. Agerholm, S.: LCF examples in HOL. The Computer Journal 38(2) (1995)
5. Altenkirch, T., Reus, B.: Monadic presentations of lambda terms using generalized inductive types. In: Flum, J., Rodríguez-Artalejo, M. (eds.) CSL 1999. LNCS, vol. 1683, pp. 453–468. Springer, Heidelberg (1999)
6. Audebaud, P., Paulin-Mohring, C.: Proofs of randomized algorithms in Coq. In: Uustalu, T. (ed.) MPC 2006. LNCS, vol. 4014, pp. 49–68. Springer, Heidelberg (2006)
7. Bartels, F., Dold, A., Pfeifer, H., Von Henke, F.W., Rueß, H.: Formalizing fixed-point theory in PVS. Technical report, Universität Ulm (1996)
8. Benton, N., Hur, C.-K.: Biorthogonality, step-indexing and compiler correctness. In: ACM International Conference on Functional Programming (2009)
9. Capretta, V.: General recursion via coinductive types. Logical Methods in Computer Science 1 (2005)
10. Coquand, T.: Infinite objects in type theory. In: Barendregt, H., Nipkow, T. (eds.) TYPES 1993, vol. 806. Springer, Heidelberg (1994)
11. Freyd, P.: Recursive types reduced to inductive types. In: IEEE Symposium on Logic in Computer Science (1990)
12. Freyd, P.: Remarks on algebraically compact categories. In: Applications of Categories in Computer Science. LMS Lecture Notes, vol. 177 (1992)
13. Joyal, A., Street, R.: The geometry of tensor calculus. Adv. in Math. 88 (1991)
14. Kahn, G.: Elements of domain theory. In: The Coq users' contributions library (1993)
15. McBride, C.: Type-preserving renaming and substitution (unpublished draft)
16. Milner, R.: Logic for computable functions: Description of a machine implementation. Technical Report STAN-CS-72-288, Stanford University (1972)
17. Moggi, E.: Notions of computation and monads. Inf. Comput. 93(1), 55–92 (1991)
18. Müller, O., Nipkow, T., von Oheimb, D., Slotosch, O.: HOLCF = HOL + LCF. J. Functional Programming 9 (1999)
19. Nipkow, T.: Winskel is (almost) right: Towards a mechanized semantics textbook. Formal Aspects of Computing 10 (1998)
20. Paulin-Mohring, C.: A constructive denotational semantics for Kahn networks in Coq. In: From Semantics to Computer Science. Essays in Honour of G Kahn (2009)
21. Petersen, K.D.: Graph model of LAMBDA in higher order logic. In: Joyce, J.J., Seger, C.-J.H. (eds.) HUG 1993. LNCS, vol. 780. Springer, Heidelberg (1994)
22. Pitts, A.M.: Computational adequacy via 'mixed' inductive definitions. In: Main, M.G., Melton, A.C., Mislove, M.W., Schmidt, D., Brookes, S.D. (eds.) MFPS 1993. LNCS, vol. 802. Springer, Heidelberg (1994)
23. Pitts, A.M.: Relational properties of domains. Inf. Comput. 127 (1996)
24. Regensburger, F.: HOLCF: Higher order logic of computable functions. In: Schubert, E.T., Alves-Foss, J., Windley, P. (eds.) HUG 1995. LNCS, vol. 971. Springer, Heidelberg (1995)
25. Reus, B.: Formalizing a variant of synthetic domain theory. J. Automated Reasoning 23 (1999)
26. Varming, C., Birkedal, L.: Higher-order separation logic in Isabelle/HOLCF. In: Mathematical Foundations of Programming Semantics (2008)

Turning Inductive into Equational Specifications

Stefan Berghofer*, Lukas Bulwahn, and Florian Haftmann**

Technische Universität München
Institut für Informatik, Boltzmannstraße 3, 85748 Garching, Germany
http://www.in.tum.de/~berghofe/
http://www.in.tum.de/~bulwahn/
http://www.in.tum.de/~haftmann/

Abstract. Inductively defined predicates are frequently used in formal specifications. Using the theorem prover *Isabelle*, we describe an approach to turn a class of systems of inductively defined predicates into a system of equations using data flow analysis; the translation is carried out *inside* the logic and resulting equations can be turned into functional program code in *SML*, *OCaml* or *Haskell* using the existing code generator of *Isabelle*. Thus we extend the scope of code generation in *Isabelle* from functional to functional-logic programs while leaving the trusted foundations of code generation itself intact.

1 Introduction

Inductively defined predicates (for short, *(inductive) predicates*) are a popular specification device in the theorem proving community. Major theory developments in the proof assistant *Isabelle/HOL* [8] make pervasive use of them, e.g. formal semantics of realistic programming language fragments [11]. From such large applications naturally the desire arises to generate *executable* prototypes from the abstract specifications. It is well-known how systems of predicates can be transformed to functional programs using *mode analysis*. The approach described in [1] for *Isabelle/HOL* works but has turned out unsatisfactorily:

- The applied transformations are not trivial but are carried out outside the *LCF* inference kernel, thus relying on a large code base to be trusted.
- Recently a lot of code generation facilities in *Isabelle/HOL* have been generalized to cover type classes and more languages than *ML*, but this has not yet been undertaken for predicates.

In our view it is high time to tackle execution of predicates again; we present a transformation from predicates to function-like equations that is not a mere re-implementation, but brings substantial improvements:

- The transformation is carried out *inside* the logic; thus the transformation is guarded by *LCF* inferences and does not increase the trusted code base.

* Supported by BMBF in the VerisoftXT project under grant 01 IS 07008 F.
** Supported by DFG project NI 491/10-1.

- The code generator itself can be fed with the function-like equations and does not need to be extended; also other tools involving equational reasoning could benefit from the transformation.
- Proposed extensions can also work *inside* the logic and do not endanger trustability.

The role of our transformation in this scenario is shown in the following picture:

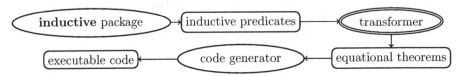

The remainder of this paper is structured as follows: we briefly review existing work in §2 and explain the preliminaries in §3. The main section (§4) explains how the translation works, followed by a discussion of further extensions (§5). Our conclusion (§6) will deal with future work.

In our presentation we use fairly standard notation, plus little *Isabelle/HOL*-specific concrete syntax.

2   Related Work

From the technical point of view, the execution of predicates has been extensively studied in the context of the programming languages *Curry* [4] and *Mercury* [10]. The central concept for executing predicates are *modes*, which describe dataflow by partitioning arguments into *input* and *output*.

We already mentioned the state-of-the-art implementation of code generation for predicates in *Isabelle/HOL* [1] which turns inductive predicates into *ML* programs extralogically using mode analysis.

Delahaye et al. provide a similar direct extraction for the *Coq* proof assistant [2]; however at most one solution is computed, multiple solutions are not enumerated.

For each of these approaches, correctness is ensured by pen-and-paper proofs. Our approach instead *animates* the correctness proof by applying it to each single predicate using the proof assistant itself; thus correctness is guaranteed by construction.

3   Preliminaries

3.1   Inductive Predicates

An inductive predicate is characterized by a collection of *introduction rules* (or *clauses*), each of which has a *conclusion* and an arbitrary number of *premises*. It corresponds to the *smallest* set closed under these clauses. As an example, consider the following predicate describing the concatenation of two lists, which can be defined in in *Isabelle/HOL* using the **inductive** command:

inductive *append* :: α *list* $\Rightarrow$ α *list* $\Rightarrow$ α *list* $\Rightarrow$ *bool* **where**
 append [] *ys ys*
 | *append xs ys zs* $\Longrightarrow$ *append* $(x \cdot xs)$ *ys* $(x \cdot zs)$

For each predicate, an *elimination* (or *case analysis*) rule is provided, which for *append* has the form

 append Xs Ys Zs $\Longrightarrow$
 $(\bigwedge ys. \; Xs = [] \Longrightarrow Ys = ys \Longrightarrow Zs = ys \Longrightarrow P) \Longrightarrow$
 $(\bigwedge xs \; ys \; zs \; x.$
  $Xs = x \cdot xs \Longrightarrow$
  $Ys = ys \Longrightarrow Zs = x \cdot zs \Longrightarrow append \; xs \; ys \; zs \Longrightarrow P) \Longrightarrow$
 P

There is also an *induction* rule, which however is not relevant in our scenario. In introduction rules, we distinguish between premises of the form $Q \; u_1 \ldots u_k$, where Q is an inductive predicate, and premises of other shapes, which we call *side conditions*. Without loss of generality, we only consider clauses without side conditions in most parts of our presentation. The general form of a clause is

$$C_i : \; Q_{i,1} \; \overline{u}_{i,1} \Longrightarrow \cdots \Longrightarrow Q_{i,n_i} \; \overline{u}_{i,n_i} \Longrightarrow P \; \overline{t}_i$$

We use $k_{i,j}$ and l to denote the *arities* of the predicates $Q_{i,j}$ and P, i.e. the length of the argument lists $\overline{u}_{i,j}$ and $\overline{t}_i$, respectively.

3.2 Code Generation

The *Isabelle* code generator views generated programs as an implementation of an equational rewrite system, e.g. the following program normalizes a list of natural numbers to its sum by equational rewriting:

sum [Suc Zero_nat, Suc Zero_nat]

```
datatype nat = Zero_nat | Suc of nat;

fun plus_nat (Suc m) n = plus_nat m (Suc n)
  | plus_nat Zero_nat n = n;

fun sum [] = Zero_nat
  | sum (m :: ms) = plus_nat m (sum ms);
```

Suc (Suc Zero_nat)

The code generator turns a set of equational theorems into a program inducing the *same* equational rewrite system. This means that any sequence of reduction steps the generated program performs on a term can be simulated in the logic:

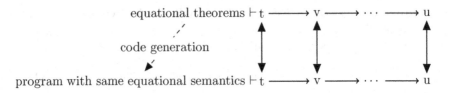

This guarantees partial correctness [3]. As a further consequence only program statements which contribute to a program's equational semantics (e.g. `fun` in ML) are correctness-critical, whereas others are not. For example, the constructors of a `datatype` in ML need only meet the syntactic characteristics of a datatype, but *not* the usual logical properties of a HOL datatype such as injectivity. This gives us some freedom in choosing datatype constructors which we will employ in §4.2.

4   Transforming Clauses to Equations

4.1   Mode Analysis

In order to execute a predicate P, its arguments are classified as *input* or *output*. For example, all three arguments of *append* could be input, meaning that the predicate just checks whether the third list is the concatenation of the first and second list. Another possibility would be to consider only the first and second argument as input, while the third one is output. In this case, the predicate actually *computes* the concatenation of the two input lists. Yet another way of using *append* would be to consider the third argument as input, while the first two arguments are output. This means that the predicate enumerates all possible ways of splitting the input list into two parts. This notion of *dataflow* is made explicit by means of *modes* [6].

Modes. For a predicate P with k arguments, we denote a particular dataflow assignment by a *mode* which is a set $M \subseteq \{1, \ldots, k\}$ such that M is exactly the set of all parameter position numbers denoting *input* parameters. A *mode assignment* for a given clause

$$Q_{i,1} \, \overline{u}_{i,1} \Longrightarrow \cdots \Longrightarrow Q_{i,n_i} \, \overline{u}_{i,n_i} \Longrightarrow P \, \overline{t}_i$$

is a list of modes $M, M_{i,1}, \ldots M_{i,n_i}$ for the predicates $P, Q_{i,1}, \ldots, Q_{i,n_i}$, where $1 \le i \le m$, $M \subseteq \{1, \ldots, l\}$ and $Q_{i,j} \subseteq \{1, \ldots, k_{i,j}\}$. Let $FV(t)$ denote the set of free variables in a term t. Given a vector of arguments $\overline{t}$ and a mode M, the projection expression $\overline{t}\langle M \rangle$ denotes the list of all arguments in $\overline{t}$ (in the order of their occurrence) whose index is in M.

Mode consistency. Given a clause

$$Q_{i,1}\,\overline{u}_{i,1} \Longrightarrow \cdots \Longrightarrow Q_{i,n_i}\,\overline{u}_{i,n_i} \Longrightarrow P\,\overline{t}_i$$

a corresponding mode assignment $M, M_{i,1}, \ldots M_{i,n_i}$ is *consistent* if there exists a chain of sets $v_0 \subseteq \cdots \subseteq v_n$ of variables generated by

1. $v_0 = FV(\overline{t}_i\langle M\rangle)$
2. $v_j = v_{j-1} \cup FV(\overline{u}_{i,j})$

such that

3. $FV(\overline{u}_{i,j}\langle M_{i,j}\rangle) \subseteq v_{j-1}$
4. $FV(\overline{t}_i) \subseteq v_n$

Consistency models the possibility of a sequential evaluation of premises in a given order, where v_j represents the known variables after the evaluation of the j-th premise:

1. initially, all variables in input arguments of P are known
2. after evaluation of the j-th premise, the set of known variables is extended by all variables in the arguments of $Q_{i,j}$,
3. when evaluating the j-th premise, all variables in the arguments of $Q_{i,j}$ have to be known,
4. finally, all variables in the arguments of P must be contained in the set of known variables.

Without loss of generality we can examine clauses under mode inference modulo reordering of premises. For side conditions R, condition 3 has to be replaced by $FV(R) \subseteq v_{j-1}$, i.e. all variables in R must be known when evaluating it. This definition yields a check whether a given clause is consistent with a particular mode assignment.

4.2   Enumerating Output Arguments of Predicates

A predicate of type $\alpha \Rightarrow bool$ is isomorphic to a set over type α; executing inductive predicates means to *enumerate* elements of the corresponding set. For this purpose we use an abstract algebra of primitive operations on such predicate enumerations. To establish an abstraction, we first define an explicit type to represent predicates:

> **datatype** $\alpha\ pred = pred\ (\alpha \Rightarrow bool)$

with a projection operator $eval :: \alpha\ pred \Rightarrow \alpha \Rightarrow bool$ satisfying

> $eval\ (pred\ f) = f$

We provide four further abstract operations on $\alpha\ pred$:

- $\bot :: \alpha\ pred$ is the empty enumeration.
- $single :: \alpha \Rightarrow \alpha\ pred$ is the singleton enumeration.
- $(\ggg) :: \alpha\ pred \Rightarrow (\alpha \Rightarrow \beta\ pred) \Rightarrow \beta\ pred$ applies a function to every element of an enumeration which itself returns an enumeration and flattens all resulting enumerations.
- $(\sqcup) :: \alpha\ pred \Rightarrow \alpha\ pred \Rightarrow \alpha\ pred$ forms the union of two enumerations.

These abstract operations, which form a *plus monad*, are used to build up the code equations of predicates (§4.3). Table 1 contains their definitions and relates them to their counterparts on sets. In order to equip these abstract operations with an executable model, we introduce an auxiliary datatype:

datatype $\alpha\ seq = Empty \mid Insert\ \alpha\ (\alpha\ pred) \mid Union\ (\alpha\ pred\ list)$

Values of type $\alpha\ seq$ are embedded into type $\alpha\ pred$ by defining:

$Seq :: (unit \Rightarrow \alpha\ seq) \Rightarrow \alpha\ pred$
$Seq\ f =$
$(case\ f\ ()\ of\ Empty \Rightarrow \bot \mid Insert\ x\ xq \Rightarrow single\ x \sqcup xq$
$\mid Union\ xqs \Rightarrow \lfloor\circ\rfloor\ xqs)$

where $\lfloor\circ\rfloor :: \alpha\ pred\ list \Rightarrow \alpha\ pred$ flattens a list of predicates into one predicate. *Seq* will serve as datatype constructor for type $\alpha\ pred$; on top of this, we prove the following code equations for our $\alpha\ pred$ algebra:

$\bot = Seq\ (\lambda u.\ Empty)$

$single\ x = Seq\ (\lambda u.\ Insert\ x\ \bot)$

$Seq\ g \ggg f =$
$Seq\ (\lambda u.\ case\ g\ ()\ of\ Empty \Rightarrow Empty$
$\qquad \mid Insert\ x\ xq \Rightarrow Union\ [f\ x,\ xq \ggg f]$
$\qquad \mid Union\ xqs \Rightarrow Union\ (map\ (\lambda x.\ x \ggg f)\ xqs))$

$Seq\ f \sqcup Seq\ g =$
$Seq\ (\lambda u.\ case\ f\ ()\ of\ Empty \Rightarrow g\ ()$
$\qquad \mid Insert\ x\ xq \Rightarrow Insert\ x\ (xq \sqcup Seq\ g)$
$\qquad \mid Union\ xqs \Rightarrow Union\ (xqs\ @\ [Seq\ g]))$

Table 1. Abstract operations for predicate enumerations

eval	$eval\ (pred\ P)\ x \longleftrightarrow P\ x$	$x \in P$
$\bot$	$\bot = pred\ (\lambda x.\ False)$	$\{\}$
$single\ x$	$single\ x = pred\ (\lambda y.\ y = x)$	$\{x\}$
$P \ggg f$	$P \ggg f = pred\ (\lambda x.\ \exists y.\ eval\ P\ y \wedge eval\ (f\ y)\ x)$	$\bigcup f\ `\ P$
$P \sqcup Q$	$P \sqcup Q = pred\ (\lambda x.\ eval\ P\ x \vee eval\ Q\ x)$	$P \cup Q$

Here $(`) :: (\alpha \Rightarrow \beta) \Rightarrow (\alpha \Rightarrow bool) \Rightarrow \beta \Rightarrow bool$ is the image operator on sets satisfying $f\ `\ A = \{y.\ \exists x \in A.\ y = f\ x\}$.

For membership tests we define a further auxiliary constant:

member :: α *seq* $\Rightarrow$ α $\Rightarrow$ *bool*
member Empty x $\longleftrightarrow$ *False*
member (Insert y yq) x $\longleftrightarrow$ $x = y \lor eval\ yq\ x$
member (Union xqs) x $\longleftrightarrow$ *list-ex* ($\lambda xq.\ eval\ xq\ x$) *xqs*

where *list-ex* :: $(\alpha \Rightarrow bool) \Rightarrow \alpha$ *list* $\Rightarrow$ *bool* is existential quantification on lists, and use it to prove the code equation

eval (Seq f) = *member (f ())*

From the point of view of the logic, this characterization of the α *pred* algebra in terms of unit abstractions might seem odd; their purpose comes to surface when translating these equations to executable code, e.g. in ML:

```
datatype 'a pred = Seq of (unit -> 'a seq)
and 'a seq = Empty | Insert of 'a * 'a pred | Union of 'a pred list;

val bot_pred : 'a pred = Seq (fn u => Empty)

fun single x = Seq (fn u => Insert (x, bot_pred));

fun bind (Seq g) f =
  Seq (fn u =>
          (case g () of Empty => Empty
            | Insert (x, xq) => Union [f x, bind xq f]
            | Union xqs => Union (map (fn x => bind x f) xqs)));

fun sup_pred (Seq f) (Seq g) =
  Seq (fn u =>
          (case f () of Empty => g ()
            | Insert (x, xq) => Insert (x, sup_pred xq (Seq g))
            | Union xqs => Union (append xqs [Seq g])));

fun eval A_ (Seq f) = member A_ (f ())
and member A_ Empty x = false
  | member A_ (Insert (y, yq)) x = eq A_ x y orelse eval A_ yq x
  | member A_ (Union xqs) x = list_ex (fn xq => eval A_ xq x) xqs;
```

In the function definitions for *eval* and *member*, the expression A_ is the dictionary for the *eq* class allowing for explicit equality checks using the overloaded constant *eq*.

In shape this follows a well-known ML technique for lazy lists: each inspection of a lazy list by means of an application f () is protected by a constructor Seq. Thus we enforce a lazy evaluation strategy for predicate enumerations even for eager languages.

4.3   Compilation Scheme for Clauses

The central idea underlying the compilation of a predicate P is to generate a function P^M for each mode M of P that, given a list of input arguments, enumerates all tuples of output arguments. The clauses of an inductive predicate can be viewed as a logic program. However, in contrast to logic programming languages like PROLOG, the execution of the functional program generated from the clauses uses *pattern matching* instead of *unification*. A precondition for the applicability of pattern matching is that the input arguments in the conclusions of the clauses, as well as the output arguments in the premises of the clauses are built up using only *datatype constructors* and variables. In the following description of the translation scheme, we will treat the pattern matching mechanism as a black box. However, our implementation uses a pattern translation algorithm due to Slind [9, §3.3], which closely resembles the techniques used in compilers for functional programming languages. The following notation will be used in our description of the translation mechanism:

$$\begin{aligned} \overline{x} &= x_1 \dots x_l & (\overline{x}) &= (x_1, \dots, x_l) \\ \overline{\tau} &= \tau_1 \dots \tau_l & \overline{\tau} \Rightarrow \sigma &= \tau_1 \Rightarrow \cdots \Rightarrow \tau_l \Rightarrow \sigma \\ \textstyle\prod \overline{\tau} &= \tau_1 \times \cdots \times \tau_l & M^- &= \{1, \dots, l\} \backslash M \end{aligned}$$

Let $P :: \overline{\tau} \Rightarrow bool$ be a predicate and $M, M_{i,1}, \dots M_{i,n_i}$ be a consistent mode assignment for the clauses C_i of P. The function P^M corresponding to mode M of P is defined as follows:

$$\begin{aligned} P^M &:: \overline{\tau}\langle M \rangle \Rightarrow (\textstyle\prod \overline{\tau}\langle M^- \rangle) \; pred \\ P^M \overline{x}\langle M \rangle &\equiv pred \; (\lambda(\overline{x}\langle M^- \rangle). \; P \; \overline{x}) \end{aligned}$$

Given the input arguments $\overline{x}\langle M \rangle :: \overline{\tau}\langle M \rangle$, function P^M returns a set of tuples of output arguments $(\overline{x}\langle M^- \rangle)$ for which $P \; \overline{x}$ holds. For modes $\{1, 2\}$ and $\{3\}$ of the introductory *append* example, the corresponding definitions are as follows:

$$\begin{aligned} append^{\{1,2\}} &:: \alpha \; list \Rightarrow \alpha \; list \Rightarrow \alpha \; list \; pred \\ append^{\{1,2\}} \; xs \; ys &= pred \; (\lambda zs. \; append \; xs \; ys \; zs) \\[4pt] append^{\{3\}} &:: \alpha \; list \Rightarrow (\alpha \; list \times \alpha \; list) \; pred \\ append^{\{3\}} \; zs &= pred \; (\lambda(xs, ys). \; append \; xs \; ys \; zs) \end{aligned}$$

The recursion equation for P^M can be obtained from the clauses characterizing P in a canonical way:

$$P^M \overline{x}\langle M \rangle = C_1 \; \overline{x}\langle M \rangle \sqcup \cdots \sqcup C_m \; \overline{x}\langle M \rangle$$

Intuitively, this means that the set of output values generated by P^M is the union of the output values generated by the clauses C_i. In order for pattern matching to work, all patterns occurring in the program must be *linear*, i.e. no variable may occur more than once. This can be achieved by renaming the free variables occurring in the terms $\overline{t}_i, \overline{u}_{i,1}, \dots, \overline{u}_{i,n_i}$, and by adding suitable equality checks to the generated program. Let $\overline{t}'_i, \overline{u}'_{i,1}, \dots, \overline{u}'_{i,n_i}$ denote these linear terms obtained by renaming the aforementioned ones, and let $\theta_i = \{\overline{y}_i \mapsto \overline{z}_i\}, \theta_{i,1} = \{\overline{y}_{i,1} \mapsto \overline{z}_{i,1}\}, \dots, \theta_{i,n_i} = \{\overline{y}_{i,n_i} \mapsto \overline{z}_{i,n_i}\}$ be substitutions such that $\theta_i(\overline{t}'_i) = t_i, \theta_{i,1}(\overline{u}'_{i,1}) = \overline{u}_{i,1}, \dots, \theta_{i,n_i}(\overline{u}'_{i,n_i}) = \overline{u}_{i,n_i}$, and $(dom(\theta_i) \cup dom(\theta_{i,1}) \cup$

$\cdots \cup dom(\theta_{i,n_i})) \cap FV(C_i) = \emptyset$. The expressions C_i corresponding to the clauses can then be defined by

$$C_i \,\overline{x}\langle M\rangle \equiv$$
$$single\ (\overline{x}\langle M\rangle) \ggg (\lambda a_0.\ case\ a_0\ of$$
$$(\overline{t}'_i\langle M\rangle) \Rightarrow if\ \overline{y}_i \neq \overline{z}_i\ then \perp else$$
$$Q^{M_{i,1}}_{i,1}\ (\overline{u}_{i,1}\langle M_{i,1}\rangle) \ggg (\lambda a_1.\ case\ a_1\ of$$
$$(\overline{u}'_{i,1}\langle M^{-}_{i,1}\rangle) \Rightarrow if\ \overline{y}_{i,1} \neq \overline{z}_{i,1}\ then \perp else$$

$$\ddots$$

$$Q^{M_{i,n_i}}_{i,n_i}\ (\overline{u}_{i,n_i}\langle M_{i,n_i}\rangle) \ggg (\lambda a_{n_i}.\ case\ a_{n_i}\ of$$
$$(\overline{u}'_{i,n_i}\langle M^{-}_{i,n_i}\rangle) \Rightarrow if\ \overline{y}_{i,n_i} \neq \overline{z}_{i,n_i}\ then \perp else$$
$$single\ (\overline{t}_i\langle M^{-}\rangle)$$
$$|\ \_ \Rightarrow \perp)$$
$$|\ \_ \Rightarrow \perp)$$
$$|\ \_ \Rightarrow \perp)$$

Here, $M^{-}_{i,1} = \{1, \ldots, k_{i,1}\}\backslash M_{i,1}, \ldots, M^{-}_{i,n_i} = \{1, \ldots, k_{i,n_i}\}\backslash M_{i,n_i}$ denote the sets of indices of *output arguments* corresponding to the respective modes. As an example, we give the recursive equations for *append* on modes $\{1,2\}$ and $\{3\}$:

$$append^{\{1,2\}}\ xs\ ys =$$
$$single\ (xs,\ ys) \ggg (\lambda a.\ case\ a\ of$$
$$([],\ zs) \Rightarrow single\ zs$$
$$|\ (z \cdot zs,\ ws) \Rightarrow \perp) \sqcup$$
$$single\ (xs,\ ys) \ggg (\lambda b.\ case\ b\ of$$
$$([],\ zs) \Rightarrow \perp$$
$$|\ (z \cdot zs,\ ws) \Rightarrow append^{\{1,2\}}\ zs\ ws \ggg (\lambda vs.\ single\ (z \cdot vs)))$$

$$append^{\{3\}}\ xs =$$
$$single\ xs \ggg (\lambda ys.\ single\ ([],\ ys)) \sqcup$$
$$single\ xs \ggg (\lambda a.\ case\ a\ of$$
$$[] \Rightarrow \perp$$
$$|\ z \cdot zs \Rightarrow append^{\{3\}}\ zs \ggg (\lambda b.\ case\ b\ of$$
$$(ws,\ vs) \Rightarrow single\ (z \cdot ws,\ vs)))$$

Side conditions can be embedded into this translation scheme using the function

$$if\text{-}pred :: \alpha$$
$$ifpred\ b = (if\ b\ then\ single\ ()\ else \perp)$$

that maps *False* and *True* to the empty sequence and the singleton sequence containing only the unit element, respectively.

4.4 Proof of Recursion Equations

We will now describe how to prove the recursion equation for P^M given in the previous section using the definition of P^M, as well as the introduction and elimination rules for P. We will also need introduction and elimination rules for the operators on type *pred*, which we show in Table 2. From the definition of P^M, we can easily derive the introduction rule

$$P \, \overline{x} \Longrightarrow eval \; (P^M \; \overline{x}\langle M \rangle) \; (\overline{x}\langle M^- \rangle)$$

and the elimination rule

$$eval \; (P^M \; \overline{x}\langle M \rangle) \; (\overline{x}\langle M^- \rangle) \Longrightarrow P \, \overline{x}$$

By extensionality (rule $=_I$), proving

$$P^M \overline{x}\langle M \rangle = \mathcal{C}_1 \; \overline{x}\langle M \rangle \sqcup \cdots \sqcup \mathcal{C}_m \; \overline{x}\langle M \rangle$$

amounts to showing that

(1) $\bigwedge x. \; eval \; (P^M \overline{x}\langle M \rangle) \; x \Longrightarrow eval \; (\mathcal{C}_1 \; \overline{x}\langle M \rangle \sqcup \cdots \sqcup \mathcal{C}_m \; \overline{x}\langle M \rangle) \; x$
(2) $\bigwedge x. \; eval \; (\mathcal{C}_1 \; \overline{x}\langle M \rangle \sqcup \cdots \sqcup \mathcal{C}_m \; \overline{x}\langle M \rangle) \; x \Longrightarrow eval \; (P^M \overline{x}\langle M \rangle) \; x$

where $x :: \prod \overline{\tau}\langle M^- \rangle$. The variable x can be expanded to a tuple of variables:

(1) $\bigwedge \overline{x}\langle M^- \rangle. \; eval \; (P^M \overline{x}\langle M \rangle) \; (\overline{x}\langle M^- \rangle) \Longrightarrow$
     $eval \; (\mathcal{C}_1 \; \overline{x}\langle M \rangle \sqcup \cdots \sqcup \mathcal{C}_m \; \overline{x}\langle M \rangle) \; (\overline{x}\langle M^- \rangle)$
(2) $\bigwedge \overline{x}\langle M^- \rangle. \; eval \; (\mathcal{C}_1 \; \overline{x}\langle M \rangle \sqcup \cdots \sqcup \mathcal{C}_m \; \overline{x}\langle M \rangle) \; (\overline{x}\langle M^- \rangle) \Longrightarrow$
     $eval \; (P^M \overline{x}\langle M \rangle) \; (\overline{x}\langle M^- \rangle)$

Proof of (1). From $eval \; (P^M \overline{x}\langle M \rangle) \; (\overline{x}\langle M^- \rangle)$, we get $P \, \overline{x}$ using the elimination rule for P^M. Applying the elimination rule for P

$$P \, \overline{x} \Longrightarrow \mathcal{E}_1 \; \overline{x} \Longrightarrow \cdots \Longrightarrow \mathcal{E}_m \; \overline{x} \Longrightarrow R$$
$$\mathcal{E}_i \; \overline{x} \equiv \bigwedge \overline{b}_i. \; \overline{x} = \overline{t}_i \Longrightarrow Q_{i,1} \; \overline{u}_{i,1} \Longrightarrow \cdots \Longrightarrow Q_{i,n_i} \; \overline{u}_{i,n_i} \Longrightarrow R$$

yields m proof obligations, each of which corresponds to an introduction rule. Note that $\overline{b}_i$ consists of the free variables of $\overline{u}_{i,j}$ and $\overline{t}_i$. For the ith introduction

Table 2. Introduction and elimination rules for operators on *pred*

$\perp_E$	$eval \perp x \Longrightarrow R$
$single_I$	$eval \; (single \; x) \; x$
$single_E$	$eval \; (single \; x) \; y \Longrightarrow (y = x \Longrightarrow R) \Longrightarrow R$
$\ggg_I$	$eval \; P \; x \Longrightarrow eval \; (Q \; x) \; y \Longrightarrow eval \; (P \ggg Q) \; y$
$\ggg_E$	$eval \; (P \ggg Q) \; y \Longrightarrow (\bigwedge x. \; eval \; P \; x \Longrightarrow eval \; (Q \; x) \; y \Longrightarrow R) \Longrightarrow R$
$\sqcup_{I1}$	$eval \; A \; x \Longrightarrow eval \; (A \sqcup B) \; x$
$\sqcup_{I2}$	$eval \; B \; x \Longrightarrow eval \; (A \sqcup B) \; x$
$\sqcup_E$	$eval \; (A \sqcup B) \; x \Longrightarrow (eval \; A \; x \Longrightarrow R) \Longrightarrow (eval \; B \; x \Longrightarrow R) \Longrightarrow R$
$ifpred_I$	$P \Longrightarrow eval \; (ifpred \; P) \; ()$
$ifpred_E$	$eval \; (ifpred \; b) \; x \Longrightarrow (b \Longrightarrow x = () \Longrightarrow R) \Longrightarrow R$
$=_I$	$(\bigwedge x. \; eval \; A \; x \Longrightarrow eval \; B \; x) \Longrightarrow (\bigwedge x. \; eval \; B \; x \Longrightarrow eval \; A \; x) \Longrightarrow A = B$

rule, we have to prove $eval\ (\mathcal{C}_1\ \overline{x}\langle M\rangle \sqcup \cdots \sqcup \mathcal{C}_m\ \overline{x}\langle M\rangle)\ (\overline{x}\langle M^-\rangle)$ from the assumptions $\overline{x} = \overline{t}_i$ and $Q_{i,1}\ \overline{u}_{i,1}, \ldots, Q_{i,n_i}\ \overline{u}_{i,n_i}$. By applying the rules $\sqcup_{I1}$ and $\sqcup_{I2}$ in a suitable order, we select the $\mathcal{C}_i$ corresponding to the ith introduction rule, which leaves us with the proof obligation $eval\ (\mathcal{C}_i\ \overline{t}_i\langle M\rangle)\ (\overline{t}_i\langle M^-\rangle)$. By the definition of $\mathcal{C}_i$ and the rule $\ggg_I$, this gives rise to the two proof obligations

$(1.i)$ $eval\ (single\ (\overline{t}_i\langle M\rangle))\ (\overline{t}_i\langle M\rangle)$

$(1.ii)$ $eval\ (case\ \overline{t}_i\langle M\rangle\ of$
$\qquad\qquad (\overline{t}_i'\langle M\rangle) \Rightarrow if\ \overline{y}_i \neq \overline{z}_i\ then\ \bot\ else$
$\qquad\qquad\qquad Q_{i,1}^{M_{i,1}}\ (\overline{u}_{i,1}\langle M_{i,1}\rangle) \ggg (\lambda a_1.\ case\ a_1\ of\ \ldots)$
$\qquad\qquad |\ \_ \Rightarrow \bot)\ \overline{t}_i\langle M^-\rangle$

Goal $(1.i)$ is easily proved using $single_I$. Concerning goal $(1.ii)$, note that $(\overline{t}_i'\langle M\rangle)$ matches $(\overline{t}_i\langle M\rangle)$, so we have to consider the first branch of the $case$ expression. Due to the definition of $\overline{t}_i'$, we also know that $\overline{y}_i = \overline{z}_i$, which means that we have to consider the $else$ branch of the if clause. This leads to the new goal

$$eval\ (Q_{i,1}^{M_{i,1}}\ (\overline{u}_{i,1}\langle M_{i,1}\rangle) \ggg (\lambda a_1.\ case\ a_1\ of\ \ldots))\ \overline{t}_i\langle M^-\rangle$$

that, by applying rule $\ggg_I$, can be split up into the two goals

$(1.iii)$ $eval\ (Q_{i,1}^{M_{i,1}}\ (\overline{u}_{i,1}\langle M_{i,1}\rangle))\ (\overline{u}_{i,1}\langle M_{i,1}^-\rangle)$

$(1.iv)$ $eval\ (case\ \overline{u}_{i,1}\langle M_{i,1}^-\rangle\ of$
$\qquad\qquad (\overline{u}_{i,1}'\langle M_{i,1}^-\rangle) \Rightarrow if\ \overline{y}_{i,1} \neq \overline{z}_{i,1}\ then\ \bot\ else\ \ldots$
$\qquad\qquad |\ \_ \Rightarrow \bot)\ \overline{t}_i\langle M^-\rangle$

Goal $(1.iii)$ follows from the assumption $Q_{i,1}\ \overline{u}_{i,1}$ using the introduction rule for $Q_{i,1}^{M_{i,1}}$, while goal $(1.iv)$ can be solved in a similar way as goal $(1.ii)$. Repeating this proof scheme for $Q_{i,2}^{M_{i,2}}, \ldots, Q_{i,n_i}^{M_{i,n_i}}$ finally leads us to a goal of the form

$$eval\ (single\ (\overline{t}_i\langle M^-\rangle))\ (\overline{t}_i\langle M^-\rangle)$$

which is trivially solvable using $single_I$.

Proof of (2). The proof of this direction is dual to the previous one: rather than splitting up the conclusion into simpler formulae, we now perform forward inferences that transform complex premises into simpler ones. Eliminating $eval\ (\mathcal{C}_1\ \overline{x}\langle M\rangle \sqcup \cdots \sqcup \mathcal{C}_m\ \overline{x}\langle M\rangle)\ (\overline{x}\langle M^-\rangle)$ using rule $\sqcup_E$ leaves us with m proof obligations of the form

$$eval\ (\mathcal{C}_i\ \overline{x}\langle M\rangle)\ (\overline{x}\langle M^-\rangle) \Longrightarrow eval\ (P^M \overline{x}\langle M\rangle)\ (\overline{x}\langle M^-\rangle)$$

By unfolding the definition of $\mathcal{C}_i$ and applying rule $\ggg_E$ to the premise of the above implication, we obtain a_0 such that

$(2.i)$ $eval\ (single\ (\overline{x}\langle M\rangle))\ a_0$

$(2.ii)$ $eval\ (case\ a_0\ of$
$\qquad\qquad (\overline{t}_i'\langle M\rangle) \Rightarrow if\ \overline{y}_i \neq \overline{z}_i\ then\ \bot\ else$
$\qquad\qquad\qquad Q_{i,1}^{M_{i,1}}\ (\overline{u}_{i,1}\langle M_{i,1}\rangle) \ggg (\lambda a_1.\ case\ a_1\ of\ \ldots)$
$\qquad\qquad |\ \_ \Rightarrow \bot)\ \overline{x}\langle M^-\rangle$

From $(2.i)$, we get $\overline{x}\langle M\rangle = a_0$ by rule $single_E$. Since a_0 must be an element of a datatype, we can analyze its shape by applying suitable case splitting rules. Of the generated cases only one case is non-trivial. In the trivial cases, a_0 does not match $(\overline{t}'_i\langle M\rangle)$, so the *case* expression evaluates to $\bot$, and the goal can be solved using $\bot_E$. In the non-trivial case, we have that $a_0 = (\overline{t}'_i\langle M\rangle)$. Splitting up the *if* expression yields two cases. In the *then* case, the whole expression evaluates to $\bot$, so the goal is again provable using $\bot_E$. In the *else* branch, we have that $\overline{y}_i = \overline{z}_i$, and hence $a_0 = (\overline{t}_i\langle M\rangle)$ by definition of $\overline{t}'_i$, which also implies $\overline{x}\langle M\rangle = \overline{t}_i\langle M\rangle$. Assumption $(2.ii)$ can thus be rewritten to

$$eval\ (Q_{i,1}^{M_{i,1}}\ (\overline{u}_{i,1}\langle M_{i,1}\rangle)) \ggeq (\lambda a_1.\ case\ a_1\ of\ \ldots)) \ \overline{x}\langle M^-\rangle$$

By another application of $\ggeq_E$, we obtain a_1 such that

$(2.iii)$ $eval\ (Q_{i,1}^{M_{i,1}}\ (\overline{u}_{i,1}\langle M_{i,1}\rangle))\ a_1$

$(2.iv)$ $eval\ (case\ a_1\ of$
$\qquad (\overline{u}'_{i,1}\langle M_{i,1}^-\rangle)) \Rightarrow if\ \overline{y}_{i,1} \neq \overline{z}_{i,1}\ then\ \bot\ else\ \ldots$
$\qquad |\ \_ \Rightarrow \bot)\ \overline{x}\langle M^-\rangle$

The assumption $(2.iv)$ is treated in a similar way as $(2.ii)$. A case analysis over a_1 reveals that the only non-trivial case is the one where $a_1 = (\overline{u}'_{i,1}\langle M_{i,1}^-\rangle)$. The only non-trivial branch of the *if* expression is the *else* branch, where $\overline{y}_{i,1} = \overline{z}_{i,1}$. Hence, by definition of $\overline{u}'_{i,1}$, it follows that $a_1 = (\overline{u}_{i,1}\langle M_{i,1}^-\rangle)$, which entitles us to rewrite $(2.iii)$ to $eval\ (Q_{i,1}^{M_{i,1}}\ (\overline{u}_{i,1}\langle M_{i,1}\rangle))\ (\overline{u}_{i,1}\langle M_{i,1}^-\rangle)$, from which we can deduce $Q_{i,1}\ \overline{u}_{i,1}$ by applying the elimination rule for $Q_{i,1}^{M_{i,1}}$. By repeating this kind of reasoning for $Q_{i,2}^{M_{i,2}}$, $\ldots$, $Q_{i,n_i}^{M_{i,n_i}}$, we also obtain that $Q_{i,2}\ \overline{u}_{i,2}$, $\ldots$, $Q_{i,n_i}\ \overline{u}_{i,n_i}$ holds. Furthermore, after the complete decomposition of $(2.iv)$, we end up with an assumption of the form

$$eval\ (single\ (\overline{t}_i\langle M^-\rangle))\ (\overline{x}\langle M^-\rangle)$$

from which we can deduce $\overline{t}_i\langle M^-\rangle = \overline{x}\langle M^-\rangle$ by an application of $single_E$. Thus, using the equations gained from $(2.i)$ and $(2.ii)$, the conclusion of the implication we set out to prove can be rephrased as

$$eval\ (P^M\ \overline{t}_i\langle M\rangle)\ (\overline{t}_i\langle M^-\rangle)$$

Thanks to the introduction rule for P^M, it suffices to prove $P\ \overline{t}_i$, which can easily be done using the introduction rule

$$Q_{i,1}\ \overline{u}_{i,1} \Longrightarrow \cdots \Longrightarrow Q_{i,n_i}\ \overline{u}_{i,n_i} \Longrightarrow P\ \overline{t}_i$$

together with the previous results.

4.5   Animating Equations

We have shown in detail how to derive executable equations from the specification of a predicate P for a consistent mode M. The results are always enumerations of type α *pred*. We discuss briefly how to get access to the enumerated values of type α proper.

Membership tests. The type constructor *pred* can be stripped using explicit membership tests. For example, we could define a suffix predicate using *append*:

$$\textit{is-suffix zs ys} \longleftrightarrow (\exists\, xs.\ \textit{append xs ys zs})$$

Using the definition of $\textit{append}^{\{2,3\}}$ this can be reformulated as

$$\textit{is-suffix zs ys} \longleftrightarrow (\exists\, xs.\ \textit{eval}\ (\textit{append}^{\{2,3\}}\ ys\ zs)\ xs)$$

from which follows

$$\textit{is-suffix zs ys} \longleftrightarrow \textit{eval}\ (\textit{append}^{\{2,3\}}\ ys\ zs \ggeq (\lambda\text{-}.\ \textit{single}\ ()))\ ()$$

using introduction and elimination rules for *op* $\gg=$ and *single*. This equation then is directly executable.

Enumeration queries. When developing inductive specifications it is often desirable to check early whether the specification behaves as expected by enumerating one or more solutions which satisfy the specification. In our framework this cannot be expressed inside the logic: values of type α *pred* are set-like, whereas each concrete enumeration imposes a certain order on elements which is not reflected in the logic. However it can be done directly on the generated code, e.g. in ML using

```
fun nexts [] = NONE
  | nexts (xq :: xqs) = case next xq
      of NONE => nexts xqs
       | SOME (x, xq) => SOME (x, Seq (fn () => Union (xq :: xqs)))
and next (Seq f) = case f ()
    of Empty => NONE
     | Insert (x, xq) => SOME (x, xq)
     | Union xqs => nexts xqs;
```

Wrapped up in a suitable user interface this allows to interactively enumerate solutions fitting to inductive predicates.

5   Extensions to the Base Framework

5.1   Higher-Order Modes

A useful extension of the framework presented in §4.3 is to allow inductive predicates that take other predicates as arguments. A standard example for such a predicate is the *reflexive transitive closure* taking a predicate of type $\alpha \Rightarrow \alpha \Rightarrow bool$ as an argument, and returning a predicate of the same type:

inductive $rtc :: (\alpha \Rightarrow \alpha \Rightarrow bool) \Rightarrow \alpha \Rightarrow \alpha \Rightarrow bool$
  for $r :: \alpha \Rightarrow \alpha \Rightarrow bool$ **where**
    $rtc\ r\ x\ x$
  $|\ r\ x\ y \Longrightarrow rtc\ r\ y\ z \Longrightarrow rtc\ r\ x\ z$

In addition to its two *arguments* of type α, *rtc* also has a *parameter* r that stays fixed throughout the definition. The general form of a mode for a higher-order predicate P with k arguments and parameters $r_1, \ldots, r_\rho$ with arities $k_1, \ldots, k_\rho$ is $(M_1, \ldots, M_\rho, M)$, where $M_i \subseteq \{1, \ldots, k_i\}$ (for $1 \leq i \leq \rho$) and $M \subseteq \{1, \ldots, k\}$. Intuitively, this mode means that $P\ r_1\ \cdots\ r_\rho$ has mode M, provided that r_i has mode M_i. The possible modes for *rtc* are $(\{\}, \{1\})$, $(\{\}, \{2\})$, $(\{\}, \{1,2\})$, $(\{1\}, \{1\})$, $(\{2\}, \{2\})$, $(\{1\}, \{1,2\})$, and $(\{2\}, \{1,2\})$. The general definition of the function corresponding to the mode $(M_1, \ldots, M_\rho, M)$ of a predicate P is

$$
\begin{aligned}
&P^{(M_1,\ldots,M_\rho,M)} \ :: \\
&\qquad (\overline{\tau}_1\langle M_1\rangle \Rightarrow (\textstyle\prod \overline{\tau}_1\langle M_1^-\rangle)\ pred) \Rightarrow \cdots \Rightarrow \\
&\qquad (\overline{\tau}_\rho\langle M_\rho\rangle \Rightarrow (\textstyle\prod \overline{\tau}_\rho\langle M_\rho^-\rangle)\ pred) \Rightarrow (\textstyle\prod \overline{\tau}\langle M^-\rangle)\ pred \\
&P^{(M_1,\ldots,M_\rho,M)}\ s_1\ \ldots\ s_\rho\ \overline{x}\langle M\rangle \equiv pred\ (\lambda(\overline{x}\langle M^-\rangle).\ P \\
&\qquad (\lambda\overline{x}_1.\ eval\ (s_1\ \overline{x}_1\langle M_1\rangle)\ (\overline{x}_1\langle M_1^-\rangle))\ \cdots \\
&\qquad (\lambda\overline{x}_\rho.\ eval\ (s_1\ \overline{x}_\rho\langle M_\rho\rangle)\ (\overline{x}_\rho\langle M_\rho^-\rangle))\ \overline{x})
\end{aligned}
$$

Since P expects predicates as parameters, but s_i are functions returning sets, these have to be converted back to predicates using *eval* before passing them to P. For *rtc*, the definitions of the functions corresponding to the modes $(\{1\}, \{1\})$ and $(\{2\}, \{2\})$ are

$$
\begin{aligned}
&rtc^{(\{1\},\{1\})} \ :: (\alpha \Rightarrow \alpha\ pred) \Rightarrow \alpha \Rightarrow \alpha\ pred \\
&rtc^{(\{1\},\{1\})}\ s\ x \equiv pred\ (\lambda y.\ rtc\ (\lambda x'\ y'.\ eval\ (s\ x')\ y')\ x\ y)
\end{aligned}
$$

$$
\begin{aligned}
&rtc^{(\{2\},\{2\})} \ :: (\alpha \Rightarrow \alpha\ pred) \Rightarrow \alpha \Rightarrow \alpha\ pred \\
&rtc^{(\{2\},\{2\})}\ s\ y \equiv pred\ (\lambda x.\ rtc\ (\lambda x'\ y'.\ eval\ (s\ y')\ x')\ x\ y)
\end{aligned}
$$

The corresponding recursion equations have the form

$$
\begin{aligned}
&rtc^{(\{1\},\{1\})}\ r\ x = \\
&\quad single\ x \ggg (\lambda x.\ single\ x) \sqcup \\
&\quad single\ x \ggg (\lambda x.\ r\ x \ggg (\lambda y.\ rtc^{(\{1\},\{1\})}\ r\ y \ggg (\lambda z.\ single\ z)))
\end{aligned}
$$

$$
\begin{aligned}
&rtc^{(\{2\},\{2\})}\ r\ y = \\
&\quad single\ y \ggg (\lambda x.\ single\ x) \sqcup \\
&\quad single\ y \ggg (\lambda z.\ rtc^{(\{2\},\{2\})}\ r\ z \ggg (\lambda y.\ r\ y \ggg (\lambda x.\ single\ x)))
\end{aligned}
$$

5.2   Mixing Predicates and Functions

When mixing predicates and functions, mode analysis treats functions as predicates where all arguments are *input*. This can restrict the number of consistent mode assignments considerably.

The following mutually inductive predicates model a grammar generating all words containing equally many *a*s and *b*s. This example, which is originally due to Hopcroft and Ullman, can be found in the Isabelle tutorial by Nipkow [8].

inductive
$S :: alfa\ list \Rightarrow bool$ **and**
$A :: alfa\ list \Rightarrow bool$ **and** $B :: alfa\ list \Rightarrow bool$

where

$$S\ []$$
$$|\ A\ w \implies S\ (b \cdot w)$$
$$|\ B\ w \implies S\ (a \cdot w)$$
$$|\ S\ w \implies A\ (a \cdot w)$$
$$|\ A\ v \implies A\ w \implies A\ (b \cdot v\ @\ w)$$
$$|\ S\ w \implies B\ (b \cdot w)$$
$$|\ B\ v \implies B\ w \implies B\ (a \cdot v\ @\ w)$$

By choosing mode $\{\}$ for the above predicates (i.e. their arguments are all output), we can enumerate all elements of the set S containing equally many as and bs. However, the above predicates cannot easily be used with mode $\{1\}$, i.e. for *checking* whether a given word is generated by the grammar. This is because of the rules with the conclusions $A\ (b \cdot v\ @\ w)$ and $B\ (a \cdot v\ @\ w)$. Since the append *function* (denoted by $@$) is not a constructor, we cannot do pattern matching on the argument. However, the problematic rules can be rephrased as

$$append\ v\ w\ vw \implies A\ v \implies A\ w \implies A\ (b \cdot vw)$$
$$append\ v\ w\ vw \implies B\ v \implies B\ w \implies B\ (a \cdot vw)$$

The problematic expression $v\ @\ w$ in the conclusion has been replaced by a new variable vw. The fact that vw is the result of appending the two lists v and w is now expressed using the *append* predicate from §3. In order to check whether a given word can be generated using these rules, *append* first enumerates all ways of decomposing the given list vw into two sublists v and w, and then recursively checks whether these words can be generated by the grammar.

6 Conclusion and Future Work

We have presented a *definitional* translation for inductive predicates to equations which can be turned into executable code using existing code generation infrastructure in *Isabelle/HOL*. This is a fundamental contribution to extend the scope of code generation from functional to functional-logic programs embedded into *Isabelle/HOL* without compromising the trusted implementation of the code generator itself. We have applied our translation to two larger case studies, the μJava semantics by Nipkow, von Oheimb and Pusch [7] and the ς-calculus by Henrio and Kammüller [5], resulting in simple interpreters for these two programming languages. Further experiments suggest the following extensions:

- Successful mode inference does not guarantee termination. Like in PROLOG, the order of premises in introduction rules can influence termination. Using termination analysis built-in in *Isabelle/HOL*, we can guess which modes lead to terminating functions. The mode analysis can use this and prefer terminating modes over possibly non-terminating ones.
- Rephrasing recursive functions to inductive predicates, as we apply it in §5.2, possibly results in more modes for the mode analysis. But applying the

transformation blindly could lead to unnecessarily complicated equations. The mode analysis should be extended to infer modes using the transformation only when required.

- The executable model for enumerations we have presented is sometimes inappropriate: it performs depth-first search which can lead to a non-terminating search in an irrelevant but infinite branch of the search tree. It has to be figured out how alternative search strategies (e.g. iterative depth-first search) can provide a solution for this.

We plan to integrate our procedure into the next *Isabelle* release.

References

1. Berghofer, S., Nipkow, T.: Executing higher order logic. In: Callaghan, P., Luo, Z., McKinna, J., Pollack, R. (eds.) TYPES 2000. LNCS, vol. 2277, p. 24. Springer, Heidelberg (2002)
2. Delahaye, D., Dubois, C., Étienne, J.F.: Extracting purely functional contents from logical inductive types. In: Schneider, K., Brandt, J. (eds.) TPHOLs 2007. LNCS, vol. 4732, pp. 70–85. Springer, Heidelberg (2007)
3. Haftmann, F., Nipkow, T.: A code generator framework for Isabelle/HOL. Tech. Rep. 364/07, Department of Computer Science, University of Kaiserslautern (2007)
4. Hanus, M.: A unified computation model for functional and logic programming. In: Proc. 24th ACM Symposium on Principles of Programming Languages (POPL 1997), pp. 80–93 (1997)
5. Henrio, L., Kammüller, F.: A mechanized model of the theory of objects. In: Bonsangue, M.M., Johnsen, E.B. (eds.) FMOODS 2007. LNCS, vol. 4468, pp. 190–205. Springer, Heidelberg (2007)
6. Mellish, C.S.: The automatic generation of mode declarations for prolog programs. Tech. Rep. 163, Department of Artificial Intelligence (1981)
7. Nipkow, T., von Oheimb, D., Pusch, C.: μJava: Embedding a programming language in a theorem prover. In: Bauer, F., Steinbrüggen, R. (eds.) Foundations of Secure Computation. Proc. Int. Summer School Marktoberdorf 1999, pp. 117–144. IOS Press, Amsterdam (2000)
8. Nipkow, T., Paulson, L.C., Wenzel, M.: Isabelle/HOL. LNCS, vol. 2283. Springer, Heidelberg (2002)
9. Slind, K.: Reasoning about terminating functional programs. Ph.D. thesis, Institut für Informatik, TU München (1999)
10. Somogyi, Z., Henderson, F.J., Conway, T.C.: Mercury: an efficient purely declarative logic programming language. In: Proceedings of the Australian Computer Science Conference, pp. 499–512 (1995)
11. Wasserrab, D., Nipkow, T., Snelting, G., Tip, F.: An operational semantics and type safety proof for multiple inheritance in C++. In: OOPSLA 2006: Proceedings of the 21st annual ACM SIGPLAN conference on Object-oriented programming languages, systems, and applications, pp. 345–362. ACM Press, New York (2006)

Formalizing the Logic-Automaton Connection

Stefan Berghofer* and Markus Reiter

Technische Universität München
Institut für Informatik, Boltzmannstraße 3, 85748 Garching, Germany

Abstract. This paper presents a formalization of a library for automata on bit strings in the theorem prover Isabelle/HOL. It forms the basis of a reflection-based decision procedure for Presburger arithmetic, which is efficiently executable thanks to Isabelle's code generator. With this work, we therefore provide a mechanized proof of the well-known connection between logic and automata theory.

1 Introduction

Although higher-order logic (HOL) is undecidable in general, there are many decidable logics such as *Presburger arithmetic* or the *Weak Second-order theory of One Successor* (WS1S) that can be embedded into HOL. Since HOL can be viewed as a logic containing a functional programming language, an interesting approach for implementing a decision procedure for such a decidable logic in a theorem prover based on HOL is to write *and verify* the decision procedure as a recursive function in HOL itself. This approach, which is called *reflection* [7], has been used in proof assistants based on type theory for quite a long time. For example, Boutin [4] has used reflection to implement a decision procedure for abelian rings in Coq. Recently, reflection has also gained considerable attention in the Isabelle/HOL community. Chaieb and Nipkow have used this technique to verify various quantifier elimination procedures for dense linear orders, real and integer linear arithmetic, as well as Presburger arithmetic [5,12]. While the decision procedures by Chaieb and Nipkow are based on *algebraic* methods like Cooper's algorithm, there are also *semantic* methods, as implemented e.g. in the MONA tool [8] for deciding WS1S formulae. In order to check the validity of a formula, MONA translates it to an automaton on bitstrings and then checks whether it has accepting states. Basin and Friedrich [1] have connected MONA to Isabelle/HOL using an *oracle-based* approach, i.e. they simply trust the answer of the tool. As a motivation for their design decision, they write:

> Hooking an 'oracle' to a theorem prover is risky business. The oracle could be buggy [. . .]. The only way to avoid a buggy oracle is to reconstruct a proof in the theorem prover based on output from the oracle, or perhaps verify the oracle itself. For a semantics based decision procedure, proof reconstruction is not a realistic option: one would have to formalize the entire automata-theoretic machinery within HOL [. . .].

* Supported by BMBF in the VerisoftXT project under grant 01 IS 07008 F.

S. Berghofer et al. (Eds.): TPHOLs 2009, LNCS 5674, pp. 147–163, 2009.

In this paper, we show that verifying decision procedures based on automata in HOL is not as unrealistic as it may seem. We develop a library for automata on bitstrings, including operations like forming the product of two automata, projection, and determinization of nondeterministic automata, which we then use to build a decision procedure for Presburger arithmetic. The procedure can easily be changed to cover WS1S, by just exchanging the automata for atomic formulae. The specification of the decision procedure is completely executable, and efficient ML code can be generated from it using Isabelle's code generator [2]. To the best of our knowledge, this is the first formalization of an automata-based decision procedure for Presburger arithmetic in a theorem prover.

The paper is structured as follows. In §2, we introduce basic concepts such as Presburger arithmetic, automata theory, bit vectors, and BDDs. The library for automata is described in §3. The actual decision procedure together with its correctness proof is presented in §4, and §5 draws some conclusions. Due to lack of space, we will not discuss any of the proofs in detail. However, the interested reader can find the complete formalization on the web[1].

2   Basic Definitions

2.1   Presburger Arithmetic

Formulae of Presburger arithmetic are represented by the following datatype:

datatype *pf = Eq (int list) int | Le (int list) int | And pf pf | Or pf pf*
| Imp pf pf | Forall pf | Exist pf | Neg pf

The atomic formulae are *Diophantine (in)equations Eq ks l* and *Le ks l*, where *ks* are the (integer) coefficients and *l* is the right-hand side. Variables are encoded using de Bruijn indices, meaning that the ith coefficient in *ks* belongs to the variable with index i. Thus, the well-known *stamp problem*

$$\forall x \geq 8. \; \exists y \; z. \; 3 * y + 5 * z = x$$

can be encoded by

Forall (Imp (Le [−1] −8) (Exist (Exist (Eq [5, 3, −1] 0))))

Like Boudet and Comon [3], we only consider variables ranging over the natural numbers. The left-hand side of a Diophantine (in)equation can be evaluated using the function

eval-dioph :: int list ⇒ nat list ⇒ int
*eval-dioph (k · ks) (x · xs) = k * int x + eval-dioph ks xs*
eval-dioph [] xs = 0
eval-dioph ks [] = 0

---

[1] http://www.in.tum.de/~berghofe/papers/automata

where xs is a valuation, $x \cdot xs$ denotes the '*Cons*' operator, and int coerces a natural number to an integer. A Presburger formula can be evaluated by

$eval\text{-}pf :: pf \Rightarrow nat\ list \Rightarrow bool$
$eval\text{-}pf\ (Eq\ ks\ l)\ xs = (eval\text{-}dioph\ ks\ xs = l)$
$eval\text{-}pf\ (Le\ ks\ l)\ xs = (eval\text{-}dioph\ ks\ xs \leq l)$
$eval\text{-}pf\ (Neg\ p)\ xs = (\neg\ eval\text{-}pf\ p\ xs)$
$eval\text{-}pf\ (And\ p\ q)\ xs = (eval\text{-}pf\ p\ xs \wedge eval\text{-}pf\ q\ xs)$
$eval\text{-}pf\ (Or\ p\ q)\ xs = (eval\text{-}pf\ p\ xs \vee eval\text{-}pf\ q\ xs)$
$eval\text{-}pf\ (Imp\ p\ q)\ xs = (eval\text{-}pf\ p\ xs \longrightarrow eval\text{-}pf\ q\ xs)$
$eval\text{-}pf\ (Forall\ p)\ xs = (\forall x.\ eval\text{-}pf\ p\ (x \cdot xs))$
$eval\text{-}pf\ (Exist\ p)\ xs = (\exists x.\ eval\text{-}pf\ p\ (x \cdot xs))$

2.2   Abstract Automata

The abstract framework for automata used in this paper is quite similar to the one used by Nipkow [11]. The purpose of this framework is to factor out all properties that deterministic and nondeterministic automata have in common. Automata are characterized by a transition function tr of type $\sigma \Rightarrow \alpha \Rightarrow \sigma$, where σ and α denote the types of *states* and *input symbols*, respectively. Transition functions can be extended to *words*, i.e. lists of symbols in a canonical way:

$steps :: (\sigma \Rightarrow \alpha \Rightarrow \sigma) \Rightarrow \sigma \Rightarrow \alpha\ list \Rightarrow \sigma$
$steps\ tr\ q\ [] = q$
$steps\ tr\ q\ (a \cdot as) = steps\ tr\ (tr\ q\ a)\ as$

The *reachability* of a state q from a state p via a word as is defined by

$reach :: (\sigma \Rightarrow \alpha \Rightarrow \sigma) \Rightarrow \sigma \Rightarrow \alpha\ list \Rightarrow \sigma \Rightarrow bool$
$reach\ tr\ p\ as\ q \equiv q = steps\ tr\ p\ as$

Another characteristic property of an automaton is its set of accepting states. Given a predicate P denoting the accepting states, an automaton is said to accept a word as iff from a starting state s we reach an accepting state via as:

$accepts :: (\sigma \Rightarrow \alpha \Rightarrow \sigma) \Rightarrow (\sigma \Rightarrow bool) \Rightarrow \sigma \Rightarrow \alpha\ list \Rightarrow bool$
$accepts\ tr\ P\ s\ as \equiv P\ (steps\ tr\ s\ as)$

2.3   Bit Vectors and BDDs

The automata used in the formalization of our decision procedure for Presburger arithmetic are of a specific kind: the input symbols of an automaton corresponding to a formula with n free variables $x_0, \ldots, x_{n-1}$ are bit lists of length n.

$$\begin{matrix} x_0 \\ \vdots \\ x_{n-1} \end{matrix} \quad \left[\begin{bmatrix} b_{0,0} \\ \vdots \\ b_{n-1,0} \end{bmatrix} \begin{bmatrix} b_{0,1} \\ \vdots \\ b_{n-1,1} \end{bmatrix} \begin{bmatrix} b_{0,2} \\ \vdots \\ b_{n-1,2} \end{bmatrix} \cdots \begin{bmatrix} b_{0,m-1} \\ \vdots \\ b_{n-1,m-1} \end{bmatrix} \right]$$

The rows in the above word are interpreted as *natural numbers*, where the left-most column, i.e. the first symbol in the list corresponds to the *least significant bit*. Therefore, the value of variable x_i is $\sum_{j=0}^{m-1} b_{i,j} 2^j$. The list of values of n variables denoted by a word can be computed recursively as follows:

nats-of-boolss :: *nat* $\Rightarrow$ *bool list list* $\Rightarrow$ *nat list*
nats-of-boolss n $[]$ = *replicate* n 0
nats-of-boolss n $(bs \cdot bss)$ =
    map $(\lambda(b, x).\ \textit{nat-of-bool}\ b + 2 * x)$ $(\textit{zip}\ bs\ (\textit{nats-of-boolss}\ n\ bss))$

where *zip* $[b_0, b_1, \ldots]$ $[x_0, x_1, \ldots]$ yields $[(b_0, x_0), (b_1, x_1), \ldots]$, *replicate* n x denotes the list $[x, \ldots, x]$ of length n, and *nat-of-bool* maps *False* and *True* to 0 and 1, respectively. We can insert a bit vector in the ith row of a word by

insertll :: *nat* $\Rightarrow$ α *list* $\Rightarrow$ α *list list* $\Rightarrow$ α *list list*
insertll i $[]$ $[]$ = $[]$
insertll i $(a \cdot as)$ $(bs \cdot bss)$ = *insertl* i a bs $\cdot$ *insertll* i as bss

where *insertl* i a bs inserts a into list bs at position i. The interaction between *nats-of-boolss* and *insertll* can be characterized by the following theorem:

If $\forall bs \in bss.$ *is-alph* n bs *and* $|bs| = |bss|$ *and* $i \leq n$ *then*
nats-of-boolss $(Suc\ n)$ $(insertll\ i\ bs\ bss)$ =
insertl i $(\textit{nat-of-bools}\ bs)$ $(\textit{nats-of-boolss}\ n\ bss)$.

Here, *is-alph* n xs means that xs is a valid symbol, i.e. the length of xs is equal to the number of variables n, $|bs|$ and $|bss|$ denote the lengths of bs and bss, respectively, and $bs \in bss$ means that bs is a member of the list bss. Moreover, *nat-of-bools* is similar to *nats-of-boolss*, with the difference that it works on a single row vector instead of a list of column vectors:

nat-of-bools :: *bool list* $\Rightarrow$ *nat*
nat-of-bools $[]$ = 0
nat-of-bools $(b \cdot bs)$ = *nat-of-bool* $b + 2 *$ *nat-of-bools* bs

Since the input symbols of our automata are bit vectors, it would be rather inefficient to just represent the transition function for a given state as an association list relating bit vectors to successor states. For such a list, the lookup operation would be exponential in the number of variables. When implementing the MONA tool, Klarlund [8] already observed that representing the transition function as a BDD is more efficient. BDDs are represented by the datatype[2]

datatype α *bdd* = *Leaf* α | *Branch* $(\alpha\ \textit{bdd})$ $(\alpha\ \textit{bdd})$

The functions *bdd-map* :: $(\alpha \Rightarrow \beta) \Rightarrow \alpha\ \textit{bdd} \Rightarrow \beta\ \textit{bdd}$ and *bdd-all* :: $(\alpha \Rightarrow \textit{bool})$ $\Rightarrow \alpha\ \textit{bdd} \Rightarrow \textit{bool}$ can be defined in a canonical way. The lookup operation, whose runtime is linear in the length of the input vector, has the definition

---

[2] Since the leaves are not just 0 or 1, Klarlund [8] calls this a *multi-terminal* BDD.

bdd-lookup :: α *bdd* $\Rightarrow$ *bool list* $\Rightarrow$ α
bdd-lookup (*Leaf x*) *bs* = *x*
bdd-lookup (*Branch l r*) (*b* · *bs*) = *bdd-lookup* (if *b* then *r* else *l*) *bs*

This operation only returns meaningful results if the height of the BDD is less or equal to the length of the bit vector. We write *bddh n bdd* to mean that the height of *bdd* is less or equal to *n*. Two BDDs can be combined using a binary operator *f* as follows:

bdd-binop :: (α $\Rightarrow$ β $\Rightarrow$ γ) $\Rightarrow$ α *bdd* $\Rightarrow$ β *bdd* $\Rightarrow$ γ *bdd*
bdd-binop f (*Leaf x*) (*Leaf y*) = *Leaf* (*f x y*)
bdd-binop f (*Branch l r*) (*Leaf y*) =
  Branch (*bdd-binop f l* (*Leaf y*)) (*bdd-binop f r* (*Leaf y*))
bdd-binop f (*Leaf x*) (*Branch l r*) =
  Branch (*bdd-binop f* (*Leaf x*) *l*) (*bdd-binop f* (*Leaf x*) *r*)
bdd-binop f (*Branch* l_1 r_1) (*Branch* l_2 r_2) =
  Branch (*bdd-binop f* l_1 l_2) (*bdd-binop f* r_1 r_2)

If the two BDDs have different heights, the shorter one is expanded on the fly. The following theorem states that *bdd-binop* yields a BDD corresponding to the pointwise application of *f* to the functions represented by the argument BDDs:

If bddh |*bs*| *l and bddh* |*bs*| *r then*
bdd-lookup (*bdd-binop f l r*) *bs* = *f* (*bdd-lookup l bs*) (*bdd-lookup r bs*).

2.4   Deterministic Automata

We represent deterministic finite automata (DFAs) by pairs of type *nat bdd list* × *bool list*, where the first and second component denotes the transition function and the set of accepting states, respectively. The states of a DFA are simply natural numbers. Note that we do not mention the start state in the representation of the DFA, since it will always be 0. Not all pairs of the above type are well-formed DFAs. The two lists must have the same length, and all leaves of the BDDs in the list representing the transition function must be valid states, i.e. be smaller than the length of the two lists. Moreover, the heights of all BDDs must be less or equal to the number of variables *n*, and the set of states must be nonempty. These conditions are captured by the following definition:

dfa-is-node :: *dfa* $\Rightarrow$ *nat* $\Rightarrow$ *bool*
dfa-is-node A $\equiv$ λq. *q* < |*fst A*|

wf-dfa :: *dfa* $\Rightarrow$ *nat* $\Rightarrow$ *bool*
wf-dfa A n $\equiv$
  ($\forall$ *bdd*$\in$*fst A*. *bddh n bdd*) $\wedge$
  ($\forall$ *bdd*$\in$*fst A*. *bdd-all* (*dfa-is-node A*) *bdd*) $\wedge$ |*snd A*| = |*fst A*| $\wedge$ 0 < |*fst A*|

Moreover, the transition function and acceptance condition can be defined by

$dfa\text{-}trans :: dfa \Rightarrow nat \Rightarrow bool\ list \Rightarrow nat$
$dfa\text{-}trans\ A\ q\ bs \equiv bdd\text{-}lookup\ (fst\ A)_{[q]}\ bs$

$dfa\text{-}accepting :: dfa \Rightarrow nat \Rightarrow bool$
$dfa\text{-}accepting\ A\ q \equiv (snd\ A)_{[q]}$

where $xs_{[i]}$ denotes the ith element of list xs. Finally, using the generic functions from §2.2, we can produce variants of these functions tailored to DFAs:

$dfa\text{-}steps\ A\quad \equiv steps\ (dfa\text{-}trans\ A)$
$dfa\text{-}accepts\ A \equiv accepts\ (dfa\text{-}trans\ A)\ (dfa\text{-}accepting\ A)\ 0$
$dfa\text{-}reach\ A\quad \equiv reach\ (dfa\text{-}trans\ A)$

2.5   Nondeterministic Automata

Nondeterministic finite automata (NFAs) are represented by pairs of type *bool list bdd list* × *bool list*. While the second component representing the accepting states is the same as for DFAs, the transition table is now a list of BDDs mapping a state and an input symbol to a finite set of successor states, which we represent as a bit vector. In order for the transition table to be well-formed, the length of the bit vectors representing the sets must be equal to the number of states of the automaton, which coincides with the length of the transition table. This well-formedness condition for the bit vectors is expressed by the predicate

$nfa\text{-}is\text{-}node :: nfa \Rightarrow bool\ list \Rightarrow bool$
$nfa\text{-}is\text{-}node\ A \equiv \lambda qs.\ |qs| = |fst\ A|$

The definition of *wf-nfa* can be obtained from the one of *wf-dfa* by just replacing *dfa-is-node* by *nfa-is-node*. Due to its "asymmetric" type, a transition function of type *nat* ⇒ *bool list* ⇒ *bool list* would be incompatible with the abstract functions from §2.2. We therefore lift the function to work on finite sets of natural numbers rather that just single natural numbers. This is accomplished by

$subsetbdd :: bool\ list\ bdd\ list \Rightarrow bool\ list \Rightarrow bool\ list\ bdd \Rightarrow bool\ list\ bdd$
$subsetbdd\ []\ []\ bdd = bdd$
$subsetbdd\ (bdd' \cdot bdds)\ (b \cdot bs)\ bdd =$
$\quad (\text{if } b \text{ then } subsetbdd\ bdds\ bs\ (bdd\text{-}binop\ bv\text{-}or\ bdd\ bdd')$
$\quad \text{else } subsetbdd\ bdds\ bs\ bdd)$

where *bv-or* is the bit-wise or operation on bit vectors, i.e. the union of two finite sets. Using this operation, *subsetbdd* combines all BDDs in the first list, for which the corresponding bit in the second list is *True*. The third argument of *subsetbdd* serves as an accumulator and is initialized with a BDD consisting of only one Leaf containing the empty set, which is the neutral element of *bv-or*:

$nfa\text{-}emptybdd :: nat \Rightarrow bool\ list\ bdd$
$nfa\text{-}emptybdd\ n \equiv Leaf\ (replicate\ n\ False)$

Using *subsetbdd*, the transition function for NFAs can now be defined as follows:

nfa-trans :: *nfa* $\Rightarrow$ *bool list* $\Rightarrow$ *bool list* $\Rightarrow$ *bool list*
nfa-trans A qs bs $\equiv$ *bdd-lookup* (*subsetbdd* (*fst A*) *qs* (*nfa-emptybdd* |*qs*|)) *bs*

A set of states is accepting iff at least one of the states in the set is accepting:

nfa-accepting' :: *bool list* $\Rightarrow$ *bool list* $\Rightarrow$ *bool*
nfa-accepting' [] *bs* = *False*
nfa-accepting' (*a* · *as*) [] = *False*
nfa-accepting' (*a* · *as*) (*b* · *bs*) = (*a* $\wedge$ *b* $\vee$ *nfa-accepting' as bs*)

nfa-accepting :: *nfa* $\Rightarrow$ *bool list* $\Rightarrow$ *bool*
nfa-accepting A $\equiv$ *nfa-accepting'* (*snd A*)

As in the case of DFAs, we can now instantiate the generic functions from §2.2. In order to check whether we can reach an accepting state from the start state, we apply *accepts* to the finite set containing only the state 0.

nfa-startnode :: *nfa* $\Rightarrow$ *bool list*
nfa-startnode A $\equiv$ *replicate* |*fst A*| *False*[0 := *True*]

nfa-steps A $\equiv$ *steps* (*nfa-trans A*)
nfa-accepts A $\equiv$ *accepts* (*nfa-trans A*) (*nfa-accepting A*) (*nfa-startnode A*)
nfa-reach A $\equiv$ *reach* (*nfa-trans A*)

where $xs[i := y]$ denotes the replacement of the ith element of xs by y.

2.6 Depth First Search

The efficiency of the automata constructions presented in this paper crucially depends on the fact that the generated automata only contain *reachable* states. When implemented in a naive way, the construction of a product DFA from two DFAs having m and n states will lead to a DFA with $m \cdot n$ states, while the construction of a DFA from an NFA with n states will result in a DFA having 2^n states, many of which are unreachable. By using a depth-first search algorithm (DFS) for the generation of the automata, we can make sure that all of their states are reachable. In order to simplify the implementation of the automata constructions, as well as their correctness proofs, the DFS algorithm is factored out into a generic function, whose properties can be proved once and for all. The DFS algorithm is based on a representation of *graphs*, as well as a data structure for storing the nodes that have already been visited. Our version of DFS, which generalizes earlier work by Nishihara and Minamide [10,13], is designed as an abstract module using the *locale* mechanism of Isabelle, thus allowing the operations on the graph and the node store to be implemented in different ways depending on the application at hand. The module is parameterized by a type α of graph nodes, a type β representing the node store, and the functions

$$succs :: \alpha \Rightarrow \alpha \; list \qquad ins :: \alpha \Rightarrow \beta \Rightarrow \beta \qquad empt :: \beta$$
$$is\text{-}node :: \alpha \Rightarrow bool \qquad memb :: \alpha \Rightarrow \beta \Rightarrow bool \qquad invariant :: \beta \Rightarrow bool$$

where $succs$ returns the list of successors of a node, and the predicate $is\text{-}node$ describes the (finite) set of nodes. Moreover, $ins\ x\ S$, $memb\ x\ S$ and $empt$ correspond to $\{x\} \cup S$, $x \in S$ and $\emptyset$ on sets. The node store must also satisfy an additional $invariant$. Using Isabelle's infrastructure for the definition of functions by well-founded recursion [9], the DFS function can be defined as follows[3]:

$$dfs :: \beta \Rightarrow \alpha \; list \Rightarrow \beta$$
$$dfs\ S\ [] = S$$
$$dfs\ S\ (x \cdot xs) = (\textit{if } memb\ x\ S \textit{ then } dfs\ S\ xs \textit{ else } dfs\ (ins\ x\ S)\ (succs\ x\ @\ xs))$$

Note that this function is *partial*, since it may loop when instantiated with ins, $memb$ and $empt$ operators not behaving like their counterparts on sets, or when applied to a list of start values not being valid nodes. However, since dfs is *tail recursive*, Isabelle's function definition package can derive the above equations without preconditions, which is crucial for the executability of dfs. The central property of dfs is that it computes the transitive closure of the successor relation:

$$\textit{If } is\text{-}node\ y \textit{ and } is\text{-}node\ x \textit{ then}$$
$$memb\ y\ (dfs\ empt\ [x]) = ((x, y) \in (succsr\ succs)^*).$$

where $succsr$ turns a successor *function* into a *relation*:

$$succsr :: (\gamma \Rightarrow \delta \; list) \Rightarrow \gamma \times \delta \Rightarrow bool$$
$$succsr\ succs \equiv \{(x, y) \mid y \in succs\ x\}$$

3   Automata Construction

In this section, we will describe all automata constructions that are used to recursively build automata from formulae in Presburger arithmetic. The simplest one is the *complement*, which we describe in §3.1. It will be used to model *negation*. The *product automaton* construction described in §3.2 corresponds to binary operators such as $\vee$, $\wedge$, and $\longrightarrow$, whereas the more intricate *projection* construction shown in §3.3 is used to deal with existential quantifiers. Finally, §3.4 illustrates the construction of automata corresponding to atomic formulae.

3.1   Complement

The complement construction is straightforward. We only exchange the accepting and the non-accepting states, and leave the transition function unchanged:

$$negate\text{-}dfa :: dfa \Rightarrow dfa$$
$$negate\text{-}dfa \equiv \lambda(t, a).\ (t, map\ Not\ a)$$

---

[3] We use $dfs\ S\ xs$ as an abbreviation for $gen\text{-}dfs\ succs\ ins\ memb\ S\ xs$.

$prod\text{-}succs :: dfa \Rightarrow dfa \Rightarrow nat \times nat \Rightarrow (nat \times nat)\ list$
$prod\text{-}succs\ A\ B \equiv \lambda(i, j).\ add\text{-}leaves\ (bdd\text{-}binop\ Pair\ (fst\ A)_{[i]}\ (fst\ B)_{[j]})\ []$

$prod\text{-}ins :: nat \times nat$
$\qquad\qquad \Rightarrow nat\ option\ list\ list \times (nat \times nat)\ list$
$\qquad\qquad \Rightarrow nat\ option\ list\ list \times (nat \times nat)\ list$
$prod\text{-}ins \equiv$
$\lambda(i, j)\ (tab, ps).\ (tab[i := tab_{[i]}[j := Some\ |ps|]],\ ps\ @\ [(i, j)])$

$prod\text{-}memb :: nat \times nat \Rightarrow nat\ option\ list\ list \times (nat \times nat)\ list \Rightarrow bool$
$prod\text{-}memb \equiv \lambda(i, j)\ (tab, ps).\ tab_{[i][j]} \neq None$

$prod\text{-}empt :: dfa \Rightarrow dfa \Rightarrow nat\ option\ list\ list \times (nat \times nat)\ list$
$prod\text{-}empt\ A\ B \equiv (replicate\ |fst\ A|\ (replicate\ |fst\ B|\ None),\ [])$

$prod\text{-}dfs :: dfa \Rightarrow dfa \Rightarrow nat \times nat \Rightarrow nat\ option\ list\ list \times (nat \times nat)\ list$
$prod\text{-}dfs\ A\ B\ x \equiv$
$gen\text{-}dfs\ (prod\text{-}succs\ A\ B)\ prod\text{-}ins\ prod\text{-}memb\ (prod\text{-}empt\ A\ B)\ [x]$

$binop\text{-}dfa :: (bool \Rightarrow bool \Rightarrow bool) \Rightarrow dfa \Rightarrow dfa \Rightarrow dfa$
$binop\text{-}dfa\ f\ A\ B \equiv$
$\mathbf{let}\ (tab, ps) = prod\text{-}dfs\ A\ B\ (0, 0)$
$\mathbf{in}\ (map\ (\lambda(i, j).\ bdd\text{-}binop\ (\lambda k\ l.\ the\ tab_{[k][l]})\ (fst\ A)_{[i]}\ (fst\ B)_{[j]})\ ps,$
$\qquad map\ (\lambda(i, j).\ f\ (snd\ A)_{[i]}\ (snd\ B)_{[j]})\ ps)$

Fig. 1. Definition of the product automaton

A well-formed DFA A will accept a word bss iff it is *not* accepted by the DFA produced by *negate-dfa*:

If wf-dfa A n and $\forall bs \in bss.$ is-alph n bs then
dfa-accepts (negate-dfa A) bss = ($\neg$ dfa-accepts A bss).

3.2 Product Automaton

Given a binary logical operator $f :: bool \Rightarrow bool \Rightarrow bool$, the product automaton construction is used to build a DFA corresponding to the formula $f\ P\ Q$ from DFAs A and B corresponding to the formulae P and Q, respectively. As suggested by its name, the state space of the product automaton corresponds to the cartesian product of the state spaces of the DFAs A and B. However, as already mentioned in §2.6, not all of the elements of the cartesian product constitute reachable states. We therefore need an algorithm for computing the reachable states of the resulting DFA. Moreover, since the automata framework described in §2.4–2.5 relies on the states to be encoded as natural numbers, we also need to produce a mapping from $nat \times nat$ to nat. All of this can be achieved just by instantiating the abstract DFS framework with suitable functions, as shown in Fig. 1. In this construction, the store containing the visited states is a pair *nat option list list* $\times$ (*nat* $\times$ *nat*) *list*, where the first component is a matrix denoting a partial map from *nat* × *nat* to *nat*. The second component of the store is a list containing all visited states (i, j). It can be viewed as a map from *nat* to *nat* × *nat*, which is the inverse of the aforementioned map. In order to compute the list of successor states of a state (i, j), *prod-succs* combines the BDDs representing the transition tables of state i of A, and of state j of B using the *Pair* operator, and then collects all leaves of the resulting BDD. The operation *prod-ins* for inserting a state into the store updates the entry at position (i, j) of the matrix *tab* with the number of visited states, and appends (i, j) to the list *ps* of visited states. By definition of DFS, this operation is guaranteed to be applied only if the state (i, j) has not been visited yet, i.e. the corresponding entry in the matrix is *None* and (i, j) is not contained in the list *ps*. We now produce a specific version of DFS called *prod-dfs* by instantiating the generic function from §2.6, and using the list containing just one pair of states as a start value. By induction on *gen-dfs*, we can prove that the matrix and the list computed by *prod-dfs* encodes a bijection between the reachable states (i, j) of the product automaton, and natural numbers k corresponding to the states of the resulting DFA, where k is smaller than the number of reachable states:

If prod-is-node A B x then
$((fst\ (prod\text{-}dfs\ A\ B\ x))_{[i][j]} = Some\ k \wedge dfa\text{-}is\text{-}node\ A\ i \wedge dfa\text{-}is\text{-}node\ B\ j) =$
$(k < |snd\ (prod\text{-}dfs\ A\ B\ x)| \wedge (snd\ (prod\text{-}dfs\ A\ B\ x))_{[k]} = (i, j)).$

The start state x must satisfy a well-formedness condition *prod-is-node*, meaning that its two components must be valid states of A and B. Using this result, as well as the fact that *prod-dfs* computes the transitive closure, we can then show by induction on *bss* that a state m is reachable in the resulting automaton via a sequence of bit vectors *bss* iff the corresponding states s_1 and s_2 are reachable via *bss* in the automata A and B, respectively:

If $\forall bs \in bss.\ is\text{-}alph\ n\ bs\ then$
$(\exists m.\ dfa\text{-}reach\ (binop\text{-}dfa\ f\ A\ B)\ 0\ bss\ m \wedge$
    $(fst\ (prod\text{-}dfs\ A\ B\ (0, 0)))_{[s_1][s_2]} = Some\ m \wedge$
    $dfa\text{-}is\text{-}node\ A\ s_1 \wedge dfa\text{-}is\text{-}node\ B\ s_2) =$
$(dfa\text{-}reach\ A\ 0\ bss\ s_1 \wedge dfa\text{-}reach\ B\ 0\ bss\ s_2).$

Finally, *bdd-binop* produces the resulting product automaton by combining the transition tables of A and B using the mapping from *nat* × *nat* to *nat* computed by *prod-dfs*, and by applying f to the acceptance conditions of A and B. Using the previous theorem, we can prove the correctness statement for this construction:

If wf-dfa A n and wf-dfa B n and $\forall bs \in bss.\ is\text{-}alph\ n\ bs\ then$
$dfa\text{-}accepts\ (binop\text{-}dfa\ f\ A\ B)\ bss = f\ (dfa\text{-}accepts\ A\ bss)\ (dfa\text{-}accepts\ B\ bss).$

3.3   Projection

Using the terminology from §2.3, the automaton for $\exists x.\ P$ can be obtained from the one for P by projecting away the row corresponding to the variable x. Since this operation yields an NFA, it is advantageous to first translate the DFA for P into an NFA, which can easily be done by replacing all the leaves in the transition table by singleton sets, and leaving the set of accepting states unchanged. The correctness of this operation called *nfa-of-dfa* is expressed by

If wf-dfa A n and ∀ bs∈bss. is-alph n bs then
nfa-accepts (nfa-of-dfa A) bss = dfa-accepts A bss.

Given a BDD representing the transition table of a particular state of an NFA, we can project away the ith variable by combining the two children BDDs of the branches at depth i using the *bv-or* operation:

quantify-bdd :: nat ⇒ bool list bdd ⇒ bool list bdd
quantify-bdd i (Leaf q) = Leaf q
quantify-bdd 0 (Branch l r) = bdd-binop bv-or l r
quantify-bdd (Suc i) (Branch l r) =
  Branch (quantify-bdd i l) (quantify-bdd i r)

To produce the NFA corresponding to the quantified formula, we just map this operation over the transition table:

quantify-nfa :: nat ⇒ nfa ⇒ nfa
quantify-nfa i ≡ λ(bdds, as). (map (quantify-bdd i) bdds, as)

Due to its type, we could apply this function repeatedly to quantify over several variables in one go. The correctness of this construction is summarized by

If wf-nfa A (Suc n) and i ≤ n and ∀ bs∈bss. is-alph n bs then
nfa-accepts (quantify-nfa i A) bss =
(∃ bs. nfa-accepts A (insertll i bs bss) ∧ |bs| = |bss|).

This means that the new NFA accepts a list *bss* of column vectors iff the original NFA accepts the list obtained from *bss* by inserting a suitable row vector *bs* representing the existential witness. Matters are complicated by the additional requirement that the word accepted by the new NFA must have the same length as the witness. This requirement can be satisfied by appending zero vectors to the end of *bss*, which does not change its interpretation. Since the other constructions (in particular the complement) only work on DFAs, we turn the obtained NFA into a DFA by applying the usual *subset construction*. The central idea is that each set of states produced by *nfa-steps* can be viewed as a state of a new DFA. As mentioned in §2.6, not all of these sets are reachable from the initial state of the NFA. Similar to the *product* construction, the algorithm for computing the reachable sets shown in Fig. 2 is an instance of the general DFS framework. The node store is now a pair of type *nat option bdd × bool list list*, where the first component is a BDD representing a partial map from finite sets (encoded as bit vectors) to natural numbers, and the second component is the list of visited states representing the inverse map. To insert new entries into a BDD, we use

$bddinsert :: \alpha \ bdd \Rightarrow bool \ list \Rightarrow \alpha \Rightarrow \alpha \ bdd$
$bddinsert \ (Leaf \ a) \ [] \ x = Leaf \ x$
$bddinsert \ (Leaf \ a) \ (w \cdot ws) \ x =$
  $(if \ w \ then \ Branch \ (Leaf \ a) \ (bddinsert \ (Leaf \ a) \ ws \ x)$
   $else \ Branch \ (bddinsert \ (Leaf \ a) \ ws \ x) \ (Leaf \ a))$
$bddinsert \ (Branch \ l \ r) \ (w \cdot ws) \ x =$
  $(if \ w \ then \ Branch \ l \ (bddinsert \ r \ ws \ x) \ else \ Branch \ (bddinsert \ l \ ws \ x) \ r)$

The computation of successor states in *subset-succs* and the transition relation in *det-nfa* closely resembles the definition of *nfa-trans* from §2.5. Using the fact that *subset-dfs* computes a bijection between finite sets and natural numbers, we can prove the correctness theorem for *det-nfa*:

If wf-nfa A n and $\forall bs \in bss.$ *is-alph n bs then*
dfa-accepts (det-nfa A) bss = nfa-accepts A bss.

Recall that the automaton produced by *quantify-nfa* will only accept words with a sufficient number of trailing zero column vectors. To get a DFA that also accepts words without trailing zeros, we mark all states as accepting from which an accepting state can be reached by reading only zeros. This construction, which is sometimes referred to as the *right quotient*, can be characterized as follows:

If wf-dfa A n and $\forall bs \in bss.$ *is-alph n bs then*
dfa-accepts (rquot A n) bss = $(\exists m.$ *dfa-accepts A (bss @ zeros m n)).*

where *zeros m n* produces a word consisting of m zero vectors of size n.

3.4   Diophantine (In)Equations

We now come to the construction of DFAs for atomic formulae, namely Diophantine (in)equations. For this purpose, we use a method due to Boudet and Comon [3]. The key observation is that xs is a solution of a Diophantine equation iff it is a solution modulo 2 and the quotient of xs and 2 is a solution of another equation with the same coefficients, but with a different right-hand side:

$(eval\text{-}dioph \ ks \ xs = l) =$
$(eval\text{-}dioph \ ks \ (map \ (\lambda x. \ x \ mod \ 2) \ xs) \ mod \ 2 = l \ mod \ 2 \ \wedge$
 $eval\text{-}dioph \ ks \ (map \ (\lambda x. \ x \ div \ 2) \ xs) =$
 $(l - eval\text{-}dioph \ ks \ (map \ (\lambda x. \ x \ mod \ 2) \ xs)) \ div \ 2)$

In other words, the states of the DFA accepting the solutions of the equation correspond to the right-hand sides reachable from the initial right-hand side l, which will again be computed using the DFS algorithm. To ensure termination of DFS, it is crucial to prove that the reachable right-hand sides m are bounded:

If $|m| \leq max \ |l| \ (\sum k \leftarrow ks. \ |k|)$ *then*
$|(m - eval\text{-}dioph \ ks \ (map \ (\lambda x. \ x \ mod \ 2) \ xs)) \ div \ 2| \leq max \ |l| \ (\sum k \leftarrow ks. \ |k|).$

---

subset-succs :: *nfa* ⇒ *bool list* ⇒ *bool list list*
subset-succs A qs ≡ *add-leaves* (*subsetbdd* (*fst A*) *qs* (*nfa-emptybdd* |*qs*|)) []

subset-ins :: *bool list*
              ⇒ *nat option bdd* × *bool list list*
                 ⇒ *nat option bdd* × *bool list list*
subset-ins qs ≡ λ(*bdd, qss*). (*bddinsert bdd qs* (*Some* |*qss*|), *qss* @ [*qs*])

subset-memb :: *bool list* ⇒ *nat option bdd* × *bool list list* ⇒ *bool*
subset-memb qs ≡ λ(*bdd, qss*). *bdd-lookup bdd qs* ≠ *None*

subset-empt :: *nat option bdd* × *bool list list*
subset-empt ≡ (*Leaf None*, [])

subset-dfs :: *nfa* ⇒ *bool list* ⇒ *nat option bdd* × *bool list list*
subset-dfs A x ≡
gen-dfs (*subset-succs A*) *subset-ins subset-memb subset-empt* [*x*]

det-nfa :: *nfa* ⇒ *dfa*
det-nfa A ≡
let (*bdd, qss*) = *subset-dfs A* (*nfa-startnode A*)
in (*map* (λ*qs*. *bdd-map* (λ*qs*. *the* (*bdd-lookup bdd qs*))
                (*subsetbdd* (*fst A*) *qs* (*nfa-emptybdd* |*qs*|)))
    qss,
    map (*nfa-accepting A*) *qss*)

Fig. 2. Definition of the subset construction

---

eq-dfa :: *nat* ⇒ *int list* ⇒ *int* ⇒ *dfa*
eq-dfa n ks l ≡
let (*is, js*) = *dioph-dfs n ks l*
in (*map* (λ*j*. *make-bdd*
         (λ*xs*. **if** *eval-dioph ks xs* mod 2 = *j* mod 2
             then *the is*$_{[int\text{-}to\text{-}nat\text{-}bij}\, ((j\, -\, eval\text{-}dioph\; ks\; xs)\; div\; 2)]}$ **else** |*js*|)
         n [])
    js @
    [*Leaf* |*js*|],
    map (λ*j*. *j* = 0) *js* @ [*False*])

Fig. 3. Definition of the automata for Diophantine equations

By instantiating the abstract *gen-dfs* function, we obtain a function

dioph-dfs :: *nat* ⇒ *int list* ⇒ *int* ⇒ *nat option list* × *int list*

that, given the number of variables, the coefficients, and the right-hand side, computes a bijection between reachable right-hand sides and natural numbers:

$((fst\ (dioph\text{-}dfs\ n\ ks\ l))_{[int\text{-}to\text{-}nat\text{-}bij\ m]} = Some\ k\ \wedge$
$|m| \leq max\ |l|\ (\sum k{\leftarrow}ks.\ |k|)) =$
$(k < |snd\ (dioph\text{-}dfs\ n\ ks\ l)|\ \wedge\ (snd\ (dioph\text{-}dfs\ n\ ks\ l))_{[k]} = m)$

The first component of the pair returned by *dioph-dfs* can be viewed as a partial map from integers to natural numbers, where *int-to-nat-bij* maps negative and non-negative integers to odd and even list indices, respectively. As shown in Fig. 3, the transition table of the DFA is constructed by *eq-dfa* as follows: if the current state corresponds to the right-hand side j, and the DFA reads a bit vector xs satisfying the equation modulo 2, then the DFA goes to the state corresponding to the new right-hand side $(j - eval\text{-}dioph\ ks\ xs)\ div\ 2$, otherwise it goes to an error state, which is the last state in the table. To produce a BDD containing the successor states for all bit vectors of length n, we use the function

$make\text{-}bdd :: (nat\ list \Rightarrow \alpha) \Rightarrow nat \Rightarrow nat\ list \Rightarrow \alpha\ bdd$
$make\text{-}bdd\ f\ 0\ xs = Leaf\ (f\ xs)$
$make\text{-}bdd\ f\ (Suc\ n)\ xs =$
  $Branch\ (make\text{-}bdd\ f\ n\ (xs\ @\ [0]))\ (make\text{-}bdd\ f\ n\ (xs\ @\ [1]))$

The key property of *eq-dfa* states that for every right-hand side m reachable from l, the state reachable from m via a word *bss* is accepting iff the list of natural numbers denoted by *bss* satisfies the equation with right-hand side m:

If $(l,\ m) \in (succsr\ (dioph\text{-}succs\ n\ ks))^*$ and $\forall bs{\in}bss.\ is\text{-}alph\ n\ bs$ then
$dfa\text{-}accepting\ (eq\text{-}dfa\ n\ ks\ l)$
  $(dfa\text{-}steps\ (eq\text{-}dfa\ n\ ks\ l)\ (the\ (fst\ (dioph\text{-}dfs\ n\ ks\ l))_{[int\text{-}to\text{-}nat\text{-}bij\ m]})\ bss) =$
  $(eval\text{-}dioph\ ks\ (nats\text{-}of\text{-}boolss\ n\ bss) = m).$

Here, *dioph-succs* n ks returns a list of up to 2^n successor states reachable from a given state by reading a single column vector of size n. The proof of the above property is by induction on *bss*, where the equation given at the beginning of this section is used in the induction step. The correctness property of *eq-dfa* can then be obtained from this result as a simple corollary:

If $\forall bs{\in}bss.\ is\text{-}alph\ n\ bs$ then
$dfa\text{-}accepts\ (eq\text{-}dfa\ n\ ks\ l)\ bss = (eval\text{-}dioph\ ks\ (nats\text{-}of\text{-}boolss\ n\ bss) = l).$

Diophantine inequations can be treated in a similar way.

4   The Decision Procedure

We now have all the machinery in place to write a decision procedure for Presburger arithmetic. A formula can be transformed into a DFA by the following function:

dfa-of-pf :: *nat* ⇒ *pf* ⇒ *dfa*
dfa-of-pf n (*Eq ks l*) = *eq-dfa n ks l*
dfa-of-pf n (*Le ks l*) = *ineq-dfa n ks l*
dfa-of-pf n (*Neg p*) = *negate-dfa* (*dfa-of-pf n p*)
dfa-of-pf n (*And p q*) = *binop-dfa* (∧) (*dfa-of-pf n p*) (*dfa-of-pf n q*)
dfa-of-pf n (*Or p q*) = *binop-dfa* (∨) (*dfa-of-pf n p*) (*dfa-of-pf n q*)
dfa-of-pf n (*Imp p q*) = *binop-dfa* (⟶) (*dfa-of-pf n p*) (*dfa-of-pf n q*)
dfa-of-pf n (*Exist p*) =
  rquot (*det-nfa* (*quantify-nfa* 0 (*nfa-of-dfa* (*dfa-of-pf* (*Suc n*) *p*)))) *n*
dfa-of-pf n (*Forall p*) = *dfa-of-pf n* (*Neg* (*Exist* (*Neg p*)))

By structural induction on formulae, we can show the correctness theorem

If ∀ *bs*∈*bss*. *is-alph n bs then*
dfa-accepts (*dfa-of-pf n p*) *bss* = *eval-pf p* (*nats-of-boolss n bss*).

Note that a closed formula is valid iff the start state of the resulting DFA is accepting, which can easily be seen by letting $n = 0$ and $bss = []$. Most cases of the induction can be proved by a straightforward application of the correctness results from §3. Unsurprisingly, the only complicated case is the one for the existential quantifier, which we will now examine in more detail. In this case, the left-hand side of the correctness theorem is

dfa-accepts
  (*rquot* (*det-nfa* (*quantify-nfa* 0 (*nfa-of-dfa* (*dfa-of-pf* (*Suc n*) *p*)))) *n*) *bss*

which, according to the correctness statement for *rquot*, is equivalent to

∃ *m*. *dfa-accepts* (*det-nfa* (*quantify-nfa* 0 (*nfa-of-dfa* (*dfa-of-pf* (*Suc n*) *p*))))
(*bss* @ *zeros m n*)

By correctness of *det-nfa*, *quantify-nfa* and *nfa-of-dfa*, this is the same as

∃ *m bs*. *dfa-accepts* (*dfa-of-pf* (*Suc n*) *p*) (*insertll* 0 *bs* (*bss* @ *zeros m n*)) ∧
    |*bs*| = |*bss*| + |*zeros m n*|

Using the induction hypothesis, this can be rewritten to

∃ *m bs*. *eval-pf p* (*nats-of-boolss* (*Suc n*) (*insertll* 0 *bs* (*bss* @ *zeros m n*))) ∧
    |*bs*| = |*bss*| + |*zeros m n*|

According to the properties of *nats-of-boolss* from §2.3, this can be recast as

∃ *m bs*. *eval-pf p* (*nat-of-bools bs* · *nats-of-boolss n bss*) ∧ |*bs*| = |*bss*| + *m*

which is obviously equivalent to the right-hand side of the correctness theorem

∃ *x*. *eval-pf p* (*x* · *nats-of-boolss n bss*)

since we can easily produce suitable instantiations for *m* and *bs* from *x*.

5   Conclusion

First experiments with the algorithm presented in §4 show that it can compete quite well with the standard decision procedure for Presburger arithmetic available in Isabelle. Even without minimization, the DFA for the stamp problem from §2.1 has only 6 states, and can be constructed in less than a second. The following table shows the size of the DFAs (i.e. the number of states) for all subformulae of the stamp problem. Thanks to the DFS algorithm, they are much smaller than the DFAs that one would have obtained using a naive construction:

		$Exist$	$Exist$	Eq [5, 3, −1] 0
$Forall$	Imp	13	9	9
6	15		Le [−1] −8	
			5	

The next step is to formalize a minimization algorithm, e.g. along the lines of Constable et al. [6]. We also intend to explore other ways of constructing DFAs for Diophantine equations, such as the approach by Wolper and Boigelot [15], which is more complicated than the one shown in §3.4, but can directly deal with variables over the integers rather than just natural numbers. To improve the performance of the decision procedure on large formulae, we would also like to investigate possible optimizations of the simple representation of BDDs presented in §2.3. Verma [14] describes a formalization of reduced ordered BDDs with *sharing* in Coq. To model sharing, Verma's formalization is based on a *memory* for storing BDDs. Due to their dependence on the *memory*, algorithms using this kind of BDDs are no longer *purely functional*, which makes reasoning about them substantially more challenging. Finally, we also plan to extend our decision procedure to cover WS1S, and use it to tackle some of the circuit verification problems described by Basin and Friedrich [1].

Acknowledgements. We would like to thank Tobias Nipkow for suggesting this project and for numerous comments, Clemens Ballarin and Markus Wenzel for answering questions concerning locales, and Alex Krauss for help with well-founded recursion and induction schemes.

References

1. Basin, D., Friedrich, S.: Combining WS1S and HOL. In: Gabbay, D., de Rijke, M. (eds.) Frontiers of Combining Systems 2. Studies in Logic and Computation, vol. 7, pp. 39–56. Research Studies Press/Wiley (2000)
2. Berghofer, S., Nipkow, T.: Executing higher order logic. In: Callaghan, P., Luo, Z., McKinna, J., Pollack, R. (eds.) TYPES 2000. LNCS, vol. 2277, p. 24. Springer, Heidelberg (2002)
3. Boudet, A., Comon, H.: Diophantine equations, Presburger arithmetic and finite automata. In: Kirchner, H. (ed.) CAAP 1996. LNCS, vol. 1059, pp. 30–43. Springer, Heidelberg (1996)

4. Boutin, S.: Using reflection to build efficient and certified decision procedures. In: Ito, T., Abadi, M. (eds.) TACS 1997. LNCS, vol. 1281, pp. 515–529. Springer, Heidelberg (1997)

5. Chaieb, A., Nipkow, T.: Proof synthesis and reflection for linear arithmetic. Journal of Automated Reasoning 41, 33–59 (2008)

6. Constable, R.L., Jackson, P.B., Naumov, P., Uribe, J.: Constructively formalizing automata theory. In: Plotkin, G., Stirling, C., Tofte, M. (eds.) Proof, Language, and Interaction: Essays in Honor of Robin Milner. MIT Press, Cambridge (2000)

7. Harrison, J.: Metatheory and reflection in theorem proving: A survey and critique. Technical Report CRC-053, SRI Cambridge (1995),
   http://www.cl.cam.ac.uk/users/jrh/papers/reflect.dvi.gz

8. Klarlund, N.: Mona & Fido: The logic-automaton connection in practice. In: Nielsen, M. (ed.) CSL 1997. LNCS, vol. 1414, pp. 311–326. Springer, Heidelberg (1998)

9. Krauss, A.: Partial recursive functions in higher-order logic. In: Furbach, U., Shankar, N. (eds.) IJCAR 2006. LNCS, vol. 4130, pp. 589–603. Springer, Heidelberg (2006)

10. Minamide, Y.: Verified decision procedures on context-free grammars. In: Schneider, K., Brandt, J. (eds.) TPHOLs 2007. LNCS, vol. 4732, pp. 173–188. Springer, Heidelberg (2007)

11. Nipkow, T.: Verified lexical analysis. In: Grundy, J., Newey, M. (eds.) TPHOLs 1998. LNCS, vol. 1479, pp. 1–15. Springer, Heidelberg (1998)

12. Nipkow, T.: Linear quantifier elimination. In: Armando, A., Baumgartner, P., Dowek, G. (eds.) IJCAR 2008. LNCS, vol. 5195, pp. 18–33. Springer, Heidelberg (2008)

13. Nishihara, T., Minamide, Y.: Depth first search. In: Klein, G., Nipkow, T., Paulson, L. (eds.) The Archive of Formal Proofs,
    http://afp.sf.net/entries/Depth-First-Search.shtml (June 2004); Formal proof development

14. Verma, K.N., Goubault-Larrecq, J., Prasad, S., Arun-Kumar, S.: Reflecting BDDs in Coq. In: He, J., Sato, M. (eds.) ASIAN 2000. LNCS, vol. 1961, pp. 162–181. Springer, Heidelberg (2000)

15. Wolper, P., Boigelot, B.: On the construction of automata from linear arithmetic constraints. In: Schwartzbach, M.I., Graf, S. (eds.) TACAS 2000. LNCS, vol. 1785, pp. 1–19. Springer, Heidelberg (2000)

Extended First-Order Logic

Chad E. Brown and Gert Smolka

Saarland University, Saarbrücken, Germany

Abstract. We consider the EFO fragment of simple type theory, which restricts quantification and equality to base types but retains lambda abstractions and higher-order variables. We show that this fragment enjoys the characteristic properties of first-order logic: complete proof systems, compactness, and countable models. We obtain these results with an analytic tableau system and a concomitant model existence lemma. All results are with respect to standard models. The tableau system is well-suited for proof search and yields decision procedures for substantial fragments of EFO.

1 Introduction

First-order logic can be considered as a natural fragment of Church's type theory [1]. In this paper we exhibit a larger fragment of type theory, called EFO, that still enjoys the characteristic properties of first-order logic: complete proof systems, compactness, and countable models. EFO restricts quantification and equality to base types but retains lambda abstractions and higher-order variables. Like type theory, EFO has a type o of truth values and admits functions that take truth values to individuals. Such functions are not available in first-order logic. A typical example is a conditional $C : o\iota\iota\iota$ taking a truth value and two individuals as arguments and returning one of the individuals. Here is a valid EFO formula that specifies the conditional and states one of its properties:

$$(\forall xy.\ C\bot xy = y \ \wedge\ C\top xy = x) \ \rightarrow\ C(x{=}y)xy = y$$

The starting point for EFO is an analytic tableau system derived from Brown's Henkin-complete cut-free one-sided sequent calculus for extensional type theory [2]. The tableau system is well-suited for proof search and yields decision procedures and the finite model property for three substantial fragments of EFO: lambda-free formulas (e.g., $pa \rightarrow pb \rightarrow p(a{\wedge}b)$), Bernays-Schönfinkel-Ramsey formulas [3], and equations between pure lambda terms (terms not involving type o). The decidability and finite model results are mostly known, but it is remarkable that we obtain them with a single tableau system.

The proofs of the main results follow the usual development of first-order logic [4,5], which applies the abstract consistency technique to a model existence lemma for the tableau system (Hintikka's Lemma). Due to the presence of higher-order variables and lambda abstractions, the proof of the EFO model existence lemma is much harder than it is for first-order logic. We employ the possible-values technique [6], which has been used in [2] to obtain Henkin models, and

S. Berghofer et al. (Eds.): TPHOLs 2009, LNCS 5674, pp. 164–179, 2009.

in [7] to obtain standard models. We generalize the model existence theorem such that we can obtain countable models using the abstract consistency technique.

In a preceding paper [7], we develop a tableau-based decision procedure for the quantifier- and lambda-free fragment of EFO and introduce the possible-values-based construction of standard models. In this paper we extend the model construction to first-order quantification and lambda abstraction. We introduce a novel subterm restriction for the universal quantifier and employ an abstract normalization operator, both essential for proof search and decision procedures.

Due to space limitations we have to omit some proofs. They can be found in the full paper at www.ps.uni-sb.de/Papers.

2   Basic Definitions

Types (σ, τ, μ) are obtained with the grammar $\tau ::= o \mid \iota \mid \tau\tau$. The elements of o are the two truth values, ι is interpreted as a nonempty set, and a function type $\sigma\tau$ is interpreted as the set of all total functions from σ to τ. For simplicity, we provide only one sort ι. Everything generalizes to countably many sorts.

We distinguish between two kinds of *names*, called *constants* and *variables*. Every name comes with a type. We assume that there are only countably many names, and that for every type there are infinitely many variables of this type. If not said otherwise, the letter a ranges over names, c over constants, and x and y over variables.

Terms (s, t, u, v) are obtained with the grammar $t ::= a \mid tt \mid \lambda x.t$ where an application st is only admitted if $s : \tau\mu$ and $t : \tau$ for some types τ and μ. Terms of type o are called *formulas*. A term is *lambda-free* if it does not contain a subterm that is a lambda abstraction. We use $\mathcal{N}s$ to denote the set of all names that have a free occurrence in the term s.

We assume that $\bot : o$, $\neg : oo$, $\wedge : ooo$, $=_\sigma : \sigma\sigma o$, and $\forall_\sigma : (\sigma o)o$ are constants for all types σ. We write $\forall x.s$ for $\forall_\sigma(\lambda x.s)$. An *interpretation* is a function $\mathcal{I}$ that is defined on all types and all names and satisfies the following conditions:

- $\mathcal{I}o = \{0, 1\}$
- $\mathcal{I}(\sigma\tau)$ is the set of all total functions from $\mathcal{I}\sigma$ to $\mathcal{I}\tau$
- $\mathcal{I}\bot = 0$
- $\mathcal{I}(\neg)$, $\mathcal{I}(\wedge)$, $\mathcal{I}(=_\sigma)$, and $\mathcal{I}(\forall_\sigma)$ are the standard interpretations of the respective logical constants.

We write $\hat{\mathcal{I}}s$ for the value the term s evaluates to under the interpretation $\mathcal{I}$. We say that an interpretation $\mathcal{I}$ is *countable* [*finite*] if $\mathcal{I}\iota$ is countable [finite]. An interpretation $\mathcal{I}$ is a *model* of a set A of formulas if $\hat{\mathcal{I}}s = 1$ for every formula $s \in A$. A set of formulas is *satisfiable* if it has a model.

The constants $\bot$, $\neg$, $\wedge$, $=_\iota$, and $\forall_\iota$ are called *EFO constants*. An *EFO term* is a term that contains no other constants but EFO constants. We write EFO_σ for the set of all EFO terms of type σ. For simplicity, we work with a restricted set of EFO constants. Everything generalizes to the remaining propositional constants, the identity $=_o$, and the existential quantifier $\exists_\iota$.

3   Normalization

We assume a *normalization operator* [] that provides for lambda conversion. The normalization operator [] must be a type preserving total function from terms to terms. We call $[s]$ the *normal form of s* and say that s is *normal* if $[s] = s$.

There are several possibilities for the normalization operator []: β-, long β-, or $\beta\eta$-normal form, all possibly with standardized bound variables [8]. We will not commit to a particular operator but state explicitly the properties we require for our results. To start, we require the following properties:

N1 $[[s]] = [s]$
N2 $[[s]t] = [st]$
N3 $[as_1 \ldots s_n] = a[s_1] \ldots [s_n]$ if the type of $as_1 \ldots s_n$ is o or ι
N4 $\hat{\mathcal{I}}[s] = \hat{\mathcal{I}}s$

Note that a ranges over names and $\mathcal{I}$ ranges over interpretations. N3 also applies for $n = 0$.

We need further properties of the normalization operator that can only be expressed with substitutions. A *substitution* is a type preserving partial function from variables to terms. If θ is a substitution, x is a variable, and s is a term that has the same type as x, we use θ_s^x to denote the substitution that agrees everywhere with θ but possibly on x where it yields s. We assume that every substitution θ can be extended to a type preserving total function $\hat{\theta}$ from terms to terms such that the following conditions hold:

S1 $\hat{\theta}a = $ if $a \in \mathrm{Dom}\,\theta$ then θa else a
S2 $\hat{\theta}(st) = (\hat{\theta}s)(\hat{\theta}t)$
S3 $[(\hat{\theta}(\lambda x.s))t] = [\widehat{\theta_t^x}s]$
S4 $[\hat{\emptyset}s] = [s]$
S5 $\mathcal{N}[s] \subseteq \mathcal{N}s$ and $\mathcal{N}(\hat{\theta}s) \subseteq \bigcup\{\mathcal{N}(\hat{\theta}a) \mid a \in \mathcal{N}s\}$

Note that a ranges over names and that $\emptyset$ (the empty set) is the substitution that is undefined on every variable.

4   Tableau System

The results of this paper originate with the tableau system $\mathcal{T}$ shown in Figure 1. The rules in the first two lines of Figure 1 are the familiar rules from first-order logic. The rules in the third and fourth line deal with embedded formulas. The *mating rule* $\mathcal{T}_{\mathrm{MAT}}$ decomposes complementary atomic formulas by introducing disequations that confront corresponding subterms. Disequations can be further decomposed with $\mathcal{T}_{\mathrm{DEC}}$. Embedded formulas are eventually raised to the top level by Rule $\mathcal{T}_{\mathrm{BE}}$, which incorporates Boolean extensionality. Rule $\mathcal{T}_{\mathrm{FE}}$ incorporates functional extensionality. It reduces disequations at functional types to disequations at lower types. The confrontation rule $\mathcal{T}_{\mathrm{CON}}$ deals with positive equations at type ι. A discussion of the confrontation rule can be found in [7]. The tableau rules are such that they add normal formulas if they are applied to normal formulas.

$$\mathcal{T}_\neg \ \frac{s\,,\,\neg s}{\bot} \qquad \mathcal{T}_{\neg\neg} \ \frac{\neg\neg s}{s} \qquad \mathcal{T}_\wedge \ \frac{s \wedge t}{s\,,\,t} \qquad \mathcal{T}_{\neg\wedge} \ \frac{\neg(s \wedge t)}{\neg s \mid \neg t}$$

$$\mathcal{T}_\forall \ \frac{\forall_\iota s}{[st]} \ t : \iota \qquad\qquad \mathcal{T}_{\neg\forall} \ \frac{\neg\forall_\iota s}{\neg[sx]} \ x : \iota \ \text{fresh}$$

$$\mathcal{T}_{\text{MAT}} \ \frac{xs_1 \dots s_n\,,\,\neg xt_1 \dots t_n}{s_1 \neq t_1 \mid \dots \mid s_n \neq t_n} \qquad \mathcal{T}_{\text{DEC}} \ \frac{xs_1 \dots s_n \neq_\iota xt_1 \dots t_n}{s_1 \neq t_1 \mid \dots \mid s_n \neq t_n}$$

$$\mathcal{T}_{\neq} \ \frac{s \neq s}{\bot} \qquad \mathcal{T}_{\text{BE}} \ \frac{s \neq_o t}{s\,,\,\neg t \mid \neg s\,,\,t} \qquad \mathcal{T}_{\text{FE}} \ \frac{s \neq_{\sigma\tau} t}{[sx] \neq [tx]} \ x : \sigma \ \text{fresh}$$

$$\mathcal{T}_{\text{CON}} \ \frac{s =_\iota t\,,\,u \neq_\iota v}{s \neq u, t \neq u \mid s \neq v, t \neq v}$$

Fig. 1. Tableau system $\mathcal{T}$

Example 4.1. The following tableau refutes the formula $pf \wedge \neg p(\lambda x.\neg\neg fx)$ where $p : (\iota o)o$ and $f : \iota o$.

$$pf \wedge \neg p(\lambda x.\neg\neg fx)$$
$$pf,\ \neg p(\lambda x.\neg\neg fx)$$
$$f \neq (\lambda x.\neg\neg fx)$$
$$fx \neq \neg\neg fx$$

$fx,\ \neg\neg\neg fx$	$\neg fx,\ \neg\neg fx$
$\neg fx$	$\bot$
$\bot$	

The rules used are $\mathcal{T}_\wedge$, $\mathcal{T}_{\text{MAT}}$, $\mathcal{T}_{\text{FE}}$, $\mathcal{T}_{\text{BE}}$, $\mathcal{T}_{\neg\neg}$, and $\mathcal{T}_\neg$.   □

5   Evidence

A *quasi-EFO formula* is a disequation $s \neq_\sigma t$ such that s and t are EFO terms and $\sigma \neq \iota$. Note that the rules $\mathcal{T}_{\text{MAT}}$ and $\mathcal{T}_{\text{DEC}}$ may yield quasi-EFO formulas when they are applied to EFO formulas. A *branch* is a set of normal formulas s such that s is either EFO or quasi-EFO.

A term $s : \iota$ is *discriminating* in a branch A if A contains a disequation $s \neq t$ or $t \neq s$ for some term t. We use $\mathcal{D}A$ to denote the set of all terms that are discriminating in a branch A.

A branch E is *evident* if it satisfies the *evidence conditions* in Figure 2. The evidence conditions correspond to the tableau rules and are designed such that a branch that is closed under the tableau rules and does not contain $\bot$ is evident. Note that the evidence conditions require less than the tableau rules:

$\mathcal{E}_\perp$ $\perp$ is not in E.

$\mathcal{E}_\neg$ If $\neg x$ is in E, then x is not in E.

$\mathcal{E}_{\neg\neg}$ If $\neg\neg s$ is in E, then s is in E.

$\mathcal{E}_\wedge$ If $s \wedge t$ is in E, then s and t are in E.

$\mathcal{E}_{\neg\wedge}$ If $\neg(s \wedge t)$ is in E, then $\neg s$ or $\neg t$ is in E.

$\mathcal{E}_\forall$ If $\forall_\iota s$ is in E, then $[st]$ is in E for all $t \in \mathcal{D}E$,
      and $[st]$ is in E for some $t \in \mathrm{EFO}_\iota$.

$\mathcal{E}_{\neg\forall}$ If $\neg\forall_\iota s$ is in E, then $\neg[st]$ is in E for some $t \in \mathrm{EFO}_\iota$.

$\mathcal{E}_{\mathrm{MAT}}$ If $xs_1 \ldots s_n$ and $\neg xt_1 \ldots t_n$ are in E where $n \geq 1$,
      then $s_i \neq t_i$ is in E for some $i \in \{1, \ldots, n\}$.

$\mathcal{E}_{\mathrm{DEC}}$ If $xs_1 \ldots s_n \neq_\iota xt_1 \ldots t_n$ is in E where $n \geq 1$,
      then $s_i \neq t_i$ is in E for some $i \in \{1, \ldots, n\}$.

$\mathcal{E}_\neq$ If $s \neq_\iota t$ is in E, then s and t are different.

$\mathcal{E}_{\mathrm{BE}}$ If $s \neq_o t$ is in E, then either s and $\neg t$ are in E or $\neg s$ and t are in E.

$\mathcal{E}_{\mathrm{FE}}$ If $s \neq_{\sigma\tau} t$ is in E, then $[sx] \neq [tx]$ is in E for some variable x.

$\mathcal{E}_{\mathrm{CON}}$ If $s =_\iota t$ and $u \neq_\iota v$ are in E,
      then either $s \neq u$ and $t \neq u$ are in E or $s \neq v$ and $t \neq v$ are in E.

Fig. 2. Evidence conditions

1. $\mathcal{E}_\neg$ is restricted to variables.
2. $\mathcal{E}_\forall$ requires less instances than $\mathcal{T}_\forall$ admits.
3. $\mathcal{E}_{\neg\forall}$ admits all EFO terms as witnesses.
4. $\mathcal{E}_\neq$ is restricted to type ι.

In §7 we will show that every evident branch is satisfiable. In §9 we will prove
the completeness of a tableau system $\mathcal{R}$ that restricts the rule $\mathcal{T}_\forall$ as suggested
by the evidence condition $\mathcal{E}_\forall$.

Example 5.1. Let $p : (\iota\iota)o$. The following branch is evident.

$$p(\lambda xy.x), \quad \neg p(\lambda xy.y), \quad (\lambda xy.x) \neq (\lambda xy.y), \quad (\lambda y.x) \neq (\lambda y.y), \quad x \neq y \qquad \square$$

6 Carriers

A carrier for an evident branch E consists of a set D and a relation $\rhd_\iota \subseteq \mathrm{EFO}_\iota \times D$
such that certain conditions are satisfied. We will show that every evident branch
has carriers, and that for every carrier $(D, \rhd_\iota)$ for an evident branch E we can
obtain a model $\mathcal{I}$ of E such that $\mathcal{I}\iota = D$ and $s \rhd_\iota \hat{\mathcal{I}}s$ for all $s \in \mathrm{EFO}_\iota$. We call
$\rhd_\iota$ a *possible-values* relation and read $s \rhd_\iota a$ as s *can be* a. Given $s \rhd_\iota a$, we say
that a is a *possible value* for s.

We assume that some evident branch E is given. We say that a set $T \subseteq \mathrm{EFO}_\iota$ is *compatible* if there are no terms $s, t \in T$ such that $([s] {\neq} [t]) \in E$. We write $s {\,\sharp\,} t$ if E contains the disequation $s {\neq} t$ or $t {\neq} s$.

Let a non-empty set D and a relation $\rhd_\iota \subseteq \mathrm{EFO}_\iota \times D$ be given. For $T \subseteq \mathrm{EFO}_\iota$ and $a \in D$ we write $T \rhd_\iota a$ if $t \rhd_\iota a$ for every $t \in T$. For all terms $s, t \in \mathrm{EFO}_\iota$, all values $a, b \in D$, and every set $T \subseteq \mathrm{EFO}_\iota$ we require the following properties:

B1 $s \rhd_\iota a$ iff $[s] \rhd_\iota a$.
B2 T compatible iff $T \rhd_\iota a$ for some $a \in D$.
B3 If $(s {=_\iota} t) \in E$ and $s \rhd_\iota a$ and $t \rhd_\iota b$, then $a = b$.
B4 For every $a \in D$ either $t \rhd_\iota a$ for some $t \in \mathcal{D}E$ or $t \rhd_\iota a$ for every $t \in \mathrm{EFO}_\iota$.

Given an evident branch E, a *carrier for E* is a pair $(D, \rhd_\iota)$ as specified above.

6.1   Quotient-Based Carriers

A branch A is *complete* if for all $s, t \in \mathrm{EFO}_\iota$ either $[s {=} t]$ is in A or $[s {\neq} t]$ is in A. We will show that complete evident branches have countable carriers that can be obtained as quotients of EFO_ι with respect to the equations contained in the branch.

Let E be a complete evident branch in the following. We write $s \sim t$ if s and t are EFO terms of type ι and $[s {=_\iota} t] \in E$. We define $\tilde{s} := \{\, t \mid t \sim s \,\}$ for $s \in \mathrm{EFO}_\iota$.

Proposition 6.1. *For all $s, t \in \mathrm{EFO}_\iota$: $s \nsim t$ iff $[s {\neq} t] \in E$.*

Proposition 6.2. $\sim$ *is an equivalence relation on EFO_ι.*

Proposition 6.3. *Let $T \subseteq \mathrm{EFO}_\iota$. Then T is compatible iff $s \sim t$ for all $s, t \in T$.*

Proof. By definition and N3, T is compatible if $[s {\neq} t] \notin E$ for all $s, t \in T$. The claim follows with Proposition 6.1.   □

Lemma 6.4. *Every complete evident branch has a countable carrier.*

Proof. Let E be a complete evident branch. We define:

$$D := \{\, \tilde{s} \mid s \in \mathrm{EFO}_\iota \,\}$$
$$s \rhd_\iota \tilde{t} :\Longleftrightarrow s \sim t$$

We will show that $(D, \rhd_\iota)$ is a carrier for E. Note that $\rhd_\iota$ is well-defined since $\sim$ is an equivalence relation. D is countable since EFO_ι is countable.

B1. We have to show that $s \sim t$ iff $[s] \sim t$. This follows with N3 and N1 since $s \sim t$ iff $[s {=} t] \in E$ and $[s] \sim t$ iff $[[s] {=} t] \in E$.

B2. If T is empty, B2 holds vacuously. Otherwise, let $t \in T$. Then T is compatible iff $s \sim t$ for all $s \in T$ by Propositions 6.3 and 6.2. Hence T is compatible iff $s \rhd_\iota \tilde{t}$ for all $s \in T$. The claim follows.

B3. Let $s=_\iota t$ in E and $s \triangleright_\iota \tilde{u}$ and $t \triangleright_\iota \tilde{v}$. Since $s=t$ is normal, we have $s \sim t$. By definition of $\triangleright_\iota$ we have $s \sim u$ and $t \sim v$. Hence $\tilde{u} = \tilde{v}$ since $\sim$ is an equivalence relation.

B4. If $\mathcal{D}E$ is empty, then $s \triangleright_\iota \tilde{t}$ for all $s,t \in \text{EFO}_\iota$ and hence the claim holds. Otherwise, let $\mathcal{D}E$ be nonempty. We show the claim by contradiction. Suppose there is a term $t \in \text{EFO}_\iota$ such that $s \not\triangleright_\iota \tilde{t}$ for all $s \in \mathcal{D}E$. Then $[s \neq t] \in E$ for all $s \in \mathcal{D}E$ by Proposition 6.1. Since $\mathcal{D}E$ is nonempty, we have $[t] \in \mathcal{D}E$ by N3. Thus $([t] \neq [t]) \in E$ by N3. Contradiction by $\mathcal{E}_{\neq}$. □

6.2   Discriminant-Based Carriers

We will now show that every evident branch has a carrier. Let an evident branch E be given. We will call a term *discriminating* if it is discriminating in E. A *discriminant* is a maximal set a of discriminating terms such that there is no disequation $s \neq t \in E$ such that $s, t \in a$. We will construct a carrier for E whose values are the discriminants.

Example 6.5. Suppose $E = \{x \neq y, x \neq z, y \neq z\}$ and $x, y, z : \iota$. Then there are 3 discriminants: $\{x\}, \{y\}, \{z\}$. □

Example 6.6. Suppose $E = \{a_n \neq_\iota b_n \mid n \in \mathbb{N}\}$ where the a_n and b_n are pairwise distinct constants. Then E is evident and there are uncountably many discriminants. □

Proposition 6.7. *If E contains exactly n disequations at ι, then there are at most 2^n discriminants. If E contains no disequation at ι, then $\emptyset$ is the only discriminant.*

Proposition 6.8. *Let a and b be different discriminants. Then:*

1. *a and b are separated by a disequation in E, that is, there exist terms $s \in a$ and $t \in b$ such that $s \sharp t$.*
2. *a and b are not connected by an equation in E, that is, there exist no terms $s \in a$ and $t \in b$ such that $(s=t) \in E$.*

Proof. The first claim follows by contradiction. Suppose there are no terms $s \in a$ and $t \in b$ such that $s \sharp t$. Let $s \in a$. Then $s \in b$ since b is a maximal compatible set of discriminating terms. Thus $a \subseteq b$ and hence $a = b$ since a is maximal. Contradiction.

The second claim also follows by contradiction. Suppose there is an equation $(s_1=s_2) \in E$ such that $s_1 \in a$ and $s_2 \in b$. By the first claim we have terms $s \in a$ and $t \in b$ such that $s \sharp t$. By $\mathcal{E}_{\text{CON}}$ we have $s_1 \sharp s$ or $s_2 \sharp t$. Contradiction since a and b are discriminants. □

Lemma 6.9. *Every (finite) evident branch has a (finite) carrier.*

Proof. Let E be an evident branch. We define:

$$D := \text{set of all discriminants}$$
$$s \vartriangleright_\iota a :\Longleftrightarrow ([s] \text{ discriminating} \Longrightarrow [s] \in a)$$

We will show that $(D, \vartriangleright_\iota)$ is a carrier for E. By Proposition 6.7 we know that D is finite if E is finite.

B1. Holds by N1.

For the remaining carrier conditions we distinguish two cases. If $\mathcal{D}E = \emptyset$, then $\emptyset$ is the only discriminant and B2, B3, and B4 are easily verified. Otherwise, let $\mathcal{D}E \neq \emptyset$.

B2⇒. Let T be compatible. Then there exists a discriminant a that contains all the discriminating terms in $\{ [t] \mid t \in T \}$. The claim follows since $T \vartriangleright a$.

B2⇐. By contradiction. Suppose $T \vartriangleright a$ and T is not compatible. Then there are terms $s, t \in T$ such that $([s] \neq [t]) \in E$. Thus $[s]$ and $[t]$ cannot be both in a. This contradicts $s, t \in T \vartriangleright a$ since $[s]$ and $[t]$ are discriminating.

B3. Let $(s \dot{=} t) \in E$ and $s \vartriangleright_\iota a$ and $t \vartriangleright_\iota b$. We show $a = b$. Since there are discriminating terms, E contains at least one disequation at type ι, and hence s and t are discriminating by $\mathcal{E}_{\text{CON}}$. By N3 s and t are normal and hence $s \in a$ and $t \in b$. Now $a = b$ by Proposition 6.8 (2).

B4. Since there are discriminating terms, we know by $\mathcal{E}_{\neq}$ that every discriminant contains at least one discriminating term. Since discriminating terms are normal, we have the claim. $\qquad\square$

7   Model Existence

We will now show that every evident branch has a model.

Lemma 7.1 (Model Existence). *Let $(D, \vartriangleright_\iota)$ be a carrier for an evident branch E. Then E has a model $\mathcal{I}$ such that $\mathcal{I}\iota = D$.*

We start the proof of Lemma 7.1. Let $(D, \vartriangleright_\iota)$ be a carrier for an evident branch E. For the rest of the proof we only consider interpretations $\mathcal{I}$ such that $\mathcal{I}\iota = D$.

7.1   Possible Values

To obtain a model of E, we need suitable values for all variables. We address this problem by defining possible-values relations $\vartriangleright_\sigma \subseteq \text{EFO}_\sigma \times \mathcal{I}\sigma$ for all types $\sigma \neq \iota$:

$$s \vartriangleright_o 0 :\Longleftrightarrow [s] \notin E$$
$$s \vartriangleright_o 1 :\Longleftrightarrow \neg[s] \notin E$$
$$s \vartriangleright_{\sigma\tau} f :\Longleftrightarrow st \vartriangleright_\tau fa \text{ whenever } t \vartriangleright_\sigma a$$

Note that we already have a possible-values relation for ι and that the definition of the possible values relations for functional types is by induction on types. Also note that if s is an EFO formula such that $[s] \notin E$ and $\neg[s] \notin E$, then both 0 and 1 are possible values for s. We will show that every EFO term has a possible value and that we obtain a model of E if we define $\mathcal{I}x$ as a possible value for x for every variable x.

Proposition 7.2. *Let* $s \in \mathrm{EFO}_\sigma$ *and* $a \in \mathcal{I}\sigma$. *Then* $s \triangleright_\sigma a \iff [s] \triangleright_\sigma a$.

Proof. By induction on σ. For o the claim follows with N1. For ι the claim follows with B1. Let $\sigma = \tau\mu$.

Suppose $s \triangleright_\sigma a$. Let $t \triangleright_\tau b$. Then $st \triangleright_\mu ab$. By inductive hypothesis $[st] \triangleright_\mu ab$. Thus $[[s]t] \triangleright_\mu ab$ by N2. By inductive hypothesis $[s]t \triangleright_\mu ab$. Hence $[s] \triangleright_\sigma a$.

Suppose $[s] \triangleright_\sigma a$. Let $t \triangleright_\tau b$. Then $[s]t \triangleright_\mu ab$. By inductive hypothesis $[[s]t] \triangleright_\mu ab$. Thus $[st] \triangleright_\mu ab$ by N2. By inductive hypothesis $st \triangleright_\mu ab$. Hence $s \triangleright_\sigma a$.   □

Lemma 7.3. *For every EFO constant* c: $c \triangleright \mathcal{I}c$.

Proof. $c = \bot$. The claim follows by $\mathcal{E}_\bot$ and N3.

$c = \neg$. Assume $s \triangleright_o a$. We show $\neg s \triangleright \mathcal{I}(\neg)a$ by contradiction. Suppose $\neg s \not\triangleright \mathcal{I}(\neg)a$. Case analysis.

$\quad a = 0$. Then $[s] \notin E$ and $\neg[\neg s] \in E$. Thus $\neg\neg[s] \in E$ by N3. Hence $[s] \in E$ by $\mathcal{E}_{\neg\neg}$. Contradiction.

$\quad a = 1$. Then $\neg[s] \notin E$ and $[\neg s] \in E$. Contradiction by N3.

$c = \wedge$. Assume $s \triangleright_o a$ and $t \triangleright_o b$. We show $s \wedge t \triangleright \mathcal{I}(\wedge)ab$ by contradiction. Suppose $s \wedge t \not\triangleright \mathcal{I}(\wedge)ab$. Case analysis.

$\quad a = b = 1$. Then $\neg[s], \neg[t] \notin E$ and $\neg[s \wedge t] \in E$. Contradiction by N3 and $\mathcal{E}_{\neg\wedge}$.

$\quad a = 0$ **or** $b = 0$. Then $[s] \notin E$ or $[t] \notin E$, and $[s \wedge t] \in E$. Contradiction by N3 and $\mathcal{E}_\wedge$.

$c = (=_\iota)$. Assume $s \triangleright_\iota a$ and $t \triangleright_\iota b$. We show $(s{=}t) \triangleright \mathcal{I}(=_\iota)ab$ by contradiction. Suppose $(s{=}t) \not\triangleright \mathcal{I}(=_\iota)ab$. Case analysis.

$\quad a = b$. Then $\neg[s{=}t] \in E$ and $s, t \triangleright_\iota a$. By B2 $\{s, t\}$ is compatible. Contradiction by N3.

$\quad a \neq b$. Then $([s]{=}[t]) \in E$ by N3. Hence $a = b$ by B1 and B3. Contradiction.

$c = \forall_\iota$. Assume $s \triangleright_{\iota o} f$. We show $\forall_\iota s \triangleright_o \mathcal{I}\forall_\iota f$ by contradiction. Suppose $\forall_\iota s \not\triangleright_o \mathcal{I}\forall_\iota f$. Case analysis.

$\quad \mathcal{I}\forall_\iota f = 0$. Then $\forall_\iota[s] \in E$ by N3 and $fa = 0$ for some value a. By $\mathcal{E}_\forall$ and B4 there exists a term t such that $[[s]t] \in E$ and $t \triangleright_\iota a$. Thus $st \triangleright fa = 0$ and hence $[st] \notin E$. Contradiction by N2.

$\quad \mathcal{I}\forall_\iota f = 1$. Then $\neg\forall_\iota[s] \in E$ by N3. By $\mathcal{E}_{\neg\forall}$ we have $\neg[[s]t] \in E$ for some term $t \in \mathrm{EFO}_\iota$. By $\mathcal{E}_{\neq}$ and B2 we have $t \triangleright a$ for some value a. Now $st \triangleright fa = 1$. Thus $\neg[st] \notin E$. Contradiction by N2.   □

We call an interpretation $\mathcal{I}$ *admissible* if it satisfies $x \triangleright \mathcal{I}x$ for every variable x. We will show that admissible interpretations exist and that every admissible interpretation is a model of E.

Lemma 7.4 (Admissibility). *Let* $\mathcal{I}$ *be admissible and* θ *be a substitution such that* $\theta x \triangleright \mathcal{I}x$ *for all* $x \in \mathrm{Dom}\,\theta$. *Then* $\hat{\theta}s \triangleright \hat{\mathcal{I}}s$ *for every EFO term* s.

Proof. By induction on s. Let s be an EFO term. By assumption, θx is EFO for all $x \in \text{Dom}\,\theta$. Hence $\hat{\theta}s$ is EFO by S5. Case analysis.

$s = a$. If $a \in \text{Dom}\,\theta$, the claim holds by assumption. If $a \notin \text{Dom}\,\theta$, then $\hat{\theta}s = a$ by S1. If a is a constant, the claim holds by Lemma 7.3. If a is a variable, the claim holds by assumption.

$s = tu$. Then $\hat{\theta}s = (\hat{\theta}t)(\hat{\theta}u)$ by S2. Now $\hat{\theta}t \triangleright \hat{\mathcal{I}}t$ and $\hat{\theta}u \triangleright \hat{\mathcal{I}}u$ by the inductive hypothesis. Now $\hat{\theta}s = (\hat{\theta}t)(\hat{\theta}u) \triangleright (\hat{\mathcal{I}}t)(\hat{\mathcal{I}}u) = \hat{\mathcal{I}}s$.

$s = \lambda x.t$ and $x : \sigma$. Moreover, let $u \triangleright_\sigma a$. We show $(\hat{\theta}s)u \triangleright (\hat{\mathcal{I}}s)a$. By Proposition 7.2 it suffices to show $[(\hat{\theta}s)u] \triangleright (\hat{\mathcal{I}}s)a$. We have $[(\hat{\theta}s)u] = [\widehat{\theta^x_u}t]$ by S3 and $(\hat{\mathcal{I}}s)a = \widehat{\mathcal{I}^x_a}t$ where $\mathcal{I}^x_a$ denotes the interpretation that agrees everywhere with $\mathcal{I}$ but possibly on x where it yields a. By inductive hypothesis we have $\widehat{\theta^x_u}t \triangleright \widehat{\mathcal{I}^x_a}t$. The claim follows with Proposition 7.2. □

7.2 Compatibility

It remains to show that there is an admissible interpretation and that every admissible interpretation is a model of E. For this purpose we define compatibility relations $\|_\sigma \subseteq \text{EFO}_\sigma \times \text{EFO}_\sigma$ for all types:

$$
\begin{aligned}
s \parallel_o t &\;:\Longleftrightarrow\; \{[s], \neg[t]\} \not\subseteq E \text{ and } \{\neg[s], [t]\} \not\subseteq E \\
s \parallel_\iota t &\;:\Longleftrightarrow\; \text{not } [s]\sharp[t] \\
s \parallel_{\sigma\tau} t &\;:\Longleftrightarrow\; su \parallel_\tau tv \text{ whenever } u \parallel_\sigma v
\end{aligned}
$$

Note that the definition of the compatibility relations for functional types is by induction on types. We say that s and t are *compatible* if $s \parallel t$. A set T of equi-typed terms is *compatible* if $s \parallel t$ for all terms $s, t \in T$. If $T \subseteq \text{EFO}_\sigma$, we write $T \triangleright a$ if a is a common possible value for all terms $s \in T$. We will show that a set of equi-typed terms is compatible if and only if all its terms have a common possible value.

The compatibility relations are reflexive. We first show $x \parallel x$ for all variables x. For the induction to go through we strengthen the hypothesis.

Lemma 7.5 (Reflexivity). *For every type σ and all EFO terms s, t, $xs_1 \ldots s_n$, $xt_1 \ldots t_n$ of type σ with $n \geq 0$:*

1. *Not both $s \parallel_\sigma t$ and $[s]\sharp[t]$.*
2. *Either $xs_1 \ldots s_n \parallel_\sigma xt_1 \ldots t_n$ or $[s_i]\sharp[t_i]$ for some $i \in \{1, \ldots, n\}$.*

Proof. By mutual induction on σ. A similar proof can be found in [7]. □

Lemma 7.6 (Common Value). *Let $T \subseteq \text{EFO}_\sigma$. Then T is compatible if and only if there exists a value a such that $T \triangleright_\sigma a$.*

Proof. By induction on σ. A similar proof can be found in [7]. □

Lemma 7.7. *Every admissible interpretation is a model of E.*

Proof. Let $\mathcal{I}$ be an admissible interpretation and $s \in E$. We show $\hat{\mathcal{I}}s = 1$. Case analysis.

Suppose s is a normal EFO term. Then $s = [s] = [\hat{\emptyset}s]$ by S4 and $s \not\rhd 0$. Moreover, $\hat{\emptyset}s \rhd \hat{\mathcal{I}}s$ by Lemma 7.4 and $s \rhd \hat{\mathcal{I}}s$ by Lemma 7.2. Hence $\hat{\mathcal{I}}s = 1$.

Suppose $s = (t \neq u)$ where t and u are normal EFO terms. Then $t = [t] = [\hat{\emptyset}t]$ and $u = [u] = [\hat{\emptyset}u]$ by S4. We prove the claim by contradiction. Suppose $\hat{\mathcal{I}}s = 0$. Then $\hat{\mathcal{I}}t = \hat{\mathcal{I}}u$. Thus $\hat{\emptyset}t, \hat{\emptyset}u \rhd \hat{\mathcal{I}}t$ by Lemma 7.4 and $t, u \rhd \hat{\mathcal{I}}t$ by Lemma 7.2. Hence $t \parallel u$ by Lemma 7.6. Thus not $[t] \sharp [u]$ by Lemma 7.5 (1). Contradiction since $([t] \neq [u]) \in E$. $\square$

We can now prove Lemma 7.1. By Lemma 7.5 (2) we know $x \parallel x$ for every variable x. Hence there exists an admissible interpretation $\mathcal{I}$ by Lemma 7.6. By Lemma 7.7 we know that $\mathcal{I}$ is a model of E. This finishes the proof of Lemma 7.1.

Theorem 7.8 (Finite Model Existence)
Every finite evident branch has a finite model.

Proof. Follows with Lemmas 6.9 and 7.1. $\square$

Lemma 7.9 (Model Existence). *Let E be an evident branch. Then E has a model. Moreover, E has a countable model if E is complete.*

Proof. Follows with Lemmas 6.9, 7.1, and 6.4. $\square$

8   Abstract Consistency

To obtain our main results, we boost the model existence lemma with the abstract consistency technique. Everything works out smoothly.

An *abstract consistency class* is a set Γ of branches such that every branch $A \in \Gamma$ satisfies the conditions in Figure 3. An abstract consistency class Γ is *complete* if for every $A \in \Gamma$ and all $s, t \in \text{EFO}_\iota$ either $A \cup \{[s{=}t]\}$ is in Γ or $A \cup \{[s{\neq}t]\}$ is in Γ.

Lemma 8.1 (Extension Lemma). *Let Γ be an abstract consistency class and $A \in \Gamma$. Then there exists an evident branch E such that $A \subseteq E$. Moreover, if Γ is complete, a complete evident branch E exists such that $A \subseteq E$.*

Proof. Let $u_0, u_1, u_2, \ldots$ be an enumeration of all formulas that can occur on a branch. We construct a sequence $A_0 \subseteq A_1 \subseteq A_2 \subseteq \cdots$ of branches such that every $A_n \in \Gamma$. Let $A_0 := A$. We define A_{n+1} by cases. If there is no $B \in \Gamma$ such that $A_n \cup \{u_n\} \subseteq B$, then let $A_{n+1} := A_n$. Otherwise, choose some $B \in \Gamma$ such that $A_n \cup \{u_n\} \subseteq B$. We consider four subcases.

1. If u_n is of the form $\forall_\iota s$, then choose A_{n+1} to be $B \cup \{[st]\} \in \Gamma$ for some $t \in \text{EFO}_\iota$. This is possible since Γ satisfies $\mathcal{C}_\forall$.
2. If u_n is of the form $\neg\forall_\iota s$, then choose A_{n+1} to be $B \cup \{\neg[st]\} \in \Gamma$ for some $t \in \text{EFO}_\iota$. This is possible since Γ satisfies $\mathcal{C}_{\neg\forall}$.

$\mathcal{C}_\perp$ $\perp$ is not in A.

$\mathcal{C}_\neg$ If $\neg x$ is in A, then x is not in A.

$\mathcal{C}_{\neg\neg}$ If $\neg\neg s$ is in A, then $A \cup \{s\}$ is in Γ.

$\mathcal{C}_\wedge$ If $s \wedge t$ is in A, then $A \cup \{s, t\}$ is in Γ.

$\mathcal{C}_{\neg\wedge}$ If $\neg(s \wedge t)$ is in A, then $A \cup \{\neg s\}$ or $A \cup \{\neg t\}$ is in Γ.

$\mathcal{C}_\forall$ If $\forall_\iota s$ is in A, then $A \cup \{[st]\}$ is in Γ for all $t \in \mathcal{D}A$,
 and $A \cup \{[st]\}$ is in Γ for some $t \in \mathrm{EFO}_\iota$

$\mathcal{C}_{\neg\forall}$ If $\neg\forall_\iota s$ is in A, then $A \cup \{\neg[st]\}$ is in Γ for some $t \in \mathrm{EFO}_\iota$.

$\mathcal{C}_{\mathrm{MAT}}$ If $xs_1 \ldots s_n$ is in A and $\neg xt_1 \ldots t_n$ is in A where $n \geq 1$,
 then $A \cup \{s_i \neq t_i\}$ is in Γ for some $i \in \{1, \ldots, n\}$.

$\mathcal{C}_{\mathrm{DEC}}$ If $xs_1 \ldots s_n \neq_\iota xt_1 \ldots t_n$ is in A where $n \geq 1$,
 then $A \cup \{s_i \neq t_i\}$ is in Γ for some $i \in \{1, \ldots, n\}$.

$\mathcal{C}_\neq$ If $s \neq_\iota t$ is in A, then s and t are different.

$\mathcal{C}_{\mathrm{BE}}$ If $s \neq_o t$ is in A, then either $A \cup \{s, \neg t\}$ or $A \cup \{\neg s, t\}$ is in Γ.

$\mathcal{C}_{\mathrm{FE}}$ If $s \neq_{\sigma\tau} t$ is in A, then $A \cup \{[sx] \neq [tx]\}$ is in Γ for some variable x.

$\mathcal{C}_{\mathrm{CON}}$ If $s =_\iota t$ and $u \neq_\iota v$ are in A,
 then either $A \cup \{s \neq u, t \neq u\}$ or $A \cup \{s \neq v, t \neq v\}$ is in Γ.

Fig. 3. Abstract consistency conditions (must hold for every $A \in \Gamma$)

3. If u_n is of the form $s \neq_{\sigma\tau} t$, then choose A_{n+1} to be $B \cup \{[sx] \neq [tx]\} \in \Gamma$
 for some variable x. This is possible since Γ satisfies $\mathcal{C}_{\mathrm{FE}}$.

4. If u_n has none of these forms, then let A_{n+1} be B.

Let $E := \bigcup_{n \in \mathbb{N}} A_n$. Note that $\mathcal{D}E = \bigcup_{n \in \mathbb{N}} \mathcal{D}A_n$. It is not difficult to verify that E is
evident. For space reasons we will only show that $\mathcal{E}_{\neg\forall}$ is satisfied. Assume $\neg\forall_\iota s$
is in E. Let n be such that $u_n = \neg\forall_\iota s$. Let $r \geq n$ be such that $\neg\forall_\iota s$ is in A_r.
Hence $\neg[st] \in A_{n+1} \subseteq E$ for some t.

It remains to show that E is complete if Γ is complete. Let Γ be complete
and $s, t \in \mathrm{EFO}_\iota$. We show that $[s = t]$ or $[s \neq t]$ is in E. Let m, n be such that
$u_m = [s{=}t]$ and $u_n = [s{\neq}t]$. We consider $m < n$, the case $m > n$ is symmetric.
If $[s{=}t] \in A_n$, we have $[s{=}t] \in E$. If $[s{=}t] \notin A_n$, then $A_n \cup \{[s{=}t]\}$ is not in Γ.
Hence $A_n \cup \{[s \neq t]\}$ is in Γ since Γ is complete. Hence $[s \neq t] \in A_{n+1} \subseteq E$. $\square$

9 Completeness

We will now show that the tableau system $\mathcal{T}$ is complete. In fact, we will show
the completeness of a tableau system $\mathcal{R}$ that is obtained from $\mathcal{T}$ by restricting
the applicability of some of the rules. We consider $\mathcal{R}$ since it provides for more
focused proof search and also yields a decision procedure for three substantial

fragments of EFO. $\mathcal{R}$ is obtained from $\mathcal{T}$ by restricting the applicability of the rules $\mathcal{T}_\forall$, $\mathcal{T}_{\neg\forall}$, and $\mathcal{T}_{FE}$ as follows:

- $\mathcal{T}_\forall$ can only be applied to $\forall_\iota s \in A$ with a term $t \in \text{EFO}_\iota$ if either $t \in \mathcal{D}A$ or the following conditions are satisfied:
  1. $\mathcal{D}A = \emptyset$ and t is a variable.
  2. $t \in \mathcal{N}A$ or $\mathcal{N}A = \emptyset$.
  3. There is no $u \in \text{EFO}_\iota$ such that $[su] \in A$.
- $\mathcal{T}_{\neg\forall}$ can only be applied to $\neg\forall_\iota s \in A$ if there is no $t \in \text{EFO}_\iota$ such that $\neg[st] \in A$.
- $\mathcal{T}_{FE}$ can only be applied to an equation $(s=_{\sigma\tau}t) \in A$ if there is no variable $x : \sigma$ such that $([sx]=[tx]) \in A$.

We use $\mathcal{R}_\forall$, $\mathcal{R}_{\neg\forall}$, and $\mathcal{R}_{FE}$ to refer to the restrictions of $\mathcal{T}_\forall$, $\mathcal{T}_{\neg\forall}$, and $\mathcal{T}_{FE}$, respectively. Note that $\mathcal{R}_\forall$ provides a novel subterm restriction that may be useful for proof search. We say a branch A is *refutable* if it can be refuted with $\mathcal{R}$. Let $\Gamma_\mathcal{T}$ be the set of all finite branches that are not refutable.

Lemma 9.1. $\Gamma_\mathcal{T}$ *is an abstract consistency class.*

Proof. For space limitations we only verify $\mathcal{C}_{\neg\forall}$. Let $\neg\forall_\iota s \in A \in \Gamma_\mathcal{T}$. Suppose $A \cup \{\neg[st]\} \notin \Gamma_\mathcal{T}$ for every $t \in \text{EFO}_\iota$. Then $A \cup \{\neg[st]\}$ is refutable for every $t \in \text{EFO}_\iota$. Hence A is refutable using $\mathcal{T}_{\neg\forall}$ and the finiteness of A. Contradiction. $\square$

Theorem 9.2 (Completeness)
$\mathcal{T}$ *and* $\mathcal{R}$ *can refute every unsatisfiable finite branch.*

Proof. It suffice to show the claim for $\mathcal{R}$. We prove the claim by contradiction. Let A be an unsatisfiable finite branch that is not refutable. Then $A \in \Gamma_\mathcal{T}$ and hence A is satisfiable by Lemmas 9.1, 8.1, and 7.9. $\square$

10   Compactness and Countable Models

A branch A is *sufficiently pure* if for every type σ there are infinitely many variables of type σ that do not occur in any formula of A. Let Γ_C be the set of all sufficiently pure branches A such that every finite subset of A is satisfiable. We write $\subseteq_f$ for the finite subset relation.

Lemma 10.1. Γ_C *is a complete abstract consistency class.*

Proof. Omitted for space limitations. Not difficult. $\square$

Theorem 10.2 (Compactness)
A branch is satisfiable if each of its finite subsets is satisfiable.

Proof. Let A be a branch such that every finite subset of A is satisfiable. Without loss of generality we assume A is sufficiently pure. Then $A \in \Gamma_C$. Hence A is satisfiable by Lemmas 10.1, 8.1, and 7.9. $\square$

Theorem 10.3 (Countable Models)
Every satisfiable branch has a countable model.

Proof. Let A be a satisfiable branch. Without loss of generality we assume that A is sufficiently pure. Hence $A \in \Gamma_C$. By Lemmas 10.1 and 8.1 we have a complete evident set E such that $A \subseteq E$. By Lemma 7.9 we have a countable model for E and hence for A. □

Theorem 10.4 (Countable Model Existence)
Every evident branch has a countable model.

Proof. Let E be an evident branch. By Lemma 7.9 we know that E is satisfiable. By Theorem 10.3 we know that E has a countable model. □

11   Decidability

The tableau system $\mathcal{R}$ defined in § 9 yields a procedure that decides the satisfiability of three substantial fragments of EFO. Starting with the initial branch, the procedure applies tableau rules until it reaches a branch that contains $\bot$ or cannot be extended with the tableau rules. The procedure returns "satisfiable" if it arrives at a terminal branch that does not contain $\bot$, and "unsatisfiable" if it finds a refutation. There are branches on which the procedure does not terminate (e.g., $\{\forall_\iota x.\ fx{\neq}x\}$). We first establish the partial correctness of the procedure.

Proposition 11.1 (Verification Soundness). *Let A be a finite branch that does not contain $\bot$ and cannot be extended with $\mathcal{R}$. Then A is evident and has a finite model.*

Proposition 11.2 (Refutation Soundness)
Every refutable branch is unsatisfiable.

For the termination of the procedure we consider the relation $A \rightarrow A'$ that holds if A and A' are branches such that $\bot \notin A \subsetneq A'$ and A' can be obtained from A by applying a rule of $\mathcal{R}$. We say that $\mathcal{R}$ *terminates* on a set Δ of branches if there is no infinite derivation $A \rightarrow A' \rightarrow A'' \rightarrow \cdots$ such that $A \in \Delta$.

Proposition 11.3. *Let $\mathcal{R}$ terminate on a set Δ of finite branches. Then satisfiability of the branches in Δ is decidable and every satisfiable branch in Δ has a finite model.*

Proof. Follows with Propositions 11.2 and 11.1 and Theorem 7.8. □

The decision procedure depends on the normalization operator employed with $\mathcal{R}$. A normalization operator that yields β-normal forms provides for all termination results proven in this section. Note that the tableau system applies the normalization operator only to applications st where s and t are both normal and t has type ι if it is not a variable. Hence at most one β-reduction is needed for normalization if s and t are β-normal. Moreover, no α-renaming is needed if the bound variables are chosen differently from the free variables. For clarity, we continue to work with an abstract normalization operator and state a further condition:

N5 The least relation $\succ$ on terms such that

1. $as_1 \ldots s_n \succ s_i$ if $i \in \{1, \ldots, n\}$
2. $s \succ [sx]$ if $s : \sigma\tau$ and $x : \sigma$

terminates on normal terms.

A type is *pure* if it does not contain o. A term is *pure* if the type of every name occurring in it (bound or unbound) is pure. An equation $s = t$ or disequation $s \neq t$ is *pure* if s and t are pure terms.

Proposition 11.4 (Pure Termination). *Let the normalization operator satisfy N5. Then $\mathcal{R}$ terminates on finite branches containing only pure disequations.*

Proof. Let $A \to A_1 \to A_2 \to \cdots$ be a possibly infinite derivation that issues from a finite branch containing only pure disequations. Then no other rules but possibly $\mathcal{T}_{\mathrm{DEC}}$, $\mathcal{R}_{\mathrm{FE}}$, and $\mathcal{T}_{\neq}$ apply and thus no A_i contains a formula that is not $\bot$ or a pure disequations (using S5). Using N5 it follows that the derivation is finite. $\square$

We now know that the validity of pure equations is decidable, and that the invalidity of pure equations can be demonstrated with finite interpretations (Proposition 11.1). Both results are well-known [9,10], but it is remarkable that we obtain them with different proofs and as a byproduct.

It is well-known that satisfiability of Bernays-Schönfinkel-Ramsey formulas (first-order $\exists^*\forall^*$-prenex formulas without functions) is decidable and the fragment has the finite model property [3]. We reobtain this result by showing that $\mathcal{R}$ terminates for the respective fragment. We call a type *BSR* if it is ι or o or has the form $\iota \ldots \iota o$. We call an EFO formula s *BSR* if it satisfies two conditions:

1. The type of every variable that occurs in s is BSR.
2. $\forall_\iota$ does not occur below a negation in s.

For simplicity, our BSR formulas don't provide for outer existential quantification. We need one more condition for the normalization operator:

N6 If $s : \iota o$ is BSR and $x : \iota$, then $[sx]$ is BSR.

Proposition 11.5 (BSR Termination). *Let the normalization operator satisfy N5 and N6. Then $\mathcal{R}$ terminates on finite branches containing only BSR formulas.*

Proof. Let $A \to A_1 \to A_2 \to \cdots$ be a possibly infinite derivation that issues from a finite branch containing only BSR formulas. Then $\mathcal{R}_{\neg\forall}$ and $\mathcal{R}_{\mathrm{FE}}$ are not applicable and all A_i contain only BSR formulas (using N6). Furthermore, at most one new variable is introduced. Since all terms of type ι are variables, there is only a finite supply. Using N5 it follows that the derivation is finite. $\square$

In [7] we study lambda- and quantifier-free EFO and show that the concomitant subsystem of $\mathcal{R}$ terminates on finite branches. The result extends to lambda-free branches containing quantifiers (e.g., $\{\forall_\iota f\}$).

Proposition 11.6 (Lambda-Free Termination). *Let the normalization operator satisfy* $[s] = s$ *for every lambda-free EFO term* s. *Then* $\mathcal{R}$ *terminates on finite lambda-free branches.*

Proof. An application of $\mathcal{R}_{\mathrm{FE}}$ disables a disequation $s \neq_{\sigma\tau} t$ and introduces new subterms as follows: a variable $x : \sigma$, two terms $sx : \tau$ and $tx : \tau$, and two formulas $sx = tx$ and $sx \neq tx$. Since the types of the new subterms are smaller than the type of s and t, and the new subterms introduced by the other rules always have type o or ι, no derivation can employ $\mathcal{R}_{\mathrm{FE}}$ infinitely often.

Let $A \to A_1 \to A_2 \to \cdots$ be a possibly infinite derivation that issues from a finite lambda-free branch and does not employ $\mathcal{R}_{\mathrm{FE}}$. It suffices to show that the derivation is finite. Observe that no new subterms of the form $\forall_\iota s$ are introduced. Hence only finitely many new subterms of type ι are introduced. Consequently, only finitely many new subterms of type o are introduced. Hence the derivation is finite. □

12 Conclusion

In this paper we have shown that the EFO fragment of Church's type theory enjoys the characteristic properties of first-order logic. We have devised a complete tableau system that comes with a new treatment of equality (confrontation) and a novel subterm restriction for the universal quantifier (discriminating terms). The tableau system decides lambda-free formulas, Bernays-Schönfinkel-Ramsey formulas, and equations between pure lambda terms.

References

1. Andrews, P.B.: Classical type theory. In: Robinson, A., Voronkov, A. (eds.) Handbook of Automated Reasoning, vol. 2, pp. 965–1007. Elsevier Science, Amsterdam (2001)
2. Brown, C.E.: Automated Reasoning in Higher-Order Logic: Set Comprehension and Extensionality in Church's Type Theory. College Publications (2007)
3. Börger, E., Grädel, E., Gurevich, Y.: The Classical Decision Problem. Springer, Heidelberg (1997)
4. Smullyan, R.M.: First-Order Logic. Springer, Heidelberg (1968)
5. Fitting, M.: First-Order Logic and Automated Theorem Proving. Springer, Heidelberg (1996)
6. Prawitz, D.: Hauptsatz for higher order logic. J. Symb. Log. 33, 452–457 (1968)
7. Brown, C.E., Smolka, G.: Terminating tableaux for the basic fragment of simple type theory. In: Giese, M., Waaler, A. (eds.) TABLEAUX 2009. LNCS (LNAI), vol. 5607, pp. 138–151. Springer, Heidelberg (2009)
8. Hindley, J.R.: Basic Simple Type Theory. Cambridge Tracts in Theoretical Computer Science, vol. 42. Cambridge University Press, Cambridge (1997)
9. Friedman, H.: Equality between functionals. In: Parikh, R. (ed.) Proc. Logic Colloquium 1972-73. Lectures Notes in Mathematics, vol. 453, pp. 22–37. Springer, Heidelberg (1975)
10. Statman, R.: Completeness, invariance and lambda-definability. J. Symb. Log. 47(1), 17–26 (1982)

Formalising Observer Theory for Environment-Sensitive Bisimulation

Jeremy E. Dawson and Alwen Tiu

Logic and Computation Group
College of Engineering and Computer Science
Australian National University
Canberra ACT 0200, Australia
http://users.rsise.anu.edu.au/~jeremy/
http://users.rsise.anu.edu.au/~tiu/

Abstract. We consider a formalisation of a notion of observer (or intruder) theories, commonly used in symbolic analysis of security protocols. An observer theory describes the knowledge and capabilities of an observer, and can be given a formal account using deductive systems, such as those used in various "environment-sensitive" bisimulation for process calculi, e.g., the spi-calculus. Two notions are critical to the correctness of such formalisations and the effectiveness of symbolic techniques based on them: decidability of message deduction by the observer and consistency of a given observer theory. We consider a formalisation, in Isabelle/HOL, of both notions based on an encoding of observer theories as pairs of symbolic traces. This encoding has recently been used in a theory of open bisimulation for the spi-calculus. We machine-checked some important properties, including decidability of observer deduction and consistency, and some key steps which are crucial to the automation of open bisimulation checking for the spi-calculus, and highlight some novelty in our Isabelle/HOL formalisations of decidability proofs.

1 Introduction

In most symbolic techniques for reasoning about security protocols, certain assumptions are often made concerning the capability of an intruder that tries to compromise the protocols. A well-known model of intruder is the so-called Dolev-Yao model [10], which assumes perfect crytography. We consider here a formal account of Dolev-Yao intruder model, formalised as some sort of deduction system. This deductive formulation is used in formalisations of various "environment-sensitive" bisimulations (see e.g., [6]) for process calculi designed for modeling security protocols, such as the spi-calculus [3]. An environment-sensitive bisimulation is a bisimulation relation which is indexed by a structure representing the intruder's knowledge, which we call an *observer theory*.

An important line of work related to the spi-calculus, or process calculi in general, is that of automating bisimulation checking. The transition semantics of these calculi often involve processes with infinite branching (e.g., transitions for

S. Berghofer et al. (Eds.): TPHOLs 2009, LNCS 5674, pp. 180–195, 2009.

input-prefixed processes in the π-calculus [12]), and therefore a *symbolic method* is needed to deal with potential infinite branches lazily. The resulting bisimulation, called *symbolic bisimulation*, has been developed for the spi-calculus [7]. The work reported in [7] is, however, only aimed at finding an effective approximation of environment-sensitive bisimulation, and there has been no metatheory developed for this symbolic bisimulation so far. A recent work by the second author [14] attempts just that: to establish a symbolic bisimulation that has good metatheory, in particular, a symbolic bisimulation which is also a congruence. The latter is also called *open bisimulation* [13]. One important part of the formulation of open bisimulation for the spi-calculus is a symbolic representation of observer theories, which needs to satisfy certain consistency properties, in addition to closure under a certain notion of "respectful substitutions", as typical in formulations of open bisimulation.

A large part of the work on open bisimulation in [14] deals with establishing properties of observer theories and their symbolic counterparts. This paper is essentially about formally verifying the results of [14] concerning properties of (symbolic) observer theories in Isabelle/HOL. In particular, it is concerned with proving decidability of the deduction system for observer theory, correctness of a finite characterisation of consistency of observer theories (hence decidability of consistency of observer theories), and preservation of consistency under respectful substitutions. Additionally, we also verify some key steps towards a decision procedure for checking consistency of symbolic observer theories, which is needed in automation of open bisimulation. A substantial formalisation work described here concerns decidability proofs. Such proofs are difficult to formalise in Isabelle/HOL, as noted in [17], due to the fact that Isabelle/HOL is based on classical logic. We essentially follow [17] in that decidability in this case can be inferred straightforwardly by inspection on the way we define total functions corresponding to the decidability problems in question. That is, we show, by meta-level inspection, that the definitions of the functions do not introduce any infinite aspect and are therefore are finitely computable.

There is a recent work [11] in formalising the spi-calculus and a notion of environment-sensitive bisimulation (called the *hedged bisimulation* [8]) in Isabelle/HOL. However, this notion of bisimulation is a "concrete" bisimulation (as opposed to symbolic), which means that the structure of observer theories is less involved and much easier to deal with compared to its symbolic counterpart. Our work on observer theories is mostly orthogonal to their work, and it can eventually be integrated into their formalisation to provide a completely formalised open bisimulation for the spi-calculus. Such an integration may not be too difficult, given that much of their work, e.g., formalisation of the operational semantics of the spi-calculus, can be reused without modifications.

We assume that the reader is familar with Isabelle proof assistant, its object logic HOL and logical frameworks in general. In the remainder of this section we briefly describe relevant Isabelle notations used throughout the paper. In Section 2, we give an overview of observer theories and an intuition behind them. We also give a brief description of two problems that will be the focus of subsequent sections, namely, those that concern decidability of consistency checking for (symbolic) observer theories. In Section 3 we consider formalisation of a notion of theory reduction and decidability of consistency checking for observer theories. In Section 4 we discuss a symbolic representation of observer theories using pairs of symbolic traces [5], called *bi-traces*, their consistency requirements and a notion of *respectful substitutions*. We prove a key lemma which relates a symbolic technique for trace refinement [5] to bi-traces, and discuss how this may lead to a decision procedure for testing bi-trace consistency. Section 5 concludes.

Isabelle notation. The Isabelle codes for the results of this paper can be found at `http://users.rsise.anu.edu.au/~jeremy/isabelle/2005/spi/`. In the statement of lemma or theorem, a name given in `typewriter` font indicates the name of the relevant theorem in our Isabelle development. We show selected theorems and definitions in the text, and more in the Appendix. A version of the paper, including the Appendix, is in `http://users.rsise.anu.edu.au/~jeremy/pubs/spi/fotesb/`. So now we indicate some key points of the Isabelle notation.

- A name preceded by ? indicates a variable: other names are entities which have been defined as part of the theory
- Conclusion β depending on assumptions α_i is `[| `α_1`;`α_2`; ... ;`α_n` |]` `==>` β
- $\forall$, $\exists$ are written as `ALL`, `EX`
- $\subseteq$, $\supseteq$, $\in$ are written as `<=`, `>=`, `:`

2   Observer Theory

An observer theory describes the knowledge accumulated by an observer in its interaction with a process (in the form of messages sent over networks), and its capability in analyzing and synthesizing messages. Since messages can be encrypted, and the encryption key may be unknown to the observer, it is not always the case that the observer can decompose all messages sent over the networks. In the presence of an active intruder, the traditional notion of bisimulation is not fine grained enough to prove interesting equivalence of protocols. A notion of bisimulation in which the knowledge and capability of the intruder is taken into account is often called an *environment-sensitive bisimulation.*

Messages are expressions formed from names, pairing constructor, e.g., $\langle M, N \rangle$, and symmetric encryption, e.g., $\{M\}_K$, where K is the encryption key and M is the message being encrypted. Note that we restrict to pairing and encryption to simplify discussion; there is no difficulty in extending the set of messages to include other constructors, including asymmetric encryption, natural numbers, etc. For technical reasons, we shall distinguish two kinds of names: *flexible names* and *rigid names*. We shall refer to flexible names as simply names. Names will be denoted with lower-case letters, e.g., a, x, y, etc., and rigid names will be denoted with bold letters, e.g., **a**, **b**, etc. We let $\mathcal{N}$ denote the set of names and $\mathcal{N}^=$ denote the set of pairs (x, x) of the same name. A name is really just a variable, i.e., a site for substitutions, and rigid names are just constants.

This slightly different terminology is to conform with a "tradition" in name-passing process calculi where names are sometimes confused with variables (see e.g., [13]). In the context of open bisimulation for the spi-calculus [14], names stand for undetermined messages which can be synthesized by the observer.

There are two aspects of an observer theory which are relevant to bisimulation methods for protocols verification (for a more detailed discussion, see, e.g., [2]):

- *Message analysis and synthesis*: This is often formalised as a deduction system with judgments of the form $\Sigma \vdash M$, where Σ is a set of messages and M is a message. The intuitive meaning is that the observer can derive M given Σ. The deduction system is given in Figure 1 using sequent calculus. The usual formulation is based on natural deduction, but there is an easy correspondence between the two presentations (see [16] for details). One can derive, for example, that $\Sigma \vdash M$ holds if $\Sigma \vdash \{M\}_K$ and $\Sigma \vdash K$ hold, i.e., if the observer can derive $\{M\}_K$ and the key K, then it can derive M.
- *Indistinguishability of messages*: This notion arises when an observer tries to differentiate two processes based on the messages output by the processes. In the presence of encryption, indistinguishability does not simply mean syntactic equality. The judgment of interest in this case takes the form $\Gamma \vdash M \leftrightarrow N$ where Γ is a finite set of pairs of messages. It means, intuitively, that the observer cannot distinguish between M and N, given the *indistinguishability assumption* Γ. We shall not go into detailed discussion on this notion of indistinguishability; it has been discussed extensively in the literature [2, 6, 8, 14]. Instead we give a proof system for message indistinguishability (or message equivalence) in Figure 2.

Note that there are some minor differences between the inference rules in Figure 1 and Figure 2 and those given in [14]. That is, the "principal" message pairs for the rules (pl) and (el) in [14], $(\langle M_a, M_b \rangle, \langle N_a, N_b \rangle)$ and $(\{M_p\}_{M_k}, \{N_p\}_{N_k})$, are also in the premises. We proved that the alternative system is equivalent and that, in both systems, weakening on the left of $\vdash$ is admissible: see Appendix A.1.

We note that, by a cut-admissibility-like result, it is possible to further remove (M_k, N_k) from the second premise of (el): see Appendix A.2.

Subsequent results in this paper are concerned mainly with the above notion of indistinguishability. We therefore identify an *observer theory* with its underlying indistinguishability assumptions (i.e., Γ in the second item above). Hence, from now on, an observer theory (or theory) is a just finite set of pairs of messages, and will be denoted with Γ. Given a theory Γ, we write $\pi_1(\Gamma)$ to denote the set obtained by projecting on the first components of the pairs in Γ. The set $\pi_2(\Gamma)$ is defined analogously.

Observer theory consistency: An important notion in the theory of environment sensitive bisimulation is that of *consistency* of an observer theory. This amounts to the requirement that any observation (i.e., any "destructive" operations related to constructors of the messages, e.g., projection, decryption) that is applicable to the first projection of the theory is also applicable to the second projection. For example, the theory $\{(\{\mathbf{a}\}_\mathbf{b}, \{\mathbf{c}\}_\mathbf{d}), (\mathbf{b}, \mathbf{c})\}$ is not consistent, since on the first projection (i.e., the set $\{\{\mathbf{a}\}_\mathbf{b}, \mathbf{b}\}$), one can decrypt the first message

$$\frac{x \in \mathcal{N}}{\Sigma \vdash x} \ (var) \qquad \frac{}{\Sigma, M \vdash M} \ (id) \qquad \frac{\Sigma \vdash M \quad \Sigma \vdash N}{\Sigma \vdash \langle M, N \rangle} \ (pr)$$

$$\frac{\Sigma \vdash M \quad \Sigma \vdash N}{\Sigma \vdash \{M\}_N} \ (er) \qquad \frac{\Sigma, M, N \vdash R}{\Sigma, \langle M, N \rangle \vdash R} \ (pl) \qquad \frac{\Sigma \vdash N \quad \Sigma, M, N \vdash R}{\Sigma, \{M\}_N \vdash R} \ (el)$$

Fig. 1. A proof system for message synthesis

$$\frac{x \in \mathcal{N}}{\Gamma \vdash x \leftrightarrow x} \ (var) \qquad \frac{(M, N) \in \Gamma}{\Gamma \vdash M \leftrightarrow N} \ (id)$$

$$\frac{\Gamma \vdash M_a \leftrightarrow N_a \quad \Gamma \vdash M_b \leftrightarrow N_b}{\Gamma \vdash \langle M_a, M_b \rangle \leftrightarrow \langle N_a, N_b \rangle} \ (pr) \qquad \frac{\Gamma \vdash M_p \leftrightarrow N_p \quad \Gamma \vdash M_k \leftrightarrow N_k}{\Gamma \vdash \{M_p\}_{M_k} \leftrightarrow \{N_p\}_{N_k}} \ (er)$$

$$\frac{\Gamma, (M_a, N_a), (M_b, N_b) \vdash M \leftrightarrow N}{\Gamma, (\langle M_a, M_b \rangle, \langle N_a, N_b \rangle) \vdash M \leftrightarrow N} \ (pl)$$

$$\frac{\Gamma \vdash M_k \leftrightarrow N_k \quad \Gamma, (M_p, N_p), (M_k, N_k) \vdash M \leftrightarrow N}{\Gamma, (\{M_p\}_{M_k}, \{N_p\}_{N_k}) \vdash M \leftrightarrow N} \ (el)$$

Fig. 2. A proof system for deducing message equivalence

$\{a\}_b$ using the second message **b**, but the same operation cannot be done on the second projection. The formal definition of consistency involves checking all message pairs (M, N) such that $\Gamma \vdash M \leftrightarrow N$ is derivable for certain similarity of observations. The first part of this paper is about verifying that this infinite quantification is not necessary. This involves showing that for every theory Γ, there is a corresponding *reduced theory* that is equivalent, but for which consistency checking requires only checking finitely many message pairs.

Symbolic observer theory: The definition of open bisimulation for name-passing calculi, such as the π-calculus, typically includes closure under a certain notion of *respectful substitutions* [13]. In the π-calculus, this notion of respectfulness is defined w.r.t. to a notion of *distinction* among names, i.e., an irreflexive relation on names which forbids identification of certain names. In the case of the spi-calculus, things get more complicated because the bisimulation relation is indexed by an observer theory, not just a simple distinction on names. We need to define a symbolic representation of observer theories, and an appropriate notion of consistency for the symbolic theories. These are addressed in [14] via a structure called *bi-traces*. A bi-trace is essentially a list of pairs of messages. It can be seen as a pair of symbolic traces, in the sense of [5]. The order of the message pairs in the list indicates the order of their creation (i.e., by the intruder or by the processes themselves). Names in a bi-trace indicate undetermined messages, which are open to instantiations. Therefore the notion of consistency of bi-traces needs to take into account these possible instantiations. Consider the following sequence of message pairs: $(\mathbf{a}, \mathbf{d}), (\{\mathbf{a}\}_\mathbf{b}, \{\mathbf{d}\}_\mathbf{k}), (\{\mathbf{c}\}_{\{x\}_\mathbf{b}}, \{\mathbf{k}\}_\mathbf{l})$. Considered as a theory, it is consistent, since none of the encryption keys are known to the observer. However, if we allow x to be instantiated to a, then the resulting theory $\{(\mathbf{a}, \mathbf{d}), (\{\mathbf{a}\}_\mathbf{b}, \{\mathbf{d}\}_\mathbf{k}), (\{\mathbf{c}\}_{\{\mathbf{a}\}_\mathbf{b}}, \{\mathbf{k}\}_\mathbf{l})\}$ is inconsistent, since on the first projection, $\{a\}_b$ can be used as a key to decrypt $\{c\}_{\{a\}_b}$, while in the second projection, no decryption is possible. Therefore to check consistency of a bi-trace, one needs to consider potentially infinitely many instances of the bi-trace. Section 4 shows some key steps to simplify consistency checking for bi-traces.

3   Observer Theory Reduction and Consistency

We now discuss our formalisation of observer theory and its consistency properties in Isabelle/HOL.

The datatype for messages is represented in Isabelle/HOL as follows.

```
datatype msg = Name nat | Rigid nat | Mpair msg msg | Enc msg msg
```

A observer theory, as already noted, is a finite set of pairs of messages. In Isabelle, we just use a set of pairs, so the finiteness condition appears in the Isabelle statements of many theorems. The judgment $\Gamma \vdash M \leftrightarrow N$ is represented by $(\Gamma, (M, N))$, or, equivalently in Isabelle, (Γ, M, N).

In Isabelle we define, inductively, a set of sequents indist which is the set of sequents derivable in the proof system for message equivalence (Figure 2). Subsequently we found it helpful to define the corresponding set of rules explicitly, calling them indpsc. The rules for message synthesis, given in Figure 1, are just a projection to one component of the rule set indpsc; we call this projection smpsc. It is straightforward to extend the notion of a projection on rule sets, so we can define the rules for message synthesis as simply smpsc $= \pi_1($indpsc$)$. The formal expression in Isabelle is more complex: see Appendix A.3. Likewise, we write $pair(X)$ to turn each message M into the pair (M, M) in a theory, sequent, rule or bi-trace X.

The following lemma relates message synthesis and message equivalence. Lemma 1(d) depends on theory consistency, to be introduced later.

Lemma 1. *(a)* (smpsc_alt) *Rule* $R \in smpsc$ *iff* $pair(R) \in indpsc$
(b) (slice_derrec_smpsc_empty) *if* $\Gamma \vdash M \leftrightarrow N$ *then* $\pi_1(\Gamma) \vdash M$
(c) (derrec_smpsc_eq) $\Sigma \vdash M$ *if and only if* $pair(\Sigma) \vdash M \leftrightarrow M$
(d) (smpsc_ex_indpsc_der) *if* $\pi_1(\Gamma) \vdash M$ *and* Γ *is consistent, then there exists* N *such that* $\Gamma \vdash M \leftrightarrow N$

3.1   Decidability of $\vdash$ and Computability of Theory Reduction

The first step towards deciding theory consistency is to define a notion of *theory reduction*. Its purpose is to extract a "kernel" of the theory with no redundancy, that is, no pairs in the kernel are derivable from the others. We need to establish the decidability of $\vdash$, and then termination of the theory reduction. In [14], Tiu observes that $\Gamma \vdash M \leftrightarrow N$ is decidable, because the right rules (working upwards) make the right-hand side messages smaller, and the left rules saturate the antecedent theory with more pairs of smaller messages. Hence for a given end sequent, there are only finitely many possible sequents which can appear in any proof of the sequent. Some results relevant to this argument for decidability are presented in Appendix A.4. Here we present an alternative proof for the decidability of $\vdash$ and termination of theory reduction.

Tiu [14, Definition 4] defines a reduction relation of observer theories:

$$\Gamma, (\langle M_a, M_b \rangle, \langle N_a, N_b \rangle) \longrightarrow \Gamma, (M_a, N_a), (M_b, N_b)$$
$$\Gamma, (\{M_p\}_{M_k}, \{N_p\}_{N_k}) \longrightarrow \Gamma, (M_p, N_p), (M_k, N_k)$$
$$\text{if } \Gamma, (\{M_p\}_{M_k}, \{N_p\}_{N_k}) \vdash M_k \leftrightarrow N_k$$

We assume that Γ does not contain $(\langle M_a, M_b \rangle, \langle N_a, N_b \rangle)$ and $(\{M_p\}_{M_k}, \{N_p\}_{N_k})$ respectively (otherwise reduction would not terminate). This reduction relation is terminating and confluent, and so every theory Γ reduces to a unique normal form $\Gamma{\Downarrow}$. It also preserves the entailment $\vdash$.

Lemma 2. (a) [15, Lemma 15] (red_nc) If $\Gamma \longrightarrow \Gamma'$ then $\Gamma \vdash M \leftrightarrow N$ if and only if $\Gamma' \vdash M \leftrightarrow N$
(b) (nf_nc) Assuming that $\Gamma{\Downarrow}$ exists, $\Gamma \vdash M \leftrightarrow N$ if and only if $\Gamma{\Downarrow} \vdash M \leftrightarrow N$

It is easy to show that $\longrightarrow$ is well-founded, since the sum of the sizes of [the first member of each of] the message pairs reduces each time. Confluence is reasonably easy to see since the side condition for the second rule is of the form $\Gamma' \vdash M_k \leftrightarrow N_k$ where Γ' is exactly the theory being reduced, and, from Lemma 2, this condition (for a particular M_k, N_k) will continue to hold, or not, when other reductions have changed Γ'. Actually, proving confluence in Isabelle was not so easy, and we describe the difficulty and our proof in Appendix A.6. Then it is a standard result, and easy in Isabelle, that confluence and termination give normal forms.

Theorem 3 (nf_oth_red). Any theory Γ has a $\longrightarrow$-normal form $\Gamma{\Downarrow}$.

A different reduction relation. As a result of Lemma 2, to decide whether $\Gamma \vdash M \leftrightarrow N$ one might calculate $\Gamma{\Downarrow}$ and determine whether $\Gamma{\Downarrow} \vdash M \leftrightarrow N$, which is easier (see Lemma 5). However to calculate $\Gamma{\Downarrow}$ requires determining whether $\Gamma \vdash M_k \leftrightarrow N_k$, so the decidability of this procedure is not obvious.

We defined an alternative version, $\longrightarrow'$, of the reduction relation, by changing the condition in the second rule, so our new relation is:

$$\Gamma, (\langle M_a, M_b \rangle, \langle N_a, N_b \rangle) \longrightarrow' \Gamma, (M_a, N_a), (M_b, N_b)$$
$$\Gamma, (\{M_p\}_{M_k}, \{N_p\}_{N_k}) \longrightarrow' \Gamma, (M_p, N_p), (M_k, N_k) \quad \text{if } \Gamma \vdash M_k \leftrightarrow N_k$$

This definition does not give the same relation, but we are able to show that the two relations have the same normal forms. Using this reduction relation, the procedure to decide whether $\Gamma \vdash M \leftrightarrow N$ is: calculate $\Gamma{\Downarrow}$ and determine whether $\Gamma{\Downarrow} \vdash M \leftrightarrow N$. Calculating $\Gamma{\Downarrow}$ requires deciding questions of the form $\Gamma' \vdash M_k \leftrightarrow N_k$, where Γ' is smaller than Γ (because a pair $(\{M_p\}_{M_k}, \{N_p\}_{N_k})$ is omitted). Thus this procedure terminates.

Note that Lemma 2 also holds for $\longrightarrow'$ since $\longrightarrow' \subseteq \longrightarrow$.

To show the two relations have the same normal forms, we first show (in Theorem 4(b)) that if Γ is $\longrightarrow$-reducible, then it is $\longrightarrow'$-reducible, even though the same reduction may not be available.

Theorem 4. (a) (red_alt_lem) If $\Gamma \vdash M_k \leftrightarrow N_k$ then either

$\Gamma \setminus \{(\{M_p\}_{M_k}, \{N_p\}_{N_k})\} \vdash M_k \leftrightarrow N_k$ or there exists Γ' such that $\Gamma \longrightarrow' \Gamma'$,

(b) (oth_red_alt_lem) If $\Gamma \longrightarrow \Delta$ then there exists Δ' such that $\Gamma \longrightarrow' \Delta'$

(c) (rsmin_or_alt) If Γ is $\longrightarrow'$-minimal (i.e., cannot be reduced further) then Γ is $\longrightarrow$-minimal

(d) (nf_acc_alt) $\Gamma \longrightarrow' \Gamma{\Downarrow}$ (where $\Gamma{\Downarrow}$ is the $\longrightarrow$-normal form of Γ)

(e) (nf_alt, nf_same) $\Gamma{\Downarrow}$ is also the $\longrightarrow'$-normal form of Γ

Proof. We show a proof of (a) here. We prove a stronger result namely: If $\Gamma \vdash M \leftrightarrow N$ and $size\, M \leq size\, Q_k$ then either $\Gamma' = \Gamma \setminus \{(\{Q_p\}_{Q_k}, \{R_p\}_{R_k})\} \vdash M \leftrightarrow N$ or there exists Δ such that $\Gamma \longrightarrow' \Delta$.

We prove it by induction on the derivation of $\Gamma \vdash M \leftrightarrow N$. If the derivation is by the (var) rule, ie, $(M, N) = (x, x)$, then clearly $\Gamma' \vdash M \leftrightarrow N$ by the (var) rule. If the derivation is by the (id) rule, ie, $(M, N) \in \Gamma$, then the size condition shows that $(M, N) \in \Gamma'$, and so $\Gamma' \vdash M \leftrightarrow N$ by the (id) rule.

If the derivation is by either of the right rules (pr) or (er), then we have $\Gamma \vdash M' \leftrightarrow N'$ and $\Gamma \vdash M'' \leftrightarrow N''$, according to the rule used, with M' and M'' smaller than M. Then, unless $\Gamma \longrightarrow' \Delta$ for some Δ, we have by induction $\Gamma' \vdash M' \leftrightarrow N'$ and $\Gamma' \vdash M'' \leftrightarrow N''$, whence, by the same right rule, $\Gamma' \vdash M \leftrightarrow N$.

If the derivation is by the left rule (pl), then $\Gamma \longrightarrow' \Delta$ for some Δ.

If the derivation is by the left rule (el), then we apply the inductive hypothesis to the *first* premise of the rule. Let the "principal" message pair for the rule be $(\{M_p\}_{M_k}, \{N_p\}_{N_k})$, so the first premise is $\Gamma \vdash M_k \leftrightarrow N_k$. Note that we apply the inductive hypothesis to a possibly *different* pair of encrypts in Γ, namely $(\{M_p\}_{M_k}, \{N_p\}_{N_k})$ instead of $(\{Q_p\}_{Q_k}, \{R_p\}_{R_k})$.

By induction, either $\Gamma \longrightarrow' \Delta$ for some Δ or (since $size\, M_k \leq size\, M_k$), we have $\Gamma \setminus \{(\{M_p\}_{M_k}, \{N_p\}_{N_k})\} \vdash M_k \leftrightarrow N_k$. Then we have $\Gamma \longrightarrow' \Delta$, as required, where $\Delta = \Gamma \setminus \{(\{M_p\}_{M_k}, \{N_p\}_{N_k})\}, (M_p, N_p), (M_k, N_k)$.   □

Since the process of reducing a theory essentially replaces pairs of compound messages with more pairs of simpler messages, this suggests that to show that $\Gamma \vdash M \leftrightarrow N$ for a reduced Γ, one need only use the rules which build up pairs of compound messages on the right. That is, one would use the right rules (pr) and (er), but not the left rules (pl) and (el). Let us define $\Gamma \vdash_r M \leftrightarrow N$ to mean that $\Gamma \vdash M \leftrightarrow N$ can be derived using the rules $(var), (id), (pr)$ and (er) of Figure 2. We call the set of these rules indpsc_virt.

We define a function is_der_virt which shows how to test $\Gamma \vdash_r M \leftrightarrow N$, and, in Lemma 5(b), prove that it does this. It terminates because at each recursive call, the size of M gets smaller. When we define a function in this way, Isabelle requires termination to be *proved* (usually it can do this automatically). Then *inspection* of the function definition shows that, assuming the theory oth is finite, the function is finitely computable. We discuss this idea further later. We also define a simpler function `is_der_virt_red`, as an alternative to `is_der_virt`, which gives the same result when Γ is reduced, see Appendix A.13.

```
recdef "is_der_virt" "measure (%(oth, M, N). size M)"
  "is_der_virt (oth, Name i, Name j) = ((Name i, Name j) : oth | i = j)"
  "is_der_virt (oth, Mpair Ma Mb, Mpair Na Nb) =
    ((Mpair Ma Mb, Mpair Na Nb) : oth |
      is_der_virt (oth, Ma, Na) & is_der_virt (oth, Mb, Nb))"
  "is_der_virt (oth, Enc Mp Mk, Enc Np Nk) =
    ((Enc Mp Mk, Enc Np Nk) : oth |
      is_der_virt (oth, Mp, Np) & is_der_virt (oth, Mk, Nk))"
  "is_der_virt (oth, M, N) = ((M, N) : oth)"
```

Lemma 5. *(a)* (nf_no_left) *If Γ is reduced and $\Gamma \vdash M \leftrightarrow N$ then $\Gamma \vdash_r M \leftrightarrow N$*
(b) (virt_dec) *$\Gamma \vdash_r M \leftrightarrow N$ if and only if* `is_der_virt` *$(\Gamma, (M, N))$*

We can now define a function `reduce` which computes a $\longrightarrow'$-normal form.

```
recdef (permissive) "reduce" "measure (setsum (size o fst))"
  "reduce S = (if infinite S then S else
    let P = (%x. x : Mpairs <*> Mpairs & x : S) ;
        Q = (%x. (if x : Encs <*> Encs & x : S then
            is_der_virt (reduce (S - {x}), keys x) else False))
    in if Ex P then reduce (red_pair (Eps P) (S - {Eps P}))
      else if Ex Q then reduce (red_enc (Eps Q) (S - {Eps Q}))
      else S)"
```

To explain this: $P (M, N)$ means $(M, N) \in S$ and M, N are both pairs; $Q (M, N)$ means $(M, N) \in S$ and M, N are both encrypts, say $\{M_p\}_{M_k}, \{N_p\}_{N_k}$, where $S \setminus \{(M, N)\} \vdash_r (M_k, N_k)$; `red_pair` and `red_enc` do a single step reduction based on the message pairs or encrypts given as their argument, `Ex P` means $\exists x. \ P \ x$, and `Eps P` means some x satisfying P, if such exists. Thus the function selects arbitrarily a pair of message pairs or encrypts suitable for a single reduction step, performs that step, and then reduces the result.

The expression `measure (setsum (size o fst))` is the termination measure, the sum of the sizes of the first member of each message pair in a theory. The function `reduce` is recursive, and necessarily terminates since at each iteration this measure function, applied to the argument, is smaller. However this function definition is sufficiently complicated that Isabelle cannot automatically prove that it terminates — thus the notation (`permissive`) in the definition.

Isabelle produces a complex definition dependent on conditions that if we change a theory by applying `red_pair` or `red_enc`, or by deleting a pair, then we get a theory which is smaller according to the measure function. Since in the HOL logic of Isabelle all functions are total, we have a function `reduce` in any event; we need to prove the conditions to prove that `reduce` conforms to the definition given above. We then get Theorem 6(a) and (b), which show how to test $\Gamma \vdash M \leftrightarrow N$ as a manifestly finitely computable function. We also prove a useful characterisation of $\Gamma \Downarrow$.

Theorem 6. *(a)* (`reduce_nf`, `reduce_nf_alt`) *reduce* $\Gamma = \Gamma\Downarrow$
(b) (`virt_reduce`, `idvr_reduce`) $\Gamma \vdash M \leftrightarrow N$ *if and only if* `is_der_virt`
$(\Gamma\Downarrow, (M, N))$, *equivalently, if and only if* `is_der_virt_red` $(\Gamma\Downarrow, (M, N))$
(c) (`reduce_alt`) *For* $(M, N) \notin \mathcal{N}^=$, $(M, N) \in \Gamma\Downarrow \setminus \mathcal{N}^=$ *iff*
  (i) $\Gamma \vdash M \leftrightarrow N$,
  (ii) M *and* N *are not both pairs, and*
  (iii) if $M = \{M_p\}_{M_k}, N = \{N_p\}_{N_k}$, *then* $\Gamma \nvdash M_k \leftrightarrow N_k$

As Urban et al point out in [17] formalising decidability — or computability — is difficult. It would require formalising the computation process, as distinct from simply defining the quantity to be computed. However, as is done in [17, §3.4], it is possible to define a quantity in a certain way which makes it reasonable to assert that it is computable. This is what we have aimed to do in defining the function **reduce**. It specifies the computation to be performed (with a caveat mentioned later). Isabelle requires us to show that this computation is terminating, and we have shown that it produces the $\longrightarrow'$-normal form. To ensure termination, we needed to base the definition of **reduce** on $\longrightarrow'$, not on $\longrightarrow$, but by Theorem 4(e), $\longrightarrow'$ and $\longrightarrow$ have the same normal forms.

Certain terminating functions are not obviously computable, for example $f\ x = (\exists y.\ P\ y)$ (even where P is computable). So our definition of **reduce** requires inspection to ensure that it contains nothing which makes it not computable. It does contain existential quantifiers, but they are in essence quantification over a finite set. The only problem is the subterms `Eps P` and `Eps Q`, that is $\epsilon x.\ P\ x$ and $\epsilon x.\ Q\ x$. These mean "some x satisfying P" (similarly Q). In Isabelle's logic, this means *some* x, but we have no knowledge of which one (and so we cannot perform precisely this computation). But our proofs went through without any knowledge of which x is denoted by $\epsilon x.\ P\ x$. Therefore it would be safe to implement a computation which makes any choice of $\epsilon x.\ P\ x$, and we can safely assert that our proofs would still hold for that computation.[1] That is, in general we assert that if a function involving $\epsilon x.\ P\ x$ can be proven to have some property, then a function which replaces $\epsilon x.\ P\ x$ by some other choice of x (satisfying P if possible) would also have that property. Based on this assertion we say that our definition of **reduce** shows that the $\longrightarrow$-normal form is computable, and so that $\Gamma \vdash M \leftrightarrow N$ is decidable.

We found that although the definition of **reduce** gives the function in a computable form, many proofs are much easier using the characterisation as the normal form. For example Lemma 2(b) is much easier using Lemma 2(a) than using the definition of **reduce**. We found this with some other results, such as: if Γ is finite, then so is $\Gamma\Downarrow$, and if Γ consists of identical pairs then so does $\Gamma\Downarrow$.

Since also Γ and $\Gamma\Downarrow$ entail the same message pairs, it is reasonable to ask which theories, other than those with the same normal form as Γ, entail the same message pairs as Γ. Now it is clear, due to the (*var*) rule, that deleting (x, x) from a theory does not change the set of entailed message pairs or the

---

[1] In general a repeated choice must be made consistently; the HOL logic *does* imply $\epsilon x.\ P\ x = \epsilon x.\ P\ x$. This point clearly won't arise for the **reduce** function.

reductions available. However we find that the condition is that theories entail the same pairs iff their normal forms are equal, modulo $\mathcal{N}^=$.

We could further change $\longrightarrow'$ by deleting the (M_k, N_k) from the second rule. Lemma 2(a) holds for this new relation. For further discussion see Appendix A.7.

Theorem 7. (a) (rsmin_names) Γ is reduced if and only if $\Gamma \setminus \mathcal{N}^=$ is reduced
(b) [15, Lemma 8] (name_equivd) $\Gamma \vdash M \leftrightarrow N$ if and only if $\Gamma \setminus \mathcal{N}^= \vdash M \leftrightarrow N$
(c) (nf_equiv_der) Theories Γ_1 and Γ_2 entail the same message pairs if and only if $\Gamma_1{\Downarrow} \setminus \mathcal{N}^= = \Gamma_2{\Downarrow} \setminus \mathcal{N}^=$

3.2   Theory Consistency

Definition 8. [15, Definition 11] A theory Γ is consistent if for every M and N, if $\Gamma \vdash M \leftrightarrow N$ then the following hold:

(a) M and N are of the same type of expressions, i.e., M is a pair (an encrypted message, a (rigid) name) if and only if N is.
(b) If $M = \{M_p\}_{M_k}$ and $N = \{N_p\}_{N_k}$ then $\pi_1(\Gamma) \vdash M_k$ implies $\Gamma \vdash M_k \leftrightarrow N_k$ and $\pi_2(\Gamma) \vdash N_k$ implies $\Gamma \vdash M_k \leftrightarrow N_k$.
(c) For any R, $\Gamma \vdash M \leftrightarrow R$ implies $R = N$ and $\Gamma \vdash R \leftrightarrow N$ implies $R = M$.

This definition of consistency involves infinite quantification. We want to eliminate this quantification by finding a finite characterisation on *reduced theories*. But first, let us define another equivalent notion of consistency, which is simpler for verification, as it does not use the deduction system for message synthesis.

Definition 9. A theory Γ satisfies the predicate thy_cons if for every M and N, if $\Gamma \vdash M \leftrightarrow N$ then the following hold:

(a) M and N are of the same type of expressions, i.e., as in Definition 8(a)
(b) for every M, N', M_p, N_p if $\Gamma \vdash M' \leftrightarrow N'$ or $\Gamma \vdash \{M_p\}_{M'} \leftrightarrow \{N_p\}_{N'}$, then $M' = M$ iff $N' = N$

Lemma 10. (a) (thy_cons_equiv) Γ is consistent iff it satisfies Definition 9
(b) (thy_cons_equivd) Γ is consistent if and only if $\Gamma \setminus \mathcal{N}^=$ is consistent
(c) [15, Lemma 19] (nf_cons) Γ is consistent if and only if $\Gamma{\Downarrow}$ is consistent
(d) (cons_der_same) If Γ_1 and Γ_2 entail the same message pairs then Γ_1 is consistent if and only if Γ_2 is consistent

Tiu [15, Proposition 20] gives a characterisation of consistency (reproduced below in Proposition 11) which is finitely checkable. In Definition 12 we define a predicate thy_cons_red which is somewhat similar. In Theorem 13 we show that, for a reduced theory, that our thy_cons_red is equivalent to consistency and to the conditions in Proposition 11. Decidability of consistency then follows from decidability of $\vdash$, and termination of normal form computation.

Proposition 11. [15, Proposition 20] A theory Γ is consistent if and only if $\Gamma{\Downarrow}$ satisfies the following conditions: if $(M, N) \in \Gamma{\Downarrow}$ then

(a) M and N are of the same type of expressions, in particular, if $M = x$, for some name x, then $N = x$ and vice versa,

(b) if $M = \{M_p\}_{M_k}$ and $N = \{N_p\}_{N_k}$ then $\pi_1(\Gamma\Downarrow) \nvdash M_k$ and $\pi_2(\Gamma\Downarrow) \nvdash N_k$.

(c) for any $(U, V) \in \Gamma\Downarrow$, $U = M$ if and only if $V = N$.

Definition 12. *A theory Γ satisfies the predicate* **thy_cons_red** *if*

(a) for all $(M, N) \in \Gamma$, M and N satisfy Proposition 11(a)

(b) for all (M, N) and $(M', N') \in \Gamma$, $M' = M$ iff $N' = N$

(c) for all $(\{M_p\}_{M_k}, \{N_p\}_{N_k}) \in \Gamma$, for all M, N such that $\Gamma \vdash M \leftrightarrow N$, $M \neq M_k$ and $N \neq N_k$

Theorem 13. *(a) (*`tc_red_iff`*) Γ is consistent iff $\Gamma\Downarrow$ satisfies* **thy_cons_red**
(b) (`thy_cons_red_equiv`*) $\Gamma\Downarrow$ satisifes Proposition 11(a) to (c) iff it satisfies* **thy_cons_red***, ie, Definition 12(a) to (c)*

4 Respectful Substitutions and Bi-trace Consistency

We now consider a symbolic representation of observer theories from [14], given below. We denote with $fn(M)$ the set of names in M. This notation is extended straightforwardly to pairs of messages, lists of (pairs of) messages, etc.

Definition 14. *A* bi-trace *is a list of message pairs marked with i (indicating input) or o (output), i.e., elements in a bi-trace have the form $(M, N)^i$ or $(M, N)^o$. Bi-traces are ranged over by h. We denote with $\pi_1(h)$ the list obtained from h by taking the first component of the pairs in h. The list $\pi_2(h)$ is defined analogously. Bi-traces are subject to the following restriction: if $h = h_1.(M, N)^o.h_2$ then $fn(M, N) \subseteq fn(h_1)$. We write $\{h\}$ to denote the set of message pairs obtained from h by forgetting the marking and the order.*

Names in a bi-trace represent *symbolic values* which are input by a process at some point. This explains the requirement that the free names of an output pair in a bi-trace must appear before the output pair. We express this restriction on name occurrences by defining a predicate `validbt` on lists of marked message pairs, and we do not mention it in the statement of each result, although it does appear in their statements in Isabelle. In our Isabelle representation the list is reversed, so that the latest message pair is the first in the list. The theory $\{h\}$ obtained from a bi-trace h is represented by `oth_of` h. Likewise for a list s of marked messages (which can be seen as a symbolic trace [5]), we can define the set $\{s\}$ of messages by forgetting the annotations and ordering.

A substitution pair $\vec{\theta} = (\theta_1, \theta_2)$ replaces free names $x \in \mathcal{N}$ by messages, using substitutions $\theta_1(\theta_2)$ for the first (second) component of each pair. For a bi-trace h, $\vec{\theta}$ respects h, or is h-respectful [15, Definition 34], if for every free name x in an input pair $(M, N)^i$, $\{h'\}\vec{\theta} \vdash x\theta_1 \leftrightarrow x\theta_2$, where h' is the part of h preceding $(M, N)^i$. This is expressed in Isabelle by $h \in$ `bt_resp` $\vec{\theta}$.

Definition 15. *[15, Definition 35] The set of* consistent bi-traces *are defined inductively (on the length of bi-traces) as follows:*

(a) The empty bi-trace is consistent.
(b) If h is a consistent bi-trace then $h.(M, N)^i$ is also a consistent bi-trace, provided that $h \vdash M \leftrightarrow N$.
(c) If h is a consistent bi-trace, then $h' = h.(M, N)^o$ is a consistent bi-trace, provided that for every h-respectful substitution pair $\vec{\theta}$, if $h\vec{\theta}$ is a consistent bi-trace then $\{h'\vec{\theta}\}$ is a consistent theory.

Given Lemma 16(c) below, it may appear that leaving out the underlined words of Definition 15 would make no difference. This minor fact can indeed be proved formally: details are given in Appendix A.16.

The following are significant lemmas from [15] which we proved in Isabelle. As an illustration of the value of automated theorem proving, we found that the original proof of (b) in a draft of [15] contained an error (which was easily fixed).

Lemma 16. *(a) [15, Lemma 24] (subst_indist) Let $\Gamma \vdash M \leftrightarrow N$ and let $\vec{\theta} = (\theta_1, \theta_2)$ be a substitution pair such that for every free name x in Γ, M or N, $\Gamma\vec{\theta} \vdash \theta_1(x) \leftrightarrow \theta_2(x)$. Then $\Gamma\vec{\theta} \vdash M\theta_1 \leftrightarrow N\theta_2$.*

(b) [15, Lemma 40] (bt_resp_comp) Let h be a consistent bi-trace, let $\vec{\theta} = (\theta_1, \theta_2)$ be an h-respectful substitution pair, and let $\vec{\gamma} = (\gamma_1, \gamma_2)$ be an $h\vec{\theta}$-respectful substitution pair. Then $\vec{\theta} \circ \vec{\gamma}$ is also h-respectful.

(c) [15, Lemma 41] (cons_subs_bt) If h is a consistent bi-trace and $\vec{\theta} = (\theta_1, \theta_2)$ respects h, then $h\vec{\theta}$ is also a consistent bi-trace.

Respectfulness of a substitution relative to a theory. Testing consistency of bi-traces involves testing whether a theory Γ is consistent after applying any respectful substitution pair $\vec{\theta}$ to it. We will present some results that (under certain conditions) if we reduce $\{h\}$ first, and then apply an h-respectful substitution, then the result is a reduced theory, to which the simpler test for consistency, thy_cons_red, applies.

The complication here is that reduction applies to a theory whereas the definition of bi-trace consistency crucially involves the ordering of the pairs of messages. We overcome this by devising the notion, thy_strl_resp, of a substitution being respectful with respect to an (unordered) theory and an ordered list of sets of variable names. Importantly, this property holds for $\{h\}$ where $\vec{\theta}$ is h-respectful, and it is preserved by reducing a theory. We use this to prove some later results involving $\{h\}\Downarrow$ and h-respectful substitutions, such as Theorem 17. Details are in Appendix A.19.

Simplifying testing consistency after substitution. Recall that a theory Γ is consistent if and only if $\Gamma\Downarrow$ is consistent (Lemma 10(c)), and if and only if $\Gamma \backslash \mathcal{N}^=$ is consistent (Lemma 10(b)). Thus, to determine whether Γ is consistent, one may calculate $\Gamma\Downarrow$ or $\Gamma\Downarrow \backslash \mathcal{N}^=$ (which is reduced, by Lemma 7(a)), and use the function thy_cons_red (by virtue of Theorem 13(a)). Therefore, the naive approach to testing bi-trace consistency is to apply θ to Γ and then reduce the result, and delete pairs $(x, x) \in \mathcal{N}^=$. We can derive results which permit a simpler approach.

Theorem 17. *Let h be a bi-trace, and let $\Gamma = \{h\}$. Let $\vec{\theta}$ be an h-respectful substitution pair, and denote its action on Γ by $\vec{\theta}$ also.*

(a) (nf_subst_nf_Ne) $\Gamma\vec{\theta}\Downarrow \setminus \mathcal{N}^= = (\Gamma\Downarrow \setminus \mathcal{N}^=)\vec{\theta}\Downarrow \setminus \mathcal{N}^=$

(b) (subst_nf_Ne_tc) $\Gamma\vec{\theta}$ *is consistent if and only if* $(\Gamma\Downarrow \setminus \mathcal{N}^=)\vec{\theta}$ *is consistent*

This, given a bi-trace h and a respectful substitution pair $\vec{\theta}$, if one wants to test whether $\Gamma\vec{\theta} = \{h\vec{\theta}\}$ is consistent, it makes no difference to the consistency of the resulting theory if one reduces the theory and deletes pairs (x, x) before substituting. This means that we need only consider substitution in a theory which is reduced and has pairs (x, x) removed.

If we disallow encryption where keys are themselves pairs or encrypts, then further simplification is possible. Thus we will require that keys are atomic (free names or rigid names, Name n or Rigid n), both initially and after substitution.

Theorem 18 (subs_not_red_ka). *Let Γ be reduced, consistent and have atomic keys. Then* $(\Gamma \setminus \mathcal{N}^=)\vec{\theta}$ *is reduced.*

Thus, if keys are atomic, the effect of Theorem 18 is to simplify the consistency test thus: to test the consistency of the substituted theory $\Gamma\vec{\theta}$, one reduces Γ to $\Gamma\Downarrow$ and deletes pairs (x, x) to get $\Gamma' = \Gamma\Downarrow \setminus \mathcal{N}^=$. One then considers substitution pairs $\vec{\theta}$ of Γ', knowing that any $\Gamma'\vec{\theta}$ is reduced and so the simpler criterion for theory consistency, thy_cons_red, applies to it. Thus we get:

Theorem 19. *Let h be a bi-trace, and let $\Gamma = \{h\}$, where Γ is consistent with atomic keys. Let $\vec{\theta}$ be an h-respectful substitution pair, and write $\Gamma\vec{\theta} = \{h\vec{\theta}\}$.*

(a) (nfs_comm) $\Gamma\vec{\theta}\Downarrow \setminus \mathcal{N}^= = (\Gamma\Downarrow \setminus \mathcal{N}^=)\vec{\theta} \setminus \mathcal{N}^=$

(b) (nfs_comm_tc) $\Gamma\vec{\theta}$ *is consistent iff* thy_cons_red *holds of* $(\Gamma\Downarrow \setminus \mathcal{N}^=)\vec{\theta} \setminus \mathcal{N}^=$

Unique Completion of a Respectful Substitution. A bi-trace can be projected into the fist or second component of each pair, giving lists of marked messages. We can equally project the definition of a respectful substitution pair, so that for a list s of marked messages, substitution θ_i respects s, $s \in$ sm_resp θ_i, iff for every free name x in an input message M^i, $\{s'\}\theta_i \vdash x\theta_i$, where $\vdash$ is here the message synthesis relation, and $\{s'\}$ is the set of marked messages prior to M^i. Given h, whose projections are s_1, s_2, if $\vec{\theta}$ respects h then clearly θ_i respects s_i (proved as bt_sm_resp, see Appendix A.20). Conversely, given θ_i which respects s_i ($i = 1$ or 2), can we complete θ_i to an h-respectful pair $\vec{\theta}$?

Theorem 20 (subst_exists, subst_unique). *Given a consistent bi-trace h whose projections to a single message trace are s_1 and s_2, and a substitution θ_1 which respects s_1, there exists θ_2 such that $\vec{\theta} = (\theta_1, \theta_2)$ respects h, and θ_2 is "unique" in the sense that any two such θ_2 act the same on names in $\pi_2(h)$.*

We defined a function which, given θ_1 in this situation, returns θ_2. First we defined a function match_rc1 which, given a theory Γ and a message M, "attempts" to determine a message N such that $\Gamma \vdash M \leftrightarrow N$. By Theorem 20 such a message is unique if Γ is consistent.

The definition of `match_rc1` (Appendix A.24) follows that of `is_der_virt_red` (Appendix A.13), so Theorem 21(a) holds whether or not Γ is actually reduced.

It will be seen that it involves testing for membership of a finite set, and corresponding uses of the ϵ operator, (as in the case of `reduce`, as discussed earlier). Therefore we assert that `match_rc1` is finitely computable.

The return type of `match_rc1` is *message* `option`, which is `Some` *res* if the result *res* is successfully found, or `None` to indicate failure.

Theorem 21. *(a)* (`match_rc1_iff_idvr`) *If Γ satisfies* `thy_cons_red`, *then*
    `is_der_virt_red` (Γ, M, N) *iff* `match_rc1` $\Gamma\ M = Some\ N$
(b) (`match_rc1_indist`) *If Γ is consistent, then*
    $\Gamma \vdash M \leftrightarrow N$ *iff* `match_rc1` $\Gamma{\Downarrow}\ M = Some\ N$

Then we defined a function `second_sub` which uses `match_rc1` to find the appropriate value of $x\theta_2$ for each new x which appears in the bi-trace, and we proved that `second_sub` does in fact compute the θ_2 of Theorem 20. See Appendix A.26 for the definition of `second_sub` and this result. The function `second_sub` tests membership of a finite set, and uses `reduce` and `match_rc1`, so we assert that `second_sub` is also finitely computable.

5   Conclusions and Further Work

We have modelled observer theories and bi-traces in the Isabelle theorem prover, and have confirmed, by proofs in Isabelle, the results of a considerable part of [14]. This work constitutes a significant step formalising open bisimulation for the spi-calculus in Isabelle/HOL, and ultimately towards a logical framework for proving process equivalence.

We discussed the issue of showing finite computability in Isabelle/HOL, using a mixed formal/informal argument, and building upon the discussion in Urban et al [17]. We defined a function `reduce` in Isabelle, and showed that it computes $\Gamma{\Downarrow}$. Isabelle required us to show that the function terminates. We asserted, with relevant discussion, that inspection shows that the definition does not introduce any infinite aspect into the computation and so asserted that therefore the function is finitely computable. Similarly, we provided a finitely computable function `is_der_virt` and proved that it tests $\Gamma \vdash M \leftrightarrow N$ for a reduced theory Γ.

We then considered bi-traces and bi-trace consistency. The problem here is that, to test bi-trace consistency, it is necessary to test whether $\Gamma\theta$ is consistent for all θ satisfying certain conditions. We proved a number of lemmas which simplify this task, and appear to lead to a finitely computable algorithm for this. In particular, our result on the unique completion of respectful substitutions that relates symbolic trace and bi-trace opens up the possibility to use symbolic trace refinement algorithm [5] to compute a notion of *bi-trace refinement*, which will be useful for bi-trace consistency checking.

Another approach to representating observer theories is to use equational theories, instead of deduction rules, e.g., as in the applied-pi calculus [1]. In this setting, the notion of consistency of a theory is replaced by the notion of *static equivalence* between knowledge of observers [1]. Baudet has shown that static equivalence between two symbolic theories is decidable [4], for a class of theories called subterm-convergent theories (which subsumes the Dolev-Yao model of intruder). It will be interesting to work out the precise correspondence between static equivalence and our notion of bi-trace consistency, as such correspondence may transfer proof techniques from one approach to the other.

Acknowledgment. We thank the anonymous referees for their comments on an earlier draft. This work is supported by the Australian Research Council through the Discovery Projects funding scheme (project number DP0880549).

References

1. Abadi, M., Fournet, C.: Mobile values, new names, and secure communication. In: POPL, pp. 104–115 (2001)
2. Abadi, M., Gordon, A.D.: A bisimulation method for cryptographic protocols. Nord. J. Comput. 5(4), 267–303 (1998)
3. Abadi, M., Gordon, A.D.: A calculus for cryptographic protocols: The spi calculus. Information and Computation 148(1), 1–70 (1999)
4. Baudet, M.: Sécurité des protocoles cryptographiques: aspects logiques et calculatoires. PhD thesis, École Normale Supérieure de Cachan, France (2007)
5. Boreale, M.: Symbolic trace analysis of cryptographic protocols. In: Orejas, F., Spirakis, P.G., van Leeuwen, J. (eds.) ICALP 2001. LNCS, vol. 2076, pp. 667–681. Springer, Heidelberg (2001)
6. Boreale, M., De Nicola, R., Pugliese, R.: Proof techniques for cryptographic processes. SIAM J. Comput. 31(3), 947–986 (2001)
7. Borgström, J., Briais, S., Nestmann, U.: Symbolic bisimulation in the spi calculus. In: Gardner, P., Yoshida, N. (eds.) CONCUR 2004. LNCS, vol. 3170, pp. 161–176. Springer, Heidelberg (2004)
8. Borgström, J., Nestmann, U.: On bisimulations for the spi calculus. Mathematical Structures in Computer Science 15(3), 487–552 (2005)
9. Dawson, J.E., Goré, R.: Formalising cut-admissibility for provability logic (submitted, 2009)
10. Dolev, D., Yao, A.: On the security of public-key protocols. IEEE Transactions on Information Theory 2(29) (1983)
11. Kahsai, T., Miculan, M.: Implementing spi calculus using nominal techniques. In: Beckmann, A., Dimitracopoulos, C., Löwe, B. (eds.) CiE 2008. LNCS, vol. 5028, pp. 294–305. Springer, Heidelberg (2008)
12. Milner, R., Parrow, J., Walker, D.: A calculus of mobile processes, Part II. Information and Computation, 41–77 (1992)
13. Sangiorgi, D.: A theory of bisimulation for the pi-calculus. Acta Inf. 33(1), 69–97 (1996)
14. Tiu, A.: A trace based bisimulation for the spi calculus: An extended abstract. In: Shao, Z. (ed.) APLAS 2007. LNCS, vol. 4807, pp. 367–382. Springer, Heidelberg (2007)
15. Tiu, A.: A trace based bisimulation for the spi calculus. Preprint (2009), http://arxiv.org/pdf/0901.2166v1
16. Tiu, A., Goré., R.: A proof theoretic analysis of intruder theories. In: Proceedings of RTA 2009 (to appear, 2009)
17. Urban, C., Cheney, J., Berghofer, S.: Mechanizing the metatheory of LF. In: LICS, pp. 45–56. IEEE Computer Society, Los Alamitos (2008)

Formal Certification of a Resource-Aware Language Implementation*

Javier de Dios and Ricardo Peña

Universidad Complutense de Madrid, Spain
C/ Prof. José García Santesmases s/n. 28040 Madrid
Tel.: 91 394 7627; Fax: 91 394 7529
jdcastro@aventia.com, ricardo@sip.ucm.es

Abstract. The paper presents the development, by using the proof assistant Isabelle/HOL, of a compiler back-end translating from a functional source language to the bytecode language of an abstract machine. The Haskell code of the compiler is extracted from the Isabelle/HOL specification and this tool is also used for proving the correctness of the implementation. The main correctness theorem not only ensures functional semantics preservation but also resource consumption preservation: the heap and stacks figures predicted by the semantics are confirmed in the translation to the abstract machine.

The language and the development belong to a wider Proof Carrying Code framework in which formal compiler-generated certificates about memory consumption are sought for.

Keywords: compiler verification, functional languages, memory management.

1 Introduction

The first-order functional language *Safe* has been developed in the last few years as a research platform for analysing and formally certifying two properties of programs related to memory management: absence of dangling pointers and having an upper bound to memory consumption.

Two features make *Safe* different from conventional functional languages: (a) the memory management system does not need a garbage collector; and (b) the programmer may ask for explicit destruction of memory cells, so that they could be reused by the program. These characteristics, together with the above certified properties, make *Safe* useful for programming small devices where memory requirements are rather strict and where garbage collectors are a burden both in space and in service availability.

The *Safe* compiler is equipped with a battery of static analyses which infer such properties [15,16,17,22]. These analyses are carried out on an intermediate language called *Core-Safe* (explained in Sec. 2.1), obtained after type-checking and desugaring the source language called *Full-Safe*. The back-end comprises two more phases:

1. A translation from *Core-Safe* to the bytecode language of an imperative abstract machine of our own, called the *Safe Virtual Machine* (SVM). We call this bytecode language *Safe-Imp* and it is explained in Sec. 2.4.

---

* Work partially funded by the projects TIN2008-06622-C03-01/TIN (STAMP), and S-0505/ TIC/ 0407 (PROMESAS).

S. Berghofer et al. (Eds.): TPHOLs 2009, LNCS 5674, pp. 196–211, 2009.
© Springer-Verlag Berlin Heidelberg 2009

2. A translation from *Safe-Imp* to the bytecode language of the *Java Virtual Machine* (JVM) [13].

We have proved our analyses correct and are currently generating Isabelle/HOL [20] scripts which, given a *Core-Safe* program and the annotations produced by the analyses, will mechanically certify that the program satisfies the properties inferred by the analyses. The idea we are trying to implement, consistently with the *Proof Carrying Code* (PCC) paradigm [18], is sending the code generated by the compiler together with the Isabelle/HOL scripts to a hypothetical code consumer who, using another Isabelle/HOL system and a database of previously proved theorems, will check the property and consequently trust the code. The annotations consist of special types in the case of the absence of dangling pointers property, and will consist of some polynomials when the space consumption analysis is finished. At this point of the development we were confronted with two alternatives:

- Either to translate the properties obtained at *Core-Safe* level to the level of the JVM bytecode, by following for instance some of the ideas of [2].
- Or to provide the certificates at the *Core-Safe* level. Then the consumer should trust that our back-end does not destroy the *Core-Safe* properties, or better, we should provide evidence that these properties are preserved.

The first alternative was not very appealing in our case. Differently to [2], where the certificate transformation is carried on at the same intermediate language, here the distance between our *Core-Safe* language and the target language is very large: the first one is functional and the second one is a kind of assembly language; new structures such as the frames stack, the operand stack, or the program counter are present in the second but not in the first; we have built a complete memory management runtime system on top of the JVM in order to avoid its built-in garbage collector, etc. The translated certificate should provide invariants and properties for all these structures. Even if all this work were done, the size of the certificates and the time needed to check them would very probably be huge. The figures reported in [26] for JVM bytecode-level certificates seem to confirm this assertion.

The second alternative has other drawbacks. One of them is that the *Core-Safe* program must be part of the certificate, because the consumer must be able to relate the properties stated at source level with the low-level code being executed. Providing the source code is not allowed in some PCC scenarios. The second drawback is that the back-end should be formally verified, and both the translation algorithm, and the theorem proving its correctness must be in the consumer database. We have chosen this second alternative because smaller certificates can be expected, but also because we feel that proving the translation correct once for all programs is more reasonable in our case than checking this correctness again and again for every translated program.

Machine-assisted compiler certification has been developed by several authors in the last few years. In Sec. 6 we review some of these works. For the certification being really trusty, the code running in the compiler's back-end should be *exactly the same* which has been proved correct by the proof-assistant. Fortunately, modern proof-assistants such as Coq [4] and Isabelle/HOL provide code extraction facilities which deliver code written in some wider-use languages such as Caml or Haskell. Of course, one must trust the translation done by the proof-assistant.

In this paper we present the certification of the first pass explained above (*Core-Safe* to *Safe-Imp*). The second pass (*Safe-Imp* to JVM bytecode) is currently being completed. The reader can find a preliminary version of it in [21].

The main improvement of this work with respect to previous efforts in compiler certification is that we prove, not only the preservation of functional semantics, but also the preservation of the resource consumption properties. As it is asserted in [11], this property can be lost as a consequence of some compiler optimisations. For instance, some auxiliary variables not present in the source may appear during the translation. In our framework, it is essential that memory consumption is preserved during the translation, since we are trying to certify exactly this property. To this aim, we introduce at *Core-Safe* level a *resource-aware* semantics and then prove that this semantics is preserved in the translation to the abstract machine.

With the aim of facilitating the understanding of the paper, and also avoiding descending to many low level details, we have made available the Isabelle/HOL scripts at `http://dalila.sip.ucm.es/safe/theories`. We recommend the reader to consult this site while reading in order to match the concepts described here with its definition in Isabelle/HOL. The paper is structured as follows: after this introduction, in Sec. 2 we motivate our *Safe* language and then present the syntax and semantics of the source and target languages. Then, Sec. 3 explains the translation and gives a small example of the generated code. Sections 2 and 3 contain large portions of material already published in [14,16]. We felt that this material was needed in order to understand the certification process. Sec. 4 is devoted to explaining the main correctness theorem and a number of auxiliary predicates and relations needed in order to state it. Sec. 5 summarises the lessons learnt, and finally a Related Work section closes the paper.

2   The Source and Target Languages

2.1   Full-Safe and Core-Safe

Safe is a first-order polymorphic functional language with a syntax similar to that of (first-order) Haskell, and with some facilities to manage memory. The memory model is based on heap regions where data structures are built. A region is a collection of cells and a cell stores exactly one constructor application. However, in *Full-Safe* regions are implicit. These are inferred [15] when *Full-Safe* is desugared into *Core-Safe*. The allocation and deallocation of regions are bound to function invocations: a *working region* is allocated when entering the call and deallocated when exiting it. All data structures allocated in this region are lost.

Inside a function, data structures may be built but they can also be destroyed by using a *destructive pattern matching*, denoted by the symbol !, which deallocates the cell corresponding to the outermost constructor. Using recursion the recursive spine of the whole data structure may be deallocated. As an example, we show an append function destroying the first list's spine, while keeping its elements in order to build the result:

```
appendD [] !     ys  = ys
appendD (x:xs)! ys  = x : appendD xs ys
```

This appending needs constant (in fact, zero) additional heap space, while the usual version needs linear additional heap space. The fact that the first list is lost

$$prog \rightarrow \overline{data_i}^n; \overline{dec_j}^m; e \qquad \{Core\text{-}Safe\ program\}$$
$$data \rightarrow \textbf{data}\ T\ \overline{\alpha_i}^n\ @\ \overline{\rho_j}^m = \overline{C_k\ \overline{t_{ks}}^{n_k}\ @\ \rho_m}^l \quad \{recursive,\ polymorphic\ data\ type\}$$
$$dec \rightarrow f\ \overline{x_i}^n\ @\ \overline{r_j}^l = e \qquad \{recursive,\ polymorphic\ function\}$$
$$e \rightarrow a \qquad \{atom:\ literal\ c\ or\ variable\ x\}$$
$$| \ x\ @\ r \qquad \{copy\ data\ structure\ x\ into\ region\ r\}$$
$$| \ x! \qquad \{reuse\ data\ structure\ x\}$$
$$| \ a_1 \oplus a_2 \qquad \{primitive\ operator\ application\}$$
$$| \ f\ \overline{a_i}^n\ @\ \overline{r_j}^l \qquad \{function\ application\}$$
$$| \ \textbf{let}\ x_1 = be\ \textbf{in}\ e \qquad \{non\text{-}recursive,\ monomorphic\}$$
$$| \ \textbf{case}\ x\ \textbf{of}\ \overline{alt_i}^n \qquad \{read\text{-}only\ case\}$$
$$| \ \textbf{case!}\ x\ \textbf{of}\ \overline{alt_i}^n \qquad \{destructive\ case\}$$
$$alt \rightarrow C\ \overline{x_i}^n \rightarrow e \qquad \{case\ alternative\}$$
$$be \rightarrow C\ \overline{a_i}^n\ @\ r \qquad \{constructor\ application\}$$
$$| \ e$$

Fig. 1. *Core-Safe* syntax

is reflected, by using the symbol ! in the type inferred for the function *appendD* :: $\forall a \rho_1 \rho_2 . [a]!@\rho_1 \rightarrow [a]@\rho_2 \rightarrow \rho_2 \rightarrow [a]@\rho_2$, where ρ_1 and ρ_2 are polymorphic types denoting the regions where the input and output lists should live. In this case, due to the sharing between the second list and the result, these latter lists should live in the same region. Another possibility is to destroy part of a data structure and to *reuse* the rest in the result, as in the following destructive *split* function:

```
splitD 0 zs!     = ([], zs!)
splitD n []!     = ([], [])
splitD n (y:ys)! = (y:ys1, ys2) where (ys1, ys2) = splitD (n-1) ys
```

The righthand side *zs!* expresses reusing the remaining list. The inferred type is:

$$splitD :: \forall a \rho_1 \rho_2 \rho_3 . \ Int \rightarrow [a]!@\rho_2 \rightarrow \rho_1 \rightarrow \rho_2 \rightarrow \rho_3 \rightarrow ([a]@\rho_1, [a]@\rho_2)@\rho_3$$

Notice that the regions used to build the result appear as additional arguments. The data structures which are not part of the function's result are inferred to be built in the local working region, which we call *self*, and they die at function termination. As an example, the tuples produced by the internal calls to `splitD` are allocated in their respective *self* regions and do not consume memory in the caller regions. The type of these internal calls is $Int \rightarrow [a]!@\rho_2 \rightarrow \rho_1 \rightarrow \rho_2 \rightarrow \rho_{self} \rightarrow ([a]@\rho_1, [a]@\rho_2)@\rho_{self}$, which is different from the external type because we allow polymorphic recursion on region types. More information about *Safe* and its type system can be found at [16].

The *Safe* front-end desugars *Full-Safe* and produces a bare-bones functional language called *Core-Safe*. The transformation starts with region inference and follows with Hindley-Milner type inference, desugaring pattern matching into **case** expressions, **where** clauses into **let** expressions, collapsing several function-defining equations into a single one, and some other transformations.

In Fig. 1 we show *Core-Safe*'s syntax, which is defined in Isabelle/HOL as a collection of datatypes. A program *prog* is a sequence of possibly recursive polymorphic data and function definitions followed by a main expression *e* whose value is the program result. The abbreviation $\overline{x_i}^n$ stands for $x_1 \cdots x_n$. Destructive pattern matching is desugared into **case!** expressions. Constructor applications are only allowed in **let** bindings. Only atoms are used in applications, and only variables are used in **case/case!** discriminants, copy and reuse expressions. Region arguments are explicit in constructor and function applications and in the copy expression. Function definitions have additional region arguments $\overline{r_j}^l$ where the function is allowed to build data structures. In the function's body only the r_j and its working region *self* may be used.

2.2 Core-Safe Semantics

In Figure 2 we show the resource-aware big-step semantics of *Core-Safe* expressions. A judgement of the form $E \vdash h, k, td, e \Downarrow h', k, v, r$ means that the expression e is successfully reduced to normal form v under runtime environment E and heap h with $k + 1$ regions, ranging from 0 to k, and that a final heap h' with $k + 1$ regions is produced as a side effect. Arguments k can be considered as attributes of their respective heaps. We highlight them in order to emphasise that the evaluation starts and ends with the same number of regions, and also to show when regions are allocated and deallocated. A value v is either a constant or a heap pointer. The argument td and the result r have to do with resource consumption and will be explained later. The semantics can be understood disregarding them. Moreover, forgetting about resource consumption produces a valid value semantics for the language.

A runtime environment E maps program variables to values and region variables to actual region identifiers which consist of natural numbers. As region allocation/deallocation are done at function invocation/return time, the live regions are organised in a region stack. A region identifier is just its offset from the bottom of this stack. We adopt the convention that for all E, if c is a con-

$$E \vdash h, k, td, c \Downarrow h, k, c, ([\,]_k, 0, 1) \; [Lit]$$

$$E[x \mapsto v] \vdash h, k, td, x \Downarrow h, k, v, ([\,]_k, 0, 1) \; [Var]$$

$$\frac{j \leq k \quad (h', p') = copy(h, p, j) \quad m = size(h, p)}{E[x \mapsto p, r \mapsto j] \vdash h, k, td, x@r \Downarrow h', k, p', ([j \mapsto m], m, 2)} \; [Var_2]$$

$$\frac{fresh(q)}{E[x \mapsto p] \vdash h \uplus [p \mapsto w], k, td, x! \Downarrow h \uplus [q \mapsto w], k, q, ([\,]_k, 0, 1)} \; [Var_3]$$

$$\frac{c = c_1 \oplus c_2}{E[a_1 \mapsto c_1, a_2 \mapsto c_2] \vdash h, k, td, a_1 \oplus a_2 \Downarrow h, k, c, ([\,]_k, 0, 2)} \; [Primop]$$

$$\frac{(f \; \overline{x_i}^n @ \overline{r_j}^l = e) \in \Sigma \qquad [x_i \mapsto E(a_i)^n, \overline{r_j \mapsto E(r'_j)}^l, self \mapsto k+1] \vdash h, k+1, n+l, e \Downarrow h', k+1, v, (\delta, m, s)}{E \vdash h, k, td, f \; \overline{a_i}^n @ \overline{r'_j}^l \Downarrow h'|_k, k, v, (\delta|_k, m, \max\{n+l, s+n+l - td\})} \; [App]$$

$$\frac{E \vdash h, k, 0, e_1 \Downarrow h', k, v_1, (\delta_1, m_1, s_1) \qquad E \cup [x_1 \mapsto v_1] \vdash h', k, td+1, e_2 \Downarrow h'', k, v, (\delta_2, m_2, s_2)}{E \vdash h, k, td, \text{let } x_1 = e_1 \text{ in } e_2 \Downarrow h'', k, v, (\delta_1 + \delta_2, \max\{m_1, |\delta_1| + m_2\}, \max\{2 + s_1, 1 + s_2\})} \; [Let_1]$$

$$\frac{j \leq k \quad fresh(p) \quad E \cup [x_1 \mapsto p] \vdash h \uplus [p \mapsto (j, C \; \overline{v_i}^n)], k, td+1, e_2 \Downarrow h', k, v, (\delta, m, s)}{E[\overline{a_i \mapsto v_i}^n, r \mapsto j] \vdash h, k, td, \text{let } x_1 = C \; \overline{a_i}^n @ r \text{ in } e_2 \Downarrow h', k, v, (\delta + [j \mapsto 1], m+1, s+1)} \; [Let_2]$$

$$\frac{E[x \mapsto p] \quad h[p \mapsto (j, C_r \; \overline{v_i}^n)] \quad E \cup [\overline{x_{r_i} \mapsto v_i}^{n_r}] \vdash h, k, td+n_r, e_r \Downarrow h', k, v, (\delta, m, s)}{E \vdash h, k, td, \text{case } x \text{ of } \overline{C_i \; \overline{x_{ij}}^{n_i} \rightarrow e_i}^n \Downarrow h', k, v, (\delta, m, s+n_r)} \; [Case]$$

$$\frac{E[x \mapsto p] \quad h^+ = h \uplus [p \mapsto (j, C_r \; \overline{v_i}^n)] \quad E \cup [\overline{x_{r_i} \mapsto v_i}^{n_r}] \vdash h, k, td+n_r, e_r \Downarrow h', k, v, (\delta, m, s)}{E \vdash h^+, k, td, \text{case! } x \text{ of } \overline{C_i \; \overline{x_{ij}}^{n_i} \rightarrow e_i}^n \Downarrow h', k, v, (\delta + [j \mapsto -1], \max\{0, m-1\}, s+n_r)} \; [Case!]$$

Fig. 2. Resource-Aware Big-Step Operational Semantics of *Core-Safe* expressions

stant, then $E(c) = c$. A heap h is a finite mapping from fresh variables p to constructor cells w of the form $(j, C\,\overline{v_i}^n)$, meaning that the cell resides in region j. By $h[p \mapsto w]$ we denote a heap h where the binding $[p \mapsto w]$ is highlighted, while $h \uplus [p \mapsto w]$ denotes the disjoint union of heap h with the binding $[p \mapsto w]$. By $h \mid_k$ we denote the heap obtained by deleting from h those bindings living in regions greater than k, and by $dom(h)$, the set $\{p \mid [p \mapsto w] \in h\}$.

The semantics of a program is the semantics of the main expression in an environment Σ, which is the set containing all the function and data declarations. Rules Lit and Var_1 just say that basic values and heap pointers are normal forms. Rule Var_2 executes a runtime system $copy$ function copying the recursive part of the data structure pointed to by p, and living in a region j', into a (possibly different) region j. In rule Var_3, the binding $[p \mapsto w]$ in the heap is deleted and a fresh binding $[q \mapsto w]$ to cell w is added. This action may create dangling pointers in the live heap, as some cells may contain free occurrences of p. Rule App shows when a new region is allocated. Notice that the body of the function is executed in a heap with $k + 2$ regions. The formal identifier $self$ is bound to the newly created region $k + 1$ so that the function body may create cells in this region or pass this region as an argument to other functions. Before returning from the function, all cells created in region $k + 1$ are deleted. Rules Let_1, Let_2, and $Case$ are the usual ones for an eager language, while rule $Case!$ expresses what happens in a destructive pattern matching: the binding of the discriminant variable disappears from the heap.

This semantics is defined in Isabelle/HOL as an inductive relation. The environment E is split into a pair (E_1, E_2) separating program variables from region arguments bindings. These and the heap are modelled as partial functions. Even though all functions are total in Isabelle/HOL, a partial function, denoted $'a \rightharpoonup {}'b$, can be easily defined as the total function $'a \Rightarrow {}'b$ $option$, where $f\,x = None$ represents that f is not defined at x.

2.3   Resource Consumption

The semantics relates the evaluation of an expression e to a resource vector $r = (\delta, m, s)$ obtained as a side effect. The first component is a partial function $\delta : \mathbb{N} \rightharpoonup \mathbb{Z}$ giving for each region k in scope the signed difference between the cells in the final and initial heaps. A positive difference means that new cells have been created in this region. A negative one means that some cells have been destroyed. By $dom(\delta)$ we denote the subset of $\mathbb{N}$ in which δ is defined. By $|\delta|$ we mean the sum $\sum_{n \in dom(\delta)} \delta(n)$ giving the total balance of cells. The remaining components m and s respectively give the $minimum$ number of fresh cells in the heap and of words in the stack needed to successfully evaluate e. When e is the main expression, these figures give us the total memory needs of the $Safe$ program. The additional argument td is the number of bindings in E which can be discarded when a normal form is reached or at function invocation. It coincides with the value returned by the function $topDepth$ of Sec. 3. As we will see there, the runtime environment E is kept in the evaluation stack and (part of) this environment is discarded by the abstract machine in those situations. By $[]_k$ we denote the function $\lambda n.0$ if $0 \le n \le k$, and $\lambda n.\bot$ otherwise. By $\delta_1 + \delta_2$ we denote the function:

$$(\delta_1 + \delta_2)(x) = \begin{cases} \delta_1(x) + \delta_2(x) & \text{if } x \in dom(\delta_1) \cap dom(\delta_2) \\ \delta_i(x) & \text{if } x \in dom(\delta_i) - dom(\delta_{3-i}),\ i \in \{1, 2\} \\ \bot & \text{otherwise} \end{cases}$$

Function *size* in rule Var_2 gives the size of the recursive spine of a data structure:

$$size(h[p \mapsto (j, C \ \overline{v_i^n})], p) = 1 + \sum_{i \in RecPos(C)} size(h, v_i)$$

where *RecPos* returns the recursive argument positions of a given constructor. In rule *App*, by $\delta|_k$ we mean a function like δ but undefined for values greater than k. The computation of these resource consumption figures takes into account how the translation will transform, and the abstract machine will execute, the corresponding expression. For instance, in rule *App* the number $\max\{n + l, s + n + l - td\}$ of fresh stack words takes into account that the first $n + l$ words are needed to store the actual arguments in the stack, then the current environment of length td is discarded, and then the function body is evaluated. In rule Let_1, a *continuation* (2 words, see Sec. 2.4) is stacked before evaluating e_1, and this evaluation leaves a value in the stack before evaluating e_2. Hence, the computation $\max\{2 + s_1, 1 + s_2\}$.

2.4   Safe-Imp Syntax and Semantics

Safe-Imp is the bytecode language of the SVM. Its syntax and semantics is depicted in Figure 3. A configuration of the SVM consists of the five components $(is, (h, k), k_0, S, cs)$, where is is the current instruction sequence, (h, k) is the current heap, k being its topmost region, S is a stack and cs is the code store where the instruction sequences resulting from the compilation of program fragments are kept. A code store is a partial function from *code labels*, denoted $p, q, \ldots$, to bytecode lists. The component k_0 is a low watermark in the heap registering which one must be the topmost region when a normal form is reached (see the semantics of the *DECREGION* instruction). The property $k_0 \leq k$ is an invariant of the execution. By $b, b_i, \ldots$ we denote heap pointers or any other item stored in the stack. The stack contains three kinds of objects: values, regions and continuations.

$$\begin{aligned} so &\rightarrow v \mid j \mid (k, p) & \{\text{stack object}\} \\ S &\rightarrow so \ list & \{\text{stack}\} \end{aligned}$$

The semantics of the *Safe-Imp* instructions is shown in terms of configuration transitions. By C_r^m we denote the data constructor which is the r-th in its **data** definition out of a total of m data constructors, and by $S!j$, the j-th element of the stack S counting from the top and starting at 0. A more complete view on how this machine has been derived from the semantics can be found at [14]. For the purpose of this paper, a short summary of the instructions follows.

Instruction *DECREGION* deletes from the heap all the regions, if any, between the current topmost region k and region k_0, excluding the latter. Each region can be deallocated with a time cost in $O(1)$ due to its implementation as a linked list (see [21] for details). Instruction *POPCONT* pops a continuation from the stack or stops the execution if there is none. Instruction *PUSHCONT* pushes a continuation. It will be used in the translation of a **let**.

Instructions *COPY* and *REUSE* just mimic the semantics given to the corresponding expressions. Instruction *CALL* jumps to a new instruction sequence and creates a new region. Function calls are always tail recursive, so there is no

Initial configuration $\Rightarrow$ Final configuration	Condition
$(DECREGION : is, (h,k), k_0, S, cs)$ $\Rightarrow (is, (h\|_{k_0}, k_0), k_0, S, cs)$	$k \geq k_0$
$([POPCONT], (h,k), k, b:(k_0,p):S, cs[p \mapsto is])$ $\Rightarrow (is, (h,k), k_0, b:S, cs)$	
$(PUSHCONT\ p : is, (h,k), k, S, cs[p \mapsto is'])$ $\Rightarrow (is, (h,k), k, (k_0,p):S, cs)$	
$(COPY : is, (h[b \mapsto (l, C\,\overline{v_i}^n)], k), k_0, b:j:S, cs)$ $\Rightarrow (is, (h',k), k_0, b':S, cs)$	$(h',b') = copy(h,b,j)$ $j \leq k$
$(REUSE : is, (h \uplus [b \mapsto w], k), k_0, b:S, cs)$ $\Rightarrow (is, (h \uplus [b' \mapsto w], k), k_0, b':S, cs)$	$fresh(b', h \uplus [b \mapsto w])$
$([CALL\ p], (h,k), k_0, S, cs[p \mapsto is])$ $\Rightarrow (is, (h,k+1), k_0, S, cs)$	
$(PRIMOP \oplus : is, (h,k), k_0, c_1:c_2:S, cs)$ $\Rightarrow (is, (h,k), k_0, c:S, cs)$	$c = c_1 \oplus c_2$
$([MATCH\ l\ \overline{p_j}^m], (h[S!l \mapsto (j, C_r^n\,\overline{v_i}^n)], k), k_0, S, cs[\overline{p_j \mapsto is_j}^m])$ $\Rightarrow (is_r, (h,k), k_0, \overline{v_i}^n:S, cs)$	
$([MATCH!\ l\ \overline{p_j}^m], (h \uplus [S!l \mapsto (j, C_r^n\,\overline{v_i}^n)], k), k_0, S, cs[\overline{p_j \mapsto is_j}^m])$ $\Rightarrow (is_r, (h,k), k_0, \overline{v_i}^n:S, cs)$	
$(BUILDENV\ \overline{K_i}^n : is, (h,k), k_0, S, cs)$ $\Rightarrow (is, (h,k), k_0, \overline{Item_k(K_i)}^n:S, cs)$	(1)
$(BUILDCLS\ C_r^m\ \overline{K_i}^n\ K : is, (h,k), k_0, S, cs)$ $\Rightarrow (is, (h \uplus [b \mapsto (Item_k(K), C_r^m\,\overline{Item_k(K_i)}^n)], k), k_0, b:S, cs)$	$Item_k(K) \leq k,\ fresh(b,h)$ (1)
$(SLIDE\ m\ n : is, (h,k), k_0, \overline{b_i}^m : \overline{b_i'}^n : S, cs)$ $\Rightarrow (is, (h,k), k_0, \overline{b_i}^m : S, cs)$	

$$(1)\quad Item_k(K) \stackrel{\text{def}}{=} \begin{cases} S!j & \text{if } K = j \in \mathbb{N} \\ c & \text{if } K = c \\ k & \text{if } K = self \end{cases}$$

Fig. 3. The *Safe* Virtual Machine (SVM)

need for a return instruction. Instruction *MATCH* does a jump depending on the constructor of the matched cell. The list of code labels $\overline{p_j}^m$ corresponds to the compilation of a set of **case** alternatives. Instruction *MATCH!* additionally destroys the matched cell. The following invariant is ensured by the translation: For every instruction sequence in the code store cs, instruction i is the last one if and only if it belongs to the set $\{POPCONT, CALL, MATCH, MATCH!\}$.

Instruction *BUILDENV* creates a portion of the environment on top of the stack: If a key K is a natural number j, the item $S!j$ is copied and pushed on the stack; if it is a basic constant c, it is directly pushed on the stack; if it is the identifier *self*, then the topmost region number k is pushed. Instruction *BUILDCLS* allocates a fresh cell and fills it with a constructor application. It uses the same conventions as *BUILDENV*. Finally, instruction *SLIDE* removes some parts of the stack and it is used to remove environment fragments.

We have defined this semantics in Isabelle/HOL as the function:

$$execSVM :: SafeImpProg \Rightarrow SVMState \Rightarrow (SVMState, SVMState)\ Either$$

where *Either* is a sum type and *SVMState* denotes a configuration $((h,k), k_0, pc, S)$ with the code store removed and the current instruction sequence replaced by a program counter $pc = (p,i)$. The code store cs is a read-only component and has been included in the type *SafeImpProg*. The current instruction can be retrieved by accessing the i-th element of the sequence $(cs\ p)$. If the result of $execSVM\ P\ s_1$ is *Left* s_1, this means that s_1 is a final state. Otherwise, it returns *Right* s_2.

3   The Translation

The translation splits the runtime environment (E_1, E_2) of the semantics into two: a compile-time one ρ mapping program variables to stack offsets, and the actual runtime environment contained in the stack. As this grows dynamically, numbers are assigned to the variables from the bottom of the environment. In this way, if the environment occupies the top m positions of the stack and $\rho[x \mapsto 1]$, then $S!(m-1)$ will contain the runtime value of x.

An expression let $x_1 = e_1$ in e_2 will be translated by pushing to the stack a continuation for e_2, and then executing the translation of e_1. A continuation consists of a pair (k_0, p) where p points to the translation of e_2 and k_0 is the lower watermark associated to e_2. It is saved in the stack because the lower watermark of e_1 is different (see the semantics of $PUSHCONT$). As e_1 and e_2 share most of their runtime environments, the continuation is treated as a barrier below which the environment must not be deleted while e_2 has not reached its normal form. So, the whole compile-time environment ρ consists of a list of smaller environments $[\delta_1, \ldots, \delta_n]$, mimicking the stack layout. Each individual block i consists of a triple (δ_i, l_i, n_i) with an environment δ_i mapping variables to numbers in the range $(1 \ldots m_i)$, a block length $l_i = m_i + n_i$, and an indicator $n_i = 2$ for all the blocks except for the first one, whose value is $n_1 = 0$. We are assuming that a continuation needs two words in the stack and that the remaining items need one word.

The offset with respect to the top of the stack of a variable x defined in the block k, denoted $\rho\, x$, is computed as follows: $\rho\, x \stackrel{\text{def}}{=} (\sum_{i=1}^{k} l_i) - \delta_k\, x$. Only the top environment may be extended with new bindings. There are three operations on compile-time environments:

1. $((\delta, m, 0) : \rho) + \{\overline{x_i \mapsto j_i}^n\} \stackrel{\text{def}}{=} (\delta \cup \{\overline{x_i \mapsto m + j_i}^n\}, m + n, 0) : \rho$.
2. $((\delta, m, 0) : \rho)^{+\!\!+} \stackrel{\text{def}}{=} (\{\}, 0, 0) : (\delta, m + 2, 2) : \rho$.
3. $topDepth\ ((\delta, m, 0) : \rho) \stackrel{\text{def}}{=} m$. Undefined otherwise.

The first one extends the top environment with n new bindings, while the second closes the top environment with a 2-indicator and then opens a new one.

Using these conventions, in Figure 4 we show an idealised version of the translation function trE taking a *Core-Safe* expression and a compile-time environment, and giving as a result a list of SVM instructions and a code store. There, $NormalForm\ \rho$ is the following list:

$$NormalForm\ \rho \stackrel{\text{def}}{=} [SLIDE\ 1\ (topDepth\ \rho), DECREGION, POPCONT]$$

The whole program translation is done by Isabelle/HOL function $trProg$ which first translates each function definition by using function trF, and then the main expression by using trE. The source file is guaranteed to define a function before its use. The translation accumulates an environment $funm$ mapping every function name to the initial bytecode sequence of its definition. The main part of $trProg$ is:

```
trProg (datas, defs, e) = (
   let ...
       ((p, funm, contm), codes) = mapAccumL trF (1, empty, []) defs;
       cs = concat codes
   in ... cs ...)
```

$$trE\ c\ \rho \qquad\qquad = (BUILDENV\ [c] : NormalForm\ \rho,\ \{\})$$
$$trE\ x\ \rho \qquad\qquad = (BUILDENV\ [\rho\ x] : NormalForm\ \rho,\ \{\})$$
$$trE\ (x@r)\ \rho \qquad\quad = (BUILDENV\ [\rho\ x, \rho\ r] : COPY : NormalForm\ \rho,\ \{\})$$
$$trE\ (x!)\ \rho \qquad\qquad = (BUILDENV\ [\rho\ x] : REUSE : NormalForm\ \rho,\ \{\})$$
$$trE\ (a_1 \oplus a_2)\ \rho \qquad = (BUILDENV\ [\rho\ a_1, \rho\ a_2] : PRIMOP : NormalForm\ \rho,\ \{\})$$
$$trE\ (f\ \overline{a_i}^n\ @\ \overline{s_j}^m)\ \rho = ([BUILDENV\ [\overline{\rho\ a_i}^n, \overline{\rho\ s_j}^m], SLIDE\ (n+m)\ (topDepth\ \rho), CALL\ p], cs')$$
$$\text{where}\quad (f\ \overline{x_i}^n\ @\ \overline{r_j}^m = e) \in defs$$
$$cs' = \{p \mapsto is\} \cup cs$$
$$(is, cs) = trE\ e\ [(\{\ \overline{r_j \mapsto m-j+1}^m, \overline{x_i \mapsto n-i+m+1}^n\}, n+m, 0)]$$

$$trE\ (\textbf{let}\ x_1 = C_l^m\ \overline{a_i}^n @ s\ \textbf{in}\ e)\ \rho = (BUILDCLS\ C_l^m\ [\overline{(\rho\ a_i)}^n]\ (\rho\ s) : is,\ cs)$$
$$\text{where}\quad (is,\ cs) = trE\ e\ (\rho + \{x_1 \mapsto 1\})$$

$$trE\ (\textbf{let}\ x_1 = e_1\ \textbf{in}\ e_2)\ \rho \qquad = (PUSHCONT\ p : is_1,\ cs_1 \cup cs_2 \cup \{p \mapsto is_2\})$$
$$\text{where}\quad (is_1, cs_1) = trE\ e_1\ \rho^{+}$$
$$(is_2, cs_2) = trE\ e_2\ (\rho + \{x_1 \mapsto 1\})$$

$$trE\ (\textbf{case}\ x\ \textbf{of}\ \overline{alt_i}^n)\ \rho \qquad = ([MATCH\ (\rho\ x)\ \overline{p_i}^n],\ \{\overline{p_i \mapsto is_i}\} \cup (\bigcup_{i=1}^n cs_i))$$
$$\text{where}\quad (is_i, cs_i) = trA\ alt_i\ \rho,\ 1 \le i \le n$$

$$trE\ (\textbf{case!}\ x\ \textbf{of}\ \overline{alt_i}^n)\ \rho \qquad = ([MATCH!\ (\rho\ x)\ \overline{p_i}^n],\ \{\overline{p_i \mapsto is_i}\} \cup (\bigcup_{i=1}^n cs_i))$$
$$\text{where}\quad (is_i, cs_i) = trA\ alt_i\ \rho,\ 1 \le i \le n$$

$$trA\ (C\ \overline{x_i}^n \to e)\ \rho \qquad = trE\ e\ (\rho + \{\overline{x_i \mapsto n-i+1}^n\})$$

Fig. 4. Translation from *Core-Safe* expressions to *Safe-Imp* bytecode instructions

$$P_1 \mapsto [BUILDCLS\ Nil_0^2\ [\ ]\ self,\ BUILDENV\ [0, 0, self],\ SLIDE\ 3\ 1,\ CALL\ P_2]$$
$$P_2 \mapsto [MATCH!\ 0\ [P_3, P_4]]$$
$$P_3 \mapsto [BUILDENV\ [1],\ SLIDE\ 1\ 3,\ DECREGION,\ POPCONT]$$
$$P_4 \mapsto [PUSHCONT\ P_5,\ BUILDENV\ [3,\ 5,\ 6],\ SLIDE\ 3\ 0,\ CALL\ P_2]$$
$$P_5 \mapsto [BUILDCLS\ Cons_1^2\ [1,\ 0]\ 5,\ BUILDENV\ [0],\ SLIDE\ 1\ 6,\ DECREGION,\ POPCONT]$$

Fig. 5. Imperative code for the *Core-Safe* `appendD` program

where *cs* is the code store resulting from the compilation, and *mapAccumL* is a higher-order function, combining *map* and *foldl*, defined to Isabelle/HOL by copying its definition from the Haskell library (`http://dalila.sip.ucm.es/safe/theories` for more details).

In Figure 5 we show the code store generated for the following *Core-Safe* program with the `appendD` function of Sec. 2.1:

$$appendD\ xs\ ys\ @\ r = \textbf{case!}\ xs\ \textbf{of}$$
$$[\ ] \qquad \rightarrow \quad ys$$
$$x : xx \quad \rightarrow \quad \textbf{let}\ yy = appendD\ xx\ ys\ @\ r\ \textbf{in}$$
$$\textbf{let}\ zz = x : yy\ @\ r\ \textbf{in}\ zz;$$
$$\textbf{let}\ l = [\ ]\ @\ self\ \textbf{in}\quad append\ l\ l\ @\ self$$

4   Formal Verification

The above infrastructure allows us to state and prove the main theorem expressing that the pair translation-abstract machine is sound and complete with respect to the resource-aware semantics. First, we make note that both the semantics and the SVM machine rules are syntax driven, and that their computations are deterministic (up to fresh names generation for the heap). So, we only need to prove that everything done by the semantics can be emulated by the machine, and that termination of the machine implies termination of the semantics (for the corresponding expression.)

First we define in Isabelle/HOL the following equivalence relation between runtime environments in the semantics and in the machine:

Definition 1. *We say that the environment* $E = (E_1, E_2)$ *and the pair* (ρ, S) *are equivalent, denoted* $(E_1, E_2) \bowtie (\rho, S)$, *if dom* $E - \{self\} = dom \ \rho$, *and* $\forall x \in dom \ E_1 \ . \ E_1(x) = S!(\rho \ x)$, *and* $\forall r \in dom \ E_2 - \{self\} \ . \ E_2(r) = S!(\rho \ r)$.

Then we define an inductive relation expressing the evolution of the SVM machine up to some intermediate points corresponding to the end of the evaluation of sub-expressions:

inductive
   $execSVMBalanced :: [SafeImpProg, SVMState, nat \ list, SVMState \ list, nat \ list] \Rightarrow bool$
                            $(\ \_ \vdash \_ , \ \_ \ \text{-}svm \rightarrow \ \_ , \ \_)$
where
   init: $P \vdash s, n\#ns \ \text{-}svm \rightarrow [s], n\#ns$
 | *step:* $\llbracket P \vdash s, n\#ns \ \text{-}svm \rightarrow s'\#ss, m\#ms;$
         $execSVM \ P \ s' = Right \ s'';$
         $m' = nat \ (diffStack \ s'' \ s' \ m);$
         $m' \geq 0;$
         $ms' = (if \ pushcont \ (instrSVM \ P \ s') \ then \ 0\#m\#ms$
                $else \ if \ popcont \ (instrSVM \ P \ s') \land ms=m''\#ms'' \ then \ (Suc \ m'')\#ms''$
                $else \ m'\#ms)\rrbracket \Longrightarrow$
         $P \vdash s, n\#ns \ \text{-}svm \rightarrow s''\#s'\#ss, ms'$

$P \vdash s, n\#ns \ -svm \rightarrow ss, 1\#ns$ represents a 'balanced' execution of the SVM corresponding to the evaluation of a source expression. Its meaning is that the *Safe-Imp* program P evolves by starting at state s and passing through all the states in the list ss (s is the last state of the list ss, and the sequence progresses towards the head of the list), with the stack decreasing at most by n positions. Should the top instruction of the current state create a smaller stack, then the machine stops at that state. The symbol $\#$ in Isabelle/HOL is the *cons* constructor for lists.

Next, we define what resource consumption means at the machine level. Given a forwards state sequence $ss = s_0 \cdots s_r$ starting at s_0 with heap h_0 and stack S_0, *maxFreshCells ss* gives the highest non-negative difference in cells between the heaps in ss and the heap h_0. Likewise, *maxFreshWords ss* gives the maximum number of fresh words created in the stack during the sequence ss with respect to S_0. Finally, *diff k h h'* gives for each region j, $0 \leq j \leq k$, the signed difference in cells between h' and h.

From the input list ds of *Core-Safe* definitions, we define the set *definedFuns ds* of the function names defined there. Also, given an expression e, *closureCalled e ds* is an inductive set giving the names of the functions reached from e by direct or indirect invocation. By $cs \sqsubseteq cs'$ we mean that the code store cs' extends the code store cs with new bindings.

Finally, we show the correctness lemma of the semantics with respect to the machine, as it has been stated and proved in Isabelle/HOL:

lemma *correctness:*
   $E \ h , k , td , e \Downarrow h' , k , v , r \longrightarrow$
   $(closureCalled \ e \ defs \subseteq definedFuns \ defs$
   $\land \ ((p, funm, contm), codes) = mapAccumL \ trF \ (1, empty, []) \ defs$
   $\land \ cs = concat \ codes$

$\wedge\ P = ((cs,\ contm),p,ct,st)$
$\wedge\ finite\ (dom\ h)$
$\longrightarrow\ (\forall\ rho\ S\ S'\ k0\ s0\ p'\ q\ ls\ is\ is'\ cs1\ j.$
$\qquad (q,\ ls,\ is,\ cs1) = trE\ p'\ funm\ fname\ rho\ e$
$\qquad \wedge\ (append\ cs1\ [(q,is',fname)]) \sqsubseteq\ cs$
$\qquad \wedge\ drop\ j\ is' = is$
$\qquad \wedge\ E \bowtie (rho,S)$
$\qquad \wedge\ td = topDepth\ rho$
$\qquad \wedge\ k0 \leq k$
$\qquad \wedge\ S' = drop\ td\ S$
$\qquad \wedge\ s0 = ((h,\ k),\ k0,\ (q,\ j),\ S)$
$\qquad \longrightarrow\ (\exists s\ ss\ q'\ i\ \delta\ m\ w.$
$\qquad\qquad\qquad P \vdash s0\ ,\ td\#tds\ \text{-}svm \longrightarrow s\ \#\ ss\ ,\ 1\#tds$
$\qquad\qquad \wedge\ s = ((h',\ k) \downarrow k0,\ k0,\ (q',\ i),\ Val\ v\ \#\ S')$
$\qquad\qquad \wedge\ fst\ (the\ (map\_of\ cs\ q'))!i = POPCONT$
$\qquad\qquad \wedge\ r = (\delta,m,w)$
$\qquad\qquad \wedge\ \delta = diff\ k\ (h,k)\ (h',k)$
$\qquad\qquad \wedge\ m = maxFreshCells\ (rev\ (s\#ss))$
$\qquad\qquad \wedge\ w = maxFreshWords\ (rev\ (s\#ss)))))$

The premises state that the arbitrary expression e is evaluated to a value v according to the *Core-Safe* semantics, that it is translated in the context of a closed *Core-Safe* program *defs* having a definition for every function reached from e, and that the instruction sequence *is* and the partial code store *cs1* are the result of the translation. Then, the execution of this sequence by the SVM starting at an appropriate state *s0* in the context of the translated program P, will reach a stopping state s having the same heap (h', k) as the one obtained in the semantics, and the same value v on top of the stack. Moreover, the memory (δ, m, w) consumed by the machine, both in the heap and in the stack, is as predicted by the semantics.

The proof is done by induction on the $\Downarrow$ relation, and with the help of a number of auxiliary lemmas, some of them stating properties of the translation and some others stating properties of the evaluation. We classify them into the following groups:

Lemmas on the evolution of the SVM. This group takes care of the first three conclusions, i.e. $P \vdash s0\ ,\ td\#tds\ \text{-}svm \longrightarrow s\ \#\ ss\ ,\ 1\#tds$ and the next two ones, and there is one or more lemmas for every syntactic form of e.

Lemmas on the equivalence of runtime environments. They are devoted to proving that the relation $(E_1, E_2) \bowtie (\rho, S)$ is preserved across evaluation. For instance, if $e \equiv f\ \overline{a_i}^n\ @\ \overline{r_j'}^l$, being f defined by the equation $f\ \overline{x_i}^n\ @\ \overline{r_j}^l = e_f$, we prove that the equivalence of the environments local to f still hold. Formally:

$\qquad (E_1, E_2) \bowtie (\rho, S)$
$\quad \wedge \quad \rho' = [(\{x_i \mapsto n - i + l + 1\}^n,\ \overline{r_j \mapsto l - j + 1}^l, \},n + l, 0)]$
$\quad \wedge \quad (E_1', E_2') = ([x_i \mapsto E(a_i)]^n, [r_j \mapsto E(r_j')]^l, self \mapsto k + 1])$
$\quad \wedge \quad S' = \overline{S!(\rho\ a_i)}^n\ @\ \overline{S!(\rho\ r_j')}^l\ @\ drop\ td\ S$
$\quad \Longrightarrow (E_1', E_2') \bowtie (\rho', S')$

Lemmas on cells charged to the heap. This group takes care of the last but two conclusion $\delta = diff\ k\ (h,k)\ (h',k)$, and there is one or more lemmas for every syntactic form of e. For instance, if $e \equiv$ **let** $x_1 = e_1$ **in** e_2, then the main lemma has essentially this form:

$$\delta_1 = \mathit{diff}\ k\ (h,k)\ (h'|_k,k)$$
$$\wedge\quad \delta_2 = \mathit{diff}\ k\ (h'|_k,k)\ (h'',k)$$
$$\implies \delta_1 + \delta_2 = \mathit{diff}\ k\ (h,k)\ (h'',k)$$

where (h,k), $(h'|_k,k)$, and (h'',k) are respectively the initial heap, and the heaps after the evaluation of e_1 and e_2.

Lemmas on fresh cells needed in the heap. This group takes care of the last but one conclusion $m = \mathit{maxFreshCells}\ (\mathit{rev}\ (s\#ss))$. If $e \equiv$ **let** $x_1 = e_1$ **in** e_2, then the main lemma has essentially this form:

$$\delta_1 = \mathit{diff}\ k\ (h,k)\ (h'|_k,k)$$
$$\wedge\quad m_1 = \mathit{maxFreshCells}\ (\mathit{rev}\ (s_1\#ss_1))$$
$$\wedge\quad m_2 = \mathit{maxFreshCells}\ (\mathit{rev}\ (s_2\#ss_2))$$
$$\implies \max\ m_1\ (m_2 + |\delta_1|) = \mathit{maxFreshCells}\ (\mathit{rev}\ (s_2\#ss_2\ @\ s_1\#ss_1\ @\ [s_0]))$$

where s_0, s_1, and s_2 are respectively the initial state of the SVM, and the states after the evaluation of e_1 and e_2.

Lemmas on fresh words needed in the stack. This group takes care of the last conclusion $w = \mathit{maxFreshWords}\ (\mathit{rev}\ (s\#ss))$. If $e \equiv f\ \overline{a_i}^n\ @\ \overline{r_j}^l$, then the main lemma has essentially this form:

$$w = \mathit{maxFreshWords}\ (\mathit{rev}\ (s\#ss))$$
$$\implies \max\ (n+l)\ (w+n+l-td) = \mathit{maxFreshWords}\ (\mathit{rev}\ (s\#ss\ @\ [s_2,s_1,s_0]))$$

where s_0, s_1, s_2 are respectively the initial state of the application, and the states after the execution of *BUILDENV* and *SLIDE*, and $s\#ss$ is the state sequence of the body of f.

That termination of the SVM implies the existence of a derivation in the semantics for the corresponding expression has not been proved for the moment.

5   Discussion

On the use of Isabelle/HOL. The complete specification in Isabelle/HOL of the syntax and semantics of our languages, of the translation functions, the theorems and the proofs, represent almost one person-year of effort. Including comments, about 7000 lines of Isabelle/HOL scripts have been written, and about 200 lemmas proved.

Isabelle/HOL gives enough facilities for defining recursive and higher-order functions. These are written in much the same way as a programmer would do in ML or Haskell. We have not found special restrictions in this respect. The only 'difficulty' is that it is not possible to write potentially non-terminating functions. One must provide a termination proof when Isabelle/HOL cannot find one. Providing such a proof is not always easy because the argument depends on some other properties such as 'there are no cycles in the heap', which are not so easy to prove. Fortunately in these cases we have expressed the same ideas using inductive relations.

Isabelle/HOL also provides inductive n-relations, transitive closures as well as ordinary first-order logic. This has made it easy to express our properties with almost the same concepts one would use in hand-written proofs. Partial functions have also been very useful in modelling programming language structures such as environments, heaps, and the like. Being able to quantify these objects in Higher-Order Logic has been essential for stating and proving the theorems.

Assessing how 'easy' it has been to conduct the proofs is another question. Part of the difficulties were related to our lack of experience in using Isabelle/HOL. The learning process was rather slow at the beginning. A second inconvenience is that proof assistants (as it must be) do not take anything for granted. Trivial facts that nobody cares to formalise in a hand-written proof, must be painfully stated and proved before they can be used. We have sparingly used the automatic proving commands such as simp_all, auto, etc., in part because they do 'too many' things, and frequently one does not recognise a lemma after using them. Also, we wanted the proof and to relate the proof to our hand-written version. As a consequence, it is very possible that our scripts are longer than needed. Finally, having programs and predicates 'living' together in a theorem has been an experience not always easy to deal with.

On the quality of the extracted code. The Haskell code extracted from the Isabelle/HOL definitions reaches 700 lines, and has undergone some changes before becoming operative in the compiler. One of these changes has been a trivial coercion between the Isabelle/HOL types *nat* and *int* and the Haskell type Int. The most important one has been the replacement of the Isabelle/HOL type $\rightharpoonup$ representing a partial function, heavily used for specifying our compile-time environments, by a highly trusty table type of the Haskell library. The code generated for $\rightharpoonup$ was just a λ-abstraction needing linear time in order to find the value associated to a key. This would lead to a quadratic compile time. Our table is implemented as a balanced tree and has also been used in other phases of the compiler. With this, the efficiency of the code generation phase is in $O(n \log n)$ for a single *Core-Safe* function of size n, and about linear with the number of functions of the input.

6   Related Work

Using some form of formal verification to ensure the correctness of compilers has been a hot topic for many years. An annotated bibliography covering up to 2003 can be found at [6]. Most of the papers reflected there propose techniques whose validity is established by formal proofs made and read by humans.

Using machine-assisted proofs for compilers starts around the seventies, with an intensificaton at the end of the nineties. For instance, [19] uses a constraint solver to asses the validity of the GNU C compiler translations. They do not try to prove the compiler correct but instead to *validate its output* by comparing it with the corresponding input. This technique was originally proposed in [23]. A more recent experiment in compiler validation is [12]. In this case the source is the term language of HOL and the target is assembly language of the ARM processor. The compiler generates for each source, the object file and a proof showing that the semantics of the source is preserved. The last two stages of the compilation are in fact formally verified, while validation of the output is used in the previous phases.

More closely related to our work are [1] which certifies the translation of a Lisp subset to a stack language by using PVS, and [25] which uses Isabelle/HOL to formalise the translation from a small subset of Java (called μ-Java) to a stripped version of the Java Virtual Machine (17 bytecode instructions). Both specify the translation functions, and prove correctness theorems similar to ours. The latter work can be considered as a first attempt on Java, and it was considerably extended by Klein, Nipkow, Berghofer, and Strecker himself in [8,9,3]. Only [3] claims that the extraction facilities of Isabelle/HOL have been used to produce an actually running Java compiler. The main emphasis is on formalisation of Java and JVM features and on creating an infrastructure on which other authors could verify properties of Java or Java bytecode programs.

A realistic C compiler for programming embedded systems has been built and verified in [5,10,11]. The source is a small C subset called *Cminor* to which C is informally translated, and the target is Power PC assembly language. The compiler runs through six intermediate languages for which the semantics are defined and the translation pass verified. The authors use the Coq proof-assistant and its extraction facilities to produce Caml code. They provide figures witnessing that the compile times obtained are competitive whith those of *gcc* running with level-2 optimisations activated. This is perhaps the biggest project on machine-assisted compiler verification done up to now.

Less related work are [7] and the MRG project [24], where certificates in Isabelle/HOL about heap consumption, based on special types inferred by the compiler, are produced. Two EU projects, EmBounded (http://www.embounded.org) and Mobius (http://mobius.inria.fr) have continued this work on certification and proof carrying code, the first one for the functional language Hume, and the second one for Java and the JVM.

As we have said in Sec. 1, the motivation for verifying the *Safe* back-end arises in a different context. We have approached this development because we found it shorter than translating the *Core-Safe* properties to certificates at the level of the JVM. Also, we expected the size of our certificates to be considerably smaller than the ones obtained with the other approach. We have improved on previous work by complementing functional correctness with a proof of resource consumption preservation.

References

1. Dold, A., Vialard, V.: A Mechanically Verified Compiling Specification for a Lisp Compiler. In: Hariharan, R., Mukund, M., Vinay, V. (eds.) FSTTCS 2001. LNCS, vol. 2245, pp. 144–155. Springer, Heidelberg (2001)
2. Barthe, G., Grégoire, B., Kunz, C., Rezk, T.: Certificate Translation for Optimizing Compilers. In: Yi, K. (ed.) SAS 2006. LNCS, vol. 4134, pp. 301–317. Springer, Heidelberg (2006)
3. Berghofer, S., Strecker, M.: Extracting a formally verified, fully executable compiler from a proof assistant. In: Proc. Compiler Optimization Meets Compiler Verification, COCV 2003. ENTCS, pp. 33–50 (2003)
4. Bertot, Y., Casteran, P.: Interactive Theorem Proving and Program Development Coq'Art: The Calculus of Inductive Constructions. Texts in Theoretical Computer Science. An EATCS Series. Springer, Heidelberg (2004)
5. Blazy, S., Dargaye, Z., Leroy, X.: Formal verification of a C compiler front-end. In: Misra, J., Nipkow, T., Sekerinski, E. (eds.) FM 2006. LNCS, vol. 4085, pp. 460–475. Springer, Heidelberg (2006)
6. Dave, M.A.: Compiler verification: a bibliography. SIGSOFT Software Engineering Notes 28(6), 2 (2003)
7. Hofmann, M., Jost, S.: Static prediction of heap space usage for first-order functional programs. In: Proc. 30th ACM Symp. on Principles of Programming Languages, POPL 2003, pp. 185–197. ACM Press, New York (2003)

8. Klein, G., Nipkow, T.: Verified Bytecode Verifiers. Theoretical Computer Science 298, 583–626 (2003)
9. Klein, G., Nipkow, T.: A Machine-Checked Model for a Java-Like Language, Virtual Machine and Compiler. ACM Transactions on Programming Languages and Systems 28(4), 619–695 (2006)
10. Leroy, X.: Formal certification of a compiler back-end, or: programming a compiler with a proof assistant. In: Principles of Programming Languages, POPL 2006, pp. 42–54. ACM Press, New York (2006)
11. Leroy, X.: A formally verified compiler back-end, July 2008, p. 79 (submitted, 2008)
12. Li, G., Owens, S., Slind, K.: Structure of a Proof-Producing Compiler for a Subset of Higher Order Logic. In: De Nicola, R. (ed.) ESOP 2007. LNCS, vol. 4421, pp. 205–219. Springer, Heidelberg (2007)
13. Lindholm, T., Yellin, F.: The Java Virtual Machine Sepecification, 2nd edn. The Java Series. Addison-Wesley, Reading (1999)
14. Montenegro, M., Peña, R., Segura, C.: A Resource-Aware Semantics and Abstract Machine for a Functional Language with Explicit Deallocation. In: Workshop on Functional and (Constraint) Logic Programming, WFLP 2008, Siena, Italy, July 2008, pp. 47–61 (2008) (to appear in ENTCS)
15. Montenegro, M., Peña, R., Segura, C.: A Simple Region Inference Algorithm for a First-Order Functional Language. In: Trends in Functional Programming, TFP 2008, Nijmegen (The Netherlands), May 2008, pp. 194–208 (2008)
16. Montenegro, M., Peña, R., Segura, C.: A Type System for Safe Memory Management and its Proof of Correctness. In: Nadathur, G. (ed.) PPDP 1999. LNCS, vol. 1702, pp. 152–162. Springer, Heidelberg (1999)
17. Montenegro, M., Peña, R., Segura, C.: An Inference Algorithm for Guaranteeing Safe Destruction. In: LOPSTR 2008. LNCS, vol. 5438, pp. 135–151. Springer, Heidelberg (2009)
18. Necula, G.C.: Proof-Carrying Code. In: ACM SIGPLAN-SIGACT Principles of Programming Languages, POPL 1997, pp. 106–119. ACM Press, New York (1997)
19. Necula, G.C.: Translation validation for an optimizing compiler. SIGPLAN Notices 35(5), 83–94 (2000)
20. Nipkow, T., Paulson, L., Wenzel, M.: Isabelle/HOL. A Proof Assistant for Higher-Order Logic. LNCS, vol. 2283. Springer, Heidelberg (2002)
21. Peña, R., Rupérez, D.: A Certified Implementation of a Functional Virtual Machine on top of the Java Virtual Machine. In: Jornadas sobre Programación y Lenguajes, PROLE 2008, Gijón, Spain, October 2008, pp. 131–140 (2008)
22. Peña, R., Segura, C., Montenegro, M.: A Sharing Analysis for SAFE. In: Selected Papers of the 7th Symp. on Trends in Functional Programming, TFP 2006, pp. 109–128 (2007) (Intellect)
23. Pnueli, A., Siegel, M., Singerman, E.: Translation Validation. In: Steffen, B. (ed.) TACAS 1998. LNCS, vol. 1384, pp. 151–166. Springer, Heidelberg (1998)
24. Sannela, D., Hofmann, M.: Mobile Resources Guarantees. EU Open FET project, IST 2001-33149 2001-2005, http://www.dcs.ed.ac.uk/home/mrg
25. Strecker, M.: Formal Verification of a Java Compiler in Isabelle. In: Voronkov, A. (ed.) CADE 2002. LNCS, vol. 2392, pp. 63–77. Springer, Heidelberg (2002)
26. Wildmoser, M.: Verified Proof Carrying Code. Ph.D. thesis, Institut für Informatik, Technical University Munchen (2005)

A Certified Data Race Analysis for a Java-like Language*

Frédéric Dabrowski and David Pichardie

INRIA, Centre Rennes - Bretagne Atlantique, Rennes, France

Abstract. A fundamental issue in multithreaded programming is detecting *data races*. A program is said to be well synchronised if it does not contain data races w.r.t. an interleaving semantics. Formally ensuring this property is central, because the JAVA Memory Model then guarantees that one can safely reason on the interleaved semantics of the program. In this work we formalise in the COQ proof assistant a JAVA bytecode data race analyser based on the conditional must-not alias analysis of Naik and Aiken. The formalisation includes a context-sensitive points-to analysis and an instrumented semantics that counts method calls and loop iterations. Our JAVA-like language handles objects, virtual method calls, thread spawning and lock and unlock operations for threads synchronisation.

1 Introduction

A fundamental issue in multithreaded programming is *data races*, i.e., the situation where two threads access a memory location, and at least one of them changes its value, without proper synchronisation. Such situations can lead to unexpected behaviours, sometimes with damaging consequences [14, 20]. The semantics of programs with multiple threads of control is described by architecture-dependent *memory models* [1, 10] which define admissible executions, taking into account optimisations such as caching and code reordering. Unfortunately, these models are generally not *sequentially consistent*, i.e., it might not be possible to describe every execution of a program as the *serialization*, or *interleaving*, of the actions performed by its threads. Although common memory models impose restrictions on admissible executions, these are still beyond intuition: writes can be seen out of order and reads can be speculative and return values from the future.

Reasoning directly on memory models is possible but hard, counter-intuitive and probably infeasible to the average programmer. As a matter of fact, the *interleaving semantics* is generally assumed in most formal developments in compilation, static analysis and so on. Hopefully, under certain conditions, the interleaving semantics can be turned into a correct approximation of admissible behaviors. Here, we focus on programs expressed in JAVA, which comes with its own, relieved from architecture specific details, memory model. Although the JAVA memory model [15, 21] does not guarantee sequential consistency for all programs, race free programs are guaranteed to be sequentially consistent. Moreover, it enjoys a major property, so called, *the datarace free guarantee*. This property states that a program whose all sequentially consistent

---

* Work partially supported by EU project MOBIUS, and by the ANR-SETI-06-010 grant.

S. Berghofer et al. (Eds.): TPHOLs 2009, LNCS 5674, pp. 212–227, 2009.

executions are race free, only admit sequentially consistent executions. In other words, proving that a program is race free can be done on a simple interleaving semantics; and doing so guarantees the correctness of the interleaving semantics for that program. It is worth noticing that data race freedom is important, not only because it guarantees semantic correctness, but also because it is at the basis of a higher level property called atomicity. The possibility to reason sequentially about atomic sections is a key feature in analysing multithreaded programs. Designing tools, either static or dynamic, aiming at proving datarace freeness is thus a fundamental matter.

This paper takes root in the european MOBIUS project[1] where several program verification techniques have been machine checked with respect to a formal semantics of the sequential JAVA bytecode language. The project has also investigated several verification techniques for multithreaded JAVA but we need a formal guarantee that reasoning on interleaving semantics is safe. While a JAVA memory model's formalisation has been done in COQ [9] and a machine-checked proof of the data race free guarantee has been given in [2] we try to complete the picture formally proving data race freeness. We study how such a machine-checked formalisation can be done for the race detection analysis recently proposed by Naik and Aiken [16–18].

The general architecture of our development is sketched in Figure 1. We formalise four static analyses : a context-sensitive points-to analysis, a must-lock analysis, a conditional must-not alias analysis based on disjoint reachability and a must-not thread escape analysis. In order to ensure the data-race freeness of the program, these analyses are used to refine, in several stages an initial over-approximation of the set of potential races of a program, with the objective to obtain an empty set at the very last stage. Each analysis is mechanically proved correct with respect to an operational semantics. However, we consider three variants of semantics. While the first one is a standard small-step semantics, the second one attaches context information to each reference and frame. This instrumentation makes the soundness proof of the points-to analysis easier. The last semantics handles more instrumentation in order to count method calls and loop iterations. Each instrumentation is proved correct with respect to the semantics just above it. The notion of safe instrumentation is formalised through a standard simulation diagram.

The main contributions of our work are as follows.

- Naik and Aiken have proposed one of the most powerful data race analysis of the area. Their analyser relies on several stages that remove pairs of potential races. Most of these layers have been described informally. The most technical one has been partially proved correct with pencil and paper for a sequential While language [17]. We formalise their work in COQ for a realistic bytecode language with unstructured control flow, operand stack, objects, virtual method calls and lock and unlock operations for threads synchronization.
- Our formalisation is an open framework with three layers of semantics. We formalise and prove correct four static analyses on top of these semantics. We expect our framework to be sufficiently flexible to allow easy integration of new certified blocks for potential race pruning.

[1] http://mobius.inria.fr

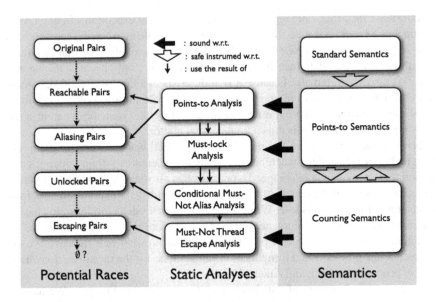

Fig. 1. Architecture of the development

2   A Challenging Example Program

Figure 2 presents an example of a source program for which it is challenging to formally prove race freeness. This example is adapted from the running example given by Naik and Aiken [17] in a While language syntax. The program starts in method main by creating in a first loop, a simple linked list l and then launches a bunch of threads of class T that all share the list l in their field data. Each thread, launched in this way, chooses non deterministically a cell m of the list and then updates m.val.f, using a lock on m.

Figure 3 presents the potential races computed for this example. A data race is described by a triplet (i, f, j) where i and j denote the program points in conflicts and f denotes the accessed field. The first over-approximation, the set of *Original Pairs* is simply obtained by typing: thanks to JAVA's strong typing, a pair of accesses may be involved in a race only if both access the same field and at least one access is a write. For each other approximation, a marked indicates a potential race. The program is in fact data race free but the size of the set of *Original Pairs* (13 pairs here) illustrates the difficulty of statically demonstrating it.

Following [18], a first approximation of races computes the field accesses that are reachable from the entry point of the program and removes also pairs where both accesses are taken by the main thread (*Reachable pairs*). Some triplets may also be removed with an alias analysis that shows that two potential conflicting accesses are not in alias (*Aliasing pairs*). Both sets rely on the points-to analysis presented in Section 4. Among the remaining potential races, several triplets can be disabled by a *must-not thread escape* analysis that predicts a memory access only concerns a reference which

```
class Main() {                    class A{};
   void main() {                  class List{ T val; List next; };
      List l = null;
      while (*) {                  class T extends java.lang.Thread {
         List temp = new List;       A f;
1:       temp.val = new T;           List data;
2:       temp.val.f = new A;         void run(){
3:       temp.next = l;                while(*) {
         l = temp; };             6:      List m = this.data;
      while (*) {                  7:      while (*) { m = m.next; }
         T t = new T;                      synchronized(m)
4:       t.data = l;               8:         { m.val.f = new A; }};
         t.start();                     }
5:       t.f = null; }}};          };
```

Fig. 2. A challenging example program

Original	Reachable	Aliasing	Unlocked	Escaping
$(1, \text{val}, 1), (1, \text{val}, 2), (2, \text{f}, 2), (3, \text{next}, 3),$ $(4, \text{data}, 4)$		✓	✓	
$(2, \text{f}, 5)$			✓	
$(5, \text{f}, 5)$		✓	✓	✓
$(4, \text{data}, 6), (3, \text{next}, 7), (1, \text{val}, 8), (2, \text{f}, 8)$	✓	✓	✓	
$(5, \text{f}, 8)$	✓		✓	✓
$(8, \text{f}, 8)$	✓	✓		✓

Fig. 3. Potential race pairs in the example program

is local to a thread at the current point (*Escaping pairs*). The last potential race $(8, \text{f}, 8)$ requires the most attention since several threads of class T are updating fields f in parallel. These writes are safe because they are guarded by a synchronization on an object which is the only ancestor of the write target in the heap. Such reasoning relies on the fact that if locks guarding two accesses are different then so are the targeted memory locations. The main difficulty comes when several objects allocated at the same program point, e.g. within a loop, may point to the same object. This last triplet is removed by the *conditional must not alias* presented in Section 5.

3   Standard Semantics

The previous example can be compiled into a bytecode language whose syntax is given below. The instruction set allows to manipulate objects, call virtual methods, start threads and lock (or unlock) objects for threads synchronization.

Compared to real JAVA, we discard all numerical manipulations because they are not relevant to our purpose. Static fields, static methods and arrays are not managed here but they are nevertheless source of several potential data races in JAVA programs. Naik's approach [18] for these layers is similar to the technique developed for objects. We estimate that adding these language features would not bring new difficulties that we have not covered yet in the current work. At last, as Naik and Aiken did before us, we only cover synchronization by locks without join, wait and interruption mechanisms. Our approach is sound in presence of such statements, but doesn't take into account the potential races they could prevent. The last missing feature is the JAVA's exception mechanism. Exceptions complicate the control flow of a JAVA program. We expect that handling this mechanism would increase the amount of formal proof but will not require new proof techniques. This is left for further work.

Program syntax. A program is a set of classes, coming with a *Lookup* function matching signatures and program points (allocation sites denoting class names) to methods.

$$\mathbb{C}_{id} \supseteq \{c_{id}, \ldots\} \qquad \mathbb{F} \supseteq \{f, g, h, \ldots\} \qquad \mathbb{M}_{id} \supseteq \{m_{id}, \ldots\}$$
$$\mathbb{V} \supseteq \{x, y, z, \ldots\} \qquad \mathbb{M}_{sig} = \mathbb{M}_{id} \times \mathbb{C}_{id}^n \times (\mathbb{C}_{id} \cup \{\texttt{void}\})$$

$$\mathbb{M} \ni \{\texttt{sig} \in \mathbb{M}_{sig}; \ \texttt{body} \in \mathbb{N} \rightharpoonup inst\}$$
$$\mathbb{C} \ni \{\texttt{name} \in \mathbb{C}_{id}; \ \texttt{fields} \subseteq \mathbb{F}; \ \texttt{methods} \subseteq \mathbb{M}\}$$

$$inst ::= \texttt{aconstnull} \mid \texttt{new } c_{id} \mid \texttt{aload } x \mid \texttt{astore } x \mid \texttt{getfield } f \mid \texttt{putfield } f$$
$$\mid \texttt{areturn} \mid \texttt{return} \mid \texttt{invokevirtual } m_{id} : (c_{id}^n) \, rtype \qquad (n \geq 0)$$
$$\mid \texttt{monitorenter} \mid \texttt{monitorexit} \mid \texttt{start} \mid \texttt{ifnd } \ell \mid \texttt{goto } \ell$$

Semantics Domain. The dynamic semantics of our language is defined over states as a labelled transition system. States and labels, or events, are defined in Figure 4, where $\rightarrow$ stands for total functions and $\rightharpoonup$ stands for partial functions. We distinguish *location* and *memory location* sets. The set of locations is kept abstract in this presentation. In this section, a memory location is itself a location ($\mathbb{L} = \mathbb{O}$). This redundancy will be useful when defining new instrumented semantics where memory locations will carry more information (Sections 4 and 5). In a state (L, σ, μ), L maps memory locations (that identify threads) to call stacks, σ denotes the heap that associates memory locations to objects (c_{id}, map) with c_{id} a class name and map a map from fields to values. We note $\texttt{class}(\sigma, l)$ for $fst(\sigma(l))$ when $l \in dom(\sigma)$. A locking state μ associates with every location ℓ a pair (ℓ', n) if ℓ is locked n times by ℓ' and the constant $\texttt{free}$ if ℓ is not held by any thread. An event $(\ell, ?_f^{ppt}, \ell')$ (resp. $(\ell, !_f^{ppt}, \ell')$) denotes a read (resp. a write) of a field f, performed by the thread ℓ over the memory location ℓ', at a program point ppt. An event τ denotes a silent action.

Transition system. Labelled transitions have the form $st \xrightarrow{e} st'$ (when e is τ we simply omit it). They rely on the usual interleaving semantics, as expressed in the rule below.

$$\frac{L\,\ell = cs \qquad L; \ell \vdash (cs, \sigma, \mu) \xrightarrow{e} (L', \sigma', \mu')}{(L, \sigma, \mu) \xrightarrow{e} (L', \sigma', \mu')}$$

Reductions of the shape $L; \ell \vdash (cs, \sigma, \mu) \xrightarrow{e} (L', \sigma', \mu')$ are defined in Figure 5. Intuitively, such a reduction expresses that in state (L, σ, μ), reducing the thread defined

$$
\begin{array}{rll}
\mathbb{L} \ni \ell & & \text{(location)} \\
\mathbb{O} = \mathbb{L} \ni \ell & & \text{(memory location)} \\
\mathbb{O}_\perp \ni v & ::= \ell \mid \texttt{Null} & \text{(value)} \\
s & ::= v :: s \mid \varepsilon & \text{(operand stack)} \\
Var \to \mathbb{O}_\perp \ni \rho & & \text{(local variables)} \\
\mathbb{O} \rightharpoonup \mathbb{C}_{id} \times (\mathbb{F} \to \mathbb{O}_\perp) \ni \sigma & & \text{(heap)} \\
PPT = \mathbb{M} \times \mathbb{N} \ni ppt & ::= (m, i) & \text{(program point)} \\
CS \ni cs & ::= (m, i, s, \rho) :: cs \mid \square & \text{(call stack)} \\
\mathbb{O} \rightharpoonup CS \ni L & & \text{(thread call stacks)} \\
\mathbb{O} \to ((\mathbb{O} \times \mathbb{N}^*) \cup \{\texttt{free}\}) \ni \mu & & \text{(locking state)} \\
st & ::= (L, \sigma, \mu) & \text{(state)} \\
e & ::= \tau \mid (\ell, ?_f^{ppt}, \ell') \mid (\ell, !_f^{ppt}, \ell') & \text{(event)}
\end{array}
$$

Fig. 4. States and actions

$$
\begin{array}{l}
\sigma[\ell.f \leftarrow v]\,\ell\,f = v \\
\sigma[\ell.f \leftarrow v]\,\ell'\,f = \sigma\,\ell'\,f \text{ if } \ell' \neq \ell \\
\sigma[\ell.f \leftarrow v]\,\ell\,f' = \sigma\,\ell\,f' \text{ if } f' \neq f
\end{array}
\qquad
\begin{array}{l}
(acquire\ \ell\ \ell'\ \mu)\ \ell' = \begin{cases} (\ell, 1) & \text{if } \mu(\ell') = \texttt{free} \\ (\ell, n+1) & \text{if } \mu(\ell') = (\ell, n) \end{cases} \\
(acquire\ \ell\ \ell'\ \mu)\ \ell'' = \mu\ \ell'' \text{ if } \ell'' \neq \ell'
\end{array}
$$

(a) Notations

$$
\texttt{getfield } f;\ell;ppt \vdash (i, \ell'::s, \rho, \sigma) \xrightarrow{\ell?_f^{ppt}\ell'}_1 (i+1, (\sigma\ \ell'\ f)::s, \rho, \sigma) \quad \text{if } \ell' \in dom(\sigma)
$$

$$
\texttt{putfield } f;\ell;ppt \vdash (i, v::\ell'::s, \rho, \sigma) \xrightarrow{\ell!_f^{ppt}\ell'}_1 (i+1, s, \rho, \sigma[\ell'.f \mapsto v]) \quad \text{if } \ell' \in dom(\sigma)
$$

$$
\texttt{new } c_{id};\ell;ppt \vdash (i, s, \rho, \sigma) \longrightarrow_1 (i+1, \ell'::s, \rho, \sigma[\ell' \mapsto new(c_{id})]) \quad \text{where } \ell' \notin dom(\sigma)
$$

$$
\frac{(m.\textbf{body})\ i;\ell;(m,i) \vdash (i, s, \rho, \sigma) \xrightarrow{e}_1 (i', s', \rho', \sigma') \quad L' = L[\ell \mapsto (m, i', s', \rho')::cs]}{L;\ell \vdash ((m, i, s, \rho)::cs, \sigma, \mu) \xrightarrow{e} (L', \sigma', \mu)} \quad (1)
$$

$$
\frac{\begin{array}{c}(m.\textbf{body})\ i = \texttt{invokevirtual } m_{id} : (c_{id}^n)\ rtype \\ s = v_n :: \ldots :: v_1 :: \ell' :: s' \\ Lookup\ (m_{id} : (c_{id}^n) rtype)\ \texttt{class}(\sigma, \ell') = m_1 \\ \rho_1 = [0 \mapsto \ell', 1 \mapsto v_1, \ldots, n \mapsto v_n] \\ L' = L[\ell \mapsto (m_1, 0, \varepsilon, \rho_1)::(m, i+1, s', \rho)::cs]\end{array}}{L;\ell \vdash ((m, i, s, \rho)::cs, \sigma, \mu) \to (L', \sigma, \mu)}
\qquad
\frac{\begin{array}{c}(m.\textbf{body})\ i = \texttt{start} \quad \neg(\ell' \in dom(L)) \\ Lookup\ (run : ()\textbf{void})\ \texttt{class}(\sigma, \ell') = m_1 \\ \rho_1 = [0 \mapsto \ell'] \\ L' = L[\ell \mapsto (m, i+1, s', \rho)::cs, \\ \ell' \mapsto (m_1, 0, \varepsilon, \rho_1)::\square]\end{array}}{L;\ell \vdash ((m, i, \ell'::s', \rho)::cs, \sigma, \mu) \to (L', \sigma, \mu)}
$$

$$
\frac{\begin{array}{c}(m.\textbf{body})\ i = \texttt{monitorenter} \\ \mu\ \ell' \in \{\texttt{free}, (\ell, n)\} \\ \mu' = acquire\ \ell\ \ell'\ \mu \\ L' = L[\ell \mapsto (m, i+1, s, \rho)::cs]\end{array}}{L;\ell \vdash ((m, i, \ell'::s, \rho)::cs, \sigma, \mu) \to (L', \sigma, \mu')}
\qquad
\frac{\begin{array}{c}(m.\textbf{body})\ i = \texttt{monitorexit} \\ \mu = acquire\ \ell\ \ell'\ \mu' \\ L' = L[\ell \mapsto (m, i+1, s, \rho)::cs]\end{array}}{L;\ell \vdash ((m, i, \ell'::s, \rho)::cs, \sigma, \mu) \to (L', \sigma, \mu')}
$$

(b) Reduction rules

Fig. 5. Standard Dynamic Semantics

by the memory location ℓ and the call stack cs, by a non deterministic choice, produces the new state (L', σ', μ'). For the sake of readability, we rely on an auxiliary relation of the shape $instr; \ell; ppt \vdash (i, s, \rho, \sigma) \xrightarrow{e}_1 (i', s', \rho', \sigma')$ for reduction of intra-procedural instructions. In Figure 5, we consider only putfield, getfield and new. Reductions for instructions are standard and produce a τ event. The notation $\sigma[\ell.f \leftarrow v]$ for field update, where $\ell \in dom(\sigma)$, is defined in Figure 5(a). It does not change the class of an object. The reduction of a new instruction pushes a fresh address onto the operand stacks and allocates a new object in the heap. The notation $\sigma[\ell \leftarrow new(cid)]$, where $\neg(\ell \in dom(\sigma))$, denotes the heap σ with a new object, at location ℓ, of class cid and with all fields equals to Null. The auxiliary relation is embedded into the semantics by rule (1). Method invocation relies on the $lookup$ function for method resolution and generates a new frame. Thread spawning is similar to method invocation. However, the new frame is put on top of an empty call stack. We omit the reduction rules for return and areturn, those rules are standard and produce a τ event. For monitorenter and monitorexit we use a partial function $acquire$ defined in Figure 5(a). Intuitively, $acquire\ \ell\ \ell'\ \mu$ results from thread ℓ locking object ℓ' in μ.

We write $RState(P)$ for the set of states that contains the initial state of a program P, that we do not describe here for conciseness concerns, and that is closed by reduction. A data race is a tuple (ppt_1, f, ppt_2) such that $Race(P, ppt_1, f, ppt_2)$ holds.

$$\frac{st \in RState(P) \quad st \xrightarrow{\ell_1 !_f^{ppt_1} \ell_0} st_1 \quad st \xrightarrow{\ell_2 \mathcal{R} \ell_0} st_2 \quad \mathcal{R} \in \{?_f^{ppt_2}, !_f^{ppt_2}\} \quad \ell_1 \neq \ell_2}{Race(P, ppt_1, f, ppt_2)}$$

The ultimate goal of our certified analyser is to guarantee *Data Race Freeness*, i.e. for all $ppt_1, ppt_2 \in PPT$ and $f \in \mathbb{F}, \neg Race(P, ppt_1, f, ppt_2)$.

4   Points-to Semantics

Naik and Aiken make intensive use of points-to analysis in their work. Points-to analysis computes a finite abstraction of the memory where locations are abstracted by their allocation site. The analysis can be made context sensitive if allocation sites are distinguished wrt. the calling context of the method where the allocation occurs.

Many static analyses use this kind of information to have a conservative approximation of the call graph and the heap of a program. Such analyses implicitly reason on instrumented semantics that directly manipulates informations on allocation sites while a standard semantics only keeps track of the class given to a reference during its allocation. In this section we formalise such an intermediate semantics.

This *points-to semantics* takes the form of a COQ module functor

Module PointsToSem (C:CONTEXT). ... **End** PointsToSem.

parameterised by an abstract notion of context which captures a large variety of points-to contexts. Figure 6 presents this notion.

A context is given by two abstract types pcontext and mcontext[2] for pointer contexts and method contexts. Function make_new_context is used to create a new

---

[2] mcontext is noted *Context* in Section 5.

Module Type CONTEXT.

   Parameter pcontext : **Set**. *(* pointer context *)*
   Parameter mcontext : **Set**. *(* method context *)*

   Parameter make_new_context :
    method → line → classId → mcontext → pcontext.
   Parameter make_call_context :
    method → line → mcontext → pcontext → mcontext.
   Parameter get_class : program → pcontext → option classId.

   Parameter class_make_new_context : ∀ p m i cid c,
    body m i = Some (New cid) →
    get_class p (make_new_context m i cid c) = Some cid.

   Parameter init_mcontext : mcontext.
   Parameter init_pcontext : pcontext.

   Parameter eq_pcontext : ∀ c1 c2:pcontext, {c1=c2}+{c1<>c2}.
   Parameter eq_mcontext : ∀ c1 c2:mcontext, {c1=c2}+{c1<>c2}.

End CONTEXT.

Fig. 6. The Module Type of Points-to Contexts

pointer context (make_new_context m i cid c) when an allocation of an object of class cid is performed at line i of a method m, called in a context c. We create a new method context (make_call_context m i c p) when building the calling context of a method called on an object of context p, at line i of a method m, itself called in a context c. At last, (get_class prog p) allows to retrieve the class given to an object allocated in a context p. The hypothesis class_make_new_context ensures consistency between get_class and make_new_context.

The simplest instantiation of this semantics takes class name as pointer contexts and uses a singleton type for method context. A more interesting instantiation is k-objects sensitivity: contexts are sequences of at most k allocation sites $(m, i) \in \mathbb{M} \times \mathbb{N}$. When creating an object at site (m, i) in a context c, we attach to this object a pointer context $(m, i) \oplus_k c$ defined by $(m, i) \cdot c'$ if $|c|=k$, $c = c' \cdot (m', i')$ and $(m, i) \cdot c$ if $|c| < k$, without any change to the current method context. When calling a method on an object we build a new frame with the same method context as the pointer context of the object.

The definition of this semantics is similar to the standard semantics described in the previous section, except that memory location are now couples of the form (ℓ, p) with ℓ a location and p a pointer context. We found convenient for our formalisation to deeply instrument the heap that is now a partial function from memory location of the form (ℓ, p) to objects. This allows us to state a property about the context of a location without mentioning the current heap (in contrast to the class of a location in the previous standard semantics). The second change concerns frames that are now of the form (m, i, c, s, ρ) with c being the method context of the current frame.

In order to reason on this semantics and its different instantiations we give to this module a module type POINTSTO_SEM such that for all modules of type CONTEXT, PointsToSem(C) : POINTSTO_SEM.

Several invariants are proved on this semantics, for example that if any memory locations (ℓ, p_1) and (ℓ, p_2) are in the domain of a heap reachable from a initial state, then $p_1 = p_2$.

```
Module PointsToSemInv (S:POINTSTO_SEM). ...
  Lemma reachable_wf_heap : ∀ p st,
    reachable p st →
    match st with (L,sigma,mu) ⇒
      ∀ l p1 p2, sigma (l,p1)<>None → sigma (l,p2)<>None → p1=p2
    end.
  Proof. ... Qed.
End PointsToSemInv.
\vspace*{-3mm}
```

Safe Instrumentation. The analyses we formalise on top of this points-to semantics are meant for proving absence of race. To transfer such a semantic statement in terms of the standard semantics, we prove simulation diagrams between the transitions systems of the standard and the points-to semantics. Such diagram then allows us to prove that each standard race corresponds to a points-to race.

```
Module SemEquivProp (S:POINTSTO_SEM).
  Lemma race_equiv :
    ∀ p ppt ppt', Standard.race p ppt ppt' → S.race p ppt ppt'.
  Proof. ... Qed.
End SemEquivProp.
\vspace*{-2.5mm}
```

Points-to Analysis. A generic context-sensitive analysis is specified as a set of constraints attached to a program. The analysis is flow-insensitive for heap and flow-sensitive for local variables and operand stacks. Its result is given by four functions

```
PtL: mcontext → method → line → var → (pcontext → Prop).
PtS: mcontext → method → line → list (pcontext → Prop).
PtR: mcontext → method → (pcontext → Prop).
PtF: pcontext → field → (pcontext → Prop).
```

that attach pointer context properties to local variables (PtL), operand stack (PtS) and method returns (PtR). PtF is the flow-insensitive abstraction of the heap.

This analysis is parameterized by a notion of context and proved correct with respect to a suitable points-to semantics. The final theorem says that if (PtL, PtS, PtR, PtF) is a solution of the constraints system then it correctly approximates any reachable states. The notion of correct approximation expresses, for example, that all memory locations (ℓ, p) in the local variables ρ or operand stack s of a reachable frame (m, i, c, s, ρ) is such that p is in the points-to set attached to PtL or PtS for the corresponding flow position (m, i, c).

Must-Lock Analysis. Fine lock analysis requires to statically understand which locks are definitely held when a given program point is reached. For this purpose we specify and prove correct a flow sensitive must-lock analysis that computes the following informations:

```
Locks:    method → line → mcontext → (var → Prop).
Symbolic: method → line → list expr.
```

At each flow position (m, i, c), Locks computes an under-approximation of the local variables that are currently held by the thread reaching this position. The specification of Locks depends on the points-to information PtL computed before. This is a useful information for the monitorexit instruction because the unlocking of a variable x can only cancel the lock information of the variables that may be in alias with x. Symbolic is a flow sensitive abstraction of the operand stack that manipulate symbolic expressions. Such expressions are path expressions of the form x, $x.f$, etc... Lock analysis only requires variable expressions but more complex expressions are useful for the conditional must lock analysis given in Section 5.

Removing False Potential Races. The previous points-to analysis supports the first two stages of the race analyser of Naik et al [18]. The first stage prunes the so called *ReachablePairs*. It only keeps in *OriginalsPairs* the accesses that may be reachable from a start() call site that is itself reachable from the main method, according to the points-to information. Moreover, it discards pairs where each accesses are performed by the main thread because there is only one thread of this kind.

The next stage keeps only the so called *AliasingPairs* using the fact that a conflicting access can only occur on references that may alias. In the example of the Figure 2, the potential race $(5, \mathtt{f}, 8)$ is cancelled because the points-to information of t and m.val are disjoints.

For each stage we formally prove that all these sets over-approximate the set of real races wrt. the points-to semantics.

5   Counting Semantics

The next two stages of our analysis require a deeper instrumentation. We introduce a new semantics with instrumentation for counting method calls and loop iterations. This semantics builds on top of the points-to semantics and uses k-contexts. All developments of this section were formalized in COQ. However, for the sake of conciseness we introduce them in a paper style. In addition to the allocation site (m, i) and the calling context c of an allocation, this semantics captures counting information. More precisely, it records that the allocation occurred after the nth iteration of flow edge $\mathcal{L}(m, i)$ in the kth call to m in context c. Given a program P, the function $\mathcal{L} \in M \times N \to Flow$, for $Flow = \mathbb{N} \times \mathbb{N}$, must satisfy $Safe_P(\mathcal{L})$ as defined below:

$$Safe_P(\mathcal{L}) \equiv \forall m, i, c_{id}.\ (m.\mathrm{body})\ i = \mathtt{new}\ c_{id} \Rightarrow$$
$$\forall n > 0, j_1, \ldots, j_n.\ leadsTo(m, i, i, j_1 \cdot \ldots \cdot j_n) \Rightarrow$$
$$\exists k < n.(j_k, j_{k+1}) = \mathcal{L}(m, i)$$

where $leadsTo(m, i, j, j_1 \cdot \ldots \cdot j_n)$ states that $j_1 \cdot \ldots \cdot j_n$ is a path from i to j in the control flow graph of m. Intuitively, the semantics counts, and records, iteration of all flow edges while $\mathcal{L}$ maps every allocation site to a flow edge, typically a loop entry. Obviously, the function defined by $\mathcal{L}(m, i) = (i, i+1)$ is safe. However, for the purpose of static analysis we need to observe that two allocations occurred within the same loop and, thus, we need a less strict definition (but still strict enough to discriminate between different allocations occurring at the same site). The function $\mathcal{L}$ might be provided by the compiler or computed afterwards with standard techniques. For example, in the bytecode version of our running example, mapping the three allocation sites of the first loop with the control flow edge of the first one is safe. We update the semantic domain as follows:

$$
\begin{array}{llll}
mVect = \mathbb{M} \times Context \to \mathbb{N} & \ni \omega & \text{(method vector)} \\
lVect = \mathbb{M} \times Context \times Flow \to \mathbb{N} \ni \pi & & \text{(iteration vector)} \\
CP & \ni cp ::= \langle m, i, c, \omega, \pi \rangle & \text{(code pointer)} \\
\mathbb{O} = \mathbb{L} \times CP & \ni \bar{\ell} & \text{(memory location)} \\
CS & \ni cs ::= (cp, s, \rho) :: cs \mid \square & \text{(call stack)} \\
& st ::= (L, \sigma, \mu, \omega_g) & \text{state}
\end{array}
$$

A frame holding the code pointer $\langle m, i, c, \omega, \pi \rangle$ is the $\omega(m, c)^{th}$ call to method m in context c (a k-context) since the execution began and, so far, it has performed $\pi(m, c, \phi)$ steps through edge ϕ of its control flow graph. In a state $(L, \sigma, \mu, \omega_g)$, ω_g is a global method vector used as a shared call counter by all threads.

Below, we sketch the extended transition system by giving rules for allocation and method invocation.

$$
\frac{
\begin{array}{c}
(m.\text{body}) \, i = \text{new } cid \qquad \forall cp. \neg((\ell', cp) \in dom(\sigma)) \\
L' = L[\bar{\ell} \mapsto (\langle m, i+1, c, \omega, \pi' \rangle, \bar{\ell}' :: s, \rho) :: cs] \\
\pi' = \pi[(m, c, (i, i+1)) \mapsto \pi(m, c, (i, i+1)) + 1] \qquad \bar{\ell}' = (\ell', \langle m, i, c, \omega, \pi \rangle)
\end{array}
}{
L; \bar{\ell} \vdash ((\langle m, i, c, \omega, \pi \rangle, s, \rho) :: cs, \sigma, \mu, \omega_g) \to (L', \sigma[\bar{\ell}' \mapsto new(cid)], \mu, \omega_g)
}
$$

$$
\frac{
\begin{array}{c}
(m.\text{body}) \, i = \text{invokevirtual } m_{id} : (cid^n) \, rtype \\
s = v_n :: \ldots :: v_1 :: \bar{\ell}' :: s' \qquad \bar{\ell}' = (a_0, \langle m_0, i_0, c_0, \omega_0, \pi_0 \rangle) \\
Lookup \, (m_{id} : (cid^n) rtype) \, \text{class}(\sigma, \bar{\ell}') = m_1 \\
c_1 = (m_0, i_0) \oplus_k c_0 \\
\pi' = \pi[(m, c, (i, i+1)) \mapsto \pi(m, c, (i, i+1)) + 1] \\
\omega_1 = \omega[(m_1, c_1) \mapsto \omega_g(m_1, c_1) + 1] \qquad \pi_1 = \pi'[(m_1, c_1, ., .) \mapsto 0] \\
\rho_1 = [0 \mapsto \bar{\ell}', 1 \mapsto v_1, \ldots, n \mapsto v_n] \qquad \omega_g' = \omega_g[(m_1, c_1) \mapsto \omega_g(m_1, c_1) + 1] \\
L' = L[\bar{\ell} \mapsto (\langle m_1, 0, c_1, \omega_1, \pi_1 \rangle, \varepsilon, \rho_1) :: (\langle m, i+1, c, \omega, \pi' \rangle, s', \rho) :: cs]
\end{array}
}{
L; \ell \vdash ((\langle m, i, c, \omega, \pi \rangle, s, \rho) :: cs, \sigma, \mu, \omega_g) \to (L', \sigma, \mu, \omega_g')
}
$$

For allocation, we simply annotate the new memory location with the current code pointer and record the current move. For method invocation, the caller records the current move. The new frame receives a copy of the vectors of the caller (after the call) where the current call is recorded and the iteration vector corresponding to this call is reseted. Except for thread spawning, omitted rules simply record the current move.

Thread spawning is similar to method invocation except that the new frame receives fresh vectors rather than copies of the caller's vectors.

Safe Instrumentation. As we did between the standard and the points-to semantics we prove a diagram simulation between the points-to semantics and the counting semantics. Is ensures that all *points-to races* correspond to a *counting race*. However, in order to use the soundness theorem of the must-lock analysis we also need to prove a bisimulation diagram. It ensures that all states that are reachable in the counting semantics correspond to a reachable state in the points-to semantics. It allows us to transfer the soundness result of the must-lock analysis in terms of the counting semantics.

Semantics invariants. Proposition 1 states that our instrumentation discriminates between memory locations allocated at the same program point. As expected, to discriminate between memory locations allocated at program point (m, i) in context c, it is sufficient to check the values of $\omega(m, c)$ and $\pi(m, c, \mathcal{L}(m, i))$.

Proposition 1. *Given a program P, if $(L, \sigma, \mu, \omega_g)$ is a reachable state of P and if $Safe_P(\mathcal{L})$ holds then, for all $\bar{\ell}_1, \bar{\ell}_2 \in dom(\sigma)$, we have*

$$\left(\begin{array}{c} \bar{\ell}_1 = (a_1, \langle m, i_1, c, \omega_1, \pi_1 \rangle) \wedge \bar{\ell}_2 = (a_2, \langle m, i_2, c, \omega_2, \pi_2 \rangle) \wedge \\ \omega_1(m, c), \pi_1(m, c, \mathcal{L}(m, i)) = \omega_2(m, c), \pi_2(m, c, \mathcal{L}(m, i)) \end{array} \right) \Rightarrow \bar{\ell}_1 = \bar{\ell}_2$$

Proving Proposition 1 requires stronger semantics invariants. Intuitively, when reaching an allocation site, no memory location in the heap domain should claim to have been allocated at the current iteration. More formally, we have proved that any reachable state is well-formed. A state $(L, \sigma, \mu, \omega_g)$ is said to be well formed if for any frame $(\langle m, i, c, \omega, \pi \rangle, s, \rho)$ in L and for any memory location $(\ell, \langle m_0, i_0, c_0, \omega_0, \pi_0 \rangle)$ in the heap domain we have

$$\omega(m, c) \leq \omega_g(m, c) \qquad \omega(m_0, c_0) \leq \omega_g(m_0, c_0)$$
$$localCoherency(\mathcal{L}, (\ell, \langle m_0, i_0, c_0, \omega_0, \pi_0 \rangle), \langle m, i, c, \omega, \pi \rangle)$$
$$\omega(m, c) \neq \omega'(m, c) \text{ for all distinct frame } (\langle m, i', c, \omega', \pi' \rangle, s', \rho') \text{ in } L$$

where $localCoherency(\mathcal{L}, (\ell, \langle m_0, i_0, c_0, \omega_0, \pi_0 \rangle), \langle m, i, c, \omega, \pi \rangle)$ stands for

$$(m, c) = (m_0, c_0) \Rightarrow \omega(m, c) = \omega_0(m, c) \Rightarrow$$
$$\pi_0(m, c, \mathcal{L}(m, i_0)) \leq \pi(m, c, \mathcal{L}(m, i_0)) \wedge$$
$$\left(\begin{array}{c} \pi_0(m, c, \mathcal{L}(m, i_0)) = \pi(m, c, \mathcal{L}(m, i_0)) \Rightarrow \\ \left(\begin{array}{c} i_0 \neq i \qquad \wedge \\ leadsTo(m, i, i_0, j_1 \cdots \cdot j_n) \Rightarrow \exists k < n, (j_k, j_{k+1}) = \mathcal{L}(m, c) \end{array} \right) \end{array} \right)$$

Type And Effect System. We have formalized a type and effect system which captures the fact that some components of vectors of a memory location are equals to the same components of vectors of : (1) the current frame when the memory location is in local variables or in the stack of the frame or (2) of another memory location pointing to it in the heap. By lack of space, we cannot describe the type and effect system here. Intuitively, we perform a points-to analysis where allocation sites are decorated with masks which tell us which components of vectors of the abstracted memory location match the same components in vectors of a given code pointer (depending on whether we consider (1) or (2)). Formally, an abstract location $\tau \in \mathbb{T}$ in our extended points-to analysis is a pair (A, F) where A is a set of allocation sites and F maps every element of A to a pair Ω, Π of abstract vectors. Abstract vectors are defined by $\Omega \in MVect = \mathbb{M} \times Context \to \{1, \top\}$ and $\Pi \in LVect = \mathbb{M} \times Context \times Flow \to \{1, \top\}$.

Our analysis computes a pair $(\mathcal{A}, \Sigma)$ where $\mathcal{A}$ provides flow-sensitive points-to information with respect to local variables and Σ provides flow-insensitive points-to information with respect to the heap. The decoration of a points-to information acts as a mask. For a memory location held by a local variable, it tells us which components of its vectors (those set to **1**) match those of the current frame. When a memory location points-to another one in the heap, it tells us which components of their respective vectors are equal.

Below we present the last stages we use for potential race pruning. For each stage the result stated by proposition 1 is crucial. Indeed, given an abstract location with allocation site (m, i, c), they rely on a property stating that whenever the decoration states that $\Omega(m, c) = \Pi(m, c, \mathcal{L}(m, i)) = 1$, the abstraction describes a unique concrete location. This property results from the combination of the abstraction relation defined in our type system and of Proposition 1.

Must Not Escape Analysis. We use the flow sensitive element $\mathcal{A}$ of the previous type and effect system to check that, at some program point, an object allocated by a thread is still local to that thread (or has not escaped yet, i.e. it is not reachable from others). More precisely, the type and effect systems is used to guarantee that, at some program point, the last object allocated by a thread at a given allocation site is still local to that thread. In particular, our analysis proves that an access performed at point 4 in our running example, is on the last object of type T allocated by the main thread (which is is local, although at each loop iteration, the new object eventually escapes the main thread). On the opposite, the pair $(5, \mathtt{f}, 8)$ cannot be removed by this analysis since the location has already escaped the main thread at point 5. This pair is removed by the aliasing analysis. Our Escape analysis improves on that of Naik and Aiken which does not distinguish among several allocations performed at the same site.

Conditional Must Not Alias Analysis. The flow-insensitive element Σ of the previous type and effect system is used to define an under-approximation DR_Σ of the notion of *disjoint reachability*. Given a finite set of heaps $\{\sigma_1, \ldots, \sigma_n\}$ and a set of allocation sites H, the disjoint reachability set $DR_{\{\sigma_1, \ldots, \sigma_n\}}(H)$ is the set of allocation sites h such that whenever an object o allocated at site h may be reachable by one or more field dereferences for some heap in $\{\sigma_1, \ldots, \sigma_n\}$, from objects o_1 and o_2 allocated at any sites in H then $o_1 = o_2$. It allows to remove the last potential race of our running example. For each potential conflict between two program points i_1 and i_2, we first compute the set May_1 and May_2 of sites that the corresponding targeted objects may points-to, using the previous points-to analysis. Then we use the must-lock analysis to compute the sets $Must_1$ and $Must_2$ of allocation sites such that for any h in $Must_1$ (resp. $Must_2$), there must exists a lock l currently held at point i_1 (resp. i_2) and allocated at h. The current targeted object must furthermore be reachable from l with respect to the heap history that leads to the current point. This last property is ensured by the path expressions that are computed with a symbolic operand stack during the must-lock analysis. At last, we remove the potential race if and only if $Must_1 \neq \emptyset$, $Must_2 \neq \emptyset$ and

$$May_1 \cap May_2 \subseteq DR_\Sigma(Must_1 \cup Must_2)$$

We formally prove that any potential race that succeeds this last check is not a real race.

6   Related Work

Static race detection. Most works on static race detection follow the lock based approach, as opposed with event ordering based approaches. This approach imposes that every pair of concurrent accesses to the same memory location are guarded by a common lock and is usually enforced by means of a type and effect discipline.

Early work [5] proposes an analysis for a λ-calculus extended with support for shared memory and multiple threads. Each allocation comes in the text of the program with an annotation specifying which lock protects the new memory location and the type and effect system checks that this lock is held whenever it is accessed. More precisely, the annotation refers to a lexically scoped lock definition, thus insuring unicity. To overcome the limitation imposed by the lexical scope of locks, existential types are proposed as a solution to encapsulate an expression with the locks required for its evaluation. This approach was limited in that it was only able to consider programs where all accesses are guarded, even when no concurrent access is possible. Moreover, it imposed the use of specific constructions to manage existential types.

A step toward treatment of realistic languages was made in [7] which considers the JAVA language and supports various common synchronization patterns, classes with internal synchronization, classes that require client-side synchronization and thread-local classes. Aside from additional synchronization patterns, the approach is similar to the previous one and requires annotations on fields (the lock protecting the field) and method declarations (locks that must be held at invocation time). However, the object-oriented nature of the JAVA language is used as a more natural mean for encapsulation. Fields of an object must be protected by a lock (an object in JAVA) accessible from this object. For example, $x.f$ may be protected by $x.g.h$ where g and h are final fields (otherwise, two concurrent accesses to $x.f$ guarded by $x.g.h$ could use different locks). Client-side synchronization and thread-local classes are respectively handled by classes parametrized by locks and a simple form of escape analysis. A similar approach, using ownership types to ensure encapsulation, was taken in [3, 4].

The analysis we consider here is that of [17, 18]. Thanks to the disjoint reachability property and to an heavy use of points-to analysis, it is more precise and captures more idioms than those above. Points-to analysis also makes it more costly but it has been proved that such analyses are tractable thanks to BDD based resolution techniques [22].

Machine checked formalisation for multithreaded JAVA. There is a growing interest in machine checked semantics proof. Leroy [13] develops a certified compiler from Cminor (a C-like imperative language) to PowerPC assembly code in COQ, but only in a sequential setting. Hobor *et al.* [8] define a modular operational semantics for Concurrent C minor and prove the soundness of a concurrent separation logic w.r.t. it, in COQ. Several formalisation of the sequential JVM and its type system have been performed (notably the work of Klein and Nipkow [11]), but few have investigated its multithreaded extension. Petri and Huisman [19] propose a realistic formalization of multithreaded JAVA bytecode in COQ, BICOLANO MT that extends the sequential semantics considered in the MOBIUS project. Lochbihler extends the JAVA source model of Klein and Nipkow [11] with an interleaving semantics and prove type safety. Aspinall and Sevcik formalise the JAVA *data race free guarantee* theorem that ensures that data races free program can only have sequentially consistent behaviors. The work of Petri and Huisman [9] follows a similar approach. The only machine checked proof of a data race analyser we are aware of is the work of Lammich and Müller-Olm [12]. Their formalisation is done at the level of an abtract semantics of a flowgraph-based program model. They formalise a locking analyses with an alias analysis technique simpler than the one used by Naik and Aiken.

7   Conclusions and Future Work

In this paper, we have presented a formalisation of a JAVA bytecode data race analysis based on four advanced static analyses: a context-sensitive points-to analysis, a must-lock analysis, a must-not thread escape analysis and a conditional must-not-alias analysis. Our soundness proofs for these analyses rely on three layers of semantics which have been formally linked together with simulation (and sometimes bisimulation) proofs. The corresponding COQ development has required a little more than 15.000 lines of code. It is available on-line at http://www.irisa.fr/lande/datarace.

This is already a big achievement and as far as we know, one of the first attempt to formally prove data race freeness. However the current specification is not executable. Our analyses are only specified as sets of constraints on logical domains as (pcontext → **Prop**). We are currently working on the implementation part, starting by an Ocaml prototype to mechanically check the example given in Section 2. Then we will have to implement in COQ the abstract domains and the transfer functions of each analysis, following the methodology proposed in our previous work [6]. Thanks to the work we have presented in this paper, these transfer functions will not have to be proved sound with respect to an operational semantics. It is sufficient (and far easier) to prove that they refine correctly the logical specification we have developed here. We plan to only formalise a result checker and check with it the result given by the untrusted analyser written in Ocaml. Extracting an efficient checker is a challenging task here because state-of-the-art points-to analysis implementations rely on such complex symbolic techniques as BDD [22].

Acknowledgment. We thank Thomas Jensen and the anonymous TPHOLs reviewers for their helpful comments.

References

1. AMD. Amd64 architecture programmer's manual volume 2: System programming. Technical Report 24593 (2007)
2. Aspinall, D., Sevcík, J.: Formalising java's data race free guarantee. In: Schneider, K., Brandt, J. (eds.) TPHOLs 2007. LNCS, vol. 4732, pp. 22–37. Springer, Heidelberg (2007)
3. Boyapati, C., Lee, R., Rinard, M.: Ownership types for safe programming: preventing data races and deadlocks. In: ACM Press (ed.) Proc. of OOPSLA 2002, New York, NY, USA, pp. 211–230 (2002)
4. Boyapati, C., Rinard, M.: A parameterized type system for race-free Java programs. In: ACM Press (ed.) Proc. of OOPSLA 2001, New York, NY, USA, pp. 56–69 (2001)
5. Flanagan, C., Abadi, M.: Types for safe locking. In: Swierstra, S.D. (ed.) ESOP 1999. LNCS, vol. 1576, pp. 91–108. Springer, Heidelberg (1999)
6. Cachera, D., Jensen, T., Pichardie, D., Rusu, V.: Extracting a Data Flow Analyser in Constructive Logic. Theoretical Computer Science 342(1), 56–78 (2005)
7. Flanagan, C., Freund, S.N.: Type-based race detection for java. In: Proc. of PLDI 2000, pp. 219–232. ACM Press, New York (2000)
8. Hobor, A., Appel, A.W., Zappa Nardelli, F.: Oracle semantics for concurrent separation logic. In: Drossopoulou, S. (ed.) ESOP 2008. LNCS, vol. 4960, pp. 353–367. Springer, Heidelberg (2008)
9. Huisman, M., Petri, G.: The Java memory model: a formal explanation. In: Verification and Analysis of Multi-threaded Java-like Programs, VAMP (2007) (to appear)
10. Intel. Intel 64 architecture memory ordering white paper. Technical Report SKU 318147-001 (2007)
11. Klein, G., Nipkow, T.: A machine-checked model for a Java-like language, virtual machine and compiler. ACM Transactions on Programming Languages and Systems 28(4), 619–695 (2006)
12. Lammich, P., Müller-Olm, M.: Formalization of conflict analysis of programs with procedures, thread creation, and monitors. In: The Archive of Formal Proofs (2007)
13. Leroy, X.: Formal certification of a compiler back-end, or: programming a compiler with a proof assistant. In: Proc. of POPL 2006, pp. 42–54. ACM Press, New York (2006)
14. Leveson, N.G.: Safeware: system safety and computers. ACM, NY (1995)
15. Manson, J., Pugh, W., Adve, S.V.: The Java Memory Model. In: Proc. of POPL 2005, pp. 378–391. ACM Press, New York (2005)
16. Naik, M.: Effective Static Data Race Detection For Java. PhD thesis, Standford University (2008)
17. Naik, M., Aiken, A.: Conditional must not aliasing for static race detection. In: Proc. of POPL 2007, pp. 327–338. ACM Press, New York (2007)
18. Naik, M., Aiken, A., Whaley, J.: Effective static race detection for java. In: Proc. of PLDI 2006, pp. 308–319. ACM Press, New York (2006)
19. Petri, G., Huisman, M.: BicolanoMT: a formalization of multi-threaded Java at bytecode level. In: Bytecode 2008. Electronic Notes in Theoretical Computer Science (2008)
20. Poulsen, K.: Tracking the blackout bug (2004)
21. Sun Microsystems, Inc. JSR 133 Expert Group, Java Memory Model and Thread Specification Revision (2004)
22. Whaley, J., Lam, M.S.: Cloning-based context-sensitive pointer alias analysis using binary decision diagrams. In: Proc. of PLDI 2004, pp. 131–144. ACM, New York (2004)

Formal Analysis of Optical Waveguides in HOL

Osman Hasan, Sanaz Khan Afshar, and Sofiène Tahar

Dept. of Electrical & Computer Engineering, Concordia University,
1455 de Maisonneuve W., Montreal, Quebec, H3G 1M8, Canada
{o_hasan,s_khanaf,tahar}@ece.concordia.ca

Abstract. Optical systems are becoming increasingly important as they
tend to resolve many bottlenecks in the present age communications and
electronics. Some common examples include their usage to meet high
capacity link demands in communication systems and to overcome the
performance limitations of metal interconnect in silicon chips. Though,
the inability to efficiently analyze optical systems using traditional analysis approaches, due to the continuous nature of optics, somewhat limits their application, specially in safety-critical applications. In order to
overcome this limitation, we propose to formally analyze optical systems
using a higher-order-logic theorem prover (HOL). As a first step in this
endeavor, we formally analyze eigenvalues for planar optical waveguides,
which are some of the most fundamental components in optical devices.
For the formalization, we have utilized the mathematical concepts of differentiation of piecewise functions and one-sided limits of functions. In
order to illustrate the practical effectiveness of our results, we present
the formal analysis of a planar asymmetric waveguide.

1 Introduction

Optical systems are increasingly being used these days, mainly because of their
ability to provide high capacity communication links, in applications ranging
from ubiquitous internet and mobile communications, to not so commonly used
but more advanced scientific domains, such as optical integrated circuits, bio-
photonics and laser material processing. The correctness of operation for these
optical systems is usually very important due to the financial or safety critical
nature of their applications. Therefore, quite a significant portion of the design
time of an optical system is spent on analyzing the designs so that functional
errors can be caught prior to the production of the actual devices. Calculus plays
a significant role in such analysis. Nonliner differential equations with transcendental components are used to model the electric and magnetic field components
of the electromagnetic light waves. The optical components are characterized by
their refractive indices and then the effects of passing electromagnetic waves of
visible and infrared frequencies through these mediums are analyzed to ensure
that the desired reflection and refraction patterns are obtained.

The analysis of optical systems has so far been mainly conducted by using
paper-and-pencil based proof methods [18]. Such traditional techniques are usually very tedious and always have some risk of an erroneous analysis due to the

S. Berghofer et al. (Eds.): TPHOLs 2009, LNCS 5674, pp. 228–243, 2009.
© Springer-Verlag Berlin Heidelberg 2009

complex nature of the present age optical systems coupled with the human-error factor. The advent of fast and inexpensive computational power in the last two decades opened up avenues for using computers in the domain of optical system analysis. Nowadays, computer based simulation approaches and computer algebra systems are quite frequently used to validate the optical system analysis results obtained earlier via paper-and-pencil proof methods. In computer simulation, complex electromagnetic wave models can be constructed and then their behaviors in an optical medium of known refractive index can be analyzed. But, computer simulation cannot provide 100% precise results since the fundamental idea in this approach is to approximately answer a query by analyzing a large number of samples. Similarly, computer algebra systems, which even though are considered to be semi-formal and are very efficient in mathematical computations, also fail to guarantee correctness of results because they are constructed using extremely complicated algorithms, which are quite likely to contain bugs. Thus, these traditional techniques should not be relied upon for the analysis of optical systems, especially when they are used in safety critical areas, such as medicine, transportation and military, where inaccuracies in the analysis may even result in the loss of human lives.

In the past couple of decades, formal methods have been successfully used for the precise analysis of a verity of hardware and software systems. The rigorous exercise of developing a mathematical model for the given system and analyzing this model using mathematical reasoning usually increases the chances for catching subtle but critical design errors that are often ignored by traditional techniques like simulation. Given the sophistication of the present age optical systems and their extensive usage in safety critical applications, there is a dire need of using formal methods in this domain. However, due to the continuous nature of the analysis and the involvement of transcendental functions, automatic state-based approaches, like model checking, cannot be used in this domain. On the other hand, we believe that higher-order-logic theorem proving offers a promising solution for conducting formal analysis of optical systems. The main reason being the highly expressiveness nature of higher-order logic, which can be leveraged upon to essentially model any system that can be expressed in a closed mathematical form. In fact, most of the classical mathematical theories behind elementary calculus, such as differentiation, limit, etc., and transcendental functions, which are the most fundamental tools for analyzing optical systems, have been formalized in higher-order logic [6]. Though, to the best of our knowledge, formal analysis of optical devices is a novelty that has not been presented in the open literature so far using any technique, including theorem proving.

In this paper, as a first step towards using a higher-order-logic theorem prover for analyzing optical systems, we present the formal analysis of planar optical waveguides operating in the *transverse electric* (TE) mode, i.e., a mode when electric field is transverse to the plane of incidence. A waveguide can be defined as an optical structure that allows the confinement of electromagnetic light waves within its boundaries by *total internal reflection* (TIR). It is considered to be one of the most fundamental components of any optical system. Some of the optical systems that heavily rely on optical waveguides, include fiber-optic communications links, fiber lasers and amplifiers for high-power applications, as well as all optical integrated circuits. A planar waveguide, which we mainly analyze in this paper, is a relatively simple but widely used structure for light confinement. It is well accepted in the optics literature that the one-dimensional analysis of this simple planar waveguide is directly applicable to many real problems and the whole concept forms a foundation for more complex optical structures [18].

In order to formally describe the behavior of the planar waveguide, we model the electric and magnetic field equations, which govern the passage of light waves through a planar waveguide, in higher-order logic. The formalization is relatively simple because in the TE mode there is no $y - axis$ dependance, which allows us to describe the electromagnetic fields as a small subset of Maxwell Equations. Based on these formal definitions, we present the verification of the eignevalue equation for a planar waveguide in the TE mode. This equation plays a vital role in designing planar waveguides, as it provides the relationship between the wavelength of light waves that need to be transmitted through a planar waveguide and the planar waveguide's physical parameters, such as refractive indices and dimensions. In this formalization and verification, we required the mathematical concepts of differentiation of piecewise functions and one-sided limits. We built upon Harrison's real analysis theories [6] for this purpose, which include the higher-order-logic formalization of differentiation and limits. We also present some new definitions that allow us to reason about the differentiation of piecewise functions and one-sided limits with minimal reasoning efforts. Finally, in order to illustrate the effectiveness of the formally verified eigenvalue equation in designing real-world optical systems, we present the analysis of a planar dielectric structure [18]. All the work described in this paper is done using the HOL theorem prover [4]. The main motivations behind this choice include the past familiarity with HOL along with the availability of Harrison's real analysis theories [6], which forms the fundamental core of our work.

The remainder of this paper is organized as follows. Section 2 gives a review of related work. In Section 3, we provide a brief introduction about planar waveguides along with their corresponding electromagnetic field equations and eigenvalues. In Section 4, we present the formalization of the electromagnetic fields for a planar waveguide. We utilize this formalization to verify the eigenvalue equation in Section 5. The analysis of a planar dielectric structure is presented in Section 6. Finally, Section 7 concludes the paper.

2   Related Work

The continuous advancement of optical devices towards increased functionality and performance comes with the challenge of developing analysis tools that are able to keep up with the growing level of sophistication. Even though, there is a significant amount of research going on in this important area of analyzing optical systems but, to the best of our knowledge, none of the available optical analysis tools are based on formal methods and the work presented in this paper is the first one of its kind. In this section, we present a brief overview of the state-of-the-art informal techniques used for optical system analysis.

The most commonly used computer based techniques for optical system analysis are based on simulation and numerical methods. Some examples include the analysis of integrated optical devices [20], optical switches [16] and biosensors [23]. Optical systems are continuous systems and thus the first step in their simulation based analysis is to construct a discrete model of the given system [5]. Once the system is discretized, the electromagnetic wave equations are solved by numerical methods. Finite difference methods are the most commonly used numerical approaches applied on wave equations. Finite difference methods applied to the time domain discretized wave equations are referred to as the Finite Difference Time Domain (FDTD) methods [21] and to the frequency domain discretized wave equations as the Finite Difference Frequency Domain (FDFD) methods [19]. Solving equations with numerical methods itself imposes an additional form of error on solutions of the problem. Besides inaccuracies, another major disadvantage, associated with the numerical methods and simulation based approaches, is the tremendous amount of CPU time and memory requirements for attaining reasonable analysis results [10]. In [9,13], the authors argued different methodologies to break the structure into smaller components to improve the memory consumption and speed of the FDTD methods. Similarly, some enhancements for the FDFD method are proposed in [22,12]. There is extensive effort on this subject and although there are some improvements but the inherent nature of numerical and simulation based methods fails all these effort to bring 100% accuracy in the analysis, which can be achieved by the proposed higher-order-logic theorem proving based approach.

Computer algebra systems incorporate a wide variety of symbolic techniques for the manipulation of calculus problems. Based on these capabilities, they have been also tried in the area of optical system analysis. For example, the analysis of planar waveguides using Mathematica [14], which is a widely used computer algebra system, is presented in [3]. With the growing interest in optical system analysis, a dedicated optical analysis package *Optica* [17] has been very recently released for Mathematica. Optica performs symbolic modeling of optical systems, diffraction, interference, and Gaussian beam propagation calculations and is general enough to handle many complex optical systems in a semi-formal manner. Computer algebra systems have also been found to be very useful for evaluating eigenvalues for transcendental equations. This feature has been extensively used along with the paper-and-pencil based analytical approaches. The idea here is to verify the eigenvalue equation by hand and then feed that equation to a computer algebra system to get the desired eigenvalues [18]. Despite all these advantages, the analysis results from computer algebra systems cannot be termed as 100% precise due to the many approximations and heuristics used for automation and reducing memory constraints. Another source of inaccuracy is the presence of unverified huge symbolic manipulation algorithms in their core, which are quite likely to contain bugs. The proposed theorem proving based approach overcomes these limitations but at the cost of significant user interaction.

3   Planar Waveguides

Planar waveguides are basically optical structures in which optical radiation
propagates in a single dimension. Planar waveguides have become the key elements in the modern high speed optical networks and have been shown to
provide a very promising solution to overcome performance limitations of metal
interconnect in silicon chips.

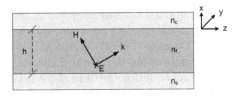

Fig. 1. Planar Waveguide Structure

The planar waveguide, shown in Figure 1, is considered to be infinite in extent
in two dimensions, lets say the yz plane, but finite in the x direction. It consists of
a thin dielectric film surrounded by materials of different refractive indices. The
refractive index of a medium is usually defined as the ratio between the phase
velocity of the light wave in a reference medium to the phase velocity in the
medium itself and is a widely used characteristic for optical devices. In Figure 1,
n_c, n_s, and n_f represent the refractive indices of the cover region, the substrate
region, and the film, which is assumed to be of thickness h, respectively. The
refractive index profile of a planar waveguide can be summarized as follows:

$$n(x) = \begin{cases} n_c & x > 0 \\ n_f & -h < x < 0 \\ n_s & x < -h \end{cases} \tag{1}$$

The most important concept in optical waveguides is that of *total internal reflection* (TIR). When a wave crosses a boundary between materials with different
refractive indices, it is usually partially refracted at the boundary surface, and
partially reflected. TIR happens when there is no refraction. Since, the objective
of waveguides is to guide waves with minimum loss, ideally we want to ensure TIR
for the waves that we want the waveguide to guide. TIR is ensured only when the
following two conditions are satisfied. Firstly, the refractive index of the trans-
mitting medium must be greater than its surroundings, $n_{medium} > n_{surrounding}$
and secondly, the angle of incidence of the wave at the medium is greater than
a particular angle, which is usually referred to as the *critical angle*. The value of
the critical angle also depends on the relative refractive index of the two materials

of the boundary. Thus, the distribution of refractive indices of the waveguides characterize the behavior of the waveguide and restricts the type of waves which the waveguide can guide.

Like all other waveguides, the planar waveguide also needs to provide the TIR conditions for the waves, which are required to be transmitted through them. The first condition is satisfied by choosing n_f to be greater than both n_s and n_c. The second condition, on the other hand, is dependent on the angle of incidence of the wave on the boundary of the waveguide and thus involves the characteristics of the wave itself, which makes it more challenging to ensure.

Basically, light is an electromagnetic disturbance propagated through the field according to electromagnetic laws. Thus, propagation of light waves through a medium can be characterized by their electromagnetic fields. Based on Maxwell equations [11], which completely describe the behavior of light waves, it is not necessary to solve electromagnetic problems for each and every field component. It is well known that for a planar waveguide, it suffices to consider two possible electric field polarizations, *transverse electric* (TE) or *transverse magnetic* (TM) [18]. In the TE mode, the electric field is transverse to the direction of propagation and it has no longitudinal component along the z-axis. Thus, the $y - axis$ component of the electric field E_y is sufficient to completely characterize the planar waveguide. Similarly, in the TM mode, magnetic field has no longitudinal components along the z-axis and solving the system only for the $y - axis$ component of the magnetic field H_y will provide us with the remaining electric field components. In this paper, we focus on the TE mode, though the TM mode can also be analyzed in a similar way.

Based on the above discussion, the electric and magnetic field amplitudes in the TE mode for the three regions, with different refractive indices, of the planar waveguide are given as follows [18]:

$$E_y(x) = \begin{cases} Ae^{\gamma_c x} & x > 0 \\ B\cos(\kappa_f) + C\sin(\kappa_f x) & -h < x < 0 \\ De^{\gamma_s(x+h)} & x < -h \end{cases} \tag{2}$$

$$H_z = \frac{j}{\omega\mu_0}\frac{\partial E_y}{\partial x} \tag{3}$$

where A, B, C, and D are amplitude coefficients, γ_c and γ_s are *attenuation coefficients* of the cover and substrate, respectively, κ_f is the *transverse component* of the wavevector $k = \frac{2\pi}{\lambda}$ in the guiding film, ω is the angular frequency of light and μ is the permeability of the medium. Some of these parameters can be further defined as follows:

$$\gamma_c = \sqrt{\beta^2 - k_0^2 n_c^2} \tag{4}$$

$$\gamma_s = \sqrt{\beta^2 - k_0^2 n_s^2} \tag{5}$$

$$\kappa_f = \sqrt{k_0^2 n_f^2 - \beta^2} \tag{6}$$

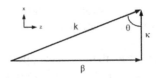

Fig. 2. Longitudinal (β) and Transverse (κ) Components of Wavevector k

where k_0 is the vacuum wavevector, such that $k_0 = \frac{k}{n}$ with n being the refractive index of the medium, and β and κ are the longitudinal and transverse components of the wavevector k, respectively, inside the film, as depicted in Figure 2. The angle θ, is the required angle of incidence of the wave.

This completes the mathematical model of the light wave in a planar waveguide, which leads us back to the original question of finding the angle of incidence θ of the wave to ensure TIR. β is the most interesting vector in this regard. It summarizes two of the very important characteristics of a wave in a medium. Firstly, because it is the longitudinal component of the wavevector, β contains the information about the wavelength of the wave. Secondly, it contains the propagation direction of the wave within the medium, which consequently gives us the angle of incidence θ. Now, in order to ensure the second condition for TIR, we need to find the corresponding βs. These specific values of βs are nominated to be the *eigenvalue of waveguides* since they contain all the information that is required to describe the behavior of the wave and the waveguide.

The electric and magnetic field equations (2) and (3) can be utilized along with their well-known continuous nature [18] to verify the following useful relationship, which is usually termed as the *eigenvalue equation* for β.

$$\tan(h\kappa_f) = \frac{\gamma_c + \gamma_s}{\kappa_f \left(1 - \frac{\gamma_c \gamma_s}{\kappa_f^2}\right)} \tag{7}$$

The good thing about this relationship is that it contains β along with all the physical characteristics of the planar waveguide, such as refractive indices and height. Thus, it can be used to evaluate the value of β in terms of the planar waveguide parameters. This way, we can tune these parameters in such a way that an appropriate value of β is attained that satisfies the second condition for TIR, i.e., $\sin^{-1}(\frac{\lambda\beta}{2\pi}) < critical\_angle$. All the values of β that satisfy the above conditions are usually termed as the TE modes in the planar waveguide.

In this paper, we present the higher-order-logic formalization of the electric and magnetic field equations for the planar wave guide, given in Equations (2) and (3), respectively. Then, based on these formal definitions, we present the formal verification of the eigenvalue equation, given in Equation (7). As outlined above, it is one of the most important relationships used for the analysis of planar waveguides, which makes its formal verification in a higher-order-logic theorem prover a significant step towards using them for conducting formal optical systems analysis.

4   Formalization of Electromagnetic Fields

In this section, we present the higher-order-logic formalization of the electric and magnetic fields for a planar waveguide in the TE mode. We also verify an expression for the magnetic field by differentiating the electric field expression.

The electric field, given in Equation (2), is a piecewise function, i.e., a function whose values are defined differently on disjoint subsets of its domain. Reasoning about the derivatives of piecewise functions in a theorem prover is a tedious task as it involves rewriting based on the classical definitions of differentiation and limit due to their domain dependant values. In order to facilitate such reasoning, we propose to formally define piecewise linear functions in terms of the *Heaviside step function* [1], which is sometimes also referred to as the *unit step function*. A Heaviside step function is a discontinuous, relatively simple piecewise, real-valued function that returns 1 for all strictly positive arguments, 0 for strictly negative arguments and its value at point 0 is usually $\frac{1}{2}$.

$$H(x) = \begin{cases} 0 & x < 0; \\ \frac{1}{2} & x = 0; \\ 1 & 0 < x. \end{cases} \tag{8}$$

By defining piecewise functions based on the Heaviside step function, we make their domain dependance implicit. It allows us to simply reason about the derivatives of piecewise functions using the differentiation properties of sum and products of functions. This way, we need to reason about the derivative of only one piecewise function, i.e., the Heaviside step function, from scratch and build upon these results to reason about the derivatives of all kinds of piecewise functions without utilizing the classical definitions. In this paper, we apply this approach to reason about the derivative of the electric field expression, given in Equation (2). The first step in this regard is to formalize the Heaviside step function as the following higher-order-logic function.

Definition 1: *Heaviside Step Function*
   ⊢ ∀ x. h_step x = if x=0 then $\frac{1}{2}$ else (if x<0 then 0 else 1)

Next, we formally verify that the derivative of h_step function for all values of its argument x, except 0, is equal to 0.

Theorem 1: *Derivative of Heaviside Step Function*
   ⊢ ∀ x. ¬(x = 0) ⇒ (deriv h_step x = 0)

where the HOL function deriv represents the derivative function [6] that accepts a real-valued function f and a differentiating variable x and returns df/dx. The proof of the above theorem is based on the classical definitions of differentiation and limit along with some simple arithmetic reasoning.

Now, the electric field of a planar waveguide, given in Equation (2), can be expressed in higher-order logic as the following function.

Definition 2: *Electric Field for the Planar Waveguide in TE mode*

$\vdash \forall$ b k n. gamma b k n = $\sqrt{b^2 - k^2 n^2}$

$\vdash \forall$ b k n. kappa b k n = $\sqrt{k^2 n^2 - b^2}$

$\vdash \forall$ A B C D n_c n_s n_f k_0 b h x.

E_field A B C D n_c n_s n_f k_0 b h x =

A $e^{(-(\text{gamma b k\_0 n\_c}))x}$ (h_step x) +

(B cos((kappa b k_0 n_f) x) + C sin((kappa b k_0 n_f) x))

(h_step (-x)) +

(D $e^{(-(\text{gamma b k\_0 n\_s}))(x+h)}$ -

(B cos((kappa b k_0 n_f) x) + C sin((kappa b k_0 n_f) x)))

(h_step (-x - h))

The function E_field accepts the four amplitude coefficients A, B, C and D, the three refractive indices for the planar waveguide n_c, n_s and n_f, corresponding to the cover, substrate and the film regions, respectively, the vacuum wave vector k_0, the longitudinal component of the wave vector b, the height of the waveguide h and the variable x for the *x-axis*. It uses the function gamma to obtain the two attenuation coefficients in the cover and substrate as (gamma b k_0 n_c) and (gamma b k_0 n_s), respectively, and the function kappa to model the transverse component of k in the guiding film as (kappa b k_0 n_f). It also utilizes the Heaviside step function h_step thrice with appropriate arguments to model the three sub domains of the piecewise electric field for the planar waveguide, described by the above parameters, according to Equation (2). It is important to note that, rather than having the undefined values for the boundaries $x = 0$ and $x = -h$, as is the case in Equation (2), our formal definition assigns fixed values to these points. But, since we will be analyzing the amplitude coefficients under the continuity of electric and magnetic fields, these point values do not alter our results as will be seen in the next section.

Next, we formalize the magnetic field expression for the planar waveguide, given in Equation (3), using the functional definition of the derivative, deriv, given in [6], as follows.

Definition 3: *Magnetic Field for the Planar Waveguide*

$\vdash \forall$ omega mu A B C D n_c n_s n_f k_0 b h x.

H_field omega mu A B C D n_c n_s n_f k_0 b h x =

$\frac{1}{\text{omega mu}}$ deriv (λx. E_field A B C D n_c n_s n_f k_0 b h x) x

The function H_field accepts the frequency omega and the permeability of the medium mu besides the same parameters that have been used for defining the electric field of the planar waveguide in Definition 2. We have removed the *imaginary unit* part from the original definition, given in Equation (3), in the above definition for simplicity as our analysis is based on the amplitudes or absolute values of electric and magnetic fields and thus requires the real portion of the corresponding complex numbers only. However, if need arises, the imaginary part can be included in the analysis as well by utilizing the higher-order-logic formalization of complex numbers [7].

Definitions 2 and 3 can now be used to formally verify a relation for the magnetic field in a planar waveguide as follows:

Theorem 2: *Expression for the Magnetic Field*

⊢ ∀ omega mu A B C D n_c n_s n_f k_0 b h x.
   ¬(x = 0) ∧ ¬(x = -h) ⇒
      H_field omega mu A B C D n_c n_s n_f k_0 b h x = $\frac{1}{\text{omega mu}}$ (
      (-(gamma b k_0 n_c)) A e$^{(-(\text{gamma b k\_0 n\_s}))x}$ (h_step x) +
      (kappa b k_0 n_f)
      (-B sin((kappa b k_0 n_f) x) + C cos((kappa b k_0 n_f) x))
         (h_step (-x)) +
      (gamma b k_0 n_s) (D e$^{((\text{gamma b k\_0 n\_s}))(x+h)}$ -
      (kappa b k_0 n_f)
      (-B sin((kappa b k_0 n_f) x) + C cos((kappa b k_0 n_f) x)))
         (h_step (-x - h)))

This theorem can be verified by proving the derivatives of the three expressions found in the definition of the electric field and the derivative of the Heaviside step function, given in Theorem 2, along with basic differentiation properties of a product and sum of functions, formally verified in [6].

5   Verification of the Eigenvalue Equation

In this section, we build upon the formal definitions of electromagnetic field relations, formalized in the previous section, to formally verify the eigenvalue equation for the planar waveguide in the TE mode, given in Equation (7).

The main idea behind our analysis is to leverage upon the continuous nature of the electric and magnetic field functions. Like all other continuous functions, a continuous piecewise function f also approaches the value $f(x0)$ at any point $x = x0$ in its domain. This condition, when applied to the boundary points $x = 0$ and $x = -h$ of our piecewise functions for the electric and magnetic fields, E_field and H_field, respectively, yields very interesting results that allow us to express the amplitude coefficients B, C and D in terms of the amplitude coefficient A and then finally utilize these relationships to verify Equation (7).

Due to the piecewise nature of our electric and magnetic field functions, the above reasoning is based on the mathematical concept of right and left hand limits, sometime referred to as one-sided limits, which are the limits of a real-valued function taken as a point in their domain is approached from the right and from the left hand side of the real axis, respectively [2]. Therefore, the first step towards the formal verification of Equation (7) is the formalization of right hand limit in higher-order logic using its classical definition as follows:

Definition 4: *Limit from the Right*

⊢ ∀ f y0 x0. right_lim f y0 x0 =
   ∀ e. 0 < e ⇒ ∃d. 0 < d ∧
      ∀x. 0 < x - x0 ∧ x - x0 < d ⇒ abs(f x - y0) < e

The abs function is the HOL function for the absolute value of a real number. According to the above definition, the limit of a real valued function $f(x)$, as x tends to $x0$ from the right is $y0$, if for all strictly positive values e, there exists a number d such that for all x satisfying $x0 < x < x0 + d$, we have $|f(x) - y0| < e$. Similarly, the left hand limit can be formalized as follows:

Definition 5: *Limit from the Left*

⊢ ∀ f y0 x0. left_lim f y0 x0 =
    ∀ e. 0 < e ⇒ ∃d. 0 < d ∧
        ∀x. -d < x - x0 ∧ x - x0 < 0 ⇒ abs(f x - y0) < e

If the normal limit of a function exists at a point and is equal to $y0$ then both the right and left limits for that function are also well-defined for the same point and are both equal to $y0$. This is an important result for our analysis and thus we formally verify it in the HOL theorem prover as the following theorem.

Theorem 4: *Limit Implies Limit from the Right and Left*

⊢ ∀ f y0 x0. (f→y0)x0 ⇒ right_lim f y0 x0 ∧ left_lim f y0 x0

The assumption of the above theorem (f → y0)x0 represents the formalization of the normal limit of a function [6] and is True only if the function f approaches $y0$ at point $x = x0$. The proof of Theorem 4 is basically a re-writing of the definitions involved along with the properties of the absolute function. We also verified the uniqueness of both right and left hand limits as follows.

Theorem 5: *Limit from the Right is Unique*

⊢ ∀f y1 y2 x0. right_lim f y1 x0 ∧ right_lim f y2 x0 ⇒(y1=y2)

Theorem 6: *Limit from the Left is Unique*

⊢ ∀f y1 y2 x0. left_lim f y1 x0 ∧ left_lim f y2 x0 ⇒(y1=y2)

The proof of Theorem 5 is by contradiction, as it is not possible that a real-valued function gets as near as possible to two unequal points in its range for the same argument. We proceed with the proof by first assuming that $\neg(y1 = y2)$ and then rewriting the statement of Theorem 5 with the definition of the function right_lim. Next, the two assumptions are specialized for $e = \frac{|y1-y2|}{2}$ case. Now, the same x is chosen for both the assumptions in such a way that the conditions on x, i.e., $x0 < x < x0 + d$, for both of the assumptions are satisfied. One such x is $\frac{min\ d1\ d2}{2} + x0$, where $d1$ and $d2$ are the $d's$ for the two assumptions, respectively, and the function min returns the minimum value out of its two real number arguments. Thus, for such an x, the two given assumptions imply that $|fx - y1| < \frac{|y1-y2|}{2}$ and $|fx - y2| < \frac{|y1-y2|}{2}$, which leads to a contradiction in both of the cases when $y1 < y2$ and $y2 < y1$. Hence, our assumption $\neg(y1 = y2)$ cannot be True and $y1$ must be equal to $y2$, which concludes the proof of Theorem 5. Theorem 6 is also verified using similar reasoning.

    The above infrastructure can now be utilized to formally verify the mathematical relationships between the amplitude coefficients. The relationship between the amplitude coefficients B and A can be formally stated as follows:

Theorem 7: $B = A$

$\vdash \forall$ A B C D n_c n_s n_f k_0 b h x. 0 < h $\land$
   ($\forall$x. (λx. E_field A B C D n_c n_s n_f k_0 b h x) contl x)
   $\Rightarrow$ (B = A)

The first assumption ensures that h is always greater than 0 and is valid since h represents the height of the waveguide. Whereas, the HOL predicate (f contl x) [6], used in the above theorem, represents the relational form of a continuous function definition, which is True when the limit of the real-valued function f exists for all points x on the real line and is equal to $f(x)$. Thus, the corresponding assumption, in the above theorem, ensures that the function E_field is continuous on the $x - axis$ and its limit at the boundary points $x = 0$ and $x = -h$ is equal to the value of the function E_field at $x = 0$ and $x = -h$.

In order to verify Theorem 7, consider the boundary point $x = 0$, for which the value of the function E_field becomes $\frac{A+B}{2}$, according to Definition 2. Now, based on Theorem 5, the limit from the right at $x = 0$ for the function E_field is also going to be $\frac{A+B}{2}$. Next, we verified, using Definition 4 along with the properties of the exponential function [6], that the limit from the right for the function E_field at point $x = 0$ is in fact equal to A. The uniqueness of the right limit property, verified in Theorem 5, can now be used to verify that A must be equal to $\frac{A+B}{2}$ as they both represent the limit from the right for the same function at the same point. This result can be easily used to discharge our proof goal $A = B$, which concludes the proof for Theorem 7.

Next, we apply similar reasoning as above with the magnetic field relation for the planar waveguide, verified in Theorem 2, at point $x = 0$ to verify the following relationship between the amplitude coefficients C and A.

Theorem 8: $C = -A\frac{\gamma_c}{\kappa_f}$

$\vdash \forall$ omega mu A B C D n_c n_s n_f k_0 b h x. (0 < h)$\land$(0 < mu)$\land$
   (0 < omega)$\land$(b < k_0 n_f)$\land$(k_0 n_s < b)$\land$(0 < n_s)$\land$(0 < k_0)$\land$
   ($\forall$x.λx.H_field omega mu A B C D n_c n_s n_f k_0 b h x) contl x)
   $\Rightarrow$ (C = $-$A$\frac{(gamma\ b\ k\_0\ n\_c)}{(kappa\ b\ k\_0\ n\_f)}$)

The additional assumptions besides, $0 < h$, used in the above theorem, ensure that the values of the functions gamma and kappa are positive real numbers and do not attain an imaginary complex number value, according to their definitions, given in Section 3. Again based on the continuity of the magnetic field H_field assumption, we know that its limit at point 0 is equal to the value of H_field at $x = 0$, say H_0. It is important to note that the value of H_0 cannot be obtained from the expression for the H_field, given in Theorem 2. Therefore, we cannot reason about its precise value but based on the continuity of H_field, we do know that it exists. This implies that the limit from right and left for this function would be also equal to H_0, according to Theorem 4. Next, we verified that limits from right and left for the magnetic field function H_field, given in Theorem 2, at point $x = 0$ are $\frac{-A(gamma\ b\ k\_0\ n\_c)}{om\ mu}$ and $\frac{C(kappa\ b\ k\_0\ n\_f)}{om\ mu}$ using Definitions 4 and 5, respectively. This leads to the verification of Theorem 8, since we already know that these two limit values are equal to H_0, using the uniqueness of limits from right and left, verified in Theorems 5 and 6.

Now, using similar reasoning as above and applying continuity of E_field and H_field at $x = -h$, we verified the following two relations to express the amplitude coefficient D in terms of the amplitude coefficients B and C.

Theorem 9: $D = B\cos(\kappa_f h) - C\sin(\kappa_f h)$

⊢ ∀ A B C D n_c n_s n_f k_0 b h x. 0 < h ∧
         (∀x. (λx. E_field A B C D n_c n_s n_f k_0 b h x) contl x)
⇒ (D = B(cos((kappa b k_0 n_f) h)) - C(sin((kappa b k_0 n_f) h)))

Theorem 10: $D = \kappa_f \dfrac{B\sin(\kappa_f h) + C\cos(\kappa_f h)}{\gamma_s}$

⊢ ∀ omega mu A B C D n_c n_s n_f k_0 b h x. (0 < h) ∧
(0 < mu)∧(0 < omega)∧(k_0 n_s < b)∧(0 < n_s)∧(0 < k_0) ∧
(∀x.(λx.H_field omega mu A B C D n_c n_s n_f k_0 b h x)contl x)
⇒(D=(kappa b k_0 n_f) $\frac{(B(sin((kappa\ b\ k\_0\ n\_f)\ h))+C(cos((kappa\ b\ k\_0\ n\_f)\ h))))}{(gamma\ b\ k\_0\ n\_s)}$

The above theorems allows us to reach an alternate expression for E_field in terms of only A, which is the amplitude of the electric field at $x = 0$. This relationship is very useful for plotting the mode profiles of guided modes [18].

The right-hand sides of the conclusions of Theorems 9 and 10 can now equated together, since both are equal to D, and the amplitudes coefficients B and C can be expressed in terms of A, using Theorems 7 and 8, respectively, to formally verify the desired relationship for evaluating the eigenvalues of the planar waveguide, given in Equation (7), as the following theorem.

Theorem 11: *Eigenvalue Equation*

⊢ ∀ omega mu A B C D n_c n_s n_f k_0 b h x. (0 < A) ∧
(0 < h) ∧ (0 < mu) ∧ (0 < omega) ∧
(b < k_0 n_f) ∧ (k_0 n_s < b) ∧ (0 < n_s) ∧ (0 < k_0) ∧
¬((kappa b k_0 n_f)² = (gamma b k_0 n_c) (gamma b k_0 n_s)) ∧
(∀x.(λx.E_field A B C D n_c n_s n_f k_0 b h x)contl x) ∧
(∀x.(λx.H_field omega mu A B C D n_c n_s n_f k_0 b h x)contl x)
    ⇒ (tan((kappa b k_0 n_f) h) =

$\dfrac{(gamma\ b\ k\_0\ n\_c)+(gamma\ b\ k\_0\ n\_s)}{(kappa\ b\ k\_0\ n\_f)\left[1-\frac{(gamma\ b\ k\_0\ n\_c)(gamma\ b\ k\_0\ n\_s)}{(kappa\ b\ k\_0\ n\_f)^2}\right]}$)

Due to the inherent soundness of the theorem proving approach, our verification results exactly matched the paper-and-pencil analysis counterparts for the eigenvalue equation, as conducted in [18], and thus can be termed as 100% precise. Interestingly, the assumption ¬((kappa b k_0 n_f)² = (gamma b k_0 n_c) (gamma b k_0 n_s)), without which the eigenvalues are undefined, was found to be missing in [18]. This fact clearly demonstrates the strength of formal methods based analysis as it allowed us to highlight this corner case, which if ignored could lead to the invalidation of the whole eigenvalue analysis.

The verification results, given in this section, heavily relied upon real analysis and thus the useful theorems available in the HOL real analysis theories [6] proved to be a great asset in this exercise. The verification task took around 2500 lines of HOL code and approximately 100 man-hours.

6   Application: Planar Asymmetric Waveguide

In this section, we demonstrate the effectiveness of Theorem 11 in analyzing the eigenvalues of a planar asymmetric waveguide [18]. The waveguide is characterized by a guiding index n_f of 1.50, the substrate index n_s of 1.45 and the cover index n_c of 1.40. The thickness of the guiding layer h is $5\mu m$. The goal is to determine the allowable values of β for this structure, assuming that the wavelength λ of $1\mu m$ is used to excite the waveguide.

In order to obtain the allowable values of β from Theorem 11, we rewrite it with the definition of the function gamma, replace the term (kappa b k_0 n_f) with k_f and express the variable b, which represents β in our Theorems, in terms of k_f as $\sqrt{(k\_0)^2(n\_f)^2 - (k\_f)^2}$, using the definition of kappa, to obtain the following alternate relationship.

Theorem 12: *Alternate form of Eigenvalue Equation*
$\vdash \forall$ omega mu A B C D n_c n_s n_f k_0 b h x. $(0 < $ A$) \wedge$
$(0 < $ h$) \wedge (0 < $ mu$) \wedge (0 < $ omega$) \wedge$
$($b $<$ k_0 n_f$) \wedge ($k_0 n_s $<$ b$) \wedge (0 < $ n_s$) \wedge (0 < $ k_0$) \wedge$
$\neg(($kappa b k_0 n_f$)^2 = ($gamma b k_0 n_c$) ($gamma b k_0 n_s$)) \wedge$
$(\forall$x.$(\lambda$x.E_field A B C D n_c n_s n_f k_0 b h x$)$contl x$) \wedge$
$(\forall$x.$(\lambda$x.H_field omega mu A B C D n_c n_s n_f k_0 b h x$)$contl x$)$
$\Rightarrow \left(\tan(k\_f\ h) = \right.$

$$\frac{\sqrt{((k\_0)^2(n\_f)^2 - (k\_f)^2)-(k\_0)^2(n\_c)^2}+\sqrt{((k\_0)^2(n\_f)^2 - (k\_f)^2)-(k\_0)^2(n\_s)^2}}{k\_f\left[1-\frac{(\sqrt{((k\_0)^2(n\_f)^2 - (k\_f)^2)-(k\_0)^2(n\_c)^2})(\sqrt{((k\_0)^2(n\_f)^2 - (k\_f)^2)-(k\_0)^2(n\_s)^2})}{(k\_f)^2}\right]}\Bigg)$$

All the quantities in the conclusion of the above theorem are known except k_f, since k_0 can be expressed in terms of the wavelength that is used to excite the waveguide, as outlined in Section 3. Though, getting a closed form solution for k_f is not possible from the above equation. Therefore, we propose to use a computer algebra system to solve for the value of k_f. Using Mathematica, the first four eigenvalues of k_f were found to be 5497.16, 10963.2, 16351 and 21545 cm^{-1}. These values can then be used to calculate the desired eigenvalues for b according to the following relationship b $= \sqrt{(k\_0)^2(n\_f)^2 - (k\_f)^2}$, and were found to be 94087, 93608, 92819 and 91752 cm^{-1}.

Hypothetically the above analysis can be divided into two parts. The first part covers the analysis starting from the electromagnetic wave equations, with the given parameters, up to the point where we obtain the alternate form of eigenvalue equation, given in Theorem 12. The second part is concerned with the actual computation of eigenvalues from Theorem 12. The first part of the above analysis was completely formal and thus 100% precise, since it was done using the HOL theorem prover. The proof script for this theorem was less than 100 lines long, which clearly demonstrates the effectiveness of our work, as it was mainly due the the availability of Theorem 11 that we were able to tackle this kind of a verification problem with such a minimal effort. The second part of the analysis cannot be handled in HOL, because of the involvement of a transcendental equation for which a closed form solution for k_f does not exist. For this part, we utilized Mathematica and obtained the desired eigenvalues. To the best of our knowledge, no other approach based on simulation, numerical methods or computer algebra systems, can provide 100% precision and soundness in the results like the proposed approach for the first part of the analysis. Whereas, in the second part, we have used a computer algebra system, which is the best option available, in terms of precision, for this kind of analysis. Other approaches used for the second part include graphical or numerical methods, which cannot compete with computer algebra systems in precision. Thus, as far as the whole analysis is concerned, the proposed method offers the most precise solution.

7   Conclusions

This paper presents the formal analysis of planar optical waveguides using a higher-order-logic theorem prover. Planar optical waveguides are simple, yet widely used optical structures and not only find their applications in wave guiding, but also in coupling, switching, splitting, multiplexing and de-multiplexing of optical signals. Hence, their formal analysis paves the way to the formal analysis of many other optical systems as well. Since the analysis is done in a theorem prover, the results can be termed as 100% precise, which is a novelty that cannot be achieved by any other computer based optical analysis framework.

We mainly present the formalization of the electromagnetic field equations for a planar waveguide in the TE mode. These definitions are then utilized to formally reason about the eigenvalue equation, which plays a vital role in the design of planar waveguides for various engineering and other scientific domains. To illustrate the effectiveness and utilization of the formally verified eigenvalue equation, we used to reason about the eigenvalues of a planar asymmetric waveguide. To the best of our knowledge, this is the first time that a formal approach has been proposed for the analysis of optical systems.

The successful handling of the planar waveguide analysis clearly demonstrates the effectiveness and applicability of higher-order-logic theorem proving for analyzing optical systems. Some of the interesting future directions in this novel domain include the verification of the eigenvalue equation for the planar waveguide in the TM mode, which is very similar to the analysis presented in this paper, and the analysis of couplers that represent two or more optical devices linked together with an optical coupling relation, which can be done by building on top of the results presented in this paper along with formalizing the couple mode theory [8] in higher-order logic. Besides these, many saftey-critical planar waveguide applications can be formally analyzed including biosensors [23] or medical imaging [15] by building on top of our results.

References

1. Abramowitz, M., Stegun, I.A.: Handbook of Mathematical Functions with Formulas, Graphs, and Mathematical Tables. Dover, New York (1972)
2. Anderson, J.A.: Real Analysis. Gordon and Breach Science Publishers, Reading (1969)
3. Costa, J., Pereira, D., Giarola, A.J.: Analysis of Optical Waveguides using Mathematica. In: Microwave and Optoelectronics Conference, pp. 91–95 (1997)
4. Gordon, M.J.C., Melham, T.F.: Introduction to HOL: A Theorem Proving Environment for Higher-Order Logic. Cambridge Press, Cambridge (1993)
5. Hafner, C.: The Generalized Multipole Technique for Computational Electromagnetics. Artech House, Boston (1990)
6. Harrison, J.: Theorem Proving with the Real Numbers. Springer, Heidelberg (1998)
7. Harrison, J.: Formalizing Basic Complex Analysis. In: From Insight to Proof: Festschrift in Honour of Andrzej Trybulec. Studies in Logic, Grammar and Rhetoric, vol. 10, pp. 151–165. University of Białystok (2007)
8. Haus, H., Huang, W., Kawakami, S., Whitaker, N.: Coupled-mode Theory of Optical Waveguides. Lightwave Technology 5(1), 16–23 (1987)
9. Hayes, P.R., O'Keefe, M.T., Woodward, P.R., Gopinath, A.: Higher-order-compact Time Domain Numerical Simulation of Optical Waveguides. Optical and Quantum Electronics 31(9-10), 813–826 (1999)
10. Heinbockel, J.H.: Numerical Methods For Scientific Computing. Trafford (2004)
11. Jackson, J.D.: Classical Electrodynamics. John Wiley & Sons, Inc., Chichester (1998)
12. Johnson, S.G., Joannopoulos, J.D.: Block-iterative Frequency Domain Methods for Maxwell's Equations in a Planewave Basis. Optics Express 8(3), 173–190 (2001)
13. Liu, Y., Sarris, C.D.: Fast Time-Domain Simulation of Optical Waveguide Structures with a Multilevel Dynamically Adaptive Mesh Refinement FDTD Approach. Journal of Lightwave Technology 24(8), 3235–3247 (2006)
14. Mathematica (2009), http://www.wolfram.com
15. Moore, E.D., Sullivan, A.C., McLeod, R.: Three-dimensional Waveguide Arrays via Projection Lithography into a Moving Photopolymer. Organic 3D Photonics Materials and Devices II 7053, 309–316 (2008)
16. Ntogari, G., Tsipouridou, D., Kriezis, E.E.: A Numerical Study of Optical Switches and Modulators based on Ferroelectric Liquid Crystals. Journal of Optics A: Pure and Applied Optics 7(1), 82–87 (2005)
17. Optica (2009), http://www.opticasoftware.com/
18. Pollock, C.R.: Fundamentals of Optoelectronics. Tom Casson (1995)
19. Rumpf, R.C.: Design and Optimization of Nano-Optical Elements by Coupling Fabrication to Optical Behavior. PhD thesis, University of Central Florida, Orlando, Florida (2006)
20. Schmidt, F., Zschiedrich, L.: Adaptive Numerical Methods for Problems of Integrated Optics. In: Integrated Optics: Devices, Materials, and Technologies VII, vol. 4987, pp. 83–94 (2003)
21. Yee, K.: Numerical Solution of Inital Boundary Value Problems involving Maxwell Equations in Isotropic Media. IEEE Transactions on Antennas and Propagation 14(3), 302–307 (1966)
22. Yin, L., Hong, W.: Domain Decomposition Method: A Direct Solution of Maxwell Equations. In: Antennas and Propagation, pp. 1290–1293 (1999)
23. Zhian, L., Wang, Y., Allbritton, N., Li, G.P., Bachman, M.: Labelfree Biosensor by Protein Grating Coupler on Planar Optical Waveguides. Optics Letters 33(15), 1735–1737 (2008)

The HOL-Omega Logic

Peter V. Homeier

U. S. Department of Defense
palantir@trustworthytools.com
http://www.trustworthytools.com

Abstract. A new logic is posited for the widely used HOL theorem prover, as an extension of the existing higher order logic of the HOL4 system. The logic is extended to three levels, adding kinds to the existing levels of types and terms. New types include type operator variables and universal types as in System F. Impredicativity is avoided through the stratification of types by ranks according to the depth of universal types. The new system, called *HOL-Omega* or HOL_ω, is a merging of HOL4, HOL2P[11], and major aspects of System F_ω from chapter 30 of [10]. This document presents the abstract syntax and semantics for the kinds, types, and terms of the logic, as well as the new fundamental axioms and rules of inference. As the new logic is constructed according to the design principles of the LCF approach, the soundness of the entire system depends critically and solely on the soundness of this core.

1 Introduction

The HOL theorem prover [3] has had a wide influence in the field of mechanical theorem proving. Despite appearing in 1988 as one of the first tools in the field, HOL has enjoyed wide acceptance around the world, and continues to be used for many substantial projects, for example Anthony Fox's model of the ARM processor. HOL's influence is seen in that three other major theorem provers, HOL Light, ProofPower, and Isabelle/HOL, have used essentially the same logic.

One of the main reasons for HOL's influence has been that the actual logic implemented in the tool, higher order logic based on Church's simple theory of types, turns out to be both easy to work with and expressive enough to be able to support most models of hardware and software that people have wished to investigate. There are theorem provers with more powerful logics, and ones with less powerful logics, but it seems that classical higher order logic fortuitously found a "sweet-spot," balancing strong expressivity with nimble ease of use.

However, despite HOL's value, it has been recognized that there are some useful concepts beyond the power of higher order logic to state. An example is the practical device of monads. Monads are particularly useful in modelling, for example, realistic computations involving state or exceptions, as a shallow embedding in a logic which itself is strictly functional, without state or exceptions.

Individual monads can and have been expressed in HOL, and used to reduce the complexity of proofs about such real-world computations.

S. Berghofer et al. (Eds.): TPHOLs 2009, LNCS 5674, pp. 244–259, 2009.

However, stating the general properties of all monads, and proving results about the class of all monads, has not been possible. The following shows why.

Let M be a postfix unary type operator that maps a type α to a type $\alpha\,M$, *unit* a prefix unary term operator of type $\alpha \to \alpha\,M$, and $\gg=$ an infix binary term operator of type $\alpha\,M \to (\alpha \to \beta\,M) \to \beta\,M$, where $k\,a \gg= h$ is $(k\,a) \gg= h$. Then M together with *unit* and $\gg=$ is a monad iff the following properties hold:

left unit:	$unit\,a \gg= k \;=\; k\,a$
right unit:	$m \gg= unit \;=\; m$
associativity:	$m \gg= (\lambda a.\ k\,a \gg= h) \;=\; (m \gg= k) \gg= h$

There are two problems with this definition in higher order logic. First, while higher order logic includes type operator *constants* like `list` and `option`, it does not support type operator *variables* like M above.

But even if it did, consider the associativity property above. There are four occurrences of $\gg=$ in that property. Among these four instances are three distinct types. Unfortunately, in higher order logic, within a single expression a variable may only have a single type. So this property would not type-check.

This is annoying because if $\gg=$ were a *constant* instead of a variable, these different instances of its basic type would be supported. What we need is a way to give $\gg=$ a single type which can then be specialized for each of $\gg=$'s four instances to produce the three distinct types required.

One way is to introduce *universal types*, as in System F [10]. A universal type is written $\forall\alpha.\sigma$, where α is a type variable and σ is a type expression, possibly including α. Such occurrences of α are bound by the universal quantification.

In addition, System F introduces abstractions of types over terms, written as $\lambda{:}\alpha.t$, where α is a type variable and t is a term. This yields a term, whose type is a universal type. Specifically, if t has type σ, then $\lambda{:}\alpha.t$ has type $\forall\alpha.\sigma$.

Given such an abstraction t, it is specialized for a particular type by $t[{:}\sigma{:}]$. This gives rise to a new form of beta-reduction on term-type applications, where $(\lambda{:}\alpha.t)[{:}\sigma{:}]$ reduces to $t[\sigma/\alpha]$. For convenience, we write $t[{:}\alpha,\beta{:}]$ for $(t[{:}\alpha{:}])[{:}\beta{:}]$.

Given these new forms, we can express the types of *unit* and $\gg=$ as

$$unit :\ \forall\alpha.\ \alpha \to \alpha\,M$$
$$\gg= :\ \forall\alpha\,\beta.\ \alpha\,M \to (\alpha \to \beta\,M) \to \beta\,M$$

and the three monad properties as

$$unit[{:}\alpha{:}]\,a\ (\gg={=}[{:}\alpha,\beta{:}])\,k \;=\; k\,a$$
$$m\ (\gg={=}[{:}\alpha,\alpha{:}])\ (unit[{:}\alpha{:}]) \;=\; m$$
$$m\ (\gg={=}[{:}\alpha,\gamma{:}])\ (\lambda a.\ k\,a\ (\gg={=}[{:}\beta,\gamma{:}])\,h) \;=\; (m\ (\gg={=}[{:}\alpha,\beta{:}])\,k)\ (\gg={=}[{:}\beta,\gamma{:}])\,h$$

What we have done here is take manual control of the typing. Since the normal HOL parametric polymorphism was inadequate, we have added facilities for type abstraction and instantiation of terms. This allows the single type of a variable to be specialized for different occurrences within the same expression.

Given the existing polymorphism in HOL, in practice universal types are needed only rarely; but when they are needed, they are absolutely essential.

In related work, as early as 1993 Tom Melham advocated adding quantification over type variables [8]. HOL-Omega includes such quantification, defining it using abstraction over type variables. Norbert Völker's HOL2P [11], a direct ancestor of this work, supports universal types quantifying over types of rank 0. HOL2P is approximately the same as HOL-Omega, but without kinds, curried type operators, or ranks > 1. Benjamin C. Pierce [10] describes a variety of programming languages with advanced type systems. HOL-Omega is similar to his system F_ω of chapter 30, but avoids F_ω's impredicativity. HOL-Omega does not include dependent types, such as found in the calculus of constructions.

In the remainder of this paper, we describe the core logic of the HOL-Omega system, and some additions to the core. In Section 2, we present the abstract syntax of HOL-Omega. Section 3 describes the set-theoretic semantics for the logic. Section 4 gives the new core rules of inference and axioms. Section 5 covers additional type and term definitions on top of the core. Section 6 presents a number of examples using the expanded logic, and in Section 7 we conclude.

2   Syntax of the HOL-Omega Logic

For reasons of space, we assume the reader is familiar with the types, terms, axioms, and rules of inference of the HOL logic, as described in [3,4,5,6]. This section presents the abstract syntax of the new HOL-Omega logic.

In HOL-Omega, the syntax consists of ranks, kinds, types, and terms.

2.1   Ranks

Ranks are natural numbers indicating the depth of universal type quantification present or permitted in a type. We use the variable r to range over ranks.

rank ::= natural

The purpose of ranks is to avoid impredicativity, which is inconsistent with HOL [2]. However, a naïve interpretation has been found to be too constrictive. For example, the HOL identity operator I has type $\alpha \to \alpha$, where α has rank 0. However, it is entirely natural to expect to apply I to values of higher ranks, and to expect I to function as the identity function on those higher-rank values. To have an infinite set of identity functions, one for each rank, would be absurd.

Inspired by new set theory, John Matthews suggested the idea of considering all ranks as being formed as a sum of a variable and a natural, where there is only one rank variable, z, ranging over naturals. This reflects the intuition that if a mathematical development was properly constructed at one rank, it could as easily have been constructed at the rank one higher, consistently at each step of the development. Only one rank variable is necessary to capture this intuition, representing the finite number of ranks that the entire development is promoted.

If there is only one rank variable, it may be understood to be always present without being explicitly modeled. Thus rank 2 signifies $z+2$. A rank substitution

θ_r indicates the single mapping $z \mapsto z + \theta_r$, so applying θ_r to $z + n$ yields $z + (n + \theta_r)$. A rank r' is an instance of r if $r' = r[\theta_r]$ for some θ_r, i.e., if $r' \geq r$.

2.2 Kinds

HOL-Omega introduces kinds as a new level in the logic, not present in HOL. Kinds control the proper formation of types just as types do for terms.

There are three varieties of kinds, namely the base kind (the kind of proper types), kind variables, and arrow kinds (the kinds of type operators).

$$
\begin{array}{llll}
\text{kind} ::= & \mathbf{ty} & \textit{(base kind)} \\
& | & \kappa & \textit{(kind variable)} \\
& | & k_1 \Rightarrow k_2 & \textit{(arrow kind)}
\end{array}
$$

We use the variable k to range over kinds, and κ to range over kind variables. The arrow kind $k_1 \Rightarrow k_2$ has domain k_1 and range k_2. Arrow kinds are also called *higher kinds*, meaning higher than the base kind. A kind k' is an instance of k if $k' = k[\theta_k]$ for some substitution θ_k, a mapping from kind variables to kinds.

2.3 Types

Replacing HOL's two varieties of types, HOL-Omega has five: type variables, type constants, type applications, type abstractions, and universal types.

$$
\begin{array}{l}
\text{type-variable} ::= \text{name} \times \text{kind} \times \text{rank} \\
\text{type-constant} ::= \text{name} \times \text{kind} \times \text{rank} \quad \textit{(instance of kind in env.)}
\end{array}
$$

$$
\begin{array}{llll}
\text{type} ::= & \alpha & \textit{(type-variable)} \\
& | & \tau & \textit{(type-constant)} \\
& | & \sigma_{arg}\, \sigma_{opr} & \textit{(type application, postfix syntax)} \\
& | & \lambda\alpha.\, \sigma & \textit{(type abstraction)} \\
& | & \forall\alpha.\, \sigma & \textit{(universal type)}
\end{array}
$$

We will use α to range over type variables, τ to range over type constants, and σ to range over types. Type constants must have kinds which are instances of the environment's kind for that type constant name.

$\begin{array}{c}\textit{Kinding:}\\[2pt] \boxed{\sigma \ : \ k}\end{array}$	$\dfrac{}{\alpha \ : \ \text{kind of } \alpha}$ $\dfrac{}{\tau \ : \ \text{kind of } \tau}$	$\dfrac{\sigma_{opr} : k_1 \Rightarrow k_2, \ \ \sigma_{arg} : k_1}{\sigma_{arg}\,\sigma_{opr} : k_2}$	$\dfrac{\alpha \ : \ k_1, \ \ \sigma \ : \ k_2}{\lambda\alpha.\sigma \ : \ k_1 \Rightarrow k_2}$ $\dfrac{\alpha \ : \ k, \ \ \sigma \ : \ \mathbf{ty}}{\forall\alpha.\sigma \ : \ \mathbf{ty}}$
$\begin{array}{c}\textit{Ranking:}\\[2pt] \boxed{\sigma :\leq r}\end{array}$	$\dfrac{}{\alpha :\leq \text{rank of } \alpha}$ $\dfrac{}{\tau :\leq \text{rank of } \tau}$	$\dfrac{\sigma_{opr} :\leq r_2, \ \ \sigma_{arg} :\leq r_1}{\sigma_{arg}\,\sigma_{opr} :\leq \max(r_1,r_2)}$ $\dfrac{\sigma :\leq r, \ \ r \leq r'}{\sigma :\leq r'}$	$\dfrac{\alpha :\leq r_1, \ \ \sigma :\leq r_2}{\lambda\alpha.\sigma :\leq \max(r_1,r_2)}$ $\dfrac{\alpha :\leq r_1, \ \ \sigma :\leq r_2}{\forall\alpha.\sigma :\leq \max(r_1+1,r_2)}$
$\begin{array}{c}\textit{Typing:}\\[2pt] \boxed{t \ : \ \sigma}\end{array}$	$\dfrac{}{x \ : \ \text{type of } x}$ $\dfrac{}{c \ : \ \text{type of } c}$	$\dfrac{t_{opr} : \sigma_1 \rightarrow \sigma_2, \ \ t_{arg} : \sigma_1}{t_{opr}\, t_{arg} : \sigma_2}$ $\dfrac{t : \forall\alpha{:}k{:}{\leq}r.\,\sigma', \ \ \sigma : k :\leq r}{t\,[{:}\sigma{:}] \ : \ \sigma'[\sigma/\alpha]}$	$\dfrac{x : \sigma_1, \ \ t : \sigma_2}{\lambda x.t : \sigma_1 \rightarrow \sigma_2}$ $\dfrac{\alpha : k :\leq r, \ \ t : \sigma}{\lambda{:}\alpha.t : \forall\alpha.\sigma}$

Existing types of HOL are fully supported in HOL-Omega. HOL type variables are represented as HOL-Omega type variables of kind **ty** and rank 0. HOL type applications of a type constant to a list of type arguments are represented in HOL-Omega as a curried type constant applied to the arguments in sequence, as $(\alpha_1, ..., \alpha_n)\tau = \alpha_n \ (... \ (\alpha_1 \ \tau)...)$.

We write $\sigma : k :\leq r$ to say that type σ has kind k and rank r.

Proper types are types of kind **ty**; only these types can be the type of a term.

In a type application of a type operator to an argument, the operator must have an arrow kind, and the domain of the kind of the operator must equal the kind of the argument. If so, the kind of the result of the type application will be the range of the kind of the operator. Also, the body of a universal type must have the base kind. These restrictions ensure types are well-kinded.

In both universal types and type abstractions, the type variable is bound over the type body. This binding structure introduces the notions of alpha and beta equivalence, as direct analogs of the corresponding notions for terms. In fact, types are identified up to alpha-beta equivalence. The following denote the same type: $\lambda\alpha.\alpha$, $\lambda\beta.\beta$, $\lambda\beta.\beta(\lambda\alpha.\alpha)$, $\gamma(\lambda\alpha.\lambda\beta.\beta)$. Beta reduction is of the form $\sigma_2(\lambda\alpha.\sigma_1) = \sigma_1[\sigma_2/\alpha]$, where $\sigma_1[\sigma_2/\alpha]$ is the result of substituting σ_2 for all free occurrences of α in σ_1, with bound type variables in σ_1 renamed as necessary.

A type σ' is an instance of σ if $\sigma'=\sigma[\theta_r][\theta_k][\theta_\sigma]$ for some rank, kind, and type substitutions $\theta_r \in \mathsf{N}$, θ_k mapping kind variables to kinds, and θ_σ mapping type variables to types. The substitutions are applied in sequence, with θ_r first.

When matching two types, the matching is higher order, so the pattern $\alpha \rightarrow \alpha \ \mu$ (where $\mu : \mathbf{ty} \Rightarrow \mathbf{ty}$) matches $\beta \rightarrow \beta$, yielding $[\alpha \mapsto \beta, \mu \mapsto \lambda\alpha.\alpha]$.

The primeval environment contains the type constants `bool`, `ind`, and `fun` as in HOL, where `bool` : **ty**, `ind` : **ty**, and `fun` : **ty** $\Rightarrow$ **ty** $\Rightarrow$ **ty**, and all three have rank 0. `fun` is usually written as the binary infix type operator $\rightarrow$, and for a function type $\sigma_1 \rightarrow \sigma_2$, we say that the domain is σ_1 and the range is σ_2. Also, for a universal type $\forall\alpha.\sigma$, we say that the domain is α and the range is σ.

2.4 Terms

HOL-Omega adds to the existing four varieties of terms two new varieties, namely term-type applications and type-term abstractions. We use x to range over term variables, c over term constants, and t over terms.

variable ::= name × type
constant ::= name × type (*an instance of type stored in environment*)

term ::= x *(variable)*
    | c *(constant)*
    | $t_{opr} \ t_{arg}$ *(application, prefix syntax)*
    | $\lambda x. \ t$ *(abstraction)*
    | $t \ [:\sigma:]$ *(term-type application)*
    | $\lambda{:}\alpha. \ t$ *(type-term abstraction)*

In applications $t_1 \, t_2$, the domain of the type of t_1 must equal the type of t_2.

As in System F, in abstractions of a type variable over a term $\lambda{:}\alpha.t$, the type variable α must not occur freely in the type of any free variable of the term t.

There are three important restrictions on term-type applications $(t \, [{:}\,\sigma\,{:}])$.

1. The type of the term t must be a universal type, say $\forall\alpha.\sigma'$.
2. The kind of α must match the kind of the type argument σ.
3. The rank of α must contain ($\geq$) the rank of the type argument σ.

The first and second restrictions ensure terms are well-typed and well-kinded.

The third restriction is necessary to avoid impredicativity, for a simpler set-theoretic model. This restriction means that the type argument is validly one of the types over which the universal type quantifies. On this key restriction, the consistency of HOL-Omega rests.

3 Semantics of the HOL-Omega Logic

3.1 A Universe for HOL-Omega Kinds, Types, and Terms

We give the ZFC semantics of HOL-Omega kinds, types, and terms in terms of a universe $\mathcal{U}$, which is fixed set of sets of sets. This development draws heavily from Pitts[6] and Völker[11]. We construct $\mathcal{U}$ as a result of first constructing sequences of sets $\mathcal{U}_0, \mathcal{U}_1, \mathcal{U}_2, ...$, and $\mathcal{T}_0, \mathcal{T}_1, \mathcal{T}_2, ...$, where $\mathcal{U}_i$ and $\mathcal{T}_i$ will only involve types of rank $\leq i$. Kinds will be modeled as elements K of $\mathcal{U}$, types will be modeled as elements T of $K \in \mathcal{U}$, and terms will be modeled as elements E of $T \in \mathcal{T} \in \mathcal{U}$.

There exist $\mathcal{U}_i$ and $\mathcal{T}_i$ for $i = 0, 1, 2, ...$, satisfying the following properties:

Inhab. Each element of $\mathcal{U}_i$ is a non-empty set of non-empty sets.

Typ. $\mathcal{U}_i$ contains a distinguished element $\mathcal{T}_i$.

Arrow. If $K \in \mathcal{U}_i$ and $L \in \mathcal{U}_i$, then $K{\to}L \in \mathcal{U}_i$, where $X{\to}Y$ is the set-theoretic (total) function space from the set X to the set Y.

Clos. $\mathcal{U}_i$ has no elements except those by **Typ** or **Arrow**.

Ext. $\mathcal{T}_{i+1}$ extends $\mathcal{T}_i$: $\mathcal{T}_i \subseteq \mathcal{T}_{i+1}$.

Sub. If $X \in \mathcal{T}_i$ and $\emptyset \neq Y \subseteq X$, then $Y \in \mathcal{T}_i$.

Fun. If $X \in \mathcal{T}_i$ and $Y \in \mathcal{T}_i$, then $X{\to}Y \in \mathcal{T}_i$.

Univ. If $K \in \mathcal{U}_i$ and $f : K \to \mathcal{T}_{i+1}$, then $\prod_{X \in K} fX \in \mathcal{T}_{i+1}$. The set theoretic product $\prod_{X \in K} fX$ is the set of all functions $g : K \to \bigcup_{X \in K} fX$ such that for all $X \in K$, $gX \in fX$.

Bool. $\mathcal{T}_0$ contains a distinguished 2-element set B = {**true**, **false**}.

Infty. $\mathcal{T}_0$ contains a distinguished infinite set I.

AllTyp. $\mathcal{T}$ is defined to be $\bigcup_{i \in \mathbb{N}} \mathcal{T}_i$.

AllArr. $\mathcal{U}$ is the closure of $\{\mathcal{T}\}$ under set theoretic function space creation.

Choice. There are distinguished elements $\text{chty}_i \in \prod_{K \in \mathcal{U}_i} K$ and $\text{ch} \in \prod_{X \in \mathcal{T}} X$. For all i and for all $K \in \mathcal{U}_i$, K is nonempty and $\text{chty}_i(K) \in K$ is an example of this, and for all $X \in \mathcal{T}$, X is nonempty and $\text{ch}(X) \in X$ is an example of this.

The system consisting of the above properties is consistent. The following construction is from William Schneeburger. Let $\mathcal{U}_i$ be the closure of $\{\mathcal{T}_i\}$ under

Arrow. Given $\mathcal{T}_i$ and $\mathcal{U}_i$, we can construct $\mathcal{T}_{i+1}$ by iteration over the ordinals [9]. Let $\mathcal{S}_0 = \mathcal{T}_i$. For all ordinals α, let $\mathcal{S}_{\alpha+1}$ be the closure under **Sub** and **Fun** of

$$\mathcal{S}_\alpha \cup \{\textstyle\prod_{X \in K} f\,X \mid K \in \mathcal{U}_i \wedge f : K \to \mathcal{S}_\alpha\}.$$

For limit ordinals λ, let $\mathcal{S}_\lambda = \bigcup_{\alpha < \lambda} \mathcal{S}_\alpha$, which is closed under **Sub** and **Fun**. Let $\mathfrak{n} = |\bigcup_{K \in \mathcal{U}_i} K|$. Then $|K| \le \mathfrak{n}$ for all $K \in \mathcal{U}_i$. Let $\mathfrak{m} = \mathfrak{n}^+$, the least cardinal $> \mathfrak{n}$. Then $\mathfrak{m}$ is a regular cardinal [9, p. 146] $> |K|$ for all $K \in \mathcal{U}_i$. Then we define $\mathcal{T}_{i+1} = \mathcal{S}_\mathfrak{m}$, which is sufficiently large by the following theorem.

Theorem 1. $\mathcal{S}_\mathfrak{m}$ *is closed under* **Univ** *(as well as* **Sub** *and* **Fun***)*.

Proof. Suppose $K \in \mathcal{U}_i$ and $f : K \to \mathcal{S}_\mathfrak{m}$. $\mathcal{S}_\mathfrak{m} = \bigcup_{\alpha < \mathfrak{m}} \mathcal{S}_\alpha$, so for each $X \in K$ define $\gamma_X =$ the smallest α s.t. $f\,X \in \mathcal{S}_\alpha$, thus $\gamma_X < \mathfrak{m}$. Define $\Gamma = \{\gamma_X \mid X \in K\}$. Then $\Gamma \subseteq \mathfrak{m}$, and $|K| < \mathfrak{m}$ so $|\Gamma| < \mathfrak{m}$ thus $\bigcup \Gamma < \mathfrak{m}$ since $\mathfrak{m}$ is regular. The image of $f \subseteq \mathcal{S}_{\bigcup \Gamma}$, so by the definition of $\mathcal{S}_{\alpha+1}$, $\prod_{X \in K} f\,X \in \mathcal{S}_{(\bigcup \Gamma)+1} \subseteq \mathcal{S}_\mathfrak{m}$.   □

3.2   Constraining Kinds and Types to a Particular Rank

The function $\_\Downarrow r$ transforms an element K of $\mathcal{U}$ into an element of $\mathcal{U}_r$:

$$T \Downarrow r = T_r$$
$$(K_1 \to K_2) \Downarrow r = K_1 \Downarrow r \to K_2 \Downarrow r$$

We need to map some elements $T \in K \in \mathcal{U}$ down to the corresponding elements in $K \Downarrow r \in \mathcal{U}_r$, when T is consistent with a type of rank r. Not all T can be so mapped; we define the subset of K that can, and the mapping, as follows.

We define the subset $K|r \subseteq K \in \mathcal{U}$ as the elements consistent with rank r, and the function $\_\downarrow r$ which transforms an element T of $K|r$ into one of $K \Downarrow r$, mutually recursively on the structure of K:

$$T|r = T_r$$
$$(K_1 \to K_2)|r = \{f \mid f \in K_1 \to K_2 \wedge$$
$$\forall (x, y) \in f.\ (x \in K_1|r \Rightarrow y \in K_2|r) \wedge$$
$$f \downarrow r \text{ is a function}\}$$

If $T \in T|r$,   then $T \downarrow r = T$
If $T \in (K_1 \to K_2)|r$, then $T \downarrow r = \{(x \downarrow r,\ y \downarrow r) \mid (x, y) \in T \wedge x \in K_1|r\}$

If $K = K_1 \to K_2$, by the definition of $T \downarrow r$, $T \downarrow r \subseteq K_1 \Downarrow r \times K_2 \Downarrow r$, and by $T \in K|r$, $T \downarrow r$ is a function, so $T \downarrow r \in K_1 \Downarrow r \to K_2 \Downarrow r = (K_1 \to K_2) \Downarrow r = K \Downarrow r$.

We can define $\_\Uparrow r : \mathcal{U}_r \to \mathcal{U}$ and $\_\uparrow r : K \Downarrow r \to K|r$ as the inverses of $\_\Downarrow r$ and $\_\downarrow r$, so that $(K \Uparrow r) \Downarrow r = K$ for all $K \in \mathcal{U}_r$ and $(T \uparrow r) \downarrow r = T$ for all $T \in K \in \mathcal{U}_r$.

$$T_r \Uparrow r = T$$
$$(K_1 \to K_2) \Uparrow r = K_1 \Uparrow r \to K_2 \Uparrow r$$

If $T \in T \Downarrow r$,   then $T \uparrow r = T$
If $T \in (K_1 \to K_2) \Downarrow r$, then $T \uparrow r = \lambda(x \in K_1).\ \text{if } x \in K_1|r \text{ then } (T(x \downarrow r)) \uparrow r$
$$\text{else } \text{chtype}(K_2, r)$$

$$\text{where } \text{chtype}(K, r) = (\text{chty}_r(K \Downarrow r)) \uparrow r$$

3.3 Semantics of Ranks and Kinds

As mentioned earlier, ranks syntactically appear as natural numbers r, but are actually combined with the hidden single rank variable z as $z + r$. A rank environment $\zeta \in \mathbb{N}$ gives the value of z. The semantics of ranks is then $[\![r]\!]_\zeta = \zeta + r$.

A kind environment ξ is a mapping from kind variables to elements of $\mathcal{U}$.

The semantics of kinds $[\![k]\!]_\xi$ is defined by recursion over the structure of k:

$$[\![\mathbf{ty}]\!]_\xi = \mathcal{T}$$
$$[\![\kappa]\!]_\xi = \xi\,\kappa$$
$$[\![k_1 \Rightarrow k_2]\!]_\xi = [\![k_1]\!]_\xi \to [\![k_2]\!]_\xi$$

3.4 Semantics of Types

We will distinguish $\mathbf{bool}$, $\mathbf{ind}$, and function types $\sigma_1 \to \sigma_2$ as special cases, in order to ensure a standard model. We assume a model M that takes a rank and kind environment (ζ, ξ) and gives a valuation of each type constant τ of kind k and rank r as an element of $[\![k]\!]_\xi \mid [\![r]\!]_\zeta$. For clarity we omit the decoration $[\![\_]\!]_M$.

A type environment ρ takes a rank and a kind environment (ζ, ξ) to a mapping of each type variable α of kind k and rank r to a value $T \in [\![k]\!]_\xi \mid [\![r]\!]_\zeta$.

$[\![\sigma]\!]_{\zeta,\xi,\rho}$ is defined by recursion over the structure of σ:

$$[\![\mathbf{bool}]\!]_{\zeta,\xi,\rho} = \mathrm{B}$$
$$[\![\mathbf{ind}]\!]_{\zeta,\xi,\rho} = \mathrm{I}$$
$$[\![\sigma_1 \to \sigma_2]\!]_{\zeta,\xi,\rho} = [\![\sigma_1]\!]_{\zeta,\xi,\rho} \to [\![\sigma_2]\!]_{\zeta,\xi,\rho}$$
$$[\![\tau]\!]_{\zeta,\xi,\rho} = M\,(\zeta, \xi)\,\tau$$
$$[\![\alpha]\!]_{\zeta,\xi,\rho} = \rho\,(\zeta, \xi)\,\alpha$$
$$[\![\sigma_{arg}\,\sigma_{opr}]\!]_{\zeta,\xi,\rho} = [\![\sigma_{opr}]\!]_{\zeta,\xi,\rho}\,[\![\sigma_{arg}]\!]_{\zeta,\xi,\rho}$$
$$[\![\lambda(\alpha : k :\leq r).\,\sigma]\!]_{\zeta,\xi,\rho} = \lambda T \in [\![k]\!]_\xi . \begin{cases} [\![\sigma]\!]_{\zeta,\xi,\rho[\alpha \mapsto T]} & \text{if } T \in [\![k]\!]_\xi \mid [\![r]\!]_\zeta \\ \mathrm{chtype}([\![k_\sigma]\!]_\xi, [\![r_\sigma]\!]_\zeta) & \text{otherwise} \end{cases}$$
$$[\![\forall(\alpha : k :\leq r).\,\sigma]\!]_{\zeta,\xi,\rho} = \prod_{T \in [\![k]\!]_\xi \Downarrow [\![r]\!]_\zeta}[\![\sigma]\!]_{\zeta,\xi,\rho[\alpha \mapsto T \uparrow [\![r]\!]_\zeta]}$$

where for $[\![\lambda(\alpha : k :\leq r).\,\sigma]\!]_{\zeta,\xi,\rho}$, if T has rank larger than the variable α, an arbitrary type of the kind k_σ and rank r_σ of σ is returned, essentially as an error.

By induction over the structure of types, it can be demonstrated that the semantics of types is consistent with the semantics of kinds and ranks, i.e.,

$$[\![\sigma : k :\leq r]\!]_{\zeta,\xi,\rho} \in [\![k]\!]_\xi \mid [\![r]\!]_\zeta.$$

3.5 Semantics of Terms

In addition to the type mapping described above, the model M is assumed, given a triple of a rank, kind, and type environments, to provide a valuation of each term constant c of type σ as an element of $[\![\sigma]\!]_{\zeta,\xi,\rho}$. A term environment μ takes a triple of a rank, kind, and type environments to a mapping of each term variable x of type σ to a value v which is an element of $[\![\sigma]\!]_{\zeta,\xi,\rho}$.

$[\![t]\!]_{\zeta,\xi,\rho,\mu}$ is defined by recursion over the structure of t:

$$[\![c]\!]_{\zeta,\xi,\rho,\mu} = M\ (\zeta,\xi,\rho)\ c$$
$$[\![x]\!]_{\zeta,\xi,\rho,\mu} = \mu\ (\zeta,\xi,\rho)\ x$$
$$[\![t_1\ t_2]\!]_{\zeta,\xi,\rho,\mu} = [\![t_1]\!]_{\zeta,\xi,\rho,\mu}\ [\![t_2]\!]_{\zeta,\xi,\rho,\mu}$$
$$[\![\lambda(x:\sigma).\,t]\!]_{\zeta,\xi,\rho,\mu} = \lambda v \in [\![\sigma]\!]_{\zeta,\xi,\rho}.\,[\![t]\!]_{\zeta,\xi,\rho,\mu[x\mapsto v]}$$
$$[\![\lambda{:}(\alpha:k:\leq r).\,t]\!]_{\zeta,\xi,\rho,\mu} = \lambda T \in [\![k]\!]_\xi {\Downarrow} [\![r]\!]_\zeta.\,[\![t]\!]_{\zeta,\xi,\rho[\alpha\mapsto T\uparrow[\![r]\!]_\zeta],\mu}$$
$$[\![t\,[{:}\sigma{:}]]\!]_{\zeta,\xi,\rho,\mu} = [\![t]\!]_{\zeta,\xi,\rho,\mu}\,([\![\sigma]\!]_{\zeta,\xi,\rho} \downarrow [\![r]\!]_\zeta)$$

where for $t\,[{:}\,\sigma\,{:}]$, the type of t must have the form $\forall\alpha.\sigma'$, and r is the rank of α.

4   Primitive Rules of Inference of the HOL-Omega Logic

HOL-Omega includes all of the axioms and rules of inference of HOL, reinterpreting them in light of the expanded sets of types and terms, and extends them with the following new rules of inference, directed at the new varieties of terms.

– Rule INST_TYPE is revised; it says that consistently and properly substituting types for free type variables throughout a theorem yields a theorem.

$$\frac{\Gamma \vdash t}{\Gamma[\sigma_1,\ \dots\ ,\sigma_n/\alpha_1,\ \dots\ ,\alpha_n] \vdash t[\sigma_1,\ \dots\ ,\sigma_n/\alpha_1,\ \dots\ ,\alpha_n]} \qquad \text{(INST\_TYPE)}$$

– Rule INST_KIND says that consistently substituting kinds for kind variables throughout a theorem yields a theorem.

$$\frac{\Gamma \vdash t}{\Gamma[k_1,\ \dots\ ,k_n/\kappa_1,\ \dots\ ,\kappa_n] \vdash t[k_1,\ \dots\ ,k_n/\kappa_1,\ \dots\ ,\kappa_n]} \qquad \text{(INST\_KIND)}$$

– Rule INST_RANK says that consistently incrementing by $n \geq 0$ the rank of all type variables throughout a theorem yields a theorem. z is the rank variable.

$$\frac{\Gamma \vdash t}{\Gamma[(z+n)/z] \vdash t[(z+n)/z]} \qquad \text{(INST\_RANK)}$$

– Rule TY_ABS says that if two terms are equal, then their type abstractions are equal, where α is not free in Γ.

$$\frac{\Gamma \vdash t_1 = t_2}{\Gamma \vdash (\lambda{:}\alpha.t_1) = (\lambda{:}\alpha.t_2)} \qquad \text{(TY\_ABS)}$$

– Rule TY_BETA_CONV describes the equality of type beta-conversion, where $t[\sigma/\alpha]$ denotes the result of substituting σ for free occurrences of α in t.

$$\frac{}{\vdash (\lambda{:}\alpha.t)[{:}\sigma{:}] = t[\sigma/\alpha]} \qquad \text{(TY\_BETA\_CONV)}$$

HOL-Omega adds one new axiom.

- Axiom TY_ETA_AX says type eta reduction is valid.

$$\vdash (\lambda{:}\alpha{:}\kappa.\ t[{:}\alpha{:}]) = t \qquad\qquad (\text{TY\_ETA\_AX})$$

To ensure the soundness of the HOL-Omega logic, all of the axioms and rules of inference need to have their semantic interpretations proven sound within set theory for all rank, kind, type, and term environments. This has not yet been formally done, but it is a priority for future work. When this is accomplished, by the LCF approach, all theorems proven within HOL-Omega will be sound.

5   Additional Type and Term Definitions

Of course the core of any system is only a point from which to begin. This section describes new type abbreviations and term constants not in HOL, defined as conservative extensions of the core logic of HOL-Omega.

5.1   New Type Abbreviations

HOL-Omega introduces the type abbreviations

$$
\begin{aligned}
\mathsf{I} &= \lambda(\alpha : {'}k).\ \alpha \\
\mathsf{K} &= \lambda(\alpha : {'}k)\ (\beta : {'}l).\ \alpha \\
\mathsf{S} &= \lambda(\alpha : {'}k \Rightarrow {'}l \Rightarrow {'}m)\ (\beta : {'}k \Rightarrow {'}l)\ (\gamma : {'}k).\ \gamma\ \beta\ (\gamma\ \alpha) \\
\mathsf{o} &= \lambda({'}f : {'}k \Rightarrow {'}l)\ ({'}g : {'}l \Rightarrow {'}m)\ (\alpha : {'}k).\ \alpha\ {'}f\ {'}g
\end{aligned}
$$

The use of kind variables ${'}k$, ${'}l$, and ${'}m$ makes these type abbreviations applicable as type operators to types with arrow kinds. o is an infix type operator, written as ${'}f \circ {'}g = \lambda\alpha.\ \alpha\ {'}f\ {'}g$. These are reminiscent of the term combinators, e.g. $\mathsf{I} = \lambda(x{:}\alpha).x$, $\mathsf{K} = \lambda(x{:}\alpha)(y{:}\beta).x$, and $(g : \beta \to \gamma) \circ (f : \alpha \to \beta) = \lambda x.\ g\ (f\ x)$.

In HOL2P, both the arguments and the results of type operator applications must have the base kind **ty**. In HOL-Omega, the arguments and results may themselves be type operators of higher kind, as managed by the kind structure. This is of great advantage, for example when using o to compose two type operators, neither of which is applied to any arguments yet.

5.2   New Terms

HOL-Omega provides universal and existential quantification of type variables over terms using the new type binder constants ∀: and ∃:, defined as

$$
\begin{aligned}
\forall{:} &= \lambda P.\ (P = (\lambda{:}\alpha{:}\kappa.\ \mathsf{T})) \\
\exists{:} &= \lambda P.\ (P \neq (\lambda{:}\alpha{:}\kappa.\ \mathsf{F}))
\end{aligned}
$$

To ease readability, the following forms are also supported:

$$
\begin{aligned}
\forall{:}\alpha{:}\kappa.\ P &= \forall{:}\ (\lambda\alpha{:}\kappa.\ P) & \forall{:}\alpha_1{:}\kappa_1\ \alpha_2{:}\kappa_2\ \dots\ .\ P &= \forall{:}\alpha_1{:}\kappa_1.\ \forall{:}\alpha_2{:}\kappa_2.\ \dots\ P \\
\exists{:}\alpha{:}\kappa.\ P &= \exists{:}\ (\lambda\alpha{:}\kappa.\ P) & \exists{:}\alpha_1{:}\kappa_1\ \alpha_2{:}\kappa_2\ \dots\ .\ P &= \exists{:}\alpha_1{:}\kappa_1.\ \exists{:}\alpha_2{:}\kappa_2.\ \dots\ P
\end{aligned}
$$

6   Examples

The HOL-Omega logic makes it straightforward to express many concepts from category theory, such as functors and natural transformations. Much of the first two examples below is ported from HOL2P [11]; the main difference is that the higher-order type abbreviations and type inference of HOL-Omega allow a more pleasing presentation. We focus on the category **Type** whose objects are the proper types of the HOL-Omega logic, and whose arrows are the (total) term functions from one type to another. The source and target of an arrow are the domain and range of the type of the function. The identity arrows are the identity functions on each type. The composition of arrows is normal functional composition. The customary check that the target of one arrow is the source of the other is accomplished automatically by the strong typing of the logic.

6.1   Functors

Functors map objects to objects and arrows to arrows. In the category **Type**, the first mapping is represented as a type $'F$ of kind $\textbf{ty} \Rightarrow \textbf{ty}$, and the second as a function of the type $'F$ functor, where functor is the type abbreviation

$$\mathsf{functor} = \lambda'F. \ \forall \alpha \ \beta. \ (\alpha \to \beta) \to (\alpha \ 'F \to \beta \ 'F).$$

To be a functor, a function of this type must satisfy the following predicate:

$functor \ (F : 'F \ \mathsf{functor}) =$
$\quad (\forall \!:\! \alpha. \ F \ (\mathtt{I} : \alpha \to \alpha) = \mathtt{I}) \ \wedge$ *Identity*
$\quad (\forall \!:\! \alpha \ \beta \ \gamma. \ \forall (f : \alpha \to \beta)(g : \beta \to \gamma). \ F \ (g \circ f) = F \ g \circ F \ f \,)$ *Composition*

where $g \circ f = \lambda x. \ g \ (f \ x)$. This is actually an abbreviated version; the parser and type inference fill in the necessary type applications, so the full version is

$functor \ (F : 'F \ \mathsf{functor}) =$
$\quad (\forall \!:\! \alpha. \ F \ [\!:\!\alpha, \alpha\!:\!] \ (\mathtt{I} : \alpha \to \alpha) = \mathtt{I}) \ \wedge$ *Identity*
$\quad (\forall \!:\! \alpha \ \beta \ \gamma. \ \forall (f : \alpha \to \beta)(g : \beta \to \gamma).$ *Composition*
$\qquad F \ [\!:\!\alpha, \gamma\!:\!] \ (g \circ f) = F \ [\!:\!\beta, \gamma\!:\!] \ g \circ F \ [\!:\!\alpha, \beta\!:\!] \ f \,)$

In what follows, these type applications will normally be omitted for clarity.

In HOL, $\mathtt{list} : \textbf{ty} \Rightarrow \textbf{ty}$ is the type of finite lists. It is defined as a recursive datatype with two constructors, $[] : \alpha \ \mathtt{list}$ and $:: \ : \alpha \to \alpha \ \mathtt{list} \to \alpha \ \mathtt{list}$. $::$ is infix. The function $\mathtt{MAP} : (\alpha \to \beta) \to (\alpha \ \mathtt{list} \to \beta \ \mathtt{list})$ is defined by

$$\begin{aligned}\mathtt{MAP} \ f \ [] \quad &= \ [] \\ \mathtt{MAP} \ f \ (x :: xs) &= \ f \ x :: \mathtt{MAP} \ f \ xs\end{aligned}$$

Then $\mathtt{MAP}$ can be proven to be a functor: $\vdash functor \ ((\lambda \!:\! \alpha \ \beta. \ \mathtt{MAP}) : \mathtt{list} \ \mathsf{functor})$. A simple functor is the identity function $\mathtt{I}$: $\vdash functor \ ((\lambda \!:\! \alpha \ \beta. \ \mathtt{I}) : \mathsf{I} \ \mathsf{functor})$. The composition of two functors is a functor. We overload $\circ$ to define this:

$$(G : 'G \ \mathsf{functor}) \circ (F : 'F \ \mathsf{functor}) = \lambda \!:\! \alpha \ \beta. \ G[\!:\!\alpha \ 'F, \beta \ 'F\!:\!] \circ F[\!:\!\alpha, \beta\!:\!]$$

The result has type $('F \circ 'G)$functor. As an example, $(\lambda{:}\alpha \ \beta.\ \text{MAP}) \circ (\lambda{:}\alpha \ \beta.\ \text{MAP}) = (\lambda{:}\alpha \ \beta.\ \text{MAP} \circ \text{MAP}) : (\text{list} \circ \text{list})$functor is a functor. The type composition operator $\circ$ reflects the category theory composition of two functors' mappings on objects. In HOL2P, the MAP functor composition example is expressed as:

$$\vdash \text{TYINST} \ (\theta \mapsto \lambda\alpha.\ (\alpha \ \text{list})\text{list}) \ functor \ (\lambda{:}\alpha \ \beta.\ \lambda f.\ \text{MAP} \ (\text{MAP} \ f))$$

Here the notation has been adjusted to that of this paper, for ease of comparison. TYINST is needed to manually instantiate a free type variable θ of the *functor* predicate with the type for this instance, which must be stated as a type abstraction. HOL-Omega's kinds and type inference enable a clearer statement:

$$\vdash functor \ (\lambda{:}\alpha \ \beta.\ \text{MAP} \circ \text{MAP})$$

Beyond the power of HOL2P, HOL-Omega supports quantification over functors:

$$\vdash \exists{:}'F.\ \exists(F : 'F \ \text{functor}).\ functor \ F.$$

6.2 Natural Transformations

Given functors F and G, a natural transformation maps objects A to arrows $FA \to GA$. In the category **Type**, we represent natural transformations as functions of the type $('F, 'G)$nattransf, where nattransf is the type abbreviation

$$\text{nattransf} = \lambda'F \ 'G.\ \forall\alpha.\ \alpha \ 'F \to \alpha \ 'G.$$

A natural transformation ϕ from a functor F to a functor G $(\phi : F \to G)$ must satisfy the following predicate:

$$nattransf \ (\phi : ('F, 'G)\text{nattransf}) \ (F : 'F \ \text{functor}) \ (G : 'G \ \text{functor}) = \\ \forall{:}\alpha \ \beta.\ \forall(h : \alpha \to \beta).\ G \ h \circ \phi = \phi \circ F \ h$$

Define the function INITS to take a list and return a list of all prefixes of it:

```
INITS []        = []
INITS (x :: xs) = [] :: MAP (λys. x :: ys) (INITS xs)
```

INITS can be proven to be a natural transformation from MAP to MAP $\circ$ MAP:

$$\vdash nattransf \ ((\lambda{:}\alpha.\quad \text{INITS})\qquad : (\text{list}, \text{list} \circ \text{list})\text{nattransf}) \\ ((\lambda{:}\alpha \ \beta.\ \text{MAP})\qquad : \text{list functor}) \\ ((\lambda{:}\alpha \ \beta.\ \text{MAP} \circ \text{MAP}) : (\text{list} \circ \text{list})\text{functor}).$$

The vertical composition of two natural transformations is defined as

$$(\phi_2 : ('G, 'H)\text{nattransf}) \circ (\phi_1 : ('F, 'G)\text{nattransf}) \ = \ \lambda{:}\alpha.\ \phi_2 \circ (\phi_1[{:}\alpha{:}])$$

The result of this vertical composition is a natural transformation:

$$\vdash nattransf (\quad \phi_1 \quad : ('F, 'G)\text{nattransf}) \ F \ G \ \wedge \\ nattransf (\quad \phi_2 \quad : ('G, 'H)\text{nattransf}) \ G \ H \ \Rightarrow \\ nattransf (\ \phi_2 \circ \phi_1 : ('F, 'H)\text{nattransf}) \ F \ H$$

A natural transformation may be composed with a functor in two ways, where the functor is either applied first or last. We define these, again overloading ∘:

$$(\phi : (\text{'}F, \text{'}G)\text{nattransf}) \circ (H : \text{'}H \text{ functor}) \quad = \lambda\text{:}\alpha. \; \phi \; [:\alpha \; \text{'}H\text{:}]$$
$$(H : \text{'}H \text{ functor}) \qquad \circ (\phi : (\text{'}F, \text{'}G)\text{nattransf}) = \lambda\text{:}\alpha. \; H \; (\phi \; [:\alpha\text{:}])$$

That the last of these is a natural transformation is expressed in HOL2P as

$$\vdash \textit{nattransf} \; \phi \; F \; G \; \wedge \; \textit{functor} \; H \; \Rightarrow$$
$$\text{TYINST} \; ((\theta_1 \mapsto \lambda\alpha. \; ((\alpha)\theta_1)\theta_3) \; (\theta_2 \mapsto \lambda\alpha. \; ((\alpha)\theta_2)\theta_3))$$
$$\textit{nattransf} \; (\lambda\text{:}\alpha. \; H \; \phi) \; (\lambda\text{:}\alpha \; \beta. \; H \circ F) \; (\lambda\text{:}\alpha \; \beta. \; H \circ G)$$

where in HOL-Omega, the type inference, higher kinds, and overloaded ∘ permit

$$\vdash \textit{nattransf} \; \phi \; F \; G \; \wedge \; \textit{functor} \; H \; \Rightarrow$$
$$\textit{nattransf} \; (H \circ \phi) \; (H \circ F) \; (H \circ G).$$

6.3 Monads

Wadler [12] has proposed using monads to structure functional programming. He defines a monad as a triple ($\text{'}M$, $\textit{unit}$, $\ggg=$) of a type operator $\text{'}M$ and two term operators $\textit{unit}$ and $\ggg=$ (where $\ggg=$ is an infix operator) obeying three laws. We express this definition in HOL-Omega as follows.

We define two type abbreviations unit and bind:

$$\text{unit} = \lambda\text{'}M. \; \forall\alpha. \; \alpha \; \rightarrow \alpha \; \text{'}M$$
$$\text{bind} = \lambda\text{'}M. \; \forall\alpha \; \beta. \; \alpha \; \text{'}M \rightarrow (\alpha \rightarrow \beta \; \text{'}M) \rightarrow \beta \; \text{'}M$$

We define a monad to be two term operators, $\textit{unit}$ and $\ggg=$, with a single common free type variable $\text{'}M : \textbf{ty} \Rightarrow \textbf{ty}$, satisfying a predicate of the three laws:

$$\textit{monad} \; (\textit{unit} : \text{'}M \; \text{unit}, \; \ggg= \; : \text{'}M \; \text{bind}) =$$

$(\forall\text{:}\alpha \; \beta. \quad \forall(a : \alpha)(k : \alpha \rightarrow \beta \; \text{'}M).$	*(Left unit)*
$\quad\quad\quad\quad \textit{unit} \; a \ggg= k \; = \; k \; a) \; \wedge$	
$(\forall\text{:}\alpha. \quad\quad \forall(m : \alpha \; \text{'}M).$	*(Right unit)*
$\quad\quad\quad\quad m \ggg= \textit{unit} \; = \; m) \; \wedge$	
$(\forall\text{:}\alpha \; \beta \; \gamma. \; \forall(m : \alpha \; \text{'}M)(k : \alpha \rightarrow \beta \; \text{'}M)(h : \beta \rightarrow \gamma \; \text{'}M).$	*(Associative)*
$\quad\quad\quad\quad (m \ggg= k) \ggg= h \; = \; m \ggg= (\lambda\alpha. \; k \; \alpha \ggg= h))$	

As an example, we define the $\textit{unit}$ and $\ggg=$ operations for a state monad as

$$\text{state} = \lambda\sigma \; \alpha. \; \sigma \rightarrow \alpha \times \sigma$$

$$\textit{state\_unit} = \lambda\text{:}\alpha. \; \lambda(x\text{:}\alpha) \; (s\text{:}\sigma). \; (x, s)$$
$$\textit{state\_bind} = \lambda\text{:}\alpha \; \beta. \; \lambda(w\text{:}(\sigma, \alpha)\text{state}) \; (f\text{:}\alpha \rightarrow (\sigma, \beta)\text{state}) \; (s\text{:}\sigma). \; \textbf{let} \; (x, s') = w \; s$$
$$\textbf{in} \; f \; x \; s'$$

Then we can prove these operations satisfy the $\textit{monad}$ predicate for $\text{'}M = \sigma$ state, taking advantage of the curried nature of state, where $(\sigma, \alpha)\text{state} = \alpha \; (\sigma \; \text{state})$:

$$\vdash \textit{monad} \; (\; \textit{state\_unit} : (\sigma \; \text{state})\text{unit}, \; \textit{state\_bind} : (\sigma \; \text{state})\text{bind} \;).$$

Wadler [12] also formulates an alternative definition of monads, expressed in terms of three operators, *unit, map,* and *join,* satisfying seven laws:

$$\text{map} = \lambda'M. \,\forall \alpha\, \beta.\, (\alpha \to \beta) \to (\alpha\, 'M \to \beta\, 'M)$$
$$\text{join} = \lambda'M. \,\forall \alpha.\, \alpha\, 'M\, 'M \to \alpha\, 'M$$

$umj\_monad\, (unit : 'M\ \text{unit},\ map : 'M\ \text{map},\ join : 'M\ \text{join}) =$

$(\forall{:}\alpha.$	$map\, (\text{I} : \alpha \to \alpha) = \text{I})$	$\wedge$	*(map_I)*
$(\forall{:}\alpha\, \beta\, \gamma.\ \forall(f : \alpha \to \beta)(g : \beta \to \gamma).$			*(map_o)*
	$map\, (g \circ f) = map\, g \circ map\, f)$	$\wedge$	
$(\forall{:}\alpha\, \beta.\quad \forall(f : \alpha \to \beta).$			*(map_unit)*
	$map\, f \circ unit = unit \circ f)$	$\wedge$	
$(\forall{:}\alpha\, \beta.\quad \forall(f : \alpha \to \beta).$			*(map_join)*
	$map\, f \circ join = join \circ map\, (map\, f))$	$\wedge$	
$(\forall{:}\alpha.$	$join \circ unit = (\text{I} : \alpha\, 'M \to \alpha\, 'M))$	$\wedge$	*(join_unit)*
$(\forall{:}\alpha.$	$join \circ map\, unit = (\text{I} : \alpha\, 'M \to \alpha\, 'M))$	$\wedge$	*(join_map_unit)*
$(\forall{:}\alpha.$	$join\, [{:}\alpha{:}] \circ map\, join = join \circ join)$		*(join_map_join)*

Given a monad defined using *unit* and $\gg=$, corresponding *map* and *join* operators $\text{MMAP}(unit, \gg=)$ and $\text{JOIN}(unit, \gg=)$ may be constructed automatically:

$\text{MMAP}\, (unit : 'M\ \text{unit},\ \gg= :\ 'M\ \text{bind})$
$\quad = \lambda{:}\alpha\, \beta.\, \lambda(f : \alpha \to \beta)\, (m : \alpha\, 'M).\, m \gg= (\lambda a.\, unit\, (f\, a))$
$\text{JOIN}\, (unit : 'M\ \text{unit},\ \gg= :\ 'M\ \text{bind})$
$\quad = \lambda{:}\alpha.\, \lambda(z : \alpha\, 'M\, 'M).\, z \gg= \text{I}$

Given a monad defined using *unit, map,* and *join,* the corresponding $\gg=$ operator $\text{BIND}(map, join)$ may also be constructed automatically:

$\text{BIND}\, (map : 'M\ \text{map},\ join : 'M\ \text{join})$
$\quad = \lambda{:}\alpha\, \beta.\, \lambda(m : \alpha\, 'M)\, (k : \alpha \to \beta\, 'M).\, join\, (map\, k\, m)$

E.g., for the state monad, $\quad state\_map\ =\ \text{MMAP}\, (state\_unit,\ state\_bind)$
$\qquad\qquad\qquad\qquad state\_join\ =\ \text{JOIN}\, (state\_unit,\ state\_bind)$
$\qquad\qquad\qquad\qquad state\_bind\ =\ \text{BIND}\, (state\_map,\ state\_join).$

Then it can be proven that these two definitions of a monad are equivalent.

$\vdash\ monad\, (unit : 'M\ \text{unit},\ \gg= :\ 'M\ \text{bind})\ \Leftrightarrow$
$\quad (umj\_monad\, (unit,\ \text{MMAP}(unit, \gg=),\ \text{JOIN}(unit, \gg=))\ \wedge$
$\qquad \gg=\ =\ \text{BIND}\, (\text{MMAP}(unit, \gg=),\ \text{JOIN}(unit, \gg=)))$

$\vdash\ umj\_monad(unit : 'M\ \text{unit},\ map : 'M\ \text{map},\ join : 'M\ \text{join})\ \Rightarrow$
$\quad monad(unit,\ \text{BIND}(map, join))$

Lack and Street [7] define monads as a category A, a functor $t : A \to A$, and natural transformations $\mu : t^2 \to t$ and $\eta : 1_A \to t$ satisfying three equations, as expressed by the commutative diagrams (in the functor category)

This definition can be expressed in HOL-Omega as follows:

cat_monad $(t : 'M$ functor, $\mu : ('M$ o $'M, 'M)$nattransf, $\eta : (I, 'M)$nattransf$) =$
    $functor\ t$ $\wedge$ $(t$ is a functor$)$
    $nattransf\ \mu\ (t \circ t)\ t$ $\wedge$ $(\mu$ is a natural transformation$)$
    $nattransf\ \eta\ (\lambda{:}\alpha\ \beta.\ I)\ t$ $\wedge$ $(\eta$ is a natural transformation$)$
    $(\mu \circ (t \circ \mu) = \mu \circ (\mu \circ t))$ $\wedge$ $(square\ commutes)$
    $(\mu \circ (t \circ \eta) = \lambda{:}\alpha.\ I)$ $\wedge$ $(left\ triangle\ commutes)$
    $(\mu \circ (\eta \circ t) = \lambda{:}\alpha.\ I)$ $(right\ triangle\ commutes)$.

It can be proven that this is equivalent to the *(unit, map, join)* definition:

    $\vdash \forall(unit : 'M$ unit$)\ map\ join.$
        $cat\_monad(map,\ join,\ unit) \Leftrightarrow umj\_monad(unit,\ map,\ join).$

Therefore all three definitions of monads are equivalent.

7   Conclusion

This document has presented a description of the core logic of the HOL-Omega theorem prover. This has been implemented as a variant of the HOL4 theorem prover. The implementation may be downloaded by the command

```
svn checkout https://hol.svn.sf.net/svnroot/hol/branches/HOL-Omega
```

Installation instructions are in the top directory.

This provides a practical workbench for developments in the HOL-Omega logic, integrated in a natural and consistent manner with the existing HOL4 tools and libraries that have been refined and extended over many years.

This implementation was designed with particular concern for backward compatibility. This was almost entirely achieved, which was possible only because the fundamental data types representing types and terms were originally encapsulated. This meant that the underlying representation could be changed without affecting the abstract view of types and terms by the rest of the system. Virtually all existing HOL4 code will build correctly, including the extensive libraries. The simplifiers have been upgraded, including higher-order matching of the new types and terms and automatic type beta-reduction. Algebraic types with higher kinds and ranks may be constructed using the familiar Hol_datatype tool [5]. Not all of the tools will work as expected on the new terms and types, as the revision process is ongoing, but they will function identically on the classic terms and types. So nothing of HOL4's power has been lost.

Also, the nimble ease of use of HOL has been largely preserved. For example, the type inference algorithm is a pure extension, so that all classic terms have the same types successfully inferred. Inference of most general types for all terms is not always possible, as also seen in System F, and type inference may fail even for typeable terms, but in practice a few user annotations are usually sufficient.

The system is still being developed but is currently useful. All of the examples presented have been mechanized in the `examples/HolOmega` subdirectory, along with further examples from *Algebra of Programming* [1] ported straightforwardly from HOL2P, including homomorphisms, initial algebras, catamorphisms, and the banana split theorem. While maintaining backwards compatibility with the existing HOL4 system and libraries, the additional expressivity and power of HOL-Omega makes this tool applicable to a great collection of new problems.

Acknowledgements. Norbert Völker's HOL2P [11] was an vital inspiration. Michael Norrish helped get the new branch of HOL4 established and to begin the new parsers and prettyprinters. John Matthews suggested adding the single rank variable to every rank. William Schneeburger justified an aggressive set-theoretic semantics of ranks. Mike Gordon has consistently encouraged this work. We honor his groundbreaking and seminal achievement in the original HOL system [3], without which none of this work would have been possible.

Soli Deo Gloria.

References

1. Bird, R., de Moor, O.: Algebra of Programming. Prentice Hall (1997)
2. Coquand, T.: A new paradox in type theory. In: Prawitx, D., Skyrms, B., Westerstahl, D. (eds.) Proceedings 9th Int. Congress of Logic, Methodology and Philosophy of Science, pp. 555–570. North-Holland, Amsterdam (1994)
3. Gordon, M.J.C., Melham, T.F.: Introduction to HOL. Cambridge University Press, Cambridge (1993)
4. Gordon, M.J.C., Pitts, A.M.: The HOL Logic and System. In: Bowen, J. (ed.) Towards Verified Systems, ch. 3, pp. 49–70. Elsevier Science B.V., Amsterdam (1994)
5. The HOL System DESCRIPTION (Version Kananaskis 4),
   http://downloads.sourceforge.net/hol/kananaskis-4-description.pdf
6. The HOL System LOGIC (Version Kananaskis 4),
   http://downloads.sourceforge.net/hol/kananaskis-4-logic.pdf
7. Lack, S., Street, R.: The formal theory of monads II. Journal of Pure Applied Algorithms 175, 243–265 (2002)
8. Melham, T.F.: The HOL Logic Extended with Quantification over Type Variables. Formal Methods in System Design 3(1-2), 7–24 (1993)
9. Monk, J.D.: Introduction to Set Theory. McGraw-Hill, New York (1969)
10. Pierce, B.C.: Types and Programming Languages. MIT Press, Cambridge (2002)
11. Völker, N.: HOL2P - A System of Classical Higher Order Logic with Second Order Polymorphism. In: Schneider, K., Brandt, J. (eds.) TPHOLs 2007. LNCS, vol. 4732, pp. 334–351. Springer, Heidelberg (2007)
12. Wadler, P.: Monads for functional programming. In: Jeuring, J., Meijer, E. (eds.) AFP 1995. LNCS, vol. 925. Springer, Heidelberg (1995)

A Purely Definitional Universal Domain

Brian Huffman

Portland State University
brianh@cs.pdx.edu

Abstract. Existing theorem prover tools do not adequately support reasoning about general recursive datatypes. Better support for such datatypes would facilitate reasoning about a wide variety of real-world programs, including those written in continuation-passing style, that are beyond the scope of current tools.

This paper introduces a new formalization of a universal domain that is suitable for modeling general recursive datatypes. The construction is purely definitional, introducing no new axioms. Defining recursive types in terms of this universal domain will allow a theorem prover to derive strong reasoning principles, with soundness ensured by construction.

1   Introduction

One of the main attractions of pure functional languages like Haskell is that they promise to be easy to reason about. However, that promise has not yet been fulfilled. To illustrate this point, let us define a couple of datatypes and functions, and try to prove some simple properties.

```
data Cont r a = MkCont ((a -> r) -> r)

mapCont :: (a -> b) -> Cont r a -> Cont r b
mapCont f (MkCont c) = MkCont (\k -> c (k . f))

data Resumption r a = Done a | More (Cont r (Resumption r a))

bind :: Resumption r a -> (a -> Resumption r b) -> Resumption r b
bind (Done x) f = f x
bind (More c) f = More (mapCont (\r -> bind r f) c)
```

Haskell programmers may recognize type `Cont` as a standard continuation monad. Along with the type definition is a map function `mapCont`, for which we expect the functor laws to hold. By itself, type `Cont` is not difficult to work with. None of the definitions are recursive, so they can be formalized easily in most any theorem prover. Proofs of the functor laws `mapCont id = id` and `mapCont (f . g) = mapCont f . mapCont g` are straightforward.

Things get more interesting with the next datatype definition. Monad experts might notice that type `Resumption` is basically a resumption monad transformer wrapped around a continuation monad. The function `bind` is the monadic bind

S. Berghofer et al. (Eds.): TPHOLs 2009, LNCS 5674, pp. 260–275, 2009.

operation for the `Resumption` monad; together with `Done` as the monadic unit, we should expect `bind` to satisfy the monad laws.

The first monad law follows trivially from the definition of `bind`. Instead, let's consider the second monad law (also known as the right-unit law) which states that `bind r Done = r`. How can we go about proving this, formally or otherwise?

It might be worthwhile to try case analysis on `r`, for a start. If `r` is equal to `Done x`, then from the definition of `bind` we have `bind (Done x) Done = Done x`, so the law holds in this case. Furthermore, if `r` is equal to ⊥, then from the strictness of `bind` we have `bind ⊥ Done = ⊥`, so the law also holds for ⊥. Finally, we must consider the case when `r` is equal to `More c`. Using the definition of `bind` we obtain the following:

```
bind (More c) Done = More (mapCont (\r -> bind r Done) c)
```

Now, if we could only rewrite the `bind r Done` on the right-hand side to `r`, then we could use the functor identity law for `mapCont` to simplify the entire right-hand side to `More c`. Perhaps an appropriate induction rule could help.

When doing induction over simple datatypes like lists, the inductive hypothesis simply assumes that the property being proved holds for an immediate subterm: We get to assume `P(xs)` in order to show `P(x : xs)`. This kind of inductive hypothesis will not work for type `Resumption`, because of the indirect recursion in its definition.

In fact, an induction rule for `Resumption` appropriate for our proof does exist. (The proof of the second monad law using this induction scheme is left as an exercise for the reader.)

$$
\begin{array}{l}
\texttt{admissible(P)} \\
\texttt{P(undefined)} \\
\forall \texttt{x. P(Done x)} \\
\forall \texttt{f c. } (\forall \texttt{x. P(f x)}) \longrightarrow \texttt{P(More (mapCont f c))} \\
\hline
\forall \texttt{x. P(x)}
\end{array}
\tag{1}
$$

This induction rule is rather unusual—the inductive step quantifies over a function `f`, and also mentions `mapCont`. It is probably not obvious to most readers that it is correct. How can we trust it? It would be desirable to formally prove such rules using a theorem prover.

Unfortunately, a fully mechanized semantics of general recursive datatypes does not yet exist. Various theorem provers have facilities for defining recursive datatypes, but none can properly deal with datatype definitions like the `Resumption` type introduced earlier. The non–strictly positive recursion causes the definition to be rejected by both Isabelle/HOL's datatype package and Coq's inductive definition mechanism.

Of all the currently available theorem proving tools, the Isabelle/HOLCF domain package is the closest to being able to support such datatypes. It uses the continuous function space, so it is not limited to strictly positive recursion. However, the domain package has some problems due to the fact that it generates non-trivial axioms "on the fly": For each type definition, the domain package declares the existence of the new type (without defining it), and asserts an appropriate type isomorphism and induction rule. The most obvious worry with this design is the potential for unsoundness. On the other hand, the desire to avoid unsoundness can lead to an implementation that is overly conservative.

In contrast with the current domain package, the Isabelle/HOL inductive datatype package [14] is purely definitional. It uses a parameterized universe type, of which new datatypes are defined as subsets. Induction rules are not asserted as axioms; rather, they are proved as theorems. Using a similar design for the HOLCF domain package would allow strong reasoning principles to be generated, with soundness ensured by construction.

The original contributions of this paper are as follows:

- A new construction of a universal domain that can represent a wide variety of types, including sums, products, continuous function space, powerdomains, and recursive types built from these. Universal domain elements are defined in terms of sets of natural numbers, using ideal completion—thus the construction is suitable for simply-typed, higher-order logic theorem provers.
- A formalization of this construction in the HOLCF library of the Isabelle theorem prover. The formalization is fully definitional; no new axioms are asserted.

Section 2 reviews various domain theory concepts used in the HOLCF formalization. The construction of the universal domain type itself, along with embedding and projection functions, are covered in Section 3. Section 4 describes how the universal domain can be used to define recursive types. After a discussion of related work in Section 5, conclusions and directions for future work are found in Section 6.

2   Background Concepts

This paper assumes some familiarity with basic concepts of domain theory: A *partial order* is a set with a reflexive, transitive, antisymmetric relation ($\sqsubseteq$). A *chain* is an increasing sequence indexed by the naturals; a *complete* partial order (cpo) has a least upper bound (lub) for every chain. A *pointed* cpo also has a least element, $\bot$. A *continuous* function preserves lubs of chains. An *admissible* predicate holds for the lub of a chain, if it holds for all elements of the chain.

HOLCF [13] is a library of domain theory built on top of the Isabelle/HOL theorem prover. HOLCF defines all of the standard notions listed above; it also defines standard type constructors like the continuous function space, and strict sums and products. The remainder of this section is devoted to some more specialized concepts included in HOLCF that support the formalization of the universal domain.

2.1   Embedding-Projection Pairs

Some cpos can be embedded within other cpos. The concept of an *embedding-projection pair* (often shortened to *ep-pair*) formalizes this notion. Let A and B

```
data Shrub = Node Shrub Shrub | Tip
data Tree = Branch Tree Tree | Leaf | Twig

embed :: Shrub -> Tree
embed (Node l r) = Branch (embed l) (embed r)
embed Tip = Twig

project :: Tree -> Shrub
project (Branch l r) = Node (project l) (project r)
project Leaf = undefined
project Twig = Tip

deflate :: Tree -> Tree
deflate (Branch l r) = Branch (deflate l) (deflate r)
deflate Leaf = undefined
deflate Twig = Twig
```

Fig. 1. Embedding-projection pairs and deflations in Haskell. Function `deflate` is equal to the composition of functions `embed` and `project`.

be cpos, and $e : A \to B$ and $p : B \to A$ be continuous functions. Then e and p are an ep-pair if $p \circ e = \mathrm{Id}_A$ and $e \circ p \sqsubseteq \mathrm{Id}_B$. The existence of such an ep-pair means that cpo A can be embedded within cpo B.

Figure 1 shows an example in Haskell, where the the type `Shrub` is embedded into the larger type `Tree`. If we embed a `Shrub` into type `Tree`, we can always project back out to recover the original value. In other words, for all s, we have `project (embed s) = s`. On the other hand, if we project a `Tree` out to type `Shrub`, then embed back into type `Tree`, we may or may not get back the same value we started with. If t of type `Tree` contains no `Leaf` constructors at all, then we have `embed (project t) = t`. Otherwise, we basically end up with a tree with all its leaves stripped off—each `Leaf` constructor is replaced with $\bot$.

2.2 Deflations

Cpos may contain other cpos as subsets. A *deflation*[1] is a way to encode such a sub-cpo as a continuous function. Let B be a cpo, and $d : B \to B$ be a continuous function. Then d is a deflation if $d \circ d = d \sqsubseteq \mathrm{Id}_B$. The image set of deflation $d : B \to B$ gives a sub-cpo of B.

Essentially, a deflation is a *value* that represents a *type*. For example, the function `deflate` in Fig. 1 is a deflation; its image set consists of exactly those values of type `Tree` that contain no `Leaf` constructors. Note that while the the definition of `deflate` does not mention type `Shrub` at all, its image set is isomorphic to type `Shrub`—in other words, `deflate` (a function value) is a representation of `Shrub` (a type).

---

[1] My usage of *deflation* follows Gunter [6]. Many authors use the term *projection* to refer to the same concept, but I prefer *deflation* because it avoids confusion with the second half of an ep-pair.

While types can be represented by deflations, type *constructors* (which are like functions from types to types) can be represented as functions from deflations to deflations. For example, the `map` function represents Haskell's list type constructor: While `deflate` is a deflation on type `Tree` that represents type `Shrub`, `map deflate` is a deflation on type `[Tree]` that represents type `[Shrub]`.

Deflations and ep-pairs are closely related. Given an ep-pair (e, p) from cpo A into cpo B, the composition $e \circ p$ is a deflation on B whose image set is isomorphic to A. Conversely, every deflation $d : B \to B$ also gives rise to an ep-pair. Define the cpo A to be the image set of d; also define e to be the inclusion map from A to B, and define $p = d$. Then (e, p) is an embedding-projection pair. So saying that there exists an ep-pair from A to B is equivalent to saying that there exists a deflation on B whose image set is isomorphic to A.

Finally we are ready to talk about what it means for a cpo to be a universal domain. A cpo U is universal for a class of cpos, if for every cpo D in the class, there exists an ep-pair from D into U. Equivalently, for every D there must exist a deflation on U with an image set isomorphic to D.

2.3   Algebraic and Bifinite Cpos

Lazy recursive datatypes often have infinite as well as finite values.[2] For example, we can define a datatype of recursive lazy lists of booleans:

```
data BoolList = Nil | Cons Bool BoolList
```

Finite values of type `BoolList` include total values like `Cons False Nil`, and `Cons True (Cons False Nil)`, along with partial finite values like `Cons False undefined`. On the other hand, recursive definitions can yield infinite values:

```
trues :: BoolList
trues = Cons True trues
```

One way to characterize the set of finite values is in terms of an `approx` function, defined below. The function `approx` is similar to the standard list function `take` that we all know and love, except that `approx 0` returns $\bot$ instead of `Nil`. (This makes each `approx n` into a deflation.) A value `xs` of type `BoolList` is finite if and only if there exists some `n` such that `approx n xs = xs`.

```
approx :: Int -> BoolList -> BoolList
approx 0 xs = undefined
approx n Nil = Nil
approx n (Cons x xs) = Cons x (approx (n-1) xs)
```

---

[2] The formalization actually uses the related concept of *compactness* in place of *finiteness*. A value k is compact iff $(\lambda x.\ k \not\sqsubseteq x)$ is an admissible predicate. Compactness and finiteness do not necessarily coincide; for example, in a cpo of ordinals, $\omega + 1$ is compact but not finite. In the context of recursive datatypes, however, the concepts are generally equivalent.

The function `approx` is so named because for any input value `xs` it generates a sequence of finite approximations to `xs`. For example, the first few approximations to `trues` are $\bot$, `Cons True` $\bot$, `Cons True (Cons True` $\bot$`)`, and so on. Each is finite, but the least upper bound of the sequence is the infinite value `trues`. This property of a cpo, where every infinite value can be written as the least upper bound of a chain of finite values, is called *algebraicity*. Thus `BoolList` is an *algebraic cpo*.

The sequence of deflations `approx n` is a chain of functions whose least upper bound is the identity function. In terms of image sets, we have a sequence of partial orders whose limit is the whole type `BoolList`.

A further property of `approx` which may not be immediately apparent is that for any `n`, the image of `approx n` is a finite set. This means that image sets of `approx n` yield a sequence of *finite* partial orders. As a limit of finite partial orders, we say that type `BoolList` is a *bifinite* cpo. More precisely, as a limit of *countably many* finite partial orders, `BoolList` is an *omega*-bifinite cpo.[3]

The omega-bifinites are a useful class of cpos because bifiniteness is preserved by all of the type constructors defined in HOLCF. Furthermore, all Haskell datatypes are omega-bifinite. Basically any type constructor that preserves finiteness will preserve bifiniteness as well. More details about the formalization of omega-bifinite domains in HOLCF can be found in [10].

2.4   Ideal Completion and Continuous Extensions

In an algebraic cpo the set of finite elements, together with the ordering relation on them, completely determines the structure of the entire cpo. We say that the set of finite elements forms a *basis* for the cpo, and the entire cpo is a *completion* of the basis.

Given a basis B with ordering relation ($\preceq$), we can reconstruct the whole algebraic cpo. The standard process for doing this is called *ideal completion*, and it is done by considering the set of ideals over the basis.

An *ideal* is a non-empty, downward-closed, directed set—that is, it contains an upper bound for any finite subset. A *principal ideal* is an ideal of the form $\{y.\ y \preceq x\}$ for some x, denoted $\downarrow x$. The set of all ideals over $\langle B, \preceq \rangle$ is denoted $\mathrm{Idl}(B)$; when ordered by subset inclusion, $\mathrm{Idl}(B)$ forms an algebraic cpo. The compact elements of $\mathrm{Idl}(B)$ are exactly those represented by principal ideals.

Note that the relation ($\preceq$) does not need to be antisymmetric. For x and y that are equivalent (that is, both $x \preceq y$ and $y \preceq x$) the principal ideals $\downarrow x$ and $\downarrow y$ are equal. This means that the ideal completion construction automatically takes care of quotienting by the equivalence induced by ($\preceq$).

Just as the structure of an algebraic cpo is completely determined by its basis, a continuous function from an algebraic cpo to another cpo is completely determined by its action on basis elements. This suggests a method for defining continuous functions over ideal completions: First, define a function f from basis

---

[3] "SFP domain" is another name, introduced by Plotkin [15], that is used for the same concept—the name stands for Sequence of Finite Posets.

B to cpo C such that f is monotone, i.e. $x \preceq y$ implies $f(x) \sqsubseteq f(y)$. Then we can define the *continuous extension* of f as $\widehat{f}(S) = \bigsqcup_{x \in S} f(x)$. The function $\widehat{f}$ is the unique continuous function of type $\mathrm{Idl}(B) \to C$ that agrees with f on principal ideals—that is, for all $x : B$, $\widehat{f}(\downarrow x) = f(x)$.

In the next section, all of the constructions related to the universal domain will be done in terms of basis values: The universal domain itself will be defined using ideal completion, and the embedding and projection functions will be defined as continuous extensions.

HOLCF includes a formalization of ideal completion and continuous extensions, which was created to support the definition of powerdomains [10].

3   Construction of the Universal Domain

Informally, an *omega-bifinite domain* is a cpo that can be written as the limit of a sequence of finite partial orders. This section describes how to construct a *universal* omega-bifinite domain U, along with an ep-pair from another arbitrary omega-bifinite domain D into U. The general strategy is as follows:

- From the bifinite structure of D, obtain a sequence of finite posets P_n whose limit is D.
- Following Gunter [7], decompose the sequence P_n further into a sequence of *increments* that insert new elements one at a time.
- Construct a universal domain basis that can encode any increment.
- Construct the actual universal domain U using ideal completion.
- Define the embedding and projection functions between D and U using continuous extension, in terms of their action on basis elements.

The process of constructing a sequence of increments is described in Sec. 3.1. The underlying theory is standard, so the section is primarily exposition; the original contribution here is the formalization of that work in a theorem prover. The remainder of the construction, including the basis and embedding/projection functions, is covered in Sec. 3.2 onwards; here both the theory and the formalization are original.

3.1   Building a Sequence of Increments

Any omega-bifinite domain D can be represented as the limit of a sequence of finite posets, with embedding-projection pairs between each successive pair. Figure 2 shows the first few posets from one such sequence.

In each step along the chain, each new poset P_{n+1} is larger than the previous P_n by some finite amount; the structure of P_{n+1} has P_n embedded within it, but it has some new elements as well.

An ep-pair between finite posets P and P', where P' has exactly one more element than P, is called an *increment* (terminology due to Gunter [8]). In Fig. 2, the embedding of P_1 into P_2 is an example of an increment.

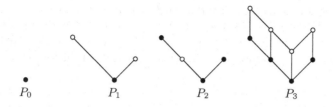

Fig. 2. A sequence of finite posets. Each P_n can be embedded into P_{n+1}; black nodes indicate the range of the embedding function.

The strategy for embedding a bifinite domain into the universal domain is built around increments. The universal domain is designed so that if a finite partial order P is representable (i.e. by a deflation), and there is an increment from P to P', then P' will also be representable.

For all embeddings from P_n to P_{n+1} that add more than one new value, we will need to decompose the single large embedding into a sequence of smaller increments. The challenge, then, is to determine in which order the new elements should be inserted. The order matters: Adding elements in the wrong order can cause problems, as shown in Fig. 3.

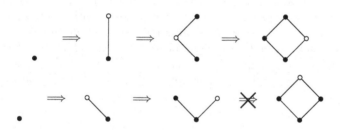

Fig. 3. The right (top) and wrong (bottom) way to order insertions. No ep-pair exists between the 3-element and 4-element posets on the bottom row.

To describe the position of a newly-inserted element, it will be helpful to invent some terminology. The set of elements *above* the new element will be known as its *superiors*. An element immediately *below* the new element will be known as its *subordinate*.

In order for the insertion of a new element to be a valid increment, it must have exactly one subordinate. The subordinate indicates the value that the increment's projection maps the new value onto.

With the four-element poset in Fig. 3, it is not possible to insert the top element last. The reason is that the element has two subordinates: If a projection function maps the new element to one, the ordering relation with the other will not be preserved. Thus a monotone projection does not exist.

A strategy for successfully avoiding such situations is to always insert maximal elements first [7, §5]. Fig. 4 shows this strategy in action. Notice that the number of superiors varies from step to step, but each inserted element always has exactly one subordinate. To maintain this invariant, the least of the four new values must be inserted last.

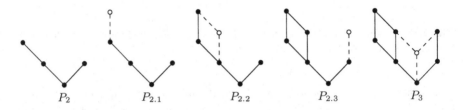

P_2 $P_{2.1}$ $P_{2.2}$ $P_{2.3}$ P_3

Fig. 4. A sequence of four increments going from P_2 to P_3. Each new node may have any number of upward edges, but only one downward edge.

Armed with this strategy, we can finally formalize the complete sequence of increments for type D. To each element x of the basis of D we must assign a sequence number $place(x)$—this numbering tells in which order to insert the values. The HOLCF formalization breaks up the definition of $place$ as follows. First, each basis value is assigned to a rank, where $rank(x) = n$ means that the basis value x first appears in the poset P_n. Equivalently, $rank(x)$ is the least n such that $approx_n(x) = x$. Then an auxiliary function pos assigns sequence numbers to values in finite sets, by repeatedly removing an arbitrary maximal element until the set is empty. Finally, $place(x)$ is defined as the sequence number of x within its (finite) rank set, plus the total size of all earlier ranks.

$$choose(A) = (\varepsilon x \in A. \ \forall y \in A. \ x \sqsubseteq y \longrightarrow x = y) \tag{2}$$

$$pos(A, x) = \begin{cases} 0, & \text{if } x = choose(A) \\ 1 + pos(A - \{choose(A)\}, x), & \text{if } x \neq choose(A) \end{cases} \tag{3}$$

$$place(x) = pos(\{y. \ rank(y) = rank(x)\}, x) + |\{y. \ rank(x) < rank(y)\}| \tag{4}$$

For the remainder of this paper, it will be sufficient to note that the $place$ function satisfies the following two properties:

- Values in earlier ranks come before values in later ranks: If $rank(x) < rank(y)$, then $place(x) < place(y)$.
- Within the same rank, larger values come first: If $rank(x) = rank(y)$ and $x \sqsubseteq y$, then $place(y) < place(x)$.

3.2   A Basis for the Universal Domain

Constructing a partial order incrementally, there are two possibilities for any newly inserted value:

- The value is the very first one (i.e. it is $\bot$)
- The value is inserted above some previous value (its subordinate), and below zero or more other previous values (its superiors)

Accordingly, we can define a datatype to describe the position of these values relative to each other. (Usage of Haskell datatype syntax is merely for convenience; this is not intended to be viewed as a lazy datatype. Here `Nat` represents the natural numbers, and `Set a` represents finite sets with elements of type a.)

```
data Basis = Bottom | Node { serial_number :: Nat
                           , subordinate :: Basis
                           , superiors :: Set Basis }
```

The above definition does not work as a datatype definition in Isabelle/HOL, because the finite set type constructor does not work with the datatype package. (Indirect recursion only works with other inductive datatypes.) But it turns out that we do not need the datatype package at all—the type `Basis` is actually isomorphic to the natural numbers. Using the bijections $\mathbb{N} \cong 1+\mathbb{N}$ and $\mathbb{N} \cong \mathbb{N} \times \mathbb{N}$ with $\mathbb{N} \cong \mathcal{P}_f(\mathbb{N})$, we can construct a bijection that lets us use $\mathbb{N}$ as the basis datatype:

$$\mathbb{N} \cong 1 + \mathbb{N} \times \mathbb{N} \times \mathcal{P}_f(\mathbb{N}) \tag{5}$$

In the remainder of this section, we will use mathematical notation to write values of the basis datatype: $\bot$ represents `Bottom`, and $\langle i, a, S \rangle$ will stand for `Node i a s`.

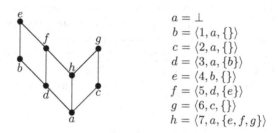

$$a = \bot$$
$$b = \langle 1, a, \{\} \rangle$$
$$c = \langle 2, a, \{\} \rangle$$
$$d = \langle 3, a, \{b\} \rangle$$
$$e = \langle 4, b, \{\} \rangle$$
$$f = \langle 5, d, \{e\} \rangle$$
$$g = \langle 6, c, \{\} \rangle$$
$$h = \langle 7, a, \{e, f, g\} \rangle$$

Fig. 5. Embedding elements of P_3 into the universal domain basis

Figure 5 shows how this system works for embedding all the elements from the poset P_3 into the basis datatype. The elements have letter names from a–h, assigned alphabetically by insertion order. In the datatype encoding of each element, the subordinate and superiors are selected from the set of previously inserted elements. Serial numbers are assigned sequentially.

The serial number is necessary to distinguish multiple values that are inserted in the same position. For example, in Fig. 5, elements b and c both have a as the subordinate, and neither has any superiors. The serial number is the only way to tell such values apart.

Note that the basis datatype seems to contain some junk—some subordinate/superiors combinations are not well formed. For example, in any valid increment, all of the superiors are positioned above the subordinate. One way to take care of this requirement would be to define a well-formedness predicate for basis elements. However, it turns out that it is possible (and indeed easier) to simply ignore any invalid elements. In the set of superiors, only those values that are above the subordinate will be considered. (This will be important to keep in mind when we define the basis ordering relation.)

There is also a possibility of multiple representations for the same value. For example, in Fig. 5 the encoding of h is given as $\langle 7, a, \{e, f, g\}\rangle$, but the representation $\langle 7, a, \{f, g\}\rangle$ would work just as well (since the sets have the same upward closure). One could consider having a well-formedness requirement for the set of superiors to be upward-closed. But this turns out not to be necessary, since the extra values do not cause problems for any of the formal proofs.

3.3   Basis Ordering Relation

To perform the ideal completion, we need to define a preorder relation on the basis. The basis value $\langle i, a, S\rangle$ should fall above a and below all the values in set S that are above a. Accordingly, we define the relation $(\preceq)$ as the smallest reflexive, transitive relation that satisfies the following two introduction rules:

$$a \preceq \langle i, a, S\rangle \tag{6}$$

$$a \preceq b \wedge b \in S \Longrightarrow \langle i, a, S\rangle \preceq b \tag{7}$$

Note that the relation $(\preceq)$ is not antisymmetric. For example, we have both $a \preceq \langle i, a, \{a\}\rangle$ and $\langle i, a, \{a\}\rangle \preceq a$. However, for ideal completion this does not matter. Basis values a and $\langle i, a, \{a\}\rangle$ generate the same principal ideal, so they will be identified as elements of the universal domain.

Also note the extra hypothesis $a \preceq b$ in Eq. (7). Because we have not banished ill-formed subordinate/superiors combinations from the basis datatype, we must explicitly consider only those elements of the set of superiors that are above the subordinate.

3.4   Building the Embedding and Projection

In the HOLCF formalization, the embedding function emb from D to U is defined using continuous extension. The first step is to define emb on basis elements, generalizing the pattern shown in Fig. 5. The definition below uses wellfounded recursion—all recursive calls to emb are on previously inserted values with smaller $place$ numbers:

$$emb(x) = \begin{cases} \bot & \text{if } x = \bot \\ \langle i, a, S\rangle & \text{otherwise} \end{cases}$$

$$\text{where } i = place(x) \tag{8}$$
$$a = emb(sub(x))$$
$$S = \{emb(y) \mid place(y) < place(x) \wedge x \sqsubseteq y\}$$

The subordinate value a is computed using a helper function sub, which is defined as $sub(x) = approx_{n-1}(x)$, where $n = rank(x)$. The ordering produced by the $place$ function ensures that no previously inserted value with the same rank as x will be below x. Therefore the previously inserted value immediately below x must be $sub(x)$, which comes from the previous rank.

In order to complete the continuous extension, it is necessary to prove that the basis embedding function is monotone. That is, we must show that for any x and y in the basis of D, $x \sqsubseteq y$ implies $emb(x) \preceq emb(y)$. The proof is by well-founded induction over the maximum of $place(x)$ and $place(y)$. There are two main cases to consider:

- Case $place(x) < place(y)$: Since $x \sqsubseteq y$, it must be the case that $rank(x) < rank(y)$. Then, using the definition of sub it can be shown that $x \sqsubseteq sub(y)$; thus by the inductive hypothesis we have $emb(x) \preceq emb(sub(y))$. Also, from Eq. (6) we have $emb(sub(y)) \preceq emb(y)$. Finally, by transitivity we have $emb(x) \preceq emb(y)$.
- Case $place(y) < place(x)$: From the definition of sub we have $sub(x) \sqsubseteq x$. By transitivity with $x \sqsubseteq y$ this implies $sub(x) \sqsubseteq y$; therefore by the inductive hypothesis we have $emb(sub(x)) \preceq emb(y)$. Also, using Eq. (8), we have that $emb(y)$ is one of the superiors of $emb(x)$. Ultimately, from Eq. (7) we have $emb(x) \preceq emb(y)$.

The projection function prj from U to D is also defined using continuous extension. The action of prj on basis elements is specified by the following recursive definition:

$$prj(a) = \begin{cases} emb^{-1}(a) & \text{if } \exists x.\ emb(x) = a \\ prj(subordinate(a)) & \text{otherwise} \end{cases} \tag{9}$$

To ensure that prj is well-defined, there are a couple of things to check. First of all, the recursion always terminates: In the worst case, repeatedly taking the subordinate of any starting value will eventually yield $\perp$, at which point the first branch will be taken since $emb(\perp) = \perp$. Secondly, note that emb^{-1} is uniquely defined, because emb is injective. Injectivity of emb is easy to prove, since each embedded value has a different serial number.

Just like with emb, we also need to prove that the basis projection function prj is monotone. That is, we must show that for any a and b in the basis of U, $a \preceq b$ implies $prj(a) \sqsubseteq prj(b)$. Remember that the basis preorder ($\preceq$) is an inductively defined relation; accordingly, the proof proceeds by induction on $a \preceq b$. Compared to the proof of monotonicity for emb, the proof for prj is relatively straightforward; details are omitted here.

Finally, we must prove that emb and prj form an ep-pair. The proof of $prj \circ emb = Id_D$ is easy: Let x be any value in the basis of D. Then using Eq. (9), we have $prj(emb(x)) = emb^{-1}(emb(x)) = x$. Since this equation is an admissible predicate on x, proving it for compact x is sufficient to show that it holds for all values in the ideal completion.

The proof of $emb \circ prj \sqsubseteq Id_U$ takes a bit more work. As a lemma, we can show that for any a in the basis of U, $prj(a)$ is always equal to $emb^{-1}(b)$ for some

$b \preceq a$ that is in the range of *emb*. Using this lemma, we then have $emb(prj(a)) = emb(emb^{-1}(b)) = b \preceq a$. Finally, using admissibility, this is sufficient to show that $emb(prj(a)) \sqsubseteq a$ for all a in U.

To summarize the results of this section: We have formalized a type U, and two polymorphic continuous functions *emb* and *prj*. For any omega-bifinite domain D, *emb* and *prj* form an ep-pair that embeds D into U. The full proof scripts are available as part of the distribution of Isabelle2009, in the theory file `src/HOLCF/Universal.thy`.

4   Algebraic Deflations and Deflation Combinators

To represent types, we need a type T consisting of all the *algebraic deflations* over U, i.e. deflations whose image sets are omega-bifinite cpos. In HOLCF, the algebraic deflations are defined using ideal completion from the set of *finite deflations*, which have finite image sets. Note that as an ideal completion, T is itself a bifinite cpo; this is important because it lets us use a fixed-point combinator to define recursive values of type T, representing recursive types.

For each of the basic type constructors in HOLCF, we can define a deflation combinator as a continuous function over type T. Using continuous extension, we start by defining the combinators on finite deflations. Below are the definitions for product and continuous function space: If D and E are finite deflations on type U, then so are $D \times_F E$ and $D \to_F E$.

$$(D \times_F E)(x) = \text{case } prj(x) \text{ of } (a, b) \to emb(D(a), E(b)) \tag{10}$$

$$(D \to_F E)(x) = emb(E \circ prj(x) \circ D) \tag{11}$$

Next, we can define combinators $(\times_T)$ and $(\to_T)$ of type $T \to T \to T$ as continuous extensions of $(\times_F)$ and $(\to_F)$. Combinators $(\oplus_T)$ and $(\otimes_T)$ for strict sums and products are defined similarly. Values $\text{unit}_T, \text{bool}_T, \text{int}_T :: T$ can also be defined to represent basic types.

The deflation combinators, together with a least fixed-point operator, can be used to define deflations for recursive types. Below are the definitions for the Cont and Resumption datatypes from the introduction:

$$\text{Cont}_T(R, A) = ((A \to_T R) \to_T R) \tag{12}$$

$$\text{Resumption}_T(R, A) = \mu D. \; A \oplus_T \text{Cont}_T(R, D) \tag{13}$$

As a recursive datatype, Resumption_T uses the fixed point operator in its definition. Also note that the definition of Resumption_T refers to Cont_T on the right-hand side—since Cont_T is a continuous function, it may be used freely within other recursive definitions. Thus it is not necessary to transform indirect recursion into mutual recursion, like the Isabelle datatype package does.

Once the deflations have been constructed, the actual Cont and Resumption types can be defined using the image sets of their respective deflations. That the Resumption type satisfies the appropriate domain isomorphism follows from the fixed-point definition. Also, a simple induction principle (a form of take induction, like what would be axiomatized by the current domain package) can be derived from the fact that Resumption$_T$ is a *least* fixed-point.

Finally, the simple take induction rule can be used to derive the higher-level induction rule shown in Eq. (1). The appearance of `mapCont` is due to the fact that (modulo some coercions to and from U) it coincides with the deflation combinator $(\lambda D.\ \text{Cont}_T(R, D))$. (This is similar to how the function `map` doubles as the deflation combinator for lists.)

5   Related Work

An early example of the purely definitional approach to defining datatypes is described by Melham, in the context of the HOL theorem prover [12]. Melham defines a type $(\alpha)\,\textit{Tree}$ of labelled trees, from which other recursive types are defined as subsets. The design is similar in spirit to the one presented in this paper—types are modeled as values, and abstract axioms that characterize each datatype are proved as theorems. The main differences are that it uses ordinary types instead of bifinite domains, and ordinary subsets instead of deflations.

The Isabelle/HOL datatype package uses a design very similar to the HOL system. The type $\alpha\,\textit{node}$, which was originally used for defining recursive types in Isabelle/HOL, was introduced by Paulson [14]; it is quite similar to the HOL system's $(\alpha)\,\textit{Tree}$ type. Gunter later extended the labelled tree type of HOL to support datatypes with arbitrary branching [9]. Berghofer and Wenzel used a similarly extended type to implement Isabelle's modern datatype package [4].

Agerholm used a variation of Melham's labelled trees to define lazy lists and other recursive domains in the HOL-CPO system [1]. Agerholm's cpo of infinite trees can represent arbitrary polynomial datatypes as subsets; however, negative recursion is not supported.

Recent work by Benton, et al. uses the colimit construction to define recursive domains in Coq [3]. Like the universal domain described in this paper, their technique can handle both positive and negative recursion. Using colimits avoids the need for a universal domain, but it requires a logic with dependent types; the construction will not work in ordinary higher-order logic.

On the theoretical side, various publications by Gunter [6,7,8] were the primary sources of ideas for my universal domain construction. The construction of the sequence of increments in Section 3 is just as described by Gunter [7, §5]. However, the use of ideal completion is original—Gunter defines the universal domain using a colimit construction instead. Given a cpo D, Gunter defines a type D^+ that can embed any increment from D to D'. The universal domain is then defined as a solution to the domain equation $D = D^+$. The construction of D^+ is similar to my `Basis` datatype, except that it is non-recursive and does not include serial numbers.

6   Conclusion and Future Work

The Isabelle/HOLCF library of domain theory now has all the basic infrastructure needed for defining general recursive datatypes without introducing axioms.

It provides a universal domain type U, into which any omega-bifinite domain can be embedded. It also provides a type T of algebraic deflations, which represent bifinite domains as values. Both are included as part of the Isabelle2009 release.

While the underlying theory is complete, the automation is not yet finished. The first area of future work is to connect the new theories to the existing domain package, so that instead of axiomatizing the type isomorphism and induction rules, the domain package can prove them from the fixed-point definitions.

The domain package will also need to be extended with automation for indirect-recursive datatypes. Such datatypes may have various possible induction rules, so this will require some design decisions about how to formulate the rules, in addition to work on automating the proofs.

Other future directions explore limitations in the current design:

- *Higher-order type constructors.* Higher-order types can be represented by deflation combinators with types like $(T \to T) \to T$. The problem is that Isabelle's type system only supports first-order types. Although, see [11] for an admittedly complicated workaround.
- *Non-regular (nested) datatypes* [5]. Deflation combinators for non-regular datatypes can be defined by taking least fixed points at type $T \to T$, rather than type T. However, since Isabelle does not support type quantification or polymorphic recursion, induction rules and recursive functions could not be defined in the normal way.
- *Higher-rank polymorphism.* This is not supported by Isabelle's type system. However, the universal domain U could be used to model such types, using the construction described by Amadio and Curien [2].
- *Generalized abstract datatypes (GADTs).* These are usually modeled in terms of some kind of type equality constraints. For example, type equality constraints are a central feature of System F_C [16], a compiler intermediate language used to represent Haskell programs. But to the extent of this author's knowledge, there is no way to model type equality constraints using deflations.

Acknowledgments. I would like to thank my advisor, John Matthews, for many encouraging discussions about HOLCF and domain theory, and also for suggesting the example used in the introduction. Thanks also to James Hook for reading drafts and providing helpful comments.

References

1. Agerholm, S.: A HOL Basis for Reasoning about Functional Programs. PhD thesis, University of Aarhus (1994)
2. Amadio, R.M., Curien, P.-L.: Domains and Lambda-Calculi. Cambridge University Press, New York (1998)
3. Benton, N., Kennedy, A., Varming, C.: Some domain theory and denotational semantics in Coq. In: Proc. 22nd International Conference on Theorem Proving in Higher Order Logics (TPHOLs 2009). LNCS, vol. 5674. Springer, Heidelberg (2009)

4. Berghofer, S., Wenzel, M.: Inductive datatypes in HOL - lessons learned in formal-logic engineering. In: Bertot, Y., Dowek, G., Hirschowitz, A., Paulin, C., Théry, L. (eds.) TPHOLs 1999. LNCS, vol. 1690, pp. 19–36. Springer, Heidelberg (1999)
5. Bird, R.S., Meertens, L.G.L.T.: Nested datatypes. In: Jeuring, J. (ed.) MPC 1998. LNCS, vol. 1422, pp. 52–67. Springer, Heidelberg (1998)
6. Gunter, C.: Profinite Solutions for Recursive Domain Equations. PhD thesis, University of Wisconsin at Madison (1985)
7. Gunter, C.A.: Universal profinite domains. Information and Computation 72(1), 1–30 (1987)
8. Gunter, C.A.: Semantics of Programming Languages: Structures and Techniques. In: Foundations of Computing. MIT Press, Cambridge (1992)
9. Gunter, E.L.: A broader class of trees for recursive type definitions for HOL. In: Joyce, J.J., Seger, C.-J.H. (eds.) HUG 1993. LNCS, vol. 780, pp. 141–154. Springer, Heidelberg (1994)
10. Huffman, B.: Reasoning with powerdomains in Isabelle/HOLCF. In: Mohamed, O.A., Muñoz, C., Tahar, S. (eds.) TPHOLs 2008. LNCS, vol. 5170, pp. 45–56. Springer, Heidelberg (2008)
11. Huffman, B., Matthews, J., White, P.: Axiomatic constructor classes in Isabelle/HOLCF. In: Hurd, J., Melham, T. (eds.) TPHOLs 2005. LNCS, vol. 3603, pp. 147–162. Springer, Heidelberg (2005)
12. Melham, T.F.: Automating recursive type definitions in higher order logic. In: Current Trends in Hardware Verification and Automated Theorem Proving, pp. 341–386. Springer, Heidelberg (1989)
13. Müller, O., Nipkow, T., von Oheimb, D., Slotosch, O.: HOLCF = HOL + LCF. Journal of Functional Programming 9, 191–223 (1999)
14. Paulson, L.C.: Mechanizing coinduction and corecursion in higher-order logic. Journal of Logic and Computation 7 (1997)
15. Plotkin, G.D.: A powerdomain construction. SIAM J. Comput. 5(3), 452–487 (1976)
16. Sulzmann, M., Chakravarty, M.M.T., Jones, S.P., Donnelly, K.: System F with type equality coercions. In: TLDI 2007: Proceedings of the 2007 ACM SIGPLAN international workshop on Types in languages design and implementation, pp. 53–66. ACM, New York (2007)

Types, Maps and Separation Logic

Rafal Kolanski and Gerwin Klein

Sydney Research Lab., NICTA* , Australia
School of Computer Science and Engineering, UNSW, Sydney, Australia
{rafal.kolanski,gerwin.klein}@nicta.com.au

Abstract. This paper presents a separation-logic framework for reasoning about low-level C code in the presence of virtual memory. We describe our abstract, generic Isabelle/HOL framework for reasoning about virtual memory in separation logic, and we instantiate this framework to a precise, formal model of ARMv6 page tables. The logic supports the usual separation logic rules, including the frame rule, and extends separation logic with additional basic predicates for mapping virtual to physical addresses. We build on earlier work to parse potentially type-unsafe, system-level C code directly into Isabelle/HOL and further instantiate the separation logic framework to C.

1 Introduction

Virtual memory is a mechanism in modern computing systems that usual programming language semantics gloss over. For the application level, the operating system (OS) is expected to provide an abstraction of plain memory and details like page faults are handled behind the scenes. While, strictly speaking, the presence of virtual memory is still observable via sharing, ignoring virtual memory is therefore defendable for the application level.

For verifying lower-level software such as the operating system itself or software for embedded devices without a complex OS layer, this is no longer true. On this layer, virtual memory plays a prominent and directly observable role. It is also the source of many defects that are frequently very frustrating to debug. A wrong, unexpected mapping from virtual to physical addresses in the machine can lead to garbled, unrecognisable data at a much later, seemingly unrelated position in the code. A wrong, non-existing mapping will lead to a page fault: if the machine attempts to read a code instruction or a data value from a virtual address without valid mapping, on most architectures, a hardware exception is raised and execution branches to the address of a registered page fault handler (which often is virtually addressed itself). Defects in the page fault handler may lead to even more obscure, non-local symptoms. The situation is complicated by the fact that these virtual-to-physical mappings are themselves encoded in

---

* NICTA is funded by the Australian Government as represented by the Department of Broadband, Communications and the Digital Economy and the Australian Research Council through the ICT Centre of Excellence program.

S. Berghofer et al. (Eds.): TPHOLs 2009, LNCS 5674, pp. 276–292, 2009.

memory, usually in a hardware-defined page table structure, and they are often manipulated through the virtual memory layer.

As an example, the completion of the very first C implementation (at the time untried and unverified) of the formally verified seL4 microkernel [8] in our group was celebrated by loading the code onto our ARMv6[1] development board and starting the boot process to generate a hello-world message. Quite expectedly, nothing at all happened. The board was unresponsive and no debug information was forthcoming. It took 3 weeks to write the C implementation following a precise specification. It took 5 weeks debugging to get it running. It turned out that the boot code had not set up the initial page table correctly, and since no page fault handler was installed, the machine just kept faulting. This was the first of a number of virtual-memory related bugs. What is worse, our verification framework for C would, at the time, not have caught any of these bugs. We have since explictly added the appropriate virtual memory proof obligations. They are derived, in part, from the work presented in this paper.

We present a framework in Isabelle/HOL for the verification of low-level C code with separation logic in the presence of virtual memory. The framework itself is abstract and generic. In earlier work [16], we described a preliminary version of it, instantiated to a hypothetical simple page table and a toy language. In that work we concentrated on showing that the logic of the framework is indeed an instance of abstract separation logic [5] and that it supports the usual separation logic reasoning, including the frame rule. Here, we concentrate on making the framework applicable to the verification of real C code. We have instantiated the framework to the high-fidelity memory model for C by Tuch et al [24] and connected it with the same C-parsing infrastructure for Isabelle/HOL that was used there. On the hardware side, we have instantiated the framework to a detailed and precise model of ARMv6 2-level hardware page tables. To our knowledge, this is the first formalisation of the ARMv6 memory translation mechanism. The resulting instantiation is a foundational, yet practical verification framework for a large subset of standard C99 [13] with the ability to reason about the effects of virtual memory when necessary and the ability to reason abstractly in the traditional separation logic style when virtual memory is not the focus.

The separation logic layer of the framework makes three additional basic predicates available: mapping from a virtual address to a value, mapping from a physical address to a value, and mapping from a virtual to a physical address. For the user of the framework, these integrate seamlessly with other separation logic formulae and they support all expected, traditional reasoning principles. Inside the framework, we invest significant effort to provide this nice abstraction, to support the frame rule, and to shield the verification user from the considerable complexity of the hardware page table layout in a modern architecture.

Our envisaged application area for this framework is low-level OS kernel code that manipulates page tables and user-level page fault handlers in microkernel

---

[1] The ARMv6 is a popular processor architecture for embedded systems, such as the iPhone or Android.

systems. To stay in the same, foundational framework, it can also be used for the remaining OS kernel without any significant reasoning overhead in a separation logic setting. Our direct application area is the verification of the seL4 microkernel [8].

The remainder of this paper is structured as follows. After introducing notation in Sect. 2, we describe in Sect. 3 an abstract type class for encoding arbitrary C types in memory. Sect. 4 describes our abstract, generic page table framework and Sect. 5 instantiates this to ARMv6. Sect. 6 integrates virtual memory into our abstract separation logic framework, first at the byte level, and then at the structured types level. Sect. 7 makes the connection to C, and, finally, Sect. 8 discusses how translation caching mechanisms can be integrated into the model.

2 Notation

This section introduces Isabelle/HOL syntax where different from standard mathematical notation.

The space of total functions is denoted by $\Rightarrow$. Type variables are written $'a$, $'b$, etc. The notation $t :: \tau$ means that HOL term t has HOL type τ.

Pairs come with the two projection functions fst $:: 'a \times 'b \Rightarrow 'a$ and snd $:: 'a \times 'b \Rightarrow 'b$. *Sets* (type $'a$ set) follow the usual mathematical convention. *Lists* support the empty list $[]$ and cons, written $x \cdot xs$. The list of natural numbers from a to (excluding) b is $[a..<b]$. We also use the standard zip and map from functional programming. The *option* type

$$\textbf{datatype } 'a \text{ option } = \text{None} \mid \text{Some } 'a$$

adjoins a new element None to a type $'a$. We use $'a$ option to model partial functions, writing $\lfloor a \rfloor$ instead of Some a and $'a \rightharpoonup 'b$ instead of $'a \Rightarrow 'b$ option. The Some constructor has an underspecified inverse called the, satisfying the $\lfloor x \rfloor$ $= x$. Lifting functions to the option type is achieved by

$$\text{option-map} = (\lambda f\, y.\ \textsf{case } y \textsf{ of None} \Rightarrow \text{None} \mid \lfloor x \rfloor \Rightarrow \lfloor f\, x \rfloor)$$

Function update is written $f(x := y)$ where $f :: 'a \Rightarrow 'b$, $x :: 'a$ and $y :: 'b$ and $f(x \mapsto y)$ stands for $f\ (x := \text{Some } y)$. *Finite integers* are represented by the type $'a$ word where $'a$ determines the word length in bits. The type supports the usual bit operations like left-shift ($<<$) and bitwise *and* ($\&\&$). The function unat converts to natural numbers (u for unsigned). Separation logic uses the concepts of disjoined maps $\perp$ and map addition $++$. They are defined below.

$$m_1 \perp m_2 \quad \equiv \textsf{dom } m_1 \cap \textsf{dom } m_2 = \emptyset$$
$$m_1 ++ m_2 \equiv \lambda x.\ \textsf{case } m_2\ x \textsf{ of None} \Rightarrow m_1\ x \mid \lfloor y \rfloor \Rightarrow \lfloor y \rfloor$$

3 Types and Value Storage

Our aim of reasoning about C programs requires a representation of the storage of C values in memory. Similarly to Tuch et al [24], we define a mem-type type class to represent these types. This section describes the abstract operations of this class and its axioms. The first such operations are serialising and restoring a value into and from bytes:

to-bytes :: $'t$::mem-type $\Rightarrow$ byte list   from-bytes (to-bytes v) = v
from-bytes :: byte list $\Rightarrow$ $'t$::mem-type

For a particular type, all values occupy the same, non-zero number of bytes in memory. We will refer to the number of these bytes as the size. The length of a type's serialisation is equal to its size. The term TYPE($'t$) of type $'t$ itself makes an Isabelle type avaiable as term.

size-of :: $'t$::mem-type itself $\Rightarrow$ nat   length (to-bytes v) = size-of TYPE($'t$)
                                                          $0 <$ size-of TYPE($'t$)

For treating types as first-class values, we require each to map to a unique tag:

type-tag :: $'t$::mem-type itself $\Rightarrow$ type-tag

In order to respect the alignment requirements of C types, mem-type instances carry alignment information. Types may only be aligned to sizes which are divisors of both the physical and virtual address space sizes:

align-of :: $'t$::mem-type itself $\Rightarrow$ nat
align-of TYPE($'a$) dvd memory-size $\wedge$ align-of TYPE($'a$) dvd addr-space-size

The model we present in this paper allows representation of all packed C types, i.e. atomic types such as int, array, and structs without padding. Tuch's work on structured C types [23] demonstrates how to extend this model to allow padding.

4   Virtual Memory

This section defines addressing and pointer conventions and describes our abstract interface to page table encodings.

4.1   Pointers and Addressing

Virtual memory is an abstraction layer on top of the physical memory in a machine. Each executing process gets its own view of physical memory, wherein each virtual address *may* be mapped to a physical address. We will henceforth refer to the function translating virtual addresses to physical ones as the *virtual map* and the application of the virtual map to a virtual address as a *lookup*.

The virtual map is partial and many-to-one — updates at one virtual address may affect values appearing at another. As in our previous work [16] memory is a partial function. Unlike our previous work [16], the work presented here is a realistic representation of physical memory and maps physical addresses to bytes. The virtual map is encoded in memory in a structure called a *page table*. Programs usually only have access to the virtual address layer, but devices may access physical memory directly. We define addresses as:

datatype ($'a$, $'p$) addr-t = Addr of $'a$

where $'a$ is the underlying address size (e.g. 32 word for 32-bit) and $'p$ is a tag: one of physical or virtual. For particular architectures, we instantiate addr-t into specific virtual and physical addresses. For the ARMv6 both virtual and physical addresses are 32-bit words, yielding the instantiations:

$$\text{vaddr} = (32 \text{ word, virtual) addr-t} \qquad \text{paddr} = (32 \text{ word, physical) addr-t}$$

ARMv6 is capable of natively addressing 8, 16 and 32 bit values in memory (corresponding to *char*, *short* and *int* in C). We have shown that these are instances of mem-type. We use addr-val $(\text{Addr } a) = a$ to extract the address.

4.2 Page Table

We now introduce our abstract interface to page table encodings. There are many such possible encodings: one-level tables, fixed multi-level tables, variable-depth guarded page tables or even just hash tables. Usually, mappings are encoded in blocks of addresses (pages, superpages, etc.), which are hardware-defined. The page table also encodes extra information such as permissions and hardware-defined flags. We generalise our previous abstract page table interface [16] slightly to accomodate multiple page sizes and briefly summarise the other definitions.

$$\text{ptable-lift} :: ('paddr \rightharpoonup 'val) \Rightarrow 'base \Rightarrow 'vaddr \rightharpoonup 'paddr$$
$$\text{ptable-trace} :: ('paddr \rightharpoonup 'val) \Rightarrow 'base \Rightarrow 'vaddr \Rightarrow 'paddr \text{ set}$$
$$\text{get-page} :: ('paddr \rightharpoonup 'val) \Rightarrow 'base \Rightarrow 'vaddr \Rightarrow 'a$$

We use ptable-lift to extract a virtual map from memory, ptable-trace to find all the physical addresses used looking up a virtual to a physical address, and get-page to find which page a virtual address is on including any machine-specific flags (such as permissions) that might be attached to it. The types $'paddr$ and $'vaddr$ represent physical and virtual pointers, while $'base$ says where we can find the page table in physical memory (e.g. the root of a two-level page table). We leave $'a$ for a generic representation of what a page is.

In order to reason about memory access in the presence of a page table, we require page table functions to conform to the rules in Fig. 1. Firstly, changing memory in areas not related to a page table lookup must not affect the lookup: if evaluation of ptable-lift and ptable-trace succeeds on smaller heap , it will also succeed on a larger one. This corresponds to the safety monotonicity property of

$$\frac{\text{ptable-lift } h_0 \ r \ vp = \lfloor p \rfloor \qquad h_0 \perp h_1}{\text{ptable-lift } (h_0 ++ h_1) \ r \ vp = \lfloor p \rfloor} \qquad \frac{\text{ptable-lift } h \ r \ vp = \lfloor p \rfloor \qquad h \perp h'}{\text{ptable-trace } (h ++ h') \ r \ vp = \text{ptable-trace } h \ r \ vp}$$

$$\frac{p \notin \text{ptable-trace } h \ r \ vp \qquad \text{ptable-lift } h \ r \ vp = \lfloor p \rfloor}{\text{ptable-trace } (h(p \mapsto v)) \ r \ vp = \text{ptable-trace } h \ r \ vp}$$

$$\frac{p \notin \text{ptable-trace } h \ r \ vp \qquad \text{ptable-lift } h \ r \ vp = \lfloor p \rfloor}{\text{ptable-lift } (h(p \mapsto v)) \ r \ vp = \lfloor p \rfloor}$$

$$\frac{\text{ptable-lift } (h_0 ++ h_1) \ r \ vp = \lfloor p \rfloor \qquad h_0 \perp h_1}{\text{ptable-lift } h_0 \ r \ vp = \lfloor p \rfloor \lor \text{ptable-lift } h_0 \ r \ vp = \text{None}}$$

Fig. 1. Abstract page table interface

separation logic [5]. Furthermore, a successful lookup must be unaffected by any heap updates outside that lookup's trace. Finally, corresponding to the frame monotonicity property [5] of separation logic, removal of information from the heap must either not affect ptable-lift or cause it to fail. Heap reduction must not return a different successful result.

5   A Formal Model of ARMv6 Page Tables

In this section, we instantiate the abstract interface described above to ARMv6 2-level page tables. We support multiple page sizes, but we omit handling of permissions — in our seL4 target setup, the ARM supervisor mode ignores permissions. Adding them would be simple.

Following ARM nomenclature [3], the first level table is called the *page directory* and the second level the *page table*. Individual *entries* at these levels are called PDEs and PTEs respectively, 32 bits wide in both cases. There is one page directory with potentially many page tables. The

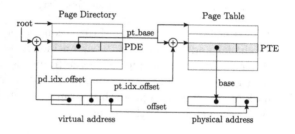

Fig. 2. ARMv6 page table lookup for SmallPage

base of the entire structure is the physical address of the page directory. Our model uses the common ARMv6 page table format where subpages are disabled. In this mode, the hardware supports mappings in four granularities: small (4Kb) and large (64Kb) pages, as well as sections (1Mb) and supersections (16Mb):

datatype page-size = SmallPage | LargePage | Section | SuperSection

Apart from invalid/reserved, a PDE either encodes the physical base address of a section, supersection or a second-level table. Within a second-level table, a valid PTE encodes the physical base address of a large or small page:

datatype pde = InvalidPDE | ReservedPDE | PageTablePDE of paddr
  | SectionPDE of paddr | SuperSectionPDE of paddr

datatype pte = InvalidPTE | LargePagePTE of paddr | SmallPagePTE of paddr

The idea of looking up a virtual address is shown in Fig. 2: figure out the base address of the appropriate structure and its size, then add the virtual address divided by that size. The get-frame function calculates the base and size:

get-frame :: heap ⇒ paddr ⇒ vaddr ⇀ (paddr × page-size)
get-frame h $root$ vp ≡
let vp-val = addr-val vp; pd-idx-$offset$ = vp-val >> 20 << 2
in case decode-pde h ($root$ + pd-idx-$offset$) of None ⇒ None
  | ⌊PageTablePDE pt-$base$⌋ ⇒ get-frame-2nd h pt-$base$ vp
  | ⌊SectionPDE $base$⌋ ⇒ ⌊($base$, Section)⌋
  | ⌊SuperSectionPDE $base$⌋ ⇒ ⌊($base$, SuperSection)⌋ | ⌊-⌋ ⇒ None

The function works by looking up a virtual address just like the ARM hardware. First, we look at the top 12 bits of the address as an index into the page directory. We then shift the index by 2 as each PDE is 4 bytes in size, add it to the base address of the page directory (*root*). We decode the PDE at this address to decide what to do next: fail on invalid/reserved, pass through the base address for sections/supersections, and go look in the second-level table in the case of a PTE pointer. We omit the definitions of decode-pde and decode-pte; they work as described in the ARMv6 manual [3]. Second-level lookup is defined similarly:

get-frame-2nd :: heap $\Rightarrow$ paddr $\Rightarrow$ vaddr $\rightharpoonup$ (paddr $\times$ page-size)
get-frame-2nd h pt-$base$ vp $\equiv$
let vp-val = addr-val vp; pt-idx-$offset$ = (vp-val >> 12) && 0xFF << 2
in case decode-pte h (pt-$base$ + pt-idx-$offset$) of None $\Rightarrow$ None
  | $\lfloor$InvalidPTE$\rfloor$ $\Rightarrow$ None | $\lfloor$LargePagePTE $base\rfloor$ $\Rightarrow$ $\lfloor(base,$ LargePage$)\rfloor$
  | $\lfloor$SmallPagePTE $base\rfloor$ $\Rightarrow$ $\lfloor(base,$ SmallPage$)\rfloor$

Starting at the physical address of the second-level table, we use the next 8 bits of the virtual address (bits 12-19) as an index, decode the PTE there, fail on invalid or return the base address of the frame along with its size.

Using get-frame, we can then implement the main lookup function ptable-lift by masking out the appropriate bits from the virtual address and adding them to the physical address of the frame:

vaddr-offset p w $\equiv$ w && mask (page-size-bits p)

ptable-lift h pt-$root$ vp $\equiv$
let vp-val = addr-val vp
in option-map ($\lambda(base,$ pg-$size)$. $base$ + vaddr-offset pg-$size$ vp-val)
  (get-frame h pt-$root$ vp)

where page-size-bits is log_2 of the page size.

Similarly, we can get the page a virtual address is on by masking out the offset bits. Also, since ARM allows multiple page sizes, the concept of a page must involve its size, instantiating the page type $'a$ to (vaddr $\times$ page-size) option:

addr-base sz w $\equiv$ w && (0xFFFFFFFF << page-size-bits sz)

get-page h $root$ vp $\equiv$
let vp-val = addr-val vp
in option-map ($\lambda(base,$ pg-$size)$. (Addr (addr-base pg-$size$ vp-val), pg-$size$))
  (get-frame h $root$ vp)

We define a sequence of n addresses starting at p as:

addr-seq p 0 = []
addr-seq p (Suc n) = p·addr-seq (p + 1) n

The final function needed to instantiate the abstract page table model from Sect. 4 is ptable-trace. The trace contains the bytes in any page directory or table entry which has successfully contributed to looking up the virtual address:

ptable-trace h $root$ vp $\equiv$
let $vp\text{-}val$ = addr-val vp; $pd\text{-}idx\text{-}offset$ = vaddr-pd-index $vp\text{-}val$ << 2;
  $pt\text{-}idx\text{-}offset$ = vaddr-pt-index $vp\text{-}val$ << 2;
  $pd\text{-}touched$ = set (addr-seq ($root$ + $pd\text{-}idx\text{-}offset$) 4);
  $pt\text{-}touched$ = $\lambda pt\text{-}base.$ set (addr-seq ($pt\text{-}base$ + $pt\text{-}idx\text{-}offset$) 4)
in case decode-pde h ($root$ + $pd\text{-}idx\text{-}offset$) of None $\Rightarrow$ $\emptyset$
  | $\lfloor$PageTablePDE $pt\text{-}base\rfloor$ $\Rightarrow$ $pd\text{-}touched$ $\cup$ $pt\text{-}touched$ $pt\text{-}base$
  | $\lfloor\text{-}\rfloor$ $\Rightarrow$ $pd\text{-}touched$

We have proved that the ptable-lift, ptable-trace and get-page functions in this section instantiate the abstract model from Sect. 4, including the axioms of Fig. 1.

6 Typed, Mapped Separation Logic

Based on our abstract page table interface of Sect. 4, we can now construct a separation logic framework for reasoning about pointer programs with types. This framework is independent of the particular page table instantiation.

Separation logic [18] is a tool for conventiently reasoning about memory and aliasing. It views memory as a partial *heap* from addresses to values, allowing for predicates which precisely state which part of the heap they hold on. At its core is the concept of separating conjunction: when the assertion $P \wedge^* Q$ holds on a heap, the heap can be split into two disjoint parts, where P holds on one part and Q on the other. Predicates which precisely define the domain of the heap they hold on allow for convenient local reasoning. This leads to the concept of local actions and the frame rule: for an action f, we can conclude $\{P \wedge^* R\} f \{Q \wedge^* R\}$ from $\{P\} f \{Q\}$ for *any* R. This expresses that the actions of f are local to the heaps described by P and Q, and therefore cannot affect any separate heap described by R. We also say that predicates *consume* parts of the heap under separating conjunction, because other predicates cannot depend on the same parts of this heap.

The basic assertion of separation logic is the maps-to arrow, holding on a heap containing only one address-value pair. From this simple assertion, more complex ones can be built. For a simple heap (paddr $\rightharpoonup$ byte) it takes the form:

$$(address \mapsto value)\ h \equiv h\ address = value \wedge \text{dom } h = \{address\}$$

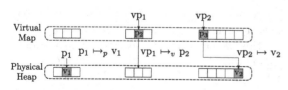

Fig. 3. The three maps-to assertions

Under separating conjunction, it consumes *address* in the heap. Tuch et al extend this basic concept all the way to reasoning about C code with structures [23].

A naive addition of virtual memory to separation logic breaks the concept of separating conjunction, the frame rule, as well as the assumption of Tuch's work of values being stored contiguously in the heap. In previous work [16], we addressed the first two in a simplified setting. In this section, we solve them in a realistic setting and extend them to reasoning about typed pointers. We introduce new maps-to arrows, as well as a new, more complex state that we use instead of a simple heap.

Our eventual goal is to be able to write the new arrows of Fig. 3 with physical or virtual addresses on the left and complex, typed C values on the right. The new arrows in Fig. 3 describe (from left to right): mappings from physical address to value, from virtual to physical address, and from virtual address to value. The next section will introduce arrows that allow raw, single bytes and explicit type information on the right. The section after that will lift this information to allow structured C types on the right.

6.1   At the Bytes Level

Following Tuch et al and our own previous work, to support both types and virtual memory, we annotate the heap with extra information, extending the state for our assertions in a first step to:

$$(\text{paddr} \rightharpoonup \text{type-data} \times \text{byte}) \times \text{ptable-base}$$

where ptable-base is any extra information needed by the virtual memory subsystem, such as the page table root (paddr in the case of ARMv6); type-data annotates which higher-level type a byte is part of. On this level it is just passed through, we will explain its purpose in Sect. 6.2.

For our maps-to assertions to be useful in separation logic, we must define which parts of the heap they consume (what their domain is). Here we run into a problem, illustrated in Fig. 4: two distinct virtual addresses map to two values via distinct physical addresses, but using the same page table entry for the lookup. Writing to one virtual pointer does not affect the value at the other, so in this sense the two maps-to predicates are separate. However, a single page table entry is involved in the lookup of both virtual

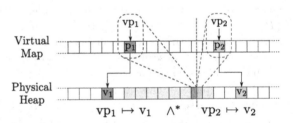

Fig. 4. Two virtual addresses resolving through the same page table entry

pointers. Under separating conjunction we can allow the entry to be consumed by either mapping or neither mapping, but not both mappings. If one consumes it, the other lacks information for a successful lookup. If neither consumes it, we lose locality: we could state the entry is separate from both mappings even though updating the entry can affect both virtual addresses!

The solution to this problem is to divide the page table entry up into two parts and share the slices between the maps-to predicates involved in the separating conjunction. This idea is similar to that of the fractional permission model of Bornat [4], with three important differences. Firstly, we do not wish to perform any explicit accounting of fractions in the most common case of the page table not being modified. Secondly, the number of virtual addresses an entry can map varies with the type of page table and the size of the mapped page. Thirdly, we want to utilise rather than recreate the proofs about partial maps and map disjunction in Isabelle/HOL. These issues are addressed by using a constant, large-enough number of slices for entries in the heap and placing them in the domain. The maximum useful number of slices is one entry mapping all virtual addresses. Thus our final state for assertions is:

fheap-state $=$ (paddr $\times$ vaddr $\rightharpoonup$ type-data $\times$ byte) $\times$ ptable-base

We refer to the first component of this state as the *typed, fragmented heap tfh*. With this new state, our physical memory maps-to predicate becomes:

$$p :\rightarrow_p v \equiv \lambda(h, r).\ (\forall vp.\ h\ (p, vp) = \lfloor v \rfloor) \wedge \text{dom}\ h = \{p\} \times \mathcal{U}$$

Like the simple maps-to predicate shown earlier, the heap at address p evaluates to value v. In the new state, it does so for all vp slices. The domain covers all slices of p, i.e. the universal set $\mathcal{U}$. This arrow works for the physical-to-value level. To define the virtual-to-physical arrow, we use our abstract page table interface. Unfortunately, this page table model knows nothing about slices and type annotations. So, to perform a lookup on vp, we derive a view of the heap *tfh* containing only slices associated with vp and discard type annotations:

h-view *tfh* $vp \equiv$ option-map snd $\circ$ ($\lambda p.\ \textit{tfh}\ (p,\ vp)$)

We can now define the virtual-to-physical arrow for mapping vp to p. It is just a ptable-lift on a heap made of slices associated with vp. The assertion consumes the vp slice of each byte used in its lookup, i.e. in ptable-trace:

vp $:\rightarrow_v p \equiv \lambda(h,r).$ let heap $=$ h-view h vp; $vmap =$ ptable-lift heap r
in $vmap\ vp = \lfloor p \rfloor \wedge$ dom $h =$ ptable-trace heap r $vp \times \{vp\}$

The virtual-to-value mapping is then just the separating conjunction of virtual-to-physical and physical-to-value.

vp $:\rightarrow v \equiv \lambda s.\ \exists p.\ (vp :\rightarrow_v p \wedge^* p :\rightarrow_p v)\ s$
$P \wedge^* Q \equiv \lambda(h, r).\ \exists h_0\ h_1.\ h_0 \perp h_1 \wedge h = h_0 ++ h_1 \wedge P\ (h_0, r) \wedge Q\ (h_1, r)$

For any of these levels, we can define the usual arrow variations [18]:

$(p :\rightarrow -)\ s \equiv \exists v.\ (p :\rightarrow v)\ s$ \qquad $(p :\hookrightarrow v)\ s \equiv (p :\rightarrow v \wedge^* \text{sep-true})\ s$
$(p :\hookrightarrow -)\ s \equiv \exists v.\ (p :\rightarrow v \wedge^* \text{sep-true})\ s$ \quad sep-true $\equiv \lambda s.\ \text{True}$

One property of this framework is that it is mostly independent of the value space, the right-hand side of the maps-to arrows. Only in the interface to the page table have we touched it at all, and then only to discard additional type information. The basic assertions we get from this section are of the form $vp :\rightarrow (b,\ t)$ where b is the byte at virtual address vp, and t is the associated type annotation.

6.2   At the Types Level

This section uses the arrows for bytes and type information we have just defined to higher-level, typed assertions for any mem-type values. We define the concept of pointers to typed values by wrapping our existing concept of addresses and adding a phantom type, like Tuch et al [24]:

datatype $('a, 'p, 't)$ ptr-t = Ptr of $('a, 'p)$ addr-t

Instantiated to the ARMv6:

$'t$ pptr = (32 word, physical, $'t$) ptr-t $'t$ vptr = (32 word, virtual, $'t$) ptr-t

Like Tuch et al we mark locations belonging to mem-type values in the heap with a type tag. The addition of virtual memory creates a new complication: if a value crosses a page boundary in virtual memory, it is not guaranteed to be contiguous at the physical level, nor even entirely loaded into memory. This means we must not only tag each byte in the heap, but also note which offset it is within the larger structure it belongs to. Our type information associated with each byte is:

type-data = type-tag × nat

We implement maps-to predicates at the typed level as a sequence of byte-level maps-to predicates, folded over separating conjunction in the usual way. For instance, we write vps $[:\rightarrow]$ vs for a sequence vps of virtual pointers mapping to a sequence of values vs. Note that these values are each of the from (b,t).

A value of type $'t$::mem-type seen in memory at either the virtual or physical level is a sequence of bytes (to-bytes) where each byte is tagged by the type-tag of $'t$ and its offset in the list:

value-seq $val \equiv$
zip (map ($\lambda seq.$ (type-tag TYPE($'t$), seq)) [0..<size-of TYPE($'t$)]) (to-bytes val)

We can now define maps-to predicates on typed pointers. Like Tuch et al [24] we employ an arbitrary guard on the pointer itself to enforce constraints such as alignment. We have not found it necessary yet to let the guard depend on the state, but this could be added easily. Compared to Tuch et al, lifting sequences of bytes to structured values is much simpler, because we already have byte-level assertions available. Between virtual and physical levels only the arrows differ.

$g \vdash p \rightarrow_p v \equiv$ ptr-seq p TYPE($'t$) $[:\rightarrow_p]$ value-seq v $\lfloor\wedge\rfloor$ ($\lambda s.\ g\ p$)
$g \vdash vp \rightarrow_v p \equiv$ ptr-seq vp TYPE($'t$) $[:\rightarrow_v]$ ptr-seq p TYPE($'t$) $\lfloor\wedge\rfloor$ ($\lambda s.\ g\ vp$)
$g \vdash vp \rightarrow v \equiv$ ptr-seq vp TYPE($'t$) $[:\rightarrow]$ value-seq v $\lfloor\wedge\rfloor$ ($\lambda s.\ g\ vp$)

where ptr-seq p $T \equiv$ addr-seq (ptr-val p) (size-of T) and addr-seq is defined in Sect. 4. Using these predicates, we can now make separation logic assertions describing the presence of typed values on the heap, visible as contiguous in either physical or virtual memory. In the common case, i.e. when not modifying the page table, our model keeps the virtual memory mechanism under the hood. We can just state, for instance, $p \rightarrow (\!|\ x = 10;\ y = 7\ |\!)$ where the right hand side is an Isabelle record of class mem-type corresponding to a C struct and the left hand side is a virtual address.

7   Connecting with C

In this section, we will connect the framework to C and define loading and storing of typed values in the program state. In the previous section, we have enriched the usual C heap with additional information: slices for specifying the domain of predicates under separating conjunction and type annotation information. We therefore need to be careful to not introduce unwanted dependencies on the additional information in the state and we need to make sure that C updates operate consistently on the extended state. We formalise load and store for virtually addressed access. Direct physical access would be similar, but simpler. In C, loading and storing are total functions. Loading from a wrongly typed or unmapped address or storing to it will produce garbage. For our intended application (seL4), we do not need to model page faults directly, but we annotate the C program with guards that make sure no page faults will occur. These annotations are added automatically during the translation into Isabelle/HOL and will produce proof obligations. Should a page-fault model be required for different applications, it is easy to add: an access to an unmapped page, instead of a guard, simply produces a branch to the page fault handler.

For a generic map h from pointers p to values, loading a mem-type value at p is merely loading its size's worth of sequential bytes starting at p (load-list-basic), making sure h contains no gaps in that range (deoption-list) and passing it to from-bytes from the type class interface.

load-list-basic h 0 p = []
load-list-basic h (Suc n) p = h $p \cdot$load-list-basic h n $(p + 1)$

deoption-list xs ≡ if None ∈ set xs then None else ⌊map the xs⌋
load-list h n p ≡ deoption-list (load-list-basic h n p)
load-value h p ≡ option-map from-bytes (load-list h (size-of TYPE($'t$)) p)

A pointer access in C is then just an application of load-value to the address-space view of memory, ignoring any read failures. We drop the additional type information that is only used in assertions, not in C, resulting in the heap type load-value expects. The as-view function is similar to h-view, but uses ptable-lift to arrive at a map from virtual addresses to values.

load-value-c s vp ≡ the (load-value (as-view s) (ptr-val vp))

As mentioned above, this function is total. The guard generated for each such access is c-guard ⊢ vp ↪ –, ensuring that the load-value-c will produce a valid result. The predicate c-guard p ensures that p is not Null and is correctly aligned for its type size.

Heap updates are similar. For a single physical address, we update all slices at that address and we leave the type annotation untouched. We can ignore entries with None, because, again, the generated guard c-guard ⊢ vp ↪ – will ensure this case does not occur. We then lift the single-byte update first to the virtual layer to provide address translation via vmap-view, and then like in Tuch et al to byte sequences to accomodate structured types.

```
void mapUserFrame(pde_t *pd, paddr_t paddr, vptr_t vptr) {
    pde_t *pdSlot; pte_t *ptSlot, *pt, pte; unsigned int ptIndex;
    pdSlot = lookupPDSlot(pd, vptr);
    ptIndex = ((unsigned int)vptr >> ARMSmallPageBits) & MASK(PT_BITS);
    pt = ptrFromPAddr(pde_coarse_ptr_get_address(pdSlot));
    ptSlot = pt + ptIndex;
    pte = pte_small_new(paddr,1,0,0,0,3,1,1,1,0);
    *ptSlot = pte;
}
```

Fig. 5. Page table code extracted from seL4

tfheap-update tfh p v $\equiv$
$\lambda ppv.$ if fst $ppv = p$ then option-map $(\lambda(td, v').\ (td, v))$ $(\mathit{tfh}\ ppv)$
        else $\mathit{tfh}\ ppv$

state-update-v s vp v $\equiv$
case vmap-view s vp of None $\Rightarrow s \mid \lfloor p \rfloor \Rightarrow$ (tfheap-update (fst s) p v, snd s)

state-update-v-list s $[\ ] = s$
state-update-v-list s $((vp, v) \cdot us) =$ state-update-v-list (state-update-v s vp v) us

c-state-update vp v s $\equiv$ state-update-v-list s (zip (ptr-seq vp TYPE$('a1)$) (to-bytes v))

For interfacing to C code, we have adapted the C parser of Tuch et al [24]. It translates a significant subset of the C99 programming language into SIMPL [19], a generic, imperative language framework in Isabelle/HOL.

As in the framework by Tuch et al we cannot prove the frame rule generically, but we can prove it automatically for each individual program. This automatic proof ultimately reduces everything to valid memory accesses and updates, based on the following rule:

$$\frac{(\text{c-guard} \vdash vp \rightarrow - \wedge^* P)\ s}{(\text{c-guard} \vdash vp \rightarrow v \wedge^* P)\ (\text{c-state-update}\ vp\ v\ s)}$$

With $P = $ sep-true, this rule becomes the state update rule by Tuch et al, corresponding to the assignment axiom in standard separation logic. The corresponding rule for memory access holds as well, of course:

$$\frac{(g \vdash vp \rightarrow v)\ s}{\text{load-value-c}\ s\ vp = v}$$

Fig. 5 shows an excerpt of typical page table manipulation code that this framework can handle. The last line of this code, for instance, would be translated into the following SIMPL statement with guard:

Guard C-Guard $\{$c-guard $'ptSlot\}$
          ($'globals :== $ heap-upd (c-state-update $'ptSlot$ $'pte$))

The heap-upd function updates the C heap (our extended state) which is merely a global variable in the semantics of the C program. The guard statement Guard throws the guard error C-Guard if the condition $\{$c-guard $'ptSlot\}$ is false, and otherwise executes the statement. In previous work [16], we have conducted a detailed case study demonstrating how page table manipulations can be verified in this framework for a simple, one-level page table. Reasoning on the C and ARM level has precisely the same structure, it just involves more detail.

8   Translation Caching

Page table lookups are expensive; they potentially involve multiple memory reads. To decrease this cost, these lookups are cached in most architectures in a translation lookaside buffer (TLB). Abstractly, the TLB can be seen as a finite, small set of virtual-to-physical mappings. They may include lookups for code instructions as well as data. It is architecture-dependent whether these are handled separately from each other or not, how large the TLBs are, and when a mapping is removed from the TLB and replaced by another. Most architectures provide assembler instructions for explicitly removing all or specific mappings from the TLB, which is called *flushing*.

Although the page table should ultimately define what a mapping is, the hardware will always first consult the TLB and ignore the contents of the page table if a TLB entry is found. When we change the page table and the TLB contains the mapping being changed, we may introduce an inconsistency. This inconsistency can be resolved by flushing the TLB such that the new page table contents will be loaded for future lookups. However, indiscriminate TLB flushes are expensive, because they will incur additional memory reads. Kernel programmers like to optimise by deferring TLB flushes as far as possible and by making them as specific as possibly.

In our model, we can add the TLB by reducing it to its safety-relevant content: whether the lookup for any specific virtual address may be inconsistent or not. What makes a TLB entry inconsistent is a change to the page table. We can turn this view around and instead keep track of inconsistent page table entries — those that have been written to since the last flush. We can reduce machinery by not caring whether a memory location currently is a page table entry or not, we just keep track of *all* locations that have been changed since the last TLB flush. If any memory read or write involves a page table entry whose location is in this set, the TLB might be inconsistent for this lookup. We can now generate guards that test for this case and require us to prove its absence.

This TLB model intergrates nicely with separating conjunction, because the set mentioned above can be implemented as an additional boolean next to the type information on the right-hand side of the maps-to arrow. Apart from the type, none of the generic framework definitions would need to change.

9   Related Work

Our work touches three main areas: separation logic, virtual memory, and C verification. For an overview on OS verification in general, see Klein [14].

Separation logic was originally conceived by O'Hearn and Reynolds et al. [12,18] and has been formalised in mechanised theorem proving systems before [25,1]. We enhance these models with the ability to reason about properties on virtual memory, adding new basic predicates, but preserving the feel and reasoning principles of separation logic.

Virtual memory formalisations have appeared in the context of OS kernel verification before [15,7,11]. Reasoning about programs *running* under virtual memory, however, especially the operating systems which control it, remains mostly unexplored. Among the exceptions is the development of the Nova micro-hypervisor [20,21]. Like our work, the Nova developers aim to use a single semantics to describe all forms of memory access which simplifies significantly in the well-behaved case. They focus on reasoning about "plain memory" in which no virtual aliasing occurs and split it into read-only and read-write regions, to permit reading the page table while in plain memory. They do not use separation logic. Our work is more abstract. We do not explicitly define "plain memory". Rather the concept emerges from the requirements and state. Tews et al also include memory-mapped devices. The necessary alignment restrictions would intergrate seamlessly into our framework via the guard mechanism. Alkassar et al. [2] have proved the correctness of a kernel page fault handler, albeit not at the separation logic level. They use a single level page table and prove that the page fault handler establishes the illusion to the user of a plain memory abstraction, swapping in pages from disk as required. We instantiate our model to an extensive, realistic model of ARMv6 2-level page tables. We are not aware of other realistic formalisations of ARM page tables; Fox [10] formalises the ARM instruction set, but omits memory translation, while Tews et al [21] formalise memory translation for IA32.

In the C verification space, we build directly on the work by Tuch et al [24,22,23] who employ separation logic to reasoning in a precise, foundational model for C memory with Isabelle/HOL infrastructure to reason about low-level, potentially type-unsafe C programs nicely and abstractly. This framework which in turn builds on Schirmer's SIMPL environment [19] is used in the verification of the seL4 microkernel [8]. We enhance the fidelity of the framework with a virtual memory layer for ARMv6 while inheriting its nice type-lifting and reasoning principles. Other work in C verification includes Key-C [17], VCC [6], and Caduceus [9]. Key-C treats only on a type-safe subset of C. VCC, which also supports concurrency, uses a memory model [6] that axiomatises a weaker version of what Tuch proves [23] and what we extend to virtual memory. Caduceus supports a large subset of C, but does not include virtual memory.

10   Conclusion and Future Work

We have presented an abstract framework for separation logic under virtual memory and have instantiated it to the C programming language as well as to ARMv6 page tables. We have shown in previous work that this framework supports one-level page tables as well as traditional separation logic reasoning, including the frame rule. We have shown here that the new instantiation supports the same basic rules for heap updates that Tuch et al provide for their C verification framework that is used in the verification of the seL4 microkernel.

Next to applying the framework to seL4 page table code in a verification case study, future work includes an Isabelle/HOL model for the translation caching mechanism that is an interesting and correctness-relevant part of most virtual memory architectures. We have sketched how the mechanism could be added to the presented model without fundamental changes. We are not aware of any other virtual memory frameworks that include TLB modelling.

The framework presented here makes the foundational verification of OS-level C code practical. It brings a significant source of errors into the realm for formal, machine-checked verification that otherwise formally verified code would ignore and fail on embarrassingly. Only when reasoning about page table modifications directly, the complexities of their encoding become visible. For reasoning on plain memory, no additional verification overhead must be paid.

Acknowledgements. We thank Thomas Sewell for commenting on a draft of this paper and Michael Norrish for help with integrating the C parser.

References

1. Affeldt, R., Marti, N.: Separation logic in Coq (2008), http://savannah.nongnu.org/projects/seplog
2. Alkassar, E., Schirmer, N., Starostin, A.: Formal pervasive verification of a paging mechanism. In: Ramakrishnan, C.R., Rehof, J. (eds.) TACAS 2008. LNCS, vol. 4963, pp. 109–123. Springer, Heidelberg (2008)
3. ARM Limited. ARM Architecture Reference Manual (June 2000)
4. Bornat, R., Calcagno, C., O'Hearn, P., Parkinson, M.: Permission accounting in separation logic. In: Proc. 32nd POPL, pp. 259–270. ACM, New York (2005)
5. Calcagno, C., O'Hearn, P.W., Yang, H.: Local action and abstract separation logic. In: Proc. 22nd LICS, pp. 366–378. IEEE Computer Society, Los Alamitos (2007)
6. Cohen, E., Moskał, M., Schulte, W., Tobies, S.: A precise yet efficient memory model for C (2008), http://research.microsoft.com/apps/pubs/default.aspx?id=77174
7. Dalinger, I., Hillebrand, M.A., Paul, W.J.: On the verification of memory management mechanisms. In: Borrione, D., Paul, W.J. (eds.) CHARME 2005. LNCS, vol. 3725, pp. 301–316. Springer, Heidelberg (2005)
8. Elphinstone, K., Klein, G., Derrin, P., Roscoe, T., Heiser, G.: Towards a practical, verified kernel. In: Proc. 11th HOTOS, pp. 117–122 (2007)
9. Filliâtre, J.-C., Marché, C.: Multi-prover verification of C programs. In: Davies, J., Schulte, W., Barnett, M. (eds.) ICFEM 2004. LNCS, vol. 3308, pp. 15–29. Springer, Heidelberg (2004)
10. Fox, A.: Formal specification and verification of ARM6. In: Basin, D., Wolff, B. (eds.) TPHOLs 2003. LNCS, vol. 2758, pp. 25–40. Springer, Heidelberg (2003)
11. Hillebrand, M.: Address Spaces and Virtual Memory: Specification, Implementation, and Correctness. PhD thesis, Saarland University, Saarbrücken (2005)
12. Ishtiaq, S.S., O'Hearn, P.W.: BI as an assertion language for mutable data structures. In: Proc. 28th POPL, pp. 14–26. ACM, New York (2001)
13. Programming languages—C, ISO/IEC 9899:1999 (1999)
14. Klein, G.: Operating system verification—An overview. Sādhanā 34(1), 27–69 (2009)
15. Klein, G., Tuch, H.: Towards verified virtual memory in L4. In: Slind, K. (ed.) TPHOLs Emerging Trends 2004, Park City, Utah, USA (2004)

16. Kolanski, R., Klein, G.: Mapped separation logic. In: Shankar, N., Woodcock, J. (eds.) VSTTE 2008. LNCS, vol. 5295, pp. 15–29. Springer, Heidelberg (2008)
17. Mürk, O., Larsson, D., Hähnle, R.: KeY-C: A tool for verification of C programs. In: Pfenning, F. (ed.) CADE 2007. LNCS, vol. 4603, pp. 385–390. Springer, Heidelberg (2007)
18. Reynolds, J.C.: Separation logic: A logic for shared mutable data structures. In: Proc. 17th IEEE Symposium on Logic in Computer Science, pp. 55–74 (2002)
19. Schirmer, N.: Verification of Sequential Imperative Programs in Isabelle/HOL. PhD thesis, Technische Universität München (2006)
20. Tews, H.: Formal methods in the Robin project: Specification and verification of the Nova microhypervisor. In: C/C++ Verification Workshop, Technical Report ICIS-R07015, Oxford, UK, July 2007, pp. 59–68. Radboud University Nijmegen (2007)
21. Tews, H., Weber, T., Völp, M.: Formal memory models for the verification of low-level operating-system code. JAR 42(2–4), 189–227 (2009)
22. Tuch, H.: Formal Memory Models for Verifying C Systems Code. PhD thesis, School Comp. Sci. & Engin., University NSW, Sydney 2052, Australia (August 2008)
23. Tuch, H.: Formal verification of C systems code: Structured types, separation logic and theorem proving. JAR 42(2–4), 125–187 (2009)
24. Tuch, H., Klein, G., Norrish, M.: Types, bytes, and separation logic. In: Hofmann, M., Felleisen, M. (eds.) POPL 2007, pp. 97–108. ACM, New York (2007)
25. Weber, T.: Towards mechanized program verification with separation logic. In: Marcinkowski, J., Tarlecki, A. (eds.) CSL 2004. LNCS, vol. 3210, pp. 250–264. Springer, Heidelberg (2004)

Acyclic Preferences and Existence of Sequential Nash Equilibria: A Formal and Constructive Equivalence

Stéphane Le Roux[*,**]

LIX, École Polytechnique, CEA, CNRS, INRIA

Abstract. In a game from game theory, a Nash equilibrium (NE) is a combination of one strategy per agent such that no agent can increase its payoff by unilaterally changing its strategy. Kuhn proved that all (tree-like) sequential games have NE. Osborne and Rubinstein abstracted over these games and Kuhn's result: they proved a sufficient condition on agents' *preferences* for all games to have NE. This paper proves a *necessary and sufficient condition*, thus accounting for the game-theoretic frameworks that were left aside. *The proof is formalised* using Coq, and contrary to usual game theory it adopts an inductive approach to trees for definitions and proofs. By rephrasing a few game-theoretic concepts, by ignoring useless ones, and by characterising the proof-theoretic strength of Kuhn's/Osborne and Rubinstein's development, this paper also *clarifies* sequential game theory. The introduction sketches these clarifications, while the rest of the paper details the formalisation.

Keywords: Coq, induction, sequential game theory, abstraction, effective generalisation.

1 Introduction

In *game theory* a few classes of games, together with related concepts, can model a wide range of real-world competitive interactions between *agents*. Game theory is applied to economics, biology, computer science, political science, *etc.*

Sequential games (*a.k.a.* games in extensive form) are a widely studied class of games. They may help model games where agents play in turn, such as Chess in [22]. Given an arbitrary set of *outcomes*, an (abstract) sequential game is a finite rooted tree where each internal node is owned by an agent and each leaf encloses an outcome. The left-hand game below involves agents a and b and outcomes oc_1, oc_2 and oc_3.

---

[*] http://www.lix.polytechnique.fr/Labo/Stephane.Leroux
[**] Anonymous referees, especially one of them, made very constructive comments.

S. Berghofer et al. (Eds.): TPHOLs 2009, LNCS 5674, pp. 293–309, 2009.
© Springer-Verlag Berlin Heidelberg 2009

Informally, a play starts at the root. If the root is a leaf, the play ends with the enclosed outcome; otherwise the root owner chooses in which child, *i.e.* subgame, the play continues. However, the concept of play is not needed. A *strategy profile* (profile for short) is a game where each internal node has chosen a child. Choices of a profile induce a unique path from the root to a leaf, which induces a unique outcome. The right-hand strategy profile above, where choices are represented by double lines, induces outcome oc_3. The left-hand profile below induces oc_1.

An agent can *convert* a profile into another one by changing its own nodes' choices. For instance agent a can convert the right-hand profile above into the left-hand one below; and below, agent b can convert the left-hand one into the right-hand one. Note that for each agent, convertibility is an equivalence relation.

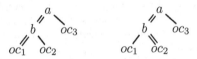

To each agent, an arbitrary binary relation over outcomes is given. It is called the agent's *preference* and it induces a relation over profiles *via* their induced outcomes. Generally speaking, a *Nash equilibrium* (NE for short) is a situation, *i.e.* a profile in the present case, that makes every agent *happy*, where an agent is happy if it cannot convert the situation into another situation that it prefers. This concept is defined in [13] and [14], and it captures the notion of NE in different types of games.

Traditional game theory involves *real-valued payoff functions* instead of abstract outcomes, *i.e.* functions mapping agents to real numbers. Agents implicitly prefer greater payoffs. The profiles below involve payoff functions where the first figure relates to agent a. The first profile is not a NE since agent a is not happy: if it played according to the profile, it would get 2, whereas by changing its choice from right to left it converts the profile into the right-hand one yielding payoff 3. The second profile below is a NE: by changing its choice agent a gets payoff 1 instead of 2 (so it is happy with the current profile), and b has no influence on the induced outcome so it is happy too. The third profile below is also a NE.

Subgame perfect equilibria [18] (SPE for short) are NE each of whose subgame is also an SPE. The second profile above is not an SPE because the subgame whose root is owned by agent b is not a NE. The third profile above is an SPE.

Kuhn [9] proved that all sequential games involving real-valued payoff functions have NE. His proof uses a recursive procedure, called *backward induction* in game theory, to build from each game an SPE (also NE). The backward induction on the left-hand game below starts by letting agent b (more generally agents at nodes closer to the leaves) maximise its payoff, in the middle picture. Then agent a (more generally agents at nodes closer to the root) maximises its payoff according to what has been chosen by b (more generally in all the subtrees).

$$
\begin{array}{ccc}
\overset{a}{\underset{\displaystyle b}{\diagup}}\;\diagdown\;2,2 & \overset{a}{\underset{\displaystyle b}{\diagup}}\;\diagdown\;2,2 & \overset{a}{\underset{\displaystyle b}{\diagup}}\;\diagdown\;2,2 \\[2pt]
1,0\quad 3,1 & 1,0\quad 3,1 & 1,0\quad 3,1
\end{array}
$$

A special case of Kuhn's theorem was formalised by Vestergaard [21] using *Coq* [2]: in this case, games are binary, *i.e.* each internal node has two choices, and payoffs are natural numbers. Instead of the usual game-theoretic approach to trees (seen as connected, acyclic graphs), Vestergaard adopts an inductive approach: a leaf is a tree, a node with two tree children is a tree. This approach is supported by Coq and is very convenient for formalisation. Also, profiles are defined directly, which shows that the traditional notion of strategy is not needed as far as [21] is concerned. See [11] for another formalisation in game theory.

Osborne and Rubinstein [16] abstracted Kuhn's proof and result. Translated in the formalism used in this paper, their result states that if agent preferences are all strict weak orders, then all sequential games with abstract outcomes have NE/SPE. (Two remarks: first, a strict weak order is equivalently an asymmetric relation whose negation is transitive, a partial order whose non-comparability is an equivalence relation, or the negation of a total preorder; second, this paper calls preference the inverse of the negation of what Osborne and Rubinstein called preference, so their result actually referred to a total preorder.)

Unfortunately they [16] do not account for, *e.g.*, *multi-criteria* games. These games involve real-valued vector payoffs to express that agents think *w.r.t.* several incommensurable dimensions, as in [19] and [3]. Agents still prefer greater payoffs, so vector $[1,3]$ is better than $[1,2]$ since $1 \geq 1$ and $3 > 2$, but $[0,3]$ and $[1,2]$ are non comparable since $0 < 1$ but $3 > 2$. The left-hand multi-criteria profile below corresponds to a backward induction since at each stage a maximal vector is chosen: $[0,3] \not< [2,2]$ and $[0,3] \not< [1,1]$. This profile induces payoff $[1,1]$ to agent a who can convert it by changing its two choices and get $[2,2] > [1,1]$: this backward induction did not yield a NE. More generally, with abstract outcomes, if agent a prefers z to x but cannot compare y with either of them, then the right-hand profile below is a backward induction but not a NE.

The following proof-theoretic characterisation of Kuhn's proof says that the result in [16] is the best possible while following Kuhn's proof structure, *i.e.* a mere backward induction. Since the preference used for multi-criteria games is not a strict weak order, this explains why the result in [16] does not account for multi-criteria games. The three propositions below are equivalent. (One proves by contraposition that the second one implies the first one, by building a game like the right-hand one above if an agent's preference is not a strict weak order.)

- The preference of each agent is a strict weak order.
- Backward induction always yields a Nash equilibrium.
- Backward induction always yields a subgame perfect equilibrium.

Nonetheless, Krieger [8] proved that every multi-criteria sequential game has a NE. His proof uses probabilities and *strategic games*, a class of games into which sequential games can be embedded. However, Krieger's result still does not account for all games with abstract outcomes and moreover his proof would not be easily formalised. Fortunately a generalisation of both [8] and [16] is given in [13] (ch. 4). Furthermore, instead of a mere sufficient condition on preferences like in [16], the following three propositions are proved equivalent.

1. The preference of each agent is acyclic.
2. Every sequential game has a Nash equilibrium.
3. Every sequential game has a subgame perfect equilibrium.

Existence of Nash equilibria in the traditional and mutlicriteria frameworks are direct corollaries of the theorem above. This is also true for other frameworks of interest as detailed in [13] (ch. 4).

The result above may be proved *via* three implications. 3) ⇒ 2) by definition. 2) ⇒ 1) is proved by contraposition: let an agent a prefer x_1 to x_0, x_2 to x_1, and so on, and x_0 to x_n. The game displayed below has no Nash equilibrium.

The main implication is 1) ⇒ 3). It may be proved as follows: first, since the preferences are acyclic, they can be linearly extended; second, following Kuhn's proof structure, for any game there exists an SPE *w.r.t.* the linear preferences; third, this SPE is also valid *w.r.t.* the *smaller* original preferences. This triple equivalence is formalised constructively in *Coq* (v8.1). In terms of proof burden, the main implication is still 1) ⇒ 3). However, the existence of a linear extension constitutes a substantial part[1] of the formal proof. In the second proof step described above one cannot merely follow Kuhn's proof structure: the definitions and proofs have to be completely rephrased (often inductively, as in [21]) and simplified in order to keep things practical and clear. Also, the formalisation is constructive: it yields an algorithm for computing equilibria, which is the main purpose of *algorithmic game theory* for various classes of games. Here in addition, the algorithm is certified since it was written in Coq.

This paragraph suggests that all the ingredients used in this generalisation were already well-known, although not yet put together: *Utility theory* prescribes embedding abstract outcomes and preferences into the real numbers and their usual total order, thus performing more than a linear extension, whereas the first proof step above is a mere linear extension; *Choice theory* uses abstract preferences and is aware of property preservation by preference inclusion, as

---

[1] The linear extension proof was also slightly modified to be part of the Coq-related CoLoR library [4].

invoked in the third proof step above. Also, [16] uses reflexive preference; Kreps [7] uses irreflexive ones; this paper assumes nothing on preferences but the way they relate to NE. Osborne and Rubinstein were most likely aware of the above-mentioned facts and techniques, but their totally preordered preferences seem utterly general and non-improvable at first glance. On contrary when considering the inverse of the negation of their preferences (*i.e.* a strict weak order), one sees that generality was just an illusion. Like natural languages, mathematical notations structure and drive our thoughts! All this suggests that in general, formalisation (and the related mindset) may not only provide a guarantee of correctness but also help build a deeper insight of the field being formalised.

Two alternative proofs of this paper's main result are mentioned below. Unlike the first proof, they cannot proceed by structural induction on trees, but strong induction on the number of internal nodes works well. One proof of 1) $\Rightarrow$ 2) is given in [13] (ch. 5). It uses only transitive closure instead of linear extension, but the proof technique suits only NE, not SPE. The (polymorphic) proof below works for both 1) $\Rightarrow$ 2) and 1) $\Rightarrow$ 3). Note that both alternative proofs show that the notion of SPE is not required to prove NE existence.

Proof. It suffices to prove it for strict weak orders. Assume a game g with $n+1$ internal nodes. (The 0 case is a leaf case.) Pick one whose children are all leaves. This node is the root of g_0 and is owned by a. Let x be an a-maximal outcome occurring in g_0. In g replace g_0 with a leaf enclosing x. This new game g' has n or less internal nodes, so there is a NE (resp. SPE) s for g'. In s replace the leaf enclosing x by a profile on g_0 where a chooses a leaf enclosing x. This yields a NE (resp. SPE) for g. (Consider happiness of agent a, then other agents.) $\quad\square$

This diversity of proofs not only proposes alternative viewpoints on the structure of sequential equilibria, but also constitutes a pool of reusable techniques for generalising the result, *e.g.*, in graphs as started in [13] (ch. 6 and 7). Nonetheless, only the first proof has been formalised.

Section 2 summarises the Coq proof for topological sorting, which was also published [12] as emerging trend in this conference. Section 3 deals with game theory, and is also meant for readers that are not too familiar with Coq.

2 Topological Sorting

The calculus of binary relations was developed by De Morgan around 1860. Then the notion of transitive closure of a binary relation (smallest transitive binary relation including a given binary relation) was defined in different manners by different people around 1890. See Pratt [17] for a historical account. In 1930, Szpilrajn [20] proved that, assuming the axiom of choice, any partial order has a linear extension, *i.e.*, is included in some total order. The proof invokes a notion close to transitive closure. In the late 1950's, The US Navy [1] designed PERT (Program Evaluation Research Task or Project Evaluation Review Techniques) for management and scheduling purposes. This tool partly consists in splitting a big project into small jobs on a chart and expressing with arrows when one job has to be done before another one can start up. In order to study the resulting directed graph, Jarnagin [15] introduced a finite and algorithmic version of Szpilrajn's result. This gave birth to the widely studied topological sorting, which spread to the industry in the early 1960's (see [10] and [5]). Some technical details and computer-oriented examples can be found in Knuth's book [6].

Section 2 summarises a few folklore results involving transitive closure and linear extension. No proof is given, but hopefully the definitions, statements, and explanations will help understand the overall structure of the development. This section requires a basic knowledge of Coq and its standard library.

In the remainder of this section, A is a *Set*; x and y have type A; R is a binary relation over A; l is a list over A; and n is a natural number. A finite "subset" of A is represented by any list involving all the elements of the subset. For the sake of readability, types will sometimes be omitted according to the above convention, even in formal statements where Coq could not infer them.

Proving constructively or computing properties about binary relations will require the following definitions about excluded middle and decidability.

Definition *eq_midex* $:= \forall \, x \, y, \, x=y \lor x\neq y$.
Definition *rel_midex* $R := \forall \, x \, y, \, R \, x \, y \lor \neg R \, x \, y$.
Definition *eq_dec* $:= \forall \, x \, y, \, \{x=y\}+\{x\neq y\}$.
Definition *rel_dec* $R := \forall \, x \, y, \, \{R \, x \, y\}+\{\neg R \, x \, y\}$.

In the Coq development similar results were proved for both excluded middle and decidability, but this section focuses on excluded middle. The main result of the section, which is invoked in section 3, says that given a middle-excluding relation (*rel_midex*), it is acyclic and equality on its domain is middle-excluding *iff* its restriction to any finite set has a middle-excluding irreflexive linear extension.

Section 2.1 gives basic new definitions about lists, relations, and finite restrictions, as well as part of the required lemmas; section 2.2 gives the definition of paths and relates it to transitive closure; section 2.3 designs increasingly complex functions leading to linear extension and the main result.

2.1   Lists, Relations, and Finite Restrictions

The lemma below will be used to extract simple, *i.e.* loop-free, paths from paths represented by lists. It says that if equality on A is middle-excluding and if an element occurs in a list over A, then the list can be decomposed into three parts: a list, one occurrence of the element, and a second list free of the element.

Lemma *In_elim_right* : *eq_midex* $\to \forall \, x \, l,$
$In \, x \, l \to \exists \, l', \exists \, l'', \, l=l'++(x::l'') \land \neg In \, x \, l''$.

The predicate *repeat_free* in *Prop* says that no element occurs more than once in a list. It is defined by recursion and used in the lemma below that will help prove that a simple path is not longer than the path from which it is extracted.

Lemma *repeat_free_incl_length* : *eq_midex* → ∀ *l l'*,
repeat_free l → *incl l l'* → *length l*≤*length l'*.

The notion of subrelation is defined below, and the lemma that follows states that a transitive relation contains its own transitive closure. (And conversely.)

Definition *sub_rel R R'* : *Prop* := ∀ *x y*, *R x y* → *R' x y*.
Lemma *transitive_sub_rel_clos_trans* : ∀ *R*,
transitive R → *sub_rel* (*clos_trans R*) *R*.

As defined below, a relation is irreflexive if no element relates to itself. So irreflexivity of a relation implies irreflexivity of its subrelations, as stated below. Together with transitive closure, irreflexivity will help state absence of cycle.

Definition *irreflexive R* : *Prop* := ∀ *x*, ¬*R x x*.
Lemma *irreflexive_preserved* : ∀ *R R'*,
sub_rel R R' → *irreflexive R'* → *irreflexive R*.

The restriction of a relation to a finite set/list is defined below. Then the predicate *is_restricted* says that the support of a relation *R* is included in the list *l*. The lemma that follows states that transitive closure preserves restriction to a given list. This will help compute finite linear extensions of finite relations.

Definition *restriction R l x y* : *Prop* := *In x l* ∧ *In y l* ∧ *R x y*.
Definition *is_restricted R l* : *Prop* := ∀ *x y*, *R x y* → *In x l* ∧ *In y l*.
Lemma *restricted_clos_trans* : ∀ *R l*,
is_restricted R l → *is_restricted* (*clos_trans R*) *l*.

2.2 Paths

Transitive closure will help guarantee acyclicity and build linear extensions. But its definition is not convenient if a witness is needed. A path is a witness, *i.e.* a list recording consecutive steps of a given relation. It is formally defined below. (Note that if relations were decidable, *is_path* could return a Boolean.) The next two lemmas state an equivalence between paths and transitive closure.

Fixpoint *is_path R x y l* {*struct l*} : *Prop* :=
match l with
— *nil* ⇒ *R x y*
— *z::l'* ⇒ *R x z* ∧ *is_path R z y l'*
end.

Lemma *clos_trans_path* : ∀ *x y*, *clos_trans R x y* → ∃ *l*, *is_path R x y l*.
Lemma *path_clos_trans* : ∀ *y l x*, *is_path R x y l* → *clos_trans R x y*.

The next lemma states that a path can be transformed into a simple path. It will help bound the computation of transitive closure for finite relations.

Lemma *path_repeat_free_length* : *eq_midex* → ∀ *y l x*, *is_path R x y l* →
∃ *l'*, ¬*In x l'* ∧ ¬*In y l'* ∧ *repeat_free l'* ∧
length l'≤ *length l* ∧ *incl l' l* ∧ *is_path R x y l'*.

The predicate *bounded_path R n* below says whether two given elements are related by a path of length n or less. It is intended to abstract over the list witness of a path while bounding its length, which transitive closure cannot do.

Inductive *bounded_path R n* : $A \to A \to Prop$:=
— *bp_intro* : $\forall\ x\ y\ l,\ length\ l \leq n \to is\_path\ R\ x\ y\ l \to bounded\_path\ R\ n\ x\ y.$

Below, two lemmas show that *bounded_path* is weaker than *clos_trans* in general, but equivalent on finite sets for some bound.

Lemma *bounded_path_clos_trans* : $\forall\ R\ n,$
sub_rel (*bounded_path R n*) (*clos_trans R*).
Lemma *clos_trans_bounded_path* : $eq\_midex \to \forall\ R\ l,$
is_restricted R l $\to$ *sub_rel* (*clos_trans R*) (*bounded_path R (length l)*) .

The first lemma below states that if a finite relation is excluding middle, so are its "bounded transitive closures". Thanks to this and the lemma above, the second lemma states that it also holds for the transitive closure.

Lemma *bounded_path_midex* : $\forall\ R\ l\ n,$
is_restricted R l $\to$ *rel_midex R* $\to$ *rel_midex* (*bounded_path R n*).
Lemma *restricted_midex_clos_trans_midex* : $eq\_midex \to \forall\ R\ l,$
rel_midex R $\to$ *is_restricted R l* $\to$ *rel_midex* (*clos_trans R*).

The following theorems state the equivalence between decidability of a relation and uniform decidability of the transitive closures of its finite restrictions. Note that decidable equality is required only for the second implication. These results remain correct when considering excluded middle instead of decidability.

Theorem *clos_trans_restriction_dec_R_dec* : $\forall\ R$
$(\forall\ l,\ rel\_dec\ (clos\_trans\ (restriction\ R\ l))) \to rel\_dec\ R.$
Theorem *R_dec_clos_trans_restriction_dec* : $eq\_dec \to \forall\ R$
rel_dec R $\to \forall\ l,\ rel\_dec\ (clos\_trans\ (restriction\ R\ l)).$

2.3   Linear Extension

This section presents a way of extending linearly an acyclic finite relation. (This is not the fastest topological sort algorithm though.) The intuitive idea is to repeat the following while it is possible: take the transitive closure of the relation and add an arc to the relation without creating 2-step cycles. Repetition ensures saturation, absence of 2-step cycle ensures acyclicity, and finiteness ensures termination. In this section this idea is implemented through several stages of increasing complexity. This not only helps describe the procedure clearly, but it also facilitate the proof of correctness by splitting it into intermediate results.

Total relations will help define linear extensions. They are defined below.

Definition *trichotomy R x y* : $Prop := R\ x\ y \vee x = y \vee R\ y\ x.$
Definition *total R l* : $Prop := \forall\ x\ y,\ In\ x\ l \to In\ y\ l \to trichotomy\ R\ x\ y.$

The definition below adds an arc to a relation if not creating 2-step cycles.

Inductive $try\_add\_arc\ R\ x\ y : A \to A \to Prop :=$
— $keep : \forall\ z\ t,\ R\ z\ t \to try\_add\_arc\ R\ x\ y\ z\ t$
— $try\_add : x{\neq}y \to \neg R\ y\ x \to try\_add\_arc\ R\ x\ y\ x\ y.$

As stated below, try_add_arc creates no cycle in strict partial orders.

Lemma $try\_add\_arc\_irrefl : eq\_midex \to \forall\ R\ x\ y,$
$transitive\ R \to irreflexive\ R \to irreflexive\ (clos\_trans\ (try\_add\_arc\ R\ x\ y)).$

While creating no cycle, the function below alternately performs a transitive closure and tries to add arcs starting at a given point and ending in a given list.

Fixpoint $try\_add\_arc\_one\_to\_many\ R\ x\ l\ \{struct\ l\} : A \to A \to Prop :=$
$match\ l\ with$
— $nil \Rightarrow R$
— $y{::}l' \Rightarrow clos\_trans\ (try\_add\_arc\ (try\_add\_arc\_one\_to\_many\ R\ x\ l')\ x\ y)$
$end.$

Preserving acyclicity, the function below alternately performs a transitive closure and tries to add all arcs starting in the first list and ending in the second.

Fixpoint $try\_add\_arc\_many\_to\_many\ R\ l'\ l\ \{struct\ l'\} : A \to A \to Prop :=$
$match\ l'\ with$
— $nil \Rightarrow R$
— $x{::}l'' \Rightarrow try\_add\_arc\_one\_to\_many\ (try\_add\_arc\_many\_to\_many\ R\ l''\ l)\ x\ l$
$end.$

While preserving acyclicity, the next function tries to add all arcs both starting and ending in a list to the restriction of the relation to that list. This function is the one that was meant informally in the beginning of section 2.3.

Definition $LETS\ R\ l : A \to A \to Prop :=$
$try\_add\_arc\_many\_to\_many\ (clos\_trans\ (restriction\ R\ l))\ l\ l.$

The five lemmas below will help state that $LETS$ constructs a middle-excluding linear extension. They rely on similar intermediate results for the intermediate functions above, but these intermediate results are not displayed.

Lemma $LETS\_transitive : \forall\ R\ l,\ transitive\ (LETS\ R\ l).$
Lemma $LETS\_restricted : \forall\ R\ l,\ is\_restricted\ (LETS\ R\ l)\ l.$
Lemma $LETS\_irrefl : eq\_midex \to \forall\ R\ l,$
$(irreflexive\ (clos\_trans\ (restriction\ R\ l)) \leftrightarrow irreflexive\ (LETS\ R\ l)).$
Lemma $LETS\_total : eq\_midex \to \forall\ R\ l,\ rel\_midex\ R \to total\ (LETS\ R\ l)\ l.$
Lemma $LETS\_midex : eq\_midex{\to}\forall\ R\ l,\ rel\_midex\ R \to rel\_midex\ (LETS\ R\ l).$

Below, a linear extension (over a list) of a relation is a strict total order (over the list) that is bigger than the original relation (restricted to the list).

Definition $linear\_extension\ R\ l\ R' := is\_restricted\ R'\ l\ \wedge$
$sub\_rel\ (restriction\ R\ l)\ R' \wedge transitive\ A\ R' \wedge irreflexive\ R' \wedge total\ R'\ l.$

Consider a middle-excluding relation. It is acyclic and equality is middle-excluding *iff* for any list there exists a decidable strict total order containing the original relation (on the list). The witness *R'* below is provided by *LETS*.

Theorem *linearly_extendable* : ∀ *R*, *rel_midex R* →
(*eq_midex* ∧ *irreflexive* (*clos_trans R*) ↔
∀ *l*, ∃ *R'*, *linear_extension R l R'* ∧ *rel_midex R'*).

3   Sequential Games

Section 3.1 presents three preliminary concepts and a lemma on lists and predicates; section 3.2 defines sequential games and strategy profiles; section 3.3 defines functions on games and profiles; section 3.4 defines the notions of preference, NE, and SPE; and section 3.5 shows that universal existence of these equilibria is equivalent to acyclicity of preferences.

3.1   Preliminaries

The function *listforall* expects a predicate on a *Set* called *A*, and returns a predicate on lists stating that all the elements in the list comply with the original predicate. This will help define, *e.g.*, the third concept below. It is recursively defined along the inductive structure of the list argument. It is typed as follows.

$$listforall : (A \rightarrow Prop) \rightarrow list\ A \rightarrow Prop$$

The function *rel_vector* expects a binary relation and two lists over *A* and states that the lists are element-wise related (which implies that they have the same length). This will help define the convertibility between two profiles. It is recursively defined along the first list argument and it is typed as follows.

$$rel\_vector : (A \rightarrow A \rightarrow Prop) \rightarrow list\ A \rightarrow list\ A \rightarrow Prop$$

Given a binary relation, the predicate *is_no_succ* returns a proposition saying that no element in a given list is the successor of a given element.

Definition *is_no_succ P x l* := *listforall* (*fun y* → ¬*P x y*) *l*.

The definition above help state lemma *Choose_and_split* that expects a decidable relation over *A* and a non-empty list over *A* and splits it into one element and two lists by choosing the first (from the head) element that is maximal among the remaining elements. The example below involves divisibility over the naturals.

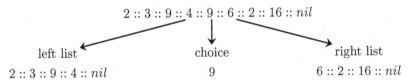

This lemma will help defined backward induction for arbitrary preferences. Note that if preferences are orders, it chooses a maximal element in the whole list.

3.2 Sequential Games and Strategy Profiles

Games and profiles are inductively defined in two steps. Graphically: if *oc* is an outcome, the left (resp. right)-hand object below is a game (resp. profile).

Below, if *a* is an agent, *g* a game, and *l* a list of games, the left-hand object is a game. If *l* is empty, *g* ensures that the internal node has as at least one child. If *s* is a profile and *r* and *t* are lists of profiles, the right-hand object is a profile where agent *a* at the root chooses the profile *s*.

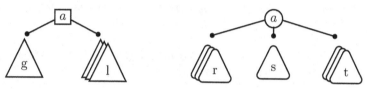

In Coq, given two sets *Outcome* and *Agent*, sequential games and strategy profiles are defined as below. (*gL* stands for game leaf and *gN* for game node.)

Variables (*Outcome* : *Set*)(*Agent* : *Set*).
Inductive *Game* : *Set* :=
| *gL* : *Outcome* → *Game*
| *gN* : *Agent* → *Game* → list *Game* → *Game*.
Inductive *Strat* : *Set* :=
| *sL* : *Outcome* → *Strat*
| *sN* : *Agent* → list *Strat* → *Strat* → list *Strat* → *Strat*.

The induction principle that Coq automatically associates to games (resp. profiles) ignores the inductive structure of lists. Mutually defining lists with games may solve this but rules out using the Coq standard library for these new lists. The principle stated below is built manually *via* a Fixpoint. There are four premises: two for the horizontal induction along empty lists and compound lists, and two for the vertical induction along leaf games and compound games.

Game_ind2 : ∀ (*P* : *Game* → *Prop*) (*Q* : *Game* → list *Game* → *Prop*),
(∀ *oc*, *P* (*gL oc*)) →
(∀ *g*, *P g* → *Q g nil*) →
(∀ *g*, *P g* → ∀ *g' l*, *Q g' l* → *Q g'* (*g* :: *l*)) →
(∀ *g l*, *Q g l* → ∀ *a*, *P* (*gN a g l*)) →
∀ *g*, *P g*

In order to prove a property ∀ *g* : *Game*, *Pg* with the induction principle *Game_ind2*, the user has to provide a predicate *Q* that is easily (yet apparently not automatically in general) derived from *P*. Also note that the induction principle for profiles requires one more premise since two lists are involved. These principles are invoked in most of the proofs of the Coq development.

3.3   Structural Definitions

This subsection presents structural definitions, *i.e.* not involving preferences, relating to games and strategy profiles.

Below, the function *UsedOutcomes* is defined recursively on the game structure: it expects a game and returns a list of the outcomes occurring in the game. It will help restrict the agent preferences to the relevant finite set prior to the topological sorting.

Fixpoint *UsedOutcomes* (g : *Game*) : *list Outcome* :=
match g with
| *gL oc* ⇒ *oc::nil*
| *gN _ g' l* ⇒ ((*fix ListUsedOutcomes* (l' : *list Game*) : *list Outcome* :=
            match l' with
            | *nil* ⇒ *nil*
            | *x::m* ⇒ (*UsedOutcomes x*)++(*ListUsedOutcomes m*)
            end) (*g'::l*))
end.

A profile induces a unique outcome, as computed by the 2-step rule ⤳ below. The intended function is also defined in Coq below, and it will help define preferences over profiles from preferences over outcomes.

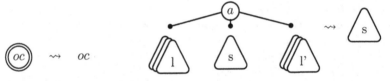

Fixpoint *InducedOutcome* (s : *Strat*) : *Outcome* :=
match s with
| *sL oc* ⇒ *oc*
| *sN a sl sc sr* ⇒ *InducedOutcome sc*
end.

The function *s2g* expects a profile and returns its underlying game by forgetting the nodes' choices. It will help state that given a game (and preferences), a well-chosen profile is a NE *for this game*. *s2g* is computed by the 3-step rule below. (The Coq definition is omitted.) Note that the two big steps are a case splitting along the structure of the first list, *i.e.*, whether it is empty or not.

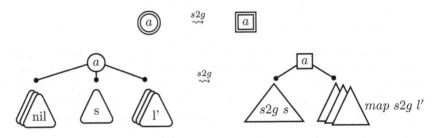

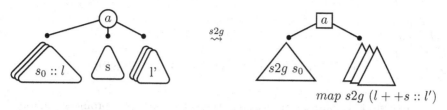

$$map \; s2g \; (l ++ s :: l')$$

Below, *Conv a s s'* means that agent *a* is able to convert *s* into *s'* by changing part of its choices. *ListConv* means component-wise convertibility of list of profiles and it is defined by mutual induction with *Conv*.

Inductive *Conv* : *Agent* → *Strat* → *Strat* → *Prop* :=
| *convLeaf* : ∀ *b oc*, *Conv b* (*sL oc*)(*sL oc*)
| *convNode* : ∀ *b a sl sl' sc sc' sr sr'*, (*length sl=length sl'* ∨ *a = b*) →
ListConv b (*sl*++(*sc::sr*)) (*sl'*++(*sc'::sr'*)) →
Conv b (*sN a sl sc sr*)(*sN a sl' sc' sr'*)

with

ListConv : *Agent* → *list Strat* → *list Strat* → *Prop* :=
| *lconvnil* : ∀ *b*, *ListConv b nil nil*
| *lconvcons* : ∀ *b s s' tl tl'*, *Conv b s s'* → *ListConv b tl tl'* →
ListConv b (*s::tl*)(*s'::tl'*).

Above, the subformula *length sl=length sl'* ∨ *a = b* ensures that only the owner of a node can change his choice at that node. *ListConv b* (*sl*++(*sc::sr*)) (*sl'*++(*sc'::sr'*)) ensures that this property holds also in the subprofiles.

A suitable 4-premise induction principle for convertibility is generated with the Coq command *Scheme*. This principle, which is not displayed, is invoked to prove that two convertible profiles have the same underlying game. Also, it is proved by induction on profiles that *Conv a* is an equivalence relation.

3.4 Concepts of Equilibrium

Below, *OcPref a* is the preference of agent *a*. It induces a preference over profiles.

Variable *OcPref* : *Agent* → *Outcome* → *Outcome* → *Prop*.
Definition *StratPref* (*a* : *Agent*)(*s s'* : *Strat*) : *Prop* :=
OcPref a (*InducedOutcome s*)(*InducedOutcome s'*).

Below, an agent is happy with a profile if it cannot convert it into a preferred profile. A Nash equilibrium (NE for short) is a profile making every agent happy.

Definition *Happy* (*s* : *Strat*)(*a* : *Agent*) : *Prop* := ∀ *s'*,
Conv a s s' → ¬*StratPref a s s'*.
Definition *Eq* (*s* : *Strat*) : *Prop* := ∀ *a*, *Happy s a*.

Instead of ¬*StratPref a s s'* above, Osborne and Rubinstein would have written *StratPref_OR a s' s*, hence the negated inversion relating the two notations.

A subgame perfect equilibrium (SPE) is a NE whose subprofiles are SPE.

Fixpoint *SPE* (*s* : *Strat*) : *Prop* := *Eq s* ∧
match s with
| *sL oc* ⇒ *True*
| *sN a sl sc sr* ⇒ (*listforall SPE sl*) ∧ *SPE sc* ∧ (*listforall SPE sr*)
end.

The following key lemma means that if the root owner of a profile *s* is happy and chooses a NE, *s* is a NE. (The converse also holds but is not needed here.)

Lemma *Eq_subEq_choice* : ∀ *a sl sc sr*,
(∀ *s s'*, *In s sl* ∨ *In s sr* → *Conv a s s'* → ¬*StratPref a sc s'*) →
Eq sc → *Eq* (*sN a sl sc sr*).

3.5   Existence of Equilibria

Assume that preferences over outcomes are decidable. This implies the same for preferences over profiles.

Hypothesis *OcPref_dec* : ∀ (*a* : *Agent*), *rel_dec* (*OcPref a*).
Lemma *StratPref_dec* : ∀ (*a* : *Agent*), *rel_dec* (*StratPref a*).

The backward induction function is defined below. Slight simplifications of the actual Coq code were made for readability purpose.

Fixpoint *BI* (*g* : *Game*) : *Strat* :=
match g with
| *gL oc* ⇒ *sL oc*
| *gN a g l* ⇒ *let* (*sl,sc,sr*):=
            Choose_and_split (*StratPref_dec a*) (*map BI l*) (*BI g*) *in*
            sN a sl sc sr
end.

It is proved (although not here) that *BI* preserves the underlying game.

Total order case. Preferences over outcomes are temporarily transitive and irreflexive. Therefore those properties also hold for preferences over profiles.

Hypothesis *OcPref_irrefl* : ∀ (*a* : *Agent*), *irreflexive* (*OcPref a*).
Hypothesis *OcPref_trans* : ∀ (*a* : *Agent*), *transitive* (*OcPref a*).
Lemma *StratPref_irrefl* : ∀ (*a* : *Agent*), *irreflexive* (*StratPref a*).
Lemma *StratPref_trans* : ∀ (*a* : *Agent*), *transitive* (*StratPref a*).

If preferences are total over a given list of outcomes, *BI* yields SPE for any sequential game using only outcomes from the list, as stated below.

Lemma *BI_SPE* : ∀ *loc* : *list Outcome*, (∀ *a* : *Agent*, *total* (*OcPref a*) *loc*) →
∀ *g* : *Game*, *incl* (*UsedOutcomes g*) *loc* → *SPE* (*BI g*).

The proof relies on the key analytical lemma *Eq_subEq_choice*. The same proof structure would work for strict weak orders and thus translates Kuhn's and

Osborne and Rubinstein's results into the abstract sequential game formalism of this paper. However, it is not even needed in the remainder.

General case. Until now equilibrium and related concepts have been defined *w.r.t.* implicit given preferences, but from now these concepts take arbitrary preferences as an explicit parameter. For instance, instead of writing *Eq s*, one shall write *Eq OcPref s* to say that *s* is a NE with respect to the family *OcPref*.

The following lemma says two things: first, the equilibrium-hood of a profile depends only on the restrictions of agent preferences to the outcomes that are used in (the underlying game of) the profile; second, removing arcs from agent preferences preserves equilibrium-hood (informally because less demanding agents are more likely to be happy). A similar result holds for SPE.

Lemma *Eq_order_inclusion* : $\forall$ *OcPref OcPref' s,*
($\forall$ *a, sub_rel* (*restriction* (*OcPref a*) (*UsedOutcomes* (*s2g s*))) (*OcPref' a*)) $\rightarrow$
Eq OcPref' s $\rightarrow$ *Eq OcPref s.*

The theorem below constitutes the main implication of the triple equivalence referred to in section 1. It generalises the results in [9], [16] and [8] to acyclic preferences. It invokes theorem *linearly_extendable* from section 2.3.

Theorem *acyclic_SPE* : $\forall$ *OcPref,* ($\forall$ *a, rel_dec* (*OcPref a*)) $\rightarrow$
($\forall$ *a, irreflexive* (*clos_trans Outcome* (*OcPref a*))) $\rightarrow$
  $\forall$ *g,* {*s : Strat* | *s2g s=g* $\land$ *SPE OcPref s*}.

4 Conclusion

This paper and the related Coq development have thus *abstracted, generalised, formalised,* and *clarified* existing game-theoretic results in [9], [16] and [8]: instead of real-valued (vector) payoffs they refer to abstract outcomes; instead of the usual order over the reals they refer to abstract preferences; instead of a mere sufficient condition on preferences they prove a necessary and sufficient condition; instead of a set-theoretic approach they use an inductive-type approach.

The difficulties that were encountered in this proof are of two sorts: first, the more general framework brings new issues, *e.g.* collecting the outcomes used in a game, topological sorting, defining backward induction for arbitrary preferences; second, the formal proof must cope with issues that were ignored in pen-and-paper proofs, *e.g.* rigorous definitions, associated proof principles, underlying game of a strategy profile. These second issues are already addressed in [21].

In the abstract-preference framework, it may sound tempting to prove *formally* the necessary and sufficient condition only for binary trees and then argue *informally* that the general case is similar or reducible to the binary case. Such an argument may contradict the idea that formalisation is a useful step towards guarantee of correctness and deeper insight: it ignores that general trees require more complex representations and induction principles and that some property holds for binary trees but not in general. Nonetheless the author believes that it would be possible to first prove the result for binary trees and then reduce *formally* general trees to binary trees. However this would require to define games, profiles, convertibility, equilibria, *etc.*, (but backward induction) for both the binary and the general settings, and it would still require the proof of topological sorting: so it would be more complex than the proof discussed in this paper.

In section 3.3 convertibility between strategy profiles was defined as an inductive type in *Prop*. This allows not even assuming decidability of agents' equality. Alternatively, assuming decidability of agents' equality would allow defining convertibility as a recursive function onto Booleans. More generally, if the Coq development were to be rephrased or generalised to graph structure instead of trees (as in [13], chap. 6 and 7), it might be interesting to discharge part of the proof burden on Coq's computational ability and keep in the proof script only the subtle reasoning. This may be achieved in part by using more recursive Boolean functions at the acceptable expense of decidability assumptions.

References

1. Anonymous. Program evaluation research task. Summary report Phase 1 and 2, U.S. Government Printing Office, Washington, D.C. (1958)
2. Berthot, Y., Castéran, P.: Interactive Theorem Proving and Program Development Coq'Art: The Calculus of Inductive Constructions. Springer, Heidelberg (2004)
3. Blackwell, D.: An analog of the minimax theorem for vector payoffs. Pacific Journal of Mathematics 6, 1–8 (1956)
4. Blanqui, F., Coupet-Grimal, S., Delobel, W., Hinderer, S., Koprowski, A.: CoLoR, a Coq Library on rewriting and termination. In: Workshop on Termination (2006)
5. Kahn, A.B.: Topological sorting of large networks. Commun. ACM 5(11), 558–562 (1962)
6. Knuth, D.E.: The Art of Computer Programming, 2nd edn., vol. 1. Addison Wesley, Reading (1973)
7. Kreps, D.M.: Notes on the Theory of Choice. Westview Press, Inc., Boulder (1988)
8. Krieger, T.: On pareto equilibria in vector-valued extensive form games. Mathematical Methods of Operations Research 58, 449–458 (2003)
9. Kuhn, H.W.: Extensive games and the problem of information. Contributions to the Theory of Games II (1953)
10. Lasser, D.J.: Topological ordering of a list of randomly-numbered elements of a network. Commun. ACM 4(4), 167–168 (1961)
11. Le Roux, S.: Non-determinism and Nash equilibria for sequential game over partial order. In: Computational Logic and Applications, CLA 2005. Discrete Mathematics & Theoretical Computer Science (2006)
12. Le Roux, S.: Acyclicity and finite linear extendability: a formal and constructive equivalence. In: Schneider, K., Brandt, J. (eds.) Theorem Proving in Higher Order Logics: Emerging Trends Proceedings, September 2007, pp. 154–169. Department of Computer Science, University of Kaiserslautern (2007)
13. Le Roux, S.: Generalisation and formalisation in game theory. Ph.d. thesis, Ecole Normale Supérieure de Lyon (January 2008)
14. Le Roux, S., Lescanne, P., Vestergaard, R.: A discrete Nash theorem with quadratic complexity and dynamic equilibria. Research report IS-RR-2006-006, JAIST (2006)

15. Jarnagin, M.P.: Automatic machine methods of testing pert networks for consistency. Technical Memorandum K-24/60, U. S. Naval Weapons Laboratory, Dahlgren, Va (1960)
16. Osborne, M.J., Rubinstein, A.: A Course in Game Theory. The MIT Press, Cambridge (1994)
17. Pratt, V.: Origins of the calculus of binary relations. In: Logic in Computer Science (1992)
18. Selten, R.: Spieltheoretische Behandlung eines Oligopolmodells mit Nachfrageträgheit. Zeitschrift für die desamte Staatswissenschaft 121 (1965)
19. Simon, H.A.: A behavioral model of rational choice. The Quarterly Journal of Economics 69(1), 99–118 (1955)
20. Szpilrajn, E.: Sur l'extension de l'ordre partiel. Fund. Math. 16 (1930)
21. Vestergaard, R.: A constructive approach to sequential Nash equilibria. Information Processing Letter 97, 46–51 (2006)
22. Zermelo, E.: Über eine Anwendung der Mengenlehre auf die Theorie des Schachspiels. In: Proceedings of the Fifth International Congress of Mathematicians, vol. 2 (1912)

Formalising FinFuns – Generating Code for Functions as Data from Isabelle/HOL

Andreas Lochbihler

Universität Karlsruhe (TH), Germany
lochbihl@ipd.info.uni-karlsruhe.de

Abstract. FinFuns are total functions that are constant except for a finite set of points, i.e. a generalisation of finite maps. We formalise them in Isabelle/HOL and present how to safely set up Isabelle's code generator such that operations like equality testing and quantification on Fin-Funs become executable. On the code output level, FinFuns are explicitly represented by constant functions and pointwise updates, similarly to associative lists. Inside the logic, they behave like ordinary functions with extensionality. Via the update/constant pattern, a recursion combinator and an induction rule for FinFuns allow for defining and reasoning about operators on FinFuns that directly become executable. We apply the approach to an executable formalisation of sets and use it for the semantics for a subset of concurrent Java.

1 Introduction

In recent years, executable formalisations, proofs by reflection [8] and automated generators for counter examples [1,5] have received much interest in the theorem proving community. All major state-of-the-art theorem provers like Coq, ACL2, PVS, HOL4 and Isabelle feature some interface to a standard (usually external) functional programming language to directly extract high-assurance code from theorems or proofs or both. Isabelle/HOL provides two code generators [3,6], which support datatypes and recursively defined functions, where Haftmann's [6] is supposed to replace Berghofer's [3]. Berghofer's, which is used to search for counter examples by default (*quickcheck*) [1], can also deal with inductively defined predicates, but not with type classes. Haftmann's additionally supports type classes and output in SML, OCaml and Haskell, but inductively defined predicates are not yet available and *quickcheck* is still experimental.

Beyond these areas, code generation is currently rather limited in Isabelle/HOL. Consequently, the everyday Isabelle user invokes the *quickcheck* facility on some conjecture and frequently encounters an error message such as "Unable to generate code for op = (λx. True)" or "No such mode [1, 2] for ...". Typically, such a message means that an assumption or conclusion involves a test on function equality (which underlies both universal and existential quantifiers) or an inductive predicate no code for which can be produced. In particular, the following restrictions curb *quickcheck*'s usefulness:

S. Berghofer et al. (Eds.): TPHOLs 2009, LNCS 5674, pp. 310–326, 2009.

- Equality on functions is only possible if the domain is finite and enumerable.
- Quantifiers are only executable if they are bounded by a finite set (e.g. $\forall x \in A.\ P\ x$).
- (Finite) sets are explicitly represented by lists, but as the set type has been merged with predicates in version Isabelle 2008, only Berghofer's code generator can work with sets properly.

The very same problems reoccur when provably correct code from a formalisation is to be extracted, although one is willing to commit more effort in adjusting the formalisation and setting up the code generator for it in that case. To apply *quickcheck* to their formalisations, end-users expect to supply little or no effort.

In the area of programming languages, states (like memories, stores, and thread pools) are usually finite, even though the identifiers (addresses, variable names, thread IDs, ...) are typically taken from an infinite pool. Such a state is most easily formalised as a (partial) function from identifiers to values. Hence, enumerating all threads or comparing two stores is not executable by default. Yet, a finite set of identifier-value pairs could easily store such state information, which is normally modified point-wisely. Explicitly using associative lists in one's formalisation, however, incurs a lot of work because one state has in general multiple representations and AC1 unification is not supported.

For such kind of data, we propose to use a new type FinFun of total functions that are constant except for finitely many points. They generalise maps, which formally are total functions of type $'a \Rightarrow 'b$ *option* that map to *None* ("undefined") almost everywhere, in two ways: First, they can replace (total) functions of arbitrary type $'a \Rightarrow 'b$. Second, their default value is not fixed to a predetermined value (like *None*). Our main technical contributions are:[1]

1. On the code level, every FinFun is represented as explicit data via two datatype constructors: constant FinFuns and pointwise update (cf. Sec. 2). *quickcheck* is set up for FinFuns and working.
2. Inside the logic, FinFuns feel very much like ordinary functions (e.g. extensionality: $f = g \longleftrightarrow (\forall x.\ f\ x = g\ x)$) and are thus easily integrated into existent formalisations. We demonstrate this in two applications (Sec. 5):
   (a) A formalisation of sets as FinFuns allows sets to be represented explicitly in the generated code.
   (b) We report on our experience in using FinFuns to represent state information for JinjaThreads [12], a semantics for a subset of concurrent Java.
3. Equality tests on, quantification over and other operators on FinFuns are all handled by Isabelle's new code generator (cf. Sec. 3).
4. All equations for code generation have passed through Isabelle's inference kernel, i.e., the trusted code base cannot be compromised by ad-hoc translations where constants in the logic are explicitly substituted by functions of the target language.
5. A recursion combinator allows to directly define functions that are recursive in an argument of FinFun type (Sec. 4).

---

[1] The FinFun formalisation is available in the Archive of Formal Proofs [13].

FinFuns are a rather restricted class of functions. To represent such functions as associative lists is common knowledge in computer science, but we focus on how to practically hide the problems that such representation issues raise during reasoning without losing the benefits of executability. In Sec. 6, we discuss which functions FinFuns can replace and which not, and compare the techniques and ideas we use with other applications. Isabelle-specific notation is defined in appendix A.

2 Type Definition and Basic Properties

To start with, we construct the new type $'a \Rightarrow_f \, 'b$ for FinFuns. This type contains all functions from $'a$ to $'b$ which map only finitely many points $a :: \, 'a$ to some value other than some constant $b :: \, 'b$, i.e. are constant except for finitely many points. We show that all elements of this type can be built from two constructors: The everywhere constant FinFun and pointwise update of a FinFun (Sec. 2.1). Code generated for operators on FinFuns will be recursive via these two kernel functions (cf. Sec. 2.2).

In Isabelle/HOL, a new type is declared by specifying a non-empty carrier set as a subset of an already existent type. The new type for FinFuns is isomorphic to the set of functions that deviate from a constant at only finitely many points:

typedef $('a,'b)$ *finfun* $= \{f::'a\Rightarrow'b \mid \exists b.\ finite\ \{a \mid f\ a \neq b\}\}$

Apart from the new type $('a,\ 'b)$ *finfun* (written $'a \Rightarrow_f \, 'b$), this introduces the set *finfun* :: $('a \Rightarrow \, 'b)$ *set* given on the right-hand side and the two bijection functions *Abs-finfun* and *Rep-finfun* between the sets *UNIV* :: $('a \Rightarrow_f \, 'b)$ *set* and *finfun* such that *Rep-finfun* is surjective and they are inverses of each other:

$$Rep\text{-}finfun\ \hat{f} \in finfun \tag{1}$$

$$Abs\text{-}finfun\ (Rep\text{-}finfun\ \hat{f}) = \hat{f} \tag{2}$$

$$f \in finfun \longrightarrow Rep\text{-}finfun\ (Abs\text{-}finfun\ f) = f \tag{3}$$

For clarity, we decorate all variable identifiers of FinFun type $'a \Rightarrow_f \, 'b$ with a hat $\hat{}$ to distinguish them from those of ordinary function type $'a \Rightarrow \, 'b$. Note that the default value b of the function, to which it does not map only finitely many points, is *not* stored in the type elements themselves. In case $'a$ is infinite, any such b is uniquely determined and would therefore be redundant. If not, $\{a \mid f\ a \neq b\}$ is finite for all $f::'a \Rightarrow \, 'b$ and $b::'b$, i.e. *finfun* = *UNIV*. Moreover, if that default value was fixed, then equality on $'a \Rightarrow_f \, 'b$ would not be as expected, cf. (5).

The function *finfun-default* $\hat{f}$ returns the default value of $\hat{f}$ for infinite domains. For finite domains, we fix it to *undefined* which is an arbitrary (but fixed) constant to represent undefinedness in Isabelle:

finfun-default $\hat{f} \equiv$ *if finite UNIV* **then** *undefined* **else** $\iota b.\ finite\ \{a \mid Rep\text{-}finfun\ \hat{f}\ a \neq b\}$

2.1   Kernel Functions for FinFuns

Having manually defined the type, we now show that every FinFun can be generated from two kernel functions similarly to a **datatype** element from its constructors: The constant function and pointwise update. For $b::'b$, let $K^f b::'a \Rightarrow_f 'b$ represent the FinFun that maps everything to b. It is defined by lifting the constant function $\lambda x::'a.\ b$ via *Abs-finfun* to the FinFun type. Similarly, pointwise update *finfun-update*, written $\_(\_ :=_f \_)$, is defined in terms of pointwise function update on ordinary functions:

$$K^f b \equiv \text{Abs-finfun } (\lambda x.\ b) \quad \text{and} \quad \hat{f}(a :=_f b) \equiv \text{Abs-finfun } ((\text{Rep-finfun } \hat{f})(a := b))$$

Note that these two kernel functions replace λ-abstraction of ordinary functions. Since the code generator will internally use these two constructors to represent FinFuns as data objects, proper λ-abstraction (via *Abs-finfun*) is not executable and is therefore deprecated. Consequently, all executable operators on FinFuns are to be defined (recursively) in terms of these two kernel functions. On the logic level, λ-abstraction is of course available via *Abs-finfun*, but it will be tedious to reason about such functions: Arbitrary λ-abstraction does not guarantee the finiteness constraint in the type definition for $'a \Rightarrow_f 'b$, hence this constraint must always be shown separately.

We can now already define what function application on $'a \Rightarrow_f 'b$ will be, namely *Rep-finfun*. To facilitate replacing ordinary functions with FinFuns in existent formalisations, we write function applications as a postfix subscript $_f$: $\hat{f}_f\ a \equiv \text{Rep-finfun } \hat{f}\ a$. This directly gives the kernel functions their semantics:

$$(K^f b)_f\ a = b \quad \text{and} \quad \hat{f}(a :=_f b)_f\ a' = (\text{if } a = a' \text{ then } b \text{ else } \hat{f}_f\ a') \quad (4)$$

Moreover, we already see that extensionality for HOL functions carries over to FinFuns, i.e. = on FinFuns does denote what it intuitively ought to:

$$\hat{f} = \hat{g} \longleftrightarrow (\forall x.\ \hat{f}_f\ x = \hat{g}_f\ x) \quad (5)$$

There are only few characteristic theorems about these two kernel functions. In particular, they are not free constructors, as e.g. the following equalities hold:

$$(K^f b)(a :=_f b) = K^f b \quad (6)$$

$$\hat{f}(a :=_f b)(a :=_f b') = \hat{f}(a :=_f b') \quad (7)$$

$$a \neq a' \longrightarrow \hat{f}(a :=_f b)(a' :=_f b') = \hat{f}(a' :=_f b')(a :=_f b) \quad (8)$$

This is natural, because FinFuns are meant to behave like ordinary functions and these equalities correspond to the standard ones for pointwise update on ordinary functions. Only $K^f\_$ is injective: $(K^f b) = (K^f b') \longleftrightarrow b = b'$. From a logician's point of view, non-free constructors are not desirable because recursion and case analysis becomes much more complicated. However, the savings in proof automation that extensionality for FinFuns permit are worth the extra effort when it comes to defining operators on FinFuns.

More importantly, these two kernel functions exhaust the type $'a \Rightarrow_f 'b$. This is most easily stated by the following induction rule, which is proven by induction on the finite set on which *Rep-finfun* $\hat{g}$ does not take the default value:

$$\frac{\forall b.\ P\ (K^f b) \qquad \forall \hat{f}\ a\ b.\ P\ \hat{f} \longrightarrow P\ \hat{f}(a :=_f b)}{P\ \hat{g}} \qquad (9)$$

Intuitively, P holds already for all FinFuns $\hat{g}$ if (i) $P\ (K^f b)$ holds for all constant FinFuns $K^f b$ and (ii) whenever $P\ \hat{f}$ holds, then $P\ \hat{f}(a :=_f b)$ holds, too. From this, a case distinction theorem is easily derived:

$$(\exists b.\ \hat{g} = (K^f b)) \vee (\exists \hat{f}\ a\ b.\ \hat{g} = \hat{f}(a :=_f b)) \qquad (10)$$

Both induction rule and case distinction theorem are weak in the sense that the $\hat{f}$ in the case for point-wise update is quantified without further constraints. Since $K^f\_$ and pointwise update are not distinct – cf. (6), proofs that do case analysis on FinFuns must always handle both cases even for constant FinFuns. Stronger induction and case analysis theorems could, however, be derived.

2.2   Representing FinFuns in the Code Generator

As mentioned above, the code generator represents FinFuns as a datatype with constant FinFun and pointwise update as (free) constructors. In Haskell, e.g., the following code is generated:

```
data Finfun a b = Finfun_update_code (Finfun a b) a b
                | Finfun_const b;
```

For efficiency reasons, we do not use *finfun-update* as a constructor for the `Finfun` datatype, as overwritten updates then would not get removed, the function's representation would keep growing. Instead, the HOL constant *finfun-update-code*, denoted $\_(\!\|\_ :=_f \_\|\!)$, is employed, which is semantically equivalent: $\hat{f}(\!\|a :=_f b\|\!) \equiv \hat{f}(a :=_f b)$. The code for *finfun-update*, however, is generated from (11) and (12):

$$(K^f b)(a :=_f b') = \textit{if } b = b' \textit{ then } K^f b \textit{ else } (K^f b)(\!\|a :=_f b'\|\!) \qquad (11)$$

$$\hat{f}(\!\|a :=_f b\|\!)(a' :=_f b') = \textit{if } a = a' \textit{ then } \hat{f}(a :=_f b') \textit{ else } \hat{f}(a' :=_f b')(\!\|a :=_f b\|\!) \qquad (12)$$

```
finfun_update :: forall a b. (Eq a, Eq b) =>
                    Finfun a b -> a -> b -> Finfun a b;
finfun_update (Finfun_update_code f a b) a' b' =
  (if eqop a a' then finfun_update f a b'
    else Finfun_update_code (finfun_update f a' b') a b);
finfun_update (Finfun_const b) a b' =
  (if eqop b b' then Finfun_const b
    else Finfun_update_code (Finfun_const b) a b');
```

where eqop is the HOL equality operator given by eqop a = (\ b -> a == b);. Hence, an update with $\_(\_:=_f \_)$ is checked against all other updates, all overwritten updates are thereby removed, and inserted only if it does not update to the default value. Using $\_(\_ :=_f \_)$ in the logic ensures that on the code level, every FinFun is stored with as few updates as possible given the fixed default value.[2]

Let, e.g., $\hat{f} = (K^f 0)(1 :=_f 5)(2 :=_f 6)$. When $\hat{f}$ is updated at 1 to 0, $\hat{f}(1 :=_f 0)$ evaluates on the code level to $(K^f 0)(2 :=_f 6)$, where all redundant updates at 1 have been removed. If the explicit code update function had been used instead, the last update would have been added to the list of updates: $\hat{f}(1 :=_f 0)$ evaluates to $(K^f 0)(1 :=_f 5)(2 :=_f 6)(1 :=_f 0)$. Exactly this problem of superfluous updates would occur if $\_(\_ :=_f \_)$ was directly used as a constructor in the exported code.

In case this optimisation is undesired, one can use *finfun-update-code* instead of *finfun-update*. Redundant updates in the representation on the code level can subsequently be deleted by invoking the *finfun-clearjunk* operator: Semantically, this is the identity function: *finfun-clearjunk* $\equiv$ *id*, but it is implemented using the following to equations that remove all redundant updates:

$$\textit{finfun-clearjunk } (K^f b) = (K^f b) \quad \text{and} \quad \textit{finfun-clearjunk } \hat{f}(a :=_f b) = \hat{f}(a :=_f b)$$

Consequently, every function that is defined recursively on FinFuns must provide two such equations for $K^f\_$ and $\_(\_ :=_f \_)$ for being executable. For function application, e.g., those from (4) are used with *finfun-update* being replaced by *finfun-update-code*.

For *quickcheck*, we have installed a sampling function that randomly creates a FinFun which has been updated at a few random points to random values. Hence, *quickcheck* can now both evaluate operators involving FinFuns and sample random values for the free variables of FinFun type in a conjecture.

3 Operators for FinFuns

In the previous section, we have shown how FinFuns are defined in Isabelle/HOL and how they are implemented in code. This section introduces more executable operators on FinFuns moving from basic ones towards executable equality.

3.1 Function Composition

The most important operation on functions and FinFuns alike – apart from application – is composition. It creates new FinFuns from old ones without losing executability: Every ordinary function $g :: 'b \Rightarrow 'c$ can be composed with a FinFun $\hat{f}$ of type $'a \Rightarrow_f 'b$ to produce another FinFun $g \circ_f \hat{f}$ of type $'a \Rightarrow_f 'c$. The operator $\circ_f$ is defined like the kernel functions via *Abs-finfun* and *Rep-finfun*:

$$g \circ_f \hat{f} \equiv \textit{Abs-finfun } (g \circ \textit{Rep-finfun } \hat{f})$$

To the code generator, two recursive equations are provided:

$$g \circ_f (K^f c) = (K^f g c) \quad \text{and} \quad g \circ_f \hat{f}(a :=_f b) = (g \circ_f \hat{f})(a :=_f g b) \tag{13}$$

---

[2] Minimal is relative to the default value in the representation (which need not coincide with *finfun-default*) – i.e. this does not include the case where changing this default value would require less updates. $(K^f 0)(\textit{True} :=_f 1)(\textit{False} :=_f 1)$ of type *bool* $\Rightarrow_f$ *nat*, e.g., is stored as $(K^f 0)(\textit{False} :=_f 1)(\textit{True} :=_f 1)$, whereas $K^f 1$ would also do.

$\circ_f$ is more versatile than composition on FinFuns only, because ordinary functions can be written directly thanks to λ abstraction. Yet, a FinFun $\hat{g}$ is equally easily composed with another FinFun $\hat{f}$ if we convert the first one back to ordinary functions: $\hat{g}_f \circ_f \hat{f}$. However, composing a FinFun with an ordinary function is not as simple. Although the definition is again straightforward:

$$\hat{f} \,{}_f\!\circ\, g \equiv \textit{Abs-finfun } (\textit{Rep-finfun } \hat{f} \circ g),$$

reasoning about ${}_f\!\circ$ is more difficult: Take, e.g., $\hat{f} = (K^f 2)(1 :=_f 1)$ and $g = (\lambda x.\ x \bmod 2)$. Then, $\hat{f} \,{}_f\!\circ\, g$ ought to be the function that maps even numbers to 2 and odd ones to 1, which is not a FinFun any more. Hence, (3) can no longer be used to reason about $\hat{f} \,{}_f\!\circ\, g$, so nothing nontrivial can be deduced about $\hat{f} \,{}_f\!\circ\, g$.

If g is injective (written $\textit{inj } g$), then $\hat{f} \,{}_f\!\circ\, g$ behaves as expected on updates:

$$\hat{f}(b :=_f c) \,{}_f\!\circ\, g = (\textbf{if } b \in \textit{range } g \textbf{ then } (\hat{f} \,{}_f\!\circ\, g)(g^{-1} b :=_f c) \textbf{ else } \hat{f} \,{}_f\!\circ\, g), \quad (14)$$

where $\textit{range } g$ denotes the range of g and g^{-1} is the inverse of g. Clearly, both $b \in \textit{range } g$ and $g^{-1} b$ are not executable for arbitrary g, so this conditional equality is not suited for code generation. If terms involving ${}_f\!\circ$ are to be executed, the above equation must be specialised to a specific g to become executable. The constant case is trivial for all g and need not be specialised: $(K^f c) \,{}_f\!\circ\, g = (K^f c)$.

This composition operator is good for reindexing the domain of a FinFun: Suppose, e.g., we need $\hat{h}_f\ x = \hat{f}_f\ (x + a)$ for some $a::int$, then $\hat{h}$ could be defined as $\hat{h} \equiv \hat{f} \,{}_f\!\circ\, g$ with $g = (\lambda x.\ x + a)$. Clearly, $\textit{inj } g$, $\textit{range } g = \textit{UNIV}$ and $g^{-1} = (\lambda x.\ x - a)$, so (14) simplifies to $\hat{f}(b :=_f c) \,{}_f\!\circ\, g = (\hat{f} \,{}_f\!\circ\, g)(b - a :=_f c)$. Unfortunately, the code generator cannot deal with such specialised recursion equations where the second parameter of $\_\ {}_f\!\circ\ \_$ is instantiated to g, so a new constant $\textit{shift } \hat{f}\ a \equiv \hat{f} \,{}_f\!\circ\, (\lambda x.\ x + a)$ must be introduced for the code generator with the recursion equations $\textit{shift } (K^f b)\ a = (K^f b)$ and $\textit{shift } \hat{f}(a' :=_f b)\ a = (\textit{shift } \hat{f}\ a)(a' - a :=_f b)$.

3.2   FinFuns and Pairs

Apart from composing FinFuns one after another, one often has to "run" FinFuns in parallel, i.e. evaluate both on the same argument and return both results as a pair. For two functions f and g, this is done by the term $\lambda x.\ (f\ x, g\ x)$. For two FinFuns $\hat{f}$ and $\hat{g}$, λ abstraction is not executable, but an appropriate operator $(\hat{f}, \hat{g})^f$ is easily defined as

$$(\hat{f}, \hat{g})^f \equiv \textit{Abs-finfun } (\lambda x.\ (\textit{Rep-finfun } \hat{f}\ x, \textit{Rep-finfun } \hat{g}\ x)).$$

This operator is most useful when two FinFuns are to be combined pointwise by some combinator h, which is then $\circ_f$-composed with this diagonal operator: Suppose, e.g., that $\hat{f}$ and $\hat{g}$ are two integer FinFuns and we need their pointwise sum, which is $(\lambda(x, y).\ x + y) \circ_f (\hat{f}, \hat{g})^f$, i.e. h is uncurried addition. The code equations are straight forward again:

$$(K^f b, K^f c)^f = K^f (b, c) \tag{15}$$

$$(K^f b, \hat{g}(a :=_f c))^f = (K^f b, \hat{g})^f (a :=_f (b, c)) \tag{16}$$

$$(\hat{f}(a :=_f b), \hat{g})^f = (\hat{f}, \hat{g})^f (a :=_f (b, \hat{g}_f\ a)) \tag{17}$$

3.3   Executable Quantifiers

Quantifiers in Isabelle/HOL are defined as higher-order functions. The universal quantifier All is defined by $All\ P \equiv P = (\lambda x.\ True)$ where P is a predicate and the binder notation $\forall x.\ P\ x$ is then just syntactic sugar for $All\ (\lambda x.\ P\ x)$. This also explains the error message of the code generator from Sec. 1. However, without λ-abstraction, there is no such nice notation for FinFuns, but the operator $\textit{finfun-All}$ for universal quantification over FinFun predicates is straightforward: $\textit{finfun-All}\ \hat{P} \equiv \forall x.\ \hat{P}_f\ x$.

Clearly, reducing universal quantification over FinFuns to All does not help with code generation, which was the main point in introducing FinFuns in the first place. However, we can exploit the explicit representation of $\hat{P}$. To that end, a more general operator $\textit{ff-All}$ of type ${}'a\ list \Rightarrow {}'a \Rightarrow_f bool \Rightarrow bool$ is necessary which ignores all points of $\hat{P}$ that are listed in the first argument:

$$\textit{ff-All}\ as\ \hat{P} \equiv \forall a.\ a \in set\ as \vee \hat{P}_f\ a$$

Clearly, $\textit{finfun-All} = \textit{ff-All}\ []$ holds. The extra list as keeps track of which points have already been updated and can be ignored in recursive calls:

$$\textit{ff-All}\ as\ (K^f b) \longleftrightarrow b \vee set\ as = UNIV \tag{18}$$

$$\textit{ff-All}\ as\ \hat{P}(\!|a :=_f b|\!) \longleftrightarrow (a \in set\ as \vee b) \wedge \textit{ff-All}\ (a \cdot as)\ \hat{P} \tag{19}$$

In the recursive case, the update a to b must either be overwritten by a previous update ($a \in set\ as$) or have b equal to $True$. Then, for the recursive call, a is added to the list as of visited points. In the constant case, either the constant is $True$ itself or all points of the domain ${}'a$ have been updated ($set\ as = UNIV$).

Via $\textit{finfun-All} = \textit{ff-All}\ []$, $\textit{finfun-All}$ is now executable, provided the test $set\ as = UNIV$ can be operationalised. Since $as::{}'a\ list$ is a (finite) list, $set\ as$ is by construction always finite. Thus, for infinite domains ${}'a$, this test always fails. Otherwise, if ${}'a$ is finite, such a test can be easily implemented.

Note that this distinction can be directly made on the basis of type information. Hence, we shift this subtle distinction into a type class such that the code automatically picks the right implementation for $set\ as = UNIV$ based on type information. Axiomatic type classes [7] allow for HOL constants being safely overloaded for different types and are correctly handled by Haftmann's code generator [6]. If the output language supports type classes like e.g. Haskell does, this feature is directly employed. Otherwise, functions in generated code are provided with an additional dictionary parameter that selects the appropriate implementation for overloaded constants at runtime.

For our purpose, we introduce a new type class $\textit{card-UNIV}$ with one parameter $\textit{card-UNIV}$ and the axiom that $\textit{card-UNIV} :: {}'a\ itself \Rightarrow nat$ returns the cardinality of ${}'a$'s universe:

$$\textit{card-UNIV}\ x = \textit{card}\ UNIV \tag{20}$$

By default, the cardinality of a type's universe is just a natural number of type nat, which itself is not related to ${}'a$ at all. Hence, $\textit{card-UNIV}$ takes an artificial parameter of type ${}'a\ itself$, where $itself$ represents types at the level of values: $TYPE({}'a)$ is the value associated with the type ${}'a$.

As every HOL type is inhabited, *card-UNIV TYPE('a)* can indeed be used to discriminate between types with finite and infinite universes by testing against 0:

$$finite \ (UNIV::{}'a \ set) \longleftrightarrow 0 < card\text{-}UNIV \ TYPE({}'a)$$

Moreover, the test *set as = UNIV* can now be written as *is-list-UNIV as* with

is-list-UNIV as ≡
let *c = card-UNIV TYPE('a)* in if *c* = 0 then *False* else |*remdups as*| = *c*

where *remdups as* removes all duplicates from the list *as*.

Note that the constraint (20) on the type class parameter *card-UNIV*, which is to be overloaded, is purely definitional. Thus, every type could be made member of the type class *card-UNIV* by instantiating *card-UNIV* to λa. *card UNIV*. However, for executability, it must be instantiated such that the code generator can generate code for it. This has been done for the standard HOL types like *unit, bool, char, nat, int,* and *'a list*, for which it is straightforward if one remembers that *card A* = 0 for all infinite sets *A*. For the type *bool*, e.g., *card-UNIV a* ≡ 2 for all *a::bool itself*. The cardinality of the universe for polymorphic type constructors like e.g. *'a × 'b* is computed by recursion on the type parameters:

$$card\text{-}UNIV \ TYPE({}'a \times {}'b) = card\text{-}UNIV \ TYPE({}'a) \cdot card\text{-}UNIV \ TYPE({}'b)$$

We have similarly instantiated *card-UNIV* for the type constructors $'a \Rightarrow 'b$, *'a option* and *'a + 'b*.

As we have the universal quantifier *finfun-All*, the executable existential quantifier is straightforward by duality: *finfun-Ex* $\hat{P}$ ≡ ¬ *finfun-All* (*Not* $\circ_f$ $\hat{P}$). As before, the pretty-print syntax ∃x. *P x* for *Ex* (λx. *P x*) in HOL cannot be transferred to FinFuns because λ-abstraction is not suited for code generation.

3.4   Executable Equality on FinFuns

Our second main goal with FinFuns, besides executable quantifiers, is executable equality tests on FinFuns. Extensionality – cf. (5) – reduces function equality to equality on every argument. However, (5) does not directly yield an implementation because it uses the universal quantifier *All* for ordinary HOL predicates, but some rewriting does the trick:

$$\hat{f} = \hat{g} \longleftrightarrow finfun\text{-}All \ ((\lambda(x, y). \ x = y) \circ_f (\hat{f}, \hat{g})^f) \tag{21}$$

By instantiating the HOL type class *eq* appropriately, the equality operator = becomes executable and in the generated code, an appropriate equality relation on the datatype is generated. In Haskell, e.g., the equality operator **==** on the type **Finfun a b** then really denotes equality like on the logic level:

```
eq_finfun :: forall a b. (FinFun.Card_UNIV a, Eq a, Eq b) ⇒
   FinFun.Finfun a b -> FinFun.Finfun a b -> Bool;
eq_finfun f g = FinFun.finfun_All
     (FinFun.finfun_comp (\ (a @ (aa, b)) -> aa == b)
                         (FinFun.finfun_Diag f g));
instance (FinFun.Card_UNIV a, Eq a, Eq b) ⇒
   Eq (FinFun.Finfun a b) where { (==) = FinFun.eq_finfun; };
```

3.5   Complexity

In this section, we briefly discuss the complexity of the above operators. We assume that equality tests require constant time. For a FinFun $\hat{f}$, let $\#\hat{f}$ denote the number of updates in its code representation. For an ordinary function g, let $\#g$ denote the complexity of evaluating g a for any a.

$K^f\_$ has constant complexity as it is a `finfun` constructor. Since $\_(\_:=_f\_)$ automatically removes redundant updates (11, 12), $\hat{f}(\_:=_f\_)$ is linear in $\#\hat{f}$, and so is application $\hat{f}_f\_$ (4). For $g \circ_f \hat{f}$, eq. (13) is recursive in $\hat{f}$ and each recursion step involves $\_(\_:=_f\_)$ and evaluating g, so the complexity is $\mathcal{O}((\#\hat{f})^2 + \#\hat{f} \cdot \#g)$.

For the product $(\hat{f}, \hat{g})^f$, we get: The base case $(K^f b, \hat{g})^f$ (15, 16) is linear in $\#\hat{g}$ and we have $\#(K^f b, \hat{g})^f = \#\hat{g}$. An update in the first parameter $(\hat{f}(a :=_f b), \hat{g})^f$ (17) executes $\hat{g}_f$ a $(\mathcal{O}(\#\hat{g}))$, the recursive call and the update $(\mathcal{O}(\#(\hat{f}, \hat{g})^f))$. Since there are $\#\hat{f}$ recursive calls and $\#(\hat{f}, \hat{g})^f \leq \#\hat{f} + \#\hat{g}$, the total complexity is bound by $\mathcal{O}(\#\hat{f} \cdot (\#\hat{f} + \#\hat{g}))$.

Since *finfun-All* is directly implemented in terms of *ff-All*, it is sufficient to analyse the latter's complexity: The base case (18) essentially executes *is-list-UNIV*. If we assume that the cardinality of the type universe is computed in constant time, *is-list-UNIV as* is bound by $\mathcal{O}(|as|^2)$ since *remdups as* takes $\mathcal{O}(|as|^2)$ steps. In case of an update (19), the updated point is checked against the list *as* $(\mathcal{O}(|as|))$ and the recursive call is executed with the list *as* being one element longer, i.e. $|as|$ grows by one for each recursive call. As there are $\#\hat{P}$ many recursive calls, *ff-All as* $\hat{P}$ has complexity $\#\hat{P} \cdot \mathcal{O}(\#\hat{P} + |as|) + \mathcal{O}((\#\hat{P} + |as|)^2) = \mathcal{O}((\#\hat{P} + |as|)^2)$. Hence, *finfun-All* $\hat{P}$ has complexity $\mathcal{O}((\#\hat{P})^2)$.

Equality on FinFuns $\hat{f}$ and $\hat{g}$ is then straightforward (21): $(\hat{f}, \hat{g})^f$ is in $\mathcal{O}(\#\hat{f} \cdot (\#\hat{f} + \#\hat{g}))$. Composing this with $\lambda(x, y).\ x = y$ takes $\mathcal{O}((\#(\hat{f}, \hat{g})^f)^2) \subseteq \mathcal{O}((\#\hat{f} + \#\hat{g})^2)$. Finally, executing *finfun-All* is quadratic in $\#((\lambda(x, y).\ x = y) \circ_f (\hat{f}, \hat{g})^f) \leq \#(\hat{f}, \hat{g})^f$. In total, $\hat{f} = \hat{g}$ has complexity $\mathcal{O}((\#\hat{f} + \#\hat{g})^2)$.

4   A Recursion Combinator

In the previous section, we have presented several operators on FinFuns that suffice for most purposes, cf. Sec. 5. However, we had to define function composition with FinFuns on either side and operations on products manually by going back to the type's carrier set *finfun* via *Rep-finfun* and *Abs-finfun*. This is not only inconvenient, but also loses the abstraction from the details of the finite set of updated points that FinFuns provide. In particular, one has to derive extra recursion equations for the code generator and prove each of them correct.

Yet, the induction rule (9) states that the recursive equations uniquely determine any function that satisfies these. Operations on FinFuns could therefore be defined by primitive recursion similarly to datatypes (cf. [2]). Alas, the two FinFun constructors are not free, so not every pair of recursive equations does indeed define a function. It might also well be the case that the equations are contradictory: For example, suppose we want to define a function *count* that counts the number of updates, i.e. *count* $(K^f c) = 0$ and *count* $\hat{f}(a :=_f b) = count$ $\hat{f} + 1$. Such a function does not exist for FinFuns in Isabelle, although it could be defined in Haskell to, e.g., compute extra-logic data such as memory consumption. Take, e.g., $\hat{f} \equiv (K^f 0)(0 :=_f 0)$. Then, $count\ \hat{f} = count\ (K^f 0) + 1 = 1$, but $\hat{f} = (K^f 0)$ by (6) and thus $count\ \hat{f} = 0$ would equally have to hold, because equality is congruent w.r.t. function application, a contradiction.

4.1 Lifting Recursion from Finite Sets to FinFuns

More abstractly, the right hand side of the recursive equations can be considered as a function: For the constant case, such a function $c::'b \Rightarrow 'c$ takes the constant value of the FinFun and evaluates to the right hand side. In the recursive case, $u::'a \Rightarrow 'b \Rightarrow 'c \Rightarrow 'c$ takes the point of the update, the new value at that point and the result of the recursive call. In this section, we define a combinator *finfun-rec* that takes c and u and defines the corresponding operator on FinFuns, similarly to the primitive recursion combinators that are automatically generated for datatypes. That is, *finfun-rec* must satisfy (22) and (23), subject to certain well-formedness conditions on c and u, which will be examined in Sec. 4.2.

$$finfun\text{-}rec\ c\ u\ (K^f b) = c\ b \qquad (22)$$

$$finfun\text{-}rec\ c\ u\ \hat{f}(a :=_f b) = u\ a\ b\ (finfun\text{-}rec\ c\ u\ \hat{f}) \qquad (23)$$

The standard means in Isabelle for defining recursive functions, namely **recdef** and the function package [10], are not suited for this task because both need a termination proof, i.e. a well-founded relation in which all recursive calls always decrease. Since K^f_ and _(_ $:=_f$ _) are not free constructors, there is no such termination order for (22) and (23). Hence, we define *finfun-rec* by recursion on the finite set of updated points using the recursion operator *fold* for finite sets:

finfun-rec $c\ u\ \hat{f} \equiv$
let $b = finfun\text{-}default\ \hat{f}$;
    $g = (\iota g.\ \hat{f} = Abs\text{-}finfun\ (map\text{-}default\ b\ g) \land finite\ (dom\ g) \land b \notin ran\ g)$
in fold $(\lambda a.\ u\ a\ (map\text{-}default\ b\ g\ a))\ (c\ b)\ (dom\ g)$

In the *let* expression, $\hat{f}$ is unpacked into its default value b (cf. Sec. 2) and a partial function $g::'a \rightharpoonup 'b$ such that $\hat{f} = Abs\text{-}finfun\ (map\text{-}default\ b\ g)$ and the finite domain of g contains only points at which $\hat{f}$ differs from its default value b, i.e. g stores precisely the updates of $\hat{f}$. Then, the update function u is folded over the finite set of points $dom\ g$ where $\hat{f}$ does not take its default value b.

All FinFun operators that we have defined in Sec. 3 via *Abs-finfun* and *Rep-finfun* can also be defined directly via *finfun-rec*. For example, the functions for $\circ_f$ directly show up in the recursive equations from (13):

$$g \circ_f \hat{f} \equiv finfun\text{-}rec\ (\lambda b.\ K^f g\ b)\ (\lambda a\ b\ \hat{f}.\ \hat{f}(a :=_f g\ b))\ \hat{f}.$$

4.2 Well-Formedness Conditions

Since all functions in HOL are total, *finfun-rec* $c\ u$ is defined for every combination of c and u. Any nontrivial property of *finfun-rec* is only provable if u is left-commutative because *fold* is unspecified for other functions. Thus, the next step is to establish conditions on the FinFun level that ensure (22) and (23). It turns out that four are sufficient:

$$u \; a \; b \; (c \; b) = c \; b \tag{24}$$

$$u \; a \; b'' \; (u \; a \; b' \; (c \; b)) = u \; a \; b'' \; (c \; b) \tag{25}$$

$$a \neq a' \longrightarrow u \; a \; b \; (u \; a' \; b' \; d) = u \; a' \; b' \; (u \; a \; b \; d) \tag{26}$$

$$\textit{finite UNIV} \longrightarrow \textit{fold} \; (\lambda a. \; u \; a \; b') \; (c \; b) \; \textit{UNIV} = c \; b' \tag{27}$$

Eq. (24), (25), and (26) naturally reflect the equalities between the constructors from (6), (7), and (8), respectively. It is sufficient to restrict overwriting updates (25) to constant FinFuns because the general case directly follows from this by induction and (26). The last equation (27) arises from the identity

$$\textit{finite UNIV} \longrightarrow \textit{fold} \; (\lambda a \; \hat{f}. \; \hat{f}(a :=_f b')) \; (K^f b) \; \textit{UNIV} = (K^f b'). \tag{28}$$

Eq. (24), (25), and (26) are sufficient for proving (23). For a FinFun operator like $\circ_f$, these constraints must be shown for specific c and u, which is usually completely automatic. Even though (27), which is required to deduce (22), must usually be proven by induction, this normally is also automatic, because for finite types $'a$, $'a \Rightarrow 'b$ and $'a \Rightarrow_f 'b$ are isomorphic via *Abs-finfun* and *Rep-finfun*.

5 Applications

In this section, we present two applications for FinFuns to demonstrate that the operations from Sec. 3 form a reasonably complete set of abstract operations.

1. They can be used to represent sets as predicates with the standard operations all being executable: membership and subset test, union, intersection, complement and bounded quantification.
2. FinFuns have been inspired by the needs of JinjaThreads [12], which is a formal semantics of multithreaded Java in Isabelle. We show how FinFuns prove essential on the way to generating an interpreter for concurrent Java.

5.1 Representing Sets with Finfuns

In Isabelle 2008, the proper type $'a \; set$ for sets has been removed in favour of predicates of type $'a \Rightarrow bool$ to eliminate redundancies in the implementation and in the library. As a consequence, Isabelle's new code generator is no longer able to generate code for sets as before: A finite set had been coded as the list of its elements. Hence, e.g. the complement operator has not been executable because the complement of a finite set might no longer be a finite set. Neither are collections of the form $\{a \mid P \; a\}$ suited for code generation.

Since FinFuns are designed for code generation, they can be used for representing sets in explicit form without explicitly introducing a set type of its own. FinFun set operations like membership and inclusion test, union, intersection and even complement are straightforward using $\circ_f$. As before, these operators are decorated with f subscripts to distinguish them from their analogues on sets:

$$\hat{f} \subseteq_f \hat{g} \equiv \textit{finfun-All} \; ((\lambda(x, y). \; x \longrightarrow y) \circ_f (\hat{f}, \hat{g})^f) \qquad -\hat{f} \equiv (\lambda b. \; \neg \; b) \circ_f \hat{f}$$

$$\hat{f} \cup_f \hat{g} \equiv (\lambda(x, y). \; x \vee y) \circ_f (\hat{f}, \hat{g})^f \qquad \hat{f} \cap_f \hat{g} \equiv (\lambda(x, y). \; x \wedge y) \circ_f (\hat{f}, \hat{g})^f$$

Obviously, these equations can be directly translated into executable code.

However, if we were to reason with them directly, most theorems about sets (as predicates) would have to be replicated for FinFuns. Although this would be straightforward, loads of redundancy would be reintroduced this way. Instead, we propose to inject FinFun sets via $_f$ into ordinary sets and use the standard operations on sets to work with them. The code generator is set up such that it preprocesses all equations for code generation and automatically replaces set operations with their FinFun equivalents by unfolding equations such as $A_f \subseteq B_f$ $\longleftrightarrow A \subseteq_f B$ and $A_f \cup B_f = (A \cup_f B)_f$. This approach works for *quickcheck*, too.

Besides the above operations, bounded quantification is also straightforward:

$$\text{finfun-Ball } \hat{A} \; P \equiv \forall x \in \hat{A}_f. \; P \; x \quad \text{and} \quad \text{finfun-Bex } \hat{A} \; P \equiv \exists x \in \hat{A}_f. \; P \; x$$

Clearly, they are not executable right away. Take, e.g., $\hat{A} = (K^f \text{ True})$, i.e. the universal set, then *finfun-Ball* $\hat{A} \; P \longleftrightarrow (\forall x. \; P \; x)$, which is undecidable if x ranges over an infinite domain. However, if we go for partial correctness, correct code can be generated: Like for the universal quantifier *finfun-All* for FinFun predicates (cf. Sec. 3.3), *ff-Ball* is introduced which takes an additional parameter xs to remember the list of points which have already been checked at previous calls.

$$\text{ff-Ball } xs \; \hat{A} \; P \equiv \forall a \in \hat{A}_f. \; a \in \text{set } xs \lor P \; a.$$

This now permits to set up recursive equations for the code generator:

$$\text{ff-Ball } xs \; (K^f b) \; P \longleftrightarrow \neg \, b \lor \text{set } xs = UNIV \lor \text{loop } (\lambda u. \; \text{ff-Ball } xs \; (K^f b) \; P)$$

$$\text{ff-Ball } xs \; \hat{A}(\!| a :=_f b |\!) \; P \longleftrightarrow (a \in \text{set } xs \lor (b \longrightarrow P \; a)) \land \text{ff-Ball } (a \cdot xs) \; \hat{A} \; P$$

In the constant case, if b is false, i.e. the set is empty, *ff-Ball* holds; similarly, if all elements of the universe have been checked already, this test is again implemented by the overloaded term *is-list-UNIV* xs (Sec. 3.3). Otherwise, one would have to check whether P holds at all points except xs, which is not computable for arbitrary P and $'a$. Thus, instead of evaluating its argument, the code for *loop* never terminates. In Isabelle, however, *loop* is simply the *unit*-lifted identity function: *loop* $f \equiv f$ (). Of course, an exception could equally be raised in place of non-termination. The bounded existential quantifier is implemented analogously.

5.2   JinjaThreads

Jinja [9] is an executable formal semantics for a large subset of Java source-code and bytecode in Isabelle/HOL. JinjaThreads [11] extends Jinja with Java's thread features on both levels. It contains a framework semantics which interleaves the individual threads whose small-step semantics is given to it as a parameter. This framework semantics takes care of all management issues related to threads: The thread pool itself, the lock state, monitor wait sets, spawning and joining a thread, etc. Individual threads communicate via the shared memory with each other and via thread actions like *Lock*, *Unlock*, *Join*, etc. with the framework semantics. At every step, the thread specifies which locks to acquire or release how many times, which thread to create or join on. In our previous work [12], this communication was modelled as a list of such actions, and a lot of pointless work went into identifying permutations of such lists which are semantically equivalent. Therefore, this has been changed such that every lock of type $'l$ now has its own list. Since only finitely many locks need to be changed in any single step, these lists are stored in a FinFun such that checking whether a step's actions are feasible in a given state is executable.

Moreover, in developing JinjaThreads, we have found that most lemmas about the framework semantics contain non-executable assumptions about the thread pool or the lock state, in particular universal quantifiers or predicates defined in terms of them. Therefore, we replaced ordinary functions that model the lock state (type $'l \Rightarrow 't \, lock$) and the thread pool (type $'t \rightharpoonup ('x, 'l) \, thread$) with FinFuns. Rewriting the existing proofs took very little effort because mostly, only fs in subscript or superscript had to be added to the proof texts because Isabelle's simplifier and classical reasoner are set up such that FinFuns indeed behave like ordinary functions.

Not to break the proofs, we did not remove the universal quantifiers in the definitions of predicates themselves, but provided simple lemmas to the code generator. For example, *locks-ok ls t las* checks whether all lock requests *las* of thread t can be met in the lock state *ls* and is defined as *locks-ok ls t las* $\equiv \forall l$. *lock-ok* $(ls_f \, l) \, t \, (las_f \, l)$, whereas the equation for code generation is

$$locks\text{-}ok \; ls \; t \; las = finfun\text{-}All \; ((\lambda(l, la). \; lock\text{-}ok \; l \; t \; la) \circ_f (ls, las)^f).$$

Unfortunately, JinjaThreads is not yet fully executable because the semantics of a single thread relies on inductive predicates. Once the code generator will handle these, we will have a certified Jinja virtual machine with concurrency to execute multithreaded Jinja programs as has been done for sequential ones [9].

6   Related Work and Conclusion

Related work. To represent (partial) functions explicitly by a list of point-value pairs is common knowledge in computer science, partial functions $'a \rightharpoonup 'b$ with finite domain have even been formalised as associative lists in the Isabelle/HOL library. However, it is cumbersome to reason with them because one single function has multiple representations, i.e. associative lists are not extensional. Coq and HOL4, e.g., also come with a formalisation of finite maps of their own and both of them fix their default value to *None*. Collins and Syme [4] have already provided a theory of partial functions with finite domain in terms of the everywhere undefined function and pointwise update. Similar to (4), (7), and (8), they axiomatize a type $('a,'b) \, fmap$ in terms of abstract operations *Empty*, *Update*, *Apply* :: $('a,'b) \, fmap \Rightarrow 'a \Rightarrow 'b$, and *Domain* and present two models: Maps $'a \rightharpoonup 'b$ with finite domain and associative lists where the order of their elements is determined with Hilbert's choice operator, but neither of these supports code generation. Moreover, equality is not extensional like ours (5), but guarded by the domains. Since these partial functions have an unspecified default value that is implicitly fixed by the codomain type and the model, they cannot be used for almost everywhere constant functions where the default value may differ from function to function. Consequently, (28) is not expressible in their setting.

Recursion over non-free kernel functions is also a well-known concept: Nipkow and Paulson [14], e.g., define a *fold* operator for finite sets which are built from the empty set and insertion of one element. However, they do not introduce a new type for finite sets, so all equations are guarded by the predicate *finite*, i.e. they cannot be leveraged by the code generator.

Nominal Isabelle [16] is used to facilitate reasoning about α-equivalent terms with binders, where the binders are non-free term constructors. The HOL type for terms is obtained by quotienting the datatype with the (free) term constructors w.r.t. α-equivalence classes. Primitive-recursive definitions must then be shown compatible with α-equivalence using a notion of freshness [17]. It is tempting to define the FinFun type universe similarly as the quotient of the datatype with constructors $K^f\_$ and $\_(\!(\_ :=_f \_)\!)$ w.r.t. the identities (6), (7), (8), and (28), because this would settle exhaustion, induction and recursion almost automatically. However, this construction is not directly possible because (28) cannot be expressed as an equality of kernel functions. Instead, we have defined the carrier set *finfun* directly in terms of the function space and established easy, sufficient (and almost necessary) conditions for recursive definitions being well-formed.

Conclusion. FinFuns generalise finite maps by continuing them with a default value in the logic, but for the code generator, they are implemented like associative lists which suffer from multiple representations for a single function. Thus, they bridge the gap between easy reasoning and these implementation issues arising from functions as data: They are as easy to use as ordinary functions. By not fixing a default value (like *None* for maps), we have been able to easily apply them to very diverse settings.

We have decided to restrict the FinFun carrier set *finfun* to functions that are constant almost everywhere. Although everything from Sec. 3 would equally work if that restriction was lifted, the induction rule (9) and recursion operator (Sec. 4) would then no longer be available, i.e. the datatype generated by the code generator would not exhaust the type in the logic. Thus, the user could not be sure that every FinFun from his formalisation can be represented as data in the generated code. Conversely, not every operator can be lifted to FinFuns: The image operator $\_ ` \_$ on sets, e.g., has no analogue on FinFun sets.

Clearly, FinFuns are a very restricted set of functions, but we have demonstrated that this lightweight formalisation is in fact useful and easy to use. In Sec. 3, we have outlined the way to executing equality on FinFuns, but we need not stop there: Other operators like e.g. currying, λ-abstraction for FinFuns $'a \Rightarrow_f 'b$ with $'a$ finite, and even the definite description operator $\iota x.\ \hat{P}_f\ x$ can all be made executable via the code generator. In terms of usability, FinFuns currently provide little support for defining new operators that can not be expressed by the existing ones: For example, recursive equations for the code generator must be stated explicitly, even if the definition explicitly uses the recursion combinator. But with some implementation effort, definitions and the code generator setup could be automated in the future.

For *quickcheck*, our implementation with at most quadratic complexity is sufficiently efficient because random FinFuns involve only a few updates. For larger applications, however, one is interested in more efficient representations. If, e.g., the domain of a FinFun is totally ordered, binary search trees are a natural option, but this requires considerable amount of work: (Balanced) binary trees must be formalised and proven correct, which could be based e.g. on [15], and all the operators that are recursive on a FinFun must be reimplemented. In practice, the user should not care about which implementation the code generator chooses, but such automation must overcome some technical restrictions, such as only one type variable for type classes or only unconditional rewrite rules for the code generator, perhaps by recurring on ad-hoc translations.

References

1. Berghofer, S., Nipkow, T.: Random testing in Isabelle/HOL. In: Proc. SEFM 2004, pp. 230–239. IEEE Computer Society, Los Alamitos (2004)
2. Berghofer, S., Wenzel, M.: Inductive datatypes in HOL – lessons learned in formal-logic engineering. In: Bertot, Y., Dowek, G., Hirschowitz, A., Paulin, C., Théry, L. (eds.) TPHOLs 1999. LNCS, vol. 1690, pp. 19–36. Springer, Heidelberg (1999)
3. Berghofer, S., Nipkow, T.: Executing higher order logic. In: Callaghan, P., Luo, Z., McKinna, J., Pollack, R. (eds.) TYPES 2000. LNCS, vol. 2277, pp. 24–40. Springer, Heidelberg (2002)
4. Collins, G., Syme, D.: A theory of finite maps. In: Schubert, E.T., Alves-Foss, J., Windley, P. (eds.) HUG 1995. LNCS, vol. 971, pp. 122–137. Springer, Heidelberg (1995)
5. Dybjer, P., Haiyan, Q., Takeyama, M.: Combining testing and proving in dependent type theory. In: Basin, D., Wolff, B. (eds.) TPHOLs 2003. LNCS, vol. 2758, pp. 188–203. Springer, Heidelberg (2003)
6. Haftmann, F., Nipkow, T.: A code generator framework for Isabelle/HOL. Technical Report 364/07, Dept. of Computer Science, University of Kaiserslautern (2007)
7. Haftmann, F., Wenzel, M.: Constructive type classes in Isabelle. In: Altenkirch, T., McBride, C. (eds.) TYPES 2006. LNCS, vol. 4502, pp. 160–174. Springer, Heidelberg (2007)
8. Harrison, J.: Metatheory and reflection in theorem proving: A survey and critique. Technical Report CRC-053, SRI International Cambridge Computer Science Research Centre (1995)
9. Klein, G., Nipkow, T.: A machine-checked model for a Java-like language, virtual machine and compiler. ACM TOPLAS 28, 619–695 (2006)
10. Krauss, A.: Partial recursive functions in higher-order logic. In: Furbach, U., Shankar, N. (eds.) IJCAR 2006. LNCS, vol. 4130, pp. 589–603. Springer, Heidelberg (2006)
11. Lochbihler, A.: Jinja with threads. The Archive of Formal Proofs. Formal proof development (2007), http://afp.sf.net/entries/JinjaThreads.shtml
12. Lochbihler, A.: Type safe nondeterminism - a formal semantics of Java threads. In: FOOL 2008 (2008)
13. Lochbihler, A.: Code generation for functions as data. The Archive of Formal Proofs. Formal proof development (2009), http://afp.sf.net/entries/FinFun.shtml
14. Nipkow, T., Paulson, L.C.: Proof pearl: Defining functions over finite sets. In: Hurd, J., Melham, T. (eds.) TPHOLs 2005. LNCS, vol. 3603, pp. 385–396. Springer, Heidelberg (2005)

15. Nipkow, T., Pusch, C.: AVL trees. The Archive of Formal Proofs. Formal proof development (2004), http://afp.sf.net/entries/AVL-Trees.shtml
16. Urban, C.: Nominal techniques in Isabelle/HOL. Journal of Automatic Reasoning 40(4), 327–356 (2008)
17. Urban, C., Berghofer, S.: A recursion combinator for nominal datatypes implemented in Isabelle/HOL. In: Furbach, U., Shankar, N. (eds.) IJCAR 2006. LNCS, vol. 4130, pp. 498–512. Springer, Heidelberg (2006)

A Notation

Isabelle/HOL formulae and propositions are close to standard mathematical notation. This subsection introduces non-standard notation, a few basic data types and their primitive operations.

Types is the set of all types which contains, in particular, the type of truth values *bool*, natural numbers *nat*, integers *int*, and the singleton type *unit* with its only element (). The space of total functions is denoted by $'a \Rightarrow 'b$. Type variables are written $'a$, $'b$, etc. The notation $t::\tau$ means that the HOL term t has type τ.

Pairs come with two projection functions *fst* and *snd*. Tuples are identified with pairs nested to the right: (a, b, c) is identical to $(a, (b, c))$ and $'a \times 'b \times 'c$ to $'a \times ('b \times 'c)$. Dually, the disjoint union of $'a$ and $'b$ is written $'a + 'b$.

Sets are represented as predicates (type $'a\ set$ is shorthand for $'a \Rightarrow bool$), but follow the usual mathematical conventions. *UNIV* :: $'a\ set$ is the set of all elements of type $'a$. The image operator $f\ {}^\backprime\ A$ applies the function f to every element of A, i.e. $f\ {}^\backprime\ A \equiv \{y \mid \exists x \in A.\ y = f\ x\}$. The predicate *finite* on sets characterises all finite sets. *card A* denotes the cardinality of the finite set A, or 0 if A is infinite. *fold f z A* folds a left-commutative[3] function $f::'a \Rightarrow 'b \Rightarrow 'b$ over a finite set A :: $'a\ set$ with initial value $z::'b$.

Lists (type $'a\ list$) come with the empty list [] and the infix constructor $\_\cdot\_$ for consing. Variable names ending in "s" usually stand for lists and $|xs|$ is the length of xs. The function *set* converts a list to the set of its elements.

Function update is defined as follows: Let $f::'a \Rightarrow 'b$, $a::'a$ and $b::'b$. Then $f(a := b) \equiv \lambda x.\ if\ x = a\ then\ b\ else\ f\ x$.

The **option** data type $'a\ option$ adjoins a new element *None* to a type $'a$. All existing elements in type $'a$ are also in $'a\ option$, but are prefixed by *Some*. For succinctness, we write $\lfloor a \rfloor$ for *Some a*. Hence, for example, *bool option* has the values *None*, $\lfloor True \rfloor$ and $\lfloor False \rfloor$.

Partial functions are modelled as functions of type $'a \Rightarrow 'b\ option$ where *None* represents undefined and $f\ x = \lfloor y \rfloor$ means x is mapped to y. Instead of $'a \Rightarrow 'b\ option$, we write $'a \rightharpoonup 'b$ and call such functions **maps**. $f(x \mapsto y)$ is shorthand for $f(x := \lfloor y \rfloor)$. The domain of f (written *dom f*) is the set of points at which f is defined, *ran f* denotes the range of f. The function *map-default b f* takes a partial function f and continues it at its undefined points with b.

The **definite description** $\iota x.\ Q\ x$ is known as Russell's ι-operator. It denotes the unique x such that $Q\ x$ holds, provided exactly one exists.

---

[3] f is left-commutative, if it satisfies $f\ x\ (f\ y\ z) = f\ y\ (f\ x\ z)$ for all x, y, and z.

Packaging Mathematical Structures

François Garillot[1], Georges Gonthier[2], Assia Mahboubi[3], and Laurence Rideau[4]

[1] Microsoft Research - INRIA Joint Centre
Francois.Garillot@inria.fr
[2] Microsoft Research Cambridge
gonthier@microsoft.com
[3] Inria Saclay and LIX, École Polytechnique
Assia.Mahboubi@inria.fr
[4] Inria Sophia-Antipolis – Méditerranée
Laurence.Rideau@inria.fr

Abstract. This paper proposes generic design patterns to define and combine algebraic structures, using dependent records, coercions and type inference, inside the CoQ system. This alternative to telescopes in particular supports multiple inheritance, maximal sharing of notations and theories, and automated structure inference. Our methodology is robust enough to handle a hierarchy comprising a broad variety of algebraic structures, from types with a choice operator to algebraically closed fields. Interfaces for the structures enjoy the convenience of a classical setting, without requiring any axiom. Finally, we present two applications of our proof techniques: a key lemma for characterising the discrete logarithm, and a matrix decomposition problem.

Keywords: Formalization of Algebra, Coercive subtyping, Type inference, CoQ, SSREFLECT.

1 Introduction

Large developments of formalized mathematics demand a careful organization. Fortunately mathematical theories are quite organized, e.g., every algebra textbook [1] describes a hierarchy of structures, from monoids and groups to rings and fields. There is a substantial literature [2,3,4,5,6,7] devoted to their formalization within formal proof systems.

In spite of this body of prior work, however, we have found it difficult to make practical use of the algebraic hierarchy in our project to formalize the Feit-Thompson Theorem in the CoQ system; this paper describes some of the problems we have faced and how they were resolved. The proof of the Feit-Thompson Theorem covers a broad range of mathematical theories, and organizing this formalization into modules is central to our research agenda. We've developed[8] an extensive set of modules for the combinatorics and set and group theory required for the "local analysis" part of the proof, which includes a rudimentary algebraic hierarchy needed to support combinatorial summations[9].

Extending this hierarchy to accommodate the linear algebra, Galois theory and representation theory needed for the "character theoretic" part of the proof has proved problematic. Specifically, we have found that well-known encodings

S. Berghofer et al. (Eds.): TPHOLs 2009, LNCS 5674, pp. 327–342, 2009.
© Springer-Verlag Berlin Heidelberg 2009

of algebraic structures using dependent types and records [2] break down in the face of complexity; we address this issue in section 2 of this paper.

Many of the cited works focused on the definition of the hierarchy rather than its use, making simplifying assumptions that would have masked the problems we encountered. For example some assume that only one or two structures are involved at any time, or that all structures are explicitly specified. The examples in section 4 show that such assumptions are impractical: they involve several different structures, often within the same expression, and some of which need to be synthesized for existing types.

We have come to realize that algebraic structures are not "modules" in the software engineering sense, but rather "interfaces". Indeed, the mathematical theory of, say, an abstract ring, is fairly thin. However, abstract rings provide an interface that allows "modules" with actual contents, such as polynomials and matrices, to be defined and, crucially, *composed*. The main function of an algebraic structure is to provide common notation for expressions and for proofs (e.g., basic lemmas) to facilitate the composition and application of these generic modules. Insisting that an interface be instantiated explicitly each time it is used negates this function, so it is critical that structures be inferred on the fly; we'll see in the next section how this can be accomplished.

Similarly, we must ensure that our algebraic interfaces are consistent with the *other* modules in our development: in particular they should integrate the existing combinatoric interfaces [8], as algebra requires equality. As described in section 3, we have therefore adapted classical algebra to our constructive combinatorics. In addition to philosophical motivations (viz., allowing constructive proof of a finitary result like the Feit-Thompson Theorem), we have practical uses for a constructive framework: it provides basic but quite useful proof automation, via the small-scale reflection methodology supported by the SSREFLECT extension to COQ [10].

Due to space constraints, we will assume some familiarity with the COQ type system [11] (dependent types and records, proof types, type inference with implicit terms and higher-order resolution) in section 2, and with the basic design choices in the Feit-Thompson Theorem development [8] (boolean reflection, concrete finite sets) in sections 3 and 4.

2   Encoding Structures

2.1   Mixins

An algebraic or combinatorial structure comprises representation types (usually only one), constants and operations on the type(s), and axioms satisfied by the operations. Within the propositions-as-types framework of COQ, the interface for all of these components can be uniformly described by a collection of dependent types: the type of operations depends on the representation type, and the statement (also a "type") of axioms depends on both the representation type and the actual operations.

For example, a path in a combinatorial graph amounts to

- a representation type T for nodes
- an edge relation $e : \mathtt{rel}\ T$

- an initial node $x_0 : T$
- the sequence $p : \mathsf{seq}\ T$ of nodes that follow x_0
- the axiom $p_P : \mathsf{path}\ e\ x_0\ p$ asserting that e holds pairwise along $x_0 :: p$.

The path "structure" is actually best left unbundled, with each component being passed as a separate argument to definitions and theorems, as there is no one-to-one relation between any of the components (there can be multiple paths with the same starting point and relation, and conversely a given sequence can be a path for different relations). Because it depends on all the other components, only the axiom p_P needs to be passed around explicitly; type inference can figure out T, e, x_0 and p from the type of p_P, so that in practice the entire path "structure" can be assimilated to p_P.

While this unbundling allows for maximal flexibility, it also induces a proliferation of arguments that is rapidly overwhelming. A typical algebraic structure, such as a ring, involves half a dozen constants and even more axioms. Moreover such structures are often nested, e.g., for the Cayley-Hamilton theorem one needs to consider the ring of polynomials over the ring of matrices over a general commutative ring. The size of the terms involved grows as C^n, where C is the number of separate components of a structure, and n is the structure nesting depth. For Cayley-Hamilton we would have $C = 15$ and $n = 3$, and thus terms large enough to make theorem proving impractical, given that algorithms in user-level tactics are more often than not nonlinear.

Thus, at the very least, related operations and axioms should be packed using CoQ's dependent records (Σ-types); we call such records *mixins*. Here is, for example, the mixin for a **Z**-module, i.e., the additive group of a vector space or a ring:

```
Module Zmodule.
Record mixin_of (M : Type) : Type := Mixin {
  zero : M; opp : M -> M; add : M -> M -> M;
  _ : associative add;  _ : commutative add;
  _ : left_id zero add;  _ : left_inverse zero opp add
}. ...
End Zmodule.
```

Here we are using a CoQ Module solely to avoid name clashes with similar mixin definitions.

Note that mixins typically provide only part of a structure; for instance a ring structure would actually comprise a representation type and three mixins: one for equality, one for the additive group, and one for the multiplicative monoid together with distributivity. A mixin can depend on another one: e.g., the ring multiplicative mixin depends on the additive one for its distributivity axioms.

Since types don't depend on mixins (it's the converse) type inference usually cannot fill in omitted mixin parameters; however, the *type class* mechanism of CoQ 8.2 [12] can do so by running ad hoc tactics after type inference.

2.2   Packed Structures

The geometric dependency of C^n on n is rather treacherous: it is quite possible to develop an extensive structure package in an abstract setting (when $n = 1$)

that will fail dramatically when used in practice for even moderate values of n. The only case when this does not occur is with $C = 1$ — when each structure is encapsulated into a single object. Thus, in addition to aesthetics, there is a strong pragmatic rationale for achieving full encapsulation.

While mixins provide some degree of packaging, it falls short of $C = 1$.

However, mixins require one object per level in the structure hierarchy. This is far from $C = 1$ because theorem proving requires deeper structure hierarchies than programming, as structures with identical operations can differ by axioms; indeed, despite our best efforts, our algebraic hierarchy is nine levels deep.

For the topmost structure in the hierarchy, encapsulation just amounts to using a dependent record to package a mixin with its representation type. For example, the top structure in our hierarchy, which describes a type with an equality comparison operation (see [8]), could be defined as follows:

```
Module Equality.
Record mixin_of (T : Type) : Type :=
  Mixin {op : rel T; _ : forall x y, reflect (x = y) (op x y)}.
Structure type : Type :=
  Pack {sort :> Type; mixin : mixin_of sort}.
End Equality.
Notation eqType := Equality.type.
Notation EqType := Equality.Pack.
Definition eq_op T := Equality.op (Equality.mixin T).
Notation "x == y" := (@eq_op _ x y).
```

CoQ provides two features that support this style of interface, Coercion and Canonical Structure. The sort :> Type declaration above makes the sort projection into a *coercion* from type to Type. This form of explicit subtyping allows any T : eqType to be used as a Type, e.g., the declaration $x : T$ is understood as x : sort T. This allows x == x to be understood as @eq_op T x x by simple first-order unification in the Hindley-Milner type inference, as @eq_op α expects arguments of type sort α.

Coercions are mostly useful for establishing generic theorems for abstract structures. A different mechanism is needed to work with specific structures and types, such as integers, permutations, polynomials, or matrices, as this calls for construing a *more* specific Type as a structure object (e.g., an eqType): coercions and more generally subtyping will not do, as they are constrained to work in the *opposite* direction.

CoQ solves this problem by using higher-order unification in combination with Canonical Structure hints. For example, assuming int is the type of signed integers, and given

```
Definition int_eqMixin := @Equality.Mixin int eqz ...
Canonical Structure int_eqType := EqType int_eqMixin.
```

CoQ will interpret 2 == 2 as @eq_op int_eqType 2 2, which is convertible to eqz 2 2. Thanks to the Canonical Structure hint, CoQ finds the solution $\alpha = $ int_eqType to the higher-order unification problem sort $\alpha \equiv_{\beta\iota\delta}$ int that arises during type inference.

2.3   Telescopes

The simplest way of packing deeper structures of a hierarchy consists in repeating the design pattern above, substituting "the parent structure" for "representation type". For instance, we could end Module Zmodule with

```
Structure zmodType : Type := Pack {sort :> eqType; _ : mixin_of sort}.
```

This makes zmodType a subtype of eqType and (transitively) of Type, and allows for the declaration of generic operator syntax $(0, x + y, -x, x - y, x * i)$, and the declaration of canonical structures such as

```
Canonical Structure int_zmodType := Zmodule.Pack int_zmodMixin.
```

Many authors [2,13,7,5] have formalized an algebraic hierarchy using such nested packed structures, which are sometimes referred to as *telescopes* [14], the term we shall use henceforth.

As the coercion of a telescope to a representation Type is obtained by transitivity, it comprises a chain of elementary coercions: given T : zmodType, the declaration $x : T$ is understood as x : Equality.sort(Zmodule.sort T). It is this explicit chain that drives the resolution of higher-order unification problems and allows structure inference for specific types. For example, the implicit α : zmodType in the term $2 + 2$ is resolved as follows: first Hindley-Milner type inference generates the constraint Equality.sort(Zmodule.sort α) $\equiv_{\beta\iota\delta}$ int. COQ then looks up the Canonical Structure int_eqType declaration associated with the pair (Equality.sort, int), reduces the constraint to Zmodule.sort $\alpha \equiv_{\beta\iota\delta}$ int_eqType which it solves using the Canonical Structure int_zmodType declaration associated with the pair (Zmodule.sort, int_eqType). Note that int_eqType is an eqType, not a Type: canonical projection values are not restricted to types.

Although this clever double use of coercion chains makes telescopes the simplest way of packing structure hierarchies, it raises several theoretical and practical issues for deep or complex hierarchies.

Perhaps the most obvious one is that telescopes are restricted to single inheritance. While multiple inheritance is rare, it does occur in classical algebra, e.g., rings can be unitary and/or commutative. possible to fake multiple inheritance by extending one base structure with the mixin of a second one (similarly to what we do in Section 3.2), provided this mixin was not inlined in the definition of the second base structure.

A more serious limitation is that the head constant of the representation type of any structure in the hierarchy is always equal to the head of the coercion chain, i.e., the Type projection of the topmost structure (Equality.sort here). This is a problem because for both efficiency and robustness, coercions and canonical projections for a type are determined by its head constant, and the topmost projection says very little about the properties of the type (e.g., only that it has equality, not that it is a ring or field).

There is also a severe efficiency issue: the complexity of COQ's term comparison algorithm is exponential in the length of the coercion chain. While this is clearly a problem specific to the current COQ implementation, it is hard and unlikely to be resolved soon, so it seems prudent to seek a design that does not run into it.

2.4   Packed Classes

We now describe a design that achieves full encapsulation of structures, like telescopes, but without the troublesome coercion chains. The key idea is to introduce an intermediate record that bundles all the mixins of a structure, but *not* the representation type; the latter is packed in a second stage, similarly to the top structure of a telescope. We call this intermediate record a *class*, by analogy with open-recursion models of objects, and Haskell type classes; hence in our design structures are represented by *packed classes*.

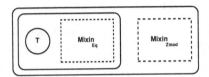
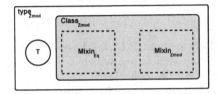

Fig. 1. Telescopes for Equality and Zmodule

Fig. 2. Packed class for Zmodule

Here is the code for the packed class for a **Z**-module:

```
Module Zmodule.
Record mixin_of (T : Type) : Type := ...
Record class_of (T : Type) : Type :=
  Class {base :> Equality.class_of T; ext :> mixin_of T}.
Structure type : Type :=
  Pack {sort :> Type; class : class_of sort; _ : Type}.
Definition unpack K (k : forall T (c : class_of T), K T c) cT :=
  let: Pack T c _ := cT return K _ (class cT) in k _ c.
Definition pack :=
  let k T c m := Pack (Class c m) T in Equality.unpack k.
Coercion eqType cT := Equality.Pack (class cT) cT.
End Zmodule.
Notation zmodType := Zmodule.type.
Notation ZmodType := Zmodule.pack.
Canonical Structure Zmodule.eqType.
```

The definitions of the `class_of` and `type` records are straightforward; `unpack` is a general dependent destructor for `cT : type` whose type is expressed in terms of `sort cT` and `class cT`. Almost all of the code is fixed by the design pattern[1]; indeed the definitions of `type` and `unpack` are literally identical for all packed classes, while usually only the name of the parent class module (here, `Equality`) changes in the definitions of `class_of` and `pack`.

Indeed, the code assumes that `Module Equality` is similarly defined. Because `Equality` is a top structure, the definitions of `class_of` and `pack` in `Equality` reduce to

---

[1] It is nevertheless impractical to use the CoQ `Module` construct to package these three fixed definitions, because of its verbose syntax and technical limitations.

```
Notation class_of := mixin_of.
Definition pack T c := @Pack T c T.
```

While Pack is the primitive constructor for type, the usual constructor is pack, whose only explicit argument is a **Z**-module mixin: it uses Equality.unpack to break the packed eqType supplied by type inference into a type and class, which it combines with the mixin to create the packed zmodType class. Note that pack ensures that the canonical Type projections of the eqType and zmodType structure are *exactly* equal.

The inconspicuous Canonical Structure Zmodule.eqType declaration is the keystone of the packed class design, because it allows CoQ's higher order unification to unify Equality.sort and Zmodule.sort. Note that, crucially, int_eqType and Zmodule.eqType int_zmodType and are *convertible*; this holds in general because Zmodule.eqType merely rearranges pieces of a zmodType. For a deeper structure, we will need to define one such conversion for each parent of the structure. This is hardly inconvenient since each definition is one line, and the convertibility property holds for any composition of such conversions.

3   Description of the Hierarchy

Figure 3 gives an account for the organization of the main structures defined in our libraries. Starred blocks denote algebraic structures that would collapse on an unstarred one in either a classical or an untyped setting. The interface for each structure supplies notation, definitions, basic theory, and generic connections with other structures (like a field being a ring).

In the following, we comment on the main design choices governing the definition of interfaces. For more details, the complete description of all the structures and their related theory, see module ssralg on http://coqfinitgroup.gforge.inria.fr/.

We do not package as interfaces all the possible combinations of the mixins we define: a structure is only packaged when it will be populated in practice. For instance integral domains and fields are defined on top of commutative rings as in standard textbooks [1], and we do not develop a theory for non commutative algebra, which hardly shares results with its commutative counterpart.

3.1   Combinatorial Structures

SubType structures. To handle mathematical objects like "the units of $\mathbf{Z}/n\mathbf{Z}$", one needs to define new types in comprehension-style, by giving a specification over an existing type. The CoQ system already provides a way to build such new types, by the means of Σ–types (dependent pairs). Unfortunately, in general, to compare two inhabitants of such a Σ-type, one needs to compare *both* components of the pairs, i.e. comparing the elements *and* comparing the related proofs.

To take advantage of the proof-irrelevance on boolean predicates when defining these new types, we use the following subType structure:

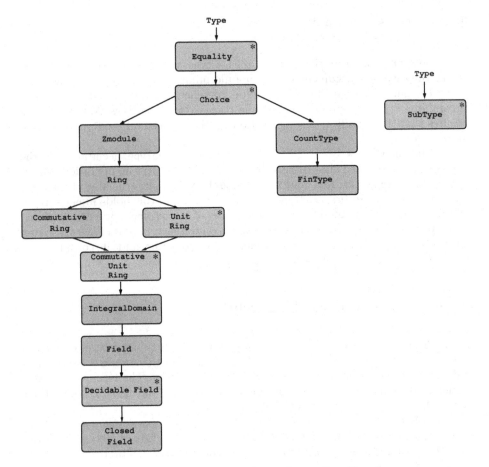

Fig. 3. The algebraic hierarchy in the ssreflect libraries

```
Structure subType (T : Type)(P : pred T): Type := SubType {
  sub_sort :> Type;
  val : sub_sort -> T;
  Sub : forall x, P x -> sub_sort;
  _ : forall K (_ : forall x Px, K (@Sub x Px)) u, K u;
  _ : forall x Px, val (@Sub x Px) = x}.
```

This interface gathers a new type sub_sort for the inhabitants of type T satisfying the boolean predicate P, with a projection val on type T, a Sub constructor, and an elimination scheme. Now, the val projection can be proved injective: to compare two elements of a subType structure on type T it is enough to compare their projections on T. A simple example of subType structure equips the type of finite ordinals:

```
Inductive ordinal (n : nat) := Ordinal m of m < n.
```

where < stands for the boolean strict order on natural numbers. Crucially, replacing a primitive CoQ Σ-type by this encoding makes it possible to *coerce* ordinal to nat. Moreover, in COQ, the definition of this inductive type automatically generates the ord_rect associated elimination scheme. We can hence easily build a (canonical) structure of subType on top of ordinal, by providing ord_rect to the SubType constructor, as the other arguments are trivial.

Types with a choice function. Our intentional, proof-irrelevant representation of finite sets was sufficient to address quotients of finite objects like finite groups [8]. However, this method does not apply to an infinite setting, where arguments like the incomplete basis theorem are pervasive.

The construction of quotients and its practice inside type theory based proofs assistants has been quite intensively studied. In classical systems like HOL, the infrastructure needed to work with quotient types is now well understood [15].

COQ provides support for Setoids [16], which is a way to define quotients by explicitly handling the involved equivalence relation, and the proved substitutive contexts. In our case, quotients have to be dependent types, the dependent parameters often being themselves (dependent) quotients. This combination of dependent types with setoids, which has proved successful in an extensional setting like NUPRL [2], is not adapted to an intentional theory like the one of COQ. Crafting and implementing a Curry-Howard based system featuring the appropriate balance between intentional and extensional type theories, as well as an internalized quotient construction is still work in progress [17].

To circumvent this limitation, we combine the structure of types with equality with a choice operator on decidable predicates in a Choice structure. This structure, at the top of the hierarchy, is embedded in every lower level algebraic structure.

To construct objects like linear bases, we need to choose *sequences* of elements. Yet a choice operator on a given type does not canonically supply a choice operator on sequences of elements of this type. This would indeed require a canonical encoding of (seq T) into T which is in general not possible: for the empty void type, (seq void) has a single inhabitant, while (seq (seq void)) is isomorphic to nat. The solution is to require a choice operator on (seq (seq T)). This leads to a canonical structure of choice for T and any (seq .. (seq T)), using a Gödel-style encoding.

Thus we arrive at the following definition for the Choice mixin and class:

```
Module Choice.
Definition xfun (T : Type) := forall P : pred T, (exists x, P x) -> T.
Definition correct (f : xfun) := forall (P : pred T) xP, P (f P xP).
Definition extensional (f : xfun) := forall P Q xP xQ,
  P =1 Q -> f P xP = f Q xQ.

Record mixin_of (T : Type) : Type := Mixin {
  xchoose : xfun T;
  xchooseP : correct xchoose;
  eq_xchoose : extensional xchoose}.

Record class_of (T : Type) : Type := Class {
  base :> Equality.class_of T; ext2 : mixin_of (seq (seq T)) }.
  ...
End Choice.
```

The xfun choice operator for boolean predicates should return a witness satisfying P, given a proof of the existence of such a witness. It is extensional with respect to both the proofs of existence and the predicates.

Countable structures. A choice structure will still not be transmitted to any desired construction (like product) over types featuring themselves a choice structure. Types with countably many inhabitants on the other side are more amenable to transmit their countability. This leads us to define a structure for these countable types, by requiring an injection pickle : T -> nat on the underlying type T.

Since the Calculus of Inductive Constructions [11] validates the axiom of countable choice, it is possible to derive a Choice structure from any countable type. However since a generic choice construction on arbitrary countable types would not always lead to the expected choice operator, we prefer to embed a Choice structure as base class for the Countable structure.

Finite types structures. The structure of types with a finite number of inhabitants is at the heart of our formalization of finite quotients [8]. The Finite mixin still corresponds to the description given in this reference, but the FinType structure now packs this mixin with a Countable base instead of an eqType. Proofs like the cardinal of the cartesian product of finite types make the most of this computational content for the enumeration. Indeed the use of (computations of) list iterators shrinks the sizes of such proofs by a factor of five compared to the abstract case.

3.2   Advanced Algebraic Structures

Commutative rings, rings with units, commutative rings with units. We package two different structures for both commutative and plain rings, as well as rings enjoying a decidable discrimination of their units. The latter structure is for instance the minimum required on a ring for a polynomial to bound the number of roots of a polynomial on that ring by the number of its roots (see lemma max_ring_poly_roots in module poly). For the ring $\mathbf{Z}/n\mathbf{Z}$, this unit predicate selects coprimes to n. For matrices, it selects those having a non-zero determinant. Its semantic and computational content can prove very efficient when developing proofs.

Yet we also want to package a structure combining the ComRing structure of commutative ring and the UnitRing deciding units, equipping for instance $\mathbf{Z}/n\mathbf{Z}$. This ComUnitRing structure has no mixin of its own:

```
Module ComUnitRing.
Record class_of (R : Type) : Type := Class {
  base1 :> ComRing.class_of R;
  ext :> UnitRing.mixin_of (Ring.Pack base1 R)}.
Coercion base2 R m := UnitRing.Class (@ext R m).
...
End ComUnitRing.
```

Its class packages the class of a ComRing structure with the mixin of a UnitRing (which reflects a natural order for further instantiation). The base1 projection coerces the ComUnitRing class to its ComRing base class. Note that this definition does *not* provide the required coercion path from a ComUnitRing class to its underlying UnitRing class, which is only provided by base2. Now the canonical structures of ComRing *and* UnitRing for a ComUnitRing structure will let the latter enjoy both theories with a correct treatment of type constraints.

Decidable fields. The DecidableField structure models fields with a decidable first order theory. One motivation for defining such a structure is our need for the decidability of the irreductibility of representation of finite groups, which is a valid but highly non trivial [18] property, pervasive in representation theory.

For this purpose, we define a reflected representation of first order formulas. The structure requires the decidability of satisfiability of atoms and their negation. Proving quantifier elimination leads to the decidability for the full first-order theory.

Closed fields. Algebraically closed fields are defined by requiring that any non constant monic polynomial has a root. Since such a structure enjoys quantifier elimination, any closed field canonically enjoys a structure of decidable field.

4 Population of the Hierarchy

The objective of this section is to give a hint of how well we meet the challenge presented in section 1: defining a concrete datatype and extending it externally with several algebraic structures that can be used in reasoning on this type. We aim at showing that this method works smoothly by going through the proofs of easy lemmas that reach across our algebraic hierarchy and manipulate a variety of structures.

4.1 Multiplicative Finite Subgroups of Fields

Motivation, notations and framework. Our first example is the well-known property that *a finite multiplicative subgroup of a field is cyclic*. When applied to F^*, the multiplicative group of non-null elements of a finite field F, it is instrumental in defining the discrete logarithm, a crucial tool for cryptography. Various textbook proofs of this result exist [19,1], prompting us to state it as:

```
1  Lemma field_mul_group_cyclic : forall (gT: finGroupType)
2      (G : {group gT}) (F : fieldType) (f : gT -> F),
3    {in G & G, {morph f : u v / u * v >-> (u * v)%R}} ->
4    {in G, forall x, f x = 1%R <-> x = 1} ->
5    cyclic G.
```

The correspondence of this lemma with its natural language counterpart becomes straightforward, once we dispense with a few notations:

%R : is a scope notation for ring operations.
{group gT} :
> The types we defined in section 3 are convenient for framing their elements in a precise algebraic setting. However, since a large proportion of the properties

we have to consider deal with relations between sets of elements sharing such an algebraic setting, we have chosen to define the corresponding set-theoretic notions, for instance sets and groups, as a selection of elements of their underlying type, as covered in [8].

`{morph f : u v / u * v >-> (u * v)%R}` :

This reads as :$\forall x, y, \ f(x * y) = (fx) *_R (fy)$.

`{in G, P}` :

If P is of the form $\forall x, \ Q(x)$ this means $\forall \ x \ \in G, \ Q(x)$. Additional & symbols (as in line 3 above) extend the notation to relativize multiple quantifiers.

The type of f along with the scope notation `%R`, allows CoQ to infer the correct interpretation for 1 and the product operator on line 3. `field_mul_group_cyclic` therefore states that any finite group G mapped to a field F by f, an injective group morphism for the multiplicative law of F, is `cyclic`.[2]

Fun with polynomials. Our proof progresses as follows: if a is an element of any such group C of order n, we already know that $a^n = 1$. C thus provides at least n distinct solutions to $X^n = 1$ in any group G it is contained in. Moreover, reading the two last lines of our goal above, it is clear that f maps injectively the roots of that equation in G to roots of $X^n = 1$ in F. Since the polynomial $X^n - 1$ has at most n roots in F, the arbitrarily chosen C is exactly the collection of roots of the equation in G.

This suffices to show that for a given n, G contains at most one cyclic group of order n. Thanks to a classic lemma ([19], 2.17), this means that G is cyclic.

An extensive development of polynomial theory on a unitary ring allows us to simply state the following definition in our proof script:

`pose P : {poly F} := ('X^n - 1)%R.`

The construction of the ring of polynomials with coefficients in a unitary ring (with F canonically unifying with such a ring) is triggered by the type annotation, and allows us to transparently use properties based on the datatype of polynomials, such as degree and root lemmas, and properties of the `Ring` structure built on the aforementioned datatype, such as having an additive inverse.

This part of the proof can therefore be quickly dispatched in CoQ. The final lemma on cyclicity works with the cardinal of a partition of G, and is a good use case for the methods developed in [9]; we complete its proof in a manner similar to the provided reference.

Importing various proof contexts inside a proof script is therefore a manageable transaction : here, we only had to provide CoQ with the type of a mapping to an appropriate unitary ring for it to infer the correct polynomial theory.

4.2   Practical Linear Algebra

Motivations. Reasoning on an algorithm that aims at solving systems of linear equations seems a good benchmark of our formalization of matrices. Indeed, the issue of representing fixed-size arrays using dependent types has pretty much become the effigy of the benefits of dependent type-checking, at least for its programatically-minded proponents.

---

[2] Unlike in [8], `cyclic` is a boolean predicate that corresponds to the usual meaning of the adjective.

However, writing functions that deal with those types implies some challenges, among which is dealing with size arguments. We want our library to simplify this task, while sharing operator symbols, and exposing structural properties of objects as soon as their shape ensures they are valid.

LUP decomposition. The LUP decomposition is a recursive function that returns, for any non-singular matrix A, three matrices P, L, U such that L is a lower-triangular matrix, U is an upper-triangular matrix, and P is a permutation matrix, and $PA = LU$.

We invite the reader to refer to ([20], 28.3) for more details about this notorious algorithm. Our implementation is strikingly similar to a tail-recursive version of its textbook parent. Its first line features a type annotation that does all of the work of dealing with matrix dimensions:

```
1   Fixpoint cormen_lup n : let M := 'M_n.+1 in M -> M * M * M :=
2     match n return let M := 'M_(1 + n) in M -> M * M * M with
3     | 0 => fun A => (1%:M, 1%:M, A)
4     | n'.+1 => fun A =>
5       let k := odflt 0 (pick [pred k | A k 0 != 0]) in
6       let A' := rswap A 0 k in
7       let Q := tperm_mx F 0 k in
8       let Schur := ((A k 0)^-1 *m: llsubmx A') *m ursubmx A' in
9       let: (P', L', U') := cormen_lup (lrsubmx A' - Schur) in
10      let P := block_mx 1 0 0 P' * Q in
11      let L := block_mx 1 0 ((A k 0)^-1 *m: (P' *m llsubmx A')) L' in
12      let U := block_mx (ulsubmx A') (ursubmx A') 0 U' in
13      (P, L, U)
14    end.
```

Here, in a fashion congruent with the philosophy of our archive, we return a value for any square matrix A, rather than just for non-singular matrices, and use the following shorthand:

`odflt 0 (pick [pred k:fT | P k])`
    returns k, an inhabitant of the finType fT such that P k if it exists, and returns 0 otherwise.

`blockmx Aul Aur All Alr`
    reads as $\left(\begin{smallmatrix} A_{ul} & A_{ur} \\ A_{ll} & A_{lr} \end{smallmatrix}\right)$.

`ulsubmx, llsubmx, ursubmx, lrsubmx`
    are auxiliary functions that use the *shape*[3] of the dependent parameter of their argument A to return respectively $A_{ul}, A_{ll}, A_{ur}, A_{lr}$ when A is as represented above. Notice we will now denote their application using the same subscript pattern.

The rest of our notations can be readily interpreted, with 1 and 0 coercing respectively to identity and null matrices of the right dimension, and (A i j) returning the appropriate $a_{i,j}$ coefficient of A through coercion.

---

[3] As crafted in line 2 above

Correctness of the algorithm. We will omit in this article some of the steps involved in proving that the LUP decomposition is correct: showing that P is a permutation matrix, for instance, involved building a theory about those matrices that correspond to a permutation map over a finite vector. But while studying the behavior of this subclass with respect to matrix operations gave some hint of the usability of our matrix library [4], it is not the part where our infrastructure shines the most.

The core of the correction lies in the following equation:

```
Lemma cormen_lup_correct : forall n A,
  let: (P, L, U) := @cormen_lup F n A in P * A = L * U.
```

Its proof proceeds by induction on the size of the matrix. Once we make sure that A' and Q (line 7) are defined coherently, it is not hard to see that we are proving is[5]:

$$
\begin{pmatrix} 1 & 0 & \cdots & 0 \\ 0 & & & \\ \vdots & & P'*A' & \\ 0 & & & \end{pmatrix} = \begin{pmatrix} 1 & & 0 & \cdots & 0 \\ a_{k,0}^{-1} \cdot P' *_m A'_{ll} & & & L' & \end{pmatrix} * \begin{pmatrix} A'_{ul} & & A'_{ur} \\ 0 & & \\ \vdots & & U' \\ 0 & & \end{pmatrix} \tag{1}
$$

where P',L',U' are provided by the induction hypothesis

$$
P' \left(A'_{lr} - a_{k,0}^{-1} \cdot A'_{ll} *_m A'_{ur} \right) = L' * U' \tag{2}
$$

Notice that we transcribe the distinction COQ does with the three product operations involved: the scalar multiplication ($\cdot$), the square matrix product ($*$), and the matrix product, accepting arbitrary sized matrices ($*_m$). Using block product expansion and a few easy lemmas allows us to transform (1) into:

$$
\begin{pmatrix} A'_{ul} & & A'_{ur} \\ P' *_m A'_{ll} & P' *_m A'_{lr} \end{pmatrix} = \begin{pmatrix} A'_{ul} & & A'_{ur} \\ a_{k,0}^{-1} \cdot P' *_m A'_{ll} *_m A'_{ul} & a_{k,0}^{-1} \cdot P' *_m A'_{ll} *_m A'_{ur} \\ & & + L' *_m U' \end{pmatrix} \tag{3}
$$

At this stage, we would like to rewrite our goal with (2) —named IHn in our script—, even though its right-hand side does not occur exactly in the equation. However, SSREFLECT has no trouble expanding the definition of the ring multiplication provided in (2) to see it exactly matches the pattern[6] - [L' *m U']IHn.

We conclude by identifying the blocks of (3) one by one. The most tedious step consists in treating the lower left block, which depends on whether we have been able to chose a non-null pivot in creating A' from A. Each alternative is resolved by case on the coefficients of that block, and it is only in that part that we use the fact that the matrix coefficients belong to a field. The complete proof is fourteen lines long.

---

[4] The theory, while expressed in a general manner, is less than ninety lines long.
[5] We will write block expressions modulo associativity and commutativity, to reduce parenthesis clutter.
[6] See [10] for details on the involved notation for the rewrite tactic.

5   Related Work

The need for packaging algebraic structures and formalizing their relative inheritance and sharing inside proof assistants is reported in literature as soon as these tools prove mature enough to allow the formalisation of significant pieces of algebra [2]. The set-theoretic Mizar Mathematical Library (MML) certainly features the largest corpus of formalized mathematics, yet covering rather different theories than the algebraic ones we presented here. Little report is available on the organization a revision of this collection of structures, apart from comments [7] on the difficulty to *maintain* it. The Isabelle/HOL system provides foundations for developing abstract algebra in a classical framework containing algebraic structures as first-class citizens of the logic and using a type-class like mechanism [6]. This library proves Sylow theorems on groups and the basic theory of rings of polynomials.

Two main algebraic hierarchies have been built using the CoQ system: the seminal abstract Algebra repository [4], covering algebraic structures from monoids to modules, and the CCorn hierarchy [5], mainly devoted to a constructive formalisation of real numbers, and including a proof of the fundamental theorem of algebra. Both are axiomatic, constructive, and setoid based. They have proved rather difficult to extend with theories like linear or multilinear algebra, and to populate with more concrete instances. In both cases, limitations mainly come from the pervasive use of setoids and the drawbacks of telescope based hierarchies pointed in section 2.

The closest work to ours is certainly the hierarchy built in Matita [21], using telescopes and a more liberal system of coercions. This hierarchy, despite including a large development in constructive analysis [22], is currently less populated than ours. For example, no counterpart of the treatment of polynomials presented in section 4 is described in the Matita system.

We are currently extending our hierarchy to extend the infrastructure to the generic theory of vector spaces and modules.

References

1. Lang, S.: Algebra. Springer, Heidelberg (2002)
2. Jackson, P.: Enhancing the Nuprl proof-development system and applying it to computational abstract algebra. PhD thesis, Cornell University (1995)
3. Betarte, G., Tasistro, A.: Formalisation of systems of algebras using dependent record types and subtyping: An example. In: Proc. 7th Nordic workshop on Programming Theory (1995)
4. Pottier, L.: User contributions in Coq, Algebra (1999),
   http://coq.inria.fr/contribs/Algebra.html
5. Geuvers, H., Pollack, R., Wiedijk, F., Zwanenburg, J.: A constructive algebraic hierarchy in Coq. Journal of Symbolic Computation 34(4), 271–286 (2002)
6. Haftmann, F., Wenzel, M.: Local theory specifications in Isabelle/Isar. In: Berardi, S., Damiani, F., de'Liguoro, U. (eds.) TYPES 2008. LNCS, vol. 5497, pp. 153–168. Springer, Heidelberg (2009)
7. Rudnicki, P., Schwarzweller, C., Trybulec, A.: Commutative algebra in the Mizar system. J. Symb. Comput. 32(1), 143–169 (2001)

8. Gonthier, G., Mahboubi, A., Rideau, L., Tassi, E., Théry, L.: A Modular Formalisation of Finite Group Theory. In: Schneider, K., Brandt, J. (eds.) TPHOLs 2007. LNCS, vol. 4732, pp. 86–101. Springer, Heidelberg (2007)
9. Bertot, Y., Gonthier, G., Ould Biha, S., Pasca, I.: Canonical big operators. In: Mohamed, O.A., Muñoz, C., Tahar, S. (eds.) TPHOLs 2008. LNCS, vol. 5170, pp. 86–101. Springer, Heidelberg (2008)
10. Gonthier, G., Mahboubi, A.: A small scale reflection extension for the Coq system. INRIA Technical report, http://hal.inria.fr/inria-00258384
11. Paulin-Mohring, C.: Définitions Inductives en Théorie des Types d'Ordre Supérieur. Habilitation à diriger les recherches, Université Claude Bernard Lyon I (1996)
12. Sozeau, M., Oury, N.: First-Class Type Classes. In: Mohamed, O.A., Muñoz, C., Tahar, S. (eds.) TPHOLs 2008. LNCS, vol. 5170, pp. 278–293. Springer, Heidelberg (2008)
13. Pollack, R.: Dependently typed records in type theory. Formal Aspects of Computing 13, 386–402 (2002)
14. Bruijn, N.G.D.: Telescopic mappings in typed lambda calculus. Information and Computation 91, 189–204 (1991)
15. Paulson, L.C.: Defining Functions on Equivalence Classes. ACM Transactions on Computational Logic 7(4), 658–675 (2006)
16. Barthe, G., Capretta, V., Pons, O.: Setoids in type theory. Journal of Functional Programming 13(2), 261–293 (2003)
17. Altenkirch, T., McBride, C., Swierstra, W.: Observational equality, now! In: Proceedings of the PLPV 2007 workshop, pp. 57–68. ACM, New York (2007)
18. Olteanu, G.: Computing the Wedderburn decomposition of group algebras by the Brauer-Witt theorem. Mathematics of Computation 76(258), 1073–1087 (2007)
19. Rotman, J.J.: An Introduction to the Theory of Groups. Springer, Heidelberg (1994)
20. Cormen, T.H., Leiserson, C.E., Rivest, R.L., Stein, C.: Introduction to Algorithms, 2nd edn. McGraw-Hill, New York (2003)
21. Sacerdoti Coen, C., Tassi, E.: Working with Mathematical Structures in Type Theory. In: Miculan, M., Scagnetto, I., Honsell, F. (eds.) TYPES 2007. LNCS, vol. 4941, pp. 157–172. Springer, Heidelberg (2008)
22. Sacerdoti Coen, C., Tassi, E.: A constructive and formal proof of Lebesgue Dominated Convergence Theorem in the interactive theorem prover Matita. Journal of Formalized Reasoning 1, 51–89 (2008)

Practical Tactics for Separation Logic

Andrew McCreight

Portland State University
mccreigh@cs.pdx.edu

Abstract. We present a comprehensive set of tactics that make it practical to use separation logic in a proof assistant. These tactics enable the verification of partial correctness properties of complex pointer-intensive programs. Our goal is to make separation logic as easy to use as the standard logic of a proof assistant. We have developed tactics for the simplification, rearranging, splitting, matching and rewriting of separation logic assertions as well as the discharging of a program verification condition using a separation logic description of the machine state. We have implemented our tactics in the Coq proof assistant, applying them to a deep embedding of Cminor, a C-like intermediate language used by Leroy's verified CompCert compiler. We have used our tactics to verify the safety and completeness of a Cheney copying garbage collector written in Cminor. Our ideas should be applicable to other substructural logics and imperative languages.

1 Introduction

Separation logic [1] is an extension of Hoare logic for reasoning about shared mutable data structures. Separation logic specifies the contents of individual cells of memory in a manner similar to linear logic [2], avoiding problems with reasoning about aliasing in a very natural fashion. For this reason, it has been successfully applied to the verification of a number of pointer-intensive applications such as garbage collectors [3,4].

However, most work on separation logic has involved paper, rather than machine-checkable, proofs. Mechanizing a proof can increase our confidence in the proof and potentially automate away some of the tedium in its construction. We would like to use separation logic in a proof assistant to verify deep properties of programs that may be hard to check fully automatically. This is difficult because the standard tactics of proof assistants such as Coq [5] cannot effectively deal with the linearity properties of separation logic. In contrast, work such as Smallfoot [6] focuses on the automated verification of lightweight specifications. We discuss other related work in Sect. 7.

In this paper, we address this problem with a suite of tools for separation-logic-based program verification of complex pointer-intensive programs. These tools are intended for the interactive verification of Cminor programs [7] in the Coq proof assistant, but should be readily adaptable to similar settings. We have chosen Cminor because it can be compiled using the CompCert verified

S. Berghofer et al. (Eds.): TPHOLs 2009, LNCS 5674, pp. 343–358, 2009.

compiler [7], allowing for some properties of source programs to be carried down to executable code. We have tested the applicability of these tools by using them to verifying the safety of a Cheney garbage collector [8], as well as a number of smaller examples.

The main contributions of this paper are a comprehensive set of tactics for reasoning about separation logic assertions (including simplification, rearranging, splitting, matching and rewriting) and a program logic and accompanying set of tactics for program verification using separation logic that strongly separate reasoning about memory from more standard reasoning. Together these tactics essentially transform Coq into a proof assistant for separation logic. The tactics are implemented in a combination of direct and reflective styles. The Coq implementation is available online from http://cs.pdx.edu/~mccreigh/ptsl/

Our tool suite has two major components. First, we have tactics for reasoning about separation logic assertions. These are focused on easing the difficulty of working with a linear-style logic within a more conventional proof assistant. These tools enable the simplification and manipulation of separation logic hypotheses and goals, as well as the discharging of goals that on paper would be trivial. These tactics are fairly modular and should be readily adaptable to other settings, from separation logic with other memory models to embeddings of linear logic in proof assistants.

The second component of our tool set is a program logic and related tactics. The program logic relates the dynamic semantics of the program to its specification. The tactics step through a procedure one statement at a time, enabling the "programmer's intuition" to guide the "logician's intuition". At each program step, there is a separation logic-based description of the current program state. A verified verification condition generator produces a precondition given the postcondition of the statement. The tactics are able to automatically solve many such steps, and update the description of the state once the current statement has been verified. Loop and branch join point annotations must be manually specified.

Organization of the Paper. We discuss the Cminor abstract machine in Sect. 2. In Sect. 3, we discuss the standard separation logic assertions we use, and the series of tactics we have created for reasoning with them. Then we discuss our program logic and its associated tactics in Sect. 4. In Sect. 5 we show how all of these pieces come together to verify the loop body of an in-place linked list reversal program. Finally, we briefly discuss our use of the previously described tactics to verify a garbage collector in Sect. 6, then discuss related work in more detail and conclude in Sect. 7.

2   Cminor

Our program tools verify programs written in Cminor, a C-like imperative language. Cminor is an intermediate language of the CompCert [7] compiler, which is a *semantics preserving* compiler from C to PowerPC assembly, giving us a

$$v ::= \mathsf{Vundef} \mid \mathsf{Vword}(w) \mid \mathsf{Vptr}(a)$$
$$e ::= v \mid \mathsf{x} \mid [e] \mid e + e \mid e \mathrel{!}= e \mid \ldots$$
$$s ::= x := e \mid [e_1] := e_2 \mid s_1; s_2 \mid \mathsf{while}(e)\ s \mid \mathsf{return}\ e$$
$$\sigma ::= (m, V)$$

Fig. 1. Cminor syntax

```
y := NULL;     // y is the part of list that has been reversed
while(x != NULL) (
   t := [x+4];  [x+4] := y;    // t is next x, store prev x
   y := x;  x := t             // advance y and x
);
return y
```

Fig. 2. In-place linked list reversal

potential path to verified machine code. We use a simplified variant of Cminor that only supports 32-bit integer and pointer values. Fig. 1 gives the syntax. We write w for 32-bit integers and a for addresses, which are always 32-bit aligned. A value v is either undefined, a 32-bit word value or a pointer. We write NULL for Vword(0). An expression e is either a value, a program variable, a memory load (*i.e.*, a dereference), or a standard arithmetic or logical operations such as addition or comparison. In this paper, a statement s is either a variable assignment, a store to memory, a sequence of statements, a while loop or a return. In the actual implementation, while loops are implemented in terms of more primitive control structures (loops, branches, blocks and exits to enclosing blocks) capable of encoding all structured control flow. Our implementation also has procedure calls. We omit discussion of these constructs, which are supported by all of the logics and tactics presented in this paper, for reasons of space.

A memory m of type *Mem* is a partial mapping from addresses to values, while a variable environment V is a partial mapping from Cminor variables x to values. A state σ is a memory plus a variable environment. In our implementation, a state also includes a stack pointer. We define two projections $\mathsf{mem}(\sigma)$ and $\mathsf{venv}(\sigma)$ to extract the memory and variable environment components of a state.

We have formally defined a standard small-step semantics for Cminor [9], omitted here for reasons of space. Expression evaluation $\mathsf{eval}(\sigma, e)$ evaluates expression e in the context of state σ and either returns $\mathsf{Some}(v)$ if execution succeeds or None if it fails. Execution of an expression will only fail if it accesses an undefined variable or an invalid memory location. A valid memory location has been allocated by some means, such as a function call, but not freed. All other degenerate cases (such as $\mathsf{Vword}(3) + \mathsf{Vundef}$) evaluate to $\mathsf{Some}(\mathsf{Vundef})$. Unlike in C, all address arithmetic is done in terms of bytes, so the second field of an object located at address a is located at address $a + 4$.

We will the program fragment shown in Fig. 2 in the remainder of this paper to demonstrate our suite of tactics. In-place list reversal takes an argument x that is a linked list and reverses the linked list in place, returning a pointer to the new head of the list. A linked list structure at address a has two fields, at a and $a + 4$.

3   Separation Logic Assertions

Imperative programs often have complex data structures. To reason about these data structures, separation logic assertions [1] describe memory by treating memory cells as a linear resource. In this section, we will describe separation logic assertions and associated tactics for Cminor, but they should be applicable to other imperative languages.

Fig. 3 gives the standard definitions of the separation logic assertions we use in this paper. We write P for propositions and T for types in the underlying logic, which in our case is the Calculus of Inductive Constructions [10] (CIC). Propositions have type $Prop$. We write A and B for separation logic assertions, implemented using a shallow embedding [11]. Each separation logic assertion is a memory predicate with type $Mem \rightarrow Prop$, so we write $A\ m$ for the proposition that memory m can be described by separation logic predicate A.

The separation logic assertion *contains*, written $v \mapsto v'$, holds on a memory m if v is an address that is the only element of the domain of m and $m(v) = v'$. The empty assertion **emp** only holds on empty memory. The trivial assertion **true** holds on every memory. The modal operator $!P$ from linear logic (also adapted to separation logic by Appel [12]) holds on a memory m if the proposition P is true and m is empty. The existential $\exists x : T.\ A$ is analogous to the standard existential operator. We omit the type T when it is clear from context, and follow common practice and write $a \mapsto -$ for $\exists x.\ a \mapsto x$.

The final and most crucial separation logic operator we will be using in this paper is *separating conjunction*, written $A * B$. This holds on a memory m if m can be split into two non-overlapping memories m_1 and m_2 such that A holds on m_1 and B holds on m_2. ($m = m_1 \uplus m_2$ holds if m is equal to $m_1 \cup m_2$ and the domains of m_1 and m_2 are disjoint.) This operator is associative and commutative, and we write $(A * B * C)$ for $(A * (B * C))$. This operator is used to specify the *frame rule*, which is written as follows in conventional Hoare logic:

$$\frac{\{A\}s\{A'\}}{\{A * B\}s\{A' * B\}}$$

$$(v \mapsto v')\ m ::= (m = \{v \rightsquigarrow v'\}) \qquad \textbf{emp}\ m ::= (m = \emptyset)$$
$$\textbf{true}\ m \quad ::= \textsf{True} \qquad\qquad\quad (!P)\ m \ ::= P \wedge \textbf{emp}\ m$$

$$(\exists x : T.\ A)\ m ::= (\exists x : T.\ A\ m)$$
$$(A * B)\ m \quad ::= \exists m_1, m_2.\ (m = m_1 \uplus m_2) \wedge A\ m_1 \wedge B\ m_2$$

Fig. 3. Definition of separation logic assertions

B describes parts of memory that s does not interact with. The frame rule is most commonly applied at procedure call sites. We have found that we do not need to manually instantiate the frame rule, thanks to our tactics and program logic.

We can use these basic operators in conjunction with Coq's standard facilities for inductive and recursive definitions to build assertions for more complex data structures. For instance, we can inductively define a separation logic assertion $\mathsf{llist}(v, l)$ that holds on a memory that consists entirely of a linked list with its head at v containing the values in the list l. A list l (at the logical level) is either empty (written nil) or contains an element X appended to the front of another list l (written $X :: l$).

$$\mathsf{llist}(a, nil) \quad ::= \ !(a = \mathsf{NULL})$$
$$\mathsf{llist}(a, v :: l') ::= \exists a'. \ a \mapsto v * (a + 4) \mapsto a' * \mathsf{llist}(a', l')$$

From this definition and basic lemmas about separation logic assertions, we can prove a number of useful properties of linked lists. For instance, if $\mathsf{llist}(v, l)$ holds on a memory and v is not NULL then the linked list is non-empty.

We can use this predicate to define part of the loop invariant for the list reversal example given in Section 2. If the variables x and y have the value v_1 and v_2, then memory m must contain two separate, non-overlapping linked lists with values l_1 and l_2. In separation logic, this is written $(\mathsf{llist}(v_1, l_1) * \mathsf{llist}(v_2, l_2)) \ m$.

3.1   Tactics

Defining the basic separation logic predicates and verifying their basic properties is not difficult, even in a mechanized setting. What can be difficult is actually constructing proofs in a proof assistant such as Coq because we are attempting to carry out linear-style reasoning in a proof assistant with a native logic that is not linear.

If A, B, C and D are regular propositions, then the proposition that $(A \land B \land C \land D)$ implies $(B \land (A \land D) \land C)$ can be easily proved in a proof assistant. The assumption and goal can be automatically decomposed into their respective components, which can in turn be easily solved.

If A, B, C and D are separation logic assertions, proving the equivalent goal, that for all m, $(A * B * C * D) \ m$ implies $(B * (A * D) * C) \ m$, is more difficult. Unfolding the definition of $*$ from Fig. 3 and breaking down the assumption in a similar way will involve large numbers of side conditions about memory equality and disjointness. While Marti et $al.$ [13] have used this approach, it throws away the abstract reasoning of separation logic.

Instead, we follow the approach of Reynolds [1] and others and reason about separation logic assertions using basic laws like associativity and commutativity. However this is not the end of our troubles. Proving the above implication requires about four applications of associativity and commutativity lemmas. This can be done manually, but becomes tedious as assertions grow larger. In real proofs, these assertions can contain more than a dozen components.

To mitigate these problems we have constructed a variety of separation logic tactics. Our goal is to make constructing proofs about separation logic predicates no more difficult than reasoning about Coq's standard logic by making the application of basic laws about separation logic assertions as simple as possible.

Now we give an example of the combined usage of our tactics. In the initial state, we have a hypothesis H that describes the current memory, and a goal on that same memory that we wish to prove. First the rewriting tactic uses a previously proved lemma about empty linked lists to simplify the assertion:

$$\frac{H:(B*A)\ m}{(A*\mathsf{llist}(v,nil)*B')\ m} \quad \rightarrow \quad \frac{H:(B*A)\ m}{(A*!(v=\mathsf{NULL})*B')\ m}$$

Next the simplification tactic extracts $!(v = \mathsf{NULL})$, creating a new subgoal $v = \mathsf{NULL}$ (not shown). Finally, a matching tactic cancels out the common parts of the hypothesis and goal, leaving a smaller proof obligation.

$$\rightarrow \quad \frac{H:(B*A)\ m}{(A*B')\ m} \quad \rightarrow \quad \frac{H':B\ m'}{B'\ m'}$$

The m' in the final proof state is fresh: it must be shown that $B\ m'$ implies $B'\ m'$ for all m'[1]. This example shows most of our tactics for separation logic assertions: simplification, rearranging, splitting, matching, and rewriting. We discuss general implementation issues, then discuss each type of tactic.

Implementation. Efficiency matters for these tools. If they are slow or produce gigantic proofs they will not be useful. Our goal is to enable the interactive development of proofs, so tactics should run quickly. Our tactics are implemented entirely in Coq's tactic language L_{tac}. To improve efficiency and reliability, some tactics are implemented reflectively [14]. Implementing a tactic reflectively means implementing it mostly in a strongly typed functional language (CIC) instead of Coq's untyped imperative tactic language. Reflective tactics can be efficient because the heavy work is done at the level of propositions, not proofs. Strong typing allows many kinds of errors in the tactics to be statically detected.

Simplification. The simplification tactic ssimpl puts a separation logic assertion into a normal form to make further manipulation simpler and to clean up assertions after other tactics have been applied. ssimpl in H simplifies the hypothesis H and ssimpl simplifies the goal. Here is an example of the simplification of a hypothesis (if a goal was being simplified, the simplification would produce three goals instead of three hypotheses):

$$\frac{H:(A*\exists x.x\mapsto v*\mathbf{emp}*\mathbf{true}*(B*\mathbf{true})*\ !P)\ m}{\cdots} \quad \rightarrow \quad \frac{\begin{array}{l}x:\mathsf{addr}\\ H':(A*x\mapsto v*B*\mathbf{true})\ m\\ H'':P\end{array}}{\cdots}$$

---

[1] The final step could be more specific (for instance, we know that m' must be a subset of m), but this is rarely useful in practice.

The basic simplifications are as follows:

1. All separation logic existentials are eliminated or introduced with Coq's meta-level existential variables (which must eventually be instantiated).
2. All modal predicates !P are removed from the assertion, and either turned into new goals or hypotheses. The tactic attempts to solve any new goals.
3. All instances of **true** are combined and moved to the right.
4. For all A, $(A * \mathbf{emp})$ and $(\mathbf{emp} * A)$ are replaced with A.
5. For all A, B and C, $((A * B) * C)$ is replaced with $(A * (B * C))$.

In addition, there are a few common cases where the simplifier can make more radical simplifications:

1. If Vundef $\mapsto v$ or $0 \mapsto v$ are present anywhere in the assertion (for any value v), the entire assertion is equivalent to False, because only addresses can be in the domain of $\mapsto$.
2. If an entire goal assertion can be simplified to **true** the simplifier tactic immediately solves the goal.
3. The tactic attempts to solve a newly simplified goal by assumption.

While most of this simplification could be implemented using rewriting, we have implemented it reflectively. The reflective tactic examines the assertion and simplifies it in a single pass, instead of the multiple passes that may be required for rewriting. In informal testing, this approach was faster than rewriting.

Rearranging. Our rearranging tactic srearr allows the separating conjunction's commutativity and associativity properties to be applied in a concise and declarative fashion. This is useful because the order and association of the components of a separation logic assertion affect the behavior of the splitting and rewriting tactics, which are described later. The rearranging tactic is invoked by srearr T, where T is a tree describing the final shape of the assertion. There is an equivalent tactic for hypotheses. The shape of the tree gives the desired *association* of the assertion while the numbering of the nodes give the desired *order*. Describing the desired term without explicitly writing it in the proof script makes the tactic less brittle. As with other tactics, srearr assumes that ssimpl has been used already, and thus is of the form $(A_1 * A_2 * ... * A_n)\ m$. The tactic fails if the ordering given is not a valid permutation.

Here is an example, invoked[2] with srearr $[7, 6, [[5, 4], 3], 1, 2]$:

$$\frac{\cdots}{(A * B * C * D * E * F * G)\ m} \quad \rightarrow \quad \frac{\cdots}{(G * F * ((E * D) * C) * A * B)\ m}$$

Permutation and reassociation are implemented as separate passes. Permutation is implemented reflectively: the tree describing the rearrangement is flattened into a list, and the assertion is also transformed into a list. For instance, the

---

[2] The actual syntax of the command is slightly different, to avoid colliding with existing syntactic definitions: `srearr .[7, 6, .[ .[5, 4], 3], 1, 2]%SL`. Coq's ability to define new syntax along with its implicit coercions makes this lighter weight than it would be otherwise.

tree $[[3,1],2]$ would become the list $[3,1,2]$ and the assertion $(A * B * C)$ would become the list $[A, B, C]$. The permutation list $[3,1,2]$ is used to reorder the assertion list $[A, B, C]$. In this example, the resulting assertion list is $[C, A, B]$. The initial assertion list is logically equivalent to the final one if the permutation list is a permutation of the indices of the assertion list. This requirement is dynamically checked by the tactic. Reassociation is implemented directly by examining the shape of the tree.

Splitting. The tactic ssplit subdivides a separation logic proof by creating a new subgoal for each corresponding part of the hypothesis and goal. This uses the standard separation logic property that if $(\forall m. \ A \ m \to A' \ m)$ and $(\forall m. \ B \ m \to B' \ m)$, then $(\forall m. \ (A * B) \ m \to (A' * B') \ m)$. Here is an example of the basic use of this tactic:

$$\frac{H : (A * B * C) \ m}{(E * F * G) \ m} \quad \longrightarrow \quad \frac{H_1 : A \ m_1}{E \ m_1} \quad \frac{H_2 : B \ m_2}{F \ m_2} \quad \frac{H_3 : C \ m_3}{G \ m_3}$$

Initially there is a goal and a hypothesis H describing memory. Afterward, there are three goals, each with a hypothesis. Memories m_1, m_2 and m_3 are freshly created, and represent disjoint subsets covering the original memory m. Splitting must be done with care as it can lead to a proof state where no further progress is possible.

The splitting tactic also has a number of special cases to solve subgoals involving $\mapsto$, and applies heuristics to try to solve address equalities that are generated. Here are two examples of this:

$$\frac{H : (a \mapsto v) \ m}{(a \mapsto v') \ m} \quad \longrightarrow \quad \frac{}{v = v'} \qquad\qquad \frac{H : (a \mapsto v) \ m}{(b \mapsto v) \ m} \quad \longrightarrow \quad \frac{}{a = b}$$

Matching. The matching tactic searchMatch cancels out matching parts of the hypothesis and goal. This matching is just syntactic equality, with the addition of some special cases for $\mapsto$. Here is an example of this tactic:

$$\frac{E : v_1 = v'_1 \quad H : (D * A * a \mapsto v_1 * B * b \mapsto v_2 * \textbf{true}) \ m}{(B' * b \mapsto - * D * a \mapsto v'_1 * A * \textbf{true}) \ m} \quad \longrightarrow \quad \frac{E : v_1 = v'_1 \quad H : (B * \textbf{true}) \ m'}{(B' * \textbf{true}) \ m'}$$

The assertion for address a is cancelled due to the equality $v_1 = v'_1$. Notice that the predicate **true**, present in both the hypothesis and goal, is *not* cancelled. If B implies $(B' * \textbf{true})$ then cancelling out **true** from goal and hypothesis will cause a provable goal to become unprovable. This is the same problem presented by the additive unit $\top$ in linear logic, which can consume any set of linear resources. We do not have this problem for matching other separation logic predicates as they generally do not have this sort of slack.

Matching is implemented by iterating over the assertions in the goal and hypothesis, looking for matches. Any matches that are found are placed in corresponding positions in the assertions, allowing the splitting tactic to carry out the actual cancellation.

Rewriting. In Coq, supporting rewriting of logically equivalent assertions must be implemented using the setoid rewriting facility. We have done this for the assertions described in this paper. By adding rewrite rules to a particular database, the user can extend our tools to support simplification of their own assertions.

At the beginning of Sect. 3.1, we gave an example of the use of the rewriting tactic. In the first proof state, the goal that contains an empty linked list that must be eliminated. Assume we have proved a theorem *arrayEmpty* having type $\forall x.\ \mathsf{array}(x, nil) \Rightarrow \mathbf{emp}$. The tactic srewrite *arrayEmpty* will change the proof state to the second proof state.

4 Program Logic

Our Cminor program logic is a verified verification condition generator (VCG). We discuss our VCG then the related tactics.

4.1 Verification Condition Generator

The VCG, stmPre, is a weakest precondition generator defined as a recursive function in Coq's logic. It takes as arguments the statement to be verified along with specifications for the various ways to exit the statement, and returns a state predicate that is the precondition for the statement. Verification requires showing that a user-specified precondition is as least as strong as the VC.

The design of the VCG is based on Appel and Blazy's program logic for Cminor [9]. Their program logic is structured as a traditional Hoare triple (though with more than three components), directly incorporates separation logic, and has more side conditions for some rules. On the other hand, our VCG is defined in terms of the operational semantics of the machine.

For the subset of Cminor we have defined in Sect. 2, the VCG only needs one specification argument, a state predicate q that is the postcondition of the current statement. The full version of the VCG that we have implemented takes other arguments giving the specifications of function calls, the precondition of program points that can be jumped to, and the post condition of the current procedure.

The definition of some of the cases of the VCG is given in Fig. 4. To simplify the presentation, only arguments required for these cases are included, leaving three arguments to stmPre: the statement, its postcondition q, and the current state σ. The precondition of a sequence of statements is simply the composition of VCs for the two statements: we generate a precondition for s', then use that as the postcondition of s.

For more complex statements, we have to handle the possibility of execution failing. To do this in a readable way, we use a Haskell-style do-notation to encode a sort of error monad. Operations that can fail, such as expression evaluation, return $\mathsf{Some}(R)$ if they succeed with result R, and None if they fail. Syntactically, the term "do $x \leftarrow M;\ N$" is similar to a let expression "let $x = M$ in N": the variable x is given the value of M, and is bound in the body N. However, the

$$\text{stmPre } (s; s') \ q \ \sigma ::= \\ \text{stmPre } s \ (\text{stmPre } s' \ q) \ \sigma$$

$$\text{stmPre } (x := e) \ q \ \sigma ::= \\ \text{do } v \leftarrow \text{eval}(\sigma, e); \\ \text{do } \sigma' \leftarrow \text{setVar}(\sigma, x, v); \\ q \ \sigma'$$

$$\text{stmPre } ([e_1] := e_2) \ q \ \sigma ::= \\ \text{do } v_1 \leftarrow \text{eval}(\sigma, e_1); \\ \text{do } v_2 \leftarrow \text{eval}(\sigma, e_2); \\ \text{do } \sigma' \leftarrow \text{storeVal } \sigma \ v_1 \ v_2; \\ q \ \sigma'$$

$$\text{stmPre } (\text{while}(e) \ s) \ q \ \sigma ::= \\ \exists I. \ I \ \sigma \wedge \forall \sigma'. \ I \ \sigma' \rightarrow \\ \text{do } v \leftarrow \text{eval}(\sigma', e); \\ (v \neq \text{Vundef}) \wedge (trueVal(v) \rightarrow \text{stmPre } s \ I \ \sigma') \wedge (v = \text{NULL} \rightarrow q \ \sigma')$$

$$(\text{do } x \leftarrow \text{Some}(v); \ P) ::= P[v/x] \qquad (\text{do } x \leftarrow \text{None}; \ P) ::= \text{False}$$

Fig. 4. Verification condition generator

reduction of this term differs, as shown in Fig. 4. In the case where $M = \text{None}$, evaluation has failed, so the entire VC becomes equivalent to False, because failure is not allowed.

The cases for variable assignment and storing to memory follow the dynamic semantics of the machine. Variable assignment attempts to evaluate the expression e, then attempts to update the value of the variable x. If it succeeds, then the postcondition q must hold on the resulting state σ'. Store works in the same way: evaluation of the two expressions and a store are attempted. For the store to succeed, v_1 must be a valid address in memory. As with assignment, the postcondition q must hold on the resulting state. With both of these statements, if any of the intermediate evaluations fail then the entire VC will end up being False, and thus impossible to prove.

The case for while loops is fairly standard. A state predicate I must be selected as a loop invariant. I must hold on the initial state σ. Furthermore, for any other states σ' such that $I \ \sigma'$ holds, it must be possible to evaluate the expression e to a value v, which cannot be Vundef. If the value v is a "true" value (*i.e.*, is either a pointer or a non-zero word value) then the precondition of the loop body s must hold, where the postcondition of the body is I. If the value is false (equal to Vword(0)) then the postcondition of the entire loop must hold.

We have mechanically verified the soundness of the verification condition generator as part of the safety of the program logic: if the program is well-formed, then we can either take another step or we have reached a valid termination state for the program.

4.2   Variable Environment Reasoning

We use separation logic to reason about memory, but define a new predicate to reason about variable environments. At any given point in a program, the contents of the variable environment V can be described by a predicate veEqv $S \ V' \ V$. S is a set of variables that are valid in V, and V' gives the values of some of the variables in V.

4.3   Tactics

The tactic vcSteps lazily unfold the VC and attempts to use the separation logic description of the state to perform symbolic execution to step through the VC. Fig. 5 shows the rough sequence of steps that vcSteps carries out automatically at an assignment. Each numbered line below the horizontal line gives an intermediate goal state as vcSteps is running. The two hypothesis V and H above the line describe the variable environment and memory of the initial state σ, and are the precondition of this statement. The first stmPre below the line is the initial VC that must be verified. The tactic unfolds the definition of the VC for a sequence (line 2), then for an assignment (line 3), then determines that the value of the variable x is a by examining the hypothesis V (line 4). The load expression now has a value as an argument, so the tactic will examine the hypothesis H to determine that address a contains value v (line 5). The relevant binding $a \mapsto v$ can occur as any subtree of the separation logic assertion. Now the do-notation can be reduced away (line 6).

Once this is done, all that remains is to actually perform the assignment. The hypothesis V proves that y is in the domain the variable file of σ, so setting the value of y to v will succeed, producing a new state σ'. The tactic simplifies the goal to step through this update, and uses V to produce a new V' that describes σ'. H can still be used as the memory of σ' is the same as the memory of σ. This results in the proof state shown in Fig. 6. The tactic can now begin to analyze the statement s in a similar manner.

This may seem like a lengthy series of steps, but it is largely invisible to the user. Breaking down statements in this manner allows the tactics to easily handle a wide variety of expressions. vcSteps will get "stuck" at various points that require user intervention, such as loops and branches where invariants must be supplied, and where the tactic cannot easily show that a memory or variable

$$V : \mathsf{veEqv}\ \{\mathsf{x}, \mathsf{y}\}\ \{(\mathsf{x} \rightsquigarrow a)\}\ (\mathsf{venv}(\sigma))$$
$$H : (A * a \mapsto v * B)\ (\mathsf{mem}(\sigma))$$

---

1) stmPre $(\mathsf{y} := [\mathsf{x}];\ s)\ P\ \sigma$
2) stmPre $(\mathsf{y} := [\mathsf{x}])\ (\mathsf{stmPre}\ s\ P)\ \sigma$
3) do $v' \leftarrow \mathsf{eval}(\sigma, [\mathsf{x}]);$ do $\sigma' \leftarrow \mathsf{setVar}(\sigma, \mathsf{y}, v');$ stmPre $s\ P\ \sigma'$
4) do $v' \leftarrow \mathsf{eval}(\sigma, [a]);$ do $\sigma' \leftarrow \mathsf{setVar}(\sigma, \mathsf{y}, v');$ stmPre $s\ P\ \sigma'$
5) do $v' \leftarrow \mathsf{Some}(v);$ do $\sigma' \leftarrow \mathsf{setVar}(\sigma, \mathsf{y}, v');$ stmPre $s\ P\ \sigma'$
6) do $\sigma' \leftarrow \mathsf{setVar}(\sigma, \mathsf{y}, v);$ stmPre $s\ P\ \sigma'$

Fig. 5. Program logic tactics: examining the state

$$V' : \mathsf{veEqv}\ \{\mathsf{x}, \mathsf{y}\}\ \{(\mathsf{x} \rightsquigarrow a), (\mathsf{y} \rightsquigarrow v)\}\ (\mathsf{venv}(\sigma'))$$
$$H : (A * a \mapsto v * B)\ (\mathsf{mem}(\sigma'))$$

$$\mathsf{stmPre}\ s\ P\ \sigma'$$

Fig. 6. Program logic tactics: updating the state

operation is safe. In the latter case, the tactics described in the previous section can be applied to manipulate the assertion to a form the program logic tactics can understand, then vcSteps can be invoked again to pick up where it left off.

In addition to the tactic for reasoning about VCs, there is a tactic veEqvSolver to automatically solve goals involving veEqv. This is straightforward, as it only needs to reason about concrete finite sets.

5   Example of Tactic Use

In this section, we demonstrate the use of our tactics by verifying a fragment of an in-place linked list reversal, given in Fig. 2. Before we can verify this program, we need to define a loop invariant $inv\ l_0\ \sigma$, where l_0 is the list of values in the initial linked list and σ is the state at a loop entry:

$$inv\ l_0\ \sigma ::= \exists v_1, v_2.\ \text{veEqv}\ \{x, y, t\}\ \{(x, v_1), (y, v_2)\}\ (\text{venv}(\sigma)) \land$$
$$\exists l_1, l_2.\ (\text{llist}(v_1, l_1) * \text{llist}(v_2, l_2))\ (\text{mem}(\sigma)) \land$$
$$\text{rev}(l_1) \mathbin{+\!\!+} l_2 = \text{rev}(l_0)$$

In the first line, the predicate veEqv requires that in the current state that at least the program variables x, y, t are valid, and that the variables x and y are equal to some values v_1 and v_2, respectively. The second line is a separation logic predicate specifying that memory contains two disjoint linked lists as seen in Sect. 3. From these two descriptions, we can deduce that the variable x contains a pointer to a linked list containing the values l_1. Finally, the invariant requires that reversing l_1 and appending l_2 results in the original list l_0. We write $\text{rev}(l)$ for the reversal of list l and $l \mathbin{+\!\!+} l'$ for appending list l' to the end of l.

To save space, we will only go over the verification of the loop body and not describe in detail the invocation of standard Coq tactics. In the loop body, we know that the loop invariant inv holds on the current state and that the value of x is not $\text{Vword}(0)$ (*i.e.*, NULL). We must show that after the loop body has executed that the loop invariant is reestablished.

Our initial proof state is thus:

$$NE : v_1 \neq \qquad V int0$$
$$L : \text{rev}(l_1) \mathbin{+\!\!+} l_2 = \text{rev}(l_0)$$
$$H : (\text{llist}(v_1, l_1) * \text{llist}(v_2, l_2))\ (\text{mem}(\sigma))$$
$$V : \text{veEqv}\ \{x, y, t\}\ \{(x, v_1), (y, v_2)\}\ (\text{venv}(\sigma))$$

$$\overline{\text{stmPre}\ (t := [x + 4]; [x + 4] := y; y := x; x := t)\ (inv\ l_0)\ \sigma}$$

Our database of rewriting rules for separation logic data structures includes the following rule: if v is not 0, then a linked list $\text{llist}(v, l)$ must have at least one element. Thus, applying our rewriting tactic to the hypothesis H triggers this rule for $v = v_1$. After applying a standard substitution tactic, we have this proof state (where everything that is unchanged is left as ...):

...

$$L : \mathsf{rev}(v :: l_1') \mathbin{+\!\!+} l_2 = \mathsf{rev}(l_0)$$
$$H' : ((v_1 \mapsto v) * (v_1 + 4 \mapsto v_1') * \mathsf{llist}(v_1', l_1') * \mathsf{llist}(v_2, l_2))\ (\mathsf{mem}(\sigma))$$

...

Now that we know that the address $v_1 + 4$ contains the value v_1', we can show that it is safe to execute the loop body. The tactic vcSteps, described in Sect. 4.3, is able to automatically step through the entire loop body, leaving an updated state description and the goal of showing that the loop invariant holds on the final state of the loop σ':

...

$$H'' : ((v_1 \mapsto v) * (v_1 + 4 \mapsto v_2) * \mathsf{llist}(v_1', l_1') * \mathsf{llist}(v_2, l_2))\ (\mathsf{mem}(\sigma'))$$
$$V' : \mathsf{veEqv}\ \{\mathsf{x}, \mathsf{y}, \mathsf{t}\}\ \{(\mathsf{x}, v_1'), (\mathsf{y}, v_1), (\mathsf{t}, v_1')\}\ (\mathsf{venv}(\sigma'))$$
$$\overline{inv\ l_0\ \sigma'}$$

Now we must instantiate two existential variables and show that they are the values of the variables x and y (and that x, y and t are valid variables). These existentials can be automatically instantiated, and the part of the goal using veEqv solved, using a few standard Coq tactics along with the veEqvSolver described in Sect. 4.3.

The remaining goal is

...

$$\exists l_3, l_4.\ (\mathsf{llist}(v_1', l_3) * \mathsf{llist}(v_1, l_4))\ (\mathsf{mem}(\sigma'))\ \wedge\ \mathsf{rev}(l_3) \mathbin{+\!\!+} l_4 = \mathsf{rev}(l_0)$$

We manually instantiate the existentials with l_1' and $(v :: l_2)$ and split the resulting conjunction using standard Coq tactics. This produces two subgoals. The second subgoal is $\mathsf{rev}(l_2) \mathbin{+\!\!+} (v :: l_1') = \mathsf{rev}(l_0)$ and can be solved using standard tactics. The first subgoal is an assertion containing $\mathsf{llist}(v_1', v :: l_2)$, which we can be simplified using standard tactics leaving the proof state

...

$$(\mathsf{llist}(v_1', l_1') * (\exists x'.(v_1 \mapsto v) * (v_1 + 4 \mapsto x') * \mathsf{llist}(x', l_2)))\ (\mathsf{mem}(\sigma'))$$

Invoking our simplification tactic ssimpl replaces the existential with a Coq meta-level existential variable "?100", leaving the goal[3]

...

$$H'' : ((v_1 \mapsto v) * (v_1 + 4 \mapsto v_2) * \mathsf{llist}(v_1', l_1') * \mathsf{llist}(v_2, l_2))\ (\mathsf{mem}(\sigma'))$$
$$\overline{(\mathsf{llist}(v_1', l_1') * (v_1 \mapsto v) * (v_1 + 4 \mapsto ?100) * \mathsf{llist}(?100, l_2))\ (\mathsf{mem}(\sigma'))}$$

This goal is immediately solved by our tactic searchMatch. The hypothesis contains $v_1 + 4 \mapsto v_2$, and the goal contains $v_1 + 4 \mapsto ?100$, so ?100 must be equal to v_2. Once ?100 is instantiated, the rest of easy to match up. Thus we have verified that the body of the loop preserves the loop invariant.

---

[3] H'' has not changed but we include it here for convenience.

6   Implementation and Application

Our tactics are implemented entirely in the Coq tactic language L_{tac}. The tool suite include about 5200 lines of libraries, such as theorems about modular arithmetic and data structures such as finite sets. Our definition of Cminor (discussed in Sect. 2) is about 4700 lines. The definition of separation logic assertions and associated lemmas (discussed in Sect. 3) are about 1100 lines, while the tactics are about 3000 lines. Finally, the program logic (discussed in Sect. 4), which includes its definition, proofs, and associated tactics, is about 2000 lines.

We have used the tactics described in this paper to verify the safety and completeness of a Cheney copying garbage collector [8] implemented in Cminor. It is *safe* because the final object heap is isomorphic to the reachable objects in the initial state and *complete* because it only copies reachable objects. This collector supports features such as scanning roots stored in stack frames, objects with an arbitrary number of fields, and precise collection via information stored in object headers. The verification took around 4700 lines of proof scripts (including the definition of the collector and all specifications), compared to 7800 lines for our previous work [4] which used more primitive tactics. The reduction in line count is despite the fact that our earlier collector did not support any of the features we listed earlier in this paragraph and did not use modular arithmetic.

There has been other work on mechanized garbage collector verification, such as Myreen *et al.* [15] who verify a Cheney collector in 2000 lines using a decompilation based approach. That publication unfortunately does not have enough detail of the collector verification to explain the difference in proof size, though it is likely due in part to greater automation. Hawblitzel *et al.* [16] used a theorem prover to automatically verify a collector that is realistic enough to be used for real C# benchmarks.

7   Related Work and Conclusion

Appel's unpublished note [12] describes Coq tactics that are very similar to ours. These tactics do not support as many direct manipulations of assertions, and his rewriting tactic appears to require manual instantiation of quantifiers. The paper describes a tactic for inversion of inductively defined separation logic predicates, which we do not support. While Appel also applies a "two-level approach" that attempts to pull things out of separation logic assertions to leverage existing proof assistant infrastructure, our approach is more aggressive about this, lifting out expression evaluation. This allows our approach to avoid reasoning about whether expressions in assertions involve memory.

We can give a rough comparison of proof sizes, using the in-place list reversal procedure. Ignoring comments and the definition of the program, by the count of wc Appel uses 200 lines and 795 words to verify this program [12]. With our tactics, ignoring comments, blank lines and the definition of the program, our verification takes 68 lines and less than 400 words.

Affeldt and Marti [13] use separation logic in a proof assistant, but unfold the definitions of the separation logic assertions to allow the use of more conventional tactics. Tuch *et al.* [17] define a mechanized program logic for reasoning about C-like memory models. They are able to verify programs using separation logic, but do not have any complex tactics for separation logic connectives.

Other work, such as Smallfoot [6], has focused on automated verification of lightweight separation logic specifications. This approach has been used as the basis for certified separation logic decisions procedures in Coq [18] and HOL [19]. Calcagno *et al.* [20] use separation logic for an efficient compositional shape analysis that is able to infer some specifications.

Still other work has focused on mechanized reasoning about imperative pointer programs outside of the context of separation logic [11,21,22] using either deep or shallow embeddings. Expressing assertions via more conventional propositions enables the use of powerful preexisting theorem provers. Another approach to program verification decompiles imperative programs into functional programs that are more amenable to analysis in a proof assistant [23,15].

The tactics we have described in this paper provide a solid foundation for the use of separation logic in a proof assistant but there is room for further automation. Integrating a Smallfoot-like decision procedure into our tactics would automate reasoning about standard data structures.

We have presented a set of separation logic tactics that allows the verification of programs using separation logic in a proof assistant. These tactics allow Coq to be used as a proof assistant for separation logic by allowing the assertions to be easily manipulated via simplification, rearranging, splitting, matching and rewriting. They also provide tactics for proving a verification condition by means of a separation logic based description of the program state. These tactics are powerful enough to verify a garbage collector.

Acknowledgments. I would like to thank Andrew Tolmach for providing extensive feedback about the tactics, and Andrew Tolmach, Jim Hook and the anonymous reviewers for providing helpful comments on this paper.

References

1. Reynolds, J.C.: Separation logic: A logic for shared mutable data structures. In: LICS 2002, Washington, DC, USA, pp. 55–74. IEEE Computer Society, Los Alamitos (2002)
2. Girard, J.Y.: Linear logic. Theoretical Computer Science 50, 1–102 (1987)
3. Birkedal, L., Torp-Smith, N., Reynolds, J.C.: Local reasoning about a copying garbage collector. In: POPL 2005, pp. 220–231. ACM Press, New York (2004)
4. McCreight, A., Shao, Z., Lin, C., Li, L.: A general framework for certifying gcs and their mutators. In: PLDI 2007, pp. 468–479. ACM, New York (2007)
5. The Coq Development Team: The Coq proof assistant, http://coq.inria.fr
6. Berdine, J., Calcagno, C., O'Hearn, P.W.: Smallfoot: Modular automatic assertion checking with separation logic. In: de Boer, F.S., Bonsangue, M.M., Graf, S., de Roever, W.-P. (eds.) FMCO 2005. LNCS, vol. 4111, pp. 115–137. Springer, Heidelberg (2006)
7. Leroy, X.: Formal certification of a compiler back-end, or: programming a compiler with a proof assistant. In: POPL 2006, pp. 42–54. ACM Press, New York (2006)

8. Cheney, C.J.: A nonrecursive list compacting algorithm. Communications of the ACM 13(11), 677–678 (1970)

9. Appel, A.W., Blazy, S.: Separation logic for small-step Cminor. In: Schneider, K., Brandt, J. (eds.) TPHOLs 2007. LNCS, vol. 4732, pp. 5–21. Springer, Heidelberg (2007)

10. Paulin-Mohring, C.: Inductive definitions in the system Coq—rules and properties. In: Bezem, M., Groote, J.F. (eds.) TLCA 1993. LNCS, vol. 664. Springer, Heidelberg (1993)

11. Wildmoser, M., Nipkow, T.: Certifying machine code safety: Shallow versus deep embedding. In: Slind, K., Bunker, A., Gopalakrishnan, G.C. (eds.) TPHOLs 2004. LNCS, vol. 3223, pp. 305–320. Springer, Heidelberg (2004)

12. Appel, A.W.: Tactics for separation logic (January 2006), http://www.cs.princeton.edu/~appel/papers/septacs.pdf

13. Marti, N., Affeldt, R., Yonezawa, A.: Formal verification of the heap manager of an os using separation logic. In: Liu, Z., He, J. (eds.) ICFEM 2006. LNCS, vol. 4260, pp. 400–419. Springer, Heidelberg (2006)

14. Boutin, S.: Using reflection to build efficient and certified decision procedures. In: Ito, T., Abadi, M. (eds.) TACS 1997. LNCS, vol. 1281, pp. 515–529. Springer, Heidelberg (1997)

15. Myreen, M.O., Slind, K., Gordon, M.J.C.: Machine-code verification for multiple architectures - an application of decompilation into logic. In: Proceedings of Formal Methods in Computer-Aided Design (FMCAD) (2008)

16. Hawblitzel, C., Petrank, E.: Automated verification of practical garbage collectors. In: POPL 2009, pp. 441–453. ACM, New York (2009)

17. Tuch, H., Klein, G., Norrish, M.: Types, bytes, and separation logic. In: POPL 2007, pp. 97–108. ACM, New York (2007)

18. Marti, N., Affeldt, R.: A certified verifier for a fragment of separation logic. In: 9th JSSST Workshop on Programming and Prog. Langs, PPL 2007 (2007)

19. Tuerk, T.: A separation logic framework in HOL. In: Otmane Ait Mohamed, C.M., Tahar, S. (eds.) TPHOLs 2008, August 2008, pp. 116–122 (2008)

20. Calcagno, C., Distefano, D., O'Hearn, P., Yang, H.: Compositional shape analysis by means of bi-abduction. In: POPL 2009, pp. 289–300. ACM, New York (2009)

21. Mehta, F., Nipkow, T.: Proving pointer programs in higher-order logic. Inf. Comput. 199(1-2), 200–227 (2005)

22. Bulwahn, L., Krauss, A., Haftmann, F., Erkök, L., Matthews, J.: Imperative functional programming with Isabelle/HOL. In: Mohamed, O.A., Muñoz, C., Tahar, S. (eds.) TPHOLs 2008. LNCS, vol. 5170, pp. 134–149. Springer, Heidelberg (2008)

23. Filliâtre, J.C., Marché, C.: The Why/Krakatoa/Caduceus platform for deductive program verification. In: Damm, W., Hermanns, H. (eds.) CAV 2007. LNCS, vol. 4590, pp. 173–177. Springer, Heidelberg (2007)

Verified LISP Implementations on ARM, x86 and PowerPC

Magnus O. Myreen and Michael J.C. Gordon

Computer Laboratory, University of Cambridge, UK

Abstract. This paper reports on a case study, which we believe is the first to produce a formally verified end-to-end implementation of a functional programming language running on commercial processors. Interpreters for the core of McCarthy's LISP 1.5 were implemented in ARM, x86 and PowerPC machine code, and proved to correctly parse, evaluate and print LISP s-expressions. The proof of evaluation required working on top of verified implementations of memory allocation and garbage collection. All proofs are mechanised in the HOL4 theorem prover.

1   Introduction

Explicit pointer manipulation is an endless source of errors in low-level programs. Functional programming languages hide pointers and thereby achieve a more abstract programming environment. The downside with functional programming (and Java/C# programming) is that the programmer has to trust automatic memory management routines built into run-time environments.

In this paper we report on a case study, which we believe is the first to produce a formally verified end-to-end implementation of a functional programming language. We have implemented, in ARM, x86 and PowerPC machine code, a program which parses, evaluates and prints LISP; and furthermore formally proved that our implementation respects a semantics of the core of LISP 1.5 [6]. Instead of assuming correctness of run-time routines, we build on a verified implementation of allocation and garbage collection.

For a flavour of what we have implemented and proved consider an example: if our implementation is supplied with the following call to `pascal-triangle`,

```
(pascal-triangle '((1)) '6)
```

it parses the string, evaluates the expression and prints a string,

```
((1 6 15 20 15 6 1)
 (1 5 10 10 5 1)
 (1 4 6 4 1)
 (1 3 3 1)
 (1 2 1)
 (1 1)
 (1))
```

where `pascal-triangle` had been supplied to it as

S. Berghofer et al. (Eds.): TPHOLs 2009, LNCS 5674, pp. 359–374, 2009.

```
(label pascal-triangle
   (lambda (rest n)
     (cond ((equal n '0) rest)
           ('t (pascal-triangle
                 (cons (pascal-next '0 (car rest)) rest) (- n '1))))))
```

with auxiliary function:

```
(label pascal-next
   (lambda (p xs)
     (cond ((atomp xs) (cons p 'nil))
           ('t (cons (+ p (car xs)) (pascal-next (car xs) (cdr xs)))))))
```

The theorem we have proved about our LISP implementation can be used to show e.g. that running `pascal-triangle` will terminate and print the first $n+1$ rows of Pascal's triangle, without a premature exit due to lack of heap space. One can use our theorem to derive sufficient conditions on the inputs to guarantee that there will be enough heap space.

We envision that our verified LISP interpreter will provide a platform on top of which formally verified software can be produced with much greater ease than at lower levels of abstraction, i.e. in languages where pointers are made explicit.

Why LISP? We chose to implement and verify a LISP interpreter since LISP has a neat definition of both syntax and semantics [12] and is still a very powerful language as one can see, for example, in the success of ACL2 [8]. By choosing LISP we avoided verifying machine code which performs static type checking.

Our proofs [14] are mechanised in the HOL4 theorem prover [19].

2 Methodology

Instead of delving into the many detailed invariants developed for our proofs, this paper will concentrate on describing the methodology we used:

▷ First, machine code for various LISP primitives, such as car, cdr, cons, was written and verified (Section 3);

  • The correctness of each code snippets is expressed as a machine-code Hoare triple [15]: $\{pre * \mathsf{pc}\ p\}\ p : code\ \{post * \mathsf{pc}\ (p + exit)\}$.

  • For cons and equal we used previously developed proof automation [15], which allows for proof reuse in between different machine languages.

▷ Second, the verified LISP primitives were input into a proof-producing compiler in such a way that the compiler can view the processors as a machine with six registers containing LISP s-expressions (Section 4);

  • The compiler [16] we use maps tail-recursive functions, defined in the logic of HOL4, down to machine code and proves that the generated code executes the original HOL4 functions.

  • Theorems describing the LISP primitives were input into the compiler, which can use them as building blocks when deriving new code/proofs.

▷ Third, LISP evaluation was defined as a (partially-specified) tail-recursive function lisp_eval, and then compiled into machine code using the compiler mentioned above (Section 5).

- LISP evaluation was defined as a tail-recursive function which only uses expressions/names for which the compiler has verified building blocks.
- lisp_eval maintains a stack and a symbol-value list.

▷ Fourth, to gain confidence that lisp_eval implements 'LISP evaluation', we proved that lisp_eval implements a semantics of LISP 1.5 [12] (Section 6).

- Our relational semantics of LISP [6] is a formalisation of a subset of McCarthy's original LISP 1.5 [12], with dynamic binding.
- The semantics abstracts the stack and certain evaluation orders.

▷ Finally, the verified LISP interpreters were sandwiched between a verified parser and printer to produce string-to-string theorems describing the behaviour of the entire implementation (Section 7).

- The parser and printer code, respectively, sets up and tears down an appropriate heap for s-expressions.

Sections 8 and 9 give quantitative data on the effort and discuss related work, respectively. Some definitions and proofs are presented in the Appendixes.

3   LISP Primitives

LISP programs are expressed in and operate over s-expressions, expressions that are either a (natural) number, a symbol or a pair of s-expressions. In HOL, s-expressions are readily modelled using a data-type with constructors:

$$\text{Num} : \mathbb{N} \rightarrow \text{SExp}$$
$$\text{Sym} : \text{string} \rightarrow \text{SExp}$$
$$\text{Dot} : \text{SExp} \rightarrow \text{SExp} \rightarrow \text{SExp}$$

LISP programs and s-expressions are conventionally written in an abbreviated string form. A few examples will illustrate the correspondence, which is given a formal definition in Appendix D.

```
(car x)   means   Dot (Sym "car") (Dot (Sym "x") (Sym "nil"))
(1 2 3)   means   Dot (Num 1) (Dot (Num 2) (Dot (Num 3) (Sym "nil")))
'f        means   Dot (Sym "quote") (Dot (Sym "f") (Sym "nil"))
(4 . 5)   means   Dot (Num 4) (Num 5)
```

Some basic LISP primitives are defined over SExp as follows:

$$\text{car } (\text{Dot } x \ y) = x$$
$$\text{cdr } (\text{Dot } x \ y) = y$$

$$\text{cons } x\, y = \text{Dot } x\, y$$

$$\text{plus (Num } m\text{) (Num } n\text{)} = \text{Num } (m + n)$$
$$\text{minus (Num } m\text{) (Num } n\text{)} = \text{Num } (m - n)$$
$$\text{times (Num } m\text{) (Num } n\text{)} = \text{Num } (m \times n)$$
$$\text{division (Num } m\text{) (Num } n\text{)} = \text{Num } (m \text{ div } n)$$
$$\text{modulus (Num } m\text{) (Num } n\text{)} = \text{Num } (m \text{ mod } n)$$

$$\text{equal } x\, y = \text{if } x = y \text{ then Sym "t" else Sym "nil"}$$
$$\text{less (Num } m\text{) (Num } n\text{)} = \text{if } m < n \text{ then Sym "t" else Sym "nil"}$$

In the definition of equal, expression $x = y$ tests standard structural equality.

3.1   Specification of Primitive Operations

Before writing and verifying the machine code implementing primitive LISP operations, a decision had to be made how to represent Num, Sym and Dot on a real machine. To keep memory usage to a minimum each Dot-pair is represented as a block of two pointers stored consecutively on the heap, each Num n is represented as a 32-bit word containing $4 \times n + 2$ (i.e. only natural numbers $0 \leq n < 2^{30}$ are representable), and each Sym s is represented as a 32-bit word containing $4 \times i + 3$, where i is the row number of symbol s in a symbol table which, in our implementation, is a linked-list kept outside of the garbage-collected heap.

Here '+2' and '+3' are used as tags to make sure that the garbage collector can distinguish Num and Sym values from proper pointers. Pointers to Dot-pairs are word-aligned, i.e. $a \text{ mod } 4 = 0$, a condition the collector tests by computing $a\ \&\ 3 = 0$, where $\&$ is bitwise-and.

This simple and small representation of SExp allows most LISP primitives from the previous section to be implemented in one or two machine instructions. For example, taking car of register 3 and storing the result in register 4 is implemented on ARM as a load instruction:

```
E5934000    ldr r4,[r3]    (* load into reg 4, memory at address reg 3 *)
```

Similarly, ARM code for performing LISP operation plus of register 3 and 4, and storing the result into register 3 is implemented by:

```
E0833004    add r3,r3,r4    (* reg 3 is assigned value reg 3 + reg 4 *)
E2433002    sub r3,r3,#2    (* reg 3 is assigned value reg 3 - 2 *)
```

The intuition here is: $(4 \times m + 2) + (4 \times n + 2) - 2 = 4 \times (m + n) + 2$.

The correctness of the above implementations of car and plus is expressed formally by the two ARM Hoare triples [15] below. Here lisp $(v_1, v_2, v_3, v_4, v_5, v_6, l)$ is an assertion, defined below, which asserts that a heap with room for l Dot-pairs is located in memory and that s-expressions $v_1...v_6$ (each of type SExp) are stored in machine registers. This lisp assertion should be understood as lifting the level of abstraction to a level where specific machine instructions make the processor seem as if it has six[1] registers containing s-expressions, of type SExp.

$$(\exists x\; y.\; \mathsf{Dot}\; x\; y = v_1)\; \Rightarrow$$
$$\{\; \mathsf{lisp}\; (v_1, v_2, v_3, v_4, v_5, v_6, l) * \mathsf{pc}\; p\; \}$$
$$p : \mathtt{E5934000}$$
$$\{\; \mathsf{lisp}\; (v_1, \mathsf{car}\; v_1, v_3, v_4, v_5, v_6, l) * \mathsf{pc}\; (p+4)\; \}$$

$$(\exists m\; n.\; \mathsf{Num}\; m = v_1 \wedge \mathsf{Num}\; n = v_2 \wedge m + n < 2^{30})\; \Rightarrow$$
$$\{\; \mathsf{lisp}\; (v_1, v_2, v_3, v_4, v_5, v_6, l) * \mathsf{pc}\; p\; \}$$
$$p : \mathtt{E0833004\; E2433002}$$
$$\{\; \mathsf{lisp}\; (\mathsf{plus}\; v_1\; v_2, v_2, v_3, v_4, v_5, v_6, l) * \mathsf{pc}\; (p+8)\; \}$$

The new assertion is defined for ARM (lisp), x86 (lisp'), and PowerPC (lisp") as maintaining a relation lisp_inv between the abstract state $v_1...v_6$ (each of type SExp) and the concrete state $x_1...x_6$ (each of type 32-bit word). The details of lisp_inv (defined in Appendix A) and the separating conjunction $*$ (explained in Myreen [14]) are unimportant for this presentation.

$$\mathsf{lisp}\; (v_1, v_2, v_3, v_4, v_5, v_6, l) =$$
$$\exists x_1\; x_2\; x_3\; x_4\; x_5\; x_6\; m_1\; m_2\; m_3\; a\; temp.\; \mathsf{m}\; m_1 * \mathsf{m}\; m_2 * \mathsf{m}\; m_3 *$$
$$\mathsf{r2}\; temp * \mathsf{r3}\; x_1 * \mathsf{r4}\; x_2 * \mathsf{r5}\; x_3 * \mathsf{r6}\; x_4 * \mathsf{r7}\; x_5 * \mathsf{r8}\; x_6 * \mathsf{r10}\; a *$$
$$\langle \mathsf{lisp\_inv}\; (v_1, v_2, v_3, v_4, v_5, v_6, l)\; (x_1, x_2, x_3, x_4, x_5, x_6, a, m_1, m_2, m_3) \rangle$$

$$\mathsf{lisp'}\; (v_1, v_2, v_3, v_4, v_5, v_6, l) =$$
$$\exists x_1\; x_2\; x_3\; x_4\; x_5\; x_6\; m_1\; m_2\; m_3\; a.\; \mathsf{m}\; m_1 * \mathsf{m}\; m_2 * \mathsf{m}\; m_3 *$$
$$\mathsf{eax}\; x_1 * \mathsf{ecx}\; x_2 * \mathsf{edx}\; x_3 * \mathsf{ebx}\; x_4 * \mathsf{esi}\; x_5 * \mathsf{edi}\; x_6 * \mathsf{ebp}\; a *$$
$$\langle \mathsf{lisp\_inv}\; (v_1, v_2, v_3, v_4, v_5, v_6, l)\; (x_1, x_2, x_3, x_4, x_5, x_6, a, m_1, m_2, m_3) \rangle$$

$$\mathsf{lisp"}\; (v_1, v_2, v_3, v_4, v_5, v_6, l) =$$
$$\exists x_1\; x_2\; x_3\; x_4\; x_5\; x_6\; m_1\; m_2\; m_3\; a\; temp.\; \mathsf{m}\; m_1 * \mathsf{m}\; m_2 * \mathsf{m}\; m_3 *$$
$$\mathsf{r2}\; temp * \mathsf{r3}\; x_1 * \mathsf{r4}\; x_2 * \mathsf{r5}\; x_3 * \mathsf{r6}\; x_4 * \mathsf{r7}\; x_5 * \mathsf{r8}\; x_6 * \mathsf{r10}\; a *$$
$$\langle \mathsf{lisp\_inv}\; (v_1, v_2, v_3, v_4, v_5, v_6, l)\; (x_1, x_2, x_3, x_4, x_5, x_6, a, m_1, m_2, m_3) \rangle$$

The following examples will use only lisp defined for ARM.

3.2   Memory Layout and Specification of 'Cons' and 'Equal'

Two LISP primitives required code longer than one or two machine instructions, namely cons and equal. Memory allocation, i.e. cons, requires an allocation procedure combined with a garbage collector. However, the top-level specification, which is explained next, hides these facts. Let size count the number of Dot-pairs in an expression.

$$\mathsf{size}\; (\mathsf{Num}\; w) = 0$$
$$\mathsf{size}\; (\mathsf{Sym}\; s) = 0$$
$$\mathsf{size}\; (\mathsf{Dot}\; x\; y) = 1 + \mathsf{size}\; x + \mathsf{size}\; y$$

---

[1] Number six was chosen since six is sufficient and suits the x86 implementation best.

The specification of cons guarantees that its implementation will always succeed as long as the number of reachable Dot-pairs is less than the capacity of the heap, i.e. less than l. This precondition under approximates pointer aliasing.

size v_1 + size v_2 + size v_3 + size v_4 + size v_5 + size $v_6 < l \Rightarrow$
{ lisp $(v_1, v_2, v_3, v_4, v_5, v_6, l) *$ pc p }
  p : E50A3018 E50A4014 E50A5010 E50A600C ... E51A8004 E51A7008
{ lisp (cons v_1 $v_2, v_2, v_3, v_4, v_5, v_6, l) *$ pc $(p + 332)$ }

The implementation of cons includes a copying collector which implements Cheney's algorithm [2]. This copying collector requires the heap to be split into two heap halves of equal size; only one of which is used for heap data at any one point in time. When a collection request is issued, all live elements from the currently used heap half are copied over to the currently unused heap half. The proof of cons is outlined in the first author's PhD thesis [14].

The fact that one half of the heap is left empty might seem to be a waste of space. However, the other heap half need not be left completely unused, as the implementation of equal can make use of it. The LISP primitive equal tests whether two s-expressions are structurally identical by traversing the expression tree as a normal recursive procedure. This recursive traversal requires a stack, but the stack can in this case be built inside the unused heap half as the garbage collector will not be called during the execution of equal. Thus, the implementation of equal uses no external stack and requires no conditions on the size of the expressions v_1 and v_2, as their depths cannot exceed the length of a heap half.

{ lisp $(v_1, v_2, v_3, v_4, v_5, v_6, l) *$ pc p }
  p : E1530004 03A0300F 0A000025 E50A4014 ... E51A7008 E51A8004
{ lisp (equal v_1 $v_2, v_2, v_3, v_4, v_5, v_6, l) *$ pc $(p + 164)$ }

4   Compiling s-Expression Functions to Machine Code

The previous sections described the theorems which state that certain machine instructions execute LISP primitives. These theorems can be used to augment the input-language understood by a proof-producing compiler that we have developed [16]. The theorems mentioned above allow the compiler to accept:

let $v_2 = $ car v_1 in ...
let $v_1 = $ plus v_1 v_2 in ...
let $v_1 = $ cons v_1 v_2 in ...
let $v_1 = $ equal v_1 v_2 in ...

Theorems for basic tests have also been proved in a similar manner, and can be provided to the compiler. For example, the following theorem shows that ARM instruction E3330003 assigns boolean value $(v_1 = $ Sym "nil") to status bit z.

{ lisp $(v_1, v_2, v_3, v_4, v_5, v_6, l) *$ pc $p * $ s }
  p : E3330003
{ lisp $(v_1, v_2, v_3, v_4, v_5, v_6, l) *$ pc $(p + 4) * $ sz $(v_1 = $ Sym "nil") $*$
  $\exists n\ c\ v.$ sn $n * $ sc $c * $ sv v }

The compiler can use such theorems to create branches on the expression assigned to status bits. The above theorem adds support for the if-statement:

$$\text{if } v_1 = \text{Sym "nil" then } \dots \text{ else } \dots$$

Once the compiler was given sufficient Hoare-triple theorems it could be used to compile functions operating over s-expressions into machine code. An example will illustrate the process. From the following function

$$\text{sumlist}(v_1, v_2, v_3) = \text{if } v_1 = \text{Sym "nil" then } (v_1, v_2, v_3) \text{ else}$$
$$\text{let } v_3 = \text{car } v_1 \text{ in}$$
$$\text{let } v_1 = \text{cdr } v_1 \text{ in}$$
$$\text{let } v_2 = \text{plus } v_2 \ v_3 \text{ in}$$
$$\text{sumlist}(v_1, v_2, v_3)$$

the compiler produces the theorem below, containing the generated ARM machine code and a precondition $\text{sumlist\_pre}(v_1, v_2, v_3)$.

$$\text{sumlist\_pre}(v_1, v_2, v_3) \Rightarrow$$
$$\{ \text{ lisp } (v_1, v_2, v_3, v_4, v_5, v_6, l) * \text{pc } p * \text{s } \}$$
$$p : \text{E3330003 0A000004 E5935000 E5934004 E0844005 E2444002 EAFFFFF8}$$
$$\{ \text{ let } (v_1, v_2, v_3) = \text{sumlist}(v_1, v_2, v_3) \text{ in}$$
$$\text{lisp } (v_1, v_2, v_3, v_4, v_5, v_6, l) * \text{pc } (p + 28) * \text{s } \}$$

The proof performed by the compiler is outlined in Appendix C, where the precondition $\text{sumlist\_pre}(v_1, v_2, v_3)$ is also defined. The automatically generated pre-functions collect side conditions that must be true for proper execution of the code, e.g. when cons is used the pre-functions collect the requirements on not exceeding the heap limit l.

5   Assembling the LISP Evaluator

LISP evaluation was defined as a large tail-recursive function lisp_eval and then compiled, to ARM, PowerPC and x86, to produce theorems of the following form. The theorem below states that the generated ARM code executes lisp_eval for inputs that do not violate any of the side conditions gathered in lisp_eval_pre.

$$\text{lisp\_eval\_pre}(v_1, v_2, v_3, v_4, v_5, v_6, l) \Rightarrow$$
$$\{ \text{ lisp } (v_1, v_2, v_3, v_4, v_5, v_6, l) * \text{pc } p * \text{s } \}$$
$$p : \text{E3360003 1A0001D1 E3A0600F E3130001 0A000009 } \dots \text{ EAFFF85D}$$
$$\{ \text{ lisp } (\text{lisp\_eval}(v_1, v_2, v_3, v_4, v_5, v_6, l)) * \text{pc } (p + 7816) * \text{s } \}$$

lisp_eval evaluates the expression stored in v_1, input v_6 is a list of symbol-value pairs against which symbols in v_1 are evaluated, inputs v_2, v_3, v_4 and v_5 are used as temporaries that are to be initialised with Sym "nil". The heap limit l had to be passed into lisp_eval due to an implementation restriction which requires lisp_eval_pre to input the same variables as lisp_eval. The side condition lisp_eval_pre uses l to state restrictions on applications of cons.

6   Evaluator Implements McCarthy's LISP 1.5

The previous sections, and Appendix C, described how a function lisp_eval was compiled down to machine code. The compiler generated some code and derived a theorem which states that the generated code correctly implements lisp_eval. However, the compiler does not (and cannot) give any evidence that lisp_eval in fact implements 'LISP evaluation'. The definition of lisp_eval is long and full of tedious details of how the intermediate stack is maintained and used, and thus it is far from obvious that lisp_eval corresponds to 'LISP evaluation'.

In order to gain confidence that the generated machine code actually implements LISP evaluation, we proved that lisp_eval implements a clean relational semantics of LISP 1.5 [6]. Our relational semantics of LISP 1.5 is defined in terms of three mutually recursive relations $\rightarrow_{eval}$, $\rightarrow_{eval\_list}$ and $\rightarrow_{app}$. Here $(fn, [arg_1; \cdots; arg_n], \rho) \rightarrow_{app} s$ means that $fn[arg_1; \cdots; arg_n] = s$ if the free variables in fn have values specified by an environment ρ; similarly $(e, \rho) \rightarrow_{eval} s$ holds if term e evaluates to s-expression s with respect to environment ρ; and $(el, \rho) \rightarrow_{eval\_list} sl$ holds if list el of expressions evaluates to list sl of expressions with respect to ρ. Here k denotes built-in function names and c constants. For details refer to Gordon [6] and Appendix A in Myreen [14].

$$\frac{\mathsf{ok\_name}\ v}{(v, \rho) \rightarrow_{eval} \rho(v)} \qquad \frac{}{(c, \rho) \rightarrow_{eval} c} \qquad \frac{}{([\,], \rho) \rightarrow_{eval} \mathtt{nil}}$$

$$\frac{(p, \rho) \rightarrow_{eval} \mathtt{nil} \land ([gl], \rho) \rightarrow_{eval} s}{([p \rightarrow e; gl], \rho) \rightarrow_{eval} s} \qquad \frac{(p, \rho) \rightarrow_{eval} x \land x \neq \mathtt{nil} \land (e, \rho) \rightarrow_{eval} s}{([p \rightarrow e; gl], \rho) \rightarrow_{eval} s}$$

$$\frac{\mathsf{can\_apply}\ k\ args}{(k, args, \rho) \rightarrow_{app} k\ args} \qquad \frac{(\rho(f), args, \rho) \rightarrow_{app} s \land \mathsf{ok\_name}\ f}{(f, args, \rho) \rightarrow_{app} s}$$

$$\frac{(e, \rho[args/vars]) \rightarrow_{eval} s}{(\lambda[[vars]; e], args, \rho) \rightarrow_{app} s} \qquad \frac{(fn, args, \rho[fn/x]) \rightarrow_{app} s}{(label[[x]; fn], args, \rho) \rightarrow_{app} s}$$

$$\frac{}{([\,], \rho) \rightarrow_{eval\_list} [\,]} \qquad \frac{(e, \rho) \rightarrow_{eval} s \land ([el], \rho) \rightarrow_{eval\_list} sl}{([e; el], \rho) \rightarrow_{eval\_list} [s; sl]}$$

We have proved that whenever the relation for LISP 1.5 evaluation $\rightarrow_{eval}$ relates expression s under environment ρ to expression r, then lisp_eval will do the same. Here t and u are translation functions, from one form of s-expressions to another. Let $\mathsf{nil} = \mathsf{Sym}\ \texttt{"nil"}$ and $\mathsf{fst}\ (x, y, \ldots) = x$.

$$\forall s\ \rho\ r.\ \ (s, \rho) \rightarrow_{eval} r\ \ \Rightarrow\ \ \mathsf{fst}\ (\mathsf{lisp\_eval}\ (t\ s, \mathsf{nil}, \mathsf{nil}, \mathsf{nil}, u\ \rho, \mathsf{nil}, l)) = t\ r$$

7   Verified Parser and Printer

Sections 4 and 5 explained how machine code was generated and proved to implement a function called lisp_eval. The precondition of the certificate theorem requires the initial state to satisfy a complex heap invariant lisp. How do we know that this precondition is not accidentally equivalent to false, making the theorem vacuously true? To remedy this shortcoming, we have verified machine code that will set-up an appropriate state from scratch.

The set-up and tear-down code includes a parser and printer that will, respectively, read in an input s-expression and print out the resulting s-expression. The development of the parser and printer started by first defining a function sexp2string which lays down how s-expressions are to be represented in string form (Appendix D). Then a function string2sexp was defined for which we proved:

$$\forall s. \; \text{sexp\_ok } s \; \Rightarrow \; \text{string2sexp (sexp2string } s) = s$$

Here sexp_ok s makes sure that s does not contain symbols that print ambiguously, e.g. Sym "", Sym "(" and Sym "2". The parsing function was defined as a composition of a lexer sexp_lex and a token parser sexp_parse (Appendix D).

$$\text{string2sexp } str = \text{car (sexp\_parse (reverse (sexp\_lex } str)) \; (\text{Sym "nil") } [])$$

Machine code was written and verified based on the high-level functions sexp_lex, sexp_parse and sexp2string. Writing these high-level definitions first was a great help when constructing the machine code (using the compiler from [16]).

The overall theorems about our LISP implementations are of the following form. If $\rightarrow_{eval}$ relates s with r under the empty environment (i.e. $(s, []) \rightarrow_{eval} r$), no illegal symbols are used (i.e. sexp_ok $(t \; s)$), running lisp_eval on t s will not run out of memory (i.e. lisp_eval_pre$(t \; s, \text{nil, nil, nil, nil, nil, } l)$), the string representation of t s is in memory (i.e. string a (sexp2string $(t \; s)$)), and there is enough space to parse t s and set up a heap of size l (i.e. enough_space $(t \; s) \; l$), then the code will execute successfully and terminate with the string representation of t r stored in memory (i.e. string a (sexp2string $(t \; r)$)). The ARM code expects the address of the input string to be in register 3, i.e. r3 a.

$\forall s \; r \; l \; p.$
  $(s, []) \rightarrow_{eval} r \wedge \text{sexp\_ok } (t \; s) \wedge \text{lisp\_eval\_pre}(t \; s, \text{nil, nil, nil, nil, nil, } l) \Rightarrow$
  $\{ \; \exists a. \; \text{r3 } a * \text{string } a \; (\text{sexp2string } (t \; s)) * \text{enough\_space } (t \; s) \; l * \text{pc } p \; \}$
  $p : \ldots$ code not shown $\ldots$
  $\{ \; \exists a. \; \text{r3 } a * \text{string } a \; (\text{sexp2string } (t \; r)) * \text{enough\_space' } (t \; s) \; l * \text{pc } (p{+}10404) \; \}$

The input needs to be in register 3 for PowerPC and the eax register for x86.

8   Quantitative Data

The idea for this project first arose approximately two years ago. Since then a decompiler [15] and compiler [16] have been developed to aid this project, which produced in total some 4,580 lines of proof automation and 16,130 lines of interactive proofs and definitions, excluding the definitions of the instruction set models [5,9,18]. Running through all of the proofs takes approximately 2.5 hours in HOL4 using PolyML.

The verified LISP implementations seem to have reasonable execution times: the `pascal-triangle` example, from Section 1, executes on a 2.4 GHz x86 processor in less than 1 millisecond and on a 67 MHz ARM processor in approximately 90 milliseconds. The PowerPC implementations have not yet been tested on real hardware. The ARM implementation is 2,601 instructions long (10,404 bytes), x86 is 3,135 instructions (9,054 bytes) and the PowerPC implementation consists of 2,929 instructions (11,716 bytes).

9   Discussion of Related Work

This project has produced trustworthy implementations of LISP. The VLISP project by Guttman et al. [7] shared our goal, but differed in many other aspects. For example, the VLISP project implemented a larger LISP dialect, namely Scheme, and emphasised rigour, not full formality:

> "The verification was intended to be rigorous, but not completely formal, much in the style of ordinary mathematical discourse. Our goal was to verify the algorithms and data types used in the implementation, not their embodiment in the code."

The VLISP project developed an implementation which translates Scheme programs into byte code that is then run on a rigorously verified interpreter. Much like our project, the VLISP project developed their interpreter in a subset of the source language: for them PreScheme, and for us, the input language of our augmented compiler, Section 4.

Work that aims to implement functional languages, in a formally verified manner, include Pike et al. [17] on a certifying compiler from Cryptol (a dialect of Haskell) to AAMP7 code; Dargaye and Leroy [4] on a certified compiler from mini-ML to PowerPC assembly; Li and Slind's work [10] on a certifying compiler from a subset of HOL4 to ARM assembly; and also Chlipala's certified compiler [3] from the lambda calculus to an invented assembly language. The above work either assumes that the environment implements run-time memory management correctly [3,4] or restricts the input language to a degree where no run-time memory management is needed [10,17]. It seems that none of the above have made use of (the now large number of) verified garbage collectors (e.g. McCreight et al. [13] have been performing correctness proofs for increasingly sophisticated garbage collectors).

The parser and printer proofs, in Section 7, involved verifying implementations of string-copy, -length, -compare etc., bearing some resemblance to pioneering work by Boyer and Yu [1] on verification of machine code. They verified Motorola MC68020 code implementing a library of string functions.

Acknowledgements. We thank Anthony Fox, Xavier Leroy and Susmit Sarkar et al. for allowing us to use their processor models for this work [5,9,18]. We also thank Thomas Tuerk, Joe Hurd, Konrad Slind and John Matthews for comments and discussions. We are grateful for funding from EPSRC, UK.

References

1. Boyer, R.S., Yu, Y.: Automated proofs of object code for a widely used micropro-
   cessor. J. ACM 43(1), 166–192 (1996)
2. Cheney, C.J.: A non-recursive list compacting algorithm. Commun. ACM 13(11),
   677–678 (1970)
3. Chlipala, A.J.: A certified type-preserving compiler from lambda calculus to assembly language. In: Programming Language Design and Implementation (PLDI),
   pp. 54–65. ACM, New York (2007)
4. Dargaye, Z., Leroy, X.: Mechanized verification of CPS transformations. In: Dershowitz, N., Voronkov, A. (eds.) LPAR 2007. LNCS, vol. 4790, pp. 211–225.
   Springer, Heidelberg (2007)
5. Fox, A.: Formal specification and verification of ARM6. In: Basin, D., Wolff, B.
   (eds.) TPHOLs 2003. LNCS, vol. 2758, pp. 25–40. Springer, Heidelberg (2003)
6. Gordon, M.: Defining a LISP interpreter in a logic of total functions. In: The ACL2
   Theorem Prover and Its Applications, ACL2 (2007)
7. Guttman, J., Ramsdell, J., Wand, M.: VLISP: A verified implementation of scheme.
   Lisp and Symbolic Computation 8(1/2), 5–32 (1995)
8. Kaufmann, M., Moore, J.S.: An ACL2 tutorial. In: Mohamed, O.A., Muñoz, C.,
   Tahar, S. (eds.) TPHOLs 2008. LNCS, vol. 5170, pp. 17–21. Springer, Heidelberg
   (2008)
9. Leroy, X.: Formal certification of a compiler back-end, or: programming a compiler
   with a proof assistant. In: Principles of Programming Languages (POPL), pp. 42–
   54. ACM Press, New York (2006)
10. Li, G., Owens, S., Slind, K.: A proof-producing software compiler for a subset of
    higher order logic. In: European Symposium on Programming (ESOP). LNCS, pp.
    205–219. Springer, Heidelberg (2007)
11. Manolios, P., Strother Moore, J.: Partial functions in ACL2. J. Autom. Reason-
    ing 31(2), 107–127 (2003)
12. McCarthy, J., Abrahams, P.W., Edwards, D.J., Hart, T.P., Levin, M.I.: LISP 1.5
    Programmer's Manual. The MIT Press, Cambridge (1966)
13. McCreight, A., Shao, Z., Lin, C., Li, L.: A general framework for certifying garbage
    collectors and their mutators. In: Ferrante, J., McKinley, K.S. (eds.) Proceedings
    of the Conference on Programming Language Design and Implementation (PLDI),
    pp. 468–479. ACM, New York (2007)
14. Myreen, M.O.: Formal verification of machine-code programs. PhD thesis, Univer-
    sity of Cambridge (2009)
15. Myreen, M.O., Slind, K., Gordon, M.J.C.: Machine-code verification for multiple
    architectures – An application of decompilation into logic. In: Formal Methods in
    Computer Aided Design (FMCAD). IEEE, Los Alamitos (2008)
16. Myreen, M.O., Slind, K., Gordon, M.J.C.: Extensible proof-producing compilation.
    In: Compiler Construction (CC). LNCS. Springer, Heidelberg (2009)
17. Pike, L., Shields, M., Matthews, J.: A verifying core for a cryptographic language
    compiler. In: Manolios, P., Wilding, M. (eds.) Proceedings of the Sixth Interna-
    tional Workshop on the ACL2 Theorem Prover and its Applications. HappyJack
    Books (2006)
18. Sarkar, S., Sewell, P., Nardelli, F.Z., Owens, S., Ridge, T., Myreen, T.B.M.O.,
    Alglave, J.: The semantics of x86-CC multiprocessor machine code. In: Principles
    of Programming Languages (POPL). ACM, New York (2009)

19. Slind, K., Norrish, M.: A brief overview of HOL4. In: Mohamed, O.A., Muñoz, C., Tahar, S. (eds.) TPHOLs 2008. LNCS, vol. 5170, pp. 28–32. Springer, Heidelberg (2008)

A   Definition of `lisp_inv` in HOL4

The definition of the main invariant of the LISP state.

```
ALIGNED a = (a && 3w = 0w)

string_mem "" (a,m,dm) = T
string_mem (STRING c s) (a,m,df) = a ∈ dm ∧
                          (m a = n2w (ORD c)) ∧ string_mem s (a+1w,m,dm)

symbol_table [] x (a,dm,m,dg,g) = (m a = 0w) ∧ a ∈ dm ∧ (x = {})
symbol_table (s::xs) x (a,dm,m,dg,g) = (s ≠ "") ∧ ¬ MEM s xs ∧
  (m a = n2w (string_size s)) ∧ {a; a+4w} ⊆ dm ∧ ((a,s) ∈ x) ∧
    let a' = a + n2w (8 + (string_size s + 3) DIV 4 * 4) in
      a < a' ∧ (m (a+4w) = a') ∧ string_mem s (a+8w,g,dg) ∧
      symbol_table xs (x - {(a,s)}) (a',dm,m,dg,g)

builtin =
  ["nil"; "t"; "quote"; "+"; "-"; "*"; "div"; "mod"; "<"; "car"; "cdr";
   "cons"; "equal"; "cond"; "atomp"; "consp"; "numberp"; "symbolp"; "lambda"]

lisp_symbol_table sym (a,dm,m,dg,g) =
  ∃syms. symbol_table (builtin ++ syms) { (b,s) | (b-a,s) ∈ sym } (a,dm,m,dg,g)

lisp_x (Num k) (a,dm,m) sym = (a = n2w (k * 4 + 2)) ∧ k < 2 ** 30
lisp_x (Sym s) (a,dm,m) sym = ALIGNED (a - 3w) ∧ (a - 3w,s) ∈ sym
lisp_x (Dot x y) (a,dm,m) sym = lisp_x x (m a,dm,m) sym ∧ a ∈ dm ∧ ALIGNED a ∧
                               lisp_x y (m (a+4w),dm,m) sym

ref_set a f = {a + 4w * n2w i | i < 2 * f + 4} ∪ {a - 4w * n2w i | i ≤ 8}

ch_active_set (a,i,e) = { a + 8w * n2w j | i ≤ j ∧ j < e }

ok_data w d = if ALIGNED w then w ∈ d else ¬(ALIGNED (w - 1w))

lisp_inv (t1,t2,t3,t4,t5,t6,l) (w1,w2,w3,w4,w5,w6,a,(dm,m),sym,(dh,h),(dg,g)) =
  ∃i u.
    let v = if u then 1 + l else 1 in
    let d = ch_active_set (a,v,i) in
      32 ≤ w2n a ∧ w2n a + 2 * 8 * l + 20 < 2 ** 32 ∧ l ≠ 0 ∧
      (m a = a + n2w (8 * i)) ∧ ALIGNED a ∧ v ≤ i ∧ i ≤ v + 1 ∧
      (m (a + 4w) = a + n2w (8 * (v + 1))) ∧
      (m (a - 28w) = if u then 0w else 1w) ∧
      (m (a - 32w) = n2w (8 * l)) ∧ (dm = ref_set a (l + l + 1)) ∧
      lisp_symbol_table sym (a + 16w * n2w l + 24w,dh,h,dg,g) ∧
      lisp_x t1 (w1,d,m) sym ∧ lisp_x t2 (w2,d,m) sym ∧ lisp_x t3 (w3,d,m) sym ∧
      lisp_x t4 (w4,d,m) sym ∧ lisp_x t5 (w5,d,m) sym ∧ lisp_x t6 (w6,d,m) sym ∧
      ∀w. w ∈ d ⟹ ok_data (m w) d ∧ ok_data (m (w + 4w)) d
```

B   Sample Verification Proof of 'Car' Primitive

The verification proofs of the primitive LISP operations build on lemmas about lisp_inv. The following lemma is used in the proof of the theorem about car described in Section 3.1. This lemma can be read as saying that, if lisp_inv relates x_1 to Dot-pair v_1, then x_1 is a word-aligned address into memory segment m, and an assignment of car v_1 to v_2 corresponds to replacing x_2 with the value of memory m at address x_1, i.e. $m(x_1)$.

$(\exists x\, y.\ \mathsf{Dot}\ x\ y = v_1)\ \wedge$
$\mathsf{lisp\_inv}\ (v_1, v_2, v_3, v_4, v_5, v_6, l)\ (x_1, x_2, x_3, x_4, x_5, x_6, a, m, m_2, m_3)\ \Rightarrow$
$(x_1\ \&\ 3 = 0) \wedge x_1 \in \mathsf{domain}\ m\ \wedge$
$\mathsf{lisp\_inv}\ (v_1, \mathsf{car}\ v_1, v_3, v_4, v_5, v_6, l)\ (x_1, m(x_1), x_3, x_4, x_5, x_6, a, m, m_2, m_3)$

One of our tools derives the following Hoare triple theorem for the ARM instruction that is to be verified: `ldr r4, [r3]` (encoded as E5934000).

$\{\mathsf{r3}\ r_3 * \mathsf{r4}\ r_4 * \mathsf{m}\ m * \mathsf{pc}\ p * \langle(r_3\ \&\ 3 = 0) \wedge r_3 \in \mathsf{domain}\ m\rangle\ \}$
$p : \mathsf{E5934000}$
$\{\mathsf{r3}\ r_3 * \mathsf{r4}\ m(r_3) * \mathsf{m}\ m * \mathsf{pc}\ (p{+}4)\ \}$

Application of the frame rule (shown in Appendix C) produces:

$\{\mathsf{r3}\ r_3 * \mathsf{r4}\ r_4 * \mathsf{m}\ m * \mathsf{pc}\ p * \langle(r_3\ \&\ 3 = 0) \wedge r_3 \in \mathsf{domain}\ m\rangle\ *$
$\mathsf{r5}\ x_3 * \mathsf{r6}\ x_4 * \mathsf{r7}\ x_5 * \mathsf{r8}\ x_6 * \mathsf{r10}\ a * \mathsf{m}\ m_2 * \mathsf{m}\ m_3 *$
$\langle\mathsf{lisp\_inv}\ (v_1, v_2, v_3, v_4, v_5, v_6, l)\ (r_3, r_4, x_3, x_4, x_5, x_6, a, m, m_2, m_3)\rangle\}$
$p : \mathsf{E5934000}$
$\{\mathsf{r3}\ r_3 * \mathsf{r4}\ m(r_3) * \mathsf{m}\ m * \mathsf{pc}\ (p{+}4) * \langle(r_3\ \&\ 3 = 0) \wedge r_3 \in \mathsf{domain}\ m\rangle\ *$
$\mathsf{r5}\ x_3 * \mathsf{r6}\ x_4 * \mathsf{r7}\ x_5 * \mathsf{r8}\ x_6 * \mathsf{r10}\ a * \mathsf{m}\ m_2 * \mathsf{m}\ m_3 *$
$\langle\mathsf{lisp\_inv}\ (v_1, v_2, v_3, v_4, v_5, v_6, l)\ (r_3, r_4, x_3, x_4, x_5, x_6, a, m, m_2, m_3)\rangle\}$

Now the postcondition can be weakened to the desired expression:

$\{\mathsf{r3}\ r_3 * \mathsf{r4}\ r_4 * \mathsf{m}\ m * \mathsf{pc}\ p * \langle(r_3\ \&\ 3 = 0) \wedge r_3 \in \mathsf{domain}\ m\rangle\ *$
$\mathsf{r5}\ x_3 * \mathsf{r6}\ x_4 * \mathsf{r7}\ x_5 * \mathsf{r8}\ x_6 * \mathsf{r10}\ a * \mathsf{m}\ m_2 * \mathsf{m}\ m_3 *$
$\langle\mathsf{lisp\_inv}\ (v_1, v_2, v_3, v_4, v_5, v_6, l)\ (r_3, r_4, x_3, x_4, x_5, x_6, a, m, m_2, m_3)\rangle\}$
$p : \mathsf{E5934000}$
$\{\ \mathsf{lisp}\ (v_1, \mathsf{car}\ v_1, v_3, v_4, v_5, v_6, l) * \mathsf{pc}\ (p + 4)\ \}$

Since variables r_3, r_4, x_3, x_4, x_5, x_6, m, m_2, m_3 do not appear in the postcondition, they can be existentially quantified in the precondition, which then strengthens as follows:

$\{\ \mathsf{lisp}\ (v_1, v_2, v_3, v_4, v_5, v_6, l) * \mathsf{pc}\ p * \langle\exists x\, y.\ \mathsf{Dot}\ x\ y = v_1\rangle\ \}$
$p : \mathsf{E5934000}$
$\{\ \mathsf{lisp}\ (v_1, \mathsf{car}\ v_1, v_3, v_4, v_5, v_6, l) * \mathsf{pc}\ (p + 4)\ \}$

The specification for car follows by moving the boolean condition:

$(\exists x\, y.\ \mathsf{Dot}\ x\ y = v_1)\ \Rightarrow$
$\{\ \mathsf{lisp}\ (v_1, v_2, v_3, v_4, v_5, v_6, l) * \mathsf{pc}\ p\ \}$
$p : \mathsf{E5934000}$
$\{\ \mathsf{lisp}\ (v_1, \mathsf{car}\ v_1, v_3, v_4, v_5, v_6, l) * \mathsf{pc}\ (p + 4)\ \}$

All of the primitive LISP operations were verified in the same manner. For the HOL4 implementation, a 50-line ML program was written to automate these proofs given the appropriate lemmas about lisp_inv.

C   Proof Performed by Compiler

Internally the compiler runs through a short proof when constructing the theorem presented in Section 4. This proof makes use of the following five proof rules derived from the definition of our machine-code Hoare triple, developed in previous work [15]. Formal definitions and detailed explanations are given in the first author's PhD thesis [14]. Here $\cup$ is simply set union.

frame:	$\{p\}\,c\,\{q\} \;\Rightarrow\; \forall r.\ \{p * r\}\,c\,\{q * r\}$
code extension:	$\{p\}\,c\,\{q\} \;\Rightarrow\; \forall d.\ \{p\}\,c \cup d\,\{q\}$
composition:	$\{p\}\,c\,\{q\} \wedge \{q\}\,d\,\{r\} \;\Rightarrow\; \{p\}\,c \cup d\,\{r\}$
move pure:	$\{p * \langle b \rangle\}\,c\,\{q\} \;=\; (b \Rightarrow \{p\}\,c\,\{q\})$
tail recursion:	$(\forall x.\ P(x) \wedge G(x) \Rightarrow \{p(x)\}\,c\,\{p(F(x))\}) \wedge$ $(\forall x.\ P(x) \wedge \neg G(x) \Rightarrow \{p(x)\}\,c\,\{q(D(x))\}) \;\Rightarrow$ $(\forall x.\ \mathsf{pre}(G,F,P)(x) \Rightarrow \{p(x)\}\,c\,\{q(\mathsf{tailrec}(G,F,D)(x))\})$

The last rule mentions tailrec and pre, which are functions that satisfy:

$$\forall x. \quad \mathsf{tailrec}(G,F,D)(x) = \text{if } G(x) \text{ then } \mathsf{tailrec}(G,F,D)(F(x)) \text{ else } D(x)$$

$$\forall x. \qquad \mathsf{pre}(G,F,P)(x) = \text{if } G(x) \text{ then } \mathsf{pre}(G,F,P)(F(x)) \wedge P(x) \text{ else } P(x)$$

Note that any tail-recursive function can be defined as an instance of tailrec, introduced using a trick by Manolios and Moore [11]. Another noteworthy feature: if $\mathsf{pre}(G,F,P)(x)$ is true then $\mathsf{tailrec}(G,F,D)$ terminates for input x.

The compiler starts its proof from the following theorems describing the test $v_1 = \mathsf{Sym}\ \texttt{"nil"}$ as well as operations car, cdr and plus.

1. $\{\ \mathsf{lisp}\ (v_1, v_2, v_3, v_4, v_5, v_6, l) * \mathsf{pc}\ p * \mathsf{s}\ \}$
   $p : \texttt{E3330003}$
   $\{\ \mathsf{lisp}\ (v_1, v_2, v_3, v_4, v_5, v_6, l) * \mathsf{pc}\ (p+4) * \mathsf{sz}\ (v_1 = \mathsf{Sym}\ \texttt{"nil"}) *$
   $\exists n\ c\ v.\ \mathsf{sn}\ n * \mathsf{sc}\ c * \mathsf{sv}\ v\ \}$

2. $(\exists x\ y.\ \mathsf{Dot}\ x\ y = v_1) \Rightarrow$
   $\{\ \mathsf{lisp}\ (v_1, v_2, v_3, v_4, v_5, v_6, l) * \mathsf{pc}\ p * \mathsf{s}\ \}$
   $p : \texttt{E5935000}$
   $\{\ \mathsf{lisp}\ (v_1, v_2, \mathsf{car}\ v_1, v_4, v_5, v_6, l) * \mathsf{pc}\ (p+4)\ \}$

3. $(\exists x\ y.\ \mathsf{Dot}\ x\ y = v_1) \Rightarrow$
   $\{\ \mathsf{lisp}\ (v_1, v_2, v_3, v_4, v_5, v_6, l) * \mathsf{pc}\ p * \mathsf{s}\ \}$
   $p : \texttt{E5933004}$
   $\{\ \mathsf{lisp}\ (\mathsf{cdr}\ v_1, v_2, v_3, v_4, v_5, v_6, l) * \mathsf{pc}\ (p+4)\ \}$

4. $(\exists m\ n.\ \mathsf{Num}\ m = v_2 \wedge \mathsf{Num}\ n = v_3 \wedge m{+}n < 2^{30}) \Rightarrow$
   $\{\ \mathsf{lisp}\ (v_1, v_2, v_3, v_4, v_5, v_6, l) * \mathsf{pc}\ p * \mathsf{s}\ \}$
   $p : \texttt{E0844005}\ \ \texttt{E2444002}$
   $\{\ \mathsf{lisp}\ (v_1, \mathsf{plus}\ v_2\ v_3, v_3, v_4, v_5, v_6, l) * \mathsf{pc}\ (p+4)\ \}$

The compiler next generates two branches to glue the code together; the branch instructions have the following specifications:

5. $\{\, \text{pc}\, p * \text{sz}\, z * \langle z \rangle\, \}\ p : \text{0A000004}\ \{\, \text{pc}\, (p+24) * \text{sz}\, z\, \}$

6. $\{\, \text{pc}\, p * \text{sz}\, z * \langle \neg z \rangle\, \}\ p : \text{0A000004}\ \{\, \text{pc}\, (p+4) * \text{sz}\, z\, \}$

7. $\{\, \text{pc}\, p\, \}\ p : \text{EAFFFFF8}\ \{\, \text{pc}\, (p-24)\, \}$

The specifications above are collapsed into theorems describing one pass through the code by composing 1,5 and 1,6,2,3,4,7, which results in:

8. $\{\, \text{lisp}\, (v_1, v_2, v_3, v_4, v_5, v_6, l) * \text{pc}\, p * s * \langle v_1 = \text{Sym "nil"} \rangle\, \}$
   $p : \text{E3330003 0A000004}$
   $\{\, \text{lisp}\, (v_1, v_2, v_3, v_4, v_5, v_6, l) * \text{pc}\, (p+28) * s\, \}$

9. $(\exists x\, y.\ \text{Dot}\, x\, y = v_1)\, \wedge$
   $(\exists m\, n.\ \text{Num}\, m = v_2 \wedge \text{Num}\, n = \text{car}\, v_1 \wedge m{+}n < 2^{30})\, \Rightarrow$
   $\{\, \text{lisp}\, (v_1, v_2, v_3, v_4, v_5, v_6, l) * \text{pc}\, p * s * \langle v_1 \neq \text{Sym "nil"} \rangle\, \}$
   $p : \text{E3330003 0A000004 E5935000 E5934004 E0844005 E2444002 EAFFFFF8}$
   $\{\, \text{lisp}\, (\text{cdr}\, v_1, \text{plus}\, v_2\, (\text{car}\, v_1, l), \text{car}\, v_1, v_4, v_5, v_6) * \text{pc}\, p * s\, \}$

Code extension is applied to theorem 8, and then the rule for introducing a tail-recursive function is applied. The compiler produces the following total-correctness specification.

10. $\text{sumlist\_pre}(v_1, v_2, v_3)\, \Rightarrow$
    $\{\, \text{lisp}\, (v_1, v_2, v_3, v_4, v_5, v_6, l) * \text{pc}\, p * s\, \}$
    $p : \text{E3330003 0A000004 E5935000 E5934004 E0844005 E2444002 EAFFFFF8}$
    $\{\, \text{let}\, (v_1, v_2, v_3) = \text{sumlist}(v_1, v_2, v_3)\, \text{in}$
    $\quad \text{lisp}\, (v_1, v_2, v_3, v_4, v_5, v_6, l) * \text{pc}\, (p+28) * s\, \}$

Here sumlist is defined as an instance of tailrec, and sumlist_pre is an instance of pre. The compiler exports sumlist_pre as the following recursive function which collects all of the side conditions that must hold for proper execution of the code:

$\text{sumlist\_pre}(v_1, v_2, v_3) =$
$\quad \text{if}\, v_1 = \text{Sym "nil"}\, \text{then true else}$
$\qquad \text{let}\, cond = (\exists x\, y.\ \text{Dot}\, x\, y = v_1)\, \text{in}$
$\qquad \text{let}\, v_3 = \text{car}\, v_1\, \text{in}$
$\qquad \text{let}\, cond = cond \wedge (\exists x\, y.\ \text{Dot}\, x\, y = v_1)\, \text{in}$
$\qquad \text{let}\, v_1 = \text{cdr}\, v_1\, \text{in}$
$\qquad \text{let}\, cond = cond \wedge (\exists m\, n.\ \text{Num}\, m = v_2 \wedge \text{Num}\, n = v_3 \wedge m{+}n < 2^{30})\, \text{in}$
$\qquad \text{let}\, v_2 = \text{plus}\, v_2\, v_3\, \text{in}$
$\qquad\quad \text{sumlist\_pre}(v_1, v_2, v_3) \wedge cond$

When the loop rule is applied above, its parameters are assigned values:

$$p = \lambda(v_1, v_2, v_3).\ \mathsf{lisp}\ (v_1, v_2, v_3, v_4, v_5, v_6, l) * \mathsf{pc}\ p * \mathsf{s}$$

$$q = \lambda(v_1, v_2, v_3).\ \mathsf{lisp}\ (v_1, v_2, v_3, v_4, v_5, v_6, l) * \mathsf{pc}\ (p + 28) * \mathsf{s}$$

$$G = \lambda(v_1, v_2, v_3).\ v_1 \neq \mathsf{Sym}\ \texttt{"nil"}$$

$$F = \lambda(v_1, v_2, v_3).\ (\mathsf{cdr}\ v_1, \mathsf{plus}\ v_2\ (\mathsf{car}\ v_1, l), \mathsf{car}\ v_1)$$

$$D = \lambda(v_1, v_2, v_3).\ (v_1, v_2, v_3)$$

$$P = \lambda(v_1, v_2, v_3).\ (v_1 \neq \mathsf{Sym}\ \texttt{"nil"}) \Rightarrow$$
$$(\exists x\ y.\ \mathsf{Dot}\ x\ y = v_1) \wedge$$
$$(\exists m\ n.\ \mathsf{Num}\ m = v_2 \wedge \mathsf{Num}\ n = \mathsf{car}\ v_1 \wedge m{+}n < 2^{30})$$

D   Definition of s-Expression Printing and Parsing

Our machine code for printing LISP s-expressions implements sexp2string.

$$\mathsf{sexp2string}\ x = \mathsf{aux}\ (x, \mathsf{T})$$
$$\mathsf{aux}\ (\mathsf{Num}\ n, b) = \mathsf{num2str}\ n$$
$$\mathsf{aux}\ (\mathsf{Sym}\ s, b) = s$$
$$\mathsf{aux}\ (\mathsf{Dot}\ x\ y, b) = \text{if isQuote } (\mathsf{Dot}\ x\ y) \wedge b \text{ then } \texttt{"'"} +\!\!+ \mathsf{aux}\ (\mathsf{car}\ y, \mathsf{T}) \text{ else}$$
$$\text{let } (a, e) = (\text{if } b \text{ then } (\texttt{"("}, \texttt{")"}) \text{ else } (\texttt{""}, \texttt{""})) \text{ in}$$
$$\text{if } y = \mathsf{Sym}\ \texttt{"nil"} \text{ then } a +\!\!+ \mathsf{aux}\ (x, \mathsf{T}) +\!\!+ e \text{ else}$$
$$\text{if isDot } y \text{ then } a +\!\!+ \mathsf{aux}\ (x, \mathsf{T}) +\!\!+ \texttt{" "} +\!\!+ \mathsf{aux}\ (y, \mathsf{F}) +\!\!+ e$$
$$\text{else } a +\!\!+ \mathsf{aux}\ (x, \mathsf{T}) +\!\!+ \texttt{" . "} +\!\!+ \mathsf{aux}\ (y, \mathsf{F}) +\!\!+ e$$
$$\mathsf{isDot}\ x = \exists y\ z.\ x = \mathsf{Dot}\ y\ z$$
$$\mathsf{isQuote}\ x = \exists y.\ x = \mathsf{Dot}\ (\mathsf{Sym}\ \texttt{"quote"})\ (\mathsf{Dot}\ y\ (\mathsf{Sym}\ \texttt{"nil"}))$$

Parsing is defined as the follows. Here reverse is normal list reversal.

$$\mathsf{string2sexp}\ str = \mathsf{car}\ (\mathsf{sexp\_parse}\ (\mathsf{reverse}\ (\mathsf{sexp\_lex}\ str))\ (\mathsf{Sym}\ \texttt{"nil"})\ [])$$

The lexing function sexp_lex splits a string into a list of strings, e.g.

$$\mathsf{sexp\_lex}\ \texttt{"(car ('23 . y))"} = [\texttt{"("}, \texttt{"car"}, \texttt{"("}, \texttt{"'"}, \texttt{"23"}, \texttt{"."}, \texttt{"y"}, \texttt{")"}, \texttt{")"}]$$

Token parsing is defined as:

$$\mathsf{sexp\_parse}\ []\ exp\ stack = exp$$
$$\mathsf{sexp\_parse}\ (\texttt{")"} :: ts)\ exp\ stack = \mathsf{sexp\_parse}\ ts\ (\mathsf{Sym}\ \texttt{"nil"})\ (exp :: stack)$$
$$\mathsf{sexp\_parse}\ (\texttt{"("} :: ts)\ exp\ stack = \mathsf{sexp\_parse}\ ts\ (\mathsf{Dot}\ exp\ (\mathsf{head}\ stack))\ (\mathsf{tail}\ stack)$$
$$\mathsf{sexp\_parse}\ (\texttt{"."} :: ts)\ exp\ stack = \mathsf{sexp\_parse}\ ts\ (\mathsf{car}\ exp)\ stack$$
$$\mathsf{sexp\_parse}\ (\texttt{"'"} :: ts)\ exp\ stack = \mathsf{sexp\_parse}\ ts\ (\mathsf{Dot}\ (\mathsf{Dot}\ (\mathsf{Sym}\ \texttt{"quote"})$$
$$(\mathsf{Dot}\ (\mathsf{car}\ exp)\ (\mathsf{Sym}\ \texttt{"nil"})))\ (\mathsf{cdr}\ exp))\ stack$$
$$\mathsf{sexp\_parse}\ (t :: ts)\ exp\ stack = \mathsf{sexp\_parse}\ ts\ (\mathsf{Dot}\ (\text{if is\_num } t \text{ then}$$
$$\mathsf{Num}\ (\mathsf{str2num}\ t) \text{ else } \mathsf{Sym}\ t)\ exp)\ stack$$

Trace-Based Coinductive Operational Semantics for While

Big-Step and Small-Step, Relational and Functional Styles

Keiko Nakata and Tarmo Uustalu

Institute of Cybernetics at Tallinn University of Technology,
Akadeemia tee 21, EE-12618 Tallinn, Estonia
{keiko,tarmo}@cs.ioc.ee

Abstract. We present four coinductive operational semantics for the While language accounting for both terminating and non-terminating program runs: big-step and small-step relational semantics and big-step and small-step functional semantics. The semantics employ traces (possibly infinite sequences of states) to record the states that program runs go through. The relational semantics relate statement-state pairs to traces, whereas the functional semantics return traces for statement-state pairs. All four semantics are equivalent. We formalize the semantics and their equivalence proofs in the constructive setting of Coq.

1 Introduction

Now and then we must program a partially recursive function whose domain of definedness we cannot decide or is undecidable, e.g., an interpreter. Reactive programs such as operating systems and data base systems are not supposed to terminate. To reason about such programs properly, we need semantics that account for both terminating and non-terminating program runs. Compilers, for example, should preserve both terminating and non-terminating behaviors of source programs [10,13]. Standard operational semantics ignore (or say too little about) non-terminating runs, so finer semantic accounts are necessary.

In this paper, we present four coinductive semantics for the While language that we claim to be both adequate for reasoning about non-terminating runs as well as well-designed. They represent four different styles of operational semantics: big-step and small-step relational and big-step and small-step functional semantics. Our semantics are based on traces, defined coinductively as possibly infinite non-empty sequences of states. What is more, the evaluation and normalization relations and functions are also coinductive/corecursive. The functional semantics are constructively possible thanks to the fact that in the trace-based setting, While becomes a total rather than partial language (every run defines a trace, even if it may be infinite). All four semantics are constructively equivalent. We have formalized our development in the Coq proof assistant, using the Ssreflect syntax extension, see http://cs.ioc.ee/~keiko/majas.tar.gz.

It might be objected against this paper that the results are unsurprising, since the semantics appear simple and enjoy all expected properties. They are simple

S. Berghofer et al. (Eds.): TPHOLs 2009, LNCS 5674, pp. 375–390, 2009.

indeed, but in the case of the two big-step semantics, this is a consequence of very careful design decisions. As a matter of fact, getting coinductive big-step semantics right is tricky, and in this situation it is really fortunate that simple solutions are available. In the paper, we discuss some of the design considerations and also show some design options that we rejected deliberately. Previous work in the literature [6,9,14] also contains some designs that are more complicated than ours or fail to have some clearly desirable properties or both. A skeptical reader may also worry that While is a toy language. We argue that While is sufficient for highlighting all important issues. In fact, our designs scale without pain to procedures and language constructs for effects such as exceptions, non-determinism and interactive input-output.

Programming and reasoning with coinductive types in type theory require taking special care about productivity. Here the type checker of Coq help us avoid mistakes by ruling out improductivity. But some limitations are imposed by the implementation. For instance, 15 years ago a type-based approach for ensuring productivity of corecursive definitions was developed [8]. This approach is more flexible than the syntactic guardedness approach [7] of Coq, but it has not been implemented. Several coding techniques have been proposed to circumvent the limitations [2,14]. In our development, we rely on syntactic productivity.

The remainder of the paper is organized as follows. We introduce traces in Section 2. We present the big-step relational semantics in Section 3, the small-step relational semantics in Section 4, and the big-step and small-step functional semantics in Sections 5 and 6, proving the equivalent along the way. We discuss related work in Section 7 to conclude in Section 8.

The language we consider is the While language, defined inductively by the following productions:

$$stmt \quad s ::= \mathsf{skip} \mid s_0; s_1 \mid x := e \mid \mathsf{if}\ e\ \mathsf{then}\ s_t\ \mathsf{else}\ s_f \mid \mathsf{while}\ e\ \mathsf{do}\ s_t$$

We assume given a supply of variables and a set of (pure) expressions, whose elements are ranged over by metavariables x and e respectively. We assume the set of values to be the integers, non-zero integers counting as truth and zero as falsity. The metavariable v ranges over values. A state, ranged over by σ, maps variables to values. The notation $\sigma[x \mapsto v]$ denotes the update of σ with v at x. We assume given an evaluation function $[\![e]\!]\sigma$, which evaluates e in the state σ. We write $\sigma \models e$ and $\sigma \not\models e$ to denote that e is true, resp. false in σ.

2   Traces

We describe the semantics of statements in terms of traces. A trace is a possibly infinite non-empty sequence of states, the sequence of all states that the run of the statement passes through, including the given initial state. We enforce non-emptiness by having the nil constructor to also take a state as an argument. Formally traces are defined coinductively by the following productions:

$$trace \quad \tau ::= \langle \sigma \rangle \mid \sigma :: \tau$$

We define bisimilarity of two traces τ, τ', written $\tau \approx \tau'$, by the coinductive interpretation of the following inference rules[1]:

$$\frac{}{\langle \sigma \rangle \approx \langle \sigma \rangle} \qquad \frac{\tau \approx \tau'}{\sigma :: \tau \approx \sigma :: \tau'}$$

Bisimilarity is reflexive, symmetric and transitive, i.e., an equivalence. The proofs are straightforward. The reader can find a gentle introduction to coinduction in Coq in [1, Ch. 13].

We want to think of bisimilar traces as equal, corresponding to quotienting the set of traces as defined above by bisimilarity. In our type-theoretic implementation, we do not "compute" the quotient. Instead, we view *trace* with bisimilarity as a setoid, i.e., a set with an equivalence relation. Accordingly, we have to make sure that all functions and predicates we define on *trace* are in fact setoid functions and predicates, i.e., insensitive to bisimilarity.

3   Big-Step Relational Semantics

The main contribution of the paper is the big-step relational semantics, presented in Fig. 1. The semantics is given by two relations $\Rightarrow$ and $\overset{*}{\Rightarrow}$, defined mutually coinductively. The $\Rightarrow$ relation relates a statement-state pair to a trace and is the evaluation relation of our interest: the proposition $(s, \sigma) \Rightarrow \tau$ expresses that running s from an initial state σ results in trace τ. It is defined by case distinction on the statement.

The auxiliary $\overset{*}{\Rightarrow}$ relation relates a statement-trace pair to a trace. Roughly the proposition $(s, \tau) \overset{*}{\Rightarrow} \tau'$ states that running s from the last state of an already accumulated trace τ results in trace τ'. The rules literally define $\overset{*}{\Rightarrow}$ as the coinductive prefix closure of $\Rightarrow$. A more precise description of $(s, \tau) \overset{*}{\Rightarrow} \tau'$ is therefore as follows. If τ is finite, then s is run from the last state of τ and τ' is obtained from τ by appending the trace produced by s. If τ is infinite (so it does not have a last state), then $(s, \tau) \overset{*}{\Rightarrow} \tau'$ is derivable for any τ' bisimilar to τ, in particular for τ. This design has the desirable consequence that, if a run of the first statement of a sequence diverges, the second statement is not run at all. Indeed, if $(s_0, \sigma) \Rightarrow \tau$ and τ is infinite, then we can derive $(s_1, \tau) \overset{*}{\Rightarrow} \tau$ and further $(s_0; s_1, \sigma) \Rightarrow \tau$. Similarly, if a run of the body of a while-loop diverges, we do not get around to retesting the guard and continuing.

A remarkable feature of the definition of $\overset{*}{\Rightarrow}$ is that it does not hinge on deciding whether the trace is finite or not, which is constructively impossible. A proof of $(s, \tau) \overset{*}{\Rightarrow} \tau'$ is simply a traversal the already accumulated trace τ: if the last element is hit, the statement is run, otherwise the traversal goes on forever.

Evaluation is a setoid predicate and it is deterministic (up to bisimulation, which is appropriate since we think of bisimilarity as equality):

Lemma 1. *For any* σ, s, τ, τ', *if* $(s, \sigma) \Rightarrow \tau$ *and* $\tau \approx \tau'$ *then* $(s, \sigma) \Rightarrow \tau'$.

---

[1] Following X. Leroy [14], we use double horizontal lines in sets of inference rules that are to be interpreted coinductively and single horizontal lines in inductive definitions.

$$\frac{}{(x := e, \sigma) \Rightarrow \sigma :: \langle \sigma[x \mapsto [\![e]\!]\sigma]\rangle} \quad \frac{}{(\mathsf{skip}, \sigma) \Rightarrow \langle \sigma \rangle} \quad \frac{(s_0, \sigma) \Rightarrow \tau \quad (s_1, \tau) \overset{*}{\Rightarrow} \tau'}{(s_0; s_1, \sigma) \Rightarrow \tau'}$$

$$\frac{\sigma \models e \quad (s_t, \sigma :: \langle \sigma \rangle) \overset{*}{\Rightarrow} \tau}{(\mathsf{if}\ e\ \mathsf{then}\ s_t\ \mathsf{else}\ s_f, \sigma) \Rightarrow \tau} \quad \frac{\sigma \not\models e \quad (s_f, \sigma :: \langle \sigma \rangle) \overset{*}{\Rightarrow} \tau}{(\mathsf{if}\ e\ \mathsf{then}\ s_t\ \mathsf{else}\ s_f, \sigma) \Rightarrow \tau}$$

$$\frac{\sigma \models e \quad (s_t, \sigma :: \langle \sigma \rangle) \overset{*}{\Rightarrow} \tau \quad (\mathsf{while}\ e\ \mathsf{do}\ s_t, \tau) \overset{*}{\Rightarrow} \tau'}{(\mathsf{while}\ e\ \mathsf{do}\ s_t, \sigma) \Rightarrow \tau'} \quad \frac{\sigma \not\models e}{(\mathsf{while}\ e\ \mathsf{do}\ s_t, \sigma) \Rightarrow \sigma :: \langle \sigma \rangle}$$

$$\frac{(s, \sigma) \Rightarrow \tau}{(s, \langle \sigma \rangle) \overset{*}{\Rightarrow} \tau} \quad \frac{(s, \tau) \overset{*}{\Rightarrow} \tau'}{(s, \sigma :: \tau) \overset{*}{\Rightarrow} \sigma :: \tau'}$$

Fig. 1. Big-step relational semantics

Lemma 2. *For any σ, s, τ and τ', if $(s, \sigma) \Rightarrow \tau$ and $(s, \sigma) \Rightarrow \tau'$ then $\tau \approx \tau'$.*

Some design decisions we have made are that skip does not grow a trace, so we have $(\mathsf{skip}, \sigma) \Rightarrow \langle \sigma \rangle$. But an assignment and testing the guard of an if- or while-statement contribute a state, i.e., constitute a small step, e.g., we have $(x := 17, \sigma) \Rightarrow \sigma :: \langle \sigma[x \mapsto 17]\rangle$, $(\mathsf{while\ false\ do\ skip}, \sigma) \Rightarrow \sigma :: \langle \sigma \rangle$ and $(\mathsf{while\ true\ do\ skip}, \sigma) \Rightarrow \sigma :: \sigma :: \sigma :: \ldots$. This is good for several reasons. First, we have that skip is the identity of sequential composition, i.e., the semantics does not distinguish s, $\mathsf{skip}; s$ and $s; \mathsf{skip}$. Second, we get a notion of small steps that fully agrees with the textbook-style small-step semantics given in the next section. The third and most important outcome is that any while-loop always progresses, because testing of the guard is a small step. Another option would be to regard testing of the guard to be instantaneous, but take leaving the loop body, or a backward jump in terms of low-level compiled code, to constitute a small step. But then we would not agree to the textbook small-step semantics.

It is not mandatory to record full states in a trace as we are doing in this paper. It would make perfect sense to record just some observable part of the intermediate states, or to only record that some states were passed through (to track ticks of the clock). Neither is (strong) bisimilarity the only interesting notion of equality of traces. Viable alternatives are various weak versions of bisimilarity (allowing collapsing finite sequences of ticks).

Discussions on alternative designs. In the rest of this section we reveal some subtleties in designing coinductive big-step semantics, by looking at several seemingly not so different but problematic alternatives that we reject[2].

Since progress of loops is not required for wellformedness of the definitions of $\Rightarrow$ and $\overset{*}{\Rightarrow}$, one might be tempted to regards guard testing to be instantaneous and modify the rules for the while-loop to take the form

$$\frac{\sigma \models e \quad (s_t, \sigma) \Rightarrow \tau \quad (\mathsf{while}\ e\ \mathsf{do}\ s_t, \tau) \overset{*}{\Rightarrow} \tau'}{(\mathsf{while}\ e\ \mathsf{do}\ s_t, \sigma) \Rightarrow \tau'} \quad \frac{\sigma \not\models e}{(\mathsf{while}\ e\ \mathsf{do}\ s_t, \sigma) \Rightarrow \langle \sigma \rangle}$$

---

[2] Our Coq development includes complete definitions of these alternative semantics.

This leads to undesirable outcomes. We can derive (while true do skip, σ) $\Rightarrow$ $\langle\sigma\rangle$, which means that the non-terminating while true do skip is considered semantically equivalent to the terminal (immediately terminating) skip. Worse, we can also derive (while true do skip; $x := 17, \sigma$) $\Rightarrow \sigma :: \langle\sigma[x \mapsto 17]\rangle$, which is even more inadequate: a sequence can continue to run after the non-termination of the first statement. Yet worse, inspecting the rules closer we discover we are also able to derive (while true do skip, σ) $\Rightarrow \tau$ for any τ! Mathematically, giving up insisting on progress in terms of growing the trace has also the consequence that the relational semantics cannot be turned into a functional one, although While should intuitively be total and deterministic. In a functional semantics, evaluation must be a trace-valued function and in a constructive setting such a function must be productive.

Another option, where assignments and test of guards are properly taken to constitute steps, could be to define $\overset{*}{\Rightarrow}$ by case distinction on the statement by rules such as

$$\frac{\tau \models^* e \quad (s_t, duplast\ \tau) \overset{*}{\Rightarrow} \tau' \quad (\text{while } e \text{ do } s_t, \tau') \overset{*}{\Rightarrow} \tau''}{(\text{while } e \text{ do } s_t, \tau) \overset{*}{\Rightarrow} \tau''} \qquad \frac{\tau \not\models^* e}{(\text{while } e \text{ do } s_t, \tau) \overset{*}{\Rightarrow} duplast\ \tau}$$

Here, $duplast\ \tau$, defined corecursively, traverses τ and duplicates its last state, if it is finite. Similarly, $\tau \models^* e$ and $\tau \not\models^* e$ traverse τ and evaluate e in the last state, if it is finite:

$$\frac{\tau \models^* e}{\sigma :: \tau \models^* e} \qquad \frac{\sigma \models e}{\langle\sigma\rangle \models^* e} \qquad \frac{\tau \not\models^* e}{\sigma :: \tau \not\models^* e} \qquad \frac{\sigma \not\models e}{\langle\sigma\rangle \not\models^* e}$$

(The rules for skip and sequence are very simple and appealing in this design.) The relation $\Rightarrow$ would then be defined uniformly by the rule

$$\frac{(s, \langle\sigma\rangle) \overset{*}{\Rightarrow} \tau}{(s, \sigma) \Rightarrow \tau}$$

It turns out that we can still derive (while true do skip, σ) $\Rightarrow \tau$ for any τ. We can even derive (while true do $x := x + 1, \sigma$) $\Rightarrow \tau$ for any τ!

The third alternative (Leroy and Grall use this technique in [14]) is most close to ours. It introduces, instead of our $\overset{*}{\Rightarrow}$ relation, an auxiliary relation $split$, defined coinductively by

$$\frac{}{split\ \langle\sigma\rangle\ \langle\sigma\rangle\ \sigma\ \langle\sigma\rangle} \qquad \frac{}{split\ (\sigma :: \tau)\ \langle\sigma\rangle\ \sigma\ (\sigma :: \tau)} \qquad \frac{split\ \tau\ \tau_0\ \sigma'\ \tau_1}{split\ (\sigma :: \tau)\ (\sigma :: \tau_0)\ \sigma'\ \tau_1}$$

so that $split\ \tau'\ \tau_0\ \sigma'\ \tau_1$ expresses that the trace τ' can be split into a concatenation of traces τ_0 and τ_1 glued together at a mid-state σ'. Then the evaluation relation is defined by replacing the uses of $\overset{*}{\Rightarrow}$ with $split$, e.g., the rule for the sequence statement would be:

$$\frac{split\ \tau'\ \tau_0\ \sigma'\ \tau_1 \quad (s_0, \sigma) \Rightarrow \tau_0 \quad (s_1, \sigma') \Rightarrow \tau_1}{(s_0; s_1, \sigma) \Rightarrow \tau'}$$

This third alternative does not cause any outright anomalies for While. But alarmingly s_1 has to be run from some (underdetermined) state within a run of s_0; s_1 even if the run of s_0 does not terminate. In a richer language with abnormal terminations, we get a serious problem: no evaluation is derived for (while true do skip); abort although the abort statement should not be reached.

4   Small-Step Relational Semantics

Devising an adequate small-step relational semantics is an easy problem compared to the one of the previous section. We can adapt the textbook inductive small-step semantics, which only accounts for terminating runs. Our semantics, given in Fig. 2, is based on a terminality predicate and one-step reduction relation. The proposition $s \downarrow$ states that s is terminal (terminates in no steps), which is possible for a sequence of skips. The proposition $(s, \sigma) \to (s', \sigma')$ states that in state σ the statement s one-step reduces to s' with the next state being σ'. These are exactly the same as one would use for an inductive semantics. The normalization relation is the terminal many-step reduction relation, defined coinductively to allow for the possibility of infinitely many steps. The proposition $(s, \sigma) \rightsquigarrow \tau$ expresses that running s from σ results in the trace τ.

Normalization is a setoid predicate and it is deterministic:

Lemma 3. *For any* s, σ, τ, τ', *if* $(s, \sigma) \rightsquigarrow \tau$ *and* $\tau \approx \tau'$ *then* $(s, \sigma) \rightsquigarrow \tau'$.

Lemma 4. *For any* s, σ, τ, τ', *if* $(s, \sigma) \rightsquigarrow \tau$ *and* $(s, \sigma) \rightsquigarrow \tau'$ *then* $\tau \approx \tau'$.

Equivalence to big-step relational semantics. Of course we expect the big-step and small-step semantics to be equivalent. We will show this in two ways: the first approach, presented in this section, directly proves the equivalence; the second approach, given in Section 6, proves the equivalence by going through

$$\frac{}{\mathsf{skip}\downarrow} \qquad \frac{s_0\downarrow \quad s_1\downarrow}{s_0; s_1\downarrow}$$

$$\frac{}{(x := e, \sigma) \to (\mathsf{skip}, \sigma[x \mapsto [\![e]\!]\sigma])} \qquad \frac{s_0\downarrow \quad (s_1, \sigma) \to (s_1', \sigma')}{(s_0; s_1, \sigma) \to (s_1', \sigma')} \qquad \frac{(s_0, \sigma) \to (s_0', \sigma')}{(s_0; s_1, \sigma) \to (s_0'; s_1, \sigma')}$$

$$\frac{\sigma \models e}{(\mathsf{if}\ e\ \mathsf{then}\ s_t\ \mathsf{else}\ s_f, \sigma) \to (s_t, \sigma)} \qquad \frac{\sigma \not\models e}{(\mathsf{if}\ e\ \mathsf{then}\ s_t\ \mathsf{else}\ s_f, \sigma) \to (s_f, \sigma)}$$

$$\frac{\sigma \models e}{(\mathsf{while}\ e\ \mathsf{do}\ s_t, \sigma) \to (s_t; \mathsf{while}\ e\ \mathsf{do}\ s_t, \sigma)} \qquad \frac{\sigma \not\models e}{(\mathsf{while}\ e\ \mathsf{do}\ s_t, \sigma) \to (\mathsf{skip}, \sigma)}$$

$$\frac{s\downarrow}{(s, \sigma) \rightsquigarrow \langle\sigma\rangle} \qquad \frac{(s, \sigma) \to (s', \sigma') \quad (s', \sigma') \rightsquigarrow \tau}{(s, \sigma) \rightsquigarrow \sigma :: \tau}$$

Fig. 2. Small-step relational semantics

functional semantics. The first approach is stronger in that it does not rely on the determinism of the semantics, thus prepares a better avenue for generalization to a language with non-determinism. (Our functional semantics deals with single-valued functions and thus the second approach relies on the determinism.)

The following lemma connects the big-step semantics with the terminality predicate and one-step reduction relation and is proved by induction.

Lemma 5. *For any s, σ and τ, if $(s, \sigma) \Rightarrow \tau$ then either $s \notdownarrow$ and $\tau = \langle \sigma \rangle$, or else there are s', σ', τ' such that $(s, \sigma) \to (s', \sigma')$ and $\tau \approx \sigma :: \tau'$ and $(s', \sigma') \Rightarrow \tau'$.*

Then correctness of the big-step semantics relative to the small-step semantics follows by coinduction:

Proposition 1. *For any s, σ and τ, if $(s, \sigma) \Rightarrow \tau$ then $(s, \sigma) \leadsto \tau$.*

The opposite direction, that the small-step semantics is correct relative to the big-step semantics, is more interesting. The proof proceeds by coinduction. At the crux is the case of the sequence statement: we are given a normalization $(s_0; s_1, \sigma) \leadsto \tau$ and the coinduction hypotheses for s_0 (resp. s_1) that enable us to deduce $(s_0, \sigma') \Rightarrow \tau'$ (resp. $(s_1, \sigma') \Rightarrow \tau'$) from $(s_0, \sigma') \leadsto \tau'$ (resp. $(s_1, \sigma') \leadsto \tau'$) for any σ', τ'. Naively, we have what we need to close the case. The assumption $(s_0; s_1, \sigma) \leadsto \tau$ ensures that τ can be split into two parts τ_0 and τ_1 such that τ_0 corresponds to running s_0 and τ_1 to running s_1. If τ_0 is finite, we can traverse τ_0 until we hit its last state, to then invoke the coinduction hypothesis on s_1. If τ_0 is infinite, we can deduce $\tau \approx \tau_0$ and $(s_1, \tau_0) \overset{*}{\Rightarrow} \tau_0$ by coinduction.

The actual proof is more involved. First we have to explicitly construct τ_0 and τ_1. This is possible by examining the proof of $(s_0; s_1, \sigma) \leadsto \tau$. Our proof defines an auxiliary function $midp$ $(s_0 \; s_1 : stmt) \; (\sigma : state) \; (\tau : trace) \; (h : (s_0; s_1, \sigma) \leadsto \tau) : trace$ by corecursion as follows. We look at the last inference in the proof h of $(s_0; s_1, \sigma) \leadsto \tau$. If $s_0; s_1$ is terminal, we return $\langle \sigma \rangle$. Otherwise we have a proof h_0 of $(s_0; s_1, \sigma) \to (s', \sigma')$ and a proof h' of $(s', \sigma') \leadsto \tau'$ for some σ', τ' such that $\tau = \sigma :: \tau'$. We look at the last inference in h_0. If s_0 is terminal, we also return $\langle \sigma \rangle$. Else it must be the case that $(s_0, \sigma) \to (s'_0, \sigma')$ for some s'_0 such that $s' = s'_0; s_1$ and we return $\sigma :: midp \; s'_0 \; s_1 \; \sigma' \; \tau' \; h'$. The corecursive call is guarded by consing σ. The following lemma is proved by coinduction.

Lemma 6. *For any s_0, s_1, σ, τ, $h : (s_0; s_1, \sigma) \leadsto \tau$, $(s_0, \sigma) \leadsto midp \; s_0 \; s_1 \; \sigma \; \tau \; h$.*

Second, we cannot decide whether τ_0 is finite as this would amount to deciding whether running s_0 from σ terminates. Our big-step semantics was carefully crafted to avoid stumbling upon this problem, by introduction of the coinductive prefix closure $\overset{*}{\Rightarrow}$ of $\Rightarrow$ to uniformly handle the cases of both the finite and infinite already accumulated trace. We need a small-step counterpart to it:

$$\frac{(s, \sigma) \leadsto \tau}{(s, \langle \sigma \rangle) \overset{*}{\leadsto} \tau} \qquad \frac{(s, \tau) \overset{*}{\leadsto} \tau'}{(s, \sigma :: \tau) \overset{*}{\leadsto} \sigma :: \tau'}$$

The proposition $(s, \tau) \overset{*}{\leadsto} \tau'$ states that running s from the last state of an already accumulated trace τ (if it has one) results in the total trace τ'. The following lemma is proved by coinduction.

$$\frac{s \notdiv}{(s,\sigma) \leadsto^{\mathrm{ind}} \sigma} \qquad \frac{(s,\sigma) \to (s',\sigma') \quad (s',\sigma') \leadsto^{\mathrm{ind}} \sigma''}{(s,\sigma) \leadsto^{\mathrm{ind}} \sigma''}$$

Fig. 3. Inductive small-step relational semantics

Lemma 7. *For any* s_0, s_1, σ, τ, $h : (s_0; s_1, \sigma) \leadsto \tau$, $(s_1, midp\ s_0\ s_1\ \sigma\ \tau\ h) \overset{*}{\leadsto} \tau$.

Only now we can finally prove that the small-step relational semantics is correct relative to the big-step relational semantics.

Proposition 2. *For any* s, σ, τ, τ', *the following two conditions hold:*

- *if* $(s,\sigma) \leadsto \tau$ *then* $(s,\sigma) \Rightarrow \tau$,
- *if* $(s,\tau) \overset{*}{\leadsto} \tau'$ *then* $(s,\tau) \overset{*}{\Rightarrow} \tau'$.

Proof. Both conditions are proved at once by mutual coinduction. We only show the first condition in the case of the sequence statement, to demonstrate how the relation $\overset{*}{\leadsto}$ helps us avoid having to decide finiteness. Suppose we have $h : (s_0; s_1, \sigma) \leadsto \tau$. By Lemmata 6 and 7, we have $(s_0, \sigma) \leadsto midp\ s_0\ s_1\ \sigma\ \tau\ h$ and $(s_1, midp\ s_0\ s_1\ \sigma\ \tau\ h) \overset{*}{\leadsto} \tau$. By invoking the coinduction hypothesis on them, we obtain $(s_0, \sigma) \Rightarrow midp\ s_0\ s_1\ \sigma\ \tau\ h$. and $(s_1, midp\ s_0\ s_1\ \sigma\ \tau\ h) \overset{*}{\Rightarrow} \tau$, from which we deduce $(s_0; s_1, \sigma) \Rightarrow \tau$.

Differences between Coq's *Prop* and *Set* force normalization to be *Set*-valued rather than *Prop*-valued, since our definition of the *trace*-valued *midp* function relies on case distinction on the proof of the given normalization proposition. Case distinction on a proof of a *Prop*-proposition is not available for constructing an element of a *Set*-set. This in turn requires the evaluation relation to also be *Set*-valued, to be comparable to normalization. A further complication is that, for technical reasons, the proofs of Lemmata 6 and 7 must rely on John Major equality [15] and the principle that two JM-equal elements of the same type are equal. Given that Coq's support for programming with (co)inductive families (in ML-style, as opposed to proving in the tactic language) is also weak (so *midp* was easily manufactured in the tactic language, but we failed to construct it in ML-style), one might wish to prove the equivalence of the big-step and small-step semantics in some altogether different way. In the subsequent sections we study functional semantics. These offer us a less direct route that is less painful in the aspects we have just described.

Adequacy relative to inductive small-step relational semantics. The textbook inductive small-step relational semantics defines normalization as the inductive terminal many-step reduction. The definition is given in Fig. 3. This normalization relation associates a state-statement pair with a state (the terminal state) rather than a trace, although a trace-based version (for an inductive concept of traces, i.e., finite sequences of states) would be obtained by a straightforward modification. For completeness of our development, we prove that the inductive and coinductive semantics agree on terminating runs.

We introduce a last-state predicate on traces inductively by the rules

$$\frac{}{\langle \sigma \rangle \downarrow \sigma} \qquad \frac{\tau \downarrow \sigma'}{\sigma :: \tau \downarrow \sigma'}$$

Proposition 3 states that the inductive semantics is correct relative to the coinductive semantics. Proposition 4 states that the coinductive semantics is correct relative to the inductive semantics for terminating runs. Both propositions are proved by induction.

Proposition 3. *For any s and σ, if $(s, \sigma) \leadsto^{\mathrm{ind}} \sigma'$ then there is τ such that $(s, \sigma) \leadsto \tau$ and $\tau \downarrow \sigma'$.*

Proposition 4. *For any s, τ, σ, σ', if $(s, \sigma) \leadsto \tau$ and $\tau \downarrow \sigma'$ then $(s, \sigma) \leadsto^{\mathrm{ind}} \sigma'$.*

The connection between our coinductive big-step semantics and the inductive big-step semantics can now be concluded from the well-known equivalence between the inductive big-step and small-step semantics. Y. Bertot has formalized the proof of this equivalence in Coq [3].

We conclude this section by citing an observation by V. Capretta [4]. The infiniteness predicate on traces is defined coinductively by the rule

$$\frac{\tau^{\uparrow}}{(\sigma :: \tau)^{\uparrow}}$$

We can prove in Coq the proposition $\forall \tau, (\neg \exists \sigma, \tau \downarrow \sigma) \to \tau^{\uparrow}$. However the proposition $\forall \tau, (\exists \sigma, \tau \downarrow \sigma) \lor \tau^{\uparrow}$ can only be proved from $\forall \tau, (\exists \sigma, \tau \downarrow \sigma) \lor \neg(\exists \sigma, \tau \downarrow \sigma)$. Constructively, this instance of the classical law of excluded middle states decidability of finiteness. For this reason, we reject what could be called sum-type semantics. For instance, a relational semantics could relate a statement-state pair to either a state for a terminating run or a special token ∞ for a non-terminating run, i.e., an element from the sum type $state + 1$, where 1 is the one-element type. Or, it could be given as the disjunction of an inductive trace-based semantics, describing terminating runs, and a coinductive trace-based semantics, describing non-terminating runs, an approach studied in [14].

5 Big-Step Functional Semantics

We now proceed to functional versions of our semantics. The standard state-based approach to While does not allow for a (constructive) functional semantics, as this would require deciding the halting problem. Working with traces has the benefit that we do not have to decide: any statement and initial state uniquely determine some trace and we do not have to know whether this trace is finite or infinite. The semantics is given in Fig. 4. The evaluation function $eval : stmt \to state \to trace$ is defined by recursion on the statement. In the cases for sequence and while it calls auxiliary functions $sequence$ and $loop$.

We first look at $loop$. It is defined together with a further auxiliary function $loopseq$ by mutual corecursion. $loop$ takes three arguments: k for evaluating the

Fixpoint eval $(s : stmt)$ $(\sigma : state)$ $\{struct\ s\} : trace :=$
 match s with
 | skip $\Rightarrow \langle \sigma \rangle$
 | $x := e \Rightarrow \sigma :: \langle \sigma[x \mapsto [\![e]\!]\sigma] \rangle$
 | $s_0; s_1 \Rightarrow$ *sequence* $(eval\ s_1)$ $(eval\ s_0\ \sigma)$
 | if e then s_t else $s_f \Rightarrow \sigma :: $ *if* $\sigma \models e$ *then eval* $s_t\ \sigma$ *else eval* $s_f\ \sigma$
 | while e do $s_t \Rightarrow \sigma :: loop\ (eval\ s_t)\ [\![e]\!]\ \sigma$
 end

CoFixpoint sequence $(k : state \rightarrow trace)$ $(\tau : trace) : trace :=$
 match τ with
 | $\langle \sigma \rangle \Rightarrow k\ \sigma$
 | $\sigma :: \tau' \Rightarrow \sigma :: $ *sequence* $k\ \tau'$
 end

CoFixpoint loop $(k : state \rightarrow trace)$ $(p : state \rightarrow bool)$ $(\sigma : state) : trace :=$
 if $p\ \sigma$ then
  match $k\ \sigma$ with
  | $\langle \sigma' \rangle \Rightarrow \sigma' :: loop\ k\ p\ \sigma'$
  | $\sigma' :: \tau \Rightarrow \sigma' :: loopseq\ k\ p\ \tau$
  end
 else $\langle \sigma \rangle$
 with loopseq $(k : state \rightarrow trace)$ $(p : state \rightarrow bool)$ $(\tau : trace) : trace :=$
 match τ with
 | $\langle \sigma \rangle \Rightarrow \sigma :: loop\ k\ p\ \sigma$
 | $\sigma :: \tau' \Rightarrow \sigma :: loopseq\ k\ p\ \tau'$
 end

Fig. 4. Big-step functional semantics

loop body from a state; p for testing the boolean guard on a state; and a state σ, which is the initial state. *loopseq* takes a trace τ, the initial trace, instead of a state, as the third argument. The two functions work as follows. *loop* takes care of repeating of the loop body, once the guard of a while loop has been evaluated. It analyzes the result and, if the guard is false, then the run of the loop terminates. If it is true, then the loop body is evaluated by calling k. *loop* then constructs the trace of the loop body by examining the result of k. If the loop body does not augment the trace, which can only happen, if the loop body is a sequence of skips, a new round of repeating the loop body is started by a corecursive call to *loop*. The corecursive call is guarded by first augmenting the trace, which corresponds to the new evaluation of the boolean guard. If the loop body augments the trace, the new round is reached by reconstruction of the trace of the current repetition with *loopseq*. On the exhaustion of this trace, *loopseq* corecursively calls *loop*, again appropriately guarded. Our choice of augmenting traces at boolean guards facilitates implementing *loop* in Coq: we exploit it to satisfy Coq's syntactic guardedness condition.

 sequence, defined by simple corecursion, is similar to *loopseq*, but does not involve repetition. It takes two arguments: k for running a statement (the second statement of a sequence) from a state and τ the already accumulated trace (resulting from running the first statement of the sequence). After reconstructing τ, *sequence* calls k on the last state of τ.

Proposition 5 proves the big-step functional semantics correct relative to the big-step relational semantics. The proof proceeds by induction on the statement and performs coinductive reasoning in the cases for sequence and while. Moreover, the way induction and coinduction hypotheses are invoked mimics the way *eval* makes recursive calls and *sequence* and *loop* make corecursive calls.

Proposition 5. *For any s, σ, $(s, \sigma) \Rightarrow eval\ s\ \sigma$.*

Proof. By induction on s. We show the interesting cases of sequence and while.

- $s = s_0; s_1$: We are given as the induction hypotheses that, for any σ, $(s_0, \sigma) \Rightarrow eval\ s_0\ \sigma$ and $(s_1, \sigma) \Rightarrow eval\ s_1\ \sigma$. We must prove $(s_0; s_1, \sigma) \Rightarrow eval\ (s_0; s_1)\ \sigma$. We do so by proving the following condition by coinduction: for any τ, $(s_1, \tau) \overset{*}{\Rightarrow} sequence\ (eval\ s_1)\ \tau$. The proof of the condition proceeds by case distinction on τ and invokes the induction hypothesis on s_1 for the case where τ is a singleton. Then we close the case by combining $(s_0, \sigma) \Rightarrow (eval\ s_0\ \sigma)$, which is obtained from the induction hypothesis on s_0, and $(s_1, (eval\ s_0\ \sigma)) \overset{*}{\Rightarrow} sequence\ (eval\ s_1)\ (eval\ s_0\ \sigma)$, which is obtained from the condition proved.
- $s = \text{while } e \text{ do } s_t$: We are given as the induction hypothesis that, for any σ, $(s_t, \sigma) \Rightarrow eval\ s_t\ \sigma$. We must prove $(\text{while } e \text{ do } s_t, \sigma) \Rightarrow eval\ (\text{while } e \text{ do } s_t)\ \sigma$. We do so by proving the following two conditions simultaneously by mutual coinduction:
  - for any σ, $(\text{while } e \text{ do } s_t, \sigma) \Rightarrow \sigma :: loop\ (eval\ s_t)\ [\![e]\!]\ \sigma$,
  - for any τ, $(\text{while } e \text{ do } s_t, \tau) \overset{*}{\Rightarrow} loopseq\ (eval\ s_t)\ [\![e]\!]\ \tau$.
  The proof of the first condition invokes the induction hypothesis on s_t. The case follows immediately from the first condition.

As an obvious corollary of Proposition 5, the relational semantics is total:

Corollary 1. *For any s, σ, there exists τ such that $(s, \sigma) \Rightarrow \tau$.*

This corollary is valuable on its own, and in fact it is one motivation for defining the functional big-step semantics. Since the conclusion is not a coinductive predicate, we cannot prove the corollary directly by coinduction.

Correctness of the big-step functional semantics relative to the big-step relational semantics is easy. In the light of Lemma 2 (determinism of the big-step relational semantics), it is an immediate consequence of Proposition 5.

Proposition 6. *For any s, σ, τ, if $(s, \sigma) \Rightarrow \tau$, then $\tau \approx eval\ s\ \sigma$.*

The fact that the corecursive functions *sequence* and *loop* must produce traces in a guarded way in the functional semantics lends further support to our definition of the big-step relational semantics. Mere wellformedness of a coinductive definition of the evaluation relation does not suffice, the rules must be tight enough to properly define the trace of a run.

Fixpoint red (*s* : *stmt*) (*σ* : *state*) {*struct s*} : *option* (*stmt* ∗ *state*) :=
  match s with
  | skip ⇒ *None*
  | *x* := *e* ⇒ *Some* (skip, *σ*[*x* ↦ $\llbracket e \rrbracket \sigma$])
  | *s*₀; *s*₁ ⇒
  match (*red s*₀ *σ*) *with*
  | *Some* (*s*₀′, *σ*′) ⇒ *Some* (*s*₀′; *s*₁, *σ*′)
  | *None* ⇒ *red s*₁ *σ*
  end
  | if *e* then *s*ₜ else *s*_f ⇒ *if* $\llbracket e \rrbracket \sigma$ *then Some* (*s*ₜ, *σ*) *else Some* (*s*_f, *σ*)
  | while *e* do *s*ₜ ⇒ *if* $\llbracket e \rrbracket \sigma$ *then Some* (*s*ₜ; while *e* do *s*ₜ, *σ*) *else Some* (skip, *σ*)
  end

CoFixpoint norm (*s* : *stmt*) (*σ* : *state*) : *trace* :=
  match red s σ with
  | *None* ⇒ ⟨*σ*⟩
  | *Some* (*s*′, *σ*′) ⇒ *σ* :: *norm s*′ *σ*′
  end

Fig. 5. Small-step functional semantics

6   Small-Step Functional Semantics

Our small-step functional semantics, defined in Fig. 5, is quite similar to the small-step relational semantics, except that it uses a function to perform one-step reductions. The option-returning one-step reduction function *red* is a functional equivalent to the jointly total and deterministic terminality predicate and one-step reduction relation of Fig. 2. It returns *None*, if the given statement is terminal; otherwise, it one-step reduces the given statement from the given state and returns the resulting statement-state pair. The normalization function *norm* calls *red* repeatedly; it is defined by corecursion (guardedness is achieved by consing the current state to the corecursive call on the next state).

The small-step functional semantics is correct relative to the small-step relational semantics:

Proposition 7. *For all s, σ,* (*s, σ*) ⤳ *norm s σ.*

That the small-step relational semantics is correct relative to the small-step functional semantics is an immediate consequence of Proposition 7 and Lemma 4 (determinism of small-step relational semantics).

Proposition 8. *For all s, σ, τ, if* (*s, σ*) ⤳ *τ, then τ* ≈ *norm s σ.*

To verify the equivalence of the small-step functional semantics to the big-step functional semantics, we first prove two auxiliary lemmas. Lemma 8 relates *norm* and *sequence*. Lemma 9 relates *norm* and *loop* together with *loopseq*. The former is proved by coinduction, the latter is by mutual coinduction.

Lemma 8. *For any s_0, s_1, σ, norm $(s_0; s_1)$ $\sigma \approx$ sequence $(norm\ s_1)$ $(norm\ s_0\ \sigma)$.*

Lemma 9. *For any e, s_t and σ, the following two conditions hold:*

- *norm (while e do s_t) $\sigma \approx \sigma :: loop$ $(norm\ s_t)$ $[\![e]\!]$ σ,*
- *for any s, norm $(s; \text{while } e \text{ do } s_t)$ $\sigma \approx loopseq$ $(norm\ s_t)$ $[\![e]\!]$ $(norm\ s\ \sigma)$.*

Equipped with these lemmata we are in the position to show the big-step and small-step functional semantics to agree up to bisimilarity.

Proposition 9. *For any s and σ, eval s $\sigma \approx$ norm s σ.*

Proof. By induction on s. We outline the proof for the main cases.

- $s = s_0; s_1$: We prove by coinduction the following condition: for any τ and τ' such that $\tau \approx \tau'$, sequence $(eval\ s_1)$ $\tau \approx$ sequence $(norm\ s_1)$ τ'. The proof of the condition uses the induction hypothesis on s_1. Then the condition, Lemma 8 and the induction hypothesis on s_0 together close the case.

- $s = \text{while } e \text{ do } s_t$: We prove the following two conditions simultaneously by mutual coinduction using the induction hypothesis:
  - for any σ, loop $(eval\ s_t)$ $[\![e]\!]$ $\sigma \approx$ loop $(norm\ s_t)$ $[\![e]\!]$ σ,
  - for any τ, τ' such that $\tau \approx \tau'$,
    loopseq $(eval\ s_t)$ $[\![e]\!]$ $\tau \approx$ loopseq $(norm\ s_t)$ $[\![e]\!]$ τ'.
  Then the first condition and Lemma 9 together close the case.

We can now prove that the small-step relational semantics correct relative to the big-step relational semantics without having to rely on dependent pattern-matching or JM equality by going through the functional semantics:

Proposition 10. *For any s, σ and τ, if $(s, \sigma) \rightsquigarrow \tau$ then $(s, \sigma) \Rightarrow \tau$.*

Proof. By Prop. 7 and Lemma 4, $\tau \approx$ norm s σ. By Prop. 9 and the transitivity of the bisimilarity relation, $\tau \approx$ eval s σ. By Prop. 5 and Lemma 1, $(s, \sigma) \Rightarrow \tau$.

7 Related Work

X. Leroy and H. Grall [14] study two approaches to big-step relational semantics for lambda-calculus, accounting for both terminating and non-terminating evaluations, fully formalized in Coq. In both approaches, evaluation relates lambda-terms to normal forms or to reduction sequences.

The first approach, inspired by Cousot and Cousot [5], uses two evaluation relations, an inductive one for terminating evaluations and a coinductive one for non-terminating evaluations. The proof of equivalence to the small-step semantics requires the use of an instance of the excluded middle, constructively amounting to deciding halting. In essence, this means adopting a sum-type solution. This has deep implications even for the big-step semantics alone: the determinism of the evaluation relation, for example, can only be shown by going through the small-step semantics.

Leroy used this approach in his work on a certified C compiler [13]. To the best of our knowledge, this was the first practical application of mechanized coinductive semantics. C is one of the most used languages for developing programs that are not supposed to terminate, such as operating systems. Hence it is important that a certified compiler for C preserves the semantics of non-terminating programs. The work on the Compcert compiler is a strong witness of the importance and practicality of mechanized coinductive semantics.

In our approach to While, we have a single evaluation relation for both terminating and non-terminating runs. The big-step semantics is equivalent to the small-step semantics constructively. Furthermore, the big-step semantics is constructively deterministic and the proof of this is without an indirection through the small-step semantics.

Leroy and Grall [14] also study a different big-step semantics where both terminating and non-terminating runs are described by a single coinductively defined evaluation relation ("coevaluation") relating lambda-terms to normal forms or reduction sequences. This semantics does not agree with the small-step semantics, since it assigns a result even to an infinite reduction sequence and continues reducing a function even after the argument diverges.

Coinductive big-step relational semantics for While similar in some aspects to Leroy and Grall's work on lambda-calculus appear in the works of Glesner [9] and Nestra [16,17]. Regardless of whether evaluation relates statement-state pairs to possibly infinite traces, possibly non-wellfounded trees of states ("fractions") or transfinite traces, these approaches have it in common that the result of a non-terminating run can be non-deterministic even for While-programs, which should be deterministic. For one technical reason or another, it becomes possible in all these semantics that after an infinite number of small steps a run reaches an under-determined limit state and continues from there. In the case of Nestra [16,17], this seems intended: he devised his non-standard "fractional" and transfinite semantics to justify a program slicing transformation that is unsound under the standard semantics. Elsewhere, the outcome appears accidental and undesired.

In our approach, we the take result of a program run to be given precisely by what can be finitely observed: we record the state of the program at every finite time instant. We never run ahead of the clock by jumping over some intermediate states (in particular, we never run ahead of the clock infinitely much) and we reject transfinite time. As a result of this design decision, the big-step semantics agrees precisely with the small-step semantics and does so even constructively.

Coinductive functional semantics similar to ours have appeared in the works of J. Rutten and V. Capretta. A difference is that instead of trace-based semantics they looked at delayed state based semantics, i.e., semantics that, for a given statement-state pair, return a possibly infinitely delayed state. Delayed states, or Burroni conaturals over states, are like conatural numbers (possibly infinite natural numbers), except that instead of the number zero their deconstruction terminates (if it does) with a state.

J. Rutten [18] gave a delayed state based coinductive small-step functional semantics for While in coalgebraic terms. The one-step reduction function is a coalgebra on statement-state pairs. The final coalgebra is given by the analysis of a delayed state into a readily available state or a unit delay of a further delayed state (the predecessor function on Burroni conaturals over states). The small-step semantics, which sends a statement-state pair to a delayed state, is given by the unique map from a coalgebra to the final coalgebra. He also discussed weak bisimilarity of delayed states. This identifies all finite delays. As he worked in classical set theory, the quotient of the set of delayed states by weak bisimilarity is isomorphic to the sum of the set of states and a one-element set. However the coalgebraic approach is not confined to the category of sets and in constructive settings the theory of weak bisimilarity is richer.

V. Capretta [4] carried out a similar project in a constructive, type-theoretic setting, focusing on combinators essential for big-step functional semantics (our *sequence* and *loop*). Central to him was the realization that the delay type constructor is a monad, more specifically a completely iterative monad (with $\langle\rangle$ the unit, *sequence* the Kleisli extension operation and *loop* the iteration operation). Similarly to Leroy and Grall, he formalized his development in Coq.

Our work is very much inspired by the designs of Rutten and Capretta. Here, however, we have replaced the delay monad by the trace monad, which is also completely iterative. Moreover, we have also considered relational semantics.

A general categorical account of small-step trace semantics has been given by I. Hasuo et al. [12].

8   Conclusion

We have devised four trace-based coinductive semantics for While in different styles of operational semantics. We were pleased to find that simple semantics covering both terminating and non-terminating program runs are possible even in the big-step relational and functional styles. The metatheory of our coinductive semantics is remarkably analogous to that of the textbook inductive semantics and on finite runs they agree. Remarkably, everything can be arranged so that in a constructive setting we never have to decide whether a trace is finite or infinite.

Acknowledgments. K. Nakata thanks X. Leroy for interesting discussions on the coinductive semantics for the Compcert compiler during her post-doctoral stay in the Gallium team at INRIA Rocquencourt. The Ssreflect library was instrumental in producing the formal development. She is also thankful for helpful advice she received via the Coq and Ssreflect mailing-lists. T. Uustalu acknowledges the many inspiring discussions on the delay monad with T. Altenkirch and V. Capretta. Both authors were supported by the Estonian Science Foundation grant no. 6940 and the EU FP6 IST integrated project no. 15905 MOBIUS.

References

1. Bertot, Y., Castéran, P.: Coq'Art: Interactive Theorem Proving and Program Development. Springer, Heidelberg (2004)
2. Bertot, Y.: Filters on coinductive streams, an application to Eratosthenes' sieve. In: Urzyczyn, P. (ed.) TLCA 2005. LNCS, vol. 3461, pp. 102–115. Springer, Heidelberg (2005)
3. Bertot, Y.: A survey of programming language semantics styles. Coq development (2007), http://www-sop.inria.fr/marelle/Yves.Bertot/proofs.html
4. Capretta, V.: General recursion via coinductive types. Logical Methods in Computer Science 1(2), 1–18 (2005)
5. Cousot, P., Cousot, R.: Inductive definitions, semantics and abstract interpretation. In: Conf. Record of 19th ACM SIGPLAN-SIGACT Symp. on Principles of Programming Languages, POPL 1992, Albuquerque, NM, pp. 83–94. ACM Press, New York (1992)
6. Cousot, P., Cousot, R.: Bi-inductive structural semantics. Inform. and Comput. 207(2), 258–283 (2009)
7. Giménez, E.: Codifying guarded definitions with recursive schemes. In: Smith, J., Dybjer, P., Nordström, B. (eds.) TYPES 1994. LNCS, vol. 996, pp. 39–59. Springer, Heidelberg (1995)
8. Giménez, E.: Structural recursive definitions in type theory. In: Larsen, K.G., Skyum, S., Winskel, G. (eds.) ICALP 1998. LNCS, vol. 1443, pp. 397–408. Springer, Heidelberg (1998)
9. Glesner, S.: A proof calculus for natural semantics based on greatest fixed point semantics. In: Knoop, J., Necula, G.C., Zimmermann, W. (eds.) Proc. of 3rd Int. Wksh. on Compiler Optimization Meets Compiler Verification, COCV 2004, Barcelona. Electron. Notes in Theor. Comput. Sci., vol. 132(1), pp. 73–93. Elsevier, Amsterdam (2005)
10. Glesner, S., Leitner, J., Blech, J.O.: Coinductive verification of program optimizations using similarity relations. In: Knoop, J., Necula, G.C., Zimmermann, W. (eds.) Proc. of 5th Int. Wksh. on Compiler Optimization Meets Compiler Verification, COCV 2006, Vienna. Electron. Notes in Theor. Comput. Sci., vol. 176(3), pp. 61–77. Elsevier, Amsterdam (2007)
11. Gonthier, G., Mahboubi, A.: A small scale reflection extension for the Coq system. Technical Report RR-6455, INRIA (2008)
12. Hasuo, I., Jacobs, B., Sokolova, A.: Generic trace semantics via coinduction. Logical Methods in Comput. Sci. 3(4), article 11(2007)
13. Leroy, X.: The Compcert verified compiler. Commented Coq development (2008), http://compcert.inria.fr/doc/
14. Leroy, X., Grall, H.: Coinductive big-step operational semantics. Inform. and Comput. 207(2), 285–305 (2009)
15. McBride, C.: Elimination with a motive. In: Callaghan, P., Luo, Z., McKinna, J., Pollack, R. (eds.) TYPES 2000. LNCS, vol. 2277, pp. 197–216. Springer, Heidelberg (2002)
16. Nestra, H.: Fractional semantic. In: Johnson, M., Vene, V. (eds.) AMAST 2006. LNCS, vol. 4019, pp. 278–292. Springer, Heidelberg (2006)
17. Nestra, H.: Transfinite semantics in the form of greatest fixpoint. J. of Logic and Algebr. Program (to appear)
18. Rutten, J.: A note on coinduction and weak bisimilarity for While programs. Theor. Inform. and Appl. 33(4–5), 393–400 (1999)

A Better x86 Memory Model: x86-TSO

Scott Owens, Susmit Sarkar, and Peter Sewell

University of Cambridge
http://www.cl.cam.ac.uk/users/pes20/weakmemory

Abstract. Real multiprocessors do not provide the sequentially consistent memory that is assumed by most work on semantics and verification. Instead, they have relaxed memory models, typically described in ambiguous prose, which lead to widespread confusion. These are prime targets for mechanized formalization. In previous work we produced a rigorous *x86-CC* model, formalizing the Intel and AMD architecture specifications of the time, but those turned out to be unsound with respect to actual hardware, as well as arguably too weak to program above. We discuss these issues and present a new *x86-TSO* model that suffers from neither problem, formalized in HOL4. We believe it is sound with respect to real processors, reflects better the vendor's intentions, and is also better suited for programming. We give two equivalent definitions of x86-TSO: an intuitive operational model based on local write buffers, and an axiomatic total store ordering model, similar to that of the SPARCv8. Both are adapted to handle x86-specific features. We have implemented the axiomatic model in our `memevents` tool, which calculates the set of all valid executions of test programs, and, for greater confidence, verify the witnesses of such executions directly, with code extracted from a third, more algorithmic, equivalent version of the definition.

1 Introduction

Most previous research on the semantics and verification of concurrent programs assumes sequential consistency: that accesses by multiple threads to a shared memory occur in a global-time linear order. Real multiprocessors, however, incorporate many performance optimisations. These are typically unobservable by single-threaded programs, but some have observable consequences for the behaviour of concurrent code. For example, on standard Intel or AMD x86 processors, given two memory locations x and y (initially holding 0), if two processors proc:0 and proc:1 respectively write 1 to x and y and then read from y and x, as in the program below, it is possible for both to read 0 *in the same execution.*

iwp2.3.a/amd4	proc:0	proc:1
poi:0	MOV [x]←$1	MOV [y]←$1
poi:1	MOV EAX←[y]	MOV EBX←[x]
Allow: 0:EAX=0 ∧ 1:EBX=0		

One can view this as a visible consequence of *write buffering*: each processor effectively has a FIFO buffer of pending memory writes (to avoid the need to

S. Berghofer et al. (Eds.): TPHOLs 2009, LNCS 5674, pp. 391–407, 2009.

block while a write completes), so the reads from y and x can occur before the writes have propagated from the buffers to main memory. Such optimisations destroy the illusion of sequential consistency, making it impossible (at this level of abstraction) to reason in terms of an intuitive notion of global time.

To describe what programmers can rely on, processor vendors document *architectures*. These are loose specifications, claimed to cover a range of past and future actual processors, which should reveal enough for effective programming, but without unduly constraining future processor designs. In practice, however, they are informal prose documents, e.g. the Intel 64 and IA-32 Architectures SDM [2] and AMD64 Architecture Programmer's Manual [1]. Informal prose is a poor medium for loose specification of subtle properties, and, as we shall see in §2, such documents are often ambiguous, are sometimes incomplete (too weak to program above), and are sometimes unsound (with respect to the actual processors). Moreover, one cannot test programs above such a vague specification (one can only run programs on particular actual processors), and one cannot use them as criteria for testing processor implementations.

Architecture specifications are, therefore, prime targets for rigorous mechanised formalisation. In previous work [19] we introduced a rigorous x86-CC model, formalised in HOL4 [11], based on the informal prose causal-consistency descriptions of the then-current Intel and AMD documentation. Unfortunately those, and hence also x86-CC, turned out to be unsound, forbidding some behaviour which actual processors exhibit.

In this paper we describe a new model, x86-TSO, also formalised in HOL4. To the best of our knowledge, x86-TSO is sound, is strong enough to program above, and is broadly in line with the vendors' intentions. We present two equivalent definitions of the model: an abstract machine, in §3.1, and an axiomatic version, in §3.2. We compensate for the main disadvantage of formalisation, that it can make specifications less widely accessible, by extensively annotating the mathematical definitions. To explore the consequences of the model, we have a hand-coded implementation in our **memevents** tool, which can explore all possible executions of litmus-test examples such as that above, and for greater confidence we have a verified execution checker extracted from the HOL4 axiomatic definition, in §4. We discuss related work in §5 and conclude in §6.

2   Many Memory Models

We begin by reviewing the informal-prose specifications of recent Intel and AMD documentation. There have been several versions, some differing radically; we contrast them with each other, and with what we know of the behaviour of actual processors.

2.1   Pre-IWP (Before Aug. 2007)

Early revisions of the Intel SDM (e.g. rev-22, Nov. 2006) gave an informal-prose model called 'processor ordering', unsupported by any examples. It is hard to give a precise interpretation of this description.

2.2   IWP/AMD64-3.14/x86-CC

In August 2007, an Intel White Paper [12] (IWP) gave a somewhat more precise model, with 8 informal-prose principles supported by 10 examples (known as litmus tests). This was incorporated, essentially unchanged, into later revisions of the Intel SDM (including rev.26–28), and AMD gave similar, though not identical, prose and tests [1]. These are essentially causal-consistency models [4]. They allow independent readers to see independent writes (by different processors to different addresses) in different orders, as below (IRIW, see also [6]), but require that, in some sense, causality is respected: *"P5. In a multiprocessor system, memory ordering obeys causality (memory ordering respects transitive visibility)"*.

amd6	proc:0	proc:1	proc:2	proc:3
poi:0	MOV [x]←$1	MOV [y]←$1	MOV EAX←[x]	MOV ECX←[y]
poi:1			MOV EBX←[y]	MOV EDX←[x]
Final: 2:EAX=1 ∧ 2:EBX=0 ∧ 3:ECX=1 ∧ 3:EDX=0				
cc : Allow; tso : Forbid				

These informal specifications were the basis for our x86-CC model, for which a key issue was giving a reasonable interpretation to this "causality". Apart from that, the informal specifications were reasonably unambiguous — but they turned out to have two serious flaws.

First, they are arguably rather weak for programmers. In particular, they admit the IRIW behaviour above but, under reasonable assumptions on the strongest x86 memory barrier, MFENCE, adding MFENCEs would not suffice to recover sequential consistency [19, §2.12]. Here the specifications seem to be much looser than the behaviour of implemented processors: to the best of our knowledge, and following some testing, IRIW is not observable in practice. It appears that some JVM implementations depend on this fact, and would not be correct if one assumed only the IWP/AMD64-3.14/x86-CC architecture [9].

Second, more seriously, they are unsound with respect to current processors. The following n6 example, due to Paul Loewenstein [14], shows a behaviour that is observable (e.g. on an Intel Core 2 duo), but that is disallowed by x86-CC, and by any interpretation we can make of IWP and AMD64-3.14.

n6	proc:0	proc:1
poi:0	MOV [x]←$1	MOV [y]←$2
poi:1	MOV EAX←[x]	MOV [x]←$2
poi:2	MOV EBX←[y]	
Final: 0:EAX=1 ∧ 0:EBX=0 ∧ [x]=1		
cc : Forbid; tso : Allow		

To see why this may be allowed by multiprocessors with FIFO write buffers, suppose that first the proc:1 write of [y]=2 is buffered, then proc:0 buffers its write of [x]=1, reads [x]=1 from its own write buffer, and reads [y]=0 from main memory, then proc:1 buffers its [x]=2 write and flushes its buffered [y]=2 and [x]=2 writes to memory, then finally proc:0 flushes its [x]=1 write to memory.

2.3   Intel SDM rev-29 (Nov. 2008)

The most recent change in the x86 vendor specifications, was in revision 29 of the Intel SDM (revision 30 is essentially identical, and we are told that there will be a future revision of the AMD specification on similar lines). This is in a similar informal-prose style to previous versions, again supported by litmus tests, but is significantly different to IWP/AMD64-3.14/x86-CC. First, the IRIW final state above is forbidden [Example 7-7, rev-29], and the previous coherence condition: *"P6. In a multiprocessor system, stores to the same location have a total order"* has been replaced by: *"P9. Any two stores are seen in a consistent order by processors other than those performing the stores".*

Second, the memory barrier instructions are now included, with *"P11. Reads cannot pass LFENCE and MFENCE instructions"* and *"P12. Writes cannot pass SFENCE and MFENCE instructions".*

Third, same-processor writes are now explicitly ordered (we regarded this as implicit in the IWP *"P2. Stores are not reordered with other stores"*): *"P10. Writes by a single processor are observed in the same order by all processors".*

This specification appears to deal with the unsoundness, admitting the n6 behaviour above, but, unfortunately, it is still problematic. The first issue is, again, how to interpret "causality" as used in P5. The second issue is one of weakness: the new P9 says *nothing* about observations of two stores by those two processors themselves (or by one of those processors and one other). Programming above a model that lacks any such guarantee would be problematic. The following n5 and n4 examples illustrate the potential difficulties. These final states were not allowed in x86-CC, and we would be surprised if they were allowed by any reasonable implementation (they are not allowed in a pure write-buffer implementation). We have not observed them on actual processors; however, rev-29 appears to allow them.

n5	proc:0	proc:1
poi:0	MOV [x]←$1	MOV [x]←$2
poi:1	MOV EAX←[x]	MOV EBX←[x]
Forbid: 0:EAX=2 ∧ 1:EBX=1		

n4	proc:0	proc:1
poi:0	MOV EAX←[x]	MOV ECX←[x]
poi:1	MOV [x]←$1	MOV [x]←$2
poi:2	MOV EBX←[x]	MOV EDX←[x]
Forbid: 0:EAX=2 ∧ 0:EBX=1∧		
1:ECX=1 ∧ 1:EDX=2		

Summarising the key litmus-test differences, we have:

	IWP/AMD64-3.14/x86-CC	rev-29	actual processors
IRIW	allowed	forbidden	not observed
n6	forbidden	allowed	observed
n4/n5	forbidden	allowed	not observed

There are also many non-differences: tests for which the behaviours coincide in all three cases. The test details are omitted here, but can be found in the extended version [16] or in [19]. They include the 9 other IWP tests, illustrating that the various load and store reorderings other than those shown in iwp2.3.a/amd4 (§1) are not possible; the AMD MFENCE tests amd5 and amd10; and several others.

3   The x86-TSO Model

Given these problems with the informal specifications, we cannot produce a useful rigorous model by formalising the "principles" they contain (as we attempted with x86-CC [19]). Instead, we have to build a reasonable model that is consistent with the given litmus tests, with observed processor behaviour, and with what we know of the needs of programmers and of the vendors intentions.

The fact that write buffering is observable (iwp2.3.a/amd4 and n6) but IRIW is not, together with the other tests that prohibit many other reorderings, strongly suggests that, apart from write buffering, all processors share the same view of memory (in contrast to x86-CC, where each processor had a separate view order). This is broadly similar to the SPARC Total Store Ordering (TSO) memory model [20,21], which is essentially an axiomatic description of the behaviour of write-buffer multiprocessors. Moreover, while the term "TSO" is not used, informal discussions suggest this matches the intention behind the rev.29 informal specification. Accordingly, we present here a rigorous x86-TSO model, with two equivalent definitions.

The first definition, in §3.1, is an abstract machine with explicit write buffers. The second definition, in §3.2, is an axiomatic model that defines valid executions in terms of memory orders and reads-from maps. In both, we deal with x86 CISC instructions with multiple memory accesses, with x86 LOCK'd instructions (CMPXCHG, LOCK;INC, etc.), with potentially non-terminating computations, and with dependencies through registers. Together with our earlier instruction semantics, x86-TSO thus defines a complete semantics of programs. The abstract machine conveys the programmer-level operational intuition behind x86-TSO, whereas the axiomatic model supports constraint-based reasoning about example programs, e.g., by our **memevents** tool in §4.

The intended scope of x86-TSO, as for the x86-CC model, covers typical user code and most kernel code: programs using coherent write-back memory, without exceptions, misaligned or mixed-size accesses, 'non-temporal' operations (e.g. MOVNTI), self-modifying code, or page-table changes.

Basic Types: Actions, Events, and Event Structures.   As in our earlier work, the action of (any particular execution of) a program is abstracted into a set of *events* (with additional data) called an *event structure*. An event represents a read or write of a particular value to a memory address, or to a register, or the execution of a fence. Our earlier work includes a definition of the set of event structures generated by an assembly language program. For any such event structure, the memory model (there x86-CC, here x86-TSO) defines what a *valid execution* is.

In more detail, each machine-code instruction may have multiple events associated with it: events are indexed by an instruction ID *iiid* that identifies which processor the event occurred on and the position in the instruction stream of the instruction it comes from (the *program order index*, or *poi*). Events also have an event ID *eiid* to identify them within an instruction (to permit multiple, otherwise identical, events). An event structure indicates when one of an instruction's events has a dependency on another event of the same instruction with an *intra _causality* relation, a partial order over the events of each instruction. An event structure also records which events occur together in a locked instruction with *atomicity* data, a set of (disjoint, non-empty) sets of events which must occur atomically together.

Expressing this in HOL, we index processors by a type proc = num, take types address and value to both be the 32-bit words, and take a location to be either a memory address or a register of a particular processor:

location = LOCATION_REG **of** proc $'reg$
         | LOCATION_MEM **of** address

The model is parameterised by a type $'reg$ of x86 registers, which one should think of as an enumeration of the names of ordinary registers EAX, EBX, etc., the instruction pointer EIP, and the status flags. To identify an instance of an instruction in an execution, we specify its processor and its program order index.

iiid = ⟨| *proc* : proc;   *poi* : num |⟩

An action is either a read or write of a value at some location, or a barrier:

dirn = R | W
barrier = LFENCE | SFENCE | MFENCE
action = ACCESS **of** dirn ($'reg$ location) value | BARRIER **of** barrier

Finally, an event has an instruction instance id, an event id (of type eiid = num, unique per iiid), and an action:

event = ⟨| *eiid* : eiid;   *iiid* : iiid;   *action* : action |⟩

An event structure E comprises a set of processors, a set of events, an intra-instruction causality relation, and a partial equivalence relation (PER) capturing sets of events which must occur atomically, all subject to some well-formedness conditions which we omit here.

event_structure = ⟨| *procs* : proc set;
                    events : ($'reg$ event)set;
                    intra_causality : ($'reg$ event)reln;
                    atomicity : ($'reg$ event)set set |⟩

Example. We show a very simple event structure below, for the program:

tso1	proc:0	proc:1
poi:0	MOV [x]←$1	MOV [x]←$2
poi:1	MOV EAX←[x]	

There are four events — the inner (blue in the on-line version) boxes. The event ids are pretty-printed alphabetically, as a,b,c,d, etc. We also show the assembly instruction that gave rise to each event, e.g. MOV [x]←$1, though that is not formally part of the event structure.

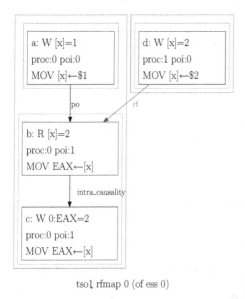

tso1 rfmap 0 (of ess 0)

Note that events contain concrete values: in this particular event structure, there are two writes of x, with values 1 and 2, a read of [x] with value 2, and a write of proc:0's EAX register with value 2. Later we show two valid executions for this program, one for this event structure and one for another (note also that some event structures may not have any valid executions). In the diagram, the instructions of each processor are clustered together, into the outermost (magenta) boxes, with program order (po) edges between them, and the events of each instruction are clustered together into the intermediate (green) boxes, with intra-causality edges as appropriate — here, in the MOV EAX←[x], the write of EAX is dependent on the read of x.

3.1   The x86-TSO Abstract Machine Memory Model

To understand our x86-TSO machine model, consider an idealised x86 multiprocessor system partitioned into two components: its memory and register state (of all its processors combined), and the rest of the system (the other parts of all the processor cores). Our abstract machine is a labelled transition system: a set of states, ranged over by s, and a transition relation $s \xrightarrow{l} s'$. An abstract machine state s models the state of the first component: the memory and register state of a multiprocessor system. The machine interacts with the rest of the system by synchronising on labels l (the interface of the abstract machine), which include register and memory reads and writes. In Fig. 1, the states s correspond to the parts of the machine shown inside of the dotted line, and the labels l correspond to the communications that traverse the dotted line boundary.

One should think of the machine as operating in parallel with the processor cores (absent their register/memory subsystems), executing their instruction streams in program order; the latter data is provided by an event structure. This partitioning does not correspond directly to the microarchitecture of any realistic x86 implementation, in which memory and registers would be implemented by separate and intricate mechanisms, including various caches. However, it is useful and sufficient for describing the programming model, which is the proper business of an architecture description. It also supports a precise correspondence with our axiomatic memory model. In more detail, the labels l are the values of

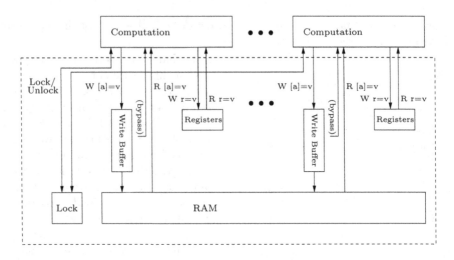

Fig. 1. The abstract machine

the HOL type:

label $=$ TAU | EVT **of** proc $('reg\ action)$ | LOCK **of** proc | UNLOCK **of** proc

- TAU, for an internal action by the machine;
- EVT p a, where a is an action, as defined above (a memory or register read or write, with its value, or a barrier), by processor p;
- LOCK p, indicating the start of a LOCK'd instruction by processor p; or
- UNLOCK p, for the end of a LOCK'd instruction by p.

(Note that there is nothing specific to any particular memory model in this interface.) The states of the x86-TSO machine are records, with fields R, giving a value for each register on each processor; M, giving a value for each shared memory location; B, modelling a write buffer for each processor, as a list of address/value pairs; and L, which is a global lock, either SOME p, if p holds the lock, or NONE. The HOL type is below.

machine_state $=(\!|\ R$: proc $\rightarrow$ $'reg$ $\rightarrow$ value option; (* per-processor registers *)
$\qquad\qquad\qquad M$: address $\rightarrow$ value option; (* main memory *)
$\qquad\qquad\qquad B$: proc $\rightarrow$ (address#value)list; (* per-processor write buffers *)
$\qquad\qquad\qquad L$: proc option(* which processor holds the lock *))$\!|$)

The behaviour of the x86-TSO machine, the transition relation $s \xrightarrow{l} s'$, is defined by the rules in Fig. 2. The rules use two auxiliary definitions: processor p is *not blocked* in machine state s if either it holds the lock or no processor does; and there are *no pending* writes in a buffer b for address a if there are no (a, v) pairs in b. Restating the rules informally:

1. p can read v from memory at address a if p is not blocked, has no buffered writes to a, and the memory does contain v at a;

Read from memory

$$\frac{\text{not\_blocked } s\ p \wedge (s.M\ a = \text{SOME}\ v) \wedge \text{no\_pending } (s.B\ p)a}{s \xrightarrow{\text{EVT } p\ (\text{ACCESS } R\ (\text{LOCATION\_MEM } a)v)} s}$$

Read from write buffer

$$\frac{\text{not\_blocked } s\ p \wedge (\exists b_1\ b_2.(s.B\ p = b_1 +\!\!+[(a, v)] +\!\!+ b_2) \wedge \text{no\_pending } b_1\ a)}{s \xrightarrow{\text{EVT } p\ (\text{ACCESS } R\ (\text{LOCATION\_MEM } a)v)} s}$$

Read from register

$$\frac{(s.R\ p\ r = \text{SOME}\ v)}{s \xrightarrow{\text{EVT } p\ (\text{ACCESS } R\ (\text{LOCATION\_REG } p\ r)v)} s}$$

Write to write buffer

$$\frac{\mathbf{T}}{s \xrightarrow{\text{EVT } p\ (\text{ACCESS } W\ (\text{LOCATION\_MEM } a)v)} s \oplus (\!| B := s.B \oplus (p \mapsto [(a, v)] +\!\!+ (s.B\ p)) |\!)}$$

Write from write buffer to memory

$$\frac{\text{not\_blocked } s\ p \wedge (s.B\ p = b +\!\!+ [(a, v)])}{s \xrightarrow{\text{TAU}} s \oplus (\!| M := s.M \oplus (a \mapsto \text{SOME}\ v); B := s.B \oplus (p \mapsto b) |\!)}$$

Write to register

$$\frac{\mathbf{T}}{s \xrightarrow{\text{EVT } p\ (\text{ACCESS } W\ (\text{LOCATION\_REG } p\ r)v)} s \oplus (\!| R := s.R \oplus (p \mapsto ((s.R\ p) \oplus (r \mapsto \text{SOME}\ v))) |\!)}$$

Barrier

$$\frac{(b = \text{MFENCE}) \implies (s.B\ p = [])}{s \xrightarrow{\text{EVT } p\ (\text{BARRIER } b)} s}$$

Lock

$$\frac{(s.L = \text{NONE}) \wedge (s.B\ p = [])}{s \xrightarrow{\text{LOCK } p} s \oplus (\!| L := \text{SOME}\ p |\!)}$$

Unlock

$$\frac{(s.L = \text{SOME}\ p) \wedge (s.B\ p = [])}{s \xrightarrow{\text{UNLOCK } p} s \oplus (\!| L := \text{NONE} |\!)}$$

Fig. 2. The x86-TSO Machine Behaviour

2. p can read v from its write buffer for address a if p is not blocked and has v as the newest write to a in its buffer;
3. p can read the stored value v from its register r at any time;
4. p can write v to its write buffer for address a at any time;
5. if p is not blocked, it can silently dequeue the oldest write from its write buffer to memory;
6. p can write value v to one of its registers r at any time;
7. if p's write buffer is empty, it can execute an MFENCE (so an MFENCE cannot proceed until all writes have been dequeued, modelling buffer flushing); LFENCE and SFENCE can occur at any time, making them no-ops;
8. if the lock is not held, and p's write buffer is empty, it can begin a LOCK'd instruction; and
9. if p holds the lock, and its write buffer is empty, it can end a LOCK'd instruction.

Consider execution paths through the machine $s_0 \xrightarrow{l_1} s_1 \xrightarrow{l_2} s_2 \cdots$ consisting of finite or infinite sequences of states and labels. We define okMpath to hold for paths through the machine that start in a valid initial state (with empty write buffers, etc.) and satisfy the following progress condition: for each memory write in the path, the corresponding TAU transition appears later on. This ensures that no write can stay in the buffer forever. (We actually formalize okMpath for the event-annotated machine described below.)

We emphasise that this is an *abstract* machine: we are concerned with its extensional behaviour: the (completed, finite or infinite) traces of labelled transitions it can perform (which should include the behaviour of real implementations), not with its internal states and the transition rules. The machine should provide a good model for programmers, but may bear little resemblance to the internal structure of implementations. Indeed, a realistic design would certainly not implement LOCK'd instructions with a global lock, and would have many other optimisations — the force of the x86-TSO model is that none of those have *programmer-visible* effects, except perhaps via performance observations. There are several variants of the machine with different degrees of locking which we conjecture are observationally equivalent. For example, one could prohibit all activity by other processors when one holds the lock, or not require write buffers to be flushed at the start of a LOCK'd instruction.

We relate the machine to event structures in two steps, which we summarise here (the HOL details can be found on-line [16]). First, we define a more intensional event-machine: we annotate each memory and register location with an event option, recording the most recent write event (if any) to that location, refine write buffers to record lists of events rather than of plain location/value pairs, and annotate labels with the relevant events. Second, we relate paths of annotated labels and event structures with a predicate okEpath that holds when the path is a suitable linearization of the event structure: there is a 1:1 correspondence between non-TAU/LOCK/UNLOCK labels of *path* and the events of E, the order of labels in *path* is consistent with program order and intra-causality, and atomic sets are properly bracketed by LOCK/UNLOCK pairs. Thus, okMpath describes paths that are valid according to the memory model, and okEpath describes those that are valid according to an event structure (that encapsulates the other aspects of processor semantics).

Theorem 1. *The annotation-erasure of the event-machine is exactly the machine presented above.* [HOL proof]

3.2   The x86-TSO Axiomatic Memory Model

Our x86-TSO axiomatic memory model is based on the SPARCv8 memory model specification [20,21], but adapted to x86 and in the same terms as our earlier x86-CC model. (Readers unfamiliar with the SPARCv8 memory model can safely ignore the SPARC-specific comments in this section.) Compared with the SPARCv8 TSO specification, we omit instruction fetches (*IF*), instruction loads (*IL*), flushes (*F*), and stbars (*S*). The first three deal exclusively with instruction memory, which we do not model, and the last is useful only under the SPARC PSO memory model. To adapt it to x86 programs, we add register and fence events, generalize to support instructions that give rise to many events (partially ordered by an intra-instruction causality relation), and generalize atomic load/store pairs to locked instructions.

An execution is permitted by our memory model if there exists an *execution witness* X for its event structure E that is a *valid execution*. An execution witness contains a *memory_order*, an *rfmap*, and an *initial_state*; the rest of this section defines when these are valid.

execution_witness =
 ⟨ memory_order : ($'reg$ event)reln;
  rfmap : ($'reg$ event)reln;
  initial_state : ($'reg$ location $\rightarrow$ value option)⟩

The memory order is a partial order that records the global ordering of memory events. It must be a total order on memory writes, and corresponds to the $\leq$ relation in SPARCv8, as constrained by the SPARCv8 **Order** condition (in figures, we use the label mo_non-po_write_write for the otherwise-unforced part of this order).

partial_order ($<_{X.memory\_order}$)(mem_accesses E)
linear_order (($<_{X.memory\_order})|_{(\text{mem\_writes } E)}$)(mem_writes E)

The initial state is a partial function from locations to values. Each read event's value must come either from the initial state or from a write event: the rfmap ('reads-from map') records which, containing (ew, er) pairs where the read er reads from the write ew. The *reads_from_map_candidates* predicate below ensures that the rfmap only relates such pairs with the same address and value. (Strictly speaking, the rfmap is unnecessary; the constraints involving it can be stated directly in terms of memory order, as SPARCv8 does. However, we find it intuitive and useful. The SPARCv8 model has no initial states.)

reads_from_map_candidates E *rfmap* =
 $\forall(ew, er) \in rfmap.(er \in$ reads $E) \wedge (ew \in$ writes $E) \wedge$
    (loc $ew =$ loc $er) \wedge ($value_of $ew =$ value_of $er)$

We lift program order from instructions to a relation po_iico E over events, taking the union of program order of instructions and intra-instruction causality. This corresponds roughly to the ; in SPARCv8. However, *intra_causality* might not relate some pairs of events in an instruction, so our po_iico E will not generally be a total order for the events of a processor.

po_strict $E =$
  $\{(e_1, e_2) \mid (e_1.iiid.proc = e_2.iiid.proc) \wedge e_1.iiid.poi < e_2.iiid.poi \wedge$
      $e_1 \in E.events \wedge e_2 \in E.events\}$
$<_{(\text{po\_iico } E)} = \text{po\_strict } E \cup E.intra\_causality$

The *check_rfmap_written* below ensures that the rfmap relates a read to the most recent preceding write. For a register read, this is the most recent write in program order. For a memory read, this is the most recent write in memory order among those that precede the read in either memory order or program order (intuitively, the first case is a read of a committed write and the second is a read from the local write buffer). The check_rfmap_written and reads_from_map_candidates predicates implement the SPARCv8 **Value** axiom above the rfmap witness data. The *check_rfmap_initial* predicate extends this to handle initial state, ensuring that any read not in the rfmap takes its value from the initial state, and that that read is not preceded by a write in memory order or program order.

previous_writes E er $<_{order} =$
  $\{ew' \mid ew' \in \text{writes } E \wedge ew' <_{order} er \wedge (\text{loc } ew' = \text{loc } er)\}$
check_rfmap_written E $X =$
  $\forall(ew, er) \in (X.rfmap).$
    if $ew \in \text{mem\_accesses } E$ **then**
        $ew \in \text{maximal\_elements (previous\_writes } E \text{ } er \text{ } (<_{X.memory\_order}) \cup$
                        $\text{previous\_writes } E \text{ } er \text{ } (<_{(\text{po\_iico } E)}))$
              $(<_{X.memory\_order})$
      else (* ew IN reg_accesses E *)
        $ew \in \text{maximal\_elements (previous\_writes } E \text{ } er \text{ } (<_{(\text{po\_iico } E)}))(<_{(\text{po\_iico } E)})$
check_rfmap_initial E $X =$
  $\forall er \in (\text{reads } E \setminus \text{range } X.rfmap).$
    $(\exists l.(\text{loc } er = \text{SOME } l) \wedge (\text{value\_of } er = X.initial\_state \text{ } l)) \wedge$
    $(\text{previous\_writes } E \text{ } er \text{ } (<_{X.memory\_order}) \cup$
      $\text{previous\_writes } E \text{ } er \text{ } (<_{(\text{po\_iico } E)})) = \{\})$

We now further constrain the memory order, to ensure that it respects the relevant parts of program order, and that the memory accesses of a LOCK'd instruction do occur atomically.

– Program order is included in memory order, for a memory read before a memory access (labelled mo_po_read_access in figures) (SPARCv8's **LoadOp**):

  $\forall er \in (\text{mem\_reads } E).\forall e \in (\text{mem\_accesses } E).$
    $er <_{(\text{po\_iico } E)} e \implies er <_{X.memory\_order} e$

- Program order is included in memory order, for a memory write before a memory write (mo_po_write_write) (the SPARCv8 **StoreStore**):

$\forall ew_1\ ew_2 \in (\text{mem\_writes}\ E).$
$ew_1 <_{(\text{po\_iico}\ E)} ew_2 \implies ew_1 <_{X.memory\_order} ew_2$

- Program order is included in memory order, for a memory write before a memory read, *if* there is an MFENCE between (mo_po_mfence). (There is no need to include fence events themselves in the memory ordering.)

$\forall ew \in (\text{mem\_writes}\ E).\forall er \in (\text{mem\_reads}\ E).\forall ef \in (\text{mfences}\ E).$
$(ew <_{(\text{po\_iico}\ E)} ef \wedge ef <_{(\text{po\_iico}\ E)} er) \implies ew <_{X.memory\_order} er$

- Program order is included in memory order, for any two memory accesses where at least one is from a LOCK'd instruction (mo_po_access/lock):

$\forall e_1\ e_2 \in (\text{mem\_accesses}\ E).\forall es \in (E.atomicity).$
$((e_1 \in es \vee e_2 \in es) \wedge e_1 <_{(\text{po\_iico}\ E)} e_2) \implies e_1 <_{X.memory\_order} e_2$

- The memory accesses of a LOCK'd instruction occur atomically in memory order (mo_atomicity), i.e., there must be no intervening memory events. Further, all program order relationships between the locked memory accesses and other memory accesses are included in the memory order (this is a generalization of the SPARCv8 **Atomicity** axiom):

$\forall es \in (E.atomicity).\forall e \in (\text{mem\_accesses}\ E \setminus es).$
$(\forall e' \in (es \cap \text{mem\_accesses}\ E).e <_{X.memory\_order} e') \vee$
$(\forall e' \in (es \cap \text{mem\_accesses}\ E).e' <_{X.memory\_order} e)$

To deal properly with infinite executions, we also require that the prefixes of the memory order are all finite, ensuring that there are no limit points, and, to ensure that each write eventually takes effect globally, there must not be an infinite set of reads unrelated to any particular write, all on the same memory location (this formalizes the SPARCv8 **Termination** axiom).

finite_prefixes $(<_{X.memory\_order})(\text{mem\_accesses}\ E)$
$\forall ew \in (\text{mem\_writes}\ E).$
finite$\{er \mid er \in E.events \wedge (\text{loc}\ er = \text{loc}\ ew) \wedge$
$er \not<_{X.memory\_order} ew \wedge ew \not<_{X.memory\_order} er\}$

A final state of a valid execution takes the last write in memory order for each memory location, together with a maximal write in program order for each register (or the initial state, if there is no such write). This is uniquely defined assuming that no instruction has multiple unrelated writes to the same register — a reasonable property for x86 instructions.

The definition of valid_execution $E\ X$ comprising the above conditions is equivalent to one in which $<_{X.memory\_order}$ is required to be a linear order, not just a partial order (again, the full details are on-line):

Theorem 2
1. *If* linear_valid_execution $E\ X$ *then* valid_execution $E\ X$.

2. *If valid_execution* E X *then there exists an* $\hat{X}$ *with a linearisation of X's memory order such that* linear_valid_execution E $\hat{X}$.   [HOL proof]

Interpreting "not reordered with".   Perhaps surprisingly, the above definition does not require that program order is included in memory order for a memory write followed by a read from the same address. The definition does imply that any such read cannot be speculated before the write (by check_rfmap_written, as that takes both $<_{(po\_iico\ E)}$ and $<_{X.memory\_order}$ into account). However, if one included a memory order edge, perhaps following a naive interpretation of the rev-29 *"P4. Reads may be reordered with older writes to different locations but not with older writes to the same location"*, then the model would be strictly stronger: the n7 example below would become forbidden, whereas it is allowed on x86-TSO. We conjecture that this would correspond to the (rather strange) machine with the Fig. 2 rules but without the read-from-write-buffer rule, in which any processor would have to flush its write buffer up to (and including) a local write before it can read from it.

n7	proc:0	proc:1	proc:2
poi:0	MOV [x]←$1	MOV [y]←$1	MOV ECX←[y]
poi:1	MOV EAX←[x]		MOV EDX←[x]
poi:2	MOV EBX←[y]		
Allow: 0:EAX=1 ∧ 0:EBX=0 ∧ 2:ECX=1 ∧ 2:EDX=0			

Examples.   We show two valid executions of the previous example program in Fig. 3. In both executions, the proc:0 W x=1 event is before the proc:1 W x=2 event in memory order (the bold mo_non-po_write_write edge). In the first execution, on the left, the proc:0 read of x reads from the most recent write in memory order (the combination of the bold mo_non-po_write_write edge and the mo_rf edge), which is the proc:1 W x=2. In the second execution, on the right, the proc:0 read of x reads from the most recent write in program order, which is the proc:0 W x=1. This example also illustrates some register events: the MOV EAX←[x] instruction gives rise to a memory read of x, followed by (in the intra-instruction causality relation) a register write of EAX.

3.3   The Machine and Axiomatic x86-TSO Models Are Equivalent

To prove that the abstract machine admits only valid executions, we define a function path_to_X from event-annotated paths that builds a linear execution witness by using the events from Tau and memory read labels in order. Thus, the memory ordering in the execution witness corresponds to the order in which events were read from and written to memory in the abstract machine.

Theorem 3. *For any well-formed event structure E and event-machine path path, if* (okEpath E path) *and* (okMpath path), *then* (path_to_X path) *is a valid execution for E.*   [HOL proof]

To prove that the abstract machine admits every valid execution, we first prove (in HOL) a lemma showing that any valid execution can be a turned into a

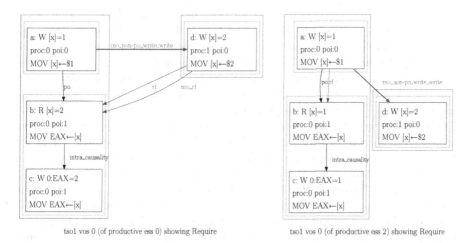

tso1 vos 0 (of productive ess 0) showing Require   tso1 vos 0 (of productive ess 2) showing Require

Fig. 3. Example valid execution witnesses (for two different event structures)

stream-like linear order over labels that satisfies several conditions (label_order in the HOL sources) describing labels in an okMpath. We then have:

Theorem 4. *For any well-formed event structure E, and valid execution X for E, there exists some event-machine path, such that* okEpath *E path and* okMpath *path, in which the memory reads and write-buffer flushes both respect* $<_{X.memory\_order}$. [hand proof, relying on the preceding lemma]

4   Verified Checker and Results

To explore the consequences of x86-TSO, we implemented the axiomatic model in our **memevents** tool, which exhaustively explores candidate execution witnesses. For greater confidence, we added to this a verified witness checker: we defined variants of event structures and execution witnesses, using lists instead of sets, wrote algorithmic versions of well_formed_event_structure and valid_execution, proved these equivalent (in the finite case) to our other definitions, extracted OCaml code from the HOL, and integrated that into **memevents**. (Obviously, this only provides assurance for positive tests, those with allowed final states.)

The **memevents** results coincide with our observations on real processors and the vendor specifications, for the 10 IWP tests, the (negated) IRIW test, the two MFENCE tests amd5 and amd10, our n2–n6, and rwc-fenced. The remaining tests (amd3, n1, n7, n8, and rwc-unfenced) are "allow" tests for which we have not observed the specified final state in practice.

5   Related Work

There is an extensive literature on relaxed memory models, but most of it does not address x86, and we are not aware of any previous model that addresses the concerns of §2. We touch here on some of the most closely related work.

There are several surveys of weak memory models, including those by Adve and Gharachorloo [3], and by Higham et al. [13]; the latter formalises a range of models, including a TSO model, in both operational and axiomatic styles, and proves equivalence results. Their axiomatic TSO model is rather closer to the operational style than ours is, and both are idealised rather than x86-specific. Burckhardt and Musuvathi [8, Appendix A] also give operational and axiomatic definitions of a TSO model and prove equivalence, but only for finite executions. Their models treat memory reads and writes and barrier events, but lack register events and locked instructions with multiple events that happen atomically. Hangel et al. [10] describe the Sun TSOtool, checking the observed behaviour of pseudo-randomly generated programs against a TSO model. Roy et al. [17] describe an efficient algorithm for checking whether an execution lies within an approximation to a TSO model, used in Intel's Random Instruction Test (RIT) generator. Boudol and Petri [7] give an operational model with hierarchical write buffers (thereby permitting IRIW behaviours), and prove sequential consistency for data-race-free (DRF) programs. Loewenstein et al. [15] describe a "golden memory model" for SPARC TSO, somewhat closer to a particular implementation microarchitecture than the abstract machine we give in §3.1, that they use for testing implementations. They argue that the additional intensional detail increases the effectiveness of simulation-based verification. Saraswat et al. [18] also define memory models in terms of local reordering, and prove a DRF theorem, but focus on high-level languages. Several groups have used proof tools to tame the intricacies of these models, including Yang et al. [22], using Prolog and SAT solvers to explore an axiomatic Itanium model, and Aspinall and Ševčík [5], who formalised and identified problems with the Java Memory Model using Isabelle/HOL.

6   Conclusion

We have described x86-TSO, a memory model for x86 processors that does not suffer from the ambiguities, weaknesses, or unsoundnesses of earlier models. Its abstract-machine definition should be intuitive for programmers, and its equivalent axiomatic definition supports the memevents exhaustive search and permits an easy comparison with related models; the similarity with SPARCv8 suggests x86-TSO is strong enough to program above. Mechanisation in HOL4 revealed a number of subtle points of detail, including some of the well-formed event structure conditions that we depend on (e.g. that instructions have no *internal* data races). We hope that this will clarify the semantics of x86 architectures.

Acknowledgements. We thank Luc Maranget for his work on memevents, and David Christie, Dave Dice, Doug Lea, Paul Loewenstein, Gil Neiger, and Francesco Zappa Nardelli for helpful remarks. We acknowledge funding from EPSRC grant EP/F036345.

References

1. AMD64 Architecture Programmer's Manual (3 vols). Advanced Micro Devices, rev. 3.14 (September 2007)
2. Intel 64 and IA-32 Architectures Software Developer's Manual (5 vols). Intel Corporation, rev. 29 (November 2008)
3. Adve, S., Gharachorloo, K.: Shared memory consistency models: A tutorial. IEEE Computer 29(12), 66–76 (1996)
4. Ahamad, M., Neiger, G., Burns, J., Kohli, P., Hutto, P.: Causal memory: Definitions, implementation, and programming. Distributed Computing 9(1), 37–49 (1995)
5. Aspinall, D., Ševčík, J.: Formalising Java's data race free guarantee. In: Schneider, K., Brandt, J. (eds.) TPHOLs 2007. LNCS, vol. 4732, pp. 22–37. Springer, Heidelberg (2007)
6. Boehm, H.-J., Adve, S.: Foundations of the C++ concurrency memory model. In: Proc. PLDI (2008)
7. Boudol, G., Petri, G.: Relaxed memory models: an operational approach. In: Proc. POPL, pp. 392–403 (2009)
8. Burckhardt, S., Musuvathi, M.: Effective program verification for relaxed memory models. Technical Report MSR-TR-2008-12, Microsoft Research (2008); Gupta, A., Malik, S. (eds.) CAV 2008. LNCS, vol. 5123, pp. 107–120. Springer, Heidelberg (2008)
9. Dice, D.: Java memory model concerns on Intel and AMD systems (January 2008), http://blogs.sun.com/dave/entry/java_memory_model_concerns_on
10. Hangal, S., Vahia, D., Manovit, C., Lu, J.-Y.J., Narayanan, S.: TSOtool: A program for verifying memory systems using the memory consistency model. In: Proc. ISCA, pp. 114–123 (2004)
11. The HOL 4 system, http://hol.sourceforge.net/
12. Intel. Intel 64 architecture memory ordering white paper. SKU 318147-001 (2007)
13. Higham, L., Kawash, J., Verwaal, N.: Defining and comparing memory consistency models. PDCS, Full version as TR #98/612/03, U. Calgary (1997)
14. Loewenstein, P.: Personal communication (November 2008)
15. Loewenstein, P.N., Chaudhry, S., Cypher, R., Manovit, C.: Multiprocessor memory model verification. In: Proc. AFM (Automated Formal Methods), FLoC workshop (August 2006), http://fm.csl.sri.com/AFM06/
16. Owens, S., Sarkar, S., Sewell, P.: A better x86 memory model: x86-TSO (extended version). Technical Report UCAM-CL-TR-745, Univ. of Cambridge (2009), Supporting material at, www.cl.cam.ac.uk/users/pes20/weakmemory/
17. Roy, A., Zeisset, S., Fleckenstein, C.J., Huang, J.C.: Fast and generalized polynomial time memory consistency verification. In: Ball, T., Jones, R.B. (eds.) CAV 2006. LNCS, vol. 4144, pp. 503–516. Springer, Heidelberg (2006)
18. Saraswat, V., Jagadeesan, R., Michael, M., von Praun, C.: A theory of memory models. In: Proc. PPoPP (2007)
19. Sarkar, S., Sewell, P., Zappa Nardelli, F., Owens, S., Ridge, T., Braibant, T., Myreen, M., Alglave, J.: The semantics of x86-CC multiprocessor machine code. In: Proc. POPL 2009 (January 2009)
20. Sindhu, P.S., Frailong, J.-M., Cekleov, M.: Formal specification of memory models. In: Scalable Shared Memory Multiprocessors, pp. 25–42. Kluwer, Dordrecht (1991)
21. SPARC International, Inc. The SPARC architecture manual, v. 8. Revision SAV080SI9308 (1992), http://www.sparc.org/standards/V8.pdf
22. Yang, Y., Gopalakrishnan, G., Lindstrom, G., Slind, K.: Nemos: A framework for axiomatic and executable specifications of memory consistency models. In: IPDPS (2004)

Formal Verification of Exact Computations Using Newton's Method

Nicolas Julien and Ioana Paşca

INRIA Sophia Antipolis
{Nicolas.Julien,Ioana.Pasca}@sophia.inria.fr

Abstract. We are interested in the verification of Newton's method. We use a formalization of the convergence and stability of the method done with the axiomatic real numbers of Coq's Standard Library in order to validate the computation with Newton's method done with a library of exact real arithmetic based on co-inductive streams. The contribution of this work is twofold. Firstly, based on Newton's method, we design and prove correct an algorithm on streams for computing the root of a real function in a lazy manner. Secondly, we prove that rounding at each step in Newton's method still yields a convergent process with an accurate correlation between the precision of the input and that of the result. An algorithm including rounding turns out to be much more efficient.

1 Introduction

The Standard Library of the Coq proof assistant [4,1] contains a formalization of real numbers based on a set of axioms. This gives the real numbers all the desired theoretical properties and makes theorem proving more agreeable and close to "pencil and paper" proofs [16]. However, this formalization has no (or little) computational meaning. During this paper we shall refer to the reals from this implementation as "axiomatic reals". We note that Coq is not a special case and proof assistants in general provide libraries with results from real analysis [5,7,8,10], but with formalizations for real numbers that are not well suited for computations. However, in a proof process, it is often the case that we are interested in computing with the real numbers (or at least approximating such computations), so a considerable effort has been invested in having libraries of exact computations for proof systems [13,15,18]. We shall refer to numbers from such implementations as "exact reals". These libraries provide verified computations for a set of operations and elementary functions on real numbers.

The results in this paper are concerned with Newton's method. Under certain conditions, this method ensures the convergence at a certain speed towards a root of the given function, the unicity of this root in a certain domain and the local stability. But, as the "paper" proof for these results depends on non-trivial theorems from analysis like the mean value theorem and concepts like continuity, derivation etc. the formal development conducted around them is based on the axiomatic reals of Coq. We would like to transfer these "theoretical" properties to the computations done with exact reals. Our work is thus conducted in two

S. Berghofer et al. (Eds.): TPHOLs 2009, LNCS 5674, pp. 408–423, 2009.

directions. On one side we are interested in proving correct Newton's method on exact reals and having algorithms that are suited for our implementation of the real numbers as co-inductive streams [13]. On the other hand we are concerned in providing appropriate theoretical results to support the correctness of the algorithms and optimizations we make.

The paper is organized as follows: in section 2 we present the theoretical results around Newton's method that have been verified with the axiomatic reals in CoQ. This section gives the formalization of well-known results in [6] and presents a new proof that was motivated by our implementation of the method on exact reals. To clarify the need for this proof, in section 3 we present a library of exact real arithmetic implemented with CoQ's co-inductive streams and we discuss how computations with Newton's method can be verified in this setting. We also design and prove correct an algorithm for computing the root of a function that is based on Newton's method and is adapted for streams. However, this algorithm is much more efficient when rounding is used during the process. The theorem we present in section 2.1 justifies this optimization, though the optimized algorithm is not completely certified. The applications of our algorithm are given in section 4.4 along with perspectives opened by the suggested improvements. We finish by discussing related work in section 5 as well as conclusions and possible extensions of our work in section 6.

2   Kantorovitch's Theorem and Related Results

Kantorovitch's theorem gives sufficient conditions for the convergence of Newton's method towards the root of a given function and establishes the unicity of this root in a certain domain. A version of this theorem as well as results concerning the speed for the convergence of the process and its stability are discussed in [6]. Preliminary results around a formalization of these theorems inside the CoQ proof assistant are described in [19]. At present all the theorems listed in this section are verified in the CoQ proof assistant. The formal proof is based on the axiomatic real numbers from CoQ's Standard Library. This choice is motivated by the concepts we needed to handle, as the library contains results from real analysis concerning convergence, continuity, derivability etc. The theorems listed bellow illustrate the type of concepts involved in the proof.

Theorem 1 (Existence). *Consider an equation $f(x) = 0$, where $f : [a, b] \rightarrow \mathbb{R}$, $a, b \in \mathbb{R}$ $f(x) \in C^{(1)}([a, b])$. Let $x^{(0)}$ be a point contained in $[a, b]$ with its closed ε-neighborhood $\overline{U_\varepsilon}(x^{(0)}) = \{|x - x^{(0)}| \leq \varepsilon\} \subset [a, b]$. If the following conditions hold:*

1. *$f'(x^{(0)}) \neq 0$ and $|\frac{1}{f'(x^{(0)})}| \leq A_0$;*
2. *$|\frac{f(x^{(0)})}{f'(x^{(0)})}| \leq B_0 \leq \frac{\varepsilon}{2}$;*
3. *$\forall x, y \in]a, b[, |f'(x) - f'(y)| \leq C|x - y|$*
4. *the constants A_0, B_0, C satisfy the inequality $\mu_0 = 2A_0 B_0 C \leq 1$.*

then, for an initial approximation $x^{(0)}$, the Newton process

$$x^{(n+1)} = x^{(n)} - \frac{f(x^{(n)})}{f'(x^{(n)})}, \quad n = 0, 1, 2, \ldots \tag{1}$$

converges and $\lim\limits_{n\to\infty} x^{(n)} = x^*$ *is a solution of the initial system, so that* $|x^* - x^{(0)}| \leq 2B_0 \leq \varepsilon$.

Theorem 2 (Uniqueness). *Under the conditions of Theorem 1 the root* x^* *of the function* f *is unique in the interval* $[x^{(0)} - 2B_0, x^{(0)} + 2B_0]$.

Theorem 3 (Speed of convergence). *Under the conditions of Theorem 1 the speed of the convergence of Newton's method is given by* $|x^{(n)} - x^*| \leq \frac{1}{2^{n-1}} \mu_0^{2^n - 1} B_0$.

Theorem 4 (Local stability). *If the conditions of Theorem 1 are satisfied and if, additionally,* $0 < \mu_0 < 1$ *and* $[x^{(0)} - \frac{2}{\mu_0}B_0, x^{(0)} + \frac{2}{\mu_0}B_0] \subset [a, b]$, *then for any initial approximation* $x'^{(0)}$ *that satisfies* $|x'^{(0)} - x^{(0)}| \leq \frac{1-\mu_0}{2\mu_0} B_0$ *the associated Newton's process converges to the root* x^*.

The convergence of the process ensures that Newton's method is indeed appropriate for determining the root of the function. The unicity of the solution in a certain domain is used in practice for isolating the roots of the function. The result on the speed of the convergence means we know a bound for the distance between a given element of the sequence and the root of the function. This represents the precision at which an element of the sequence approximates the root. In practice this theorem is used to determine the number of iterations needed in order to achieve a certain precision for the solution. The result on the stability of the process will help with efficiency issues as it allows the use of an approximation rather than an exact real.

We do not present here the proofs of the theorems, we just give a few elements of these proofs that are needed in understanding the next section. For details on the proofs we refere the reader to [6]. The central element of the proof is an induction process that establishes a set of properties for each element of the Newton sequence. The proof introduces the auxiliary sequences $\{A_n\}_{n\in\mathbb{N}}$, $\{B_n\}_{n\in\mathbb{N}}$ and $\{\mu_n\}_{n\in\mathbb{N}}$:

$$A_n = 2A_{n-1} \tag{2}$$

$$B_n = A_{n-1}B_{n-1}^2 C = \frac{1}{2}\mu_{n-1}B_{n-1} \tag{3}$$

$$\mu_n = 2A_n B_n C = \mu_{n-1}^2 \tag{4}$$

For each element of the Newton sequence, we are able to verify properties that are similar to those for $x^{(0)}$. Reasoning by induction we get the following:

- $f'(x^{(n)}) \neq 0$ and $|\frac{1}{f'(x^{(n)})}| \leq A_n$
- $|f(x^{(n)})/f'(x^{(n)})| \leq B_n \leq \frac{\varepsilon}{2^{n+1}}$
- $\mu_n \leq 1$

Notice that hypothesis 3. of Theorem 1 is a property of the function and it does not depend on the elements of Newton's sequence.

From the above relations we get the convergence, unicity and speed of convergence for the sequence.

For Theorem 4 (local stability) we prove that the new initial approximation $x'^{(0)}$ satisfies similar hypotheses as those for $x^{(0)}$. The new constants are $A' = \frac{4}{3+\mu_0} A_0$ and $B' = \frac{3+\mu_0}{4\mu_0} B_0$. This makes that $\mu' = 2A'B'C = 1$ and we can verify that

- $f'(x'^{(0)}) \neq 0$ and $|\frac{1}{f'(x'^{(0)})}| \leq A'$
- $|f(x'^{(0)})/f'(x'^{(0)})| \leq B'$
- $\mu' \leq 1$

We are thus in the hypotheses of Theorem 1 and by applying this theorem we conclude that the process converges to the same root x^*.

Notice, however, that for the new constants we get $\mu' = 1$. If we do a Newton iteration, we would get the new $\mu'' = \mu'^2 = 1$ (cf. equation (4)) and we would not be able to do an approximation again, because Theorem 4 requires $\mu'' < 1$. To correct this, we impose a finer approximation $|x_0 - x'_0| \leq \frac{(1-\mu_0)}{4\mu_0} B_0$. This new approximation yields the following formulas for the constants:

$$A' = \frac{8}{7+\mu_0} A_0 \tag{5}$$

$$B' = \frac{\mu_0^2 + 46\mu_0 + 17}{8(7+\mu_0)\mu_0} B_0 \tag{6}$$

this makes that

$$\mu' = \frac{\mu_0^2 + 46\mu_0 + 17}{(7+\mu_0)^2} < 1 \tag{7}$$

We summarize these results in:

Corollary 1. *If the conditions of Theorem 1 are satisfied and if, additionally, $0 < \mu_0 < 1$ and $[x^{(0)} - \frac{2}{\mu_0} B_0, x^{(0)} + \frac{2}{\mu_0} B_0] \subset [a, b]$, then for any initial approximation $x'^{(0)}$ that satisfies $|x'^{(0)} - x^{(0)}| \leq \frac{1-\mu_0}{4\mu_0} B_0$ the associated Newton's process converges to the root x^*.*

2.1 Newton's Method with Rounding

We now have all the necessary tools to state and prove a theorem on the behavior of Newton's method if we consider rounding at each step. The rounding we do is just good enough to ensure the convergence. This theorem is particularly interesting for computations in arbitrary or multiple precision, as it relates the number of iterations with the precision of the input and that of the result. This means that for the first iterations we need a lower precision, as we are not close to the root. We will later increase the precision of our input with the desired precision for the result.

Theorem 5. *We consider a function $f : [a, b] \to \mathbb{R}$ and an initial approximation $x^{(0)}$ satisfying the conditions in Theorem 1.*
We also consider a function $rnd : \mathbb{N} \times \mathbb{R} \to \mathbb{R}$ that models the approximation we will make at each step in the perturbed Newton sequence:

$$t^{(0)} = x^{(0)} \text{ and } t^{(n+1)} = rnd_{n+1}\left(t^{(n)} - \frac{f(t^{(n)})}{f'(t^{(n)})}\right)$$

If

1. $\forall n \forall x, x \in]a, b[\Rightarrow rnd_n(x) \in]a, b[$
2. $\frac{1}{2} \le \mu_0 < 1$
3. $[x^{(0)} - 3B_0, x^{(0)} + 3B_0] \subset [a, b]$
4. $\forall n \forall x, |x - rnd_{n+1}(x)| \le \frac{1}{3^n} R_0$, *where* $R_0 = \frac{1 - \mu_0^2}{8\mu_0} B_0$

then

a. *the sequence* $\{t^{(n)}\}_{n \in \mathbb{N}}$ *converges and* $\lim\limits_{n \to \infty} t^{(n)} = x^*$ *where* x^* *is the root of the function f given by Theorem 1*
b. $\forall n, |x^* - t^{(n)}| \le \frac{1}{2^{n-1}} B_0$

The first hypothesis makes sure that the new value will also be in the range of the function. The second and third hypotheses come from the use of the stability property of the Newton sequence (see Corollary 1). The fourth hypothesis controls the approximation we are allowed to make at each iteration. The conclusion gives us the convergence of the process to the same limit as Newton's method without approximations. Also we give an estimate of the distance from the computed value to the root at each step.

Proof. Our proof is based on those for theorems 1 - 4 and corollary 1. To give the intuition behind the proof, we decompose Newton's perturbed process $t^{(n)}$ as follows:

1. set $t^{(0)} := x^{(0)}$
2. do a Newton iteration to get $x^{(1)} := t^{(0)} - \frac{f(t^{(0)})}{f'(t^{(0)})}$
3. do an approximation of the result to get $t^{(1)} := rnd(x^{(1)})$
4. set $t^{(0)} := t^{(1)}$ and go to step 2.

Now let's look at these steps individually:

- At step 1. we start with the initial $x^{(0)}$ that satisfies the conditions in Theorem 1. This means that Newton's method from this initial point converges to the root x^* (cf. Theorem 1).
- At step 2. we consider a Newton sequence starting with $x^{(1)}$. This sequence is the same as the sequence at step 1. except that we "forget" the first element of the sequence and start with the second. It is trivial that this sequence converges to the root x^*. We note that (cf. proof of Theorem 1) we can associate the constants A_1, B_1 to the initial iteration of this sequence and get the corresponding hypotheses from Theorem 1.

o At step 3. we consider Newton's sequence starting from $t^{(1)}$. This initial point is just an approximation of the initial point of the previously considered sequence. From Corollary 1 we get the convergence of the new sequence to the same root x^*. Moreover, the proof of Corollary 1 gives us the constants A', B' associated to the initial point that also satisfy the hypotheses of Theorem 1. This means we can start the process over again.

If we take $x^{(0)}$ and then all the initial iterations of the sequences formed at step 3. we get back our perturbed Newton's sequence. But decomposing the problem as we did gives the intuition of why this sequence should converge. However, just having a set of sequences that all converge to the same root does not suffice to prove that the sequence formed with all initial iterations of these sequences will also converge to the same root. The reason is simple, the approximation at step 3. could bring us back to the initial point $x^{(0)}$ which would still yield a convergent Newton's sequence, but which would not make the new element of the perturbed sequence any closer to the root than the previous one. To get the convergence of the perturbed sequence we need to control the approximation we make. We will see in what follows that hypothesis 4. suffices to ensure the convergence of the new process.

To make the intuitive explanation more formal we consider the sequence of sequences of real numbers $\{Y_p\}_{p \in \mathbb{N}}$ defined as follows:
$Y_0^n = x^{(n)}$ is the original Newton's sequence;
Y_1 is given by
$Y_1^0 = rnd_1(x^{(1)})$;
$Y_1^{n+1} = Y_1^n - f(Y_1^n)/f'(Y_1^n)$ is the Newton's sequence associated to the initial iteration Y_1^0;
we continue in the same manner and for an arbitrary p we define Y_p as follows
$Y_{p+1}^0 = rnd_{p+1}(Y_p^1)$;
$Y_{p+1}^{n+1} = Y_{p+1}^n - f(Y_{p+1}^n)/f'(Y_{p+1}^n)$.
We notice that taking the first element in each of these sequences forms our perturbed Newton's process:

$$Y_0^0 = x^{(0)} = t^{(0)} \text{ and}$$

$$Y_{n+1}^0 = rnd_{n+1}(Y_n^0 - f(Y_n^0)/f'(Y_n^0)) = rnd_{n+1}(t^{(n)} - f(t^{(n)})/f'(t^{(n)})) = t^{(n+1)}$$

Following our plan, we now show that for each p the sequence $\{Y_p^n\}_{n \in \mathbb{N}}$ converges to x^* and ensures a certain bound in the error.

o We start with sequence $\{Y_0^n\}_{n \in \mathbb{N}}$. Since it coincides with the initial sequence, the properties from Theorem 1 are trivially satisfied. For the initial point Y_0^0 we have the associated constants A_0, B_0. Applying Theorem 1 we get that $\lim_{n \to \infty} Y_0^n = x^*$ and $|x^* - Y_0^0| \le 2B_0$.

o Before considering $\{Y_1^n\}_{n \in \mathbb{N}}$, we note that the sequence $\overline{Y}_0^n = Y_0^{n+1}$ (i.e. the previously considered sequence where we start from the second element) also satisfies the conditions, with initial point $\overline{Y}_0^0 = Y_0^1$ and constants $\overline{A}_0 = 2A_0$ and $\overline{B}_0 = A_0 B_0^2 C$. The laws for these constants are deduced from relations

(2), (3). We get that $\lim_{n\to\infty} \overline{Y}_0^n = x^*$ and $|x^* - \overline{Y}_0^0| = |x^* - Y_0^1| \leq 2\overline{B}_0 = 2(A_0 B_0^2 C)$.

○ Now we consider $\{Y_1^n\}_{n\in\mathbb{N}}$. The initial point of this sequence is $Y_1^0 = rnd_1$ $(Y_0^0 - f(Y_0^0)/f'(Y_0^0)) = rnd_1(\overline{Y}_0^0)$. We are in the situation of Corollary 1, where we have a converging sequence ($\{\overline{Y}_0^n\}_{n\in\mathbb{N}}$) and we introduce an approximation in the initial iteration. To be able to apply this corollary we need to verify $0 < \overline{\mu}_0 < 1$, $[\overline{Y}_0^0 - \frac{2}{\overline{\mu}_0}\overline{B}_0, \overline{Y}_0^0 + \frac{2}{\overline{\mu}_0}\overline{B}_0] \subset [a, b]$ and $|rnd_1(\overline{Y}_0^0) - \overline{Y}_0^0| \leq \frac{1-\overline{\mu}_0}{4\overline{\mu}_0}\overline{B}_0$. We will show later on that under our hypotheses these three conditions are indeed verified. From Corollary 1 we get the new constants according to relations (5), (6). This makes that we find ourselves again in the conditions of Theorem 1 and we can deduce that $\lim_{n\to\infty} \overline{Y}_1^n = x^*$ and

$$|x^* - Y_1^0| \leq 2B' = 2\frac{\overline{\mu}_0^2 + 46\overline{\mu}_0 + 17}{8(7+\overline{\mu}_0)\overline{\mu}_0}\overline{B}_0.$$

We are in the appropriate conditions to start this process again and explain in the same manner the properties for $\{Y_2^n\}_{n\in\mathbb{N}}$, $\{Y_3^n\}_{n\in\mathbb{N}}$, etc. The auxiliary sequences are given by the following relations:

$$A_0' = A_0 \text{ and } A_{n+1}' = \frac{8}{7 + \overline{\mu}_n(2A_n')}$$

$$B_0' = B_0 \text{ and } B_{n+1}' = \frac{\overline{\mu}_n^2 + 46\overline{\mu}_n + 17}{8(7+\overline{\mu}_n)\overline{\mu}_n}(A_n' B_n'^2 C)$$

$$\overline{\mu}_n = 2(2A_n')(A_n' B_n'^2 C)C = (2A_n' B_n' C)^2$$

we also consider

$$\mu_{n+1}' = 2A_{n+1}' B_{n+1}' C = \frac{\overline{\mu}_n^2 + 46\overline{\mu}_n + 17}{(7+\overline{\mu}_n)^2} = \frac{\mu_n'^2 + 46\mu_n'^2 + 17}{(7 + \mu_n'^2)^2}$$

$$R_n = \frac{1 - \overline{\mu}_n}{4\overline{\mu}_n}\overline{B}_n = \frac{1-\mu_n'^2}{4\mu_n'^2}(\frac{1}{2}\mu_n' B_n') = \frac{1-\mu_n'^2}{8\mu_n'}B_n'$$

Using the above reasoning steps, we get by induction that $|Y_n^0 - x^*| \leq 2B_n'$ and we also manage to show $\forall n, B_{n+1}' \leq \frac{1}{2}B_n' \leq \frac{1}{2^{n-1}}B_0$. The latter relations is deduced from the above formulas by basic manipulations. It trivially implies the convergence of the perturbed sequence to the root x^*.

We need some auxiliary results to ensure that Corollary 1 is applied in the appropriate conditions each time we make a rounding. These results are as follow:

○ $0 < \frac{1}{2} \leq \mu_0 = \mu_0' \leq \mu_n' \leq \mu_{n+1}' \leq \ldots < 1$

○ $R_{n+1} \leq \frac{1}{3}R_n \leq \ldots \leq \frac{1}{3^n}R_0 = \frac{1}{3^n}\frac{1-\overline{\mu}_0}{4\overline{\mu}_0}\overline{B}_0 = \frac{1}{3^n}\frac{1-\mu_0^2}{8\mu_0}B_0$

○ $|Y_{n+1}^0 - Y_n^0| \leq \frac{1}{2^n}B_0 + \frac{1}{3^n}R_0$

○ $[\overline{Y}_n^0 - \frac{2}{\overline{\mu}_n}\overline{B}_n, \overline{Y}_n^0 + \frac{2}{\overline{\mu}_n}\overline{B}_n] \subseteq [Y_0^0 - 3B_0, Y_0^0 + 3B_0] \subset [a, b]$

We do not discuss all the details as they are elementary reasoning steps concerning inequalities, second degree equations or geometric series. All these results have been formalized in CoQ to ensure that no steps are overlooked.

Remarks. Independent of verification of exact computations, this proof has an interest from a proof engineering point of view. We were able to come up with the proof because we had formalized theorems 1 - 4 inside a proof assistant. Such a formalization forces the user to understand the structure of the proof on one hand and to handle details with care on the other. Thus, an assisted proof is usually more structured and more detailed than a paper proof (especially in domains where automatic techniques are difficult to implement, like real analysis). For example, while on paper the auxiliary sequences $\{A_n\}_{n\in\mathbb{N}}$, $\{B_n\}_{n\in\mathbb{N}}$ appear during the proof, on the computer they are defined apart from the proof, allowing the user to better understand their importance and use similar sequences in the new proof. A proof assistant is also helpful with syntactic aspects like properly constructing the induction hypothesis and doing the bookkeeping to make sure all needed details are taken into consideration.

3   A Coq Library for Exact Real Arithmetic

The exact real library we are considering represents a real in the interval $[-1, 1]$ as a lazy infinite sequence of signed digits of an arbitrary integer base. The signed digits of a base β are the integers in $[-\beta + 1, \beta - 1]$. We denote $s_1 : : s$ the infinite sequence beginning by the digit s_1 and followed by the infinite sequence s. The real number r represented by such an infinite sequence s in base β is :

$$r = [\![s]\!]_\beta = [\![s_1 : : s_2 : : s_3 : : \ldots]\!]_\beta = \sum_{i=1}^\infty \frac{s_i}{\beta^i}.$$

A real number represented by a stream and for which we know the first digit can be written as: $r = [\![s_1 : : s]\!]_\beta = \frac{s_1 + [\![s]\!]_\beta}{\beta}$.

Having signed digits makes our representation redundant. For example we can represent $\frac{1}{3}$ as $[\![3 : : 3 : : 3 : : 3 \ldots]\!]_{10}$ but also as $[\![4 : : -7 : : 4 : : -7 \ldots]\!]_{10}$. For each digit k the set of real numbers that admit a representation beginning by this digit is: $[\frac{k-1}{\beta}, \frac{k+1}{\beta}]$. The sets associated to consecutive digits overlap with a constant magnitude of $\frac{2}{\beta}$. The main benefit of this redundancy is that we are able to design algorithms for which we can decide a possible first digit of the output. Without redundancy this is in general undecidable. Take the example of addition: $[\![0 : : 3 \ldots]\!]_{10} + [\![0 : : 6 \ldots]\!]_{10}$ may need infinite precision to decide whether the first digit is 0 or 1. In the case of signed digits we give 1 as a first digit knowing we can always go back to a smaller number by using a negative digit. We also note that in our example it was sufficient to know two digits of the input to decide the first digit of the output and this is true for addition in general.

Designing an algorithm therefore requires approximating the result to a precision that is sufficient to determine a possible first digit. Also, since our real numbers are infinite streams, the algorithms need to be designed in such a way that we are always able to provide an extra digit of the result. This is done by co-recursive calls on our co-inductive streams.

In CoQ, co-induction [9] provides a way to describe potentially infinite datatypes as our infinite sequences of digits. It offers both efficient lazy evaluation and nice proof schemes. The type of infinite sequences of objects of some type A is defined as follows

CoInductive stream (A : Set) : Set := | Cons : A → stream A → stream A.

Cons should not be understood as a way to construct an infinite stream from another since we cannot build an initial infinite stream, but as a way to decompose an infinite stream into a finite part and an infinite part that could be described again with a new Cons and so on.

Our real numbers will be streams of signed digits, so we also need to create a model for the digits. They are abstracted with respect to the base and the implementation of integers, so both the base and the type of integers can be chosen by the user. Here we will denote their type by digit.

We can define new streams using co-recursive functions, for instance the stream of 0s, which obviously represents the real number 0.

CoFixpoint zero : stream digit := Cons 0 zero.

To prove the correctness of the algorithms on streams of digits we first define a relation between these streams and the axiomatic reals of CoQ. This relation is based on the relation between the real value of a sequence, its first digit and the real value of the following sequence as noted previously. We formalize this relation as a co-inductive predicate :

CoInductive represents $(\beta : \mathbb{Z})$: stream digit → $\mathbb{R}$ → Prop :=
| rep : $\forall s\ r\ k, -\beta < k < \beta \to\quad -1 \leq r \leq 1 \to$
represents $\beta\ s\ r$ → represents β (Cons $k\ s$) $\frac{k+r}{\beta}$.

This relation also makes sure that streams only represent reals in $[-1, 1]$ and that the digits are in the set of the allowed signed digits.

The correctness of our algorithms is verified when we manage to express a represents relation between our implementation and the standard in the CoQ library. For instance the proof that the multiplication is correct is formulated in this way :

Theorem mult_correct :
$\forall x\ y\ vx\ vy$, represents x vx → represents y vy → represents $(x \otimes y)$ (vx $*$ vy).

In means that every time we have an exact real (i.e. a stream of digits) x that represents an axiomatic real vx and an y that represents a vy than our multiplication of streams x and y (here denoted $\otimes$) will represent the multiplication of axiomatic reals vx and vy.

For further details on algorithms and proofs for this library we refer the reader to [13].

4 Newton's Method on Exact Reals

4.1 Correctness of Newton's Method

We want to prove correctness of computation with Newton's method on exact reals in the same manner we proved correctness of multiplication in section 3. We code Newton's algorithm for both exact reals and axiomatic reals. For simplification we use a function g on exact reals to represent the ratio $\frac{f(x)}{f'(x)}$ of axiomatic reals.

Fixpoint EXn g ex0 n {struct n}: stream digit:= **match** n **with**
    | 0 $\Rightarrow$ ex0 | S n $\Rightarrow$ **let** exn:= (EXn g x0 n) **in** exn $\ominus$ g exn **end.**

Fixpoint Xn f f' x0 n {struct n}: R:= **match** n **with**
    | 0 $\Rightarrow$ x0 | S n $\Rightarrow$ **let** xn:= (Xn x0 f f' n) **in** xn $-$ f xn / f' xn **end.**

The relation between elements of the same rank in the two sequences:
    $\forall$ n, represents (EXn g EX0 n) (Xn X0 f f' n)
is almost trivial, if we have a represents relation for the initial iteration and for the function.

Theorem EXn_correct : $\forall$g ex0 f f' x0 n, represents ex0 x0 $\rightarrow$
($\forall$x vx, represents x vx $\rightarrow$ represents (g x) (f vx / f' vx)) $\rightarrow$
($\forall$n, $-1 \leq$Xn x0 f f' n ≤ 1) $\rightarrow$ represents (EXn g ex0 n) (Xn x0 f f' n).

The proof follows from the correction of the subtraction on streams with respect to the subtraction on axiomatic reals.

    This theorem allows us to transfer properties proved for Newton's method on axiomatic reals to the method implemented on exact reals. If we satisfy the conditions of Theorem 1 for the function f and the initial iteration $X0$, then we can compute the root of the function at an arbitrary accuracy, given by Theorem 3 (speed of convergence). From the same theorem we get the rank to which we need to compute for a given accuracy to be obtained. However, if we wanted to increase this accuracy, we would need to redo all the computation for the new rank. We want to avoid this and take advantage of the lazy evaluation characteristic for streams: we can design an algorithm that uses Newton's method to compute an arbitrary number of digits for the root of a given function, under certain conditions for this function.

4.2 An Algorithm for Exact Computation of Roots

We consider a function $f : [-1, 1] \rightarrow \mathbb{R}$ with x^* the root of f and a suitable initial approximation $x^{(0)}$ for Newton's process. We have to find a possible first digit of the result x^* in base β. For this we use make_digit which requires a precision of $\frac{\beta-2}{2\beta^2}$ of the result to make the appropriate choice of the first digit. Indeed make_digit changes the first two digits of a representation by an equivalent two digits prefix in order to ensure that each number close to a precision of $\frac{\beta-2}{2\beta^2}$ admits a representation begining by the same first new digit.

To determine the number of Newton iterations that ensures this precision we use Theorem 3 (speed of convergence), which gives us n s.t. $|x^{(n)} - x^*| \leq \frac{\beta-2}{2\beta^2}$. We choose as a first digit for x^* the first digit d_1 of a representation of $x^{(n)}$. This gives us $x^* = \frac{d_1+x_1^*}{\beta}$, where x_1^* is the number formed from the remaining digits of x^*. Since $f(x^*) = 0$, we get $f(\frac{d_1+x_1^*}{\beta}) = 0$. This means we can define a new function $f_1(x) := f(\frac{d_1+x}{\beta})$, and x_1^* is the root of f_1. Determining the second digit of x^* is equivalent to determining the first digit of x_1^*. We repeat the previous steps for function f_1 and we take as the initial approximation the remaining digits of $x^{(n)}$, given by $\overline{x}^{(n)} = \beta x^{(n)} - d_1$. Now we have a co-recursive process to produce the digits of the root of our function one by one. If we simplify our algorithm by using $g = \frac{f}{f'}$, when we transform g in g_1 we get

$$g_1(x) := \frac{f_1(x)}{f_1'(x)} = \frac{f(\frac{d_1+x}{\beta})}{\frac{1}{\beta}f'(\frac{d_1+x}{\beta})} = \beta \times g(\frac{d_1+x}{\beta})$$

For the exact real implementation in CoQ we express the algorithm on streams of digits, so we remind that for the stream $d_1 :: x$, we have $[\![d_1 :: x]\!]_\beta = \frac{d_1+[\![x]\!]_\beta}{\beta}$

CoFixpoint exact_newton (g: stream digit → stream digit) ex0 n:=
  match (make_digit (EXn g ex0 n) **with**
    |d1::x' ⇒ d1::exact_newton (fun x ⇒ ($\beta \odot$ g (d1::x))) x' n
  end.

We note that $\beta \odot$ represents a specific function provided by the library which computes efficiently the multiplication of a stream by the base. It is possible to determine a value for n that is sufficent to allow the production of a digit at each co-recursive call. This simplifies the algorithm but reduces the quadratic convergence to linear convergence.

The formal verification of this algorithm means we have to prove that the output of this algorithm represents the root of the function f. For this we use Theorems 1 - 3 (see section 2) on axiomatic reals and a version of the theorem EXn_correct (section 4.1) that links Newton's method on exact reals to Newton's method on streams. We need to show that if the initial function f satisfies the hypotheses of Theorem 1 then the function f_1 built at the co-recursive call will also satisfy these hypotheses, thus yielding a correct algorithm.

The hypotheses of Theorem 1 impose that

1. $f \in C^{(1)}(]-1,1[)$
2. $\forall x, y \in]-1,1[, |f'(x) - f'(y)| \leq C|x - y|$
3. $f'(x^{(0)}) \neq 0$ and $|\frac{1}{f'(x^{(0)})}| \leq A_0$;
4. $|\frac{f(x^{(0)})}{f'(x^{(0)})}| \leq B_0 \leq \frac{\varepsilon}{2}$;
5. $\mu_0 = 2A_0B_0C \leq 1$.

We analyze $f_1(x) := f(\frac{d_1+x}{\beta})$ for which we have $f_1'(x) = \frac{1}{\beta}f'(\frac{d_1+x}{\beta})$ and the new initial iteration $\overline{x}^{(n)} = \beta x^{(n)} - d_1$

1. the class of the function is obviously the same, so $f \in C^{(1)}(]-1, 1[)$
2. $|f'_1(x) - f'_1(y)| = |\frac{1}{\beta}f'(\frac{d_1+x}{\beta}) - \frac{1}{\beta}f'(\frac{d_1+y}{\beta})| \leq \frac{1}{\beta}C|\frac{d_1+x}{\beta} - \frac{d_1+y}{\beta}| = \frac{1}{\beta^2}C|x - y|$
3. $f'_1(\overline{x}^{(n)}) = f'(x^{(n)}) \neq 0$ and $|\frac{1}{f'_1(\overline{x}^{(n)})}| = |\beta\frac{1}{f'(x^{(n)})}| \leq \beta A_n;$
4. $|\frac{f_1(\overline{x}^{(n)})}{f'_1(\overline{x}^{(n)})}| = |\beta\frac{f(x^{(n)})}{f'(x^{(n)})}| \leq \beta B_n;$
5. $\overline{\mu}_n = 2\beta A_n \beta A_n \frac{1}{\beta^2}C = 2A_n B_n C \leq 1.$

Relations 3. - 5. are given by the proof of Theorem 1. We are now able to prove by co-induction that `represents (exact_newton g ex0 n)` x^*.

4.3   Improvements of the Algorithm

Though short, elegant and proven correct, the algorithm presented in this section is not usable in practice as it is very slow. There are two main reasons for this:

1. The certified computations from the library require a precision of the operands higher than that of the result. We saw that in the case of addition one extra digit is required, but for other operations and function this precision can be higher. When we have an expression where we perform several operations, the precision demanded for each individual operand is a lot higher than the precision of the output. In the case of Newton's method, each iteration only brings a certain amount of information, so using a higher precision will not improve the result.
2. This approach relies on the higher-order capabilities of the functional programming language: the first argument of the `exact_newton` function is itself a function that becomes more and more complex as `exact_newton` calls itself recursively. The management of this function is somehow transparent to the programmer, but it has a cost: a new closure is built at every recursive call to `exact_newton` and when the function g is called, all the closures built since the initial call have to be unraveled to obtain the operations that really need to be performed. This cost can be avoided by building directly a first order data structure.

We discuss two possible improvements of this algorithm, dealing with these two issues. For the first point the solution is simple, just use the significant digits in the stream. Determining which are these significant digits and certifying the result is still possible thanks to Theorem 5. We implement a `truncate` function that given a stream s returns the stream containing the first n digits of s and sets the rest to zero. This function represents the `rnd` function on axiomatic reals (see Theorem 5).

```
Fixpoint truncate s n {struct n} :=
  match n with | 0 ⇒ zero | S n' ⇒
    match s with | d :: s' ⇒ d :: truncate s' n' end
  end.
```

The perturbed Newton's method becomes:

Fixpoint Etn g ex0 (n : nat) {struct n} : stream digit := **match** n **with**
  | 0 ⇒ ex0 | S n' ⇒ **let** tn := (Etn g ex0 n') **in** (truncate (tn ⊖ (g tn)) (φ n'))
    end.

The function φ controls the approximation we can make at each iteration and follows the constraints imposed by Theorem 5. The exact_newton algorithm will work in the same way with this sequence as with the original method.

CoFixpoint exact_newton_rnd (g: stream digit → stream digit) ex0 n:=
  match (make_digit (Etn g ex0 n)) **with**
    |d1::x' ⇒ d1::exact_newton_rnd (fun x ⇒ (β ⊙ g (d1::x))) x' n
  end.

Though the proof for this new algorithm is not finalized yet, we feel there is no real difficulty in obtaining it as both the algorithm and the optimization we make are proven correct.

To tackle the second point in our list of possible improvements we make explicit the construction of the new function g in the co-recursive call.

CoFixpoint exact_newton_aux
  (g : stream digit → stream digit) (Xn : stream digit) k n :=
    let Xn' := make_k_digits x0 (EXn g x0 n) k **in**
    (nth k Xn') :: exact_newton_aux g Xn' (S k) n.

Definition exact_newton2 (g : stream digit → stream digit)
  (X0 : stream digit) n := exact_newton_aux g X0 0 n.

The function make_k_digits takes three arguments: two streams x and y and an integer k and produces $k + 1$ digits of y by copying the first k digits in x and computing the digit $k + 1$ by using make_digit. For the function to perform correctly we must ensure that the first k digits in y can indeed be the same as those in x. In our case this results from the theorem on the speed of convergence of Newton's method, which make sure that a certain element is close enough to the root. The way the algorithm works is that it does iterations always for the same function g. It produces digits one at a time. Once it reached enough precision to certify an extra digit, the $(k+1)$th, it gets this digit by using the function nth and it continues to compute where it left of.

This algorithm performs better than the previous one, but the optimizations performed in this case seem more difficult to prove correct. At the time of writing this paper we have the good properties on make_k_digit and the proof of correctness of the algorithm is in progress.

4.4   Applications to the Square Root

Newton's method is commonly used for the implementation of nth root function or division. We discuss the example of the square root to illustrate the behaviour of our algorithms. The square root of a positive real number a is the root of the function $f_{sqrt}(x) = x^2 - a$. The corresponding function g_{sqrt} is $\frac{f_{sqrt}(x)}{f'_{sqrt}(x)} = \frac{x}{2} - \frac{a}{2x}$. Due to restrictions about implementing the inverse function of exact reals, the library provides functions of the family $x \mapsto \frac{1}{\beta^n x}$ where $n > 0$. So we chose instead the function $f_{sqrt}(x) = \beta^2 x^2 - a$ which corresponds to $g_{sqrt}(x) = \frac{x}{2} - \frac{a}{2\beta^2 x}$. The root of this function is $\frac{\sqrt{a}}{\beta}$. So a final multiplication by the base will give the expected result. We apply the algorithm to this function g_{sqrt} and the user provide a suitable initial approximation. We prove in CoQ that the resulting function actually computes a representation of the square root function on axiomatic reals divided by the base.

Definition Ssqrt (a : stream digit) ex0 n := exact_newton (g_sqrt a) ex0 n.
Theorem sqrt_correct : $\forall$ (a : stream digit) (va : R),
    represents a va $\rightarrow$ represents (Ssqrt a) ((sqrt va)/β).

The original algorithm is slow. For example the computation of the first digit of $\sqrt{\frac{1}{2}}$ in base 2^{124} using the original algorithm blocks the system, while for the same algorithm improved with approximations we get the equivalent precision of 37 decimal digits in 12 seconds. The second algorithm exact_newton2 brings an improvement at each new digit we want to obtain making the algorithm run in average twice as fast. We should also take into consideration that using $f(x) = \frac{a}{x^2} - 1$ can improve our execution times considerably as there is only one division involved. Nevertheless, our intention here was not to implement an efficient square root, but to test the capabilities of the previously presented algorithms.

5   Related Work

This work presents different angles in the formal verification of a numerical algorithm. A lot of work is being done concerning formally verified exact real arithmetic libraries. Besides the library presented here, the development [15] for PVS and [18] also for CoQ are two of the most recent such implementations. These two libraries have computations that are verified with respect to the real analysis formalizations in PVS and C-CoRN [5], respectively. A significant part of the work presented here could be reproduced in any of these libraries. In the case of [18] the exact reals operations and functions are verified via an isomorphism between the exact reals and the C-CoRN real structure; there is also an isomorphism between C-CoRN reals and Standard Library reals (see [14]), so in theory it should be possible to verify computations by using the presented proofs and the two isomorphisms.

Concerned with exact real arithmetic and also with co-inductive aspects we mention the work of Niqui [17]. This works aims to obtain all field operations on real numbers via the Edalat-Potts algorithm for lazy exact arithmetic.

Results of the convergence of Newton's method with rounding have been proved for some special cases like the the inverse and the square root [3]. Of course, in these cases the speed of convergence is better than in the general case.

The proof of correctness of square root algorithms has been the subject of several formal developments. We mention [2] for the verification of the GMP square root algorithm, [11] for an Intel architecture square root algorithm and [20] for the verification of the square root algorithm in an IBM processor.

A general algorithm using Newton's method was developped by Hur and Davenport [12] on a different representation of exact reals but not in a formally verified setting.

6   Conclusions and Perspectives

As a case study of theorem proving in numerical analysis, this work tries to underline three aspects of such a development: how do design and formalize the necessary proofs from "paper" mathematics, how to prove correct numerical methods implemented on exact reals and provide certified computations, how to design and verify specific algorithms for an implementation of exact reals. The CoQ development can be found at http://www-sop.inria.fr/marelle/Exact_Newton.

To the best of the authors' knowledge, the result and proof of Theorem 5 are new, though the authors are not experts in numerical analysis. Using only a predetermined precision for our computation makes it that our formalization can be seen as an (imperfect) model of computation in multiple or arbitrary precision, thus validating Newton's method in such a context. The proof of Theorem 5 was motivated by the need to improve the algorithm discussed in section 4.2. The contribution of proof assistants in obtaining this proof is twofold: the proof was motivated by a formal development and the proof was constructed inside the proof assistant, following the pattern of existing proofs.

The algorithms presented in section 4 are also new. The verified algorithm, though it is not of practical use, it can serve as a model in obtaining proofs for our optimized algorithms. The next step is to put together the two proofs we presented here and get a formally verified algorithm for computing roots in a (more) efficient manner. With the results presented here we see no difficulties in obtaining this proof.

The axiomatic formalization on Newton's method contains the multivariable version of the theorems 1 - 4. This means we can solve systems of (non-)linear equations using this method. Exact real arithmetic libraries do not yet treat such cases, so it would be an interesting experiment to see to what extent the results presented here can be obtained in the multivarite setting. We note that the proof of Theorem 5 has the same structure for the multivariate case. We presented here the real case to make it easier for the reader to follow and to show the correspondence with results from section 4.2.

Acknowledgments

We thank Yves Bertot for his help and constructive suggestions.

References

1. Bertot, Y., Castéran, P.: Interactive Theorem Proving and Program Development, Coq'Art:the Calculus of Inductive Constructions. Springer, Heidelberg (2004)
2. Bertot, Y., Magaud, N., Zimmermann, P.: A Proof of GMP Square Root. J. Autom. Reasoning 29(3-4), 225–252 (2002)
3. Brent, R.P., Zimmermann, P.: Modern Computer Arithmetic (2006) (in preparation), http://www.loria.fr/zimmerma/mca/pub226.html
4. Coq development team. The Coq Proof Assistant Reference Manual, version 8.1. (2006)
5. Cruz-Filipe, L., Geuvers, H., Wiedijk, F.: C-CoRN: The Constructive Coq Repository at Nijmegen. In: Asperti, A., Bancerek, G., Trybulec, A. (eds.) MKM 2004. LNCS, vol. 3119, pp. 88–103. Springer, Heidelberg (2004)
6. Démidovitch, B., Maron, I., et al.: Éléments de calcul numérique. Mir - Moscou (1979)
7. Fleuriot, J.D.: On the mechanization of real analysis in Isabelle/HOL. In: Harrison, J., Aagaard, M. (eds.) TPHOLs 2000. LNCS, vol. 1869, pp. 146–162. Springer, Heidelberg (2000)
8. Gamboa, R., Kaufmann, M.: Nonstandard Analysis in ACL2. Journal of automated reasoning 27(4), 323–428 (2001)
9. Giménez, E.: Codifying guarded definitions with recursive schemes. In: Dybjer, P., Nordström, B., Smith, J. (eds.) TYPES 1994. LNCS, vol. 996, pp. 39–59. Springer, Heidelberg (1995)
10. Harrison, J.: Theorem Proving with the Real Numbers. Springer, Heidelberg (1998)
11. Harrison, J.: Formal verification of square root algorithms. Formal Methods in System Design 22(2), 143–153 (2003)
12. Hur, N., Davenport, J.H.: A generic root operation for exact real arithmetic. In: Blanck, J., Brattka, V., Hertling, P. (eds.) CCA 2000. LNCS, vol. 2064, pp. 82–87. Springer, Heidelberg (2001)
13. Julien, N.: Certified exact real arithmetic using co-induction in arbitrary integer base. In: Garrigue, J., Hermenegildo, M.V. (eds.) FLOPS 2008. LNCS, vol. 4989, pp. 48–63. Springer, Heidelberg (2008)
14. Kaliszyk, C., O'Connor, R.: Computing with classical real numbers. In: CoRR, abs/0809.1644 (2008)
15. Lester, D.R.: Real Number Calculations and Theorem Proving. In: Mohamed, O.A., Muñoz, C., Tahar, S. (eds.) TPHOLs 2008. LNCS, vol. 5170, pp. 215–229. Springer, Heidelberg (2008)
16. Mayero, M.: Formalisation et automatisation de preuves en analyses reelle et numerique. Ph.D thesis, Université de Paris VI (2001)
17. Niqui, M.: Coinductive formal reasoning in exact real arithmetic. Logical Methods in Computer Science 4(3-6), 1–40 (2008)
18. O'Connor, R.: Certified Exact Transcendental Real Number Computation in Coq. In: Mohamed, O.A., Muñoz, C., Tahar, S. (eds.) TPHOLs 2008. LNCS, vol. 5170, pp. 246–261. Springer, Heidelberg (2008)
19. Paşca, I.: A Formal Verification for Kantorovitch's Theorem. Journées Francophones des Langages Applicatifs, 15–29 (2008)
20. Sawada, J., Gamboa, R.: Mechanical verification of a square root algorithm using taylor's theorem. In: Aagaard, M., O'Leary, J.W. (eds.) FMCAD 2002. LNCS, vol. 2517, pp. 274–291. Springer, Heidelberg (2002)

Construction of Büchi Automata for LTL Model Checking Verified in Isabelle/HOL

Alexander Schimpf[1], Stephan Merz[2], and Jan-Georg Smaus[1]

[1] University of Freiburg, Germany
{schimpfa,smaus}@informatik.uni-freiburg.de
[2] INRIA Nancy, France
Stephan.Merz@loria.fr

Abstract. We present the implementation in Isabelle/HOL of a translation of LTL formulae into Büchi automata. In automaton-based model checking, systems are modelled as transition systems, and correctness properties stated as formulae of temporal logic are translated into corresponding automata. An LTL formula is represented by a (generalised) Büchi automaton that accepts precisely those behaviours allowed by the formula. The model checking problem is then reduced to checking language inclusion between the two automata. The automaton construction is thus an essential component of an LTL model checking algorithm. We implemented a standard translation algorithm due to Gerth *et al.* The correctness and termination of our implementation are proven in Isabelle/HOL, and executable code is generated using the Isabelle/HOL code generator.

1 Introduction

The term *model checking* [2] subsumes several algorithmic techniques for the verification of reactive and concurrent systems, in particular with respect to properties expressed as formulae of temporal logics. More specifically, the context of our work are LTL model checking algorithms based on Büchi automata [19]. In this approach, the system to be verified is modelled as a finite transition system and the property is expressed as a formula φ of *linear temporal logic* (LTL). The formula φ constrains executions, and the transition system is deemed correct (with respect to the property) is all its executions satisfy φ. After translating the formula into a Büchi automaton [1], the model checking problem can be rephrased in terms of language inclusion between the transition system (interpreted as a Büchi automaton) and the automaton representing φ or, technically more convenient, as an emptiness problem for the product of the transition system and the automaton representing $\neg\varphi$.

In this paper, we present a verified implementation in Isabelle/HOL of the classical translation algorithm due to Gerth *et al.* [7] of LTL formulae into Büchi automata.[1] The automaton translation is at the heart of automaton-based model

---

[1] The Isabelle sources on which our paper is based are available at http://www.informatik.uni-freiburg.de/~ki/papers/diplomarbeiten/LTL2LGBA.zip. Extensive documentation on Isabelle can be found at http://isabelle.in.tum.de. Throughout this paper *Isabelle* refers to *Isabelle/HOL*.

S. Berghofer et al. (Eds.): TPHOLs 2009, LNCS 5674, pp. 424–439, 2009.

checking algorithms, and an error in the design or the implementation of the translation algorithm compromises the soundness of the verdict returned by the model checker. Indeed, the original implementation of the translation proposed by Gastin and Oddoux [6] contained a flaw that went unnoticed for several years, despite wide-spread use within the Spin model checker. The purpose of our work is to demonstrate that it is feasible to obtain an executable program implementing such a translation from a formalisation in a modern interactive proof assistant. Assuming that the kernel of the proof assistant and the code generator are correct, we thus obtain a highly trustworthy implementation.

We chose the algorithm of Gerth *et al.* because it is well-known and representative of the problems that such algorithms pose. More recent algorithms such as [6] are known to behave better for larger LTL formulae, but they require additional automata-theoretic concepts, and we leave their formalisation as a worthwhile and challenging topic for future work.

The algorithm of Gerth *et al.* is based on the construction of a graph of nodes labelled with subformulae of the original formula, similar to a tableau construction [4]. Acceptance conditions on infinite runs complement the tableau and enforce "eventuality" (liveness) properties. The main theorem states that the generated automaton should accept precisely those words (system executions) that are models of the temporal formula. The correctness of the translation is by no means obvious; in fact, we already found proving the termination of the method to be quite challenging. In our formalisation, we limit ourselves to data structures and operations that are supported by the Isabelle code generator. In this way, extraction of executable code becomes straightforward, but we are limited to relatively low-level constructions.

The paper is organised as follows: In the next section, we provide some preliminaries on LTL and Büchi automata. In Sect. 3, we recall the algorithm proposed by Gerth *et al.* [7]. Section 4 presents our implementation of the algorithm. In Sect. 5, we discuss the proof of termination and correctness of this implementation. Section 6 concludes. The results presented in this paper were obtained within the Diploma Thesis of the first author [14].

2   Preliminaries

2.1   Linear Temporal Logic

Linear-time temporal logic LTL [13] is a popular formalism for expressing correctness properties about (runs of) reactive systems. It extends propositional logic by modal operators that refer to future points of time.

Definition 1. Let Prop be a finite, non-empty set of propositions. The set Φ of *LTL formulae* is inductively defined as follows:

- Prop $\subseteq \Phi$;
- if $\varphi \in \Phi$ and $\psi \in \Phi$, then $\neg\varphi \in \Phi$, $\varphi \vee \psi \in \Phi$, $X\varphi \in \Phi$ ("next φ"), and $\varphi \cup \psi \in \Phi$ ("φ until ψ").

Further logical connectives can be defined as abbreviations. In particular, we will use the propositional constants $\top$ (true) and $\bot$ (false), as well as the operators $\wedge$ and $\vee$ ("release"), which are the duals of $\vee$ and U.

The semantics of an LTL formula is defined with respect to a *(temporal) interpretation* $\xi = a_0 a_1 \ldots$, which is an ω-word[2] over 2^{Prop}, consisting of propositional interpretations $a_i \in 2^{\mathsf{Prop}}$. The set a_0 contains exactly those propositions that are true in the initial state of the temporal interpretation ξ, a_1 gives the propositions true in the second state, and so on. When $\xi = a_0 a_1 \ldots$, we write ξ_i for a_i and $\xi|_i$ for the suffix $a_i a_{i+1} \ldots$, which is itself a temporal interpretation.

Definition 2. The relation $\xi \models \varphi$ ("ξ *is a model of* φ" or "φ *holds of* ξ") is inductively defined as follows:

$$
\begin{array}{lll}
\xi \models p & \text{iff} & p \in \xi_0 \quad (p \in \mathsf{Prop}) \\
\xi \models \neg\varphi & \text{iff} & \xi \not\models \varphi \\
\xi \models \varphi \vee \psi & \text{iff} & \xi \models \varphi \text{ or } \xi \models \psi \\
\xi \models \mathsf{X}\varphi & \text{iff} & \xi|_1 \models \varphi \\
\xi \models \varphi \,\mathsf{U}\, \psi & \text{iff} & \text{there exists } i \in \mathbb{N} \text{ such that } \xi|_i \models \psi \text{ and } \xi|_j \models \varphi \text{ for all } 0 \le j < i.
\end{array}
$$

2.2   Generalised Büchi Automata

The automata-theoretic approach to LTL model checking [19] relies on translating LTL formulae φ to Büchi automata $\mathcal{A}_\varphi$ such that a word ξ is accepted by $\mathcal{A}_\varphi$ if and only if $\xi \models \varphi$. The following variant of Büchi automata underlies the algorithm by Gerth et al. [7].

Definition 3. A *generalised Büchi automaton* (GBA) $\mathcal{A}$ is tuple (Q, I, δ, F) where:

- Q is a finite set of states;
- $I \subseteq Q$ is the set of initial states;
- $\delta \subseteq Q \times Q$ is the transition relation;
- $F \subseteq 2^Q$ is the set of acceptance sets (the *acceptance family*).

An ω-word σ over Q is called *path* of $\mathcal{A}$ if $\sigma_0 \in I$ and $(\sigma_i, \sigma_{i+1}) \in \delta$ for all $i \in \mathbb{N}$. The *limit* of ω-word σ is given as $\mathrm{limit}(\sigma) := \{q \mid \exists_\infty n. \, \sigma_n = q\}$[3]. The GBA $\mathcal{A}$ *accepts* a path σ of $\mathcal{A}$ if $\mathrm{limit}(\sigma) \cap M \neq \emptyset$ holds for all $M \in F$.

Observe that Def. 3 does not mention an alphabet. Instead, it is conventional to label automaton states by sets of propositional interpretations and use these labels to define the acceptance of a temporal interpretation by a GBA. Formally, this is achieved by the following definition, where $\mathcal{D}$ is chosen as 2^{Prop}.

Definition 4. A *labelled generalised Büchi automaton* (LGBA) is given by a triple $(\mathcal{A}, \mathcal{D}, \mathcal{L})$ where:

---

[2] An ω-word over alphabet Σ is a sequence $s_0 s_1 \ldots$ where $s_i \in \Sigma$ for all $i \in \mathbb{N}$.

[3] The symbol $\exists_\infty$ means "there are infinitely many".

- $\mathcal{A} = (Q, I, \delta, F)$ is a GBA;
- $\mathcal{D}$ is a finite set of labels;
- $\mathcal{L} : Q \to 2^{\mathcal{D}}$ is the label function.

A path σ of $\mathcal{A}$ is *consistent with* an ω-word ξ over $\mathcal{D}$ if $\xi_i \in \mathcal{L}(\sigma_i)$ for all $i \in \mathbb{N}$. An LGBA *accepts* an ω-word ξ over $\mathcal{D}$ iff it (more precisely, its underlying GBA) accepts some path of $\mathcal{A}$ that is consistent with ξ.

In model checking, systems are modelled as Kripke structures, that is, finite transition systems whose states are labelled with propositional interpretations. A Kripke structure $\mathcal{K}$ is an LGBA whose underlying GBA has a trivial (empty) acceptance family, and whose label function assigns a single propositional interpretation to every state. Assuming that the LGBA $\mathcal{A}$ represents the complement of the LTL formula φ ($\mathcal{A}$ accepts precisely those executions of which φ does not hold), $\mathcal{K}$ is a model of φ if no execution is accepted by both $\mathcal{K}$ and $\mathcal{A}$, i.e. if the intersection of the languages accepted by the two automata is empty.

3   Generating an LGBA for an LTL Formula

We recall the algorithm proposed by Gerth *et al.* [7] for computing an LGBA $\mathcal{A}_{\varphi}$ (with set of labels 2^{Prop}) for an LTL formula φ such that $\mathcal{A}_{\varphi}$ accepts a temporal interpretation ξ iff $\xi \vDash \varphi$.

The construction of $\mathcal{A}_{\varphi}$ proceeds in three stages. First, one builds the graph of the underlying GBA, using a procedure similar to a tableau construction [4]. Second, the function for labelling states of the LGBA is defined. Finally, the acceptance family is determined based on the set of "until" subformulae of φ. We now describe each stage in more detail.

The first step builds a graph of nodes (which will become the automaton states) that contain subformulae of φ. Intuitively, a node "promises" that the formulae it contains hold of any temporal interpretation that has an accepting run starting at that node. The construction is essentially based on "recursion laws" of LTL such as

$$\mu \, \mathsf{U} \, \psi \; \leftrightarrow \; \psi \vee (\mu \wedge \mathsf{X}(\mu \, \mathsf{U} \, \psi)) \tag{1}$$

that are used to split a promised formula into promises for the current state and for the successor state. The initial states of the automaton will be precisely those nodes that promise φ.

Without loss of generality, we assume that φ is given in *negation normal form* (NNF), i.e. the negation symbol is only applied to propositions. Transformation to NNF is straightforward once we include the dual operators $\wedge$ and $\vee$ among the set of logical connectives, using laws such as $\neg(\varphi \vee \psi) \equiv \neg\varphi \wedge \neg\psi$.

Gerth *et al.* [7] represent each node of the graph by a record with the following fields:

- *Name*: a unique identifier of the node.

- *Incoming*: the set of names of all nodes that have an edge pointing to the current node. Using this field, the entire graph is represented as the set of its nodes.
- *New*: A set of LTL formulae promised by this node but that have not yet been processed. This set is used during the construction and is empty for all nodes of the final graph.
- *Old*: A set of LTL formulae promised by this node and that have already been processed.
- *Next*: A set of LTL formulae that all successor nodes must promise.
- *Father*: During the construction, nodes will be split. This field contains the name of the node from which the current one has been split. It is used by Gerth *et al.* solely for reasoning about the algorithm, and we will not mention it any further.

The algorithm successively moves formulae from *New* to *Old*, decomposing them, and inserting subformulae into *New* and *Next* as appropriate. When the *New* field is empty, a successor node is generated whose *New* field equals the *Next* field of the current node. The algorithm maintains a list of all nodes generated so far to avoid generating duplicate nodes; this is essential for ensuring termination of the algorithm. More formally, the algorithm is realised by the function expand whose pseudo-code is reproduced in Fig. 1. For reasons of space and clarity, we omit some parts of the code in this presentation, in particular, some of the cases for the currently considered formula η, while preserving the original line numbering.

The automaton graph is constructed by the following function call:

$$\text{expand}([\, Name \Leftarrow \textbf{new\_name}(), Incoming \Leftarrow \{\text{init}\}, \qquad (2)$$
$$New \Leftarrow \{\varphi\}, Old \Leftarrow \emptyset, \Leftarrow Next \Leftarrow \emptyset\,], \emptyset)$$

where φ is the input LTL formula and init is a reserved identifier: all nodes whose *Incoming* field contains init will be initial states of the automaton.

In the second step of the construction, we define the function labelling the nodes with sets of propositional interpretations, each represented as the set of propositions that evaluate to true. The label of a node q is defined as the set of interpretations that are compatible with $Old(q)$. Formally, let

$$Pos(q) = Old(q) \cap \mathsf{Prop} \quad \text{and} \quad Neg(q) = \{\eta \in \mathsf{Prop} \mid \neg\eta \in Old(q)\}.$$

A propositional interpretation X is compatible with q iff it satisfies all atomic propositions in $Pos(q)$ but none in $Neg(q)$. This motivates the definition

$$\mathcal{L}(q) = \{X \subseteq \mathsf{Prop} \mid X \supseteq Pos(q), \; X \cap Neg(q) = \emptyset\}. \qquad (3)$$

It remains to define the acceptance family of the LGBA. Reconsider the "recursion law" (1) for the U operator, which is implemented by lines 20–27 of the code of Fig. 1. Every node "promising" a formula $\mu \cup \psi$ has one successor promising ψ and a second successor promising μ and $X(\mu \cup \psi)$. Thus, the graph of the LGBA may contain paths such that all nodes along the path promise μ

```
 3: function expand(Node, Nodes_Set)
 4:   if New(Node)=∅ then
 5:     if ∃ND∈Nodes_Set with Old(ND)=Old(Node) and Next(ND)=Next(Node) then
 6:       Incoming(ND):=Incoming(ND)∪Incoming(Node);
 7:         return(Nodes_Set)
 8:     else return(expand([Name⇐new_name(), Incoming⇐{Name(Node)},
10:         New⇐Next(Node), Old⇐ ∅, Next⇐ ∅], {Node}∪Nodes_Set))
11:   else
12:     let η ∈New(Node);
13:     New(Node):=New(Node)\{η};
14:     case η of
15:     ¬P ⇒
18:         Old(Node):=Old(Node)∪{η};
19:         return(expand(Node, Nodes_Set));
20:     η = μ U ψ ⇒
21:       Node1:=[Name⇐new_name(), Incoming⇐Incoming(Node),
22:         New⇐New(Node)∪({μ}\ Old(Node))
23:         Old⇐Old(Node)∪{η}, Next⇐Next(Node)∪{η}];
24:       Node2:=[Name⇐new_name(), Incoming⇐Incoming(Node),
25:         New⇐New(Node)∪({ψ}\ Old(Node))
26:         Old⇐Old(Node)∪{η}, Next⇐Next(Node)];
27:         return(expand(Node2, expand(Node1, Nodes_Set)));
32: end expand
```

Fig. 1. The algorithm by Gerth *et al.* [7] (incomplete)

but no node promises ψ. Such paths are not models of $\mu \cup \psi$, which requires ψ to be true eventually, and the acceptance family is defined in order to exclude them. Formally, we define for each formula $\mu \cup \psi$ the set of nodes

$$F_{\mu U \psi} \; = \; \{q \in Q \mid \mu \cup \psi \notin Old(q) \text{ or } \psi \in Old(q)\}, \tag{4}$$

and define the acceptance family F as

$$F \; = \; \{F_{\mu U \psi} \mid \mu \cup \psi \text{ is a subformula of } \varphi\}. \tag{5}$$

4   Implementation in Isabelle

4.1   LTL Formulae

We represent LTL formulae in Isabelle as an inductive data type. For the purposes of this presentation, we restrict to NNF formulae, although our full development also includes unrestricted LTL formulae and NNF transformation. For simplicity, we represent atomic propositions as strings; alternatively, the type of propositions could be made a parameter of the data type definition.

```
datatype
  frml = LTLTrue                 ("true")
       | LTLFalse                ("false")
       | LTLProp string          ("prop'(_')")
       | LTLNProp string         ("nprop'(_')")
       | LTLAnd frml frml        ("_ and _")
       | LTLOr frml frml         ("_ or _")
       | LTLNext frml            ("X _")
       | LTLUntil frml frml      ("_ U _")
       | LTLUDual frml frml      ("_ V _")
```

The above definition includes the concrete syntax for each clause of the data type. For example, (X prop(''p'')) and prop(''q'') would be the Isabelle representation of $(Xp) \wedge q$.

We next introduce types for representing ω-words and temporal interpretations, and define the semantics of LTL formulae by a straightforward primitive recursive function definition.

```
types
   'a word = nat ⇒ 'a
   interprt = "(string set) word"
```

fun semantics :: "[interprt, frml] ⇒ bool" ("_ ⊨ _" [80,80] 80)
where

```
     "ξ ⊨ true = True"
   | "ξ ⊨ false = False"
   | "ξ ⊨ prop(q) = (q∈ξ(0))"
   | "ξ ⊨ nprop(q) = (q∉ξ(0))"
   | "ξ ⊨ φ and ψ = (ξ ⊨ φ ∧ ξ ⊨ ψ)"
   | "ξ ⊨ φ or ψ = (ξ ⊨ φ ∨ ξ ⊨ ψ)"
   | "ξ ⊨ X φ = (suffix 1 ξ ⊨ φ)"
   | "ξ ⊨ φ U ψ = (∃i. suffix i ξ ⊨ ψ ∧ (∀j<i. suffix j ξ ⊨ φ))"
   | "ξ ⊨ φ V ψ = (∀i. suffix i ξ ⊨ ψ ∨ (∃j<i. suffix j ξ ⊨ φ))"
```

4.2 Büchi Automata

We now encode GBAs and LGBAs in Isabelle, following Defs. 3 and 4. In this encoding we approximate the set Q of states by a type parameter 'q, which will later be instantiated by the type representing the nodes of the graph. Although not enforced by the definition, finiteness of the actual set of nodes will be ensured by the termination of the algorithm, which produces states one by one.

GBAs and LGBAs are naturally modelled as *records* in Isabelle. Since we aim at producing executable code, all sets that appear in the original definition are represented as lists.

```
record 'q gba =
  initial :: "'q list"
```

```
trans    :: "('q × 'q) list"
accept   :: "'q list list"
```

```
record 'q lgba =
  gbauto :: "'q gba"
  label  :: "'q ⇀ string list list"
```

The node labelling function is represented by a *partial* function (denoted by the ⇀ symbol) because it needs only be defined over actual states of the LGBA, whereas the type 'q may contain extra elements.

It remains to define the runs and the acceptance family of (L)GBAs. The following definitions are a straightforward transcription of Def. 3: the utility function set from the Isabelle library computes the set of list elements, the limit function is defined as indicated in Def. 3.

definition gba_path :: "['q gba, 'q word] ⇒ bool" **where**
  "gba_path A σ
    ≡ σ 0 ∈ set (initial A) ∧
      (∀n. ((σ n), σ (Suc n)) ∈ set (trans A))"
definition gba_accept :: "['q gba, 'q word] ⇒ bool" **where**
  "gba_accept A σ
    ≡ gba_path A σ ∧
      (∀i<length (accept A). limit σ ∩ set (accept A!i) ≠ {})"

The acceptance condition for an LGBA is defined in a similar fashion. Finally, the predicate lgba_accept characterises the language of an LGBA: a temporal interpretation ξ is accepted by the LGBA A if there exists some path σ that is accepted by the GBA underlying A and that is consistent with ξ (cf. Def. 4).

definition lgba_accept :: "['q lgba, interprt] ⇒ bool"
where
  "lgba_accept A ξ
    ≡ ∃σ. (∀i. ξ i ∈ set (map set (the (label A (σ i)))))
        ∧ gba_accept (gbauto A) σ"

The use of the function "the" in the above code is a technicality related to the fact that the node labelling function is, in principle, partial. What matters is that "the (label A (σ i))" is of type string list list.

4.3   Translation from LTL to LGBA

We now formalise in Isabelle the algorithm due to Gerth *et al.* that we have presented informally in Sect. 3. As discussed there, three elementary steps have to be addressed:

- construct the graph of the underlying GBA using the expand function;
- define the acceptance family of the GBA;

```
function (sequential) expand :: "[cnode, node list] ⇒ node list"
where
  "expand ([], n) ns
   = (if (∃nd∈set ns. set (old nd) = set (old n) ∧
                      set (next nd) = set (next n))
      then upd_nds (λn nd. set (old nd) = set (old n) ∧
                          set (next nd) = set (next n)) ns n
      else expand (next n,
                   (|name = Suc(name n),
                     incoming = [name n],
                     old = [],
                     next = []|)) (n#ns))"

| "expand ((nprop(q))#fs, n) ns
   = expand (fs, n(| old := (nprop(q))#(old n) |)) ns"

| "expand ((μ U ψ) #fs, n) ns
   = (let nds = expand (μ#fs,
                        n(| old := (μ U ψ)#(old n),
                            next := (μ U ψ)#(next n) |)) ns
      in expand (ψ#fs,
                 n(| name := ...,
                     old := (μ U ψ)#(old n) |)) nds)"
```

Fig. 2. The Isabelle implementation of expand, simplified

- compute the labelling of the states of the LGBA with sets of propositional interpretations.

The algorithm expand constructs a graph, represented as a set of nodes. In Isabelle, we again use lists instead of finite sets in order to simplify code generation. We represent node names as integers, and model a node as a record containing the fields introduced in Sect. 3. We omit the *Father* field, which is unnecessary for the construction of the graph. We also replace the field *New*, which is used only during the construction, by an extra argument to the expand function. More precisely, the first argument of the function is of type cnode, defined as a pair of a formula list and a node.

```
record node =
  name :: nat
  incoming :: "nat list"
  old :: "frml list"
  next :: "frml list"
types cnode = "frml list * node"
```

Figure 2 contains the fragment of the definition of function expand in Isabelle that corresponds to the pseudo-code shown in Fig. 1. The function upd_nds merges the `incoming` fields of the current node with those of the already constructed nodes whose `old` and `next` fields agree with those of the current node.

For the sake of presentation, the code shown in Fig. 2 is somewhat simplified with respect to our Isabelle theories: the actual definition produces a pair consisting of a list of nodes and the highest used node name, which is used in the (omitted) definition of the name of the node created in the second call to `expand` in the clause for "until" formulae. Moreover, the actual definition checks for duplicates whenever a formula is added to the `old` or `next` components of a node.

The graph for an LTL formula is computed by the function `create_graph`, which in analogy to (2) is defined as

definition `create_graph :: "frml ⇒ node list"`
where
  `"create_graph φ`
    `≡ expand ([φ], (| name = 1, incoming = [0],`
                 `old = [], next = [] |)) []"`

We now address the second problem, i.e. the computation of the acceptance family for an LTL formula and a graph represented as a list of nodes. The following function `accept_family` is a quite direct transcription of the definition of the acceptance family in (5):

definition `accept_family :: "[frml, node list] ⇒ node list list"`
where
  `"accept_cond φ ns`
    `≡ map (λη. case η of`
            `_ U ψ ⇒ [q←ns. η∈set(old q) ⟶ ψ∈set(old q)])`
    `(all_until_frmls φ)"`

where `all_until_frmls` computes the list of "until" subformulae of the argument formula, without duplicates. It is now straightforward to define a function `create_gba` that constructs a GBA (of type `node gba`) from a node list representing the graph.

It remains to compute the function labelling the nodes with sets of propositional interpretations, in order to obtain an LGBA. The following definitions implement the labelling defined by (3) in a straightforward way.

definition
  `gen_label :: "[string list list, node] ⇒ string list list"`
where
  `"gen_label lbls n`
    `≡ [xs←lbls. set (pos_props (old n)) ⊆ set xs`
             `∧ list_inter xs (neg_props (old n)) = []]"`
definition
  `create_lgba :: "frml ⇒ node lgba"`
where

```
"create_lgba φ
  ≡ (let ns = create_graph φ in
     (| gbauto = create_gba φ ns,
        label = [ns[↦]map (gen_label (list_Pow (get_props φ)))
                     ns] |))"
```

The auxiliary functions pos_props and neg_props compute the lists of positive and negative literals contained in a list of formulae; get_props computes the list of atomic propositions contained in a temporal formula.

4.4   Code Generation

We have set up our theories in such a way that they use only data types and operations supported by the code generator, except for certain tests that convert lists to sets. In order to make these tests executable, we derive some auxiliary lemmas such as

```
lemma [code inline]:
  "set xs ⊆ set ys ⟷ list_all (λx. x mem ys) xs"
lemma [code inline]:
  "set xs = set ys ⟷ set xs ⊆ set ys ∧ set ys ⊆ set xs"
```

After these preliminaries, executable code can be extracted by simply issuing the command

export_code create_lgba in OCaml file "ltl2lgba.ml"

from the Isabelle theory file. This command produces an OCaml module containing the function create_lgba and all definitions and functions on which that function depends.

In order to use this code we have manually written a parser and driver program that parses an LTL formula, calls the function create_lgba, and outputs the result. We have used this program to generate automata corresponding to formulae φ_n that are representative of the verification of liveness properties under fairness constraints[4]

$$\varphi_n \equiv \neg((\mathsf{GF}p_1 \wedge \ldots \wedge \mathsf{GF}p_n) \Longrightarrow \mathsf{G}(q \Longrightarrow \mathsf{F}r))$$

for atomic propositions p_i, q, and r.

We have compared our code with implementations of the algorithm of Gerth et al. that are available in the tools Spin (http://spinroot.com) and Wring (http://vlsi.colorado.edu). The running times (in seconds, on a dual-core notebook computer with a 2.4GHz CPU and 2GB of RAM) for translating φ_n are shown in the table on the right. However, this comparison is

	Our code	Spin	Wring
$n = 5$	30	> 1200	90
$n = 6$	540	> 1200	900

Table 1. Runtimes

---

[4] $\mathsf{F}\psi$ ("finally ψ") is an abbreviation for $\top\ \mathsf{U}\ \psi$; $\mathsf{G}\psi$ ("globally ψ") denotes $\neg\mathsf{F}\neg\psi$.

not quite fair, because the other tools go on to translate the LGBA to ordinary Büchi automata. We plan to formalise this additional (polynomial) translation in the future, but take the present results as an indication that the execution times of the implementation generated from Isabelle are not prohibitive.

We have used the LTL-to-Büchi translator testbench [17] for gaining additional confidence in our program, including the hand-written driver. As expected, our code passes all the tests.

5   Verifying the Automaton Construction

Our main motivation for implementing the algorithm in Isabelle is of course the possibility to verify the correctness of our definitions. Assuming we trust Isabelle's proof kernel and its code generator, we obtain a verified program for translating LTL formulae into LGBA. We outline the correctness proof in this section. In fact, we must address two subproblems: we prove that the function **expand** terminates on all arguments, and we show that a temporal interpretation is accepted by the resulting LGBA iff it is a model of the input formula.

5.1   Termination

HOL is a logic of total functions, and it is essential for consistency to prove that every function that we define terminates. Indeed, Isabelle inserts a termination predicate in all theorems that involve a function whose termination has not been proven. Termination of the **expand** function (cf. Fig. 2) is not obvious on first sight but, remarkably, is not discussed at all in the original paper [7].

Consider Fig. 2. A call to **expand** is of the form **expand (fs,n) ns**. Now in all cases of the definition, except the first one, some formula is removed from **fs**, suggesting a well-founded ordering based on the size of the list **fs**. (This observation is also true of the cases of the definition omitted in Fig. 2.)

However, that simple definition breaks down for the first case where argument **fs** equals **[]**. Indeed, the recursive call constructs a new node based on the contents of the **next** field of the node **n**. In this case, the termination argument must be based on the argument **ns** of the function call. The apparent difficulty here is that this list does not become shorter on recursive calls, but (potentially) longer, so it is not completely obvious how to define a well-founded order. The solution here is to find a suitable upper bound for the argument **ns**. This can be done using the fact that all the nodes that are ever constructed contain subformulae of the input formula φ in their fields **old** and **next**, the same holds for the argument **fs** of formulae to process, and no two different nodes containing the same formulae in their **old** and **next** fields are ever constructed. It follows that there are only finitely many possible nodes since there exist only finitely many distinct sets of subformulae of φ. Very roughly speaking, the well-founded order by which argument **ns** decreases is given by (LIM φ - **ns**) where LIM is a function that calculates the appropriate upper bound given an LTL formula φ. The actual definition of the upper bound, which appears in the definition of the ordering below, depends on the arguments of function expand, not the formula φ.

The two orderings are combined lexicographically, that is to say, either the argument ns decreases w.r.t. the ordering discussed above, or the ns argument stays the same and there is a decrease on the fs argument.

The termination proof is complicated further by the fact that we have a nested recursive call in the last case. This is obvious in line 27 in Fig. 1, but the let expression in Fig. 2 amounts to the same. We therefore start off by showing a partial termination property, which states that if expand terminates, then nds $\supseteq$ ns, where nds is the result computed by the inner call (see Fig. 2). This partial result is then used to show that the arguments of the outer recursive call are smaller according to the well-founded ordering explained above.

The termination order is formally defined in Isabelle as follows:

abbreviation
```
"expand_term_ord ≡
    inv_image (finite_psubset <*lex*> less_than)
               (λ(n, ns). (nds_limit n ns - (old_next_pair ' set ns),
                           size_frml_list (fst n)))"
```

We explain this definition. The termination order compares pairs of the form (n, ns) where n is a cnode and ns is a node list. This corresponds exactly to the argument types of expand. The function $\lambda(n, ns)$.... in the above definition turns (n, ns) into another pair, say (st, sz), where st is given by the old and next fields of all nodes in ns and *subtracting* those from the set of all *possible* old and next fields—i.e., st states "how far ns is from the limit". The second argument sz is simply the length of the list appearing as the first component of the pair n. To compare two pairs (n, ns) and (n', ns'), the function is used to compute the corresponding (st, sz) and (st', sz'), and those pairs are compared using a lexicographical combination of $\subseteq$ and $\leq$.

The formal termination proof takes about 500 lines of Isar proof script.

5.2   Correctness

We now address the proper correctness proof of the algorithm, whose idea is presented in the original paper [7]. We have to prove that the LGBA computed by function create_lgba φ accepts precisely those temporal structures that are a model of φ. Formally, this is expressed as the Isabelle theorem

theorem lgba_correct:
  assumes "$\forall i.\ \xi\ i \in$ Pow (set (get_props φ))"
    shows "lgba_accept (create_lgba φ) $\xi \longleftrightarrow \xi \models \varphi$".

The hypothesis of the theorem states that ξ is a temporal interpretation over 2^{Prop} where Prop is the set of atomic propositions that occur in φ (cf. Sect. 2.2).

As explained in Sect. 3, the idea of the construction is to construct nodes that "promise" certain formulae and to make sure that these promises are enforced along any path starting at that node. However, the graph construction by itself can ensure this only partly. For example, we can prove the following lemma about "until" formulae promised by a node:

```
lemma L4_2a:
  assumes "gba_path (gbauto (create_lgba φ)) σ"
      and "f U g ∈ set (old (σ 0))"
    shows "(∀i. {f, f U g} ⊆ set (old (σ i))
              ∧ g ∉ set (old (σ i)))
         ∨ (∃j. (∀i<j. {f, f U g} ⊆ set (old (σ i)))
              ∧ g ∈ set (old (σ j)))".
```

In other words, we know for any path that starts at a node promising formula $f \cup g$ that f and $f \cup g$ are promised as long as g is not promised. However, we cannot be sure that g will indeed be promised by some node along the path. We defined the acceptance family precisely in a way to make sure that such paths are non-accepting, and indeed we can prove the following stronger lemma about the *accepting* paths starting at a node promising some formula $f \cup g$:

```
lemma L4_2b:
  assumes "gba_path (gbauto (create_lgba φ)) σ"
      and "f U g ∈ set (old (σ 0))"
      and "gba_accept (gbauto (create_lgba φ)) σ"
    shows "∃j. (∀i<j. {f, f U g} ⊆ set (old (σ i)))
              ∧ g ∈ set (old (σ j))"
```

The proof of theorem `lgba_correct` above relies on similar lemmas for each temporal operator, and then proves by induction on the structure of LTL formulae that all formulae promised along an accepting path indeed hold of the corresponding suffix of the temporal interpretation. For the proof of the "if" direction of theorem `lgba_correct` we inductively construct an accepting path for any temporal interpretation satisfying a formula. The length of the overall correctness proof is about 4500 lines of Isar proof script. The effort of working out the Isabelle proofs was around four person months.

6   Conclusion

In this paper we have presented a formally verified definition of labelled generalised Büchi automata in the interactive proof assistant Isabelle. Our formalisation is based on the classical algorithm by Gerth et al. [7], and Isabelle can generate executable code from our definitions. In this way, we obtain a highly trustworthy program for a critical component of a model checking engine for LTL.

Few formalisations of similar translations have been studied in the literature. Schneider [15] presents a HOL conversion for LTL that produces a symbolic encoding of an LGBA, which can be used in connection with a symbolic (in particular BDD-based) model checker. In contrast, our implementation produces a full LGBA that can be used with explicit-state LTL model checkers. Moreover, it generates a stand-alone program that can be used independently of any particular proof assistant. The second author [11] previously presented a formalisation of weak alternating automata (WAA [12]), including a translation of LTL formulae into WAA. Due to their much richer combinatorial structure, WAA afford a rather straightforward LTL translation of linear complexity, whereas the translation into (generalised) Büchi automata is exponential. Indeed, the main contribution of [11] was the formalisation of a game-theoretic argument due to [10,18] that underlies a complementation procedure for WAA.

Since the translation of LTL formulae to Büchi automata is of exponential complexity, one cannot expect to translate large formulae. Fortunately, the formulae that express typical correctness properties of concurrent systems are quite small. Although efficiency was not of much concern to us during the development of our theories, our experiments so far indicate that the extracted program does not behave significantly worse than existing implementations of the algorithm of Gerth *et al.* Of course, several improvements to the code are possible. For example, we could represent the sets of propositional interpretations labelling the automaton states symbolically instead of through an explicit enumeration, for example using a Boolean function that checks whether an interpretation is consistent with the label. Optimisations at a lower level could be obtained by replacing the list representation of finite sets with a more efficient data structure.

More significant optimisations could be achieved by basing the construction on a different algorithm altogether. Although the construction of Gerth *et al.* is well known and widely implemented, several alternative constructions have been studied in the literature [3,16,6,5,8], and the algorithm presented in [6] is widely considered to behave best in practice. This algorithm makes use of more advanced automata-theoretic notions, including WAA and various simulation relations on WAA and Büchi automata. These concepts have wider applications than just the automata constructions used in model checkers, including the complementation of ω-automata [9] and the synthesis of concurrent systems.

Encouraged by the success we have had so far, we would indeed like to formalise the construction of [6] in future work. Our current formalisation will continue to serve as an important building block that contains essential, fundamental concepts.

References

1. Büchi, R.: On a decision method in restricted second-order arithmetic. In: Intl. Cong. Logic, Methodology, and Philosophy of Science 1960, pp. 1–12. Stanford University Press (1962)
2. Clarke, E.M., Grumberg, O., Peled, D.A.: Model Checking. MIT Press, Cambridge (2002)
3. Daniele, M., Giunchiglia, F., Vardi, M.: Improved automata generation for linear temporal logic. In: Halbwachs, N., Peled, D.A. (eds.) CAV 1999. LNCS, vol. 1633, pp. 249–260. Springer, Heidelberg (1999)

4. Fitting, M.C.: Proof Methods for Modal and Intuitionistic Logic. Synthese Library: Studies in Epistemology, Logic, Methodology and Philosophy of Science. D. Reidel, Dordrecht (1983)

5. Fritz, C.: Constructing Büchi automata from linear temporal logic using simulation relations for alternating Büchi automata. In: Ibarra, O.H., Dang, Z. (eds.) CIAA 2003. LNCS, vol. 2759, pp. 35–48. Springer, Heidelberg (2003)

6. Gastin, P., Oddoux, D.: Fast LTL to Büchi automata translation. In: Berry, G., Comon, H., Finkel, A. (eds.) CAV 2001. LNCS, vol. 2102, pp. 53–65. Springer, Heidelberg (2001)

7. Gerth, R., Peled, D., Vardi, M.Y., Wolper, P.: Simple on-the-fly automatic verification of linear temporal logic. In: Dembinski, P., Sredniawa, M. (eds.) 15th Intl. Symp. Protocol Specification, Testing, and Verification (PSTV 1996). IFIP Conference Proceedings, vol. 38, pp. 3–18. Chapman & Hall, Boca Raton (1996)

8. Gurumurthy, S., Kupferman, O., Somenzi, F., Vardi, M.Y.: On complementing nondeterministic Büchi automata. In: Geist, D., Tronci, E. (eds.) CHARME 2003. LNCS, vol. 2860, pp. 96–110. Springer, Heidelberg (2003)

9. Kupferman, O., Vardi, M.: Complementation constructions for nondeterministic automata on infinite words. In: Halbwachs, N., Zuck, L. (eds.) TACAS 2005. LNCS, vol. 3440, pp. 206–221. Springer, Heidelberg (2005)

10. Kupferman, O., Vardi, M.Y.: Weak alternating automata are not that weak. ACM Trans. Comput. Log. 2(3), 408–429 (2001)

11. Merz, S.: Weak alternating automata in Isabelle/HOL. In: Aagaard, M.D., Harrison, J. (eds.) TPHOLs 2000. LNCS, vol. 1869, pp. 424–441. Springer, Heidelberg (2000)

12. Muller, D., Saoudi, A., Schupp, P.: Weak alternating automata give a simple explanation of why most temporal and dynamic logics are decidable in exponential tim. In: 3rd IEEE Symp. Logic in Computer Science (LICS 1988), Edinburgh, Scotland, pp. 422–427. IEEE Press, Los Alamitos (1988)

13. Pnueli, A.: The temporal semantics of concurrent programs. Theoretical Computer Science 13, 45–60 (1981)

14. Schimpf, A.: Implementierung eines Verfahrens zur Erzeugung von Büchi-Automaten aus LTL-Formeln in Isabelle. Diplomarbeit, Albert-Ludwigs-Universität Freiburg (2008), http://www.informatik.uni-freiburg.de/~ki/papers/diplomarbeiten/schimpf-diplomarbeit-08.pdf

15. Schneider, K., Hoffmann, D.W.: A HOL conversion for translating linear time temporal logic to ω-automata. In: Bertot, Y., Dowek, G., Hirschowitz, A., Paulin, C., Théry, L. (eds.) TPHOLs 1999. LNCS, vol. 1690, pp. 255–272. Springer, Heidelberg (1999)

16. Somenzi, F., Bloem, R.: Efficient Büchi automata from LTL formulae. In: Halbwachs, N., Peled, D.A. (eds.) CAV 1999. LNCS, vol. 1633, pp. 257–263. Springer, Heidelberg (1999)

17. Tauriainen, H., Heljanko, K.: Testing LTL formula translation into Büchi automata. International Journal on Software Tools for Technology Transfer 4(1), 57–70 (2002), http://www.tcs.hut.fi/Software/lbtt/

18. Thomas, W.: Complementation of Büchi automata revisited. In: Rozenberg, G., Karhumäki, J. (eds.) Jewels are forever, Contributions on Theoretical Computer Science in Honor of Arto Salomaa, pp. 109–122. Springer, Heidelberg (2000)

19. Vardi, M.Y., Wolper, P.: Reasoning about infinite computations. Information and Computation 115(1), 1–37 (1994)

A Hoare Logic for the State Monad

Proof Pearl

Wouter Swierstra

Chalmers University of Technology
wouter@chalmers.se

Abstract. This pearl examines how to verify functional programs written using the state monad. It uses Coq's Program framework to provide strong specifications for the standard operations that the state monad supports, such as return and bind. By exploiting the monadic structure of such programs during the verification process, it becomes easier to prove that they satisfy their specification.

1   Introduction

Monads help structure functional programs. Yet proofs about monadic programs often start by expanding the definition of return and bind. This seems rather wasteful. If we exploit this structure when writing *programs*, why should we discard it when writing *proofs*? This pearl examines how to verify functional programs written using the state monad. It is my express aim to take advantage of the monadic structure of these programs to guide the verification process.

This pearl is a literate Coq script [15]. Most proofs have been elided from the typeset version, but a complete development is available from my homepage. Throughout this paper, I will assume that you are familiar with Coq's syntax and have some previous exposure to functional programming using monads [16].

2   The State Monad

Let me begin by motivating the state monad. Consider the following inductive data type for binary trees:

```
Inductive Tree (a : Set) : Set :=
  | Leaf : a → Tree a
  | Node : Tree a → Tree a → Tree a.
```

Now suppose we want to define a function that replaces every value stored in a leaf of such a tree with a unique integer, i.e., no two leaves in the resulting tree should share the same label.

The obvious solution, given by the relabel function below, keeps track of a natural number as it traverses the tree.

S. Berghofer et al. (Eds.): TPHOLs 2009, LNCS 5674, pp. 440–451, 2009.

Fixpoint relabel $(a : \mathsf{Set})\ (t : \mathsf{Tree}\ a)\ (s : \mathsf{nat}) : \mathsf{Tree}\ \mathsf{nat} * \mathsf{nat}$
  $:= \mathbf{match}\ t\ \mathbf{with}$
      | Leaf $\_ \Rightarrow (\mathsf{Leaf}\ s, 1 + s)$
      | Node $l\ r \Rightarrow \mathbf{let}\ (l', s') := \mathsf{relabel}\ a\ l\ s$
                     $\mathbf{in}\ \mathbf{let}\ (r', s'') := \mathsf{relabel}\ a\ r\ s'$
                     $\mathbf{in}\ (\mathsf{Node}\ l'\ r', s'')$
  end.

The relabel function uses its argument number as the new label for the leaves. To make sure that no two leaves get assigned the same number, the number returned at a leaf is incremented. In the Node case, the number is threaded through the recursive calls appropriately.

While this solution is correct, there is some room for improvement. It is all too easy to pass the wrong number to a recursive call, thereby forgetting to update the state. To preclude such errors, the *state monad* may be used to carry the number implicitly as the tree is traversed.

For some fixed type of state $s : \mathsf{Set}$, the state monad is:

Definition State $(a : \mathsf{Set}) : \mathsf{Type} := s \to a * s$.

A computation in the state monad State a takes an initial state as its argument. Using this initial state, it performs some computation yielding a pair consisting of a value of type a and a final state.

The two monadic operations, return and bind, are defined as follows:

Definition return $(a : \mathsf{Set}) : a \to \mathsf{State}\ a := \mathbf{fun}\ x\ s \Rightarrow (x, s)$.

Definition bind $(a\ b : \mathsf{Set}) : \mathsf{State}\ a \to (a \to \mathsf{State}\ b) \to \mathsf{State}\ b$
  $:= \mathbf{fun}\ c_1\ c_2\ s_1 \Rightarrow \mathbf{let}\ (x, s_2) := c_1\ s_1$
                     $\mathbf{in}\ c_2\ x\ s_2$.

The return function lifts any pure value into the state monad, leaving the state untouched. Two computations may be composed using the bind function. It passes both the state and the result arising from the first computation as arguments to the second computation.

In line with the notation used in Haskell [11], I will use a pair of infix operators to write monadic computations. Instead of bind, I will sometimes write $\gg\!\!=$, a right-associative infix operator. Secondly, I will write $c_1 \gg c_2$ instead of bind $c_1\ (\mathbf{fun}\ \_ \Rightarrow c_2)$. This operator binds two computations, discarding the intermediate result.

Besides return and bind, there are two other operations that may be used to construct computations in the state monad:

Definition get : State $s := \mathbf{fun}\ s \Rightarrow (s, s)$.

Definition put : $s \to \mathsf{State}\ \mathsf{unit} := \mathbf{fun}\ s\ \_ \Rightarrow (\mathsf{tt}, s)$.

The get function returns the current state, whereas put overwrites the current state with its argument.

We can now redefine the relabelling function to use the state monad as follows:

Fixpoint relabel (a : Set) (t : Tree a) : State nat (Tree nat)
  := **match** t **with**
     | Leaf _ $\Rightarrow$ get $\ggg$ **fun** n $\Rightarrow$
           put (S n) $\ggg$
           return (Leaf n)
     | Node l r $\Rightarrow$ relabel l $\ggg$ **fun** l' $\Rightarrow$
             relabel r $\ggg$ **fun** r' $\Rightarrow$
             return (Node l' r')
   end.

Note that the type variable s has been instantiated to nat – the state carried around by the relabelling function is a natural number. By using the state monad, we no longer need to pass around this number by hand. This definition is less error prone: all the 'plumbing' is handled by the monadic combinators.

3   The Challenge

How can we prove that this relabelling function is correct?

Before we can talk about correctness, we need to establish the specification that we expect the relabel function to satisfy. One way of formulating the desired specification is by defining the following auxiliary function that flattens a tree to a list of labels:

Fixpoint flatten (a : Set) (t : Tree a) : list a
  := **match** t **with**
     | Leaf x $\Rightarrow$ x :: nil
     | Node l r $\Rightarrow$ flatten l ++ flatten r
  end.

We will prove that for any tree t and number x, the list flatten (fst (relabel t x)) does not have any duplicates. This property does not completely characterise relabelling – we should also check that the argument tree has the same shape as the resulting tree. This is relatively straightforward to verify as the relabelling function clearly maps leaves to leaves and nodes to nodes. Proving that the resulting tree satisfies the proposed specification, however, is not so easy.

4   Decorating the State Monad

The relabel function in the previous section is simply typed. We can certainly use proof assistants such as Coq to formalise equational proofs about such functions. In this paper, however, I will take a slightly different approach.

In this paper I will use *strong specifications*, i.e., the type of the relabel function will capture information about its behaviour. Simultaneously completing the function definition and the proof that this definition satisfies its specification yields programs that are *correct by construction*. This approach to verification can be traced back to Martin-Löf [6].

To give a strong specification of the relabelling function, we decorate computations in the state monad with additional propositional information. Recall that the state monad is defined as follows:

Definition State (a : Set) : Type := $s \to a * s$.

We can refine this definition slightly: instead of accepting *any* initial state of type s, the initial state should satisfy a given precondition. Furthermore, instead of returning *any* pair, the resulting pair should satisfies a postcondition relating the initial state, resulting value, and final state. Bearing these two points in mind, we arrive at the following definition of a state monad enriched with Hoare logic [2, 3].

Definition Pre : Type := $s \to$ Prop.

Definition Post (a : Set) : Type := $s \to a \to s \to$ Prop.

Program Definition HoareState (pre : Pre) (a : Set) ($post$: Post a) : Set
:= **forall** i : { t : s | $pre\ t$ }, { (x, f) : $a * s$ | $post\ i\ x\ f$ }.

Coq uses the notation { x : a | $p\ x$ } for strong specifications. Such a specification is inhabited by a pair consisting of a value x of type a, together with a proof that x satisfies the property p.

The code presented here uses Coq's Program framework [12, 13]. Defining a function that manipulates strong specifications using the Program framework is a two stage process: once we have defined the computational fragment, we are presented with a series of proof obligations that must be fulfilled before the Program framework can generate a complete Coq term. When defining the computational fragment, we do not have to manipulate proofs, but rather focus on programming with the first components of the strong specifications. In fact, the definition of the HoareState type above already uses one aspect of the Program framework: a projection is silently inserted to extract a value of type s from the variable i in the postcondition.

Although we have defined the HoareState type, we still need to define the return and bind functions. The return function does not place any restriction on the input state; it simply returns its second argument, leaving the state intact:

Definition top : Pre := **fun** $s \Rightarrow$ True.

Program Definition return (a : Set)
: **forall** x, HoareState top a (**fun** $i\ y\ f \Rightarrow i = f \wedge y = x$)
:= **fun** $x\ s \Rightarrow (x, s)$.

This definition of the return is identical to the original definition of the state monad: we have only made its behaviour evident from its type. The Program framework automatically discharges the trivial proofs necessary to complete the definition.

The corresponding revision of bind is a bit more subtle. Recall that the bind of the state monad has the following type.

$$\text{State } a \rightarrow (a \rightarrow \text{State } b) \rightarrow \text{State } b$$

You might expect the definition of the revised bind function to have a type of the form:

$$\text{HoareState } P_1 \ a \ Q_1 \rightarrow (a \rightarrow \text{HoareState } P_2 \ b \ Q_2) \rightarrow \text{HoareState} \dots b \dots$$

Before we consider the precondition and postcondition of the resulting computation, note that we can generalise this slightly. In the above type signature, the second argument of the bind function is not dependent. We can parametrise P_2 and Q_2 by the result of the first computation:

$$\text{HoareState } P_1 \ a \ Q_1$$
$$\rightarrow (\textbf{forall } (x : a), \text{HoareState } (P_2 \ x) \ b \ (Q_2 \ x))$$
$$\rightarrow \text{HoareState} \dots b \dots$$

This generalisation allows the pre- and postconditions of the second computation to refer to the results of the first computation.

Now we need to choose a suitable precondition and postcondition for the composite computation returned by the bind function. To motivate the choice of pre- and postcondition, recall that the bind of the state monad is defined as follows:

Definition bind $(a \ b : \text{Set}) : \text{State } a \rightarrow (a \rightarrow \text{State } b) \rightarrow \text{State } b$
$:= \textbf{fun } c_1 \ c_2 \ s_1 \Rightarrow \textbf{let } (x, s_2) := c_1 \ s_1$
$\qquad \textbf{in } c_2 \ x \ s_2.$

The bind function starts by running the first computation, and subsequently feeds its result to the second computation. So clearly the precondition of the composite computation should imply the precondition of the first computation c_1 – otherwise we could not justify running c_1 with the initial state s_1. Furthermore the postcondition of the first computation should imply the precondition of the second computation – if this wasn't the case, we could not give grounds for the call to c_2. These considerations lead to the following choice of precondition for the composite computation:

$$\textbf{fun } s_1 \Rightarrow P_1 \ s_1 \wedge \textbf{forall } x \ s_2, Q_1 \ s_1 \ x \ s_2 \rightarrow P_2 \ x \ s_2$$

What about the postcondition? Recall that a postcondition is a relation between the initial state, resulting value, and the final state. We would expect the postcondition of both argument computations to hold after executing the composite computation resulting from a call to bind. This composite computation, however, cannot refer to the initial state passed to the second computation or the results of the first computation: it can only refer to its own initial state and results. To solve this we existentially quantify over the results of the first computation, yielding the below postcondition for the bind operation.

$$\textbf{fun } s_1 \ y \ s_3 \Rightarrow \textbf{exists } x, \textbf{exists } s_2, Q_1 \ s_1 \ x \ s_2 \wedge Q_2 \ x \ s_2 \ y \ s_3$$

In words, the postcondition of the composite computation states that there is an intermediate state s_2 and a value x resulting from the first computation, such that these satisfy the postcondition of the first computation Q_1. Furthermore, the postcondition of the second computation Q_2 relates these intermediate results to the final state s_3 and the final value y.

Once we have chosen the desired precondition and postcondition of bind, its definition is straightforward:

Program Definition bind : **forall** a b P_1 P_2 Q_1 Q_2,
    (HoareState P_1 a Q_1) $\rightarrow$
    (**forall** $(x : a)$, HoareState $(P_2$ $x)$ b $(Q_2$ $x))$ $\rightarrow$
    HoareState (**fun** $s_1 \Rightarrow P_1$ $s_1 \wedge$ **forall** x s_2, Q_1 s_1 x $s_2 \rightarrow P_2$ x s_2)
        b
        (**fun** s_1 y $s_3 \Rightarrow$ **exists** x, **exists** s_2, Q_1 s_1 x $s_2 \wedge Q_2$ x s_2 y s_3)
  := **fun** a b P_1 P_2 Q_1 Q_2 c_1 c_2 $s_1 \Rightarrow$
    match c_1 s_1 **with**
      $(x, s_2) \Rightarrow c_2$ x s_2
    end.

This definition does give rise to two proof obligations: the intermediate state s_2 must satisfy the precondition of the second computation c_2; the application c_2 x s_2 must satisfy the postcondition of bind. Both these obligations are fairly straightforward to prove.

Before we have another look at the relabel function, we redefine the two auxiliary functions get and put to use the HoareState type:

Program Definition get : HoareState top s (**fun** i x $f \Rightarrow i = f \wedge x = i$)
  := **fun** $s \Rightarrow (s, s)$.
Program Definition put $(x : s)$: HoareState top unit (**fun** $\_\_f \Rightarrow f = x$)
  := **fun** $\_ \Rightarrow (\text{tt}, x)$.

Both functions have the trivial precondition top. The postcondition of the get function guarantees that it will return the current state without modifying it. The postcondition of the put function declares that the final state is equal to put's argument.

5   Relabelling Revisited

Finally, we return to the original question: how can we prove that the relabel function satisfies its specification?

Using the HoareState type, we now arrive at the definition of the relabelling function presented in Figure 1. The function definition of relabel is identical to the version using the state monad in Section 3. The only novel aspect is the choice of pre- and postcondition.

As we do not need any assumptions about the initial state, we choose the trivial precondition top. The postcondition uses two auxiliary functions, size and

```
Fixpoint size (a : Set) (t : Tree a) : nat :=
  match t with
      | Leaf x ⇒ 1
      | Node l r ⇒ size  l + size  r
  end.
Fixpoint seq (x n : nat) : list nat :=
  match n with
      | 0 ⇒ nil
      | S k ⇒ x :: seq (S x) k
  end.
Program Fixpoint relabel (a : Set) (t : Tree a) :
  HoareState nat top
            (Tree nat)
            (fun i t f ⇒ f = i + size t ∧ flatten  t = seq i (size t))
  := match t with
      | Leaf x ⇒ get  ≫= fun n ⇒
                  put (n + 1) ≫
                  return (Leaf n)
      | Node l r ⇒ relabel  l ≫= fun l' ⇒
                   relabel  r ≫= fun r' ⇒
                   return (Node l' r')
  end.
```

Fig. 1. The revised definition of the relabel function

seq, and consists of two parts. First of all, the final state should be exactly size t larger than the initial state, where t refers to the resulting tree. Furthermore, when the relabelling function is given an initial state i, flattening t should yield the sequence $i, i + 1, ...i + \text{size } t$.

This definition gives rise to two proof obligations, one for each branch of the pattern match in the relabel function. In the Leaf case, the proof obligation is trivial. It is discharged automatically by the Program framework. To solve the remaining obligation, we need to apply several tactics to trigger β-reduction and introduce the assumptions. After giving the variables in the context more meaningful names, we arrive at the proof state in Figure 2.

To complete the proof, we must prove that the postcondition holds for the tree Node l r under the assumption that it holds for recursive calls to l and r. The first part of the conjunction follows immediately from the assumptions *finalRes*, *sizeR*, and *sizeL* and the associativity of addition. The second part of the conjunction is a bit more interesting. After applying the induction hypotheses, *flattenL* and *flattenR*, the remaining goal becomes:

$$================================$$
$$\text{seq } i \text{ (size } l) + \!\!\!+ \text{ seq } lState \text{ (size } r) = \text{seq } i \text{ (size } l + \text{size } r)$$

1 *subgoal*

i : nat
t : Tree nat
n : nat
l : Tree nat
$lState$: nat
$sizeL$: $lState = i +$ size l
$flattenL$: flatten $l =$ seq i (size l)
r : Tree nat
$rState$: nat
$sizeR$: $rState = lState +$ size r
$flattenR$: flatten $r =$ seq $lState$ (size r)
$finalState$: $rState = n$
$finalRes$: $t =$ Node l r
================================
$n = i +$ size $t \land$ flatten $t =$ seq i (size t)

Fig. 2. Proving the obligation of the relabelling function

To complete the proof we need to use the assumption *sizeL*. If we had chosen the obvious postcondition flatten $t =$ seq i (size t) we would not have been able to complete this proof. Once we apply *sizeL* we can use one last lemma to complete the proof:

Lemma SeqSplit : **forall** y x z, seq x $(y + z) =$ seq x y ++ seq $(x + y)$ z.

This lemma is easy to prove by induction on y.

It is interesting to note that extracting a Haskell program from this revised relabelling function yields the same extracted code as the original definition of relabel in Section 2. As Coq's extraction mechanism discards propositional information, using the HoareState type does introduce any computational overhead.

6   Wrapping It Up

Now suppose we need to show that relabel satisfies a weaker postcondition. For instance, consider the NoDup predicate on lists from the Coq standard libraries. A list satisfies the NoDup predicate if it does not contain duplicates. The predicate's definition is given below.

Inductive NoDup : list $a \rightarrow$ Prop :=
| NoDup_nil : NoDup nil
| NoDup_cons : **forall** x xs, $x \notin xs \rightarrow$ NoDup $xs \rightarrow$ NoDup $(x :: xs)$.

How can we prove that the tree resulting from a call to the relabelling function satisfies NoDup (flatten t)?

We cannot define a relabelling function that has this postcondition – the induction hypotheses are insufficient to complete the required proofs in the Node case. We can, however, weaken the postcondition and strengthen the precondition explicitly. In line with Hoare Type Theory [10, 9, 8], we call this operation do:

> **Program Definition** do $(s\ a : \mathsf{Set})\ (P_1\ P_2 : \mathsf{Pre}\ s)\ (Q_1\ Q_2 : \mathsf{Post}\ s\ a):$
> $(\textbf{forall}\ i, P_2\ i \rightarrow P_1\ i) \rightarrow (\textbf{forall}\ i\ x\ f, Q_1\ i\ x\ f \rightarrow Q_2\ i\ x\ f) \rightarrow$
>   HoareState $s\ P_1\ a\ Q_1 \rightarrow$ HoareState $s\ P_2\ a\ Q_2$
> $:= \textbf{fun}\ \_\ \_\ c \Rightarrow c.$

This function has no computational content. It merely changes the precondition and postcondition associated with a computation in the HoareState type. We can now define the final version of the relabelling function as follows:

> **Program Fixpoint** finalRelabel $(a : \mathsf{Set})\ (t : \mathsf{Tree}\ a):$
>   HoareState (top nat) (Tree nat) $(\textbf{fun}\ i\ t\ f \Rightarrow$ NoDup (flatten t))
> $:= \mathsf{do}\ \_\ \_\ (\text{relabel}\ a\ t).$

The precondition is unchanged. As a result, the first argument to the do function is trivial. To complete this definition, however, we need to prove that the postcondition can be weakened appropriately. This proof boils down to showing that the list seq i (size t) does not have any duplicates. Using one last lemma, **forall** $n\ x\ y, x < y \rightarrow \neg\mathsf{In}\ x$ (seq $y\ n$), we complete the proof.

7   Discussion

Related Work

This pearl draws inspiration from many different sources. Most notably, it is inspired by recent work on Hoare Type Theory [10, 9, 8]. Ynot, the implementation of Hoare Type Theory in Coq, postulates the existence of return, bind, and do to use Hoare logic to reason about functions that use mutable references. This paper shows how these functions may be *defined* in Coq, rather than postulated. Furthermore, the HoareState type generalises their presentation somewhat: where Hoare Type Theory has specifically been designed to reason about mutable references, this pearl shows that the HoareState type can be used to reason about *any* computation in the state monad.

The relabelling problem is taken from Hutton and Fulger [4], who give an equational proof. Their proof, however, revolves around defining an intermediate function relabel' that carries around an (infinite) list of fresh labels.

> relabel' : **forall** $a\ b$, Tree $a \rightarrow$ State (list b) (Tree b)

To prove that relabel meets the required specification, Hutton and Fulger prove various lemmas relating relabel' and relabel. It is not clear how their proof techniques can be adapted to other functions in the state monad.

Similar techniques have been used by Leroy [5] in the Compcert project. His solution, however, revolves around defining an auxiliary data type:

Inductive Res $(a : \mathsf{Set})\ (t : s) : \mathsf{Set} :=$
 | Error : Res $a\ t$
 | OK : $a \to$ **forall** $(t' : s), \mathsf{R}\ t\ t' \to$ Res $a\ t$.

Where R is some relation between states. Unfortunately, the bind of this monad yields less efficient extracted code, as it requires an additional pattern match on the Res resulting from the first computation. Using the HoareState type, it may be possible to rule out errors by strengthening the precondition, thereby eliminating the need for this additional pattern match. Furthermore, the HoareState type presented here is slightly more general as its postcondition may also refer to the result of the computation.

Similar monadic structures to the one presented here have appeared in the verification of the seL4 microkernel [1] and security protocol verification [14]. There are a few differences between these approaches and the development presented here. Firstly, the postconditions presented here are ternary relations between the initial state, result, and final state. As a result, we do not need to introduce auxiliary variables to relate intermediate results. Sprenger and Basin [14] construct a Hoare logic on top of a weakest-precondition calculus. They present a shallow embedding of a series of logical rules that describe how the return and bind behave. On the other hand, Cock et al. [1] present their rules are presented as predicate transformers, using Isabelle/HOL's verification condition generator to infer the weakest precondition of a computation. The approach taken here focuses on programming with strong specifications in type theory, where the type of a computation fixes the desired pre- and postcondition.

Further Work

I have not provided justification for the choice of pre- and postcondition of bind and return. Other choices are certainly possible. For instance, we could choose the following type for return:

forall x, HoareState top a (**fun** $i\ y\ f \Rightarrow$ True)

Clearly this is a bad choice – applying the return function will no longer yield any information about the computation. It would be interesting to investigate if the choices presented here are somehow canonical, for instance, by showing that the HoareState type forms a monad in some category of strong specifications. McKinna's thesis [7] on the categorical structure of strong specifications may form the starting point for such research.

Using the HoareState type to write larger programs will lead to larger proof obligations. For this approach to scale, it is important to provide a suitable set of custom tactics to alleviate the burden of proof. Some tactics that are already provided by the Program framework proved useful in the development presented here, but further automation might still be necessary.

Acknowledgements. I would like to thank Matthieu Sozeau for developing Coq's Program framework and for helping me to use it. Both Matthieu and Jean-Philippe Bernardy provided invaluable feedback on a draft version of this paper. I would like to thank Thorsten Altenkirch, Peter Hancock, Graham Hutton, and James McKinna for their useful suggestions. Finally, I would like to thank the anonymous referees for their extremely helpful reviews.

References

[1] Cock, D., Klein, G., Sewell, T.: Secure microkernels, state monads and scalable refinement. In: Munoz, C., Ait, O. (eds.) TPHOLs 2008. LNCS, vol. 5170, pp. 167–182. Springer, Heidelberg (2008)

[2] Floyd, R.W.: Assigning meanings to programs. Mathematical Aspects of Computer Science 19 (1967)

[3] Hoare, C.A.R.: An axiomatic basis for computer programming. Communications of the ACM 12(10), 576–580 (1969)

[4] Hutton, G., Fulger, D.: Reasoning about effects: seeing the wood through the trees. In: Proceedings of the Ninth Symposium on Trends in Functional Programming (2008)

[5] Leroy, X.: Formal certification of a compiler back-end, or: programming a compiler with a proof assistant. In: POPL 2006: 33rd Symposium on Principles of Programming Languages, pp. 42–54. ACM Press, New York (2006)

[6] Martin-Löf, P.: Constructive mathematics and computer programming. In: Proceedings of a discussion meeting of the Royal Society of London on Mathematical logic and programming languages, pp. 167–184. Prentice-Hall, Inc., Englewood Cliffs (1985)

[7] McKinna, J.: Deliverables: a categorical approach to program development in type theory. Ph.D thesis, School of Informatics at the University of Edinburgh (1992)

[8] Nanevski, A., Morrisett, G.: Dependent type theory of stateful higher-order functions. Technical Report TR-24-05, Harvard University (2005)

[9] Nanevski, A., Morrisett, G., Birkedal, L.: Polymorphism and separation in Hoare Type Theory. In: ICFP 2006: Proceedings of the Eleventh ACM SIGPLAN International Conference on Functional Programming (2006)

[10] Nanevski, A., Morrisett, G., Shinnar, A., Govereau, P., Birkedal, L.: Ynot: Reasoning with the awkward squad. In: ICFP 2008: Proceedings of the Twelfth ACM SIGPLAN International Conference on Functional Programming (2008)

[11] Peyton Jones, S. (ed.): Haskell 98 Language and Libraries: The Revised Report. Cambridge University Press, Cambridge (2003)

[12] Sozeau, M.: Subset coercions in Coq. In: Altenkirch, T., McBride, C. (eds.) TYPES 2006. LNCS, vol. 4502, pp. 237–252. Springer, Heidelberg (2007)

[13] Sozeau, M.: Un environnement pour la programmation avec types dépendants. Ph.D thesis, Université de Paris XI (2008)

[14] Sprenger, C., Basin, D.: A monad-based modeling and verification toolbox with application to security protocols. In: Schneider, K., Brandt, J. (eds.) TPHOLs 2007. LNCS, vol. 4732, pp. 302–318. Springer, Heidelberg (2007)

[15] The Coq development team. The Coq proof assistant reference manual. LogiCal Project, Version 8.2 (2008)

[16] Wadler, P.: The essence of functional programming. In: POPL 1992: Conference Record of the Nineteenth Annual ACM SIGPLAN-SIGACT Symposium on Principles of Programming Languages, pp. 1–14 (1992)

Certification of Termination Proofs Using CeTA*

René Thiemann and Christian Sternagel

Institute of Computer Science, University of Innsbruck, Austria
{rene.thiemann,christian.sternagel}@uibk.ac.at

Abstract. There are many automatic tools to prove termination of term rewrite systems, nowadays. Most of these tools use a combination of many complex termination criteria. Hence generated proofs may be of tremendous size, which makes it very tedious (if not impossible) for humans to check those proofs for correctness.

In this paper we use the theorem prover Isabelle/HOL to automatically certify termination proofs. To this end, we first formalized the required theory of term rewriting including three major termination criteria: dependency pairs, dependency graphs, and reduction pairs. Second, for each of these techniques we developed an executable check which guarantees the correct application of that technique as it occurs in the generated proofs. Moreover, if a proof is not accepted, a readable error message is displayed. Finally, we used Isabelle's code generation facilities to generate a highly efficient and certified Haskell program, CeTA, which can be used to certify termination proofs without even having Isabelle installed.

1 Introduction

Termination provers for term rewrite systems (TRSs) became more and more powerful in the last years. One reason is that a proof of termination no longer is just some reduction order which contains the rewrite relation of the TRS. Currently, most provers construct a proof in the dependency pair framework which allows to combine basic termination techniques in a flexible way. Then a termination proof is a tree where at each node a specific technique has been applied. So instead of stating the precedence of some lexicographic path order (LPO) or giving some polynomial interpretation, current termination provers return proof trees which reach sizes of several megabytes. Hence, it would be too much work to check by hand whether these trees really form a valid proof.

That we cannot blindly trust the output of termination provers is regularly demonstrated: Every now and then some tool delivers a faulty proof for some TRS. But most often this is only detected if there is some other prover giving the opposite answer on the same TRS, i.e., that it is nonterminating. To solve this problem, in the last years two systems have been developed which automatically certify or reject a generated termination proof: CiME/Coccinelle [4,6] and Rainbow/CoLoR [3] where Coccinelle and CoLoR are libraries on rewriting for

---

* This research is supported by FWF (Austrian Science Fund) project P18763.

S. Berghofer et al. (Eds.): TPHOLs 2009, LNCS 5674, pp. 452–468, 2009.

Coq (`http://coq.inria.fr`), and CiME and Rainbow are used to convert proof trees into Coq-proofs which heavily rely on the theorems within those libraries.

$$\text{proof tree} \xrightarrow{\text{CiME/Rainbow}} \texttt{proof.v} \xrightarrow{\text{Coq + Coccinelle/CoLoR}} \text{accept/failure}$$

In this paper we present a new combination, CeTA/IsaFoR, to automatically certify termination proofs. Note that the system design has two major differences in comparison to the two existing ones. First, our library IsaFoR (*Isabelle Formalization of Rewriting*, containing 173 definitions, 863 theorems, and 269 functions) is written for the theorem prover Isabelle/HOL[1] [16] and not for Coq.

Second, and more important, instead of generating for each proof tree a new Coq-proof using the auxiliary tools CiME/Rainbow, our library IsaFoR contains several executable "check"-functions (within Isabelle) for each termination technique we formalized. We have formally proven that whenever such a check is accepted, then the termination technique is applied correctly. Hence, we do not need to create an individual Isabelle-proof for each proof tree, but just call the "check"-function for checking the whole tree (which does nothing else but calling the separate checks for each termination technique occurring in the tree). This second difference has several advantages:

- In the other two systems, whenever a proof is not accepted, the user just gets a Coq-error message that some step in the generated Coq-proof failed. In contrast, our functions deliver error messages using notions of term rewriting.
- Since the analysis of the proof trees in IsaFoR is performed by executable functions, we can just apply Isabelle's code-generator [11] to create a certified Haskell program [17], CeTA, leading to the following workflow.

$$\text{IsaFoR} \xrightarrow{\text{Isabelle}} \text{Haskell program} \xrightarrow{\text{Haskell compiler}} \text{CeTA}$$
$$\text{proof tree} \xrightarrow{\text{CeTA}} \text{accept/error message}$$

Hence, to use our certifier CeTA (*Certified Termination Analysis*) you do not have to install any theorem prover, but just execute some binary. Moreover, the runtime of certification is reduced significantly. Whereas the other two approaches take more than one hour to certify all (≤ 580) proofs during the last certified termination competition, CeTA needs less than two minutes for all (786) proofs that it can handle. Note that CeTA can also be used for modular certification. Each single application of a termination technique can be certified—just call the corresponding Haskell-function.

Concerning the techniques that have been formalized, the other two systems offer techniques that are not present in IsaFoR, e.g., LPO or matrix interpretations. Nevertheless, we also feature one new technique that has not been certified

---

[1] In the remainder of this paper we just write Isabelle instead of Isabelle/HOL.

so far. Whereas currently only the initial dependency graph estimation of [1] has been certified, we integrated the most powerful estimation which does not require tree automata techniques and is based on a combination of [9,12] where the function tcap is required. Initial problems in the formalization of tcap led to the development of etcap, an equivalent but more efficient version of tcap which is also beneficial for termination provers. Replacing tcap by etcap within the termination prover T$_T$T$_2$ [14] reduced the time to estimate the dependency graph by a factor of 2. We will also explain, how to reduce the number of edges that have to be inspected when checking graph decompositions.

Another benefit of our system is its robustness. Every proof which uses weaker techniques than those formalized in IsaFoR is accepted. For example, termination provers can use the graph estimation of [1], as it is subsumed by our estimation.

The paper is structured as follows. In Sect. 2 we recapitulate the required notions and notations of term rewriting and the dependency pair framework (DP framework). Here, we also introduce our formalization of term rewriting within IsaFoR. In Sect. 3–6 we explain our certification of the four termination techniques we currently support: dependency pairs (Sect. 3), dependency graph (Sect. 4), reduction pairs (Sect. 5), and combination of proofs in the dependency pair framework (Sect. 6). However, to increase readability we abstract from our concrete Isabelle code and present the checks for the techniques on a higher level. How we achieved readable error-messages while at the same time having maintainable Isabelle proofs is the topic of Sect. 7. We conclude in Sect. 8 where we show how CeTA is created from IsaFoR and where we give experimental data.

IsaFoR, CeTA, and all details about our experiments are available at CeTA's website http://cl-informatik.uibk.ac.at/software/ceta.

2   Formalizing Term Rewriting

We assume some basic knowledge of term rewriting [2]. Variables are denoted by x, y, z, etc., function symbols by f, g, h, etc., terms by s, t, u, etc., and substitutions by σ, μ, etc. Instead of $f(t_1, \ldots, t_n)$ we write $f(t_n)$. The set of all variables occurring in term t is denoted by $\mathcal{V}\mathrm{ar}(t)$. By $\mathcal{T}(\mathcal{F}, \mathcal{V})$ we denote the set of terms over function symbols from $\mathcal{F}$ and variables from $\mathcal{V}$.

In the following we give an overview of our formalization of term rewriting in IsaFoR. Our main concern is *termination* of rewriting. This property—also known as *strong normalization*—can be stated without considering the structure of terms. Therefore it is part of our Isabelle theory AbstractRewriting. An abstract rewrite system (ARS) is represented by the type ('a×'a)set in Isabelle. Strong normalization (SN) of a given ARS $\mathcal{A}$ is equivalent to the absence of an infinite sequence of $\mathcal{A}$-steps. On the lowest level we have to link our notion of strong normalization to the notion of well-foundedness as defined in Isabelle. This is an easy lemma since the only difference is the orientation of the relation, i.e., $\mathrm{SN}(\mathcal{A}) = \mathrm{wf}(\mathcal{A}^{-1})$. At this point we can be sure that our notion of strong normalization is valid.

Now we come to the level of first-order terms (in theory `Term`):

```
datatype ('f,'v)"term" = Var 'v | Fun 'f "('f,'v)term list"
```

Many concepts related to terms are formalized in `Term`, e.g., an induction scheme for terms (as used in textbooks), substitutions, contexts, the (proper) subterm relation etc.

By restricting the elements of some ARS to terms, we reach the level of TRSs (in theory `Trs`), which in our formalization are just binary relations over terms.

Example 1. As an example, consider the following TRS, encoding rules for subtraction and division on natural numbers.

$$\mathsf{minus}(x, 0) \to x \qquad\qquad \mathsf{div}(0, \mathsf{s}(y)) \to 0$$
$$\mathsf{minus}(\mathsf{s}(x), \mathsf{s}(y)) \to \mathsf{minus}(x, y) \quad \mathsf{div}(\mathsf{s}(x), \mathsf{s}(y)) \to \mathsf{s}(\mathsf{div}(\mathsf{minus}(x, y), \mathsf{s}(y)))$$

Given a TRS $\mathcal{R}$, $(\ell, r) \in \mathcal{R}$ means that ℓ is the lhs and r the rhs of a rule in $\mathcal{R}$ (usually written as $\ell \to r \in \mathcal{R}$). The *rewrite relation* induced by a TRS $\mathcal{R}$ is denoted by $\to_{\mathcal{R}}$ and has the following definition in IsaFoR:

Definition 2. *Term s rewrites to t by $\mathcal{R}$, iff there are a context C, a substitution σ, and a rule $\ell \to r \in \mathcal{R}$ such that $s = C[\ell\sigma]$ and $t = C[r\sigma]$.*

Note that this section contains the only parts where you have to trust our formalization, i.e., you have to believe that $\mathrm{SN}(\to_{\mathcal{R}})$ as defined in IsaFoR really describes "$\mathcal{R}$ is terminating."

3   Certifying Dependency Pairs

Before we introduce dependency pairs [1] formally and give some details about our Isabelle formalization, we recapitulate the ideas that led to the final definition (including a refinement proposed by Dershowitz [7]).

For a TRS $\mathcal{R}$, strong normalization means that there is no infinite derivation $t_1 \to_{\mathcal{R}} t_2 \to_{\mathcal{R}} t_3 \to_{\mathcal{R}} \cdots$. Additionally we can concentrate on derivations, where t_1 is minimal in the sense that all its proper subterms are terminating. Such terms are called *minimal nonterminating*. The set of all minimally nonterminating terms with respect to a TRS $\mathcal{R}$ is denoted by $\mathcal{T}_{\mathcal{R}}^{\infty}$. Observe that for every term $t \in \mathcal{T}_{\mathcal{R}}^{\infty}$ there is an initial part of an infinite derivation having a specific shape: A (possibly empty) derivation taking place below the root, followed by an application of some rule $\ell \to r \in \mathcal{R}$ at the root, i.e., $t \xrightarrow{>\epsilon*}_{\mathcal{R}} \ell\sigma \xrightarrow{\epsilon}_{\mathcal{R}} r\sigma$, for some substitution σ. Furthermore, since $r\sigma$ is nonterminating, there is some subterm u of r, such that $u\sigma \in \mathcal{T}_{\mathcal{R}}^{\infty}$, i.e., $r\sigma = C[u\sigma]$. Then the same reasoning can be used to get a root reduction of $u\sigma$, ..., cf. [1].

To get rid of the additional contexts C a new TRS, $\mathrm{DP}(\mathcal{R})$, is built.

Definition 3. *The set* DP($\mathcal{R}$) *of dependency pairs of* $\mathcal{R}$ *is defined as follows: For every rule* $\ell \to r \in \mathcal{R}$, *and every subterm* u *of* r *such that* u *is not a proper subterm of* ℓ *and such that the root of* u *is defined,*[2] $\ell^\sharp \to u^\sharp$ *is contained in* DP($\mathcal{R}$). *Here* $t^\sharp$ *is the same as* t *except that the root of* t *is marked with the special symbol* $\sharp$.

Example 4. The dependency pairs for the TRS from Ex. 1 consist of the rules

$$\mathsf{M}(\mathsf{s}(x), \mathsf{s}(y)) \to \mathsf{M}(x, y) \quad \text{(MM)} \qquad \mathsf{D}(\mathsf{s}(x), \mathsf{s}(y)) \to \mathsf{M}(x, y) \qquad\qquad \text{(DM)}$$
$$\mathsf{D}(\mathsf{s}(x), \mathsf{s}(y)) \to \mathsf{D}(\mathsf{minus}(x, y), \mathsf{s}(y)) \quad \text{(DD)}$$

where we write M instead of minus$^\sharp$ and D instead of div$^\sharp$ for brevity.

Note that after switching to '$\sharp$'-terms, the derivation from above can be written as $t^\sharp \to_{\mathcal{R}}^* \ell^\sharp \sigma \to_{\mathrm{DP}(\mathcal{R})} u^\sharp \sigma$. Hence every nonterminating derivation starting at a term $t \in \mathcal{T}_{\mathcal{R}}^\infty$ can be transformed into an infinite derivation of the following shape where all $\to_{\mathrm{DP}(\mathcal{R})}$-steps are applied at the root.

$$t^\sharp \to_{\mathcal{R}}^* s_1^\sharp \to_{\mathrm{DP}(\mathcal{R})} t_1^\sharp \to_{\mathcal{R}}^* s_2^\sharp \to_{\mathrm{DP}(\mathcal{R})} t_2^\sharp \to_{\mathcal{R}}^* \cdots \tag{1}$$

Therefore, to prove termination of $\mathcal{R}$ it suffices to prove that there is no such derivation. To formalize DPs in Isabelle we modify the signature such that every function symbol now appears in a plain version and in a $\sharp$-version.

```
datatype 'f shp = Sharp 'f ("_♯") | Plain 'f ("_@")
```

Sharping a term is done via

```
fun plain :: "('f,'v)term => ('f shp,'v)term"
where "plain(Var x)   = Var x"
   | "plain(Fun f ss) = Fun f@ (map plain ss)"
```

```
fun sharp :: "('f,'v)term => ('f shp,'v)term"
where "sharp(Var x)   = Var x"
   | "sharp(Fun f ss) = Fun f♯ (map plain ss)"
```

Thus $t^\sharp$ in Def. 3 is the same as sharp(t). Since the function symbols in DP($\mathcal{R}$) are of type 'f shp and the function symbols of $\mathcal{R}$ are of type 'f, it is not possible to use the same TRS $\mathcal{R}$ in combination with DP($\mathcal{R}$). Thus, in our formalization we use the lifting ID—that just applies plain to all lhss and rhss in $\mathcal{R}$.

Considering this technicalities and omitting the initial derivation $t^\sharp \to_{\mathcal{R}}^* s_1^\sharp$ from the derivation (1), we obtain

$$s_1^\sharp \to_{\mathrm{DP}(\mathcal{R})} t_1^\sharp \to_{\mathrm{ID}(\mathcal{R})}^* s_2^\sharp \to_{\mathrm{DP}(\mathcal{R})} t_2^\sharp \to_{\mathrm{ID}(\mathcal{R})}^* \cdots$$

and hence a so called infinite (DP($\mathcal{R}$), ID($\mathcal{R}$))-*chain*. Then the corresponding *DP problem* (DP($\mathcal{R}$), ID($\mathcal{R}$)) is called to be not *finite*, cf. [8]. Notice that in IsaFoR a DP problem is just a pair of two TRSs over arbitrary signatures—similar to [8].

---

[2] A function symbol f is *defined* (w.r.t. $\mathcal{R}$) if there is some rule $f(\ldots) \to r \in \mathcal{R}$.

In IsaFoR an infinite chain[3] and finite DP problems are defined as follows.

```
fun ichain where "ichain(P,R) s t σ = (∀i.
  (s i,t i) ∈ P ∧ (t i)·(σ i) →*_R (s(i+1))·(σ(i+1))"
```

```
fun finite_dpp where
  "finite_dpp(P,R) = (¬(∃s t σ. ichain (P,R) s t σ))"
```

where '$t \cdot \sigma$' denotes the application of substitution σ to term t.

We formally established the connection between strong normalization and finiteness of the initial DP problem $(DP(\mathcal{R}), ID(\mathcal{R}))$. Although this is a well-known theorem, formalizing it in Isabelle was a major effort.

Theorem 5. $\text{wf\_trs}(\mathcal{R}) \land \text{finite\_dpp}(DP(\mathcal{R}), ID(\mathcal{R})) \longrightarrow SN(\rightarrow_{\mathcal{R}})$.

The additional premise $\text{wf\_trs}(\mathcal{R})$ ensures two well-formedness properties for $\mathcal{R}$, namely that for every $\ell \rightarrow r \in \mathcal{R}$, ℓ is not a variable and that $\mathcal{V}ar(r) \subseteq \mathcal{V}ar(\ell)$.

At this point we can obviously switch from the problem of proving $SN(\rightarrow_{\mathcal{R}})$ for some TRS $\mathcal{R}$, to the problem of proving $\text{finite\_dpp}(DP(\mathcal{R}), ID(\mathcal{R}))$, and thus enter the realm of the DP framework [8]. Here, the current technique is to apply so-called *processors* to a DP problem, in order to get a set of simpler DP problems. This is done recursively, until the leafs of the so built tree consist of DP problems with empty $\mathcal{P}$-components (and therefore are trivially finite). For this to be correct, the applied processors need to be sound, i.e., every processor *Proc* has to satisfy the implication

$$(\forall p \in Proc(\mathcal{P}, \mathcal{R}). \text{ finite\_dpp}(p)) \longrightarrow \text{finite\_dpp}(\mathcal{P}, \mathcal{R})$$

for every input. The termination techniques that will be introduced in the following sections are all such (sound) processors.

So much to the underlying formalization. Now we will present how the check in IsaFoR certifies a set of DPs $\mathcal{P}$ that was generated by some termination tool for some TRS $\mathcal{R}$. To this end, the function `checkDPs` is used.

$$\text{checkDPs}(\mathcal{P}, \mathcal{R}) = \text{checkWfTRS}(\mathcal{R}) \land \text{computeDPs}(\mathcal{R}) \subseteq \mathcal{P}$$

Here `checkWfTRS` checks the two well-formedness properties mentioned above (the difference between `wf_trs` and `checkWfTRS` is that only the latter is executable) and `computeDPs` uses Def. 3, which is currently the strongest definition of DPs. To have a robust system, the check does not require that exactly the set of DPs w.r.t. to Def. 3 is provided, but any superset is accepted. Hence we are also able to accept proofs from termination tools that use a weaker definition of $DP(\mathcal{R})$. The soundness result of `checkDPs` is formulated as follows in IsaFoR.

Theorem 6. *If* `checkDPs`$(\mathcal{P}, \mathcal{R})$ *is accepted then finiteness of* $(\mathcal{P}, ID(\mathcal{R}))$ *implies* $SN(\rightarrow_{\mathcal{R}})$.

---

[3] We also formalized *minimal* chains, but here only present chains for simplicity.

4   Certifying the Dependency Graph Processor

One important processor to prove finiteness of a DP problem is based on the *dependency graph* [1,8]. The dependency graph of a DP problem $(\mathcal{P}, \mathcal{R})$ is a directed graph $\mathcal{G} = (\mathcal{P}, E)$ where $(s \rightarrow t, u \rightarrow v) \in E$ iff $s \rightarrow t, u \rightarrow v$ is a $(\mathcal{P}, \mathcal{R})$-chain. Hence, every infinite $(\mathcal{P}, \mathcal{R})$-chain corresponds to an infinite path in $\mathcal{G}$ and thus, must end in some strongly connected component (SCC) S of $\mathcal{G}$, provided that $\mathcal{P}$ contains only finitely many DPs. Dropping the initial DPs of the chain results in an infinite $(S, \mathcal{R})$-chain.[4] Hence, if for all SCCs S of $\mathcal{G}$ the DP problem $(S, \mathcal{R})$ is finite, then $(\mathcal{P}, \mathcal{R})$ is finite. In practice, this processor allows to prove termination of each block of mutual recursive functions separately.

To certify an application of the dependency graph processor there are two main challenges. First of all, we have to certify that a valid SCC decomposition of $\mathcal{G}$ is used, a purely graph-theoretical problem. Second, we have to generate the edges of $\mathcal{G}$. Since the dependency graph $\mathcal{G}$ is in general not computable, usually estimated graphs $\mathcal{G}'$ are used which contain all edges of the real dependency graph $\mathcal{G}$. Hence, for the second problem we have to implement and certify one estimation of the dependency graph.

Notice that there are various estimations around and that the result of an SCC decomposition depends on the estimation that is used. Hence, it is not a good idea to implement the strongest estimation and then match the result of our decomposition against some given decomposition: problems arise if the termination prover used a weaker estimation and thus obtained larger SCCs.

Therefore, in the upcoming Sect. 4.1 about graph algorithms we just speak of decompositions where the components do not have to be SCCs. Moreover, we will also elaborate on how to minimize the number of tests $(s \rightarrow t, u \rightarrow v) \in E$. The reason is that in Sect. 4.2 we implemented one of the strongest dependency graph estimations where the test for an edge can become expensive. In Sect. 4.3 we finally show how to combine the results of Sections 4.1 and 4.2.

4.1   Certifying Graph Decompositions

Instead of doing an SCC decomposition of a graph within IsaFoR we base our check on the decomposition that is provided by the termination prover. Essentially, we demand that the set of components is given as a list $\langle C_1, \dots, C_k \rangle$ in topological order where the component with no incoming edges is listed first. Then we aim at certifying that every infinite path must end in some C_i. Note that the general idea of taking the topological sorted list as input was already publicly mentioned at the "Workshop on the certification of termination proofs" in 2007. In the following we present how we filled the details of this general idea.

The main idea is to ensure that all edges $(p, q) \in E$ correspond to a step forward in the list $\langle C_1, \dots, C_k \rangle$, i.e., $(p, q) \in C_i \times C_j$ where $i \leq j$. However, iterating over all edges of $\mathcal{G}$ will be costly, because it requires to perform the test $(p, q) \in E$ for all possible edges $(p, q) \in \mathcal{P} \times \mathcal{P}$. To overcome this problem we do not iterate over the edges but over $\mathcal{P}$. To be more precise, we check that

---

[4] We identify an SCC S with the set of nodes S within that SCC.

$$\forall (p,q) \in \mathcal{P} \times \mathcal{P}. \ (\exists i \leq j. \ (p,q) \in \mathcal{P}_i \times \mathcal{P}_j) \vee (p,q) \notin E \tag{2}$$

where the latter part of the disjunction is computed only on demand. Thus, only those edges have to be computed, which would contradict a valid decomposition.

Example 7. Consider the set of nodes $\mathcal{P} = \{(DD), (DM), (MM)\}$. Suppose that we have to check a decomposition of $\mathcal{P}$ into $L = \langle \{(DD)\}, \{(DM)\}, \{(MM)\} \rangle$ for some graph $\mathcal{G} = (\mathcal{P}, E)$. Then our check has to ensure that the dashed edges in the following illustration do not belong to E.

$$(DD) \ \dashleftarrow\!---\ (DM) \ \dashleftarrow\!---\ (MM)$$

It is easy to see that (2) is satisfied for every list of SCCs that is given in topological order. What is even more important, whenever there is a valid SCC decomposition of $\mathcal{G}$, then (2) is also satisfied for every subgraph. Hence, regardless of the dependency graph estimation a termination prover might have used, we accept it, as long as our estimation delivers less edges.

However, the criterion is still too relaxed, since we might cheat in the input by listing nodes twice. Consider $\mathcal{P} = \{p_1, \ldots, p_m\}$ where the corresponding graph is arbitrary and $L = \langle \{p_1\}, \ldots, \{p_m\}, \{p_1\}, \ldots, \{p_m\} \rangle$. Then trivially (2) is satisfied, because we can always take the source of edge (p_i, p_j) from the first part of L and the target from the second part of L. To prevent this kind of problem, our criterion demands that the sets C_i in L are pairwise disjoint.

Before we formally state our theorem, there is one last step to consider, namely the handling of singleton nodes which do not form an SCC on their own. Since we cannot easily infer at what position these nodes have to be inserted in the topological sorted list—this would amount to do an SCC decomposition on our own—we demand that they are contained in the list of components.[5]

To distinguish a singleton node without an edge to itself from a "real SCC", we require that the latter ones are marked. Then condition (2) is extended in a way that unmarked components may have no edge to themselves. The advantage of not marking a component is that our IsaFoR-theorem about graph decomposition states that every infinite path will end in some marked component, i.e., here the unmarked components can be ignored.

Theorem 8. *Let $L = \langle C_1, \ldots, C_k \rangle$ be a list of sets of nodes, some of them marked, let $\mathcal{G} = (\mathcal{P}, E)$ be a graph, let α be an infinite path of $\mathcal{G}$. If*

- $\forall (p,q) \in \mathcal{P} \times \mathcal{P}. \ (\exists i < j. \ (p,q) \in C_i \times C_j) \vee (\exists i. \ (p,q) \in C_i \times C_i \wedge C_i \ is \ marked) \vee (p,q) \notin E$ *and*
- $\forall i \neq j. \ C_i \cap C_j = \emptyset$

then there is some suffix β of α and some marked C_i such that all nodes of β belong to C_i.

---

[5] Note that Tarjan's SCC decomposition algorithm produces exactly this list.

Example 9. If we continue with example Ex. 7 where only components {(MM)} and {(DD)} are marked, then our check also analyzes that $\mathcal{G}$ contains no edge from (DM) to itself. If it succeeds, every infinite path will in the end only contain nodes from {(DD)} or only nodes from {(MM)}. In this way, only 4 edges of $\mathcal{G}$ have to be calculated instead of analyzing all 9 possible edges in $\mathcal{P} \times \mathcal{P}$.

4.2 Certifying Dependency Graph Estimations

What is currently missing to certify an application of the dependency graph processor, is to check, whether a singleton edge is in the dependency graph or not. Hence, we have to estimate whether the sequence $s \to t, u \to v$ is a chain, i.e., whether there are substitutions σ and μ such that $t\sigma \to_{\mathcal{R}}^* u\mu$. An obvious solution is to just look at the root symbols of t and u—if they are different there is no way that the above condition is met (since all the steps in $t\sigma \to_{\mathcal{R}}^* u\mu$ take place below the root, by construction of the dependency pairs). Although efficient and often good enough, there are more advanced estimations around.

The estimation EDG [1] first replaces via an operation cap all variables and all subterms of t which have a defined root-symbol by distinct fresh variables. Then if cap(t) and u do not unify, it is guaranteed that there is no edge.

The estimation EDG*[12] does the same check and additionally uses the reversed TRS $\mathcal{R}^{-1} = \{r \to \ell \mid \ell \to r \in \mathcal{R}\}$, i.e., it uses the fact that $t\sigma \to_{\mathcal{R}}^* u\mu$ implies $u\mu \to_{\mathcal{R}^{-1}}^* t\sigma$ and checks whether cap(u) does not unify with t. Of course in the application of cap(u) we have to take the reversed rules into account (possibly changing the set of defined symbols) and it is not applicable if $\mathcal{R}$ contains a collapsing rule $\ell \to x$ where $x \in \mathcal{V}$.

The last estimation we consider is based on a better version of cap, called tcap [9]. It only replaces subterms with defined symbols by a fresh variable, if there is a rule that unifies with the corresponding subterm.

Definition 10. *Let $\mathcal{R}$ be a TRS.*

- tcap($f(t_n)$) = $f(\text{tcap}(t_1), \ldots, \text{tcap}(t_n))$ *iff* $f(\text{tcap}(t_1), \ldots, \text{tcap}(t_n))$ *does not unify with any variable renamed left-hand side of a rule from $\mathcal{R}$*
- tcap(t) *is a fresh variable, otherwise*

To illustrate the difference between cap and tcap consider the TRS of Ex. 1 and $t = \text{div}(0,0)$. Then cap(t) = x_{fresh} since div is a defined symbol. However, tcap(t) = t since there is no division rule where the second argument is 0.

Apart from tree automata techniques, currently the most powerful estimation is the one based on tcap looking both forward as in EDG and backward as in EDG*. Hence, we aimed to implement and certify this estimation in IsaFoR.

Unfortunately, when doing so, we had a problem with the domain of variables. The problem was that although we first implemented and certified the standard unification algorithm of [15], we could not directly apply it to compute tcap. The reason is that to generate fresh variables as well as to rename variables in rules apart, we need a type of variables with an infinite domain. One solution would have been to constrain the type of variables where there is a function which delivers a fresh variable w.r.t. any given finite set of variables.

However, there is another and more efficient approach to deal with this problem than the standard approach to rename and then do unification. Our solution is to switch to another kind of terms where instead of variables there is just one special constructor "□" representing an arbitrary fresh variable. In essence, this data structure represents contexts which do not contain variables, but where multiple holes are allowed. Therefore in the following we speak of ground-contexts and use $C, D, \ldots$ to denote them.

Definition 11. *Let $[\![C]\!]$ be the equivalence class of a ground-context C where the holes are filled with arbitrary terms: $[\![C]\!] = \{C[t_1, \ldots, t_n] \mid t_1, \ldots, t_n \in \mathcal{T}(\mathcal{F}, \mathcal{V})\}$.*

Obviously, every ground-context C can be turned into a term t which only contains distinct fresh variables and vice-versa. Moreover, every unification problem between t and ℓ can be formulated as a *ground-context matching problem* between C and ℓ, which is satisfiable iff there is some μ such that $\ell\mu \in [\![C]\!]$.

Since the result of tcap is always a term which only contains distinct fresh variables, we can do the computation of tcap using the data structure of ground-contexts; it only requires an algorithm for ground-context matching. To this end we first generalize ground-context matching problems to multiple pairs (C_i, ℓ_i).

Definition 12. *A ground-context matching problem is a set of pairs $\mathcal{M} = \{(C_1, \ell_1), \ldots, (C_n, \ell_n)\}$. It is solvable iff there is some μ such that $\ell_i\mu \in [\![C_i]\!]$ for all $1 \leq i \leq n$. We sometimes abbreviate $\{(C, \ell)\}$ by (C, ℓ).*

To decide ground-context matching we devised a specialized algorithm which is similar to standard unification algorithms, but which has some advantages: it does neither require occur-checks as the unification algorithm, nor is it necessary to preprocess the left-hand sides of rules by renaming (as would be necessary for standard tcap). And instead of applying substitutions on variables, we just need a basic operation on ground-contexts called merge such that $\mathsf{merge}(C, D) = \bot$ implies $[\![C]\!] \cap [\![D]\!] = \emptyset$, and $\mathsf{merge}(C, D) = E$ implies $[\![C]\!] \cap [\![D]\!] = [\![E]\!]$.

Definition 13. *The following rules simplify a ground-context matching problem into solved form (where all terms are distinct variables) or into $\bot$.*

$$
\begin{array}{llll}
(a) & \mathcal{M} \cup \{(\Box, \ell)\} & \Rightarrow_{\mathsf{match}} \mathcal{M} & \\
(b) & \mathcal{M} \cup \{(f(\boldsymbol{D}_n), f(\boldsymbol{u}_n))\} & \Rightarrow_{\mathsf{match}} \mathcal{M} \cup \{(D_1, u_1), \ldots, (D_n, u_n)\} & \\
(c) & \mathcal{M} \cup \{(f(\boldsymbol{D}_n), g(\boldsymbol{u}_k))\} & \Rightarrow_{\mathsf{match}} \bot & \text{if } f \neq g \text{ or } n \neq k \\
(d) & \mathcal{M} \cup \{(C, x), (D, x)\} & \Rightarrow_{\mathsf{match}} \mathcal{M} \cup \{(E, x)\} & \text{if } \mathsf{merge}(C, D) = E \\
(e) & \mathcal{M} \cup \{(C, x), (D, x)\} & \Rightarrow_{\mathsf{match}} \bot & \text{if } \mathsf{merge}(C, D) = \bot
\end{array}
$$

Rules $(a\text{--}c)$ obviously preserve solvability of $\mathcal{M}$ (where $\bot$ represents an unsolvable matching problem). For Rules (d,e) we argue as follows:

$\{(C, x), (D, x)\} \cup \ldots$ is solvable iff
- there is some μ such that $x\mu \in [\![C]\!]$ and $x\mu \in [\![D]\!]$ and $\ldots$ iff
- there is some μ such that $x\mu \in [\![C]\!] \cap [\![D]\!]$ and $\ldots$ iff
- there is some μ such that $x\mu \in [\![\mathsf{merge}(C, D)]\!]$ and $\ldots$ iff
- $\{(\mathsf{merge}(C, D), x)\} \cup \ldots$ is solvable

Since every ground-context matching problem in solved form is solvable, we have devised a decision procedure. It can be implemented in two stages where the first stage just normalizes by the Rules $(a\text{--}c)$, and the second stage just applies the Rules (d,e). It remains to implement merge.

$$\mathsf{merge}(\square, C) \Rightarrow_{\mathsf{merge}} C$$
$$\mathsf{merge}(C, \square) \Rightarrow_{\mathsf{merge}} C$$
$$\mathsf{merge}(f(C_n), g(D_k)) \Rightarrow_{\mathsf{merge}} \bot \qquad \text{if } f \neq g \text{ or } n \neq k$$
$$\mathsf{merge}(f(C_n), f(D_n)) \Rightarrow_{\mathsf{merge}} f(\mathsf{merge}(C_1, D_1), \ldots, \mathsf{merge}(C_n, D_n))$$
$$f(\ldots, \bot, \ldots) \Rightarrow_{\mathsf{merge}} \bot$$

Note that our implementations of the matching algorithm and the merge function in IsaFoR are slightly different due to different data structures. For example matching problems are represented as lists of pairs, so it may occur that we have duplicates in $\mathcal{M}$. The details of our implementation can be seen in IsaFoR (theory Edg) or in the source of CeTA.

Soundness and completeness of our algorithms are proven in IsaFoR.

Theorem 14.
- *If* $\mathsf{merge}(C, D) \Rightarrow^*_{\mathsf{merge}} \bot$ *then* $[\![C]\!] \cap [\![D]\!] = \emptyset$.
- *If* $\mathsf{merge}(C, D) \Rightarrow^*_{\mathsf{merge}} E$ *then* $[\![C]\!] \cap [\![D]\!] = [\![E]\!]$.
- *If* $(C, \ell) \Rightarrow^*_{\mathsf{match}} \bot$ *then there is no* μ *such that* $\ell\mu \in [\![C]\!]$.
- *If* $(C, \ell) \Rightarrow^*_{\mathsf{match}} \mathcal{M}$ *where* $\mathcal{M}$ *is in solved form, then there exists some* μ *such that* $\ell\mu \in [\![C]\!]$.

Using $\Rightarrow_{\mathsf{match}}$, we can now easily reformulate tcap in terms of ground-context matching which results in the efficient implementation etcap.

Definition 15.
- $\mathsf{etcap}(f(t_n)) = f(\mathsf{etcap}(t_1), \ldots, \mathsf{etcap}(t_n))$ *iff* $(f(\mathsf{etcap}(t_1), \ldots, \mathsf{etcap}(t_n)), \ell) \Rightarrow^*_{\mathsf{match}} \bot$ *for all rules* $\ell \to r \in \mathcal{R}$.
- $\square$, *otherwise*

One can also reformulate the desired check to estimate the dependency graph whether $\mathsf{tcap}(t)$ does not unify with u in terms of etcap. It is the same requirement as demanding $(\mathsf{etcap}(t), u) \Rightarrow^*_{\mathsf{match}} \bot$. Again, the soundness of this estimation has been proven in IsaFoR where the second part of the theorem is a direct consequence of the first part by using the soundness of the matching algorithm.

Theorem 16. *(a) Whenever* $t\sigma \to^*_{\mathcal{R}} s$ *then* $s \in [\![\mathsf{etcap}(t)]\!]$.
(b) Whenever $t\sigma \to^*_{\mathcal{R}} u\tau$ *then* $(\mathsf{etcap}(t), u) \not\Rightarrow^*_{\mathsf{match}} \bot$ *and* $(\mathsf{etcap}(u), t) \not\Rightarrow^*_{\mathsf{match}} \bot$ *where* $\mathsf{etcap}(u)$ *is computed w.r.t. the reversed TRS* $\mathcal{R}^{-1}$.

4.3   Certifying Dependency Graph Decomposition

Eventually we can connect the results of the previous two subsections to obtain one function to check a valid application of the dependency graph processor.

$$\text{checkDepGraphProc}(\mathcal{P}, L, \mathcal{R}) = \text{checkDecomposition}(\text{checkEdg}(\mathcal{R}), \mathcal{P}, L)$$

where checkEdg just applies the criterion of Thm. 16 (b).

In IsaFoR the soundness result of our check is proven.

Theorem 17. *If* checkDepGraphProc$(\mathcal{P}, L, \mathcal{R})$ *is accepted and if for all* $\mathcal{P}' \in L$ *where* $\mathcal{P}'$ *is marked, the DP problem* $(\mathcal{P}', \mathcal{R})$ *is finite, then* $(\mathcal{P}, \mathcal{R})$ *is finite.*

To summarize, we have implemented and certified the currently best dependency graph estimation which does not use tree automata techniques. Our check-function accepts any decomposition which is based on a weaker estimation, but requires that the components are given in topological order. Since our algorithm computes edges only on demand, the number of tests for an edge is reduced considerably. For example, the five largest graphs in our experiments contain 73,100 potential edges, but our algorithm only has to consider 31,266. This reduced the number of matchings from 13 millions down to 4 millions.

Furthermore, our problem of not being able to generate fresh variables or to rename variables in rules apart led to a more efficient algorithm for tcap based on matching instead of unification: simply replacing tcap by etcap in $T_T T_2$ reduced the time for estimating the dependency graph by a factor of two.

5   Certifying the Reduction Pair Processor

One important technique to prove finiteness of a DP problem $(\mathcal{P}, \mathcal{R})$ is the so-called *reduction pair processor*. The general idea is to use a well-founded order where all rules of $\mathcal{P} \cup \mathcal{R}$ are weakly decreasing. Then we can delete all strictly decreasing rules from $\mathcal{P}$ and continue with the remaining dependency pairs.

We first state a simplified version of the reduction pair processor as it is introduced in [8], where we ignore the usable rules refinement.

Theorem 18. *If all the following properties are satisfied, then finiteness of* $(\mathcal{P} \setminus \succ, \mathcal{R})$ *implies finiteness of* $(\mathcal{P}, \mathcal{R})$.

(a) $\succ$ *is a well-founded and stable order*
(b) $\succsim$ *is a stable and monotone quasi-order*
(c) $\succsim \circ \succ \,\subseteq\, \succ$ *and* $\succ \circ \succsim \,\subseteq\, \succ$
(d) $\mathcal{P} \subseteq\, \succ \cup \succsim$ *and* $\mathcal{R} \subseteq\, \succsim$

Of course, to instantiate the reduction pair processor with a new kind of reduction pair, e.g., LPO, polynomial orders,..., we first have to prove the first three properties for that kind of reduction pairs. Since we plan to integrate many reduction pairs, but only want to write the reduction pair processor once, we tried to minimize these basic requirements such that the reduction pair processor still remains sound in total. In the end, we replaced the first three properties by:

(a) $\succ$ is a well-founded and stable *relation*
(b) $\succsim$ is a stable and monotone *relation*
(c) $\succsim \circ \succ \subseteq \succ$

In this way, for every new class of reduction pairs, we do not have to prove transitivity of $\succ$ or $\succsim$ anymore, as it would be required for Thm. 18. Currently, we just support reduction pairs based on polynomial interpretations with negative constants [13], but we plan to integrate other reduction pairs in the future.

For checking an application of a reduction pair processor we implemented a generic function `checkRedPairProc` in Isabelle, which works as follows. It takes as input two functions `checkS` and `checkNS` which have to approximate a reduction pair, i.e., whenever `checkS`(s, t) is accepted, then $s \succ t$ must hold in the corresponding reduction pair and similarly, `checkNS` has to guarantee $s \succsim t$.

Then `checkRedPairProc(checkS, checkNS, P, P', R)` works as follows:

- iterate once over P to divide P into $P_\succ$ and $P_{\not\succ}$ where the former set contains all pairs of P where `checkS` is accepted
- ensure for all $s \to t \in R \cup P_{\not\succ}$ that `checkNS`(s, t) is accepted, otherwise reject
- accept if $P_{\not\succ} \subseteq P'$, otherwise reject

The corresponding theorem in IsaFoR states that a successful application of `checkRedPairProc`$(\ldots, P, P', R)$ proves that (P, R) is finite whenever (P', R) is finite. Obviously, the first two conditions of `checkRedPairProc` ensure condition (d) of Thm. 18. Note, that it is not required that all strictly decreasing pairs are removed, i.e., our checks may be stronger than the ones that have been used in the termination provers.

6   Certifying the Whole Proof Tree

From Sect. 3–5 we have basic checks for the three techniques of applying dependency pairs (`checkDPs`), the dependency graph processor (`checkDepGraphProc`), and the reduction pair processor (`checkRedPairProc`). For representing proof trees within the DP framework we used the following data structures in IsaFoR.

```
datatype 'f RedPair = NegPolo "('f × (cint × nat list))list"

datatype ('f,'v)DPProof = ...⁶
    | PisEmpty
    | RedPairProc "'f RedPair" "('f,'v)trsL" "('f,'v)DPProof"
    | DepGraphProc "(('f,'v)DPProof option × ('f,'v)trsL)list"

datatype ('f,'v)TRSProof = ...⁶
    | DPTrans "('f shp,'v)trsL" "('f shp,'v)DPProof"
```

---

⁶ CeTA supports even more techniques, cf. CeTA's website for a complete list.

The first line fixes the format for reduction pairs, i.e., currently of (linear) polynomial interpretations where for every symbol there is one corresponding entry. E.g., the list $[(f, (-2, [0, 3]))]$ represents the interpretation where $\mathcal{P}ol(f)(x, y) = \max(-2 + 3y, 0)$ and $\mathcal{P}ol(g)(x_1, \ldots, x_n) = 1 + \Sigma_{1 \leq i \leq n} x_i$ for all $f \neq g$.

The datatype DPProof represents proof trees for DP problems. Then the check for valid DPProofs gets as input a DP problem $(\mathcal{P}, \mathcal{R})$ and a proof tree and tries to certify that $(\mathcal{P}, \mathcal{R})$ is finite. The most basic technique is the one called PisEmpty, which demands that the set $\mathcal{P}$ is empty. Then $(\mathcal{P}, \mathcal{R})$ is trivially finite.

For an application of the reduction pair processor, three inputs are required. First, the reduction pair redp, i.e., some polynomial interpretation. Second, the dependency pairs $\mathcal{P}'$ that remain after the application of the reduction pair processor. Here, the datatype trsL is an abbreviation for lists of rules. And third, a proof that the remaining DP problem $(\mathcal{P}', \mathcal{R})$ is finite. Then the checker just has to call createRedPairProc(redp, $\mathcal{P}, \mathcal{P}', \mathcal{R}$) and additionally calls itself recursively on $(\mathcal{P}', \mathcal{R})$. Here, createRedPairProc invokes checkRedPairProc where checkS and checkNS are generated from redp.

The most complex structure is the one for decomposition of the (estimated) dependency graph. Here, the topological list for the decomposition has to be provided. Moreover, for each subproblem $\mathcal{P}'$, there is an optional proof tree. Subproblems where a proof is given are interpreted as "real SCCs" whereas the ones without proof remain unmarked for the function checkDepGraphProc.

The overall function for checking proof trees for DP problems looks as follows.

```
checkDPProof(𝒫,ℛ,PisEmpty) = (𝒫 = [])
checkDPProof(𝒫,ℛ,(RedPairProc redp 𝒫' prf)) =
  createRedPairProc(redp,𝒫,𝒫',ℛ) ∧ checkDP(𝒫',ℛ,prf)
checkDPProof(𝒫,ℛ,DepGraphProc 𝒫's) =
  checkDepGraphProc(𝒫,map (λ(prf0,𝒫').(isSome prf0,𝒫')) 𝒫's, ℛ)
  ∧ ⋀(Some prf,𝒫')∈𝒫's checkDPProof(𝒫',ℛ,prf)
```

Theorem 19. *If* checkDPProof$(\mathcal{P}, \mathcal{R}, prf)$ *is accepted then* $(\mathcal{P}, \mathcal{R})$ *is finite.*

Using checkDPProof it is now easy to write the final method checkTRSProof for proving termination of a TRS, where computeID is an implementation of ID.

```
checkTRSProof(ℛ,DPTrans 𝒫 prf) =
  checkDPs(ℛ,𝒫) ∧ checkDPProof(𝒫,computeID(ℛ),prf)
```

For the external usage of CeTA we developed a well documented XML-format, cf. CeTA's website. Moreover, we implemented two XML parsers in Isabelle, one that transforms a given TRS into the internal format, and another that does the same for a given proof. The function certifyProof, finally, puts everything together. As input it takes two strings (a TRS and its proof). Then it applies the mentioned parsers and afterwards calls checkTRSProof on the result.

Theorem 20. *If* certifyProof$(\mathcal{R}, prf)$ *is accepted then* $\mathrm{SN}(\rightarrow_\mathcal{R})$.

To ensure that the parser produces the right TRS, after the parsing process it is checked that when converting the internal data-structures of the TRS back to

XML, we get the same string as the input string for the TRS (modulo white-space). This is a major benefit in comparison to the two other approaches where it can and already has happened that the uncertified components Rainbow/CiME produced a wrong proof goal from the input TRS, i.e., they created a termination proof within Coq for a different TRS than the input TRS.

7   Error Messages

To generate readable error messages, our checks do not have a Boolean return type, but a monadic one (isomorphic to 'e option). Here, None represents an accepted check whereas Some e represents a rejected check with error message e. The theory ErrorMonad contains several basic operations like >> for conjunction of checks, <- for changing the error message, and isOK for testing acceptance.

Using the error monad enables an easy integration of readable error messages. For example, the real implementation of checkTRSProof looks as follows:

```
fun checkTRSProof where "checkTRSProof R (DPTrans P prf) = (
    checkDPs R P
        <- (λs. ''error ...'' @ showTRS R @ ''...'' @ showTRS P @ s)
  >> checkDPProof P (computeID R) prf
        <- (λs. ''error below switch to dependency pairs'' @ s))"
```

However, since we do not want to adapt the proofs every time the error mes-sages are changed, we setup the Isabelle simplifier such that it hides the details of the error monad, but directly removes all the error handling and turns monadic checks via isOK(...) into Boolean ones using the following lemmas.

```
lemma "isOK(m >> n) = isOK(m) ∧ isOK(n)"
lemma "isOK(m <- s) = isOK(m)"
```

Then for example isOK(checkTRSProof R (DPTrans P prf)) directly simpli-fies to isOK(checkDPs R P) ∧ isOK(checkDPProof P (computeID R) prf).

8   Experiments and Conclusion

Isabelle's code-generator is invoked to create CeTA from IsaFoR. To compile CeTA one auxiliary hand-written Haskell file CeTA.hs is needed, which just reads two files (one for the TRS, one for the proof) and then invokes certifyProof.

We tested CeTA (version 1.03) using T_TT_2 as termination prover (TC and TC$^+$). Here, T_TT_2 uses only the techniques of this paper in the combination TC, whereas in TC$^+$ all supported techniques are tried, including usable rules and nontermi-nation. We compare to CiME/Coccinelle using AProVE [10] or CiME [5] as provers (ACC,CCC), and to Rainbow/CoLoR using AProVE or Matchbox [18] (ARC,MRC) where we take the results of the latest certified termination competition in Nov 2008[7] involving 1391 TRSs from the termination problem database.

---

[7] http://termcomp.uibk.ac.at/

We performed our experiments using a PC with a 2.0 GHz processor running Linux where both TₜT₂ and CeTA where aborted after 60 seconds. The following table summarizes our experiments and the termination competition results.

	TC	TC⁺	ACC	CCC	ARC	MRC
proved / disproved	401 / 0	572 / 214	532 / 0	531 / 0	580 / 0	458 / 0
certified	391 / 0	572 / 214	437 / 0	485 / 0	558 / 0	456 / 0
rejected	10	0	3	0	0	2
cert. timeouts	0	0	92	46	22	0
total cert. time	33s	113s	6212s	6139s	7004s	3602s

The 10 proofs that CeTA rejected are all for nonterminating TRSs which do not satisfy the variable condition. Since TC supports only polynomial orders as reduction pairs, it can handle less TRSs than the other combinations. But, there are 44 TRSs which are only solved by TC (and TC⁺), the reason being the time-limit of 60 seconds (19 TRSs), the dependency graph estimation (8 TRSs), and the polynomial order allowing negative constants (17 TRSs).

The second line clearly shows that TC⁺ (with nontermination and usable rules support) currently is the most powerful combination with 786 certified proofs. Moreover, TC⁺ can handle 214 nonterminating and 102 terminating TRSs where none of ACC, CCC, ARC, and MRC were successful. The efficiency of CeTA is also clearly visible: the average certification time in TC and TC⁺ for a single proof is by a factor of 50 faster than in the other combinations.[8]

For more details on the experiments we refer to CeTA's website.

To conclude, we presented a modular and competitive termination certifier, CeTA, which is directly created from our Isabelle library on term rewriting, IsaFoR. Its main features are that CeTA is available as a stand-alone binary, the efficiency, the dependency graph estimation, nontermination and usable rules support, the error handling, and the robustness.

As each sub-check for a termination technique can be called separately, and as our check to certify a whole termination proof just invokes these sub-checks, it seems possible to integrate other techniques (even if they are proved in a different theorem prover) as long as they are available as executable code. However, we will need a common proof format and a compatible definition.

As future work we plan to certify several other termination techniques where we already made progress in the formalization of semantic labeling and the subterm-criterion. We would further like to contribute to a common proof format.

---

[8] Note that in the experiments above, for each TRS, each combination might have certified a different proof. In an experiment where the certifiers where run on the same proofs for each TRS (using only techniques that are supported by all certifiers, i.e., EDG and linear polynomials without negative constants), CeTA was even 190 times faster than the other approaches and could certify all 358 proofs, whereas each of the other two approaches failed on more than 30 proofs due to timeouts.

References

1. Arts, T., Giesl, J.: Termination of term rewriting using dependency pairs. Theoretical Computer Science 236, 133–178 (2000)
2. Baader, F., Nipkow, T.: Term Rewriting and All That. Cambridge University Press, Cambridge (1998)
3. Blanqui, F., Delobel, W., Coupet-Grimal, S., Hinderer, S., Koprowski, A.: CoLoR, a Coq library on rewriting and termination. In: Proc. WST 2006, pp. 69–73 (2006)
4. Contejean, E., Courtieu, P., Forest, J., Pons, O., Urbain, X.: Certification of automated termination proofs. In: Konev, B., Wolter, F. (eds.) FroCos 2007. LNCS, vol. 4720, pp. 148–162. Springer, Heidelberg (2007)
5. Contejean, E., Marché, C., Monate, B., Urbain, X.: CiME, http://cime.lri.fr
6. Courtieu, P., Forest, J., Urbain, X.: Certifying a termination criterion based on graphs, without graphs. In: Mohamed, O.A., Muñoz, C., Tahar, S. (eds.) TPHOLs 2008. LNCS, vol. 5170, pp. 183–198. Springer, Heidelberg (2008)
7. Dershowitz, N.: Termination dependencies. In: Proc. WST 2003, pp. 27–30 (2003)
8. Giesl, J., Thiemann, R., Schneider-Kamp, P.: The dependency pair framework: Combining techniques for automated termination proofs. In: Baader, F., Voronkov, A. (eds.) LPAR 2004. LNCS (LNAI), vol. 3452, pp. 301–331. Springer, Heidelberg (2005)
9. Giesl, J., Thiemann, R., Schneider-Kamp, P.: Proving and disproving termination of higher-order functions. In: Gramlich, B. (ed.) FroCos 2005. LNCS (LNAI), vol. 3717, pp. 216–231. Springer, Heidelberg (2005)
10. Giesl, J., Schneider-Kamp, P., Thiemann, R.: AProVE 1.2: Automatic termination proofs in the DP framework. In: Furbach, U., Shankar, N. (eds.) IJCAR 2006. LNCS (LNAI), vol. 4130, pp. 281–286. Springer, Heidelberg (2006)
11. Haftmann, F.: Code generation from Isabelle/HOL theories (April 2009), http://isabelle.in.tum.de/doc/codegen.pdf
12. Hirokawa, N., Middeldorp, A.: Automating the dependency pair method. Information and Computation 199(1-2), 172–199 (2005)
13. Hirokawa, N., Middeldorp, A.: Tyrolean Termination Tool: Techniques and features. Information and Computation 205(4), 474–511 (2007)
14. Korp, M., Sternagel, C., Zankl, H., Middeldorp, A.: Tyrolean Termination Tool 2. In: Proc. RTA 2009. LNCS, vol. 5595, pp. 295–304 (2009)
15. Martelli, A., Montanari, U.: An efficient unification algorithm. ACM Transactions on Programming Languages and Systems 4(2), 258–282 (1982)
16. Nipkow, T., Paulson, L.C., Wenzel, M.T.: Isabelle/HOL. LNCS, vol. 2283. Springer, Heidelberg (2002)
17. Peyton Jones, S., et al.: The Haskell 98 language and libraries: The revised report. Journal of Functional Programming 13(1)–255 (2003)
18. Waldmann, J.: Matchbox: A tool for match-bounded string rewriting. In: van Oostrom, V. (ed.) RTA 2004. LNCS, vol. 3091, pp. 85–94. Springer, Heidelberg (2004)

A Formalisation of Smallfoot in HOL

Thomas Tuerk

University of Cambridge Computer Laboratory
William Gates Building, JJ Thomson Avenue, Cambridge CB3 0FD, United Kingdom
http://www.cl.cam.ac.uk

Abstract. In this paper a general framework for separation logic inside the HOL theorem prover is presented. This framework is based on Abstract Separation Logic. It contains a model of an abstract, imperative programming language as well as an abstract specification logic for this language. While the formalisation mainly follows the original definition of Abstract Separation Logic, it contains some additional features. Most noticeably is added support for procedures.

As a case study, the framework is instantiated to build a tool that is able to parse Smallfoot specifications and verify most of them completely automatically. In contrast to Smallfoot this instantiation can handle the content of data-structures as well as their shape. This enables it to verify fully functional specifications. Some noteworthy examples that have been verified are parallel mergesort and an interactive filter-function for single linked lists.

1 Motivation

Separation logic is an extension of Hoare logic that allows local reasoning [7, 9]. It is used to reason about mutable data structures in combination with low level imperative programming languages that use pointers and explicit memory management. Thanks to local reasoning, it scales better than classical Hoare logic to the verification of large programs and can easily be used to reason about parallelism. There are several implementations: Smallfoot [2], SLAyer[1] and SpaceInvader [5] are probably some of the best know examples. Moreover, there are formalisations inside theorem provers [1, 6, 10, 11].

The problem, as I see it, is that all these tools and formalisations focus on one concrete setting. They fix the programming languages, their exact semantics, the supported specifications etc. However, there are a lot of different possible design choices and the tools differ in these. I'm therefore building a general framework for separation logic in HOL that can be instantiated to a variety of different separation logics. By building such a framework, I hope to be able to concentrate on the essence of separation logic as well as keeping the formalisation clean and easy.

In this paper, the results of these efforts to build a separation logic framework in HOL are presented. The framework is based on *Abstract Separation Logic* [4],

---

[1] http://research.microsoft.com/SLAyer/

S. Berghofer et al. (Eds.): TPHOLs 2009, LNCS 5674, pp. 469–484, 2009.

an abstract, high level variant of separation logic. It consists of both an abstract, imperative programming language and an abstract specification logic for this language. Both the abstract language and the specification logic are designed to be instantiated to a concrete programming language and a concrete language for specifications.

As a case study, I instantiated this framework to build a tool similar to Smallfoot [2], one of the oldest and best documented separation logic tools. Smallfoot is able to automatically prove specifications about programs written in a simple, low-level imperative language with support for parallelism. The tool, called Holfoot, combines ideas from Abstract Separation Logic, *Variables as Resource in Hoare Logic* [8] and Smallfoot. It is able to parse Smallfoot-specifications and prove nearly all of them completely automatically inside the HOL theorem prover. In addition to Smallfoot, specifications can talk about the content of data-structures as well as their shape. Proving the resulting fully functional specifications exploits the fact that Holfoot is implemented inside HOL. All existing libraries and proof tools can be used, while a substantial amount of automation is still available to reason about the structure of the program.

Reasoning about the data-content as well as the shape of data-structures is one of the main challenges of separation logic tools at the moment. To help the communication within the community and in general to further the progress of the field, a benchmark collection called *A Heap of Problems*[2] was created. It collects interesting examples, usually with at least a natural-language description, a C-implementation and some pseudo-code. Often implementations for a specific separation logic tool are available as well.Moreover, there are proofs of the examples using different tools and techniques.

Here, I would like to highlight just two of these benchmark examples: mergesort, whose verification needs some knowledge about orderings and permutations, and filtering of a single linked list, whose iterative version uses a very complicated loop invariant. Both examples can easily be verified using Holfoot. The tool is able to reason automatically about the shape part of the problem, leaving the user to reason about properties of the data-content, i.e. about the essence of these algorithms. Fully functional specifications of simpler algorithms like reversing or copying of a single-linked-list, determining its length or a recursive filter function can even be verified completely automatically. For more examples and discussions about them, please have a lock at the *A Heap of Problems* webpage.

It took considerable effort to build this framework and instantiate it. This work cannot be presented here in detail due to space limitations. Therefore, the next section, will present a high level view on Holfoot. It is intended to give a glimpse of the features and power of this tool. Semantic foundations and implementation details are not discussed. This high level presentation of Holfoot is followed by a detailed description of the formalisation of Abstract Separation Logic in HOL. This description explains the semantic background of Holfoot. However, it is barely scratched, how the Abstract Separation Logic framework is

---

[2] http://wiki.heap-of-problems.org.

instantiated to build Holfoot. The paper ends with a section about future work and some conclusions.

2   Formalisation of Smallfoot

Smallfoot [2] is one of the oldest and best documented separation logic tools. It is able to automatically prove specifications about programs written in a simple, low-level imperative language, which is designed to resemble C. This language contains pointers, local and global variables, dynamic memory allocation/deallocation, conditional execution, while-loops and recursive procedures with call-by-value and call-by-reference arguments. Moreover, there is support for parallelism with conditional critical regions that synchronise the access to so-called resources. Smallfoot-specifications are concerned with the shape of memory. Common specifications, for example, say that some stack-variable points to a single linked list in memory. However, nothing is e. g. said about the length of the list or about its data-content.

Smallfoot comes with a selection of example specifications. There are common algorithms about single linked lists like copying, reversing or deallocating them. Another set of examples contains similar algorithms for trees. There is an implementation of mergesort, some code about queues, circular-lists, buffers and similar examples. Holfoot[3] is able to parse Smallfoot-specifications and prove most of the mentioned examples completely automatically inside the HOL theorem prover.

While some features like local variables or procedures with call-by-value arguments took some effort, and while it turned out to be useful to use explicit permission for stack-variables, it was nevertheless possible to formalise Smallfoot based on Abstract Separation Logic in a natural way. As far as I know, this is the first time Abstract Separation Logic has been used to implement a separation logic tool. The formalisation of Smallfoot illustrates that Abstract Separation Logic is powerful and flexible enough to model languages and specifications used by well-known separation logic tools. Moreover, it demonstrates that it is possible to automate reasoning in this framework. While Holfoot is slower than Smallfoot, it provides the additional assurance of a formal proof inside HOL. That this is really valuable, is underlined by the fact, that an error in Smallfoot was detected while building Holfoot. Due to a bug in its implementation, Smallfoot handles call-by-value parameters like call-by-reference ones.

However, besides a formal foundation and much higher trust in the tool, another advantage of Holfoot is, that it is straightforward to use all the libraries and proof-tools HOL provides. Smallfoot specifications talk about the shape of data-structures. The Smallfoot-specification of mergesort for example states that mergesort returns a single linked list. It does not guarantee anything about the content of this list, much less that mergesort really sorts lists. In fact, to prove a fully functional specification of mergesort, substantial knowledge about

---

[3] Holfoot as well as a collection of examples can be found in the HOL-repository.

permutations of lists, orderings and sorted lists is needed. Here, the existing infrastructure of HOL is very useful.

Once the formalisation of the features provided by Smallfoot was completed, it was straight-forward to extend it with support for the content of data-structures. This allows the verification of fully functional specifications. Holfoot is able to automatically verify fully functional specifications of simple algorithms like list-reversal, list-copy or list-length:

```
list_copy(z;c) [data_list(c,data)] {
  local x,y,w,d;
  if (c == NULL) {z=NULL;}
  else {
    z=new(); z->tl=NULL;
    x = c->dta; z->dta = x;
    w=z;
    y=c->tl;
    while (y != NULL) [
       data_lseg(c,
         ''_data1++[_cdate]'',y) *
       data_list(y,_data2) *
       data_lseg(z,_data1,w) *
       w |-> tl:0,dta:_cdate *
       ''data:num list =
         _data1 ++ _cdate::_data2''] {
      d=new(); d->tl=NULL;
      x=y->dta; d->dta=x;
      w->tl=d; w=d;
      y=y->tl;
    }
  }
} [data_list(c,data) * data_list(z,data)]
```

```
list_reverse(i;) [data_list(i,data)] {
  local p, x;
  p = NULL;
  while (i != NULL) [
       data_list(i,_idata) *
       data_list(p,_pdata) *
       ''(data:num list) =
         (REVERSE _pdata) ++ _idata''] {
    x = i->tl; i->tl = p; p = i; i = x;
  }
  i = p;
} [data_list(i,''REVERSE data'')]


list_length(r;c) [data_list(c,cdata)] {
  local t;
  if (c == NULL) {r = 0;} else {
    t = c->tl;
    list_length(r;t);
    r = r + 1;
  }
} [data_list(c,cdata) *
   r == ''LENGTH (cdata:num list)'']
```

The syntax of the this pseudo-code used by Smallfoot and Holfoot is indented to be close to C. However, there are some uncommon features: the arguments of a procedure before the semicolon are call-by-reference arguments, the others call-by-reference ones. So the argument z of list_copy is a call-by-reference argument, whereas c is a call-by-value argument. The pre- and postconditions of procedures are denoted in brackets around the procedure's body. Similarly, loops are annotated with their invariant. In specifications, a variable name that starts with an underscore denotes an existentially quantified variable. For example, data1, data2 and cdate are existentially quantified in the loop-invariant of copy. This invariant requires that data can somehow be split into these three. How it is split changes from iteration to iteration. Finally, everything within quotation marks is regarded as a HOL term. So, REVERSE or LENGTH are not part of the Smallfoot formalisation but functions from HOL's list library.

While these simple algorithms can be handled completely automatically, more complicated ones like the aforesaid mergesort need user interaction. However, even in these interactive proofs, there is a clear distinction between reasoning about the content and about the shape. While the shape can mostly be handled automatically, the user is left to reason about properties of the content. Let's consider the following specification of parallel mergesort:

```
merge(r;p,q) [data_list(p,pdata) *
    data_list(q,qdata) *
    ''(SORTED $<= pdata) /\
      (SORTED $<= qdata)''] {
  local t, q_date, p_date;
```

```
if (q == NULL) r = p;
else if(p == NULL) r = q;
else {
  p_date = p->dta;
  q_date = q->dta;
```

```
    if (q_date < p_date) {                    p->tl = t2;
        t = q; q = q->tl;                     t1->tl = r;
    } else {                                  r = t1;
        t = p; p = p->tl;                   }
    }                                       }
    merge(r;p,q);                       } [data_list(p,_pdata) *
    t->tl = r; r = t;                       data_list(r,_rdata) *
}                                           ''PERM (_pdata ++ _rdata) data'']
} [data_list(r,_rdata) *
    ''(SORTED $<= _rdata) /\          mergesort(r;p) [data_list(p,data)] {
      (PERM (pdata ++ qdata) _rdata)'']   local q,q1,p1;
                                          if (p == NULL) r = p;
split(r;p) [data_list(p,data)] {         else {
    local t1,t2;                             split(q;p);
    if (p == NULL) r = NULL;                 mergesort(q1;q) || mergesort(p1;p);
    else {                                   merge(r;p1,q1);
        t1 = p->tl;                        }
        if (t1 == NULL) r = NULL;        } [data_list(r,_rdata) *
        else {                               ''(SORTED $<= _rdata) /\
            t2 = t1->tl;                       (PERM data _rdata)'']
            split(r;t2);
```

Holfoot can automatically reduce this fully functional specification of mergesort to a small set of simple verification conditions. These verification conditions are just concerned with permutations and sorted lists. The whole structure of the program and the shape of the data-structures can be handled automatically. Some of the remaining verification conditions are very simple as for example SORTED $<= x::xs ==> SORTED xs. Others require some knowledge about permutations like PERM (x::(xs ++ ys)) l ==> PERM (x::(xs ++ y::ys)) (y::l). However, most of them can easily be handled by automated proof tools for permutations and orderings. The only remaining verification conditions are of the form

```
SORTED $<= x::xs  /\  SORTED $<= y::ys  /\ SORTED $<= l  /\
y < x  /\  PERM l (x::xs++ys)  ==> SORTED $<= y::l
```

Their proof needs a combination of properties of permutations and sorted lists. Thus, the standard proof tools fail and a tiny manual proof is required. The following proof-script is sufficient to prove the given specification of mergesort:

```
val thm = smallfoot_verbose_prove(mergesort-specification-filename,
    SMALLFOOT_VC_TAC THEN
    ASM_SIMP_TAC (arith_ss++PERM_ss) [SORTED_EQ, SORTED_DEF, transitive_def] THEN
    REPEAT STRIP_TAC THEN (
        IMP_RES_TAC PERM_MEM_EQ THEN
        FULL_SIMP_TAC list_ss [] THEN
        RES_TAC THEN ASM_SIMP_TAC arith_ss []
    ));
```

After parsing and preprocessing the specification stored in the given file, verification conditions are generated using SMALLFOOT_VC_TAC. This single call is sufficient to eliminate the whole program structure and leave just the described verification conditions. The next line calls some proof-tools for permutations and sorted lists and is able to discharge most of the verification conditions. The rest of the proof-script handles the remaining verification conditions which are all of the aforesaid form.

As this example illustrates, human interaction is often only needed to reason about the essence of an algorithms and HOL provides powerful tools to aid this reasoning. This shows the power of Holfoot and with it the flexibility and power of the whole framework.

3   Formalisation of Abstract Separation Logic

In the previous section, a high-level view of Holfoot and with it of the framework and its capabilities was presented. In this section its semantic foundations – *Abstract Separation Logic* [4] – will be explained. This explanation follows closely the HOL formalisation[4].

Abstract Separation Logic abstracts from both the concrete states and the concrete programming language. Instead of using a concrete model of memory consisting usually of a stack and a heap, Abstract Separation Logic uses an abstract set of states Σ. A partial function $\circ$, called *separation combinator*, is used to combine states.

Definition 1 (Separation Combinator). A *separation combinator* $\circ$ is a partially defined function that satisfies the following properties:

- $\circ$ is partially associative
- $\circ$ is partially commutative
- $\circ$ is cancellative, i. e.
  $\forall s_1, s_2, s_3.$ Defined$(s_1 \circ s_2)$ $\wedge$ $(s_1 \circ s_2 = s_1 \circ s_3)$ $\Longrightarrow$ $(s_2 = s_3)$ holds
- for all states s there exists a neutral element u_s with $u_s \circ s = s$

Definition 2 (Separateness, Substates). This definition of separation combinators induces notions of *separateness* (#) and *substates* ($\preceq$).

$$s_1 \# s_2 \text{ iff } s_1 \circ s_2 \text{ is defined} \qquad s_1 \preceq s_3 \text{ iff } \exists s_2.\ s_3 = s_1 \circ s_2$$

Definition 3 (*, emp). Predicates are as usual elements of the powerset of states $P(\Sigma)$. This allows to define the spatial conjunction operator $*$ of separation logic and its neutral element *emp* as follows:

$$P * Q := \{s \mid \exists p, q.\ (p \circ q = s) \wedge p \in P \wedge q \in Q\}$$
$$emp := \{u \mid \exists s.\ u \circ s = s\}$$

$*$ forms together with *emp* a commutative monoid. Other standard separation logic constructs can be defined in a natural way as well. There is a shallow embedding of the most common constructs available in the framework. Additional constructs can be added easily.

In order to instantiate the framework, one has to provide a concrete set of states Σ and a concrete separation combinator $\circ$.

---

[4] The sources can be found in the HOL - repository at Sourceforge in the subdirectory `examples/separationLogic`.

Example 4. Heaps, modelled as finite partial functions, are commonly used with separation logic. In this model, Σ is the set of all heaps and $\circ$ is given by

$$h_1 \circ h_2 = \begin{cases} h_1 \uplus h_2 & \text{iff } \text{dom}(h_1) \cap \text{dom}(h_2) = \emptyset \\ undefined & \text{otherwise} \end{cases}$$

In this setting, two heaps are disjoint ($h_1 \mathbin{\#} h_2$) iff their domains are disjoint. The combination of two separate heaps ($h_1 \circ h_2$) is their disjoint union. The empty heap is the neutral element for all heaps.

3.1   Actions

The programming language used by Abstract Separation Logic is abstract as well. Its elementary constructs are actions.

Definition 5 (Action). An *action* $act : \Sigma \to P(\Sigma)^\top$ is a function from a state to a set of states or a special failure state $\top$.

If executing an action act in a state s results in $\top$, then an error may occur during the execution of the action. Otherwise, if $act(s)$ results in a set of states S, no error can occur and executing the action will nondetermistically lead to one of the states in S. The empty set can be used to model actions that do not terminate. Actions can be combined to form new actions. The most common combination is consecutive execution:

$$(act_1; act_2)(s) = \begin{cases} \top & \text{if } act_1(s) = \top \\ \top & \text{if } \exists s'. \; s' \in act_1(s) \; \wedge \; act_2(s') = \top \\ \bigcup_{s' \in act_1(s)} act_2(s') & \text{otherwise} \end{cases}$$

Another common combination is nondeterministic choice:

$$\left(\bigsqcup_{act \in act\text{-}set} act \right)(s) = \begin{cases} \top & \text{if } \exists act \in act\text{-}set. \; act(s) = \top \\ \bigcup_{act \in act\text{-}set} act(s) & \text{otherwise} \end{cases}$$

$$act_1 + act_2 = \bigsqcup_{act \in \{act_1, \; act_2\}} act$$

Definition 6 (Semantic Hoare Triples). For predicates P, Q and an action act, a *semantic Hoare triple* $\ll P \gg act \ll Q \gg$ holds, iff for all states p that satisfy the *precondition* P the action does not fail, i.e. $\forall p \in P. \; act(p) \neq \top$, and leads to a state that satisfies the *postcondition* Q, i.e. $\forall p \in P. \; act(p) \subseteq Q$. Notice, that this describes partial correctness, since a Hoare triple is trivially satisfied, if act does not terminate, i.e. if $act(s) = \emptyset$ holds.

Local reasoning is an essential feature of separation logic. It allows to extend a specification with an arbitrary context:

$$\frac{\ll P \gg \quad act \quad \ll Q \gg}{\ll P * R \gg \quad act \quad \ll Q * R \gg}$$

In order to provide local reasoning, only those actions are considered whose specifications can be safely extended using this inference rule. These actions are called *local*.

Definition 7 (Local Actions). An action act is called *local*, iff for all states s, s_1, s_2 with $s = s_1 \circ s_2$ and $act(s_1) \neq \top$ the evaluation of the action on the extended state does not fail ($act(s) \neq \top$) and $act(s) \subseteq act(s_1) * \{s_2\}$ holds.

The *skip* action defined by $skip(s) := \{s\}$ is a simple example of a local action. Other examples are $diverge(s) := \emptyset$ or $fail(s) := \top$. Sequential composition and nondeterministic choice preserve locality. The set of local actions forms together with the following order a complete lattice.

Definition 8 (Order of Actions). $act_1 \sqsubseteq act_2$ iff act_2 allows more behaviour than act_1, i. e. iff $\forall s. (act_2(s) = \top) \lor (act_1(s) \subseteq act_2(s))$ holds. Notice that this is equivalent to $\forall P, Q. \ll P \gg act_2 \ll Q \gg \implies \ll P \gg act_1 \ll Q \gg$.

This lattice of local actions is used to define a *best local action* as an infimum of local actions in this lattice. The HOL formalisation contains the corresponding definitions and theorems. However, here the discussion of this lattice is skipped. Instead an equivalent, high level characterisation is used.

Definition 9 (Best Local Action). Given a precondition P and a postcondition Q the *best local action* $bla[P, Q]$ is the most general local action that satisfies $\ll P \gg bla[P, Q] \ll Q \gg$. This means:

- $bla[P, Q]$ is a local action
- $\ll P \gg bla[P, Q] \ll Q \gg$ holds
- $bla[P, Q]$ is more general than any local actions act with $\ll P \gg act \ll Q \gg$, i. e. $act \sqsubseteq bla[P, Q]$

One common use of the best local action bla are the materialisation and annihilation actions. $materialise(P) := bla[emp, P]$ can be used to materialise some new part of the state that satisfies the predicate P. Similarly, $annihilate(P) := bla[P, emp]$ is used to annihilate some part of the state that satisfies P. Notice, that for certain P the annihilation $annihilate(P)$ behaves unexpectedly. If there is more than one substate that satisfies P, then $annihilate(P)$ diverges. Therefore, usually just *precise* predicates are used with annihilation:

Definition 10 (Precise Predicates). A predicate P is called *precise* iff for every state there is at most one substate that satisfies P.

As shown by the examples of materialisation and annihilation, bla is useful to define local actions. Often it is however necessary to relate the pre- and postcondition. For example, the postcondition of an action that increments the value of a variable needs to refer to the old value of this variable. This leads to the following extension of best local actions:

Definition 11 (Quantified Best Local Action). Given two functions $P_{(\cdot)}$ and $Q_{(\cdot)}$ that map some argument type to predicates the *quantified best local action* (*qbla*) is the most general local action that satisfies

$$\forall arg. \ll P_{arg} \gg qbla[P_{(\cdot)}, Q_{(\cdot)}] \ll Q_{arg} \gg$$

Another useful local action is *assume*. Given a predicate, *assume* skips if the predicate holds and diverges if it does not hold. In the next section, *assume* is used in combination nondeterministic choice and Kleene star to model conditional execution and loops. In order to be a local action, the predicate has to be intuitionistic, though.

Definition 12 (Intuitionistic Predicate). A predicate P is called *intuitionistic*, iff $P * true = P$ holds. This means that iff P holds for a state s, then it holds for all superstates $s' \succeq s$ as well. The *intuitionistic negation* $\neg_i P$ holds in a state s, if P does not hold for all superstates $s' \succeq s$. P is called *decided* in a set of states S, iff $\forall s \in S. \ s \in P \ \lor \ s \in \neg_i P$ holds.

For an intuitionistic predicate P the local action $assume(P)$ can be defined as

$$assume(P)(s) = \begin{cases} \{s\} & \text{if } s \in P \\ \emptyset & \text{if } s \in \neg_i P \\ \top & \text{otherwise} \end{cases}$$

3.2 Programs

This notion of local actions is extended to an abstraction of an imperative programming language. The basic constructs of this language are local actions. Besides local actions, the language contains the usual control structures like conditional execution and while-loops. Additionally, nondeterminism, concurrency and semaphores are supported. The definition of the semantics of this language follow ideas from Brooks [3] about Concurrent Separation Logic. Programs are translated to a set of traces that capture all possible interleavings during concurrent execution. The semantics of a program is given by nondeterministic choice between the semantics of its traces. As an additional layer of abstraction proto-traces are used between programs and traces.

Definition 13 (Proto-Trace). The set of *proto-traces* PTr is inductively defined to be the smallest set with

- $act \in PTr$ for all local actions act
- $pt_1 \ ; \ pt_2 \in PTr$ (sequential composition) for $pt_1, pt_2 \in PTr$
- $pt_1 \ || \ pt_2 \in PTr$ (parallel composition) for $pt_1, pt_2 \in PTr$
- $proccall(name, arg) \in PTr$ (procedure call) for all procedure-names $name$ and all arguments arg
- $l.pt \in PTr$ (lock declaration) for a lock l and $pt \in PTr$
- $with \ l \ do \ pt \in PTr$ (critical region) for a lock l and $pt \in PTr$

Definition 14 (Program). A program is a set of proto-traces. The set of all programs is denoted by *Prog*.

Definition 15 (Atomic Action). An *atomic action* is either a local action, a check $check(act_1, act_2)$ for local actions act_1, act_2 or a lock operation $P(l)$ or $V(l)$ for a lock l.

Definition 16 (Trace). A trace is a list of atomic actions. Let ϵ denote the empty trace. The concatenation of two traces t_1, t_2 is denoted as $t_1 \cdot t_2$.

To define the traces of a program, an environment is needed that fixes the semantics of procedure calls.

Definition 17 (Procedure Environment). A *procedure environment* is a finite map $penv : procedure\text{-}names \xrightarrow{\text{fin}} arguments \to Prog$ from procedure-names to a function from procedure arguments to programs.

Definition 18 (Traces of Proto-traces). Given an procedure environment $penv$, the traces of a proto-trace t after unfolding procedures n-times with respect to $penv$ (denoted as $T_{penv}^n(t)$) are given by:

$$T_{penv}^n(act) = \{act\}$$
$$T_{penv}^n(pt_1 ; pt_2) = \{t_1 \cdot t_2 \mid t_1 \in T_{penv}^n(pt_1) \wedge t_2 \in T_{penv}^n(pt_2)\}$$
$$T_{penv}^n(pt_1 \parallel pt_2) = \bigcup_{t_1 \in T_{penv}^n(pt_1), t_2 \in T_{penv}^n(pt_2)} t_1 \; zip \; t_2$$

$$T_{penv}^n(proccall(name, arg)) = \begin{cases} \{fail\} & \text{if } name \notin dom(penv) \\ \emptyset & \text{if } name \in dom(penv) \wedge n = 0 \\ \bigcup_{pt \in penv(name, arg)} T_{penv}^{n-1}(pt) & \text{otherwise} \end{cases}$$

$$T_{penv}^n(l.pt) = \{remove\text{-}locks(l,t) \mid t \in T_{penv}^n(pt) \wedge t \text{ is } l\text{-synchronised}\}$$
$$T_{penv}^n(with \; l \; do \; pt) = \{P(l) \cdot t \cdot V(l) \mid t \in T_{penv}^n(pt)\}$$

In this definition, *remove-locks(l,t)* removes all atomic actions concerned with the lock l, i.e. $P(l)$ and $V(l)$, from the trace t. A trace is *l-synchronised*, iff the lock-actions $P(l)$ and $V(l)$ are properly aligned. Finally, the auxiliary function *zip* builds all interleavings of two traces. It is given by

$$add\text{-}check(a_1, a_2, t) = \begin{cases} check(a_1, a_2) \cdot t & \text{if } a_1 \text{ and } a_2 \text{ are local actions} \\ t & \text{otherwise} \end{cases}$$

$$\epsilon \; zip \; t = t \; zip \; \epsilon = \{t\}$$

$$(a_1; t_1) \; zip \; (a_2; t_2) = \begin{cases} add\text{-}check(a_1, a_2, t) \mid t \in \{a_1; u \mid u \in t_1 \; zip \; (a_2; t_2)\} \cup \\ \{a_2; u \mid u \in (a_1; t_1) \; zip \; t_2\} \end{cases}$$

Finally, the traces of a proto-trace pt and a program p with respect to $penv$ are defined as

$$T_{penv}(pt) = \bigcup_{n \in \mathbb{N}} T_{penv}^n(pt) \qquad T_{penv}(p) = \bigcup_{pt \in p} T_{penv}(pt)$$

It remains to define the semantics of traces. Local actions in traces are just interpreted by themselves. Checks are added to enforce race-freedom. The semantics of lock actions is however more complicated.

One central idea behind Concurrent Separation Logic is to split the state into parts for each thread and each lock: a lock protects a part of the state. If a thread holds a lock, it can access this state, otherwise it cannot. Therefore, a precise predicate called *lock invariant* is associated with each lock. This invariant abstracts the part of the state that is protected by the lock. *materialise* and *annihilate* actions are used to make this abstracted state accessible/inaccessible.

Definition 19 (Semantics of Atomic Actions). The semantics of an atomic action with respect to a *lock-environment lenv : locks* $\rightarrow P(\Sigma)$ is given by

$$[\![act]\!]_{lenv} = act$$

$$[\![check(act_1, act_2)]\!]_{lenv}(s) = \begin{cases} \{s\} & \text{if } \exists s_1, s_2.\ s = s_1 \circ s_2 \ \wedge \\ & \quad act_1(s_1) \neq \top \ \wedge \ act_2(s_2) \neq \top \\ \top & \text{otherwise} \end{cases}$$

$$[\![P(l)]\!]_{lenv} = materialise(lenv(l))$$

$$[\![V(l)]\!]_{lenv} = annihilate(lenv(l))$$

Notice, that the semantics of an atomic action is a local action.

Definition 20 (Semantics of Traces, Programs). The semantics of a trace with respect to a lock-environment is the sequential combination of the semantics of its atomic actions. The semantics of a program is given by the nondeterministic choice between the semantics of its traces.

$$[\![\epsilon]\!]_{lenv} = skip \quad [\![a \cdot t]\!]_{lenv} = [\![a]\!]_{lenv} \ ; \ [\![t]\!]_{lenv} \quad [\![prog]\!]_{(penv,lenv)} = \bigsqcup_{t \in T_{penv}(prog)} [\![t]\!]_{lenv}$$

Notice that the semantics of a program is a always a local action. This allows concepts for actions to be easily lifted to programs:

Definition 21 (Hoare triple). A *Hoare triple* $\rhd_{(penv,lenv)} \{P\} \ prog \ \{Q\}$ holds, iff $\ll P \gg [\![prog]\!]_{(penv,lenv)} \ll Q \gg$ holds. If a Hoare triple holds for all environments, it is written as $\{P\} \ prog \ \{Q\}$.

Definition 22 (Program Abstractions). A program p_2 is an abstraction of a program p_1 with respect to some environment *env* (denoted as $p_1 \sqsubseteq_{env} p_2$), iff $[\![p_1]\!]_{env} \sqsubseteq [\![p_2]\!]_{env}$ holds.

3.3 Programming Constructs

In the previous section a concept of programs has been introduced. However, these programs hardly resemble the usual programs written in imperative languages. Common constructs like loops or conditional execution are missing. However, these can be easily defined.

Every proto-trace pt can be regarded as the program $\{pt\}$. This immediately enriches the programming language with procedure calls and local actions. In particular, one can use *skip, fail, assume, diverge, bla* and *qbla* as programs. A lot of instructions can easily be defined using *bla* or *qbla*. Given some suitable definitions for a state containing a stack, one could for example define an instruction that increments a variable as $x{+}{+} = qbla[\lambda c.\ x = c, \lambda c.\ x = (c+1)]$.

The HOL-formalisation uses a shallow embedding of local actions. So, any function $f : \Sigma \to P(\Sigma)^\top$ can be used as a program. However, to enforce that just local actions are used, f is implicitly replaced by *fail*, if it is not local. The other constructs for proto-traces can be lifted to programs as well:

$$p_1 \ ; \ p_2 = \{pt_1 \ ; \ pt_2 \mid pt_1 \in p_1 \cup \{diverge\} \ \wedge \ pt_2 \in p_2 \cup \{diverge\}\}$$
$$p_1 \parallel p_2 = \{pt_1 \parallel pt_2 \mid pt_1 \in p_1 \ \wedge \ pt_2 \in p_2\}$$
$$l.p = \{l.pt \mid pt \in p\}$$
$$with\ l\ do\ p = \{with\ l\ do\ pt \mid pt \in p\}$$

Some other constructs that are not available for proto-traces can be defined using the fact that programs are just sets of proto-traces. A simple example is nondeterministic choice: $p_1 + p_2 := p_1 \cup p_2$. In combination with *assume* and sequential composition of programs, this can be used to define conditional execution:

$$if\ B\ then\ p_1\ else\ p_2 = (assume(B); p_1) \ + \ (assume(\neg_i B); p_2)$$

This definition of conditional execution might seem weird. Remember however, that the framework is just interested in partial correctness. Therefore, it is fine to nondetermistically choose between paths and then diverge, if the wrong choice has been made.

Loops can be defined in a similar manner. However, to define loops, Kleene star is needed:

$$p^0 = skip \qquad p^{n+1} = p \ ; \ p^n \qquad p^* = \bigcup_{n \in \mathbb{N}} p^n$$
$$while\ B\ do\ p = (assume(B); p)^* \ ; \ assume(\neg_i B)$$

This time, one chooses nondetermistically, how often one needs to go around the loop. If the wrong number of iterations is picked, the trace is aborted by one of the *assume* statements.

Notice the definition of Kleene star. It is represented as a shallow embedding in HOL. Moreover, it uses nondeterministic choice over an infinite set of proto-traces. This simple example illustrates how flexible and powerful the combination of shallow and deep embeddings is. Depending on the needs of a concrete instantiation this power and flexibility can be used to define more constructs.

3.4   Inference Rules

Using the semantics of Abstract Separation Logic as presented above, one can deduce high-level inference rules. These inferences are used to verify specification on a high-level of abstraction instead of breaking every proof down to the semantic foundations. Some important inference rules, that are valid in Abstract Separation Logic are:

$$\frac{P_2 \Rightarrow P_1 \qquad Q_1 \Rightarrow Q_2 \qquad \rhd_{env} \{P_1\}p\{Q_1\}}{\rhd_{env} \{P_2\} \ p\{Q_2\}} \qquad\qquad \frac{p_1 \sqsubseteq p_2 \qquad \rhd_{env} \{P\}p_2\{Q\}}{\rhd_{env} \{P\} \ p_1\{Q\}}$$

$$\frac{\rhd_{env} \{P\} \ p \ \{Q\}}{\rhd_{env} \{P * R\} \ p \ \{Q * R\}} \qquad\qquad \frac{\rhd_{env} \{P\} \ p_1 \ \{Q\} \qquad \rhd_{env} \{Q\} \ p_2 \ \{R\}}{\rhd_{env} \{P\} \ p_1 \ ; \ p_2 \ \{R\}}$$

$$\frac{B \text{ is decided in } P}{\rhd_{env} \{P\} \ assume(B) \ \{P \ \wedge \ B\}} \qquad\qquad \frac{}{\rhd_{env} \{P_{arg}\} \ qbla[P_{(\cdot)}, Q_{(\cdot)}]\{Q_{arg}\}}$$

$$\frac{\rhd_{env} \{P\} \ p \ \{P\}}{\rhd_{env} \{P\} \ p^* \ \{P\}} \qquad\qquad \frac{\rhd_{env} \{P\} \ p_1 \ \{Q\} \qquad \rhd_{env} \{P\} \ p_2 \ \{Q\}}{\rhd_{env} \{P\} \ p_1 + p_2 \ \{Q\}}$$

$$\frac{name \in dom(penv) \qquad \rhd_{(penv, lenv)} \{P\} \ penv(name, arg) \ \{Q\}}{\rhd_{(penv, lenv)} \{P\} \ \text{proccall}(name, arg)\{Q\}} \qquad\qquad \frac{B \text{ is decided in } P \qquad \rhd_{env} \{B \wedge P\} \ p_1 \ \{Q\} \qquad \rhd_{env} \{\neg_i B \wedge P\} \ p_2 \ \{Q\}}{\rhd_{env} \{P\} \ \text{if } B \text{ then } p_1 \text{ else } p_2 \ \{Q\}}$$

$$\frac{B \text{ is decided in } P \qquad \rhd_{env} \{B \wedge P\} \ p \ \{P\}}{\rhd_{env} \{P\} \ \text{while} \ B \ \text{do} \ p \ \{\neg_i B \wedge P\}} \qquad\qquad \frac{\rhd_{env} \{P_1\} \ p_1 \ \{Q_1\} \qquad \rhd_{env} \{P_2\} \ p_2 \ \{Q_2\}}{\rhd_{env} \{P_1 * P_2\} \ p_1 \ || \ p_2 \ \{Q_1 * Q_2\}}$$

$$\frac{lenv(l) = r \qquad \rhd_{(penv, lenv)} \{P\} \ p \ \{Q\}}{\rhd_{(penv, lenv)} \{P * r\} \ l.p \ \{Q * r\}} \qquad\qquad \frac{lenv(l) = r \qquad \rhd_{(penv, lenv)} \{P * r\} \ p \ \{Q * r\}}{\rhd_{(penv, lenv)} \{P\} \ \text{with } l \ \text{do } p\{Q\}}$$

These inference rules are very useful. However, the reader might notice, that there is a problem with recursive functions. The inference rule that handles procedure-calls replaces the call with the definition of the procedure. This is fine for non-recursive functions. However, an implicit induction is needed for recursive functions.

Definition 23 (Procedure Specification). A *procedure specification* consists of a lock-environment *lenv*, a procedure-environment *penv* and specification functions $P_{(\cdot, \cdot, \cdot)}$, $Q_{(\cdot, \cdot, \cdot)}$. It holds, iff all procedures satisfy their specification in the given environment:

$$\forall f \in dom(penv), \text{arg}, x. \ \rhd_{(penv, lenv)} \{P_{(f, \text{arg}, x)}\}\text{proccall}(name, arg)\{Q_{(f, \text{arg}, x)}\}$$

To prove that a procedure specification holds, it is sufficient to show that assuming that all procedures satisfy their specification, their bodies satisfy the specification. One does not need to show that possible recursions terminate, since Abstract Separation Logic just talks about partial correctness.

There is tool-support in the HOL-formalisation to handle procedure specifications. To prove a procedure specification, it is sufficient to prove the specifications of all procedure bodies, where a procedure call *proccall*(*name, arg*) has been replaced by $qbla[P_{(name,\text{arg},\cdot)}, Q_{(name,\text{arg},\cdot)}]$. This means that the resulting Hoare triples do not contain procedure calls any more. Therefore, the resulting Hoare triples do not depend on the procedure environment.

Making the Hoare triples independent from the lock-environment as well is not necessary, but often useful. The lock-operations *l.p* and *with l do p* can be eliminated by introducing *annihilate*(*lenv*(*l*)) and *materialise*(*lenv*(*l*)) at appropriate places in *p*. This moves the knowledge about lock invariants from the environment to the program itself, making the environment redundant.

Loops can be eliminated in a similar manner. Given a loop-invariant $I_{(\cdot)}$ such that for all x an intuitionistic predicate B is decided in I_x, a while-loop *while B do p* can be abstracted by $qbla[I_{(\cdot)}, I_{(\cdot)} \wedge \neg_i B]$, if it can be proved that the body of the loop really satisfies the invariant.

After these preprocessing steps, one usually just needs to reason about programs consisting of local actions and conditional-execution, for which the presented inference rules are very useful.

3.5   Holfoot

The instantiation of the Abstract Separation Logic framework to Holfoot consists of two steps. First, the framework is instantiated to use a stack that maps variables to permissions and values. This instantiation is based on ideas from *Variables as Resource in Hoare Logic* [8]. The concrete type of the variables and values is not specified. Similarly, the stack is just a part of an abstract state. Nevertheless, this instantiation is sufficient to reason about pure expressions, assignments, local variables, etc. In a second step, this setting is instantiated to Holfoot.

Holfoot represents stack-variables with strings and uses natural numbers as values. Furthermore the abstract component of the state is instantiated to a heap from locations (represented by natural numbers without zero) and tags (represented as strings) to values (natural numbers). Using this concrete representation of a state, it is easy to define actions on these states. For example, the field-lookup action v = e->t is defined as

```
val holfoot_field_lookup_action_def = Define '
  (holfoot_field_lookup_action v e t) ((st,h):holfoot_state)) =
    if (~(var_res_sl___has_write_permission v st) \/
        IS_NONE (e st)) then NONE else
    let loc = (THE loc_opt) in (
      if (~(loc IN FDOM h) \/ (loc = 0)) then NONE else
      SOME {var_res_ext_state_var_update v ((h ' loc) t) (st,h)})';
```

This action fails, if there is no write permission on the variable v or if the expression e fails to be evaluated in the current state (for example because a read permission on a variable it uses is missing). Otherwise, it checks whether the location pointed to by e is in the heap. If it is, the value of v is updated by the value found in the heap at that location indexed by tag **t**. Otherwise, i. e. if the location is not in the heap, the action fails.

Similarly to actions, it is straightforward to define predicates. For example, **e1 |-> L** is defined as:

```
val holfoot_ap_points_to_def = Define '
  holfoot_ap_points_to e1 L = \(st,h):holfoot_state.
    let loc_opt = (e1 st) in (IS_SOME (loc_opt) /\
    let (loc = THE loc_opt) in (~(loc = 0) /\ ((FDOM h)= {loc}) /\
    (FEVERY (\(tag,exp). IS_SOME (exp st) /\ (THE (exp st) = (h' loc) tag)) L))';
```

This definition of **|->** is used to define predicates for single linked lists. The data content of these lists is represented by lists of natural numbers. Therefore HOL's list libraries can be used to reason about the data content.

Since actions and predicates are shallowly embedded, it is easy to extend Holfoot with new actions and predicates. Moreover, the automation has been designed with extensions in mind.

4   Conclusion and Future Work

The main contribution of this work is the formalisation of Abstract Separation Logic and demonstrating that Abstract Separation Logic is powerful and flexible enough to be used as a basis for separation logic tools. The formalisation of Abstract Separation Logic contains some minor extensions like the addition of procedures. However, it mainly follows the original definitions [4].

The Smallfoot case study demonstrates the potential of Abstract Separation Logic. However, it is interesting in its own right as well. The detected bug in Smallfoot shows that high-assurance implementations of even comparatively simple tools like Smallfoot are important. Moreover, Holfoot is one of the very few separation logic tools that can reason about the content of data-structures as well as the shape. Combining separation logic with reasoning about data-content is currently one of the main challenges for separation logic tools. As the example of parallel mergesort demonstrates, Holfoot can answer this challenge by combining the power of the interactive prover HOL with the automation separation logic provides.

In the future, I will try to improve the level of automation. Moreover, I plan to add a concept of arrays to Holfoot. This will put my claim that Holfoot is easily extensible to a test, since it requires adding new actions for allocating / deallocating blocks of the heap as well as adding predicates for arrays. However, the main purpose of adding arrays is reasoning about pointer arithmetic. It will be interesting to see, how HOL can help to verify algorithms that use pointer-arithmetic and how much automation is possible.

Acknowledgements

I would like to thank Matthew Parkinson, Mike Gordon, Alexey Gotsman, Magnus Myreen and Viktor Vafeiadis for a lot of discussions, comments and criticism.

References

[1] Appel, A.W., Blazy, S.: Separation logic for small-step Cminor. In: Schneider, K., Brandt, J. (eds.) TPHOLs 2007. LNCS, vol. 4732, pp. 5–21. Springer, Heidelberg (2007)

[2] Berdine, J., Calcagno, C., O'Hearn, P.W.: Smallfoot: Modular automatic assertion checking with separation logic. In: de Boer, F.S., Bonsangue, M.M., Graf, S., de Roever, W.-P. (eds.) FMCO 2005. LNCS, vol. 4111, pp. 115–137. Springer, Heidelberg (2006)

[3] Brookes, S.: A semantics for concurrent separation logic. Theor. Comput. Sci. 375(1-3), 227–270 (2007)

[4] Calcagno, C., O'Hearn, P.W., Yang, H.: Local action and abstract separation logic. In: LICS 2007: Proceedings of the 22nd Annual IEEE Symposium on Logic in Computer Science, Washington, DC, USA, pp. 366–378. IEEE Computer Society, Los Alamitos (2007)

[5] Distefano, D., O'Hearn, P.W., Yang, H.: A local shape analysis based on separation logic. In: Hermanns, H., Palsberg, J. (eds.) TACAS 2006. LNCS, vol. 3920, pp. 287–302. Springer, Heidelberg (2006)

[6] Marti, N., Affeldt, R., Yonezawa, A.: Towards formal verification of memory properties using separation logic. In: 22nd Workshop of the Japan Society for Software Science and Technology, Tohoku University, Sendai, Japan, September 13–15. Japan Society for Software Science and Technology (2005)

[7] O'Hearn, P.W., Reynolds, J.C., Yang, H.: Local reasoning about programs that alter data structures. In: Fribourg, L. (ed.) CSL 2001 and EACSL 2001. LNCS, vol. 2142, pp. 1–19. Springer, Heidelberg (2001)

[8] Parkinson, M., Bornat, R., Calcagno, C.: Variables as resource in hoare logics. In: LICS 2006: Proceedings of the 21st Annual IEEE Symposium on Logic in Computer Science, Washington, DC, USA, pp. 137–146. IEEE Computer Society, Los Alamitos (2006)

[9] Reynolds, J.C.: Separation logic: A logic for shared mutable data structures. In: LICS 2002: Proceedings of the 17th Annual IEEE Symposium on Logic in Computer Science, Washington, DC, USA, pp. 55–74. IEEE Computer Society, Los Alamitos (2002)

[10] Tuch, H., Klein, G., Norrish, M.: Types, bytes, and separation logic. In: POPL 2007: Proceedings of the 34th annual ACM SIGPLAN-SIGACT symposium on Principles of programming languages, pp. 97–108. ACM, New York (2007)

[11] Weber, T.: Towards mechanized program verification with separation logic. In: Marcinkowski, J., Tarlecki, A. (eds.) CSL 2004. LNCS, vol. 3210, pp. 250–264. Springer, Heidelberg (2004)

Liveness Reasoning with Isabelle/HOL[*]

Jinshuang Wang[1,2], Huabing Yang[1], and Xingyuan Zhang[1]

[1] PLA University of Science and Technology, Nanjing 210007, China
[2] State Key Lab for Novel Software Technology, Nanjing University, Nanjing 210093, China
{wangjinshuang,xingyuanz}@gmail.com

Abstract. This paper describes an extension of Paulson's inductive protocol verification approach for liveness reasoning. The extension requires no change of the system model underlying the original inductive approach. Therefore, all the advantages, which makes Paulson's approach successful for safety reasoning are kept, while liveness reasoning becomes possible. To simplify liveness reasoning, a new fairness notion, named *Parametric Fairness* is used instead of the standard ones. A probabilistic model is established to support this new fairness notion. Experiments with small examples as well as real world communication protocols confirm the practicality of the extension. All the work has been formalized with Isabelle/HOL using Isar.

Keywords: Liveness Proof, Inductive Protocol Verification, Probabilistic Model, Parametric Fairness.

1 Introduction

Paulson's inductive approach has been used to verify safety properties of many security protocols [1, 2]. The success gives incentives to extend this approach to a general approach for protocol verification. To achieve this goal, a method for the verification of liveness properties is needed. According to Manna and Pnueli [3], temporal properties are classified into three classes: safety properties, response properties and reactivity properties. The original inductive approach only deals with safety properties. In this paper, proof rules for liveness properties (both response and reactivity) are derived under the same execution model as the original inductive approach. These liveness proof rules can be used to reduce the proof of liveness properties to the proof of safety properties, a task well solved by the original approach.

The proof rules are derived based on a new notion of fairness, *parametric fairness*, which is an adaption of the α-fairness [4,5,6] to the setting of HOL. *Parametric fairness* is properly stronger than standard fairness notions such as *weak fairness* and *strong fairness*. We will explain why the use of *parametric fairness* can deliver more liveness results through simpler proofs.

A probabilistic model is established to show the soundness of our new fairness notion. It is proved that the set of all parametrically fair execution traces is measurable and has probability 1. Accordingly, the definition of parametric fairness is reasonable.

---

[*] This research was funded by 863 Program(2007AA01Z409) and NNSFC(60373068) of China.

S. Berghofer et al. (Eds.): TPHOLs 2009, LNCS 5674, pp. 485–499, 2009.
© Springer-Verlag Berlin Heidelberg 2009

The practicability of this liveness reasoning approach has been confirmed by experiments of various sizes [7, 8, 9]. All the work has been formalized with Isabelle/HOL using Isar [10, 11], although the general approach is not necessarily confined to this particular system.

The paper is organized as the following: section 2 presents *concurrent system*, the execution model of inductive approach; section 3 gives a shallow embedding of LTL (Linear Temporal Logic); section 4 gives an informal explanation of parametric fairness; section 5 explains the liveness proof rules; section 6 establishes a probabilistic model for parametric fairness; section 7 describes liveness verification examples. section 8 discusses related works; section 9 concludes.

2   Concurrent Systems

In the inductive approach, system state only changes with the happening of events. Accordingly, it is natural to represent a system state with the list of events happening so far, arranged in reverse order. A system is concurrent because its states are nondeterministic, where a state is *nondeterministic*, if there are more than one event eligible to happen under that state. The specification of a *concurrent system* is just a specification of this eligible relation. Based on this view, formal definition of concurrent system is given in Fig. 1.

$$\sigma_i \equiv \sigma\, i$$

primrec $[\sigma]_0 = []$
$\qquad [\sigma]_{(Suc\ i)} = \sigma_i \,\#\, [\sigma]_i$

$\tau\, [cs{>}\, e \equiv (\tau, e) \in cs$

inductive-set vt :: $('a\ list \times 'a)\ set \Rightarrow 'a\ list\ set$
  for cs ::$('a\ list \times 'a)\ set$ **where**
  vt-nil [intro] : $[] \in vt\ cs\ |$
  vt-cons [intro]: $[\![\tau \in vt\ cs;\ \tau\, [cs{>}\, e]\!] \Longrightarrow (e\, \#\, \tau) \in vt\ cs$

consts derivable :: $'a \Rightarrow 'b \Rightarrow bool$ (- ⊢ - [64, 64] 50)
  fnt-valid-def: $cs \vdash \tau \equiv \tau \in vt\ cs$
  inf-valid-def: $cs \vdash \sigma \equiv \forall i.\ [\sigma]_i\, [cs{>}\, \sigma_i$

Fig. 1. Definition of Concurrent System

The type of concurrent systems is $('a\ list \times 'a)\ set$, where $'a$ is the type of events. Concurrent systems are written as cs. The expression $(\tau, e) \in cs$ means that event e is eligible to happen under state τ in concurrent system cs. The notation $(\tau, e) \in cs$ is abbreviated as $\tau\, [cs{>}\, e$. The set of reachable states is written as $vt\ cs$ and $\tau \in vt\ cs$ is abbreviated as $cs \vdash \tau$. An execution of a concurrent system is an infinite sequence of events, represented as a function with type $nat \Rightarrow 'a$. The i-th event in execution σ is abbreviated as σ_i. The prefix consisting of the first i events is abbreviated as $[\sigma]_i$. For σ to be a valid execution of cs (written as $cs \vdash \sigma$), σ_i must be eligible to happen under $[\sigma]_i$.

Infinite execution σ is called *nondeterministic* if it has infinitely many nondeterministic prefixes. It is obvious from the definition of $cs \vdash \sigma$ that σ represents the choices on which event to happen next at its nondeterministic prefixes.

3   Embedding LTL

LTL (Linear Temporal Logic) used to represent liveness properties in this paper is defined in Fig. 2. LTL formulae are written as φ, ψ, κ etc. The type of LTL formulae is defined as $'a$ tlf. The expression $(\sigma, i) \models \varphi$ means that LTL formula φ is valid at moment i of the infinite execution σ. The operator $\models$ is overloaded, so that $\sigma \models \varphi$ can be defined as the abbreviation of $(\sigma, 0) \models \varphi$. The *always* operator $\square$, *eventual* operator $\Diamond$, *next* operator $\odot$, *until* operator $\triangleright$ are defined literally. An operator $\langle \text{-} \rangle$ is defined to lift a predicate on finite executions up to a LTL formula. The temporal operator $\hookrightarrow$ is the lift of logical implication $\longrightarrow$ up to LTL level. For an event e, the term $(\!|e|\!)$ is a predicate on finite executions stating that the last happened event is e. Therefore, the expression $\langle (\!|e|\!) \rangle$ is an LTL formula saying that event e happens at the current moment.

---

types $'a$ $tlf = (nat \Rightarrow 'a) \Rightarrow nat \Rightarrow bool$

consts valid-under :: $'a \Rightarrow 'b \Rightarrow bool$ (- $\models$ - [64, 64] 50)
defs (**overloaded**) pr $\models \varphi \equiv$ let $(\sigma, i) = pr$ in pr φ σ i
defs (**overloaded**) $(\sigma{::}nat{\Rightarrow}'a) \models (\varphi{::}'a\,tlf) \equiv (\sigma{::}nat \Rightarrow 'a, (0{::}nat)) \models \varphi$

$\square\varphi \equiv \lambda\,\sigma\,i.\,\forall\,j.\,i \le j \longrightarrow (\sigma, j) \models \varphi$
$\Diamond\varphi \equiv \lambda\,\sigma\,i.\,\exists\,j.\,i \le j \wedge (\sigma, j) \models \varphi$

constdefs lift-pred :: $('a\,list \Rightarrow bool) \Rightarrow 'a\,tlf$ $(\langle\text{-}\rangle$ [65] 65)
$\langle P \rangle \equiv \lambda\,\sigma\,i.\,P\,[\![\sigma]\!]_i$

constdefs lift-imply :: $'a\,tlf \Rightarrow 'a\,tlf \Rightarrow 'a\,tlf$ (- $\hookrightarrow$ - [65, 65] 65)
$\varphi \hookrightarrow \psi \equiv \lambda\,\sigma\,i.\,\varphi\,\sigma\,i \longrightarrow \psi\,\sigma\,i$

constdefs last-is :: $'a \Rightarrow 'a\,list \Rightarrow bool$
last-is e $\tau \equiv$ (case τ of $[]$ $\Rightarrow$ False $|$ $(e' \# \tau') \Rightarrow e' = e$)

syntax -is-last :: $'a \Rightarrow ('a\,list \Rightarrow bool)$ $((\!|\text{-}|\!)$ [64] 1000)
translations $(\!|e|\!) \rightleftharpoons$ last-is e

Fig. 2. The Embedding of Linear Temporal Logic in Isabelle/HOL

4   The Notion of Parametric Fairness

In this section, *parametric fairness* is introduced as a natural extension of standard fairness notions. To show this point, the definition of parametric fairness *PF* is given in Fig. 3 as the end of a spectrum of fairness notions starting from standard ones.

A study of Fig. 3 may reveal how the definition of *PF* is obtained through incremental modifications, from the standard *weak fairness WF* to the standard *strong fairness SF* and finally through the less standard extreme fairness *EF*. Each fairness notion *?F* has a

---

constdefs WF_α :: $('a\ list \times 'a)\ set \Rightarrow 'a \Rightarrow (nat \Rightarrow 'a) \Rightarrow bool$
  $WF_\alpha\ cs\ e\ \sigma \equiv \sigma \models \Box(\lambda\,\sigma\,i.\ [\sigma]_i\ [cs> e) \longrightarrow \sigma \models \Box\Diamond(\lambda\,\sigma\,i.\ \sigma_i = e)$

constdefs WF :: $('a\ list \times 'a)\ set \Rightarrow (nat \Rightarrow 'a) \Rightarrow bool$
  $WF\ cs\ \sigma \equiv \forall\ e.\ WF_\alpha\ cs\ e\ \sigma$

constdefs SF_α :: $('a\ list \times 'a)\ set \Rightarrow 'a \Rightarrow (nat \Rightarrow 'a) \Rightarrow bool$
  $SF_\alpha\ cs\ e\ \sigma \equiv \sigma \models \Box\Diamond(\lambda\,\sigma\,i.\ [\sigma]_i\ [cs> e) \longrightarrow \sigma \models \Box\Diamond(\lambda\,\sigma\,i.\ \sigma_i = e)$

constdefs SF :: $('a\ list \times 'a)\ set \Rightarrow (nat \Rightarrow 'a) \Rightarrow bool$
  $SF\ cs\ \sigma \equiv \forall\ e.\ SF_\alpha\ cs\ e\ \sigma$

constdefs EF_α :: $('a\ list \times 'a)\ set \Rightarrow ('a\ list \Rightarrow bool) \Rightarrow ('a\ list \Rightarrow 'a) \Rightarrow 'a\ seq \Rightarrow bool$
  $EF_\alpha\ cs\ P\ E\ \sigma \equiv \sigma \models \Box\Diamond(\lambda\,\sigma\,i.\ P\ [\sigma]_i \wedge [\sigma]_i\ [cs> E\ [\sigma]_i)$
  $\qquad\qquad\qquad\longrightarrow \sigma \models \Box\Diamond(\lambda\,\sigma\,i.\ P\ [\sigma]_i \wedge \sigma_i = E\ [\sigma]_i)$

constdefs EF :: $('a\ list \times 'a)\ set \Rightarrow (nat \Rightarrow 'a) \Rightarrow bool$
  $EF\ cs\ \sigma \equiv \forall\ P\ E.\ EF_\alpha\ cs\ P\ E\ \sigma$

types $'a\ pe = ('a\ list \Rightarrow bool) \times ('a\ list \Rightarrow 'a)$

constdefs PF :: $('a\ list \times 'a)\ set \Rightarrow 'a\ pe\ list \Rightarrow (nat \Rightarrow 'a) \Rightarrow bool$
  $PF\ cs\ pel\ \sigma \equiv list\text{-}all\ (\lambda\ (P, E).\ EF_\alpha\ \ cs\ P\ E\ \sigma)\ pel$

Fig. 3. Definition of Fairness Notions

corresponding pre-version $?F_\alpha$. For example, *WF* is obtained from WF_α by quantifying over *e*.

For any nondeterministic execution σ, its progress towards desirable states depends on the choice of helpful events under the corresponding helpful states (or helpful prefixes). The association between helpful states and helpful events can be specified using (P, E)-pairs, where *P* is a predicate on states used to identify helpful states and *E* is the *choice function* used to choose the corresponding helpful events under *P*-states. A pair (P, E) is called *enabled* under state τ if $P\ \tau$ is true, and it is called *executed* under τ if $E\ \tau$ is chosen to happen.

An infinite execution σ is said to treat the pair (P, E) fairly, if (P, E) is executed infinitely often in σ, unless (P, E) is enabled for only finitely many times. In an unified view, every fairness notion is about how fair (P, E)-pairs should be treated by infinite executions. For example, *EF* requires that any expressible (P, E)-pair be fairly treated while *PF* only requires that (P, E)-pairs in the parameter *pel* be fairly treated, and this explains why it is called *parametric fairness*.

Common sense tells us that a fair execution σ should make nondeterministic choices randomly. However, standard fairness notions such as *WF* and *SF* fail to capture this intuition. Consider the concurrent system defined in Fig. 4 and Fig. 5.

Fig. 4 is a diagram of the concurrent system cs_2 formally defined in Fig. 5, where diagram states are numbered with the value of function F_2. Any *intuitively fair* execution starting from state 2 should finally get into state 0. If a fairness notion *?F* correctly captures our intuition, the following property should be valid:

$$[\![cs_2 \vdash \sigma;\ ?F\ cs_2\ \sigma]\!] \Longrightarrow \sigma \models \Box(\langle \lambda\,\tau.\ F\,\tau = 2 \rangle \hookrightarrow \Diamond\,\langle \lambda\,\tau.\ F\,\tau = 0 \rangle) \qquad (1)$$

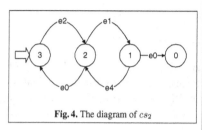

Fig. 4. The diagram of cs_2

---

datatype Evt $= e_0 \mid e_1 \mid e_2 \mid e_3 \mid e_4$

fun F_2 :: Evt list $\Rightarrow$ nat
where
$F_2 \; [] = 3 \mid$
$F_2 \; (e_0 \; \# \; \tau) = ($if $(F_2 \; \tau = 1)$ then 0 else
$\qquad\qquad\qquad$ if $(F_2 \; \tau = 2)$ then 3 else $F_2 \; \tau) \mid$
$F_2 \; (e_1 \; \# \; \tau) = ($if $(F_2 \; \tau = 2)$ then 1 else $F_2 \; \tau) \mid$
$F_2 \; (e_2 \; \# \; \tau) = ($if $(F_2 \; \tau = 3)$ then 2 else $F_2 \; \tau) \mid$
$F_2 \; (e_4 \; \# \; \tau) = ($if $(F_2 \; \tau = 1)$ then 2 else $F_2 \; \tau)$

inductive-set cs_2 :: (Evt list $\times$ Evt) set
where
$r_0 : F_2 \; \tau = 1 \Longrightarrow (\tau, e_0) \in cs_2 \mid$
$r_1 : F_2 \; \tau = 2 \Longrightarrow (\tau, e_1) \in cs_2 \mid$
$r_2 : F_2 \; \tau = 3 \Longrightarrow (\tau, e_2) \in cs_2 \mid$
$r_3 : F_2 \; \tau = 2 \Longrightarrow (\tau, e_0) \in cs_2 \mid$
$r_4 : F_2 \; \tau = 1 \Longrightarrow (\tau, e_4) \in cs_2$

Fig. 5. The definition of cs_2

Unfortunately, neither *WF* nor *SF* serves this purpose. Consider the execution $(e_2. \; e_1. \; e_4. \; e_0)^\omega$, which satisfies both *WF* cs_2 and *SF* cs_2 while violating the conclusion of (1).

The deficiency of standard fairness notions such as *SF* and *WF* is their failure to specify the association between helpful states and helpful events explicitly. For example, *SF* only requires any infinitely enabled event be executed infinitely often. Even though execution $(e_2. \; e_1. \; e_4. \; e_0)^\omega$ satisfies *SF*, it is not intuitively random in that it is still biased towards avoiding the helpful event e_0 under the corresponding helpful state *1*. Therefore, (1) is not valid if *?F* is instantiated either to *SF* or *WF*. *Extreme fairness* was proposed by Pnueli [12] to solve this problem. The *EF* is a direct expression of extreme fairness in HOL, which requires that *all* expressible (P,E)-pairs be fairly treated. Execution $(e_2. \; e_1. \; e_4. \; e_0)^\omega$ does not satisfy *EF*, because the (P, E)-pair $(\lambda\tau. \; F_2 \; \tau = 1, \; \lambda\tau. \; e_1)$ is not fairly treated. In fact, (1) is valid if *?F* is instantiated to *EF*.

Unfortunately, a direct translation of *extreme fairness* in HOL is problematic, because the universal quantification over P, E may accept any well-formed expression of the right type. Given any nondeterministic execution σ, it is possible to construct a pair (P_σ, E_σ) which is not fairly treated by σ. Accordingly, any nondeterministic execution σ is not *EF*, as confirmed by the following lemma:

$$\sigma \models \Box\Diamond(\lambda\sigma \; i. \; \{e. \; [\![\sigma]\!]_i \; [cs> e \wedge e \neq \sigma_i\} \neq \{\}) \Longrightarrow \neg \; EF \; cs \; \sigma \qquad (2)$$

The premise of (2) formally expresses that σ is nondeterministic. The construction of (P_σ, E_σ) is a diagonal one which makes a choice different from the one made by σ on every nondeterministic prefix. Detailed proof of (2) can be found in [13].

Now, since most infinite executions of a concurrent system are nondeterministic, *EF* will rule out almost all executions. If *EF* is used as the fairness premises of a liveness property, the overall statement is practically meaningless. Parametric fairness *PF* is proposed to solve the problem of *EF*. Instead of requiring all (P, E)-pairs be fairly treated, *PF* only requires (P, E)-pairs appearing in its parameter *pel* be fairly treated. Since there are only finitely many (P, E)-pairs in *pel*, most executions of a concurrent system are kept, even if every (P, E)-pair in *pel* rules out some measurement of them. Section 6 will make this argument precise by establishing a probabilistic model for *PF*.

5   Liveness Rules

According to Manna [3], response properties are of the form $\sigma \models \Box(\langle P \rangle \hookrightarrow \Diamond\langle Q \rangle)$, where $\langle P \rangle$ and $\langle Q \rangle$ are *past formulae* (in [3]'s term), obtained by lifting predicates on finite traces. The conclusion of (1) is of this form. Reactivity properties are of the form $\sigma \models (\Box\Diamond\langle P \rangle) \hookrightarrow (\Box\Diamond\langle Q \rangle)$ meaning: *if P holds infinitely often in σ, then Q holds infinitely often in σ as well.*

The proof rule for response property is the theorem *resp-rule*:

$$[\![RESP\ cs\ F\ E\ N\ P\ Q;\ cs \vdash \sigma;\ PF\ cs\ \{\!F, E, N\!\}\ \sigma]\!] \Longrightarrow \sigma \models \Box\langle P \rangle \hookrightarrow \Diamond\langle Q \rangle$$

and the proof rule for reactivity property is the theorem *react-rule*:

$$[\![REACT\ cs\ F\ E\ N\ P\ Q;\ cs \vdash \sigma;\ PF\ cs\ \{\!F, E, N\!\}\ \sigma]\!] \Longrightarrow \sigma \models \Box\Diamond\langle P \rangle \hookrightarrow \Box\Diamond\langle Q \rangle$$

The symbols used in these two theorems are given in Fig. 6.

---

consts pel-of :: ($'$a list $\Rightarrow$ nat) $\Rightarrow$ ($'$a list $\Rightarrow$ $'$a) $\Rightarrow$ nat $\Rightarrow$
$\qquad\qquad\qquad (((\,'$a list $\Rightarrow$ bool) $\times$ ($'$a list $\Rightarrow$ $'$a)) list ($\{\!$-, -, -$\!\}$ [64, 64, 64] 1000)
primrec $\{\!F, E, 0\!\} = [\,]$
$\qquad\quad \{\!F, E, (\text{Suc } n)\!\} = (\lambda\,\tau.\,F\,\tau = \text{Suc } n,\,E)\ \#\ \{\!F, E, n\!\}$

syntax -drop :: $'$a list $\Rightarrow$ nat $\Rightarrow$ $'$a list ($\lceil$-$\rceil$- [64, 64] 1000)
translations $\lceil l \rceil_n \rightleftharpoons$ drop n l

$\lceil P \longmapsto \neg Q * \rceil \equiv \lambda\,\tau.\,(\exists i \le |\tau|.\,P\,\lceil\tau\rceil_i \wedge (\forall k.\,0 < k \wedge k \le i \longrightarrow \neg Q\,\lceil\tau\rceil_k))$

locale RESP $=$
$\quad$**fixes** cs :: ($'$a list $\times$ $'$a) set **and** F :: $'$a list $\Rightarrow$ nat **and** E :: $'$a list $\Rightarrow$ $'$a
$\quad$**and** N :: nat **and** P :: $'$a list $\Rightarrow$ bool **and** Q :: $'$a list $\Rightarrow$ bool
$\quad$**assumes** mid: $[\![cs \vdash \tau;\ \lceil P \longmapsto \neg Q * \rceil\ \tau;\ \neg\,Q\ \tau]\!] \Longrightarrow 0 < F\,\tau \wedge F\,\tau < N$
$\quad$**assumes** fd: $[\![cs \vdash \tau;\ 0 < F\,\tau]\!] \Longrightarrow \tau\ [cs\!> E\,\tau \wedge F\ (E\,\tau\ \#\ \tau) < F\,\tau$

locale REACT $=$
$\quad$**fixes** cs :: ($'$a list $\times$ $'$a) set **and** F :: $'$a list $\Rightarrow$ nat **and** E :: $'$a list $\Rightarrow$ $'$a
$\quad$**and** N :: nat **and** P :: $'$a list $\Rightarrow$ bool **and** Q :: $'$a list $\Rightarrow$ bool
$\quad$**assumes** init: $[\![cs \vdash \tau;\ P\,\tau]\!] \Longrightarrow F\,\tau < N$
$\quad$**assumes** mid: $[\![cs \vdash \tau;\ F\,\tau < N;\ \neg\,Q\,\tau]\!] \Longrightarrow \tau\ [cs\!> E\,\tau \wedge F\ (E\,\tau\ \#\ \tau) < F\,\tau$

Fig. 6. Premises of Liveness Rules

Let's explain the *resp-rule* first. Premise *RESP cs F E N P Q* expresses constraints on *cs*'s state transition diagram. *RESP* is defined as a locale predicate, where assumption *mid* requires: for any state τ, after reaching P before reaching Q (characterized by

$[P \longmapsto \neg Q *] \tau$), the value of $F \tau$ is between 0 and N; assumption fd requires that if F τ is between 0 and N, event $E \tau$ must be eligible to happen under τ and the happening of it will decrease the value of function F. For any state satisfying $[P \longmapsto \neg Q *] \tau$, by repeatedly applying fd, a path leading from τ to Q can be constructed. If every pair in list $[(\lambda \tau . F \tau = n, E) \mid n \in \{1 \ldots N\}]$ is fairly treated by σ, σ will follow this path and eventually get into a Q state. This fairness requirement is expressed by the premise PF cs $\{F, E, N\}$ σ, where $\{F, E, N\}$ evaluates to $[(\lambda \tau . F \tau = n, E) \mid n \in \{1 \ldots N\}]$.

As an example, when $resp$-$rule$ is used to prove statement (1), the P is instantiated to $(\lambda \tau . F_2 = 2)$, and Q to $(\lambda \tau . F_2 = 0)$. The F is instantiated to F_2, and E to the following E_2:

constdefs E_2 :: *Evt list* $\Rightarrow$ *Evt*
  $E_2 \tau \equiv$ (*if* $(F_2 \tau = 3)$ *then* e_2 *else if* $(F_2 \tau = 2)$ *then* e_1 *else if* $(F_2 \tau = 1)$ *then* e_0 *else* e_0)

The N is instantiated to 3, so that $\{F_2, E_2, 3\}$ evaluates to $[(\lambda \tau . F_2 \tau = 3, E), (\lambda \tau . F_2 \tau = 2, E), (\lambda \tau . F_2 \tau = 1, E)]$.

Now, let's explain rule $react$-$rule$. Premise PF cs $\{F, E, N\}$ σ still has the same meaning. Since now P-states are reached infinitely often, a condition much stronger than in $resp$-$rule$, premise $REACT$ can be weaker, which only requires there exists a path leading to Q from every P state, while $RESP$ requires the existence of such a path on every $[P \longmapsto \neg Q *]$ state.

6   Probabilistic Model for PF

6.1   Some General Measure Theory

The definition of probability space given in Fig. 7 is rather standard, where U is the base set, F the measurables, Pr the measure function. The definition of measure space uses standard notions such as σ-algebra, positivity and countable additivity. In the definition of countable additivity, we use Isabelle library function $sums$, where 'f $sums$ c' stands for $\sum_{n=0}^{\omega} f(n) = c$.

Carathéodory's extension theorem [14, 15] is the standard way to construct probability space. The theorem is proved using Isabelle/HOL:

$$[algebra\ (U, F); positive\ (F, Pr); countably\text{-}additive\ (F, Pr)]$$
$$\Longrightarrow \exists P. (\forall A.\ A \in F \longrightarrow P\ A = Pr\ A) \wedge measure\text{-}space\ (U, sigma\ (U, F), P) \tag{3}$$

where the definition of $sigma$ is in Fig. 8.

6.2   Probability Space on Infinite Executions

Elements used to construct a measure space for infinite executions are defined in Fig. 7 as locale RCS, which accepts a parameter R, where $R(\tau, e)$ is the probability of choosing event e to happen under state τ. The definition of RCS is given in Fig. 9. The purpose of RCS is to define a measure space $(PA, Path, \mu)$ on infinite executions in terms of parameter R. The underlying concurrent system is given by CS. Set $N \tau$ contains all events eligible to happen under state τ. Function π is a measure function on finite executions. $Path$ is the set of valid infinite executions of CS.

consts algebra :: ('a set × 'a set set) ⇒ bool
  algebra (U, F) = (F ⊆ Pow(U) ∧ {} ∈ F ∧ (∀ a∈F. (U − a) ∈ F) ∧
                (∀ a b. a ∈ F ∧ b ∈ F ⟶ a ∪ b ∈ F))

consts sigma-algebra ::('a set × 'a set set) ⇒ bool
  sigma-algebra(U, F) = (F ⊆ Pow(U) ∧ U ∈ F ∧ (∀ a ∈ F. U − a ∈ F) ∧
                (∀ a. (∀ i::nat. a(i) ∈ F) ⟶ (⋃ i. a(i)) ∈ F))

consts positive:: ('a set set × ('a set ⇒ real)) ⇒ bool
  positive(F, Pr) = (Pr {} = 0 ∧ (∀ A. A ∈ F ⟶ 0 ≤ Pr A))

consts countably-additive:: ('a set set × ('a set ⇒ real)) ⇒ bool
  countably-additive(F, Pr) = (∀ f::(nat ⇒ 'a set). range(f) ⊆ F ∧
                (∀ m n. m ≠ n ⟶ f(m) ∩ f(n) = {}) ∧ (⋃ i. f(i)) ∈ F
            ⟶ (λn. Pr(f(n))) sums Pr (⋃ i. f(i)))

consts measure-space:: ('a set × 'a set set × ('a set ⇒ real)) ⇒ bool
  measure-space (U, F, Pr) = (sigma-algebra (U, F) ∧ positive (F, Pr) ∧
                countably-additive (F, Pr))

consts prob-space:: ('a set × 'a set set × ('a set ⇒ real)) ⇒ bool
  prob-space (U, F, Pr) = (measure-space (U, F, Pr) ∧ Pr U = 1)

Fig. 7. Definition of Probability Space

Set *palgebra-embed*($[\tau_0, \ldots, \tau_n]$) contains all infinite executions prefixed by some τ in $[\tau_0, \tau_1, \ldots, \tau_n]$, and *palgebra-embed*($[\tau_0, \ldots, \tau_n]$) is said to be supported by $[\tau_0, \ldots, \tau_n]$. The measure μ S is defined in terms of measures on supporting sets for S.

It is proved that *algebra* (*Path, PA*), *positive* (*PA, μ*) and *countably-additive* (*PA, μ*). Applying *Carathéodory's extension theorem* to these gives (*PA, Path, μ*) as a measure space on infinite executions of *CS*. It can be further derived that (*PA, Path, μ*) is a probability space.

inductive-set sigma :: ('a set × 'a set set) ⇒ 'a set set **for** M::('a set × 'a set set) **where**
  basic: (let (U, A) = M in (a ∈ A)) ⟹ a ∈ sigma M|
  empty: {} ∈ sigma M|
  complement: a ∈ sigma M ⟹ (let (U, A) = M in U − a) ∈ sigma M|
  union: (⋀i::nat. a i ∈ sigma M) ⟹ (⋃ i. a i) ∈ sigma M

Fig. 8. Definition of the sigma Operator

6.3 The Probabilistic Meaning of PF

In Fig. 10, locale *BTS* is proposed as a refinement of *RCS*. The motivation of *BTS* is to provide a low bound *bnd* for *R*. Function *P* is the probability function on execution sets. Based on *BTS*, the following theorem can be proved and it gives a meaning to *PF*:

Theorem 1. $\dfrac{[BTS\ R\ bnd;\ set\ pel \neq \{\}]}{\Longrightarrow BTS.P\ R\ \{\sigma \in RCS.Path\ R.\ PF\ \{(\tau, e).\ 0 < R\ (\tau, e)\}\ pel\ \sigma\} = 1}$

```
locale RCS =
  fixes R :: ('a list × 'a) ⇒ real
  assumes Rrange: 0 ≤ R(τ, e) ∧ R(τ, e) ≤ 1
  fixes CS :: ('a list × 'a) set
  defines CS-def: CS ≡ {(τ, e) . 0 < R(τ, e)}
  fixes N :: 'a list ⇒ 'a set
  defines N-def: N τ ≡ {e. 0 < R(τ, e)}
  assumes Rsum1: CS ⊢ τ ⟹ (∑ e ∈ (N τ). R(τ,e)) = 1
begin
fun π :: 'a list ⇒ real where
  π [] = 1 |
  π (e#τ) = R(τ, e) * π τ

definition Path:: 'a seq set where
  Path ≡ {σ. (∀ i. π [σ]_i > 0)}

definition palg-embed :: 'a list ⇒ 'a seq set where
  palg-embed τ ≡ {σ ∈ Path. [σ]_|τ| = τ}

fun palgebra-embed :: 'a list list ⇒ 'a seq set where
  palgebra-embed [] = {} |
  palgebra-embed (τ#l) = (palg-embed τ) ∪ palgebra-embed l

definition PA :: 'a seq set set where
  PA ≡ {S. ∃l. palgebra-embed l = S ∧ S ⊆ Path}

definition palg-measure :: 'a list list ⇒ real (μ_0) where
  μ_0 l ≡ (∑ τ ∈ set l. π τ)

definition palgebra-measure:: 'a list list ⇒ real  (μ_1) where
  μ_1 l ≡ inf (λr. ∃l'. palgebra-embed l = palgebra-embed l' ∧ μ_0 l' = r)

definition μ:: 'a seq set ⇒ real where
  μ S ≡ sup (λr. ∃b. μ_1 b = r ∧ (palgebra-embed b) ⊆ S)
end
```

Fig. 9. Formalization of Probabilistic Execution

Theorem 1 shows *almost all* executions are fair in the sense of *PF*. It says that the set of valid executions satisfying *PF* has probability *1*. The underlying concurrent system is $\{(\tau,e).0 < R(\tau,e)\}$, which is the expansion of *CS* in *RCS*. The probability function *BTS.P R* is the one derived from *R* using *BTS*. The set *RCS.Path R* contains all valid infinite executions of the underlying *CS*.

6.4   The Proof of Theorem 1

For the proof of theorem 1, a generalized notion of fairness *GF* is introduced in Fig. 11. *GF* is defined as a locale accepting three parameters, *cs* a concurrent system, *L* a countable set of labels, and *l* is a labeling function. A label ι is said to be *enabled* in state τ (written as *enabled* ι τ) if there exists some event *e* eligible to happen under state τ and ι belongs to $l(\tau,e)$. A label ι is said to be *taken* in state τ if ι is assigned

---

locale BTS = RCS +
  fixes bnd :: real
  assumes system-bound: $0 < $ bnd $\wedge$ bnd ≤ 1
  and bound-imply: $R(\tau, e) > 0 \Longrightarrow R(\tau, e) \geq$ bnd
  definition P::($'$a seq set $\Rightarrow$ real) **where**
  P $\equiv$ (SOME P$'$::($'$a seq set $\Rightarrow$ real).
        ($\forall$ A. A $\in$ PA $\longrightarrow$ P$'$ A $= \mu$ A) $\wedge$ measure-space (Path, sigma (Path, PA), P$'$))

Fig. 10. The definition of locale BTS

to the last execution step of τ, which is denoted by $hd(\tau)$, and the state before the last step is denoted by $tl(\tau)$. Label ι is said to be *fairly* treated by an infinite execution, if ι is taken infinite many times whenever it is enabled infinite many times in σ. An infinite execution is fair in the sense of *GF*, if all labels in set L are fairly treated. By instantiating label set L with the set of elements on *pel*(denoted by *set pel*), and label function l with function lf where lf $(\tau, e) = \{(Q, E)| (Q, E) \in (set\ pel) \wedge Q\ \tau \wedge e = (E\ \tau)\}$, it can be shown *PF* is just an instance of *GF*:

$$PF\ cs\ pel\ \sigma = GF.fair\ cs\ (set\ pel)\ (lf\ pel)\ \sigma \qquad (4)$$

It is proved that the probability of *GF* fair execution equals *1*:

$$BTS.P\ R\ \{\sigma \in RCS.Path\ R.\ GF.fair\ \{(\tau, e).\ 0 < R\ (\tau, e)\}\ L\ l\ \sigma\} = 1 \qquad (5)$$

The combination of (4) and (5) gives rise to Theorem 1. The intuition behind (4) it that (P, E)-pairs in *PF* can be seen as labels in *GF*. To prove (5), it is sufficient to prove the probability of unfair executions equals to *0*. In turn, it is sufficient to prove the probability of unfair executions with respect to any one label ι equals to *0*, because

---

locale GF =
  fixes cs:: ($'$a list $\times$ $'$a) set
  fixes L :: $'$b set
  assumes countable-L: countable L
  and non-empty-L: L $\neq$ {}
  fixes l :: $'$a list $\times$ $'$a $\Rightarrow$ $'$b set
  assumes subset-l: l$(\tau,e) \subseteq$ L
begin
  definition enabled:: $'$b $\Rightarrow$ $'$a list $\Rightarrow$ bool **where** enabled $\iota\ \tau \equiv (\exists e.\ (\iota \in l(\tau, e) \wedge (\tau, e) \in cs))$

  definition taken :: $'$b $\Rightarrow$ $'$a list $\Rightarrow$ bool **where** taken $\iota\ \tau \equiv \iota \in l(tl(\tau), hd(\tau))$

  definition fairι :: $'$b $\Rightarrow$ (nat $\Rightarrow$ $'$a) $\Rightarrow$ bool **where**
    fair$\iota\ \iota\ \sigma \equiv \sigma \models \Box\Diamond(\lambda\sigma$ i. enabled $\iota\ [\![\sigma]\!]_i) \longrightarrow \sigma \models \Box\Diamond(\lambda\sigma$ i. taken $\iota\ [\![\sigma]\!]_{Suc(i)})$

  definition fair :: (nat $\Rightarrow$ $'$a) $\Rightarrow$ bool **where** fair $\sigma \equiv \forall \iota \in$ L. fair$\iota\ \iota\ \sigma$
end

Fig. 11. A general fairness definition

the number of labels is countable. If label ι is treated unfairly by execution σ, σ must avoid ι infinite many times, with every avoidance reduce the probability by a certain proportion. This series of avoidances finally reduce the probability to *0*.

7 Liveness Proof Experiments

7.1 Elevator Control System

In [7], Yang developed a formal verification of elevator control system. The events of an elevator control system are: *Arrive p m n*: User *p* arrives at floor *m*, planning to go to floor *n*; *Enter p m n*: User *p* enters elevator at floor *m*, planning to go to floor *n*; *Exit p m n*: User *p* gets off elevator at floor *n*, *m* is the floor, where user *p* entered elevator; *Up n*: A press of the ▲-button on floor *n*; *Down n*: A press of the ▼-button on floor *n*; *To n*: A press of button *n* on elevator's control panel; *StartUp n*: Elevator starts moving upward from floor *n*; *StartDown n*: Elevator starts moving down from floor *n*; *Stop n m*: Elevator stops at floor *m*, before stopping, the elevator is moving from floor *n* to floor *m*; *Pass n m*: Elevator passes floor *n* without stopping, before passing floor *m*, the elevator is moving from floor *n* to floor *m*.

The proved liveness property for the elevator control system is:

$$\llbracket \textit{elev-cs} \vdash \sigma;$$
$$\textit{PF elev-cs}$$
$$(\{\!|F\,p\,m\,n, E\,p\,m\,n, 4*H+3|\!\} @ \{\!|FT\,p\,m\,n, ET\,p\,m\,n, 4*H+3|\!\}) \; \sigma\rrbracket$$
$$\implies \sigma \models \Box\langle(\!|\textit{Arrive } p\ m\ n|\!)\rangle \hookrightarrow \Diamond\langle(\!|\textit{Exit } p\ m\ n|\!)\rangle$$

The conclusion says: if user *p* arrives at floor *m* and wants to go to floor *n* (represented by the happening of event *Arrive p m n*), then he will eventually get there (represented by the happening of event *Exit p m n*).

The liveness proof splits into two stages. The first is to prove once user *p* arrives, it will eventually get into elevator, the second is to prove once *p* gets into elevator, it will eventually get out of elevator at its destination. Both stages use rule *resp-rule*. The definition of *cs*, *P* and *Q* is obvious. The difficult part is to find a proper definition of *F* and *E* so that premise *RESP cs F E N P Q* can be proved. We explain the finding of *F* and *E* by example of a 5-floor elevator system, the state diagram of which is shown in Fig. 12. Helpful transitions are represented using normal lines while unhelpful transitions with dotted lines.

The arrival of *p* (event *Arrive p 2 4*) brings system from state *Arrive p 2 4 ∉ arr-set τ* to *Arrive p 2 4 ∈ arr-set τ ∧ 2 ∉ up-set τ*. To call the elevator, *p* will press the ▲-button (event *Up 2*), and this will bring system into macro-state *Arrive p 2 4 ∈ arr-set τ ∧ 2 ∈ up-set τ*, which contains many sub-states characterized by the status of elevator. Status (n, m) means moving from floor *n* to *m*, while $d(n,n)$ means stopped on floor *n* in down passes, $u(n,n)$ in up passes. It can be seen that helpful transitions in Fig. 12 form a spanning tree rooted at the desirable state. Premise *RESP cs F E N P Q* is valid if *F τ* evaluates to the height of *τ* in the spanning tree and *E τ* evaluates to the helpful event leading out of *τ*. Fig. 12 shows such *F* and *E* can be defined. Details of the definition can be found in [7]. Once *F* and *E* is given, the rest of the proof is straight forward. Statistics in Table 1 may give some idea the amount of work required.

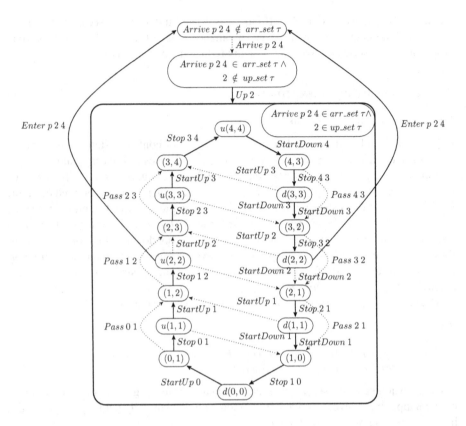

Fig. 12. The state transition diagram of elevator for *Arrive p 2 4*

Table 1. Summary of Isabelle/HOL proof scripts

Contents	Nr. of lines	Nr. of lemmas	Nr. of working days
Definitions of the elevator control system, *F*, *E*, *FT* and *ET*	≤ 200	–	*
Safety properties	579	12	2
First stage of the proof	2012	10	5
Second stage of the proof	1029	8	3

7.2 Mobile Ad Hoc Secure Routing Protocols

In mobile Ad Hoc networks, nodes keep moving around in a certain area while communicating over wireless channels. The communication is in a peer-to-peer manner, with no central control. This makes the network very susceptible to malicious attacks. Therefore, security is an important issue in Mobile Ad Hoc networks. Secure routing protocol is chosen as our verification target.

To counteract attacks, specialized procedures are used by secure routing protocols together with cryptographic mechanisms such as public key and hashing. The goal of verification is to make sure the procedures combined with cryptographic mechanisms do not collapse even in highly adversary environments.

Attackers are modeled using the same method as Paulson's [1]. The correctness of routing protocol is verified by combining it with an upper transfer layer. It is proved that the transfer protocol can deliver every user packet with the support of the underlying routing protocol. This is a typical liveness property, which says if a user packet gets into the transfer layer at one end it will eventually get out at the other end, just like an elevator passenger.

Because nodes are moving around, existing routes between nodes constantly become obsolete, in which case routing protocols have to initiate route finding procedure, this procedure may be interfered by attacks aimed at slowing down or even collapsing the network.

The verification is like the one for elevator but on a much larger scale. We have managed to find definitions of F and E, which essentially represent spanning trees covering all relevant states like the one in Fig. 12. The proof is laborious but doable, suggesting that specialized tactics may be needed.

We have verified two secure routing protocols, details can be found in [8, 9]. The work in [8,9] confirmed the practicability of our approach on one hand and the resilience of the security protocol on the other hand.

8   Related Works

Approaches for verification of concurrent systems can roughly be divided into theorem proving and model checking. The work in this paper belongs to the theorem proving category, which can deal with infinite state systems.

The purpose of this paper is to extend Paulson's inductive protocol verification approach [1, 2] to deal with general liveness properties, so that it can be used as a general protocol verification approach. The notion of PF and the corresponding proof rules were first proposed by Zhang in [13]. At the same time, Yang developed a benchmark verification for elevator control system to show the practicality of the approach [7]. Later, Yang used the approach serious to verify the liveness properties of Mobile Ad Hoc network protocols [8,9]. These works confirm the practicality of the extension.

The liveness rules in this paper relies on a novel fairness notion PF, an adaption of the α-fairness [4, 5, 16] to suit the setting of HOL. The use of PF can derive more liveness properties and usually the proofs are simpler than using the standard WF and SF.

According to Baier's work [5], PF should have a sensible probabilistic meaning. To confirm this, Wang established a probabilistic model for PF, to show that the measure of PF executions equals 1 [17]. The formalization of this probabilistic model is deeply influenced by Hurd and Stefan [14, 15], however, due to different type restrictions from [14], and the need of extension theorem which is absent from [15], most proofs have to be done from scratch. Wang's work established the soundness of PF.

9 Conclusion

The advantage of theorem proving over model checking is its ability to deal with infinite state systems. Unfortunately, relatively less work is done to verify liveness properties using theorem proving. This paper improves the situation by proposing an extension of Paulson's inductive approach for liveness verification and showing the soundness, feasibility and practicality of the approach.

The level of automation in our work is still low compared to model checking. One direction for further research is to develop specialized tactics for liveness proof. Additionally, the verification described in this paper is still at abstract model level. Another important direction for further research is to extend it to code level verification.

References

1. Paulson, L.C.: The inductive approach to verifying cryptographic protocols. Journal of Computer Security 6(1-2), 85–128 (1998)
2. Paulson, L.C.: Inductive analysis of the Internet protocol TLS. ACM Transactions on Computer and System Security 2(3), 332–351 (1999)
3. Manna, Z., Pnueli, A.: Completing the temporal picture. Theor. Comput. Sci. 83(1), 91–130 (1991)
4. Pnueli, A., Zuck, L.D.: Probabilistic verification. Information and Computation 103(1), 1–29 (1993)
5. Baier, C., Kwiatkowska, M.: On the verification of qualitative properties of probabilistic processes under fairness constraints. Information Processing Letters 66(2), 71–79 (1998)
6. Jaeger, M.: Fairness, computable fairness and randomness. In: Proc. 2nd International Workshop on Probabilistic Methods in Verification (1999)
7. Yang, H., Zhang, X., Wang, Y.: Liveness proof of an elevator control system. In: The 'Emerging Trend' of TPHOLs, Oxford University Computing Lab. PRG-RR-05-02, pp. 190–204 (2005)
8. Yang, H., Zhang, X., Wang, Y.: A correctness proof of the srp protocol. In: 20th International Parallel and Distributed Processing Symposium (IPDPS 2006), Proceedings, Rhodes Island, Greece, April 25-29 (2006)
9. Yang, H., Zhang, X., Wang, Y.: A correctness proof of the dsr protocol. In: Cao, J., Stojmenovic, I., Jia, X., Das, S.K. (eds.) MSN 2006. LNCS, vol. 4325, pp. 72–83. Springer, Heidelberg (2006)
10. Nipkow, T., Paulson, L.C., Wenzel, M.: Isabelle/HOL — A Proof Assistant for Higher-Order Logic. LNCS, vol. 2283. Springer, Heidelberg (2002)
11. Wenzel, M.: Isar - a generic interpretative approach to readable formal proof documents. In: Nipkow, T., Paulson, L.C., Wenzel, M.T. (eds.) Isabelle/HOL. LNCS, vol. 2283. Springer, Heidelberg (2002)
12. Pnueli, A.: On the extremely fair treatment of probabilistic algorithms. In: Proceedings of the fifteenth annual ACM symposium on Theory of computing, pp. 278–290. ACM Press, New York (1983)
13. Zhang, X., Yang, H., Wang, Y.: Liveness reasoning for inductive protocol verification. In: The 'Emerging Trend' of TPHOLs, Oxford University Computing Lab. PRG-RR-05-02, pp. 221–235 (2005)
14. Hurd, J.: Formal Verification of Probabilistic Algorithms. Ph.D thesis, University of Cambridge (2002)

15. Richter, S.: Formlizing integration theory with an application to probabilistic algorithms. In: Slind, K., Bunker, A., Gopalakrishnan, G.C. (eds.) TPHOLs 2004. LNCS, vol. 3223, pp. 271–286. Springer, Heidelberg (2004)

16. Pnueli, A.: On the extremely fair treatment of probabilistic algorithms. In: ACM (ed.) Proceedings of the 15th annual ACM Symposium on Theory of Computing, Boston, Massachusetts, April 25–27, pp. 278–290. ACM Press, New York (1983)

17. Wang, J., Zhang, X., Zhang, Y., Yang, H.: A probabilistic model for parametric fairness in isabelle/hol. Technical Report 364/07, Department of Computer Science, University of Kaiserslautern (2007)

Mind the Gap

A Verification Framework for Low-Level C

Simon Winwood[1,2], Gerwin Klein[1,2], Thomas Sewell[1], June Andronick[1],
David Cock[1], and Michael Norrish[1,3]

[1] NICTA*, Australia
[2] School of Computer Science and Engineering, UNSW, Sydney, Australia
[3] Computer Sciences Laboratory, ANU, Canberra, Australia
{first-name.last-name}@nicta.com.au

Abstract. This paper presents the formal Isabelle/HOL framework we
use to prove refinement between an executable, monadic specification and
the C implementation of the seL4 microkernel. We describe the refinement
framework itself, the automated tactics it supports, and the connection to
our previous C verification framework. We also report on our experience
in applying the framework to seL4. The characteristics of this microkernel
verification are the size of the target (8,700 lines of C code), the treatment
of low-level programming constructs, the focus on high performance, and
the large subset of the C programming language addressed, which includes
pointer arithmetic and type-unsafe code.

1 Introduction

The seL4 kernel [10] is a high-performance microkernel in the L4 family [18], targeted at secure, embedded devices. In verifying such a complex and large – 8,700
lines of C – piece of software, scalability and separation of concerns are of the ut-
most importance. We show how to achieve both for low-level, manually optimised,
real-world C code.

Fig. 1 shows the layers and
proofs involved in the verification
of seL4. The top layer is an abstract, operational specification of
seL4; the middle layer is an executable specification derived automatically [8, 11] from a working
Haskell prototype of the kernel;
the bottom layer is a hand-written
and hand-optimised C implementation. The aim is to connect the
three layers by formal proof in Is-
abelle/HOL [21].

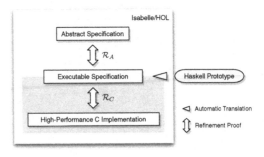

Fig. 1. Refinement steps in L4.verified

* NICTA is funded by the Australian Government as represented by the Department of
Broadband, Communications and the Digital Economy and the Australian Research
Council through the ICT Centre of Excellence program.

S. Berghofer et al. (Eds.): TPHOLs 2009, LNCS 5674, pp. 500–515, 2009.

Previously, we presented a verification framework [5] for proving refinement between the abstract and executable specifications. This paper presents the framework for the second refinement step: the formal, machine-checked proof that the high-performance C implementation of seL4 correctly implements the executable specification.

With these two refinement steps, we manage to isolate two aspects of the verification of seL4. In the first refinement step, which we call $\mathcal{R}_A$, we dealt mostly with semantic concepts such as relationships between data structures and system-global conditions for safe execution. We estimate that 80% of the effort in $\mathcal{R}_A$ was spent on such invariants. In the second refinement step, $\mathcal{R}_C$, the framework we present in this paper allows us to reduce our proof effort and to reuse the properties shown in $\mathcal{R}_A$. The first refinement step established that the kernel design works, the second closes the gap to C.

Paper Structure. We begin with an example that sketches the details of a typical kernel function. We then explain how the components of the verification framework fit together, summarising relevant details of our earlier work on the monadic, executable specification [5], and on our C semantics and memory model [25, 26, 27]. In particular, we describe the issues involved in converting the C implementation into Isabelle/HOL. The main part of the paper shows the refinement framework with its fundamental definitions, rules, and automated tactics. We demonstrate the framework's performance by reporting on our experience so far in applying it to the verification of substantial parts of the seL4 C implementation (474 out of 518 functions, 91%).

2   Example

The seL4 kernel [10] provides the following operating system kernel services: inter-process communication, threads, virtual memory, access control, and interrupt control. In this section we present a typical function, cteMove, with which we will illustrate the verification framework.

Access control in seL4 is based on *capabilities*. A capability contains an object reference along with access rights. A *capability table entry* (CTE) is a kernel data structure with two fields: a capability and an *mdbNode*. The latter is bookkeeping information and contains a pair of pointers which form a doubly linked list.

The cteMove operation, shown in Fig. 2, moves a capability table entry from *src* to *dest*. The left-hand side of the figure shows the executable specification in Isabelle/HOL, while the right-hand side shows the corresponding C code.

The first 6 lines in Fig. 2 initialise the destination entry and clear the source entry; the remainder of the function updates the pointers in the doubly linked list. During the move, the capability in the entry may be diminished in access rights. Thus, the argument *cap* is this possibly diminished capability, previously retrieved from the entry at *src*.

```
cteMove cap src dest ≡
do
    cte ← getCTE src;
    mdb ← return (cteMDBNode cte);
    updateCap dest cap;
    updateCap src NullCap;
    updateMDB dest (const mdb);
    updateMDB src (const nullMDBNode);

    updateMDB
        (mdbPrev mdb)
        (λm. m (| mdbNext := dest |));

    updateMDB
        (mdbNext mdb)
        (λm. m (| mdbPrev := dest |))
od
```

```
void cteMove (cap_t newCap,
             cte_t *srcSlot, cte_t *destSlot){
    mdb_node_t mdb; uint32_t prev_ptr, next_ptr;
    mdb = srcSlot->cteMDBNode;
    destSlot->cap = newCap;
    srcSlot->cap = cap_null_cap_new();
    destSlot->cteMDBNode = mdb;
    srcSlot->cteMDBNode = nullMDBNode;
    prev_ptr = mdb_node_get_mdbPrev(mdb);
    if(prev_ptr) mdb_node_ptr_set_mdbNext(
                    &CTE_PTR(prev_ptr)->cteMDBNode,
                    CTE_REF(destSlot));
    next_ptr = mdb_node_get_mdbNext(mdb);
    if(next_ptr) mdb_node_ptr_set_mdbPrev(
                    &CTE_PTR(next_ptr)->cteMDBNode,
                    CTE_REF(destSlot));
}
```

Fig. 2. cteMove: executable specification and C implementation

In this example, the C source code is structurally similar to the executable specification. This similarity is not accidental: the executable specification describes the low-level design with a high degree of detail. Most of the kernel functions exhibit this property. Even so, the implementation here makes a small optimisation: in the specification, updateMDB always checks that the given pointer is not NULL. In the implementation this check is done for prev_ptr and next_ptr – which may be NULL – but omitted for srcSlot and destSlot. In verifying cteMove we will have to prove these checks are not required.

3   The Executable Specification Environment

Operations in the executable specification of seL4, such as cteMove, are written in a monadic style inspired by Haskell. The type constructor $'a$ kernel is a monad representing computations returning a value of type $'a$; such values can be injected into the monad using the return :: $'a \Rightarrow 'a$ kernel operation. The composition operator, bind :: $'a$ kernel $\Rightarrow ('a \Rightarrow 'b$ kernel$) \Rightarrow 'b$ kernel, evaluates the first operation and makes the return value available to the second operation. The ubiquitous do ... od syntax seen in Fig. 2 is syntactic sugar for a sequence of operations composed using bind. There are also operations for accessing and mutating k-state, the underlying state.

The type $'a$ kernel is isomorphic to k-state $\Rightarrow ('a \times$ k-state$)$ set $\times$ bool. The motivation for, and formalisation of, this monad are detailed in earlier work [5]. In summary, we take a conventional state monad and add nondeterminism and a failure flag. Nondeterminism, required to model some interactions between kernel and hardware, is modelled by allowing a set of possible outcomes in the return type. The boolean failure flag is used to indicate unrecoverable errors and invalid assertions, and is set only by the fail :: $'a$ kernel operation. The destructors mResults and mFailed access, respectively, the set of outcomes and the failure flag of a monadic operation evaluated at a state.

The specification environment provides a verification condition generator (VCG) for judgements of the form $\{P\}\ a\ \{R\}$, and a refinement calculus for the monadic model. One feature of this calculus is that the refinement property cannot hold if the failure flag is set by the executable specification, thus $\mathcal{R}_A$ implies non-failure of the executable level. In particular, this allows all assertions in the executable specification to be taken as assumptions in the proof of $\mathcal{R}_C$.

4   Embedding C

In this section we describe our infrastructure for parsing C into Isabelle/HOL and for reasoning about the result. The seL4 kernel is implemented almost entirely in C99 [16]. Direct hardware accesses are encapsulated in machine interface functions, some of which are implemented in ARMv6 assembly. In the verification, we axiomatise the assembly functions using Hoare triples.

Fig. 3 gives an overview of the components involved. The right-hand side shows our instantiation of SIMPL [23], a generic, imperative language inside Isabelle/HOL. The SIMPL framework provides a program representation, a semantics, and a VCG. This language is generic in its expressions and state space. We instantiate both components to form C-SIMPL, with a precise C memory model and C expressions, generated by a parser. The left-hand side of Fig. 3 shows this process: the parser takes a C program and produces a C-SIMPL program.

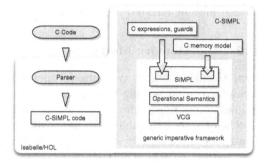

Fig. 3. C language framework

4.1   The SIMPL Framework

SIMPL provides a data type and semantics for statement forms; expressions are shallowly embedded. The following is a summary of the relevant SIMPL syntactic forms, where e represents an expression

$$c \triangleq \mathsf{SKIP} \mid \textprime v := e \mid c_1 \;;\; c_2 \mid \mathsf{IF}\ e\ \mathsf{THEN}\ c_1\ \mathsf{ELSE}\ c_2\ \mathsf{FI} \mid \mathsf{WHILE}\ e\ \mathsf{DO}\ c\ \mathsf{OD}$$
$$\mid \mathsf{TRY}\ c_1\ \mathsf{CATCH}\ c_2\ \mathsf{END} \mid \mathsf{THROW} \mid \mathsf{Call}\ f \mid \mathsf{Guard}\ F\ P\ c$$

The semantics are canonical for an imperative language. The $\mathsf{Guard}\ F\ P\ c$ statement throws the fault F if the condition P is false and executes c otherwise.

Program states in SIMPL are represented by Isabelle records. The record contains a field for each local variable in the program, and a field *globals* containing all global variables and the heap. Variables are then simply functions on the state. SIMPL includes syntactic sugar for dealing with such functions: the term $\textprime srcSlot$ refers to the local variable *srcSlot* in the current state. For example, the set of program states where *srcSlot* is NULL is described by $\{\!\!\mid \textprime srcSlot = \mathsf{NULL}\mid\!\!\}$.

The semantics are represented by judgements of the form $\Gamma \vdash \langle c, x \rangle \Rightarrow x'$ which means that executing statement c in state x terminates and results in state x'; the parameter Γ maps function names to function bodies. Both x and x' are *extended states*: for normal program states, Normal s, the semantics are as expected; abrupt termination states (Abrupt s) are propagated until a surrounding TRY ... CATCH ... END statement is reached; and Stuck and Fault u states, generated by calls to non-existent procedures and failed Guard statements respectively, are passed through unchanged. Abrupt states are generated by THROW statements and are used to implement the C statements `return`, `break`, and `continue`.

The SIMPL environment also provides a VCG for partial correctness triples; Hoare-triples are represented by judgements of the form $\Gamma \vdash_{/F} P \ c \ C,A$, where P is the precondition, C is the postcondition for normal termination, A is the postcondition for abrupt termination, and F is the set of ignored faults; if F is $\mathcal{U}$, the universal set, then all Guard statements are effectively ignored. Both A and F may be omitted if empty.

4.2   The Memory Model

Our C subset allows type-unsafe operations including casts. To achieve this soundly, the underlying heap model is a function from addresses to bytes. This allows, for example, the C function `memset`, which sets each byte in a region of the heap to a given value. We use the abbreviation $\mathcal{H}$ for the heap in the current state; the expression $\mathcal{H} \ p$ reads the object at pointer p, while $\mathcal{H}(p \mapsto v)$ updates the heap at pointer p with value v.

While this model is required for such low-level memory accesses, it is too cumbersome for routine verification. By extending the heap model with typing information and using tagged pointers we can lift bytes in the heap into Isabelle terms. Pointers, terms of type $'a$ ptr, are raw addresses wrapped by the polymorphic constructor Ptr; the phantom type $'a$ carries the type information. Pointers may be unwrapped via the ptr-val function, which simply extracts the enclosed address. Struct field addressing is also supported: the pointer $\&(p \rightarrow [f])$ refers to the field f at the address associated with pointer p. The details of this memory model are described by Tuch *et al* [27, 26].

4.3   From C to C-SIMPL

The parser translates the C kernel into a C-SIMPL program. This process generally results in a C-SIMPL program that resembles the input. Here we describe the C subset we translate, and discuss those cases where translation produces a result that is not so close to the input.

Our C Subset. As mentioned above, local variables in SIMPL are represented by record fields. It is therefore not meaningful to take their address in the framework, and so the first restriction of our C subset is that local variables may not have their addresses taken. Global variables may, however, have their addresses taken. As we translate all of the C source at once, the parser can determine exactly which globals do have their addresses taken, and these variables are then given addresses in the heap. Global variables that do not have their addresses taken are, like locals, simply fields in the program state. The restriction on local variables could be relaxed at the cost of higher reasoning overhead.

The other significant syntactic omissions in our C subset are union types, bitfields, goto statements, and switch statements that allow cases to fall-through. We handle union types and bitfields with an automatic code generator [4], described in Sect. 6, that implements these types with structs and casts. Furthermore, we do not allow function calls through function pointers and take care not to introduce a more deterministic evaluation order than C prescribes. For instance, we translate the side-effecting C expressions ++ and -- as statements.

Internal Function Calls and Automatic Modifies Proofs. SIMPL does not permit function calls within expressions. If a function call appears within an expression in the input C, we lift it out and transform it into a function call that will occur before the expression is evaluated. For example, given a global variable x, the statement z = x + f(y) becomes tmp = f(y); z = x + tmp, where tmp is a new temporary variable.

This translation is only sound when the lifted functions are side-effect free: evaluation of the functions within the original expression is linearised, making the translated code more deterministic than warranted by the C semantics. The parser thus generates a VCG "modifies" proof for each function, stating which global variables are modified by the function. Any function required to be side-effect free, but not proved as such, is flagged for the verification team's attention.

Guards and Short-Circuit Expressions. Our parser uses Guard statements to force verifiers to show that potentially illegal conditions are avoided. For example, expressions involving pointer dereferences are enclosed by guards which require the pointer to be aligned and non-zero.

Guards are statement-level constructors, so whole expressions accumulate guards for their sub-expressions. However, C's short-circuiting expression forms (&&, || and ?:) mean that sub-expressions are not always evaluated. We translate such expressions into a sequence of if-statements, linearising the evaluation of the expression. When no guards are involved, the expression in C can become a C-SIMPL expression, using normal, non-short-circuiting, boolean operators.

Example. While we have shown the C implementation in the example Fig. 2, refinement is proven between the executable specification and the imported C-SIMPL code. For instance, the assignment mdb = srcSlot->cteMDBNode in Fig. 2 is translated into the following statement in C-SIMPL

$$\text{MemGuard } \&(\text{'}srcSlot \rightarrow [\text{cteMDBNode-C}])$$
$$(\text{'}mdb := = \mathcal{H} \&(\text{'}srcSlot \rightarrow [\text{cteMDBNode-C}]))$$

The MemGuard constructor abbreviates the alignment and non-NULL conditions for pointers.

5   Refinement

Our verification goal is to prove refinement between the executable specification and the C implementation. Specifically, this means showing that the C kernel entry points for interrupts, page faults, exceptions, and system calls refine the executable specification's top-level function callKernel. We show refinement using a variation of forward simulation [7] we call *correspondence*: evaluation of corresponding functions takes related states to related states.

In previous work [5], while proving $\mathcal{R}_A$, we found it useful to divide the proof along the syntactic structure of both programs as far as possible, and then prove the resulting subgoals semantically. Splitting the proof has two main benefits: firstly, it is a convenient unit of proof reuse, as the same pairing of abstract and concrete functions recurs frequently for low-level functions; and secondly, it facilitates proof development by multiple people. One important feature of this approach is that preconditions are discovered lazily *à la* Dijkstra [9]. Rules for showing correspondence typically build preconditions from those of the premises.

In this section we describe the set of tools and techniques we developed to ease the task of proving correspondence in $\mathcal{R}_C$. First, we give our definition of correspondence, followed by a discussion of the use of the VCG. We then describe techniques for reusing proofs from $\mathcal{R}_A$ to solve proof obligations from the implementation. Next, we present our approach for handling operations with no corresponding analogue. Finally, we describe our splitting approach and sketch the proof of the example.

5.1   The Correspondence Statement

In practice, the definition of correspondence is more complex than simply linking related states, as: (1) verification typically requires preconditions to hold of the initial states; (2) we allow early returns from functions and breaks from loops; and (3) function return values must be related.

To deal with early return, we extend the semantics to lists of statements, using the judgement

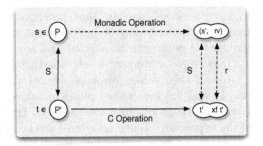

Fig. 4. Correspondence

$\Gamma \Vdash \langle c \cdot hs, s \rangle \Rightarrow x'$. The statement sequence hs is a *handler stack*; it collects the CATCH handlers which surround usages of the statements return, continue, and break. If c terminates abruptly, each statement in hs is executed in sequence until one terminates normally.

Relating the return values of functions is dealt with by annotating the correspondence statement with a *return value relation* r. Although evaluating a monadic operation results in both a new state and a return value, functions in C-SIMPL return values by updating a function-specific local variable; because local variables are fields in the state record, this is a function from the state. We thus annotate the correspondence statement with an *extraction function xf*, a function which extracts the return value from a program state.

The correspondence statement is illustrated in Fig. 4 and defined below

ccorres r xf P P' hs a c $\equiv$
  $\forall (s, t) \in \mathcal{S}.\ \forall t'.\ s \in P \wedge t \in P' \wedge \neg$ mFailed $(a\ s) \wedge \Gamma \Vdash \langle c \cdot hs, t \rangle \Rightarrow t'$
  $\longrightarrow \exists (s', rv) \in$ mResults $(a\ s)$.
    $\exists t'_N.\ t' =$ Normal $t'_N \wedge (s', t'_N) \in \mathcal{S} \wedge r\ rv\ (xf\ t'_N)$

The definition can be read as follows: given related states s and t with the preconditions P and P' respectively, if the abstract specification a does not fail when evaluated at state s, and the concrete statement c evaluates under handler stack hs in extended state t to extended state t', then the following must hold:

1. evaluating a at state s returns some value rv and new abstract state s';
2. the result of the evaluation of c is some extended state Normal t'_N, that is, not Abrupt, Fault, or Stuck;
3. states s' and t'_N are related by the state relation $\mathcal{S}$; and
4. values rv and $xf\ t'_N$ – the extraction function applied to the final state of c – are related by r, the given return value relation.

Note that a is non-deterministic: we may pick any suitable rv and s'. As mentioned in Sect. 3, the proof of $\mathcal{R}_A$ entails that the executable specification does not fail. Thus, in the definition of ccorres, we may assume $\neg$ mFailed $(a\ s)$. In practice, this means assertions and other conditions for (non-)failure in the executable specification become known facts in the proof. For example, the operation getCTE *srcSlot* in the example in Fig. 2 will fail if there is no CTE at *srcSlot*. We can therefore assume in the refinement proof that such an object exists. Of course, these facts are only free because we have already proven them in $\mathcal{R}_A$.

Example. To prove correspondence for cteMove, we must, after unfolding the function bodies, show the statement in Fig. 5. The cteMove operation has no return value, so our extraction function (xf in the definition of ccorres above) and return relation (r above) are trivial. The specification precondition (P above) is the system invariant *invs*, while the implementation precondition (P' above) relates the formal parameters *destSlot*, *srcSlot*, and *newCap* to the specification arguments *dest*, *src*, and *cap* respectively. As all functions are wrapped in a TRY ... CATCH SKIP block to handle **return** statements, the handler stack is the singleton list containing SKIP.

5.2   Proving Correspondence via the VCG

Data refinement predicates can, in general [7], be rephrased and solved as Hoare triples. We do this in our framework by using the VCG after applying the following rule

```
ccorres (λ- -. True) (λ-. ()) invs
   {| ′destSlot = Ptr dest ∧ ′srcSlot s = Ptr src ∧ ccap-relation cap ′newCap|} [SKIP]
   (do
        cte ← getCTE src;
        mdb ← return (cteMDBNode cte);            ⎫
        updateCap dest cap;                        ⎬ cteMove spec.
        ...                                        ⎭
   od)
   (MemGuard &(′srcSlot→[cteMDBNode-C])
      (′mdb :== H &(′srcSlot→[cteMDBNode-C]));     ⎫
   MemGuard &(′destSlot→[cap-C])                    ⎬ cteMove impl.
      (′globals :== H(&(′destSlot→[cap-C]) ↦ ′newCap);
   ...)                                            ⎭
```

Fig. 5. The cteMove correspondence statement

$$\frac{\forall s. \; \Gamma \vdash \{t \mid s \in P \land t \in P' \land (s,\, t) \in \mathcal{S}\} \quad c \quad \{t' \mid \exists (rv,\, s') \in \mathsf{mResults}\ (a\ s).\ (s',\, t') \in \mathcal{S} \land r\ rv\ (xf\ t')\}}{\mathsf{ccorres}\ r\ xf\ P\ P'\ hs\ a\ c}$$

In essence, this rule states that to show correspondence between a and c, for a given initial specification state s, it is sufficient to show that executing c results in normal termination where the final state is related to the result of evaluating a at s. The VCG precondition can assume that the initial states are related and satisfy the correspondence preconditions.

Use of this rule in verifying correspondence is limited by two factors. Firstly, the verification conditions produced by the VCG may be excessively large or complex. Our experience is that the output of a VCG step usually contains a separate term for every possible path through the target code, and that the complexity of these terms tends to increase with the path length. Secondly, the specification return value and result state are existential, and thus outside the range of our extensive automatic support for showing universal properties of specification fragments. Fully expanding the specification is always possible, and in the case of deterministic operations will yield a single state/return value pair, but the resulting term structure may also be large.

In the case of our example, the goal produced by the VCG has 377 lines before unfolding the specification and 800 lines afterward. Verifying such non-trivial functions is made practical by the approach described in the remainder of this section.

5.3   Local Variable Lifting

The feasibility of proving $\mathcal{R}_C$ depends heavily on proof reuse from $\mathcal{R}_A$. Consider the following rule for dealing with a guard introduced by the parser (see Sect. 4.3)

$$\frac{\mathsf{ccorres}\ r\ xf\ G\ G'\ hs\ a\ c}{\mathsf{ccorres}\ r\ xf\ (G \cap \mathsf{cte\text{-}at}'\ (\mathsf{ptr\text{-}val}\ p))\ G'\ hs\ a\ (\mathsf{MemGuard}\ (\lambda s.\ p)\ c)}$$

This states that the proof obligation introduced by MemGuard at the CTE pointer p can be discharged, assuming that there exists a CTE object on the specification side (denoted cte-at' (ptr-val p)); this rule turns a proof obligation from the implementation into an assumption of the specification. There is, however, one major problem: the pointer p cannot depend on the C state, because it is also used on the specification side.

To see why this is such a problem, recall that local variables in C-SIMPL are fields in the state record; any pointer, apart from constants, in the program will *always* refer to the state, making the above rule inapplicable; in the example, the first guard refers to the local variable ´srcSlot.

All is not lost, however: the values in local variables generally correspond to some value available in the specification. We have developed an approach that automatically replaces such local variables with new HOL variables representing their value. Proof obligations which refer to the local variable can then be solved by facts about the related value from the specification precondition. We call this process *lifting*.

Example. If we examine the preconditions to the example proof statement in Fig. 5, we note the assumption ´srcSlot = Ptr *src* and observe that *srcSlot* depends on the C state. By lifting this local variable and substituting the assumption, we get the following implementation fragment

> MemGuard &(Ptr *src*→[cteMDBNode-C])
>   (´*mdb* :== $\mathcal{H}$ &(Ptr *src*→[cteMDBNode-C]);
>   . . .

The pointer Ptr *src* no longer depends on the C state and is a value from the specification side, so the MemGuard can be removed with the above rule.

Lifting is only sound if the behaviour of the lifted code fragment is indistinguishable from that of the original code; the judgement $d' \sim d[v/f]$ states that replacing applications of the function f in statement d with value v results in the equivalent statement d'. This condition is defined as follows

$$d' \sim d[v/f] \equiv \forall t\ t'.\ f\ t = v \longrightarrow \Gamma\vdash \langle d, \text{Normal } t\rangle \Rightarrow t' = \Gamma\vdash \langle d', \text{Normal } t\rangle \Rightarrow t'$$

This states that d and d' must be semantically equivalent, assuming f has the value v in the initial state. In practice, d' depends on a locally bound HOL variable; in such cases, it will appear as $d'\ v$.

Lifting is accomplished through the following rule

$$\frac{\forall v.\ d'\ v \sim d[v/f] \qquad \forall v.\ P\ v \longrightarrow \text{ccorres } r\ xf\ G\ G'\ hs\ a\ (d'\ v)}{\text{ccorres } r\ xf\ G\ (G' \cap \{s \mid P\ (f\ s)\})\ hs\ a\ d}$$

Note that d', the lifted fragment, appears only in the assumptions; proving the first premise involves inventing a suitable candidate. We have developed tactic support for automatically calculating the lifted fragment and discharging such proof obligations, based on a set of syntax-directed proof rules.

5.4   Symbolic Execution

The specification and implementation do not always match: there may be fragments on either side that are artefacts of the particular model. In our example,
it is clear that the complex function getCTE has no direct analogue; the implementation accesses the heap directly.

In both cases we have rules to symbolically execute the code using the appropriate VCG, although we must also show that the fragment preserves the state relation. On the implementation side this case occurs frequently; in the example we
have the cap_null_cap_new, mdb_node_get_mdbNext, and mdb_node_get_mdbPrev
functions. We have developed a tactic which can symbolically execute any side-
effect free function which has a VCG specification. This tactic also takes advantage of variable lifting: the destination local variable is replaced by a new HOL
variable and we gain the assumption that the variable satisfies the function's
postcondition.

5.5   Splitting

If we examine our example, there is a clear match between most lines. Splitting allows us to take advantage of this structural similarity by considering each
match in isolation; formally, given the specification fragment do $rv \leftarrow a;\ b\ rv$
od and the implementation fragment $c;\ d$, splitting entails proving a first correspondence between a and c and a second between b and d.

In the case where we can prove that c terminates abruptly, we discard d.
Otherwise, the following rule is used

$$\frac{\begin{array}{cc} \text{ccorres } r'\ xf'\ P\ P'\ hs\ a\ c & \forall v.\ d'\ v \sim d[v/xf'] \\ \forall rv\ rv'.\ r'\ rv\ rv' \longrightarrow \text{ccorres } r\ xf\ (Q\ rv)\ (Q'\ rv\ rv')\ hs\ (b\ rv)\ (d'\ rv') \\ \{R\}\ a\ \{Q\} \qquad \Gamma \vdash_{/\mathcal{U}} R'\ c\ \{s \mid \forall rv.\ r'\ rv\ (xf'\ s) \longrightarrow s \in Q'\ rv\ (xf'\ s)\} \end{array}}{\text{ccorres } r\ xf\ (P \cap R)\ (P' \cap R')\ hs\ (\text{do } rv \leftarrow a;\ b\ rv\ \text{od})\ (c;\ d)}$$

In the second correspondence premise, d' is the result of lifting xf' in d; this
enables the proof of the second correspondence to use the result relation from
the first correspondence. To calculate the final preconditions, the rule includes
VCG premises to move the preconditions from the second correspondence across
a and c. In the C-SIMPL VCG obligation, we may ignore any guard faults as
their absence is implied by the first premise. In fact, in most cases the C-SIMPL
VCG step can be omitted altogether, because the post condition collapses to
true after simplifications.

We have developed a tactic which assists in splitting: C-SIMPL's encoding
of function calls and struct member updates requires multiple specialised rules.
The tactic symbolically executes and moves any guards if required, determines
the correct splitting rule to use, instantiates the extraction function, and lifts
the second correspondence premise.

Example. After lifting, moving the guard, and symbolically executing the getCTE
function, applying the above rule to the example proof statement in Fig. 5 gives
the following as the first proof obligation

```
ccorres cmdb-relation mdb (...) {...} [SKIP]
  (return (cteMDBNode cte))
  (´mdb :== H &(Ptr src→[cteMDBNode-C])
```

This goal, proved using the VCG approach from Sect. 5.2, states that, apart from the state correspondence, the return value from the specification side (cteMDBNode *cte*) and implementation side (H &(Ptr *src*→[cteMDBNode-C]) stored in *mdb*) are related through cmdb-relation, that is, the linked list pointers in the returned specification node are equal to those in the implementation.

5.6 Completing the Example

The proof of the example, cteMove, is 25 lines of Isabelle/HOL tactic style proof. The proof starts by weakening the preconditions (here abbreviated P and P') with new Isabelle schematic variables; this allows preconditions to be calculated on demand in the correspondence proofs.

We then lift the function arguments and proceed to prove by splitting; the leaf goals are proved as separate lemmas using the C-SIMPL VCG. Next, the correspondence preconditions are moved back through the statements using the two VCGs on specification and implementation. The final step is to solve the proof obligation generated by the initial precondition weakening: the stated preconditions (our P and P') must imply the calculated preconditions.

The lifting and splitting phase takes 9 lines, the VCG stage takes 1 line, using tactic repetition, while the final step takes 15 lines and is typically the trickiest part of any correspondence proof.

6 Experience

In this section we explore how our C subset influenced the kernel implementation and performance. We then discuss our experience in applying the framework.

We chose to implement the C kernel manually, rather than synthesising it from the executable specification. Initial investigations had shown that generated C code would not meet the performance requirements of a real-world microkernel. Message-passing (IPC) performance, even in the first hand-written version, completed after two person months, was slow, on the order of the Mach microkernel. After optimisation, this operation is now comparable to that of the modern, commercially deployed, OKL4 2.1 [22] microkernel: we measured 206 cycles for OKL4's hand-crafted assembly IPC path, and 756 cycles for its non-optimised C version on the ARMv6 Freescale i.MX31 platform. On the same hardware, our C kernel initially took over 3000 cycles, after optimisations 299. The fastest other IPC implementation for ARMv6 in C we know of is 300 cycles.

The C subset and the implementation developed in parallel, influencing each other. We extended the subset with new features such as multiple side-effect free function calls in expressions, but we also needed to make trade-offs such as for references to local variables. We avoided passing large structures on the

Table 1. Code and proof statistics. Changes for step $\mathcal{R}_C$.

	Lines			Changes	
	Haskell/C	Isabelle	Proof	Bugs	Convenience
Executable specification	5,700	13,000	117,000	8	10
Implementation	8,700	15,000	50,000[a]	97	34

[a] With 474 of 518 (91%) of the functions verified.

C stack across function boundaries. Instead, we stored these in global variables and accessed them through pointers. Whilst the typical pattern was of conflicting pressures from implementation and verification, in a few cases both sides could be neatly satisfied by a single solution. We developed a code-generation tool [4] for efficient, packed bitfields in tagged unions with a clean, uniform interface. This tool not only generates the desired code, but also the associated Isabelle/HOL proofs and specifications that integrate directly into our refinement framework. The resulting compiled code is faster, more predictable, and more compact than the bitfield code emitted by GCC on ARM.

Code and proof statistics are shown in Table 1. Of the 50,000 lines of proof in $\mathcal{R}_C$, approximately 5,000 lines are framework related, 7,700 lines are automatically generated by our bitfield tool, and the remaining 37,300 lines are handwritten. We also have about 1,000 lines of tactic code. We spent just over 2 person years in 6 months of main activity on this proof and estimate another two months until completion. We prove an average of 3–4 functions per person per week.

One important aspect of the verification effort was our ability to change both the specification and the implementation. These changes, included in Table 1, fell into two categories: true bug fixes and proof convenience changes. In the specification, bug fixes were not related to safety — the proof of $\mathcal{R}_A$ guaranteed this. Rather, they export implementation restrictions such as the number of bits used for a specific argument encoding. Although both versions were safe, refinement was only possible with the changed version. The implementation had not been intensively tested, because it was scheduled for formal verification. It had, however, been used in a number of student projects and was being ported to the x86 architecture when verification started. The former activities found 16 bugs in the ARMv6 code; the verification has so far found 97. Once the verification is complete, the only reason to change the code will be for performance and new features: C implementation defects will no longer exist.

A major aim in developing the framework presented in this paper was the avoidance of invariant proofs on the implementation. We achieved this primarily through proof reuse from $\mathcal{R}_A$: the detailed nature of the executable specification's treatment of kernel objects meant that the state relation fragment for kernel objects was quite simple; this simplicity allowed proof obligations from the implementation to be easily solved with facts from the specification. Furthermore, when new invariants were required we could prove them on the executable specification. For example, the encoding of Isabelle's option type using a default value in C (such as NULL) required us to show that these default values never occurred as valid values.

We discovered that the difficulty of verifying any given function in $\mathcal{R}_C$ was determined by the degree of difference between the function in C and its executable specification, arising either from the control structures of C or its impure memory model. Unlike the proof of $\mathcal{R}_A$, the semantic complexity of the function seems mostly irrelevant. For instance, the operation which deletes a capability — by far the most semantically complex operation in seL4 — was straightforward to verify in $\mathcal{R}_C$. On the other hand, a simpler operation which employs an indiscriminate memset over a number of objects was comparatively difficult to verify. It is interesting to note that, even here, proofs from $\mathcal{R}_A$ were useful in proving facts about the implementation.

An important consequence of the way we split up proofs is that local reasoning becomes possible. No single person needed a full, global understanding of the whole kernel implementation.

7   Related Work

Earlier work on OS verification includes PSOS [12] and UCLA Secure Unix [28]. Later, Bevier [3] describes verification of process isolation properties down to object code level, but for an idealised kernel (KIT) far simpler than modern microkernels. We use the same general approach — refinement — as KIT and UCLA Secure Unix, however the scale, techniques for each refinement step, and level of detail we treat are significantly different.

The Verisoft project [24] is working towards verifying a whole system stack, including hardware, compiler, applications, and a simplified microkernel VAMOS. The VFiasco [15] project is attempting to verify the Fiasco kernel, another variant of L4 directly on the C++ level. For a comprehensive overview of operating system verification efforts, we refer to Klein [17].

Deductive techniques to prove annotated C programs at the source code level include Key-C [20], VCC [6], and Caduceus [13], recently integrated into the Frama-C framework [14]. Key-C only focuses on a type-safe subset of C. VCC, which also supports concurrency, appears to be heavily dependent on large axiomatisations; even the memory model [6] axiomatises a weaker version of what Tuch proves [26]. Caduceus supports a large subset of C, with extensions to handle certain kinds of unions and casts [1,19]. These techniques are not directly applicable to refinement, although Caduceus has at least been used [2] to extract a formal Coq specification for verifying security and safety properties.

We directly use the SIMPL verification framework [23] from the Verisoft project, but we instantiate it differently. While Verisoft's main implementation language is fully formally defined from the ground up, with well-defined Pascal-like semantics and C-style syntax, we treat a true, large subset of C99 [16] on ARMv6 with all the realism and ugliness this implies. Our motivation for this is our desire to use standard tool-chains and compilers for real-world deployment of the kernel. Verisoft instead uses its own non-optimising compiler, which in exchange is formally verified. Another difference is the way we exploit structural similarities between our executable specification and C implementation. Verisoft uses the standard VCG-based methodology for implementation verification. Our framework allows us to transport invariant properties and Hoare-triples from our existing proof on the executable specification [5] down to the C level. This allowed us to avoid invariants on the C level, speeding up the overall proof effort significantly.

8 Conclusion

We have presented a formal framework for verifying the refinement of a large, monadic, executable specification into a low-level, manually performance-optimised C implementation. We have demonstrated that the framework performs well by applying it to the verification of the seL4 microkernel in Isabelle/HOL, and by completing a large part of this verification in a short time. The framework allows us to take advantage of the large number of invariants proved on the specification level, thus saving significant amounts of work. We were able to conduct the semantic reasoning on the more pleasant monadic, shallowly embedded specification level, and leave essentially syntactic decomposition to the C level.

We conclude that our C verification framework achieves both the scalability in terms of size, as well as the separation of concerns that is important for distributing such a large proof over multiple people.

Acknowledgements. We thank Timothy Bourke and Philip Derrin for reading and commenting on drafts of this paper.

References

1. Andronick, J.: Modélisation et Vérification Formelles de Systèmes Embarqués dans les Cartes à Microprocesseur—Plate-Forme Java Card et Système d'Exploitation. Ph.D thesis, Université Paris-Sud (March 2006)
2. Andronick, J., Chetali, B., Paulin-Mohring, C.: Formal verification of security properties of smart card embedded source code. In: Fitzgerald, J.S., Hayes, I.J., Tarlecki, A. (eds.) FM 2005. LNCS, vol. 3582, pp. 302–317. Springer, Heidelberg (2005)
3. Bevier, W.R.: Kit: A study in operating system verification. IEEE Transactions on Software Engineering 15(11), 1382–1396 (1989)
4. Cock, D.: Bitfields and tagged unions in C: Verification through automatic generation. In: Beckert, B., Klein, G. (eds.) Proc, 5th VERIFY, Sydney, Australia, August 2008. CEUR Workshop Proceedings, vol. 372, pp. 44–55 (2008)
5. Cock, D., Klein, G., Sewell, T.: Secure microkernels, state monads and scalable refinement. In: Mohamed, O.A., Muñoz, C., Tahar, S. (eds.) TPHOLs 2008. LNCS, vol. 5170, pp. 167–182. Springer, Heidelberg (2008)
6. Cohen, E., Moskal, M., Schulte, W., Tobies, S.: A precise yet efficient memory model for C (2008),
http://research.microsoft.com/apps/pubs/default.aspx?id=77174

7. de Roever, W.-P., Engelhardt, K.: Data Refinement: Model-Oriented Proof Methods and their Comparison. Cambridge Tracts in Theoretical Computer Science, vol. 47. Cambridge University Press (1998)
8. Derrin, P., Elphinstone, K., Klein, G., Cock, D., Chakravarty, M.M.T.: Running the manual: An approach to high-assurance microkernel development. In: Proc. ACM SIGPLAN Haskell WS, Portland, OR, USA (September 2006)
9. Dijkstra, E.W.: Guarded commands, nondeterminacy and formal derivation of programs. CACM 18(8), 453–457 (1975)
10. Elphinstone, K., Klein, G., Derrin, P., Roscoe, T., Heiser, G.: Towards a practical, verified kernel. In: Proc. 11th Workshop on Hot Topics in Operating Systems (2007)
11. Elphinstone, K., Klein, G., Kolanski, R.: Formalising a high-performance microkernel. In: Leino, R. (ed.) VSTTE, Microsoft Research Technical Report MSR-TR-2006-117, Seattle, USA, August 2006, pp. 1–7 (2006)
12. Feiertag, R.J., Neumann, P.G.: The foundations of a provably secure operating system (PSOS). In: AFIPS Conf. Proc., 1979 National Comp. Conf., New York, NY, USA, June 1979, pp. 329–334 (1979)
13. Filliâtre, J.-C., Marché, C.: Multi-prover verification of C programs. In: Davies, J., Schulte, W., Barnett, M. (eds.) ICFEM 2004. LNCS, vol. 3308, pp. 15–29. Springer, Heidelberg (2004)
14. Frama-C (2008), http://frama-c.cea.fr/
15. Hohmuth, M., Tews, H.: The VFiasco approach for a verified operating system. In: Proc. 2nd ECOOP-PLOS Workshop, Glasgow, UK (October 2005)
16. Programming languages—C, ISO/IEC 9899:1999 (1999)
17. Klein, G.: Operating system verification—An overview. Sādhanā 34(1), 27–69 (2009)
18. Liedtke, J.: On μ-kernel construction. In: Proc. 15th SOSP (December 1995)
19. Moy, Y.: Union and cast in deductive verification. In: Proc. C/C++ Verification Workshop, Technical Report ICIS-R07015. Radboud University Nijmegen (2007)
20. Mürk, O., Larsson, D., Hähnle, R.: KeY-C: A tool for verification of C programs. In: Pfenning, F. (ed.) CADE 2007. LNCS, vol. 4603, pp. 385–390. Springer, Heidelberg (2007)
21. Nipkow, T., Paulson, L.C., Wenzel, M.T.: Isabelle/HOL. LNCS, vol. 2283. Springer, Heidelberg (2002)
22. Open Kernel Labs. OKL4 v2.1 (2008), http://www.ok-labs.com
23. Schirmer, N.: Verification of Sequential Imperative Programs in Isabelle/HOL. Ph.D thesis, Technische Universität München (2006)
24. Schirmer, N., Hillebrand, M., Leinenbach, D., Alkassar, E., Starostin, A., Tsyban, A.: Balancing the load — leveraging a semantics stack for systems verification. JAR, special issue on Operating System Verification 42(2-4), 389–454 (2009)
25. Tuch, H.: Formal Memory Models for Verifying C Systems Code. Ph.D thesis, School Comp. Sci. & Engin., University NSW, Sydney 2052, Australia (August 2008)
26. Tuch, H.: Formal verification of C systems code: Structured types, separation logic and theorem proving. JAR, special issue on Operating System Verification 42(2–4), 125–187 (2009)
27. Tuch, H., Klein, G., Norrish, M.: Types, bytes, and separation logic. In: Hofmann, M., Felleisen, M. (eds.) Proc. 34th POPL, pp. 97–108. ACM, New York (2007)
28. Walker, B., Kemmerer, R., Popek, G.: Specification and verification of the UCLA Unix security kernel. CACM 23(2), 118–131 (1980)

Author Index